SOUTH–WESTERN

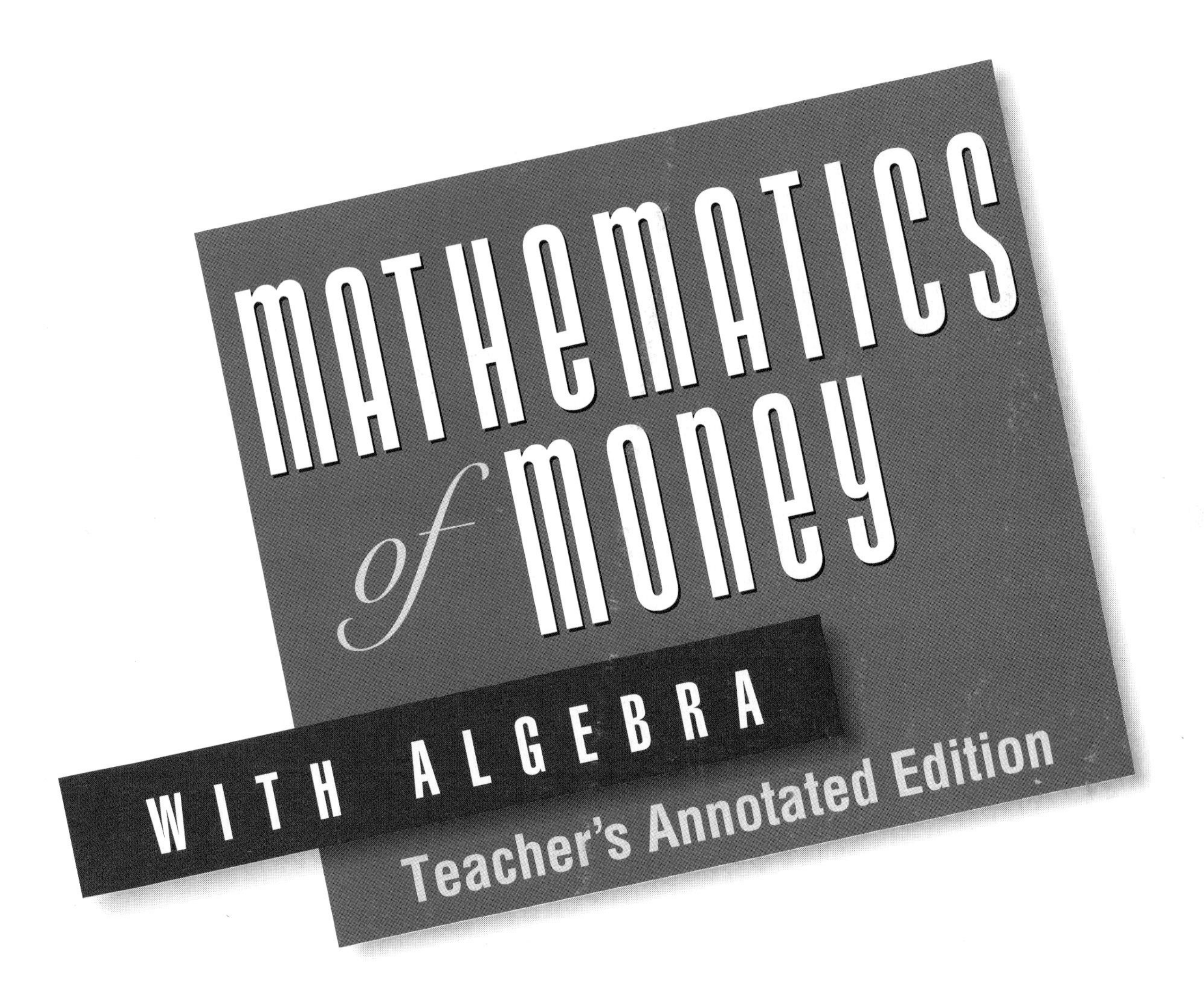

Cheryl Clayton
Richardson High School
Richardson, Texas

SOUTH-WESTERN PUBLISHING CO.

ISBN: 0–538–63472–3
2 3 4 5 6 7 8 KI 01 00 99 98 97 96 95 94
Printed in the United States of America

I(T)P
International Thomson Publishing

South-Western Publishing Co. is an ITP Company. The ITP trademark is used under license.

SOUTH—WESTERN

MATHEMATICS of MONEY

WITH ALGEBRA

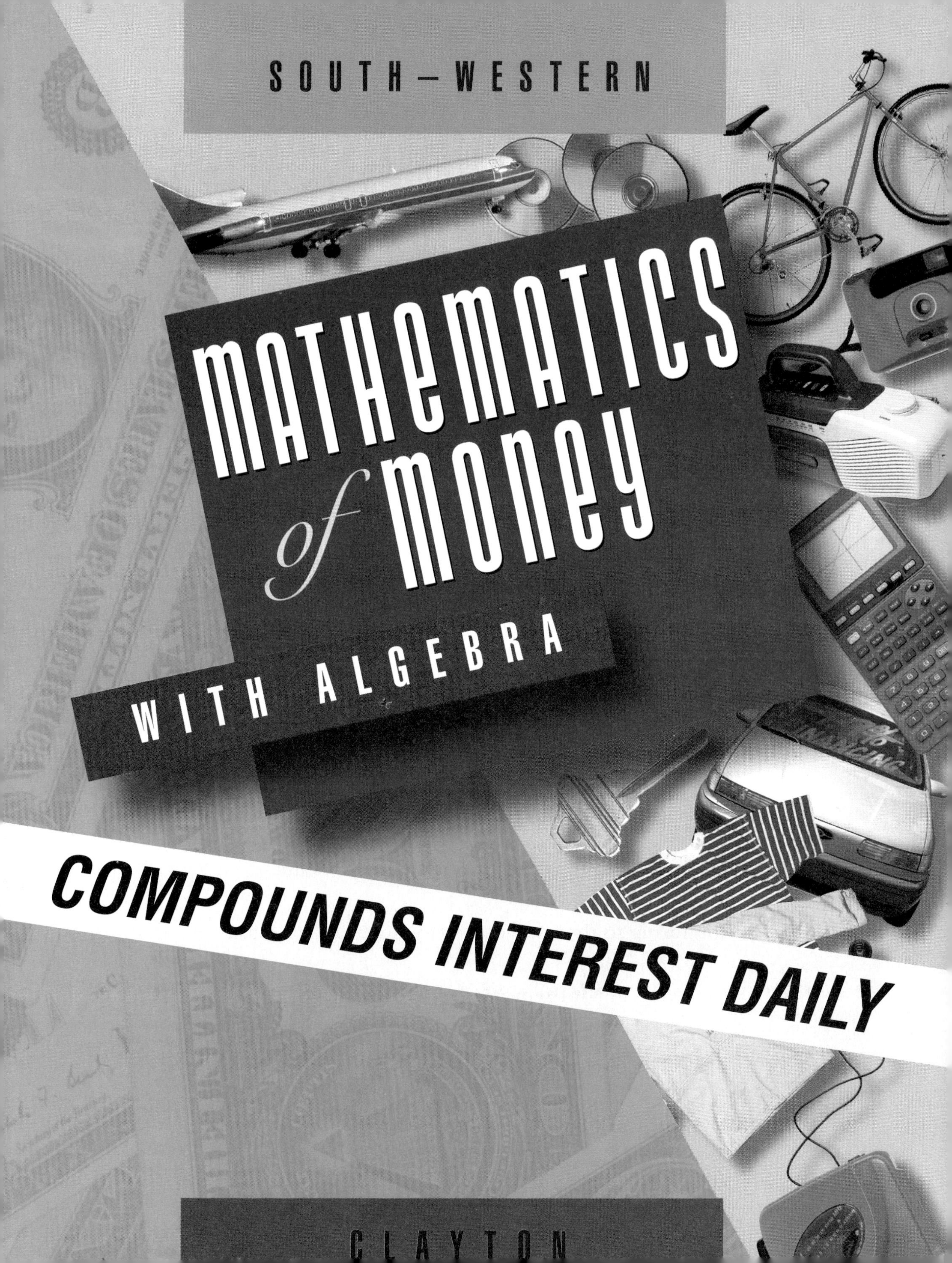

COMPOUNDS INTEREST DAILY

CLAYTON

Watch **your students' interest in** *algebra* **grow...with real-life** *money skills.*

Here's what you've been waiting for—a relevant primer for real life. In fact, if it's not relevant, we didn't include it. Now, your students won't be muttering to themselves, "Why do I have to learn this?" Everything in **Mathematics of Money with Algebra** makes sense. Everything will help them develop financial decision-making skills they can use throughout their lives.

INTERESTING CHARACTERS

Every chapter begins by involving students immediately. They'll identify with the characters—students their own age—and with the situations of the story—real problems to be solved with real math.

RNING

ER

differ from those a salaried salesperson
uld make. Betty tells us why.
Another friend, Luis, will see how and
why deductions are made from his paycheck.
His first paycheck from the grocery
store where he works comes
as a surprise to Luis. He
had expected to receive
his full hourly rate,
but he received
considerably less.

Alex, Daph[illegible]
forward to the [illegible]
have all their [illegible]
are glad to b[illegible]
lives. At t[illegible]
some of [illegible]
to them[illegible]
their c[illegible]

5-5 THE LURE OF CREDIT TERMS: THE MERCHANT PROFITS

QUICK REFERENCE

TEACHER'S RESOURCES AND ANSWER KEY
Lesson Quiz Answers 5-5, p. 56
Reteaching Activity Answers 5-5, p. 99
Enrichment Activity Answers 5-5, p. 100

EXTENSION ACTIVITIES
Reteaching Activity 5-5, p. 35
Enrichment Activity 5-5, p. 40

Maria discovered that there are methods to reduce the cost of credit. Selecting a shorter amortization period, a lower interest rate, and a larger down payment enables the borrower to reduce the total payments on a loan. In addition, prepaying a loan can often save the borrower money. Many car ads mention terms that include cash-back opportunities and significantly lower interest rates than financial institutions are offering.

Maria wonders whether she can save money on the purchase of her car by taking advantage of some of these terms. Patrick warns that merchants can increase their profits by enticing customers to choose credit terms that are favorable to the merchants.

OBJECTIVES: *In this lesson, we will help Maria to:*

- *Learn how merchants generate credit income and attract customers with special credit plans.*
- *Calculate the total financed price with lower-than-market-rate interest and rebate plans.*
- *Determine the merchants' profits with lower-than-market-rate interest and rebate plans.*
- *Observe the drawbacks associated with rent-to-own plans.*

226 CHAPTER 5 CONSUMER CREDIT

INTERESTING STORYLINES

Lesson Openers develop the scenario for the students—focusing on the decisions that will need to be made and the factors that will affect those decisions. Objectives are clearly presented.

Every chapter begins with an Algebra Refresher that will help students recall the skills they need to complete the upcoming work. There's also an Algebra Review within each lesson providing extra practice problems.

ALGEBRA *REFRESHER*

Remember that the *slope m* of a line is found by taking any two points and dividing the change in y by the change in x.

$$m = \frac{y_2 - y_1}{y_2 - x_1}$$

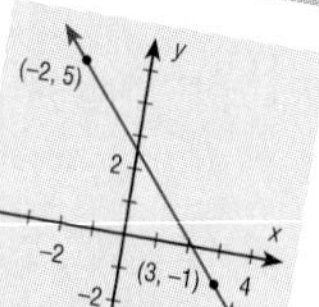

For the graph and points shown the slope m is

$$m = \frac{5-(-1)}{-2-3} = \frac{6}{-5} = -\frac{6}{5}$$

In an equation of the form $y = mx + b$, m is the slope and b is the y-intercept. Use the slope-intercept method to graph each equation.

Example $2y + 3x = 12$

$2y = -3x + 12$

$y = -\frac{3}{2}x + 6$ Solve for y.

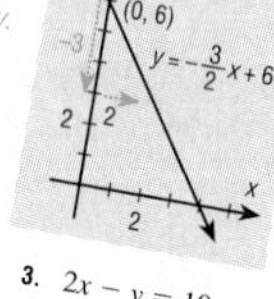

When $x = 0$, $y = 6$. The y-intercept is 6.

The slope is the coefficient of x, $-\frac{3}{2}$.

1. $y = 2x - 3$
   See Additional Answers.
2. $3y = 2x - 6$
3. $2x - y = 10$
4. $x = 10 - 5y$
5. $3y + 2x = -12$
6. $y - x = 0$

To graph a *linear inequality,* first graph the related equation, then test points to see which side of the line satisfies the inequality. To graph several inequalities, shade the regions for each one and see where they overlap. Remember to include the line of the equation in the solution for ≤ or ≥.

Find the solution of each system.

Example $y \le 2x + 3$

$y > -x + 1$ (1)

Equation (1) ha...

3. Equ...

12

OBJECTIVES: *In this lesson, we will help Hari to:*

- *Graph the cost and revenue functions.*
- *Determine the break-even point algebraically.*
- *Determine the break-even point using the graphing calculator.*

BREAK-EVEN POINT

We have seen that establishing prices is a most important step for any business. Prices and the number of products sold determine a company's revenue. If a business's total cost is greater than its revenue, then the business incurs a *loss.* If the revenue is greater than the total cost, then the business makes a *profit.* The point at which cost and revenue are equal is called the **break-even point.** This point is given numerically by the number of products sold at a given price. It is also useful to illustrate the break-even point by graphing the line for cost and the line for revenue. The break-even point is where these two lines intersect.

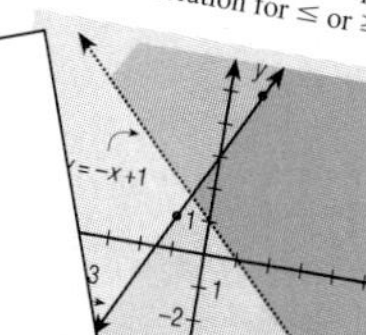

The region between the lines and above the break-even point is the profit region. The region between the lines and below the break-even point is the loss region.

Ask Yourself

1. Why are profits important to a business?
   See Additional Answers.
2. What is the break-even point?
   The point at which cost and revenue are equal.
3. What factors determine the break-even point?
   Cost, revenue, number of products sold, selling price

ADDITIONAL ANSWERS

1. Profits are needed to investigate new business opportunities and to provide reserves for emergencies.

ALGEBRA *REVIEW*

Write in slope-intercept form, and find the slope.

1. $x + y = 10$
   $y = -x + 10$; $m = -1$
2. $y = 2x - 3$
   $y = 2x - 3$; $m = 2$
3. $3y = 2x + 6$
   $y = \frac{2}{3}x + 3$; $m = \frac{2}{3}$
4. $2x - 4y = 20$
   $y = \frac{1}{2}x - 5$; $m = \frac{1}{2}$

Graph each equation.

5. $y = 2x - 5$
   See Additional Answers.
6. $2x - 3y = 12$

Find the point of intersection of each pair of equations.

7. $x + y = 8$
   $x - y = 12$
   (10, −2)
8. $y = 2x$
   $2x + 2y = 18$
   (3, 6)

Use a graphing calculator to approximate the points of intersection to the nearest hundredth.

9. $2x + 5y = 17$
   $3x - y = 21$
   (7.18, 0.53)
10. $4y - 3.7x = 16$
    $3.4x + 2y = 34.6$
    (5.07, 8.69)

All pages shown are from ***Mathematics of Money with Algebra*** *Teacher's Edition – featuring overprinted answers and side notes– a single source for planning and in-class reference.*

SKILLS FOR A LIFETIME

Here's the heart of the matter—a comprehensive skills section. Students will use their algebra skills to solve relevant situational problems. Sections include "Sharpen Your Skills," " Try Your Skills," and "Exercise Your Skills."

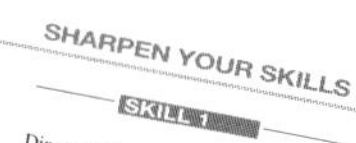

SHARPEN YOUR SKILLS

SKILL 1

Dinotech Engineering needs to hire a new engineer this year to help design some of the computer components used in the U.S. space program. The cost of the following extra taxes and fringe benefits will figure into their plans.

a. 7.65% of gross pay for Social Security and Medicare (FICA)
b. 5% of gross pay to match retirement contributions
c. $80 per month to help pay for health insurance
d. $10 per month for a life insurance policy

Dinotech operates 5 days a week, 52 weeks a year, or 260 days. Of the 260 workdays each year, the company grants its employees 10 vacation days, 6 holidays, and up to 10 sick-leave days.

EXAMPLE 1 Dinotech plans to offer the new engineer $45,000 per year as a starting salary.

QUESTIONS
1. How much must the company budget to cover the engineer's fringe benefits?
2. How much of the engineer's $45,000 salary is paid for the nonworking days?

SOLUTIONS
1. a. $0.0765 \cdot 45{,}000 = 3442.50$ — 7.65% of $45,000 for FICA
b. $0.05 \cdot 45{,}000 = 2250$ — 5% of $45,000 for the retirement plan
c. $80 \cdot 12 = 960$ — $80 per month for health insurance
d. $10 \cdot 12 = 120$ — $10 per month for life insurance

Total additional costs: $6772.50

16

CHAPTER 1 PERSONAL EARNING POWER

2. The number of nonworking days is
10 vacation days + 10 sick-leave days + 6 holidays, or 26 days
Write a proportion and solve it.

$$\frac{\text{Nonworking days}}{\text{total days}} = \frac{\text{dollars earned on nonworking days}}{\text{amount earned for entire year}}$$

$$\frac{26}{260} = \frac{x}{45{,}000}$$

$$\frac{1}{10} = \frac{x}{45{,}000} \quad \text{Simplify the left side.}$$

$$1 \cdot 45{,}000 = 10x \quad \text{Use the rule of proportions.}$$

$$\frac{45{,}000}{10} = \frac{10x}{10} \quad \text{Divide each side by 10.}$$

$$4500 = x$$

Dinotech pays $4500 for nonworking days.

COMMON ERRORS
When solving proportions using the "special rule," students should carefully keep track of the four terms of the proportion in order to pair them correctly. Some students may be aided by the language of classical geometry: "The product of the means (the middle terms) equals the product of the extremes (the first and last terms)."

SKILL 2

EXAMPLE 2 Daphne's older friend Chloe has completed her engineering training and is considering accepting a position at Aerotech. Chloe has also been offered a position at Electron Research and has described the two offers to Daphne.

	Aerotech	Electron Research
Salary	$48,000	$48,000
Medical insurance	$50/month	$60/month
Retirement	3% of gross pay	$2\frac{1}{2}$% of gross pay
Education	$1000/year	$1200/year
Travel Allowance	$120/month	$100/month

QUESTION Which of the two companies is offering Chloe more money including fringe benefits?

SOLUTION
Aerotech is offering:

Salary	$48,000	
Medical insurance: $50/month	600	$50 \cdot 12$
Retirement: 3% of 48,000	1,440	$0.03 \cdot 48{,}000$
Educational expenses	1,000	
Travel allowance	1,440	$120 \cdot 12$
Total offer	$52,480	

LESSON 1–2 EMPLOYMENT OPPORTUNITIES: SO[illegible]

17

TRY YOUR SKILLS

One week, Freda works 25 hours. Freda is single and claims no exemptions.
1. Find her gross pay if she makes $5.50 per hour. $137.50
2. Find the amount of her federal income tax withholding. $13.00
3. Find the amount of her FICA withholding. $10.52
4. Find her take-home pay. $113.98

Jackie is hired to deliver work and run errands. One week, she works 15 hours at $4.80 an hour. She is single and does not claim herself as an exemption.
5. Find her gross pay. $72.00
6. Find the amount of her federal income tax withholding. $4.00
7. What will be her FICA withholding? $5.51
8. Find her take-home pay. $62.49

GUIDED PRACTICE
The *Try Your Skills* exercises review the calculations one by one. Students should do them individually. Then go over each one explaining the process and checking for understanding.

EXERCISE YOUR SKILLS

See Additional Answers.
1. What are several qualities or factors that are needed to make a small business successful?
2. What are some of the responsibilities of a good manager?
3. Give two different meanings of the word *labor.*
4. What investment do workers make in a business?
5. Construct a payroll register for weekly deductions to record the amount for federal income tax withholding based on the tax table in the Reference Section, for FICA withholding based on 7.65% of gross pay, and for take-home pay. Kendra worked 20 hours at a rate of $5.25 an hour, with no exemptions. Paulo worked 9 hours at $8.20 an hour, with one exemption. Michael worked 29 hours at $6.10 an hour, with no exemptions. Also record the totals for each category. All of the employees are single. See Additional Answers.

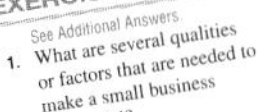

KEY TERMS
consumers
entrepreneurship
FICA
gross pay
labor
management
market economy
producers
take-home pay

ADDITIONAL ANSWERS
1. Answers may vary. (Ex.: good management, good market, quality product)
2. Answers may vary. (Ex.: making plans, empowering workers)
3. (1) efforts and skills (2) workers
4. time and labor

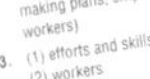

CHAPTER 4 A VENTURE INTO BUSINESS

140

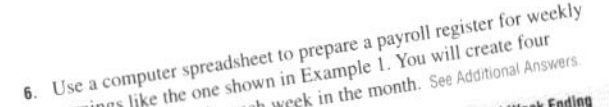

6. Use a computer spreadsheet to prepare a payroll register for weekly earnings like the one shown in Example 1. You will create four spreadsheets, one for each week in the month. See Additional Answers.

Employee	Number of Exemptions	Marital Status	Hourly Rate	Hours Worked Week Ending 4/4	4/11	4/18	4/25
Catlyn	0	S	$3.75	16	14	21	32
Sara	1	S	4.50	20	28	25	18
Joleen	0	S	8.00	26	16	16	24
Hernando	1	S	6.24	14	10	30	20

7. Use a computer spreadsheet with the headings shown below to prepare four weekly payroll registers for each of the employees in Exercise 6. Refer to the tables for Federal Income Tax Withholding in the Reference Section to find the federal income tax to be withheld. For FICA withholding, use 7.65% of gross pay.

PAYROLL REGISTER FOR WEEKLY DEDUCTIONS

Employee	Gross Pay	Income Tax Withholding	FICA Withholding	Total Deductions	Take-Home Pay

8. Use a spreadsheet program to prepare a monthly payroll summary with the headings shown below for each employee from Exercises 6 and 7. Show the weekly totals for payroll information to prepare this summary.

MONTHLY PAYROLL SUMMARY

Employee: ____________

Week	Gross Pay	Income Tax Withholding	FICA Withholding	Take-Home Pay

9. Construct a computer spreadsheet that will be a payroll register combining the information and calculations for earnings with those of deductions. Use the spreadsheet to compute gross pay, deductions, and take-home pay for the following employees. All of them are single.

Employee	Exemptions	Hourly Rate	Hours
Adikes	1	$7.85	10
Carney	1	5.30	18
Inez	0	6.78	27
Sun-Li	0	9.36	17

How would you need to adjust the FICA tax for a person who earns the following in a year.
10. More than $57,000, but less than $135,000.
11. More than $135,000.

INDEPENDENT PRACTICE

ASSIGNMENT GUIDE
Exercises 1-4 ask about the context of the lesson. A short discussion can again remind students that what they are learning is an important part of an adult's working life.
Exercises 5-9 assess students ability to produce the spreadsheets or tables and do the calculations needed for payroll.
Exercises 10-17 assess student's ability to find FICA withholding on salaries greater than $57,600.

CLOSURE
Tell students to draw a paper and pencil payroll table with books closed. Tell them to include the necessary categories and to make their own figures. See how well they recall the different categories and procedures.

LESSON QUIZ
1. A person earns $7 an hour for 40 hours and time and a half for 5 hours of overtime. What is her gross pay? (**$332.50**)
2. At 7.65%, what will be the FICA withholding amount for the person in Exercise 1 above? (**$25.44**)

ADDITIONAL ANSWERS
10. When the total gross salary has reached $57,600, only find 1.45% of the salary.
11. When the total gross salary has reached $135,000, there are no FICA taxes for the remainder of the year.

141

LESSON 4–1 EMPLOYEES ARE PART OF THE COMPANY

MATHEMATICS OF MONEY WITH ALGEBRA

COMPREHENSIVE REVIEW AND TESTING

Mathematics of Money with Algebra provides on-going review of skills learned in earlier chapters and skills just learned. Students will benefit from the continuous review.

CLOSURE

After covering the three skills, explain how equations, spreadsheets, and graphs can describe the same situation. Have students discuss the benefits and uses of each. When is a spreadsheet most useful? When is a graph the best way to present data?

LESSON QUIZ

1. A small business has variable costs of $200, a unit cost of $3.50, and produces 40 of its product. The business sells 30 of the product at a selling price of $6.00. Find the profit or loss? **(Loss, $160; Explain that a business does not always sell all of its product. Ask what it might do in trying to sell the remainder.)**
2. Graph both equations on the same axes: the cost of x items at $3.50; the sales revenue from x items at $6.00 each.

ADDITIONAL ANSWERS (EXERCISE YOUR SKILLS)

21. $c = 4n + 190$; $r = 7.5n$; $p = 3.5n - 190$

(MIXED REVIEW)

5. To: Farley Smith; For: VCR repair; Balance: $363.50
7. To: Abe's Supermarket; For: groceries; Balance: $668.73
8. No; there is a $1.40 discrepancy

Students selling book bags at $7.50 per bag have the following costs.

Fixed costs: Labor, 35 hours at $4.50 per hour
Advertising, energy, and transportation, $32.50
Variable costs: Bags, $4.00 each

21. Determine the cost, revenue, and profit functions. See Additional Answers.
22. Set up a spreadsheet to show costs, revenues, and profits for quantities of 25, 50, and 75.
23. Graph the cost function.

MIXED REVIEW

1-1 1. Tracy earns $130 for working 24 hours. Of that time, 20 hours are at her regular rate and the remainder is at $1\frac{1}{2}$ times the regular rate. What is her regular hourly rate? $5.00 per hour
1-2 2. Janice earned $29,670 last year. Her fringe benefits were 14.5% of her salary. What was the value of the fringe benefits? $4302.15
1-3 3. Louis earns a salary of $145 a week and also a commission of 4.5% on his sales. What are his total weekly earnings in a week in which his sales are $1,324? $204.58
1-3 4. A house has been sold for $190,000. The real-estate agent who sold the house earned a commission of 3.5%. The broker for whom she worked earned 2.5%. How much did each receive? agent: $6650; broker: $4750

2-2 Enter the following transactions in a check-register form. Make up your own form with the following headings: Check Number, Date, Checks/Deposits, Amount, and Balance. Under "Checks/Deposits" have two lines for each entry, the top one labeled "To:" and the bottom one labeled "For:". Find the new balance after each transaction. The starting balance is $620. See Additional Answers.

5. Check 701, February 17, to Farley Smith, $256.50 for VCR repair
6. Deposit on February 19 of $400 For: Deposit; Balance: $763.50
7. Check 702, February 19, to Abe's Supermarket, $94.77 for groceries

2-3 8. Earl's check register shows a balance of $104.56. He knows that he has a $20 check outstanding. His bank statement shows a balance of $125.96. Does his account balance? Explain your answer.
9. Tabitha sells annual memberships in an exercise club. Her boss has offered her a choice between Plan 1, no salary but a straight $50 for each membership sold, and Plan 2, a weekly base salary of $225 plus $25 for each membership sold. How many memberships would Tabitha have to sell to make the same amount of money under either plan? 9 memberships
10. Use the Rule of 72 to find how long it will take your $4550 savings account to double in value if it is growing at a rate ... per year. 11 years

160 CHAPTER 4 A VENTURE ...

CHAPTER 4 REVIEW

Find the following earnings:

	Number of Hours	Hourly Rate	Weekly Earnings
1.	38	$ 3.35	$127.30
2.	26	$ 7.50	$195.00
3.	41	22.70	930.70
4.	22	14.80	325.00

Find the weekly take-home pay for each of these single people if each claims one exemption. Use the tables for federal income tax withholding in the Reference Section.

	Gross Pay	Income Tax Withholding	FICA Withholding	Take-Home Pay
5.	$127.30	$ 5.00	$ 9.74	$112.56
6.	195.00	16.00	14.92	164.08
7.	930.70	183.00	71.20	676.50
8.	325.60	35.00	24.91	265.69

Duanita and Alisha want to make and sell stuffed kittens. Use the information given in the box to help you find the total production costs for manufacturing the amounts of toys in Exercises 9, 10, and 11.

Fixed Costs		Variable Costs	
Labor	$3.50 per hour for 30 hours	Materials	$2.25 each
		Packaging	$0.02 each
Advertising	$3.00		
Energy	$3.58		
Transportation	$5.92		

9. 20 kittens $162.90 10. 40 kittens $208.30 11. 60 kittens $253.70

Duanita and Alisha will sell stuffed kittens for $7.00 each. Use the costs in Exercises 9-11 to find the profit or loss from selling each amount.

12. 20 kittens 13. 40 kittens 14. 60 kittens

Use your results from Exercises 9–14 to make a spreadsheet that shows the following information for producing and selling 20, 40, and 60 kittens.

	Fixed Cost	Unit Cost	Number Produced	Total Cost	Unit Price	Revenue	Profit (Loss)
15.	$117.50	$2.27	20	$162.90	$7.00	$140.00	(22.90)
16.	117.50	2.27	40	208.30	7.00	280.00	71.70
17.	117.50	2.27	60	253.70	7.00	420.00	166.30

ADDITIONAL ANSWERS

12. loss: $22.90
13. profit: $71.70
14. profit: $166.30

178 CHAPTER 4 REVIEW

18. Write the cost function. $c = 2.27n + 117.50$
19. Write the revenue function. $r = 7n$
20. Write the profit function. $p = 4.73n - 117.50$
21. Graph the cost function. See Additional Answers.
22. Graph the revenue function on the same set of axes as the cost function.
23. Find the break-even point algebraically and using graphing techniques.
24. Why is it important to know how much profit (or loss) a business makes?
25. If a company does not make enough money to break even, what kinds of changes should be considered? Write a paragraph explaining the changes you would consider if your company were losing money and what you would do next if each strategy did not work.
26. After one year, the employees in Oscar's company expected to receive a raise in salary. Explain how an employee's thinking about a raise in salary might differ from the owner's thinking. What should Oscar consider before deciding whether to give raises? Why might it be better for the company and for the employees themselves if salary raises are small? When might salary raises be large?

Use the following information to do Exercises 27–30. When Duanita and Alisha were successful in the stuffed kitten business, they expanded and began to make stuffed puppies as well. Their cost for materials for the kittens continued to be $2.27, but the materials for the puppies cost $3.00. They sold the kittens for $7.00 and the puppies for $7.50. To meet the demand, they had to sell more kittens than puppies. They could not obtain materials to make more than 40 kittens per week and more than 20 puppies per week. To pay their expenses, they had to sell a total of more than 50 per week. Let x represent the stuffed kittens and y represent the stuffed puppies.

27. Write the objective function to minimize cost. $c = 2.27x + 3y$
28. Write the objective function to maximize revenue. $r = 7x + 7.5y$
29. Write the objective function to maximize profit. $p = 4.73x + 4.5y$
30. Write inequalities for the constraints. $x > y$; $x \leq 40$; $y \leq 20$; $x + y > 50$
31. Graph the constraints and shade the region that satisfies all the inequalities. See Additional Answers.
32. Find the corners of the region in Exercise 31. A(30, 20); B(40, 10); C(40, 20)
33. Find the quantities that will minimize cost within the constraints.
34. Find the quantities that will maximize revenue within the constraints.
35. Find the quantities that will maximize profit within the constraints.

ADDITIONAL ANSWERS

23. 25 kittens
24. It helps the company to analyze its costs and to plan expansion or condense production.
25. Answers may vary. Include higher prices, higher production, lower costs, and so on.
26. Answers may vary. (Ex.: The employees may not realize that getting raises affects costs, and therefore profit or loss. When profits are high, raises could be large. Small raises may insure the existence of the company.)
33. 40 kittens and 10 puppies
34. 40 kittens and 20 puppies
35. 40 kittens and 20 puppies

179 CHAPTER 4 REVIEW

CHAPTER 4 TEST

Find the weekly earnings for each. See Additional Answers.

1. 42 hours at $4.80 per hour
2. 33 hours at $17.25 per hour

Find the weekly take-home pay for each gross pay. Each person is single and claims one withholding allowance (exemption). Tables for federal income tax withholding are in the Reference Section.

3. $201.60
4. $569.25
5. $202.50
6. $844.20

Akira sells stadium cushions. Find the total production cost to manufacture the numbers of cushions indicated in Exercises 7–9.

Fixed Costs		Variable Costs	
Labor	$4.25 per hour for 30 hours	Materials	$1.40 each
Advertising	$2.00	Packaging	$0.50 each
Energy	$2.58		
Transportation	$4.42		

7. 30 cushions
8. 60 cushions
9. 90 cushions

Find the profit or loss for selling the following amounts at $5.50 each.

10. 30 cushions
11. 60 cushions
12. 90 cushions

Use your results from the previous exercises to complete the spreadsheet.

	Fixed Cost	Unit Cost	Number Produced	Total Cost	Unit Price	Revenue	Profit (Loss)
13.	$136.50	$1.90	30	$193.50	$5.50	$165.00	(28.50)
14.	136.50	1.90	60	250.50	5.50	330.00	79.50
15.	136.50	1.90	90	307.50	5.50	495.00	187.50

16. Graph the cost and revenue functions. See Additional Answers
17. Find the break-even point algebraically. 38 cushions

Jod and Jessie sell stadium cushions and caps. The cushions cost $1.90; the caps cost $2.25. They sell the cushions for $5.00 and the caps for $6.00. They can obtain no more than 100 cushions and 75 caps per week. To meet demands they have to sell a total of at least 120 of the two together. They cannot packag[e] more than 150 per week. Let x represent the cushions and y represent the caps, and find the poin[t] that will give maximum and minimum quantities.

18. Use linear programming to graph the given constraints, and find the poin[t] that will give maximum and minimum quantities. See Additional Answers.
19. What quantities will give the minimum cost? 100 cushions and 20 caps
20. What quantities will give the maximum revenue? 75 cushions and 75 cap[s]
21. What quantities will give the maximum profit? 75 cushions and 75 caps

ALTERNATIVE ASSESSMENT

Have students prepare a 5 to 10 minute presentation on a small business of their choice. The presentation should include: the variable costs; the factors that contribute to the unit costs of production; and what considerations enter into the establishment of prices. A student should also explain whether the company makes a profit and the outlook for the future.

Have students write a brief scenario about a production business of their choice. Ask them to develop cost and sales functions, and draw graphs. In addition, ask them to include the element of time, indicating how the business was threatened even though it eventually achieved break-even and made a profit.

Provide a graph with a polygonal region of the type covered in the lesson. Include the inequalities for each line. Ask students to write a minimum or maximum problem for which the given graph would be the solution.

ADDITIONAL ANSWERS

1. $201.60
2. $569.25
3. $169.18
4. $445.70
5. $170.01
6. $621.62
7. $193.50
8. $250.50
9. $307.50
10. Loss: $28.50
11. Profit: $79.50
12. Profit: $187.50

CUMULATIVE REVIEW

1-1 **1.** Your regular rate is $6.22 per hour. You also receive $1\frac{1}{2}$ times your hourly rate for any hours over 20 in one week. Last week you worked 26 hours. You also received $12.50 in tips. Find your total wages for the week. $192.88

1-1 **2.** You work according to the time schedule shown. You earn $5 per hour. Find your total earnings for the week if you receive $22 in tips that week.

Monday	Tuesday	Wednesday	Thursday	Friday	Saturday
3–6	3–7	3–6	3–6	4–9	8–12

1-2 **3.** You earn $9.30 per hour. You work 35 hours a week for 50 weeks and get 2 weeks of paid vacation as well as 18% of your base salary in fringe benefits. Find your yearly earnings including benefits. $19,972.68

1-3 **4.** You are a real-estate agent and sold a house for $110,000. You earn a 4% commission on the first $100,000 of each sale and 6% commission on the amount over $100,000. How much commission did you earn on the sale?

1-4 **5.** You earned $94.00. Your income tax withholding is $13 and your FICA withholding is 7.65% of your earnings. Find your take-home pay. $73.81

2-1 **6.** You wrote 33 checks one month. The bank charges $0.0225 for the first 20 checks that you write and $0.08 for each written check over 20. How much are the charges for that month? $1.49

2-2 **7.** Your employer made out your paycheck with your first and middle initials and full last name. Write your name as you would to endorse the check.

3-1 **8.** You save $28 per week. You want to buy a stereo system that costs $589. For how many weeks must you save before you can have the stereo system? 22 wk

3-1 **9.** Felix withdraws the interest from his savings account as soon as the interest is posted twice a year. What is the simple interest earned in $2\frac{1}{2}$ years on his $500 balance? The annual interest rate is 3.5%. $43.75

3-2 **10.** Find the value of an investment that is originally worth $1,000 after it has been compounded quarterly for 1 year at an 8.5% annual rate. $1087.75

3-3 **11.** Suppose that the reserve requirement on banking deposits is 12%. What will be the multiplier that corresponds to this requirement? $8.33

4-3 **12.** The fixed cost of producing hand-painted sweatshirts is $285 per week. The variable costs are $8 per shirt and $1 per shirt for paints. What is the total cost of producing 25 shirts in one week? $510

4-3 **13.** Use the information in Exercise 12 to find the profit or loss, given that each sweatshirt is sold for $20. Loss: $10

4-4 **14.** Write the equations for the cost function and revenue function of Exercises 12 and 13. Solve these equations to find the break-even point for the sweatshirt-selling operation. $c = 9n + 285$; $r = 20n$; 26 sweatshirts

ADDITIONAL ANSWERS

2. $132.00
4. $4600
7. Answers may vary.

PROJECTS THAT REINFORCE LEARNING

Each chapter ends with one or more related projects for the students to complete in groups and individually. Whether students form a small business or take a field trip to area banks, the projects help reinforce the relevancy of what they've learned.

PROJECT 4–1: **Retail Sales**

Work in small groups to set up imaginary companies. Select items that your company would like to sell or services that it would provide. You may use information from the chapter to plan costs or do your own research to find the costs of producing various services or products. Tell how you arrived at these costs for each product or service that your company produces.

1. Decide on the product or service for your company. Describe that product or service. Be sure to include all elements including advertising, energy use, transportation, and packaging.
2. Decide on the number of hours per week each employee (member of the group) will work. Decide the hourly wages for each person. Decide the marital status and number of exemptions for each employee.
3. Complete a payroll register showing each employee's earnings for four consecutive weeks.
4. Furnish a four-week payroll summary for each employee.
5. List your fixed and variable costs. Use a spreadsheet to show fixed costs, variable costs, number produced, total costs, average sale (or sale per unit), total sales, and profit or loss. Your spreadsheet should have entries that result in a loss and others that result in a profit.
6. Write the cost, revenue, and profit functions for your product.
7. Construct a line graph showing the break-even point for the business.
8. List several realistic constraints related to quantities and costs. Use linear programming to minimize costs or maximize sales revenue within the given constraints.

Extensions

1. After you have completed a plan for your company, get together with other groups in your class and try to sell them your product.
2. On the basis of feedback from other groups, redesign your product or service to be more useful or attractive to potential customers.
3. Listen in turn to other groups' presentations, and give them feedback about their product or service. Try to make comments that will help them in redesigning their product or rethinking their service.

COMPOUNDS YOUR STUDENTS' INTEREST DAILY

MATHEMATICS OF MONEY WITH ALGEBRA

TABLE OF CONTENTS

STUDENT EDITION
ISBN 0-538-63471-5/Stock No. MP02AA

TEACHER'S ANNOTATED EDITION
ISBN 0-538-63472-3/Stock No. MP02AW

EXTENSION ACTIVITIES
ISBN 0-538-63496-0/Stock No. MP02AD

TEACHER'S RESOURCES AND ANSWER KEY
ISBN 0-538-63497-9/Stock No. MP02AX

MICROEXAM II

IBM 3 1/2"	*ISBN 0-538-63660-2/Stock No. MP02AH88T*
Macintosh	*ISBN 0-538-63661-0/Stock No. MP02AH71T*
Apple 5 1/4"	*ISBN 0-538-63663-7/Stock No. MP02AH73T*

Call 1-800-543-7972 to order!
Or, FAX your order to 1-800-453-7882.

Contents

About the Teacher's Annotated Edition

The Teacher's Annotated Edition provides many features to help you to plan effectively. Answers to all exercises are provided, either overprinted below the exercise, in the side column, or in the Additional Answers section in the back of the book.

The following features in the side column provide a lesson plan outline.

- Focus
- Instruction
- Reteaching
- Enrichment
- Guided Practice
- Independent Practice
- Closure

The following features give additional help with instruction.

- Teaching the Lesson
- Critical Thinking
- Technology Hint
- Focus on Algebra
- Common Errors
- ESL Students
- At-Risk Students
- Interdisciplinary Investigation
- Cooperative Learning

A Lesson Quiz is provided in each lesson and specific Alternative Assessment suggestions are given beside each Chapter Test.

Time Line

The following time line suggests how a total of 164 class days may be allocated. Two days are allocated for each lesson. One day each for review, testing, cumulative review, and projects are allocated for each chapter.

Chapter	Days
1	12
2	10
3	10
4	14
5	14
6	18
7	10
8	12
9	10
10	12
11	10
12	12
13	10
14	10

OVERVIEW

Young people who are finishing their last years of high school today are becoming independent adults in a highly complex world. They must learn to function as adults in a society that is changing more rapidly than ever before. One facet of teenage life does not change. Teenagers have always wanted the freedom to choose their own friends, to make decisions independently of their parents, and to make and spend their own money. And they still do.

Young people today are entering the job market in large numbers even before they complete high school. Many are finding the realities of acquiring, handling, and spending money to be a challenging and sometimes overwhelming task. ***Mathematics of Money with Algebra*** will offer the opportunity for them to:

- Review and strengthen their algebra skills
- Extend their mathematical skills beyond Algebra 1
- Sharpen the mathematical skills they need to manage their own finances
- Become aware of the temptations they will face as they choose how to spend their money.

In ***Mathematics of Money with Algebra*** the computer and graphing calculator is used consistently to help students become informed consumers and capable money managers. Hopefully familiarity with the computer as a personal financial tool will make it easier to take advantage of the next advances when the current ones become obsolete.

Mathematics of Money with Algebra was written with a firm belief that each person must be able to communicate with language. Reading, writing, and discussing concepts orally are activities that should not be confined to an English classroom. The writing style used in ***Mathematics of Money with Algebra*** is designed so that reading about installment loans, bank accounts, and life insurance is appealing to high school students.

By introducing characters who resemble people they know, glimpses of life are given as it might be experienced by students. We observe one student who has overdrawn his bank account wondering if he can stuff twenty-dollar bills back into the Automatic Teller Machine. Another student, whose father carries too many credit cards, is embarrassed in a restaurant when the father accidentally scatters his cards on the floor. One student who has a job as a clerk in a toy store relates how a disgruntled customers threw a toy and knocked over a building block display when she told him that his credit card had been rejected.

Building upon a foundation of thorough and realistic objectives, ***Mathematics of Money with Algebra*** is designed to touch one aspect of our real lives where we will all feel it—in our pocketbooks. The teacher is encouraged to use true-to-life materials wherever possible. The exercises and most particularly the projects should be viewed as opportunities to bring into the classroom actual advertisements, listings, sales slips, articles, and so on.

The characters in Chapters 1 and 4 discover how money can be earned. They show us the differences between earning wages, earning a salary, working on commission, and operating a small business. They wonder why amounts of money are deducted from their paychecks when their actual take-home pay is less than they had expected.

In Chapters 2 and 3, the characters share with us some of their experiences using checking and savings accounts. Students learn to keep accurate records by deducting from a check register thousands of dollars of prize money for a major U.S. golf tournament. One of the characters is intrigued by the idea that because of compound interest, money that is left sitting, under proper conditions, seems to put down roots and grow new money!

In Chapters 5 and 6 our characters reveal to us the very pervasive and persuasive world of credit. Their eyes are opened to the large sums of money banks and lenders are acquiring from willing consumers who are paying high rates of interest on their credit purchases. In our world of immediate gratification, we consumers have listened to the siren's call: "Have your credit card ready." We have learned to carry a wallet full of credit cards. We have mastered the art of paper shuffling so well that we can have the money chasing its tail around in circles for years without ever making any progress in reducing our debt.

Chapters 7 and 8 show our characters handling investments, shopping for life insurance, and planning for retirement. Teenagers, who have a tendency to see their lives stretching out endlessly in front of them, are not serious retirement planners nor are they heavy purchasers of life insurance policies. The characters face situations that are carefully designed to make these topics relevant.

Students compete with each other in a quest for quick wealth by investing fictitious money in the stock market. Students also have the opportunity to participate in a long term project which requires small amounts of daily attention and careful management of detailed data.

Chapter 9 shows our characters facing their responsibilities as taxpayers and making their way through the 1040 forms. They will see a connection between how much was taken out of their paycheck and how much of that money will be kept by the IRS. One character discovers "Tax Freedom Day."

Chapters 10 and 11 give our characters the chance to calculate the costs of owning and operating their cars. One of our young drivers describes a vivid and frightening lesson on the necessity of good automobile insurance.

Several other characters give us tips on making travel plans that illustrate how family members who take trips together can return home still speaking to one another.

The housing chapters, 12 and 13, follow our graduating high school students as they move out of their parent's homes and into their own apartments. They discover the many financial and life-style decisions they must make when choosing and sharing an apartment. Parents of our characters share with them some of the responsibilities of home ownership. One character encourages his family to move to a new house so that he will no longer have to share a bedroom with his brother.

The last chapter is a culmination of the preceding ones. Students will analyze fixed and variable expenses and compare these with their monthly income, then make adjustments in whatever expenses they can. Just as many families discover each month—if there are more bills than money to pay them—someone may get left out. If that someone is not to be the grocer, another source of spending may have to be reduced. Our teenage characters will share some ideas about staying within the budget.

As our teenage students, like the characters in the book, approach adulthood and work their way through the transition, they will face many financial decisions. ***Mathematics of Money with Algebra*** offers assistance in financial planning, so that the choices the students make to establish an independent life style in their own apartment or house will be informed choices.

ASSESSMENT

Traditional Assessment

Mathematics of Money with Algebra incorporates traditional assessment in the following ways:

Pupil Text Mixed reviews, chapter reviews, chapter tests, and cumulative reviews provide many ways of assessing student work in the traditional manner. In addition, Algebra Refreshers appear at the beginning of each chapter and Algebra Reviews appear in each lesson so that teachers can continue to assess and provide needed help and remediation where indicated.

Teacher's Edition A lesson quiz is provided for each lesson so that you can assess student work on a daily basis. Also provided is a focus on algebra connections that allows you and your students an opportunity to assess and remediate algebra skills on a daily basis.

Teacher Resources and Answer Key Included are Chapter Tests (A and B) and a Year-End Test. These could be used as pre- and post-tests. Students taking pre-tests can identify weaknesses and pinpoint concepts to which they need to pay close attention while studying the chapter. Lesson quizzes are also included for monitoring of each lesson.

Alternative Assessment Methods

Assessment alternatives include the following:

Interviews Interviews can be used frequently and consist simply of talking with students in an effort to assess their thinking–a two-way communication effort.

Oral Activities You can use oral questions to help you assess more fairly those students who have difficulty reading.

Writing samples Encouraging students to record their ideas can help them organize their thinking.

Observations Teachers have always made observations about student performance as they attempt to evaluate students' insight into mathematics.

Student self-assessment It is important to involve students in the process of making informed assessment about their own work as they use it to plan ways of improving their performance in the future.

Classroom discussion Teachers can use in class discussion as an assessment tool as they assess student thinking and understanding of the "big idea."

Performance tasks Performance tasks involve presenting students with a mathematical task or project. Teachers can observe students as they complete the task or review students' solutions to assess what students have learned and how they apply this knowledge to other various situations.

Portfolios Students can use portfolios to collect representative samples of their work over a period of time. A portfolio might include records of interviews, oral activities, observations, and performance tasks, as well as student self-assessments. In many cases, it may also include homework, classwork, or traditional tests.

Variables to Consider When Selecting Assessment Methods

Open-Ended Often, you may wish to select a method that allows various responses to a problem. Other types of open-ended problems may involve multiple approaches to reach a single answer.

Duration Assessment duration may vary in length according to the time needed to complete the task successfully.

Individual or group work Some types of assessment lend themselves to individual work; others to work with a partner; still others to work in a group.

On-demand work, draft work, and perfected work Responses from on-demand work provides limited information because they are simply a student's first draft work. Perfected work, on the other hand, is the students' final product after being give the opportunity to revise, access multiple resources, and attain feedback.

Presentation and response modes Different types of responses—oral, written, demonstration, projects—should be used whenever appropriate.

Cross-content and mathematics specific Often an assessment can embed mathematics in other disciplines, such as science and literature.

Choice You may wish to have students choose samples of their best work to showcase their accomplishments in a portfolio. In doing so, students demonstrate their ability to assess their work and provide you with additional information about their work.

PERFORMANCE-BASED ASSESSMENT

Performance Standards

To implement performance-based assessment, you much first develop performance standards–one or more statements about intended levels of student performance in mathematics. These statements should include desired concepts and skills needed to perform the task in such a way that students' work can be judged on many levels of performance. This, in effect, allows students to show what they know and what they can do with what they know. *Performance benchmark* is a term used to describe an indicator of progress toward a performance standard.

In the *Assessment Standards for School Mathematics*, NCTM recommends that these six steps can be used to create performance standards:

1. Identify important "big ideas" in mathematics
2. Identify the expected knowledge of concepts and skills, know-how, effort, and context related to the "big ideas" of mathematics
3. Collect or create several examples of problem situations and questions about those problems that could be used to assess students' capabilities on this standard
4. Develop new scoring procedures to judge student performance
5. Report the results of student performance
6. Conduct an equity review of the entire process and its consequences.

What to Use in *Mathematics of Money with Algebra* Pupil Text

Lessons Each lesson uses applications that connect mathematics to many different aspects of the real world. The lessons also offer opportunities through questions and suggestions for students to expand their ideas in relation to the concepts being taught. In addition, each lesson uses many concepts and skills previously learned as a foundation in their quest to provide solutions to the "big idea."

Projects Projects included at the end of each chapter in the pupil text provide opportunities for your classroom to use performance-based assessment. Most are open-ended–allowing for multiple responses and many require different types of presentation and response modes. Many are interdisciplinary and allow students to select examples of best work to include in the portfolio. The duration of each project varies according to the difficulty and complexity of the task.

Chapter Openers Chapter openers provide an overview of the topics considered in the chapter. They can be used to help students expand their ideas in relation to the topics presented.

What to Use in *Mathematics of Money with Algebra* Teacher's Annotated Edition

Alternative Assessments Suggestions for projects and investigations appear in the notes for almost every lesson. The time involved with these projects is not as great as that involved in the pupil text project; however, these smaller projects offer an opportunity to assess students thinking, understanding, and work habits on a daily basis.

Enrichment and Reteaching Many enrichment suggestions can be used as performance-based tasks, and some reteaching suggestions might be used for those students who do not yet have the necessary skills to complete the enrichment suggestions and need positive reinforcement for those skills they can master.

Cooperative Learning Suggestions for group work occur in the notes for many lessons.

SCORING

For performance-based assessment, you can use scoring criteria, a rubric, along with some previously scored examples known as "anchors." A generic rubric is shown below. You will want to customize it for each task so that fair assessment can made in specific cases.

Generic Rubric

4 or 4+ Accomplishes or exceeds the task
This score indicates that the response by the student accomplishes the intended purpose. The student has used strategies and skills that will solve the problem; however, not all answers must be the same. A 4+ indicates a perfect response. A 4 contains only minor defects that students can easily fix without written feedback or additional teaching. Answers should be judged not on length, but rather on how effective they are when viewing the "big idea."

3 Ready for Needed Revision
Only written feedback from the teacher is needed for the student to revise the task. Additional dialogue or additional teaching is not required.

2 Partial Success
Only part of the task is completed. Students will probably need additional teaching and more discussion involving the task in order to revise their work.

1 Little Success
Even though the student tried to solve the problem, there is little success. Perhaps they didn't understand the problem or they might have used inappropriate skills and strategies. Nevertheless, additional teaching and remediation are needed.

0 No Response or Off Task
Whatever the cause the student did not complete the task successfully.

There are four basic scoring levels: 4, 3, 2, 1. There are two variations: a plus sign used to show work that was over and above what was expected and a zero used when there was no response or students were off task.

When scoring a paper, first divide responses into two categories:

those complete or ready for revision
(3 or greater)

those that indicate students need more instruction
(2 or less)

First decide which are ready for revision.	Then use the rubric to assign each task a score.	
Complete or ready for revision	+	Distinction
	4	Accomplishes the task
	3	Ready for revision
Needs further instruction	2	Partial success
	1	Little success
	0	No response or off task

MANAGEMENT

It is important to establish how many and what types of performance-based tasks should appear in every student's portfolio and communicate this clearly to your students at the beginning of the term. Criteria for an appropriate mathematics portfolio include tasks that are both open-ended and curriculum based, as well as those tasks that connect mathematics to other disciplines and to the real world and others that encourage and enable students to move beyond the task.

The number of pieces per grading period depends upon the type of work included in the portfolio as well as the length of the grading period itself.

You may wish to have students choose some of their best work based on the parameters you have devised to include in a portfolio, thus encouraging the process of self-awareness and self-assessment. Encourage students to take pride in their portfolio and continually challenge them to include their best work.

REFERENCES

Barnes, John. *How to Learn Basic Bookkeeping in Ten Easy Lessons.* New York: Harper & Row. 1978.

Beusterein, Pat ed. *Summer Employment Guide of the United States.* Cincinnati: Writer's Digest Books. 1990.

Bolles, Richard Nelson. *What Color is Your Parachute? A Practical Manual for Job-Hunters and Career Changers.* Ten Speed Press. 1985.

Cohen, William A. *The Entrepreneur and Small Business Problem Solver.* Wiley. 1983.

de Blaye, Edouard. *Fromer's Dollarwise USA.* New York: Prentice Hall. 1990.

Dorfman, Mark S. and Saul W. Adelman. *The Dow Jones-Irwin Guide to Life Insurance.* Homewood, IL: Down Jones-Irwin. 1988.

Douglas, Martha C. *Go For it: A Career-Planning Guide for Young Adults.* Ten Speed Press. 1983.

Dyer, Mary Lee. *Practical Bookkeeping for the Small Business.* Contemporary Books. 1976.

Editors of Consumer Guide. *Your Retirement.* A&W Publishers. 1981.

Estes, Jack. *Compound Interest and Annuity Tables.* New York: McGraw-Hill Paperback. 1976.

Feingold, S. Norman and Norma Reno Miller. *Emerging Careers: New Occupations for the Year 2000 and Beyond.* Garret Park Press. 1983.

Fucini, Joseph J. and Suzy Fucini. *Entrepreneurs: The Men and Women behind Famous Brand Names and How They Made It.* Boston: G.K. Hall & Co. 1986.

Goldstein, Jerome. *How to Start a Family Business and Make It Work.* Evans. 1984.

Haldane, Bernard, and others. *Job Power: The Young People's Job-Finding Guide.* Acropolis Books. 1980.

Milton, Arthur. *How Your Life Insurance Robs You.* New York: Carol Publishing Group. 1990.

Mitchell, Joyce Slayton. *See Me More Clearly: Career and Life Planning for Teens with Physical Disabilities.* Harcourt, Brace, Jovanovich. 1980.

Montgomery, Richard H. *Home Energy Audit: Your Guide to Understanding and Reducing Your Home Energy Costs.* Wiley. 1983.

Porter, Sylvia. *Sylvia Porter's New Money Book for the 80's.* New York: Doubleday & Company. 1979.

Quinn, Jane Bryant. *Everyone's Money Book.* Delacorte. 1979.

Rooney, Andy. *"Cash Standard," "Banks and Jesse James," and "Paying Bills" in Pieces of Mind.* New York: Avon Books.

Rooney, Andy. *"Places Not to Go on Vacations"* and *"Some Thoughts about Vacations,"* in *Word for Word.* New York: Berkeley Publishing Group.

Samtur, Susan J. and Tad Tuleja. *Cashing In at the Checkout: The New and Incredibly Easy System* . New York: The Stonesong Press. 1979.

Shanahan, William F. *College - Yes or No.* Arco. 1980.

Sobel, Robert and David B. Sicilia. *The Entrepreneurs: An American Experience.* Boston: Houghton Mifflin Company. 1986.

United States Bureau of Labor. *Occupational Outlook Handbook: Information on Major Occupations for Use in Guidance.* U.S. Government Printing Office. 1985.

Woelfel, Charles J. *Desktop Guide to Money, Time Interest, and Yields.* Chicago: Probus Publishing Company. 1986.

"Annual Roundup for New Car Buyers" in *Consumer Reports Buying Guide.*

Best Western International 1990 Road Atlas and Travel Guide. Best Western International, Inc.

Circular E - Employer's Tax Guide. Internal Revenue Service.

Consumer Action's Auto Insurance Guide. Consumer Action. San Francisco, CA.

Consumer Reports Money-Saving Guide to Energy in the Home. Doubleday. 1978.

Edmund's New Car Prices. Edmund's Foreign Car Prices. Edmund's Used Car Prices. Edmund Publications Corporation.

Energy Alternatives. Time-Life Books. 1982.

How to Set Up Your Own Business. American Institute of Small Business. 1984.

Investor's Information Kit. New York Stock Exchange Publication Department. 1979.

Rand McNally Road Atlas & Travel Guide. San Francisco: Rand McNally & Company.

Reader's Digest Consumer Advisor: An Action Guide to Your Rights. Pleasantville, NY: Reader's Digest Association.

Tax Guide for Teachers. Tax Guide for College Teachers. Academic Information Service, Inc.

The Federal Reserve System: Purposes and Functions. Board of Governors of the Federal Reserve, Washington, D.C. 1963.

The Language of Investing and How to Get Help When You Invest. New York Stock Exchange Big Board's basic educational pamphlets.

1989 NASDAQ Fact Book. National Association of Securities Dealers, Inc. NASD Corporate Communications. 1989.

SOUTH–WESTERN

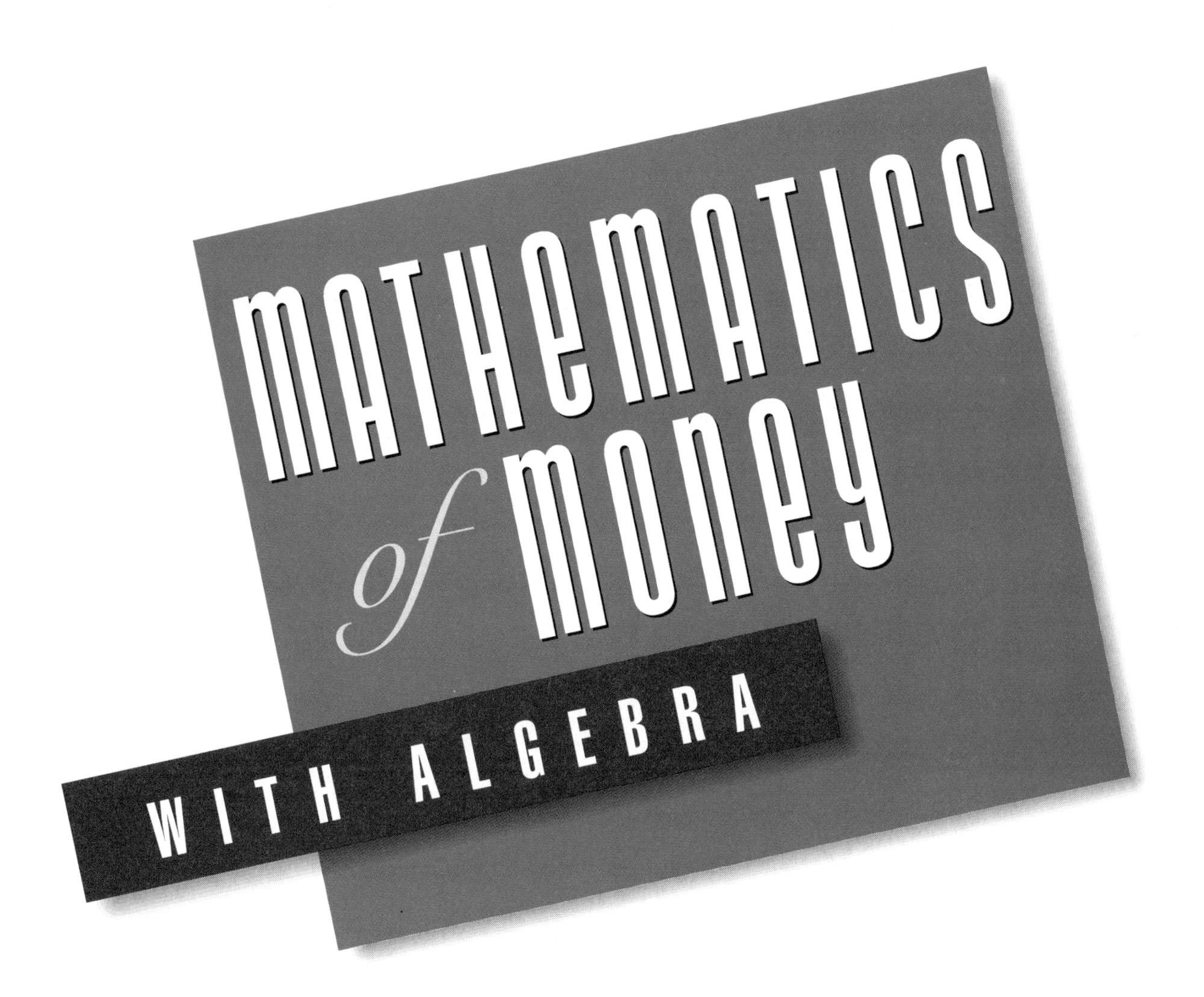

Cheryl Clayton
Richardson High School
Richardson, Texas

SOUTH-WESTERN PUBLISHING CO.

Managing Editor: Eve Lewis
Developmental Editor: Enid M. Nagel
Production Editor: Shannon O'Connor
Coordinating Editor: Patricia Matthews Boies
Associate Photo Editor/Stylist: Fred M. Middendorf
Marketing Manager: Carol Ann Dana
Design: Pronk & Associates
Production Services: PC&F, Inc.

ISBN: 0–538–63471–5
2 3 4 5 6 7 8 KI 01 00 99 98 97 96 95 94
Printed in the United States of America

I(T)P
International Thomson Publishing

ABOUT THE AUTHOR

Cheryl Clayton, M.Ed.

Since 1979, the author has taught at Richardson High School, in Richardson, Texas where she teaches the popular mathematics of money course on which this book is based. Ms. Clayton's career has included teaching fundamentals of mathematics, algebra, geometry, trigonometry, analytic geometry and consumer mathematics at the junior high and high school levels. She received a Bachelor's degree from Texas Christian University and a Master's degree in Education from North Texas State University. Ms. Clayton is married, has two sons, and lives in Dallas.

REVIEWERS

Korita M. Azopardi
Chairperson
Mathematics Department
Calallen High School
Corpus Christi, Texas

George Benedetti
Business Education Instructor
West Valley High School
Yakima, Washington

Deborah D. Carruthers
Teacher, Computer Science
and Mathematics
Ross S. Sterling High School
Baytown, Texas

Amy Davidson
Mathematics Teacher
Westfield High School
Houston, Texas

Arne Engebretsen
Chair, Mathematics and
Computer Science
Greendale High School
Greendale, Wisconsin

Bruce E. Engelhard
Mathematics Teacher
Newport Alternative School
Newport, Rhode Island

Mary Hansen
Mathematics Teacher
South View Senior High School
Hope Mills, North Carolina

Brenda Beck Lynch
Mathematics Teacher
Willis Independent
School District
Willis, Texas

Shirley D. Oliver
Mathematics Teacher
Tomball High School
Tomball, Texas

Randy L. Pippen
Mathematics Department Chair
Lisle High School
Lisle, Illinois

Charles W. Tillerson
Chair, Department of
Mathematics
Highland Park High School
Dallas, Texas

Thurman Watson
Mathematics Teacher
Robert E. Lee High School
Baytown, Texas

CONTENTS

CHAPTER 9

CHAPTER 10

CHAPTER 11

CHAPTER 12

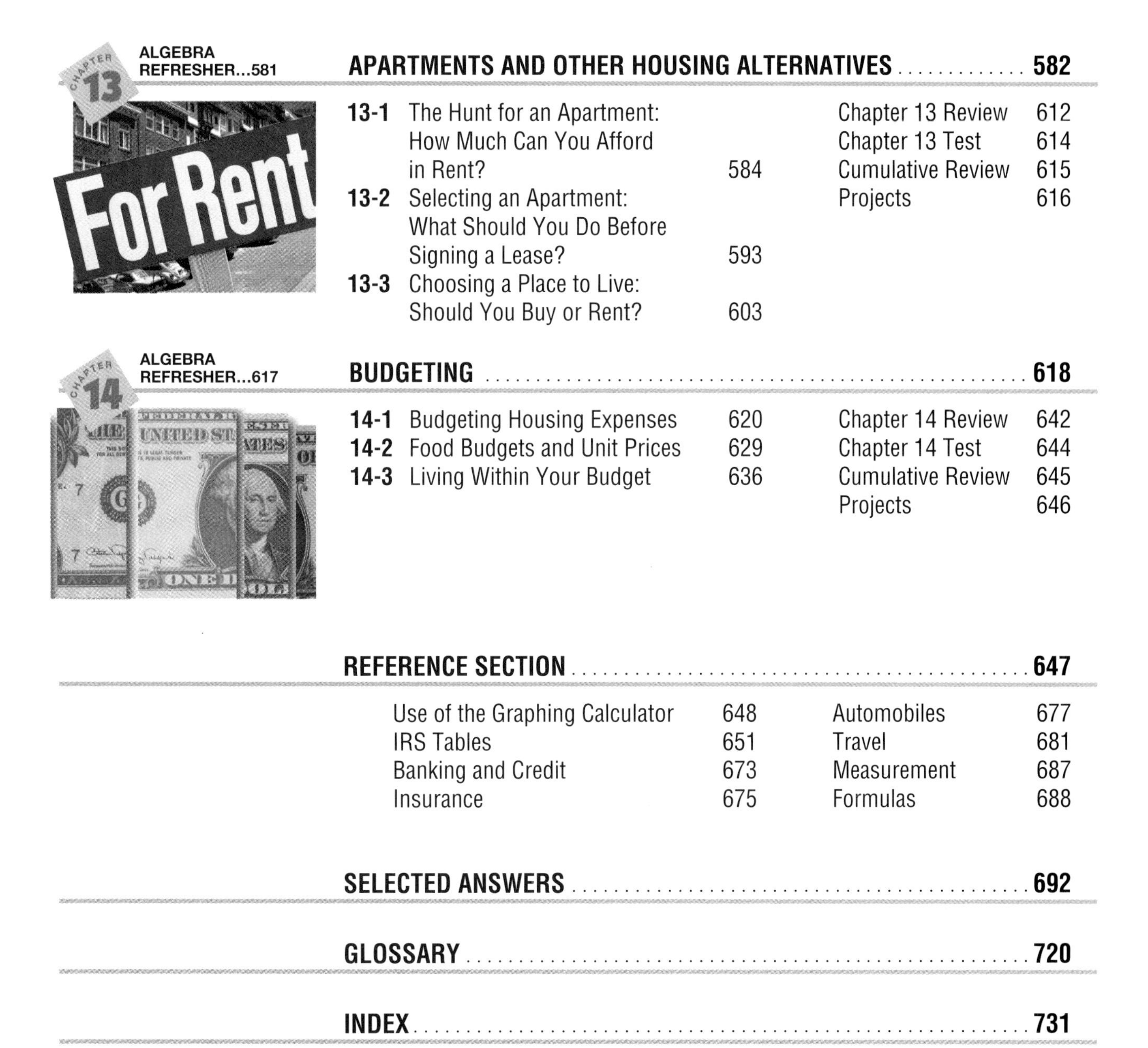

CHAPTER
13
For Rent
CHAPTER
14

LETTER TO THE STUDENT

Many of you are entering the job market even before completing high school. Some of you are finding the realities of acquiring, spending, and saving money to be a challenging and sometimes overwhelming task. ***Mathematics of Money With Algebra*** will use and extend the skills you learned in Algebra I to help you manage your own finances.

This program encourages your active involvement by constructing and applying mathematical ideas through business and consumer themes. Since decision making is a major thrust, you are encouraged to use calculators and computers as tools for learning and doing mathematics.

The readings in ***Mathematics of Money With Algebra*** give you sketches of life that might seem familiar. You, like the characters in the book, are approaching adulthood and are faced with many financial decisions. This program offers assistance in financial planning so that the choices you make to establish an independent life-style will be informed choices.

Mathematics of Money With Algebra contains information you can use now and in the future. For example, when buying a car, you are encouraged to consider the cost of maintaining a car as well as the purchase price. By including up-to-date data and common situations, this text will help you gain confidence in your mathematical abilities as you solve problems which affect you and your family every day.

In order to use ***Mathematics of Money With Algebra*** completely and effectively, you must be able to use the basic skills of algebra automatically. This text is designed to help you succeed because your Algebra I skills will be completely reviewed in the special features *Algebra Refresher* and *Algebra Review.*

Graphing calculators and computer spreadsheet applications show you how graphs can communicate concepts about data more effectively. You will have opportunities to work with other students and learn more about businesses in your community as you complete the projects at the end of each chapter.

A final word of encouragement: Your knowledge of a variety of techniques to approach and solve problems will be greatly enhanced in this course if you read carefully and enjoy the challenge of mathematical problem solving. When you value mathematics, doors open to a variety of related disciplines and careers.

ALGEBRA *REFRESHER*

Simplify each expression. When you simplify a *numerical expression,* simplify within parentheses first. Then do all multiplications and divisions from left to right followed by all additions and subtractions from left to right.

Example $6.3 + 0.4(2.7) - (6.3 - 2.7 + 1)$

$= 6.3 + 0.4(2.7) - 4.6$ Simplify within parentheses first.

$= 6.3 + 1.08 - 4.6$ Multiply before adding and subtracting.

$= 7.38 - 4.60$ Add and subtract from left to right.

$= 2.78$

1. $5.1 + (11.4 - 6)$
10.5

2. $9.7 - (12.5 - 10)$
7.2

3. $3 \bullet 4.2 - 0.1 \bullet 8$
11.8

4. $0.06 \bullet 850$
51

5. $0.06 \bullet 850 + 0.45 \bullet 730$
379.5

6. $0.765 \bullet 5760 + 0.145 \bullet 13{,}500$
6363.9

Each of the following *algebraic expressions* represents the product of 7 and a number *h*.

$7h$ $7 \bullet h$ $7(h)$ $(7)(h)$

Evaluate an expression by replacing the variable by a number. Then simplify. Evaluate each expression for $h = 6.7$ or $x = 8.6$.

Example $6.2 \div 2 + 8h$

$= 6.2 \div 2 + 8(6.7)$ Substitute 6.7 for *h*.

$= 3.1 + 53.6$ Divide and multiply first, then add.

$= 56.7$

7. $x - 3.6$
5

8. $2 \bullet h + 4$
17.4

9. $2 + 4(h)$
28.8

10. $2 + 3(10 - x)$
6.2

11. $350 \bullet h \bullet 15$
35,175

12. $(1 + x \div 100)(1.5)$
1.629

Solve each *equation* for the variable. To solve an equation, ask yourself what operation was performed on the variable. Then ask how you can undo the operation. Remember to always keep the equation in balance.

Example $7h - 316 = 594$ 316 is subtracted from 7*h*.

$7h - 316 + 316 = 594 + 316$ Add 316 to each side.

$7h + 0 = 910$ $-316 + 316 = 0$

$7h = 910$ *h* is multiplied by 7.

$\frac{7h}{7} = \frac{910}{7}$ Divide each side by 7.

$1h = 130$ $7 \div 7 = 1$

$h = 130$ $1h = h$

13. $13x = 143$
$x = 11$

14. $0.055P = 660$
$P = 12{,}000$

15. $1116 = t - 261$
$t = 1377$

16. $0.7x + 100 = 37$
$x = -90$

17. $12x - 59 = 841$
$x = 75$

18. $5.2e + 498 = 2500$
$e = 385$

CHAPTER 1

PERSONAL EARNING POWER

IN THIS CHAPTER YOU WILL MEET ALEX and his friends. These young people in their late teens are fictitious, but the decisions that they face are very real. Every year, real young people make choices similar to those that you will see Alex and his friends making.

You will see how Alex carefully considers the balance between the time and effort required in taking a part-time job and the financial rewards of working. He sees how the information that he reads in newspaper job ads translates into a paycheck for regular and overtime hours.

Daphne, a year younger than Alex and his classmates, watches her older brother Darrin make career decisions and begins to think about making her own. She learns about sources of help in exploring the world of work. She also learns how a person's income is enhanced by company benefits other than pay.

Alex's friend Betty will help us compare the consistent income of a salaried job with the variable income of a job paid by commission. A salesperson on a commission may make recommendations to the customer that differ from those a salaried salesperson would make. Betty tells us why.

Another friend, Luis, will see how and why deductions are made from his paycheck. His first paycheck from the grocery store where he works comes as a surprise to Luis. He had expected to receive his full hourly rate, but he received considerably less.

Alex, Daphne, Betty, and Luis look forward to the time when they will no longer have all their decisions made for them. They are glad to begin taking charge of their own lives. At the same time the importance of some of those decisions is becoming clear to them. Accepting the responsibility for their choices is not easy.

1–1 Salaries: Making Money Takes Time

1–2 Employment Opportunities: So Many Choices

1–3 Do More, Make More: Commissions and Payments by Item

1–4 Deductions: Who Gets What?

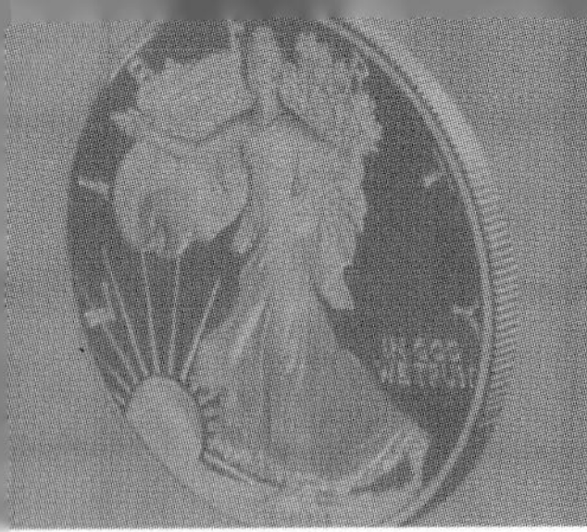

1-1 SALARIES: MAKING MONEY TAKES TIME

QUICK REFERENCE

TEACHER'S RESOURCES AND ANSWER KEY

Lesson Quiz Answers 1-1, p. 38
Reteaching Activity Answers 1-1, p.91
Enrichment Activity Answers 1-01, p. 92

EXTENSION ACTIVITIES

Reteaching Activity 1-1, p. 1
Enrichment Activity 1-1, p. 5

Alex and his classmates are entering their last year of high school. Many of his friends have already had summer jobs. Some have started saving money for college or for other schooling.

Alex's parents have been asking him what he plans to do after graduation. His girlfriend, Alice, is planning to go to State University next fall. He hates the thought of being away from Alice for a long time. They plan to get married after she gets her law degree, but that could take a while. If Alex goes to State to be with Alice, he may decide to study accounting, but he is not really handy with numbers. What Alex really enjoys is tinkering around with the van that his parents bought him when he learned to drive.

Alex could use some extra cash, since the van needs a few parts and he would like to take Alice to the Senior Prom. Alex's older sister Karen rode to the prom last year in a chauffeured limousine—Alex can just imagine what that must have cost!

The $483 that Alex saved from mowing lawns this summer is almost gone. Alex has been thinking about getting a job after school, partly to earn extra cash but also to get a taste of what the world of work will be like. If he can find a way to choose a career that he would really like, he may decide that he needs more training. Then getting a job might bring in the extra money to pay for it.

OBJECTIVES: *In this lesson, we will help Alex to:*

- *Compute the earnings from a part-time job.*
- *Compare an hourly wage with a weekly or monthly salary.*

EARNING INCOME

Alex already knows a few facts about money. He knows that his family's standard of living is determined by the amount of income his parents earn. **Income** is the money received from investments or from a person's activities at work or in business. His family, like most American families, depends on its labor for its primary income. Approximately nine out of ten American workers earn income in the form of wages or salary. A **wage** is an hourly or daily rate of pay; a **salary** is a weekly, monthly, or yearly rate of pay. Most clerical workers and people who do physical labor are wage earners. Professional and technical workers are usually salary earners. About 10% of all workers are in business for themselves and earn **self-employment income** instead of wages or salaries. These include such people as physicians, shopkeepers, writers, photographers, and farmers.

You may have wondered why a few baseball players have actually been paid over one million dollars for a season while the President of the United States earns $200,000 a year or why an actor who becomes a star earns far more than a professional accountant or teacher. Income from labor reflects social values, the monetary worth of the labor, and the demand for the labor in relation to the number of people who can do the job.

Alex began examining the classified ads in the employment section of the Sunday newspaper. He did not see any ads offering million-dollar salaries to baseball players. He did find, however, that part-time jobs available to students pay a specific amount for each hour worked. That amount is an **hourly rate.** The hours the employee is required to work each week are the employee's **regular hours.**

Some weeks an employee will be asked to work more than the required amount of hours. These extra hours are called **overtime hours** or simply overtime. The hourly rate for these overtime hours is generally greater than the regular hourly rate. Salaried employees receive the same salary each pay period whether they work the minimum number of hours required or put in extra time during weekends and evenings.

FOCUS

ALGEBRA CONNECTIONS

Ask students which, if any, of the following expressions represent an overtime rate of $1\frac{1}{2}$ times the regular rate of pay, given that r represents the regular rate.

$r + \frac{1}{2}r$ $1.5r$ $1\frac{1}{2}r$

$r + 1\frac{1}{2}$ $\frac{3}{2}r$

(all except $r + 1\frac{1}{2}$)

Ask Yourself

1. If an employee is paid by the hour and works more than the required number of hours in a week, the employee is generally paid more for the extra hours. What are the extra hours called?
   overtime
2. If an employee is paid a monthly salary, will the employee be paid more for working on the weekend?
   no
3. In which kinds of jobs do workers earn hourly or daily wages?
   Answers may vary. (Ex.: clerical, physical labor)

ALGEBRA *REVIEW*

Evaluate the expression $rh + t$ for the given values of h, r, and t.

1. $h = 5$, $r = 4.4$, $t = 7$
   29
2. $h = 8.5$, $r = 5.60$, $t = 11$
   58.6
3. $h = 19$, $r = 6$, $t = 23.50$
   137.5
4. $h = 12$, $r = 5.80$, $t = 14$
   83.6

Evaluate the expression $20r + (h - 20) \cdot 1.5r + t$ for the given values of h, r, and t.

5. $h = 21$, $r = 6.20$, $t = 34.50$
   167.8
6. $h = 27$, $r = 4.00$, $t = 16.00$
   138
7. $h = 20$, $r = 5.15$, $t = 17.50$
   120.5
8. $h = 30$, $r = 5.50$, $t = 20$
   212.5

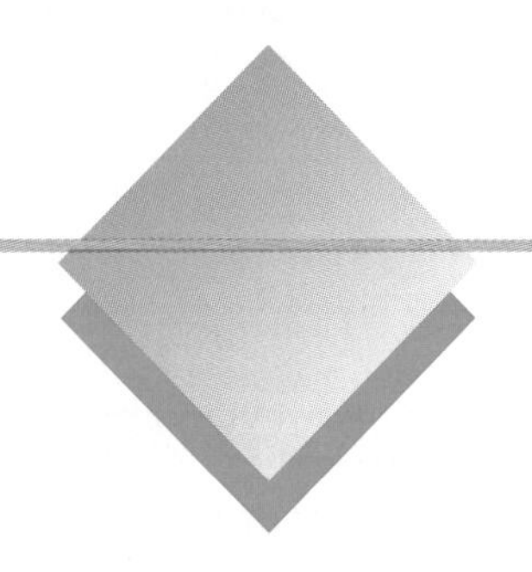

SHARPEN YOUR SKILLS

SKILL 1

Alex's friend Ed delivers pizzas. Ed receives tips from customers if the pizza is still hot and if the order is properly filled.

Ed's earnings can be expressed by the following equation.

Earnings = hourly rate • number of hours + tips

$E = rh + t$ where r = hourly rate
h = number of hours
t = tips

EXAMPLE 1 During the first week of February, Ed worked $16\frac{1}{2}$ hours and received $28.50 in tips. His hourly rate is $3.40.

QUESTION How much did Ed earn that week?

SOLUTION

$E = rh + t$ Substitute values in the equation.
$E = 3.40 \cdot 16.5 + 28.50$ $r = 3.40$, $h = 16.5$, $t = 28.50$
$E = 84.60$ Use your calculator.

Ed earned $84.60 that week.

SKILL 2

EXAMPLE 2 If Ed works more than 20 hours in one week, he receives $1\frac{1}{2}$ times his regular wage rate of $3.40 for each of the extra hours. During the second week of February there was an ice storm. Many people ordered pizza to be delivered. That week, Ed worked 26 hours and received $43.75 in tips.

QUESTION How much money did Ed make that week?

SOLUTION
Ed's earnings that week is the sum of his regular earnings, his overtime earnings, and his tips.

Overtime hours = total hours − regular hours = $h - 20$
Total earnings = regular earnings + overtime earnings + tips

Total earnings = number of regular hours • hourly rate + overtime hours • overtime rate + tips

$E = 20r + (h - 20) \cdot 1.5r + t$ The overtime rate is $1\frac{1}{2}$ or 1.5 times the regular rate.
$E = 20 \cdot 3.4 + (26 - 20) \cdot 1.5 \cdot 3.4 + 43.75$
$E = 142.35$ Use your calculator.

Ed earned $142.35 that week.

INSTRUCTION

TEACHING THE LESSON
Have the students read the opening scenario in order to understand the nature of the choices that students such as Alex face as they decide whether to continue their formal education or to concentrate on the practical question of earning a living. Before working with the equations of Skills 1 and 2, students should do the *Algebra Review* exercises that provide practice for those equations.

SKILL 3

EXAMPLE 3 Alex wondered how salaries stated by the week, month, and year compare with each other and with hourly wages. He devised a plan to find out by using his computer **spreadsheet program.**

QUESTION How much are the hourly, weekly, monthly, and yearly salaries of each of the jobs shown in these advertisements?

1.

RECEPTIONIST must have good phone skills. $6.50/h to start. Hours: 8:30am–5pm. Equal Oppty Employment. Call Monday–Thursday only.

2.

COMMUNITY WORK Start a new career with the state's oldest consumer group. Great future! No experience nec. Paid training, holidays, vacations. Up to $350 per wk.

3.

ADMINISTRATIVE ASSISTANT Social service agency seeks Admin. Asst. for data entry. Must have exp. with dBase III systems. Typing 40 wpm. Sal. $15 K + benefits.

CRITICAL THINKING

Students will be familiar with the concept of being paid "time-and-a-half" for overtime. Ask them why they think an employer might be willing to offer this option to an employee rather than hire a new employee at the regular rate.

SOLUTION

Alex knows that he could quickly compute the answers using his calculator. However, he is trying to learn how to use his computer spreadsheet program, which can do many calculations quickly if he sets up the spreadsheet properly.

Open the spreadsheet program to a new blank spreadsheet filled with blank rows and columns. The intersection of a *row* and a *column* is called a **cell.** A cell is labeled by the column and row that forms it. Begin by entering the headings Hourly, Weekly, Monthly, and Yearly in cells A1, B1, C1, and D1.

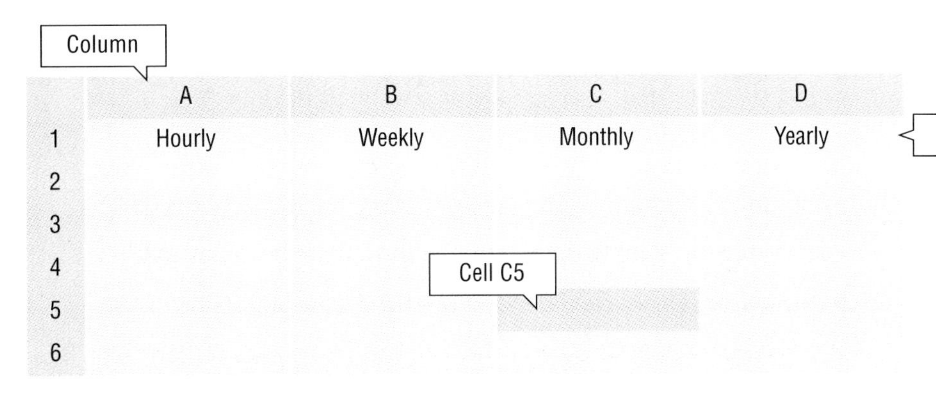

AT-RISK STUDENTS

Generate interest in the matter of actual hourly rates by having students conduct a survey among employers in the community. Have the students ask about hourly rates that employers pay and the number of hours that must be reached before an employee may be paid $1\frac{1}{2}$ times his or her regular rate.

Alex enters the salaries that he knows and makes the computer calculate the others by using formulas.

Enter $6.50, the hourly salary for the receptionist, in cell A2.

For the formula for the weekly salary, use the hourly salary, which is located in cell A2. Multiply it by 40, since there are 40 hours in a workweek.

Enter +A2*40 in cell B2.

TECHNOLOGY HINT

Skill 3 introduces the use of computer spreadsheet programs to facilitate the organization and processing of data. Point out that Alex can generate weekly, monthly, and yearly rates for as many hourly rates as he wishes starting on line 3 by just typing in the hourly rates in column A and then using the spreadsheet's COPY command to generate the appropriate results for columns B, C, and D. Point out that the spreadsheet program automatically adjusts the formulas to reflect the variables for the current row. Thus, in row 3, the COPY command will change the formula +A2*40 to +A3*40, and similarly for the other formulas.

FOCUS ON ALGEBRA

Compare the notations used in the spreadsheet formulas to those used in standard algebra, pointing out that an expression such as "B2," which specifies a location for a number, plays much the same role as an algebraic variable. *(NCTM Standard 4, p. 146; Standard 5, p. 150)*

COMMON ERRORS

Students who have not recently worked with multiple operations may need to be careful about the order of operations. For a quick check on this skill, have students do *Algebra Refresher* Exercises 1-6.

For the formula for the yearly salary, use the weekly salary, which is located in cell B2. Multiply it by 52, since there are 52 weeks in a year.

Enter +B2*52 in cell D2.

Since there are 12 months in a year, divide the yearly salary in cell D2 by 12.

Enter +D2/12 in cell C2.

40 hours = 1 workweek
52 weeks = 1 year
12 months = 1 year

Remember that * is a symbol for multiplication and / is a symbol for division.

Alex would like his answers rounded to two decimal places, since they represent money. Highlight the portion of the spreadsheet to be formatted and choose the option for fixed format of two decimal places. Although the spreadsheet will then display the numbers rounded to two decimal places, the computer will retain the numbers to many decimal places in its memory and use these values in computations.

	A	B	C	D
1	Hourly	Weekly	Monthly	Yearly
2	6.50	260.00	1126.67	13520
3		+A2*40	+D2/12	+B2*52
4				

Enter $350, the weekly salary for community work, in cell B3.

To determine the hourly salary, use the formula +B3/40 in cell A3.

To determine the yearly salary, use the formula +B3*52 in cell D3.

For the monthly salary, use the formula +D3/12 in cell C3.

Enter $15,000, the yearly salary for the administrative assistant, in cell D4.

To determine the monthly salary, use the formula +D4/12 in cell C4.

To determine the weekly salary, use the formula +D4/52 in cell B4.

For the hourly salary, use the formula +B4/40 in cell A4.

	A	B	C	D
1	Hourly	Weekly	Monthly	Yearly
2	6.50	260.00	1126.67	13520
3	8.75	350.00	1516.67	18200
4	7.21	288.46	1250.00	15000

The community work is the highest-paying job, and the receptionist is the lowest-paying job.

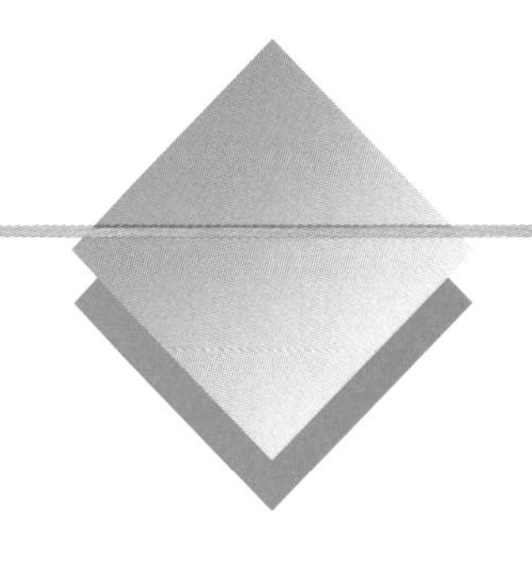

TRY YOUR SKILLS

Write an equation of the form $E = rh + t$ where r represents the regular hourly rate, h represents the number of hours worked, and t represents any tip or bonus. Then find the weekly earnings for each of the following "employees."

1. I am the tape recorder in the French language lab. I make $11.50 per hour, and I worked $29\frac{1}{2}$ hours this week. I received a bonus of $3.50 when I taught the students how to say "I love you" in French. $342.75
2. I am the button that turns on the light when you open the refrigerator door. My door was open $22\frac{1}{2}$ hours this week. My salary is $10.00 per hour, and I did not receive a bonus because I was on when my owner knocked over a bottle of juice onto his toe. $225.00

3.–4. Assume that the employees in Exercises 1 and 2 received overtime pay of $1\frac{1}{2}$ times their regular rate for all hours they worked over 20. Write an equation to represent their weekly earnings of the form

$$E = 20r + (h - 20) \cdot 1.5r + t$$

where r represents the regular hourly rate, h represents the number of hours worked, and t represents any tip or bonus. Determine the total earnings for each employee. **3.** $397.38 **4.** $237.50

Create a spreadsheet to calculate the hourly, weekly, monthly, and yearly earnings for each of the jobs described below.

40 hours = 1 workweek
52 weeks = 1 year
12 months = 1 year

5.

HEALTH CARE Better Life Health Care is looking for a respiratory therapist. Earn as much as $17.85/h.

Weekly: $714
Monthly: $3,094
Yearly: $37,128

6.

Mold Maker
Jr. MOLD MAKER We pay as high as $8.50/h after 1 year.

Weekly: $340
Monthly: $1,473.33
Yearly: $17,680

7.

Data Processing Computer Operator
Must know WordPerfect and want to learn other programs. $700/wk.

Hourly: $17.50
Monthly: $3,033.33
Yearly: $36,400

GUIDED PRACTICE

Remind students that they will sometimes need to use the box that contains information such as the length of a workweek. It is included, not so much as a prescription of a "normal" work calendar, but rather as a means of assuring some uniformity in the answers that students obtain for the exercises.

RETEACHING

Students who have difficulty with algebraic settings may be more successful with an arithmetic approach, perhaps combined with dimensional analysis (see below). Rework Example 1 without the formula, treating each term of the formula as a single operation, multiplication first followed by addition.

ENRICHMENT

Some students may benefit from an introduction to dimensional analysis. For example, the equation of Skill 1 can concentrate on the wages portion of the earnings in the form

$$\frac{\text{dollars}}{\text{hours}} \cdot \text{hours}$$

which simplifies to dollars by "canceling" the two hour(s) labels. This is a standard "trick" that scientists routinely use when dealing with physical formulas.

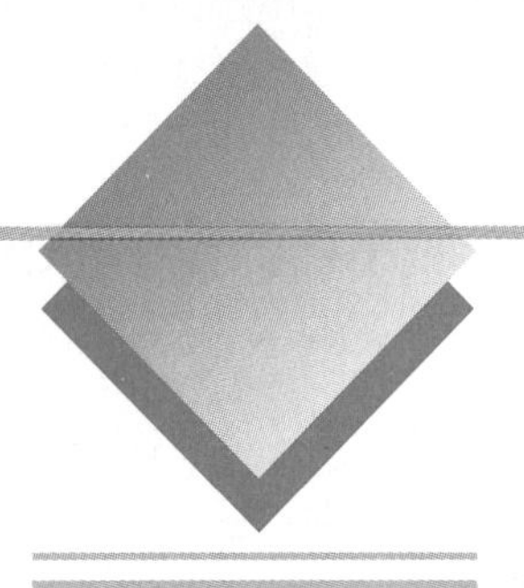

EXERCISE YOUR SKILLS

See Additional Answers.

1. What is the major difference between the way earnings are computed for salaried employees and for hourly employees?
2. Why would an employer be willing to pay extra money for overtime hours?
3. Why would salaried employees be willing to work extra hours if their salary remained the same?
4. Why do many professional baseball players earn more money each year than the President of the United States?
5. If a student takes a part-time job, what do you think is the maximum number of hours the student should be allowed to work per week? Be sure to take into consideration the time need to keep up with school work.

KEY TERMS

cell
hourly rate
income
overtime hours
regular hours
salary
self-employment income
spreadsheet program
wage

Write an equation of the form $E = rh + t$, and find the weekly earnings for each of the following "employees."

6. I am the pay telephone in the hall by the principal's office. I work all the time but got paid $12.00 per hour for 29 hours this week. I got a $25.00 bonus from a guy whose girlfriend told him through me that she had decided not to break up with him. $373.00
7. I am the gate that goes up and down at the tollbooth. I made $11.50 per hour for $29\frac{1}{2}$ hours of work this week and received a $0.50 bonus from a taxi driver who got past me before I could close after the car in front of him. $339.75
8. I dispense tokens for video games at Showbiz. I eat $5 bills and spit out 20 tokens at a time. I worked 27 hours last week at $13.00 per hour and received an extra $1.00 when the tokens got stuck and a little boy kicked me. $352.00
9. I am the coffee machine in the teachers' lounge at the high school. I work many extra hours each week keeping these folks awake, but I was paid for only $26\frac{1}{2}$ hours at $11.00 per hour. I got no bonus because I ran out of coffee during the last staff meeting. $291.50
10. I am the golf cart that Karl pulls around on the golf course. I carry all his clubs and his lucky red rabbit's foot that his brother sold to him when he lost his. I worked 31 hours this week and made $14.00 per hour and received a $20.00 bonus when Karl made his first par on the number 2 hole at Keyton Park. $454.00

INDEPENDENT PRACTICE

ASSIGNMENT GUIDE

Exercises 1-5 are open-ended questions that aim to encourage students to reflect on the world of work into which they will soon enter. Exercises 6-17 provide practice in calculating earnings obtained from both wages and tips. In particular, Exercises 12-17 show the effect that "time-and-a-half" can have on overall income. Practice in computer spreadsheets is provided by Exercises 18-26.

11. I am the button that you push when you want to ride your bicycle on the bike path across the freeway. I make the "Walk" sign come on. I make \$12.50 per hour, and I worked $31\frac{1}{2}$ hours last week. I did not receive a bonus because I am not fast enough for the speed-racers. \$393.75

12.–17. Assume that the "employees" of Exercises 6–11 received overtime pay of $1\frac{1}{2}$ times their regular rate for all hours they worked over 20. Write an equation to represent their weekly earnings of the form

$$E = 20r + (h - 20) \cdot 1.5r + t$$

where r represents the regular hourly rate, h represents the number of hours worked, and t represents any tip or bonus. Determine the total earnings for each employee. See Additional Answers.

Create a spreadsheet to calculate the hourly, weekly, monthly, and yearly earnings for each of the jobs described below.

40 hours = 1 workweek
52 weeks = 1 year
12 months = 1 year

18.

Health Care
NURSES
RNs UP TO \$35/h.

Weekly: \$1,400
Monthly: \$6,066.67
Yearly: \$72,800

19.

Maintenance
MECHANIC Center Manuf. is in need of a person with 3–5 years exp. in machine repair/punch press. \$10.70/h.

20.

Sales!
Interesting inside sales position. Start at \$5.75/h.

21.

Hairdresser
English and Spanish speaking. Guaranteed pay \$300/wk. to start. Apply in person only.

22.

General Retail Clerk
Are you getting nowhere with that same old job? Earn \$600 per week. No exp. nec.

23.

DRIVERS
Drivers for local messenger. Earn \$450 per week. Must have own car and good driving record.

24.

Pharmacist
Unlimited opportunity with growing drug store chain. Starting salary \$33,000.

Hourly: \$15.87
Weekly: \$634.62
Monthly: \$2750

25.

Executive Asst.
Investment comp. seeks experienced and motivated individual to run office. Submit resume. From \$40K.

Hourly: \$19.23
Weekly: \$769.23
Monthly: \$3333.33

26.

CHIEF CHEMIST
Will pay up to \$55,000 for the right person with at least 15 years experience.

Hourly: \$26.44
Weekly: \$1057.69
Monthly: \$4583.33

CLOSURE

Have students discuss the advantages and disadvantages of different forms of compensation, such as hourly wages, yearly salaries, and tip income. Why are some jobs paid on an hourly basis and others on a weekly basis?

LESSON QUIZ

1. Jason earns \$6.30 an hour for the first 25 hours of the week and then earns $1\frac{1}{2}$ times his regular rate. How much does he earn in 30 hours? **(\$204.75)**
2. Jason's employer pays 2 times the regular pay rate for work on Sundays. How much would Jason earn if 3 of his 30 hours were on Sunday? **(\$214.20)**

ADDITIONAL ANSWERS

12. \$427.00
13. \$394.38
14. \$397.50
15. \$327.25
16. \$531.00
17. \$465.63
19. Weekly: \$428
    Monthly: \$1,854.67
    Yearly: \$22,256
20. Weekly: \$230
    Monthly: \$996.67
    Yearly: \$11,960
21. Hourly: \$7.50
    Monthly: \$1,300
    Yearly: \$15,600
22. Hourly: \$15.00
    Monthly: \$2,600
    Yearly: \$31,200
23. Hourly: \$11.25
    Monthly: \$1,950
    Yearly: \$23,400

1-2 EMPLOYMENT OPPORTUNITIES: SO MANY CHOICES

QUICK REFERENCE

TEACHER'S RESOURCES AND ANSWER KEY
Lesson Quiz Answers 1-2, p. 38
Reteaching Activity Answers 1-2, p. 91
Enrichment Activity Answers 1-2, p. 92

EXTENSION ACTIVITIES
Reteaching Activity 1-1, p. 2
Enrichment Activity 1-1, p. 6

Daphne is one year younger than Alex. Her older brother Darrin has had trouble finding a career. Daphne and Darrin have had long talks about "life after high school," and Daphne hopes to learn from Darrin's experience.

Ever since he and his family visited Cape Canaveral when he was 11, Darrin's ambition was to become an astronaut. When he knew that his eyesight was not good enough for astronaut training, he decided to try to participate in the space program in some other way.

He went to State University to major in aeronautical engineering. About mid-October, he began to wish that he had not skimmed through the science and mathematics courses he had taken in high school. Actually, he had taken them only because his friends had.

After only a year at State he came home and enrolled in a community college. Then he dropped out of college, took a job, and moved into an apartment. He had planned to save his money and return to State, but he discovered that living in an apartment was too expensive for him. He moved back home.

Daphne is glad to have Darrin at home but is troubled by the upsetting effect that all the changes have had on the family. She would like to have a clearer picture of what is waiting "out there" before she wastes a lot of time and money. She wants to train for a career that interests her and for which she has the required skills. She also wants to have some reason to believe that she will be able to find a job at the end of her training.

OBJECTIVES: *In this lesson, we will help Daphne discover how to:*

- *Identify career fields suited to talents and interests.*
- *Investigate jobs and their educational requirements.*
- *Predict what the job market will be like in several years.*
- *Understand what benefits an employer may offer in addition to a paycheck.*

CHOOSING A CAREER

Daphne wonders why Darrin is having such a hard time choosing a career. She knows that it is probably one of the most important decisions in a person's life. What she may not know is that surveys indicate that fewer than 50% of people who work enjoy their jobs. This figure would surely be much higher if young people entering the workplace took more time to assess their abilities and interests. Fortunately, Daphne does realize that her choice of career is important, not only because it directly affects how she will be spending most of her waking hours, but also because it will have a direct impact on the overall quality of her personal life. Unlike many people, Daphne knows that she can make choices. She intends to focus on fields that not only will offer her an attractive income, but also will provide her with a sense of satisfaction and self-esteem.

The types of occupations in the United States can be counted in the thousands. Some of these require long periods of education or training. In fact, most jobs require post–high school education or training. Perhaps some workers, like Darrin, missed an opportunity to benefit from the education they were being offered. Of the 40 occupations with the largest projected job growth in the next decade, only one in four will require a college degree or specialized training, according to employment projections published by the U.S. Bureau of Labor Statistics. This agency groups occupations in 13 clusters of related jobs, as follows:

- Industrial production
- Service
- Sales
- Transportation
- Technical and repair jobs
- Social science
- Art, design, and communication
- Office
- Education
- Construction
- Scientific and technical
- Health
- Social service

FOCUS

ALGEBRA CONNECTIONS

Ask students how they would go about finding the portion of their pay that is attributable to days when they don't work. How would they go about finding this out? Have them review the idea of proportion covered in the *Algebra Review.*

ASSESS YOURSELF

With the large number of occupational choices in front of her, Daphne wonders, "Where do I begin?" If you are asking yourself that same question, start with what you know about your own interests and abilities. Do you like frequent contact with other people, or do you prefer to spend a lot of time alone? Are you a good follower, or do you prefer directing others in a work effort? Identify your personal skills and interests.

- Do you communicate effectively with others?
- Are you comfortable with computational tasks?
- Do you enjoy conducting a research project or investigation?
- Do you prefer working with your hands?
- Have others complimented you on your creative talent?
- Do you relate unusually well with other people?
- Have you shown talent in leading others in social or athletic activities?

INSTRUCTION

TEACHING THE LESSON

Have students seriously ask themselves the question posed in the section of the lesson called "Assess Yourself" in order to evaluate their abilities as honestly as they can.

After that, they will be able more effectively to cope with the research on careers that they must eventually undertake in reference books such as the *Occupational Outlook Handbook* mentioned in the lesson.

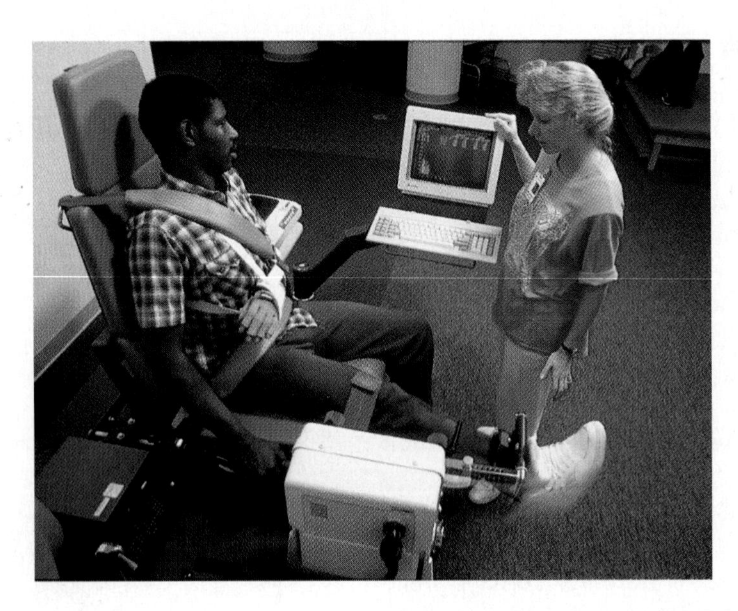

You also need to match your individual talents, interests, and goals with those demanded by various fields of work. A good place to begin your research is with the *Occupational Outlook Handbook* published by the U.S. Bureau of Labor Statistics. Many other sources are also available in school and public libraries. You may also find answers to your questions by interviewing people in fields that interest you.

EDUCATION AND INCOME

The amount and kind of education you receive greatly affects your choice of occupation and thus your current and lifetime income. People who have received postsecondary education can expect to earn more than $1,400,000 during their lifetimes. This is nearly two and a half times the $600,000 that is likely to be earned by workers with fewer than eight years of schooling and more than one and a half times as much as high school graduates. To maximize income, it pays for you to acquire additional education and training.

FOCUS ON ALGEBRA

The ability to solve a proportion is one of the fundamental mathematical skills that all citizens should have. It is highlighted in this lesson, but also underlies many of the mathematical challenges that students will encounter everywhere, from finding the best price in a supermarket to calculating whether one has enough gas to drive to grandmother's house for Thanksgiving. *(NCTM Standard 1, p. 137; Standard 5, p. 150)*

As the demand for goods and services changes, workers often have to change jobs. It is estimated that college-educated workers change jobs from four to eight times in their lifetimes. Workers with only a high school education change jobs even more frequently.

TECHNOLOGY

Job opportunities and occupational trends are being changed by technology. **Technology** enables goods and services to be produced more efficiently through the use of machines and labor-saving methods. As a percent of the total work force, the number of blue-collar workers (carpenters, plumbers, and factory workers, for example) is decreasing while the demand for white-collar workers (office workers, salespeople, and technicians, for example) is increasing. Technology requires workers to have better training to fill new types of jobs.

FRINGE BENEFITS

Wages and salary are just one kind of financial reward for work. Many companies also offer other benefits such as paid vacations, life and health insurance, free uniforms, retirement plans, stock options, and discounts on company products. These are examples of **fringe benefits.** A fringe benefit is like a hidden paycheck within a paycheck, amounting to between 10% and 25% of total employee compensation.

What do you get for this 10% or 25%? First of all, there are benefits that the company is compelled to offer by law, such as taxes to fund your state's unemployment compensation system, which provides temporary income for workers who are temporarily out of work through no fault of their own. A company must also pay for worker's compensation insurance (to cover work-related injuries or illnesses) and for part of your Social Security/Medicare tax. This last tax is called the **FICA tax**, since it is imposed as a consequence of the *Federal Insurance Contributions Act.*

Examples of nonmandatory employee benefits include paid sick leave, paid holidays, and paid vacation time. However, many companies go well beyond these commonly offered benefits. For example, some companies pay part or all of the premiums for your health insurance or life insurance. A company might also supplement funds that you put into a company retirement plan by matching your contributions with funds of its own.

Many companies encourage employees to enroll in study programs such as college courses, seminars, workshops, or training programs. Often, a company will partly or entirely subsidize education or training programs because it considers such programs to be a benefit for the company as well as the employee.

As Daphne continues her search for employment and career opportunities, she will take into account not only the fact that fringe benefits are a part of her total compensation package, but also the fact that they are a significant part of her potential employer's costs.

CRITICAL THINKING

One of the most important fringe benefits offered by employers is health insurance. Ask students why, for many workers, this benefit may be more valuable coming from an employer than if the worker had to obtain health insurance for himself or herself.

Ask Yourself

1. According to the U.S. Bureau of Labor Statistics, 40 occupations have the largest projected job growth in the next decade. How many of these will require a college degree or specialized technical training?
   1 in 4
2. You may begin considering the choice of a career by identifying your own interests and abilities. Name eight skills in which you are strong.
   Answers may vary.
3. One step in deciding on a career is to match your interests and talents with those required by various fields of work. What publication by the U.S. Bureau of Labor Statistics can you consult for this information?
   Occupational Outlook Handbook
4. How has technology affected the percent of blue-collar workers in the total work force?
   Decreasing
5. What fringe benefits would you consider to be the most important in choosing a company to work for?
   Answers may vary. (Ex.: health insurance, vacations, retirement)

ALGEBRA *REVIEW*

Solve each proportion.
To solve a *proportion,* use the *Rule of Proportions:*

If $\frac{a}{b} = \frac{c}{d}$, then $ad = bc$.

Example $\frac{x}{14} = \frac{65}{7}$

$x \cdot 7 = 14(65)$ — $x \cdot 7$ is "*ad*," and 14(65) is "*bc*."

$7x = 910$ — $x \cdot 7$ and $7x$ are equivalent expressions. Divide both sides by 7.

$x = 130$

1. $\frac{x}{6} = \frac{4}{5}$
   4.8
2. $\frac{x}{6.2} = \frac{3}{31}$
   0.6
3. $\frac{4}{5} = \frac{x}{15}$
   12
4. $\frac{2.1}{9} = \frac{x}{6}$
   1.4
5. $\frac{3.6}{9} = \frac{80}{x}$
   200
6. $\frac{132}{x} = \frac{6}{7}$
   154
7. $\frac{52}{5200} = \frac{x}{17{,}500}$
   175
8. $\frac{26}{x} = \frac{260}{34{,}000}$
   3400

SHARPEN YOUR SKILLS

Dinotech Engineering needs to hire a new engineer this year to help design some of the computer components used in the U.S. space program. The cost of the following extra taxes and fringe benefits will figure into their plans.

a. 7.65% of gross pay for Social Security and Medicare (FICA)

b. 5% of gross pay to match retirement contributions

c. $80 per month to help pay for health insurance

d. $10 per month for a life insurance policy

Dinotech operates 5 days a week, 52 weeks a year, or 260 days. Of the 260 workdays each year, the company grants its employees 10 vacation days, 6 holidays, and up to 10 sick-leave days.

EXAMPLE 1 Dinotech plans to offer the new engineer $45,000 per year as a starting salary.

QUESTIONS

1. How much must the company budget to cover the engineer's fringe benefits?
2. How much of the engineer's $45,000 salary is paid for the nonworking days?

SOLUTIONS

1. **a.** $0.0765 \cdot 45{,}000 = 3442.50$ — 7.65% of $45,000 for FICA

   b. $0.05 \cdot 45{,}000 = 2250$ — 5% of $45,000 for the retirement plan

   c. $80 \cdot 12 = 960$ — $80 per month for health insurance

   d. $10 \cdot 12 = 120$ — $10 per month for life insurance

   Total additional costs: $6772.50

2. The number of nonworking days is

   10 vacation days + 10 sick-leave days + 6 holidays, or 26 days

   Write a proportion and solve it.

$$\frac{\text{Nonworking days}}{\text{total days}} = \frac{\text{dollars earned on nonworking days}}{\text{amount earned for entire year}}$$

$$\frac{26}{260} = \frac{x}{45{,}000}$$

$$\frac{1}{10} = \frac{x}{45{,}000} \quad \text{Simplify the left side.}$$

$$1 \cdot 45{,}000 = 10x \quad \text{Use the rule of proportions.}$$

$$\frac{45{,}000}{10} = \frac{10x}{10} \quad \text{Divide each side by 10.}$$

$$4500 = x$$

   Dinotech pays $4500 for nonworking days.

COMMON ERRORS

When solving proportions using the "special rule," students should carefully keep track of the four terms of the proportion in order to pair them correctly. Some students may be aided by the language of classical geometry: "The product of the means (the middle terms) equals the product of the extremes (the first and last terms)."

SKILL 2

EXAMPLE 2 Daphne's older friend Chloe has completed her engineering training and is considering accepting a position at Aerotech. Chloe has also been offered a position at Electron Research and has described the two offers to Daphne.

	Aerotech	Electron Research
Salary	$48,000	$48,000
Medical insurance	$50/month	$60/month
Retirement	3% of gross pay	$2\frac{1}{2}$% of gross pay
Education	$1000/year	$1200/year
Travel Allowance	$120/month	$100/month

QUESTION Which of the two companies is offering Chloe more money including fringe benefits?

SOLUTION

Aerotech is offering:

Salary	$48,000	
Medical insurance: $50/month	600	50 • 12
Retirement: 3% of 48,000	1,440	0.03 • 48,000
Educational expenses	1,000	
Travel allowance	1,440	120 • 12
Total offer	$52,480	

COOPERATIVE LEARNING

The fringe-benefits options of *Try Your Skills* can be converted to spreadsheet formulas for quick calculation and tabulation. Have students work in groups to prepare spreadsheets for the job offers of Example 2. Then have the different groups compare their results.

Electron Research is offering:

Salary	\$48,000	
Medical Insurance: \$60/month	720	60 • 12
Retirement: $2\frac{1}{2}$ % of 48,000	1,200	0.025 • 48,000
Educational expenses	1,200	
Travel allowance	1,200	100 • 12
Total offer	\$52,320	

Aerotech's offer is slightly higher than Electron Research's.

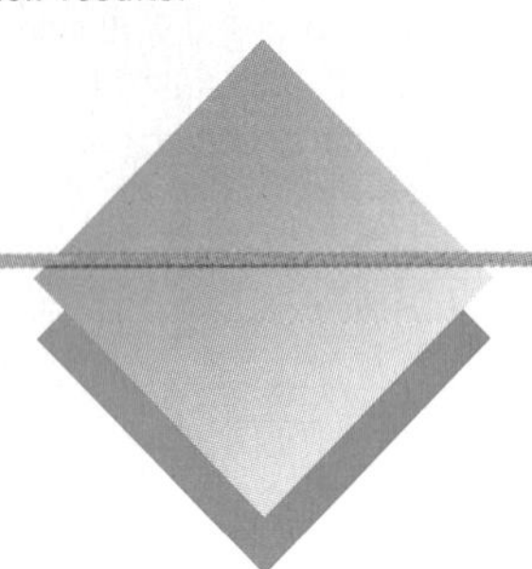

TRY YOUR SKILLS

Fast Facts is a company that copies documents and creates brochures, charts, booklets, and other types of materials for other businesses. The company's compensation committee is submitting four possible fringe benefits options to the company's board of directors. For each employee and fringe benefits option,

a. find the amount that Fast Facts must pay for extra taxes, retirement, health and life insurance, and other fringe benefits.

b. find how much salary is paid for the given number of nonworking days in a total of 260 days.

Help students to formulate some of the proportions in *Try Your Skills* Exercises 1-4, part b.

	EMPLOYEE	SALARY	FRINGE BENEFITS OPTION
1.	Copier/collater	\$12,000	7.65% of gross for FICA taxes
	a. \$1698.00		\$65/month for life and health insurance
	b. \$1200.00		26 nonworking days
2.	Clerk/receptionist	\$14,500	7.65% of gross for FICA taxes
	a. \$2179.25		\$65/month for life and health insurance
	b. \$1338.46		2% of gross for retirement
			24 nonworking days
3.	Graphics designer	\$33,000	7.65% of gross for FICA taxes
	a. \$5954.50		\$65/month for life and health insurance
	b. \$3807.69		5% of gross for retirement
			\$1000/year for training
			30 nonworking days
4.	Accountant	\$48,000	7.65% of gross for FICA taxes
	a. \$7272.00		5% of gross for retirement
	b. \$3876.92		\$100/month for life and health insurance
			15 vacation days and 6 holidays

For each of the following groups of employment opportunities,

a. calculate the total amount of money being offered, including fringe benefits.

b. determine which of the three positions offers the most money.

5.

Company	A	B	C
Salary	$37,500	$38,000	$35,800
Retirement benefits	3.5% of gross pay	3% of gross pay	5% of gross pay
Medical insurance	$55/month	$45/month	$50/month
Educational expense	$500/year	$800/year	$1000/year
Travel allowance	$2400/year	$100/month	$150/month

a. $42,372.50; $41,680; $40,990 b. A

6.

Company	A	B	C
Salary	$42,500	$41,800	$44,000
Stock options	5% of gross pay	6.5% of gross pay	4.5% of gross pay
Medical and life insurance	$100/month	$115/month	$100/month
Car lease	$250/month	$280/month	$240/month
Travel allowance	$350/month	$375/month	$300/month

a. $53,025; $53,757; $53,660 b. B

7. Mr. Grant receives a salary of $34,000 and takes vacation days and holidays for 28 of the company's 242 operating days. How much of his salary is paid for the days during which he does not work? $3933.88

EXERCISE YOUR SKILLS

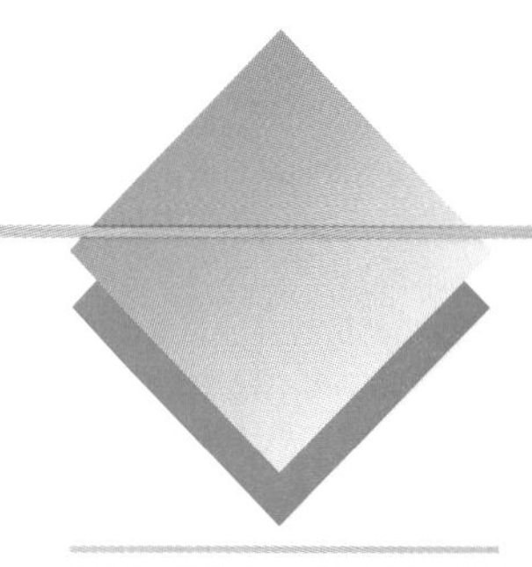

1. What are some examples of jobs that are available even though people who are looking for employment cannot find jobs? Answers may vary.
2. Why do you think that workers with a high school education change jobs more frequently than college-educated workers? Answers may vary.
3. As an aid in answering some of your questions about a particular kind of job, it may be helpful to interview someone who is already working in that field. What are six questions that you might ask in such an interview? Answers may vary. (Ex.: What do you like most about your job? What skills do you use? What training did you need?)

KEY TERMS

FICA tax
fringe benefits
technology

ASSIGNMENT GUIDE

Exercises 1-3 are discussion questions that allow students to further investigate the ways of the workplace. Exercises 4-11 help students to appreciate the viewpoint of employers who must expend money on fringe benefits that do not go directly to their employees. Exercises 12-14 provide an opportunity for students to appreciate the fact that the most attractive position is not always the one that offers the highest salary.

It is convenient to solve proportions using the special rule shown in Skill 1. However, students should be reminded that proportions can be solved using the same general methods that apply to all equations. Show students that the equation of Skill 1 can be solved by multiplying each side by the product of the denominators, that is, by "*bd*" in the formula provided in the *Algebra Review.*

For each company fringe-benefits policy described below,

a. find the amount that the company must pay for extra taxes, retirement, health and life insurance, and other fringe benefits.

b. find how much salary is paid for the given number of nonworking days in a total of 240 operating days.

	EMPLOYEE	SALARY	FRINGE-BENEFITS POLICY
4.	Computer operator	$40,000	7.65% of gross for FICA taxes
	a. $8260		5% of gross for retirement
	b. $4000		$100/month for life and health insurance
			$2000/year for training
			24 nonworking days
5.	Delivery person	$12,000	7.65% of gross for FICA taxes
	a. $1518		$50/month for health insurance
	b. $600		8 holidays and 4 sick days
6.	Office manager	$56,000	7.65% of gross for FICA taxes
	a. $17,084		$150/month for life and health insurance
	b. $4,200		$200/month per diem expense allowance
			$250/month car lease
			10% of gross in stock options
			18 nonworking days
7.	Lab technician	$28,000	7.65% of gross for FICA taxes
	a. $4542.00		$200/month for health insurance
	b. $3033.33		26 nonworking days
8.	Theatre manager	$35,000	7.65% of gross for FICA taxes
	a. $3477.50		$800/year for training
	b. $2187.50		15 vacation days and holidays
9.	Executive secretary	$42,000	7.65% of gross for FICA taxes
	a. $9273		$225/month for life and health insurance
	b. $3675		8% of gross in stock options
			15 vacation days, 6 holidays
10.	Copy Editor	$25,000	7.65% of gross for FICA taxes
	a. $4012.50		$175/month for health insurance
	b. $2083.33		10 holidays and 10 vacation days
11.	Pharmacist	$52,000	7.65% of gross for FICA taxes
	a. $10,398.00		$275/month for health insurance
	b. $4,766.67		6% of gross for retirement
			22 nonworking days

For each of the three employment opportunities **A, B,** and **C,**

a. calculate the total amount of money being offered, including fringe benefits.

b. determine which of the three positions offers the most money.

12.

Company	A	B	C
Salary	$22,000	$22,000	$22,000
Retirement benefits	2% of gross pay	3% of gross pay	2.5% of gross pay
Medical insurance	$25/month	$20/month	$40/month
Expense allowance	$200/month	$250/month	$300/month

a. $25,140; $25,900; $26,630 **b.** C

13.

Company	A	B	C
Salary	$27,000	$30,000	$23,500
Retirement benefits	4% of gross pay	2% of gross pay	5% of gross pay
Medical insurance	$100/month	none	$150/month
Expense allowance	none	$200/month	$350/month

a. $29,280; $33,000; $30,675 **b.** B

14.

Company	A	B	C
Salary	$47,000	$39,000	$35,500
Retirement benefits	none	3.5% of gross pay	1% of gross pay
Medical insurance	$50/month	$125/month	$200/month
Expense allowance	$300/month	$300/month	$500/month

a. $51,200; $45,465; $44,255 **b.** A

MIXED REVIEW

Write an equation of the form $E = rh + t$ where r represents the regular hourly rate, h represents the number of hours worked, and t represents any bonus or tip. Then find the weekly earnings for each of the following "employees."

1. I am the gremlin who arranges for you to be stopped by every traffic light between your home and your job on any day that you are running a little late. I get paid only when I actually make you late for work. This week I worked 28 hours and got paid $14.50 per hour. I get no bonuses because no one likes me. $406.00

1-1

2. I am the auto-focus camera that you took with you on your once-in-a-lifetime vacation to the South Seas. I worked for 32 hours at $10.50 an hour helping you take 125 well-focused photos out of 128. I also received a $35 bonus for one photo that earned you a prize in a local photography contest. $371.00

1-1

ENRICHMENT

Ask students whether the proportion

$$\frac{a-b}{b} = \frac{c-d}{d}$$

is true if the basic proportion

$$\frac{a}{b} = \frac{c}{d}$$

is true. As a hint, ask them the value of $\frac{b}{b}$ and of $\frac{d}{d}$.

(yes, since the given proportions simplifies

first to $\frac{a}{b} - \frac{b}{b} = \frac{c}{d} - \frac{d}{d}$

then to $\frac{a}{b} - 1 = \frac{c}{d} - 1$

and finally to the basic proportion)

CLOSURE

Ask students which kinds of fringe benefits they feel are of greatest importance to them. Which benefit would they be willing to do without if the value of the benefit could be exchanged directly for a salary increase?

LESSON QUIZ

Damien has fringe benefits that are 27% of his salary of $24,000 a year. Of the company's 250 operating days, he has 12 non-working days.

1. What is the total compensation, including fringe benefits? **($30,480)**

2. How much of his salary is paid for days during which he does not work? **($1152)**

MIXED REVIEW *(continued)*

3.–4. Assume that the "employees" of Exercises 1–2 received overtime
1-1 pay of $1\frac{1}{2}$ times their regular rate for all hours they worked over 20. Write an equation to represent their weekly earnings of the form

$$E = 20r + (h - 20) \cdot 1.5r + t$$

Where r represents the regular hourly rate, h represents the number of hours worked, and t represents any tip or bonus. Determine the total earnings for each employee. **3.** $464 **4.** $434

Create a spreadsheet to calculate the hourly, weekly, monthly, and yearly earnings for each job described below.

40 hours = 1 workweek
52 weeks = 1 year
12 months = 1 year

5. 1-1

Do you know how to sell? Start with us at $6.25/h and GROW!

Weekly: $250
Monthly: $1,083.33
Yearly: $13,000

6. Hourly: $10.50
Monthly: $1,820
Yearly: $21,840

6. 1-1

Laboratory Assistant
We are looking for a qualified worker trained in the daily operating procedures of a biotechnical laboratory. Two years experience desired but not necessary. Starting salary $420/ week.

7. 1-1

COMPUTER PROGRAMMER
Experience required. Flexible hours can be arranged. Starting salary $3,800/month.

Hourly: $21.92
Weekly: $876.92
Yearly: $45,600

1-3 DO MORE, MAKE MORE: COMMISSIONS AND PAYMENTS BY ITEM

QUICK REFERENCE

TEACHER'S RESOURCES AND ANSWER KEY

Lesson Quiz Answers 1-3, p. 40
Reteaching Activity Answers 1-3, p. 91
Enrichment Activity Answers 1-3, p. 92

EXTENSION ACTIVITIES

Reteaching Activity 1-3, p. 3
Enrichment Activity 1-3, p. 7

Betty goes to high school with Alex. She knows that Alex is trying to decide whether getting a part-time job is worth the effort and time involved. She and Alex were dating before Alex met Alice. In fact, Betty was with Alex when he was shopping around for his van.

Alex and his parents seemed to know a lot about car salespeople when they were picking out the van. At one of the lots they visited, a saleswoman kept trying to push Alex's father and mother into choosing a gorgeous, current-year model. Betty noticed that the saleswoman kept saying that it was "loaded." It had not only a very powerful engine but all the extras: "power this" and "power that," a really plush interior, and all kinds of built-in storage space. Of course, it had a "powerful" price tag also. Betty guessed that it cost more than Alex's parents wanted to spend. After all, Alex just needed transportation to and from school.

As they left the car lot and drove toward another lot, where Alex would eventually buy his van, Betty asked why the saleswoman had seemed so eager to sell the most expensive van they had. Alex's father explained, "Oh, she works on commission, Betty. You see, the more money we spend, the more she makes! Perhaps she has not stopped to realize that if she does not sell us what we want, she will not make anything from us!"

Betty was not sure she understood how a person's earnings could be affected by the price of a van. She decided to find out.

OBJECTIVES: In this lesson, we will help Betty to:

- *Determine how much a salesperson earns when his or her pay is based on a commission.*
- *Determine how much a person earns when his or her pay is based on a piece rate, or item rate.*
- *Understand that the earnings of a person who works on commission or piece rate can vary from month to month.*

EARNING COMMISSIONS

FOCUS

ALGEBRA CONNECTIONS

Ask students what algebraic expression represents the annual earnings of someone who has a weekly salary of x dollars and who also earns a commission of 10% on y dollars of sales for the year. **($52x + 0.10y$)**

People whose salaries depend on how much they sell in a given period of time are said to be paid on **commission.** These salespeople receive a percent of each amount that they bring in. For example, Jason, who works at a used-car lot, earns a 15% commission on every car he sells. Today, Jason sold a car for $6000. His earnings are 15% of $6000.

$$6000 \cdot 0.15 = 900$$

Jason earned $900. The more cars Jason sells, the more money he makes. If he sells no cars on a particular day, he earns no money that day.

Some salespeople earn a regular salary plus a commission, as illustrated in the ad below. If the headline is to be believed, the salesperson could earn $50,000 in a year, but he or she would probably have to make a lot of sales to earn that much.

Auto

$50,000+ POTENTIAL

We are seeking 2 aggressive self-motivated people to add to our highly successful team. Previous auto-sales experience helpful. We are especially interested in those with HEAVY RETAIL SALES EXP. We have an intensive in-house training program to launch you onto a successful sales career. We offer:

- Weekly salary & Comm.
- New demo
- Top medical ins. program
- Heavy advertising
- Excellent service dept. to keep your customers happy

Suppose that the weekly salary is \$300. For 52 weeks the salary will be \$15,600 (300 • 52 weeks). If the salesperson earns 10% commission on all sales and sells \$344,000 worth of cars during the year, the total earnings will be

\$15,600	\$300 • 52 = regular salary
+ 34,400	\$344,000 • 0.10 = commission
\$50,000	Total earnings for the year

If these are new cars, then between 25 and 40 of them are probably being sold a year. If the commission is less than 10%, the salesperson will have to sell more cars to make the same amount of money.

Some companies also encourage their salespeople to sell more by offering them a **graduated commission.** If they sell up to a specified amount, their commission rate will be a certain percent. If they sell more than that amount, the percent will increase.

CRITICAL THINKING

The lesson cautions students to be alert to the possibility that a salesperson is being paid "push money" by a manufacturer to induce the salesperson to steer customers away from competing brands. Ask students how they might recognize that a salesperson is being paid push money.

PIECE RATE

Instead of receiving a salary for working a certain amount of time, some people are paid a **piece rate,** or **item rate** for the number of items that they produce or sell. If you work fast or work on your own schedule, being paid a piece rate may be advantageous to you. Like commission, a piece rate is sometimes combined with a **base salary,** which you would receive regardless of how many items you produced or sold.

PUSH MONEY

Consumers are easily influenced when they lack adequate information about products that they buy. Unfortunately, in many stores, salespeople are not informed enough about a product to provide adequate information to customers. They are trained just to take a customer's order.

Another problem is the salesperson who is paid **push money,** a cash incentive provided by a *manufacturer* for selling the manufacturer's products. Push money is not the same as a commission, which is money paid by the *store* to salespeople to sell *all* the products in the store. If you notice a salesperson aggressively pushing a certain product, that salesperson may be receiving push money. Let the buyer beware!

ALGEBRA *REVIEW*

Solve for x or a.

1. $14x - 8 = 20$
   $x = 2$
2. $3.5 + 2.5a = 18$
   $a = 5.8$
3. $(3.5 + 2.5)a = 18$
   $a = 3$
4. $26x = 78$
   $x = 3$
5. $\frac{1}{26}x = 78$
   $x = 2028$
6. $300 + 0.10x = 50{,}000$
   $x = 497{,}000$

Simplify.

Example

$$2m + 3.5m - 2.7p + 8$$
$$= (2 + 3.5)m - 2.7p + 8$$
$$= 5.5m - 2.7p + 8$$

7. $9x + 3.5x$
   $12.5x$
8. $0.05x - 0.03x + 7$
   $0.02x + 7$
9. $(2.1 + 3.9)b + 4b$
   $10b$
10. $1y + 14y$
    $15y$
11. $0y + 14y + 1 \cdot 5$
    $14y + 5$
12. $y + 14y + 0 \cdot 5$
    $15y$

CLASSIFIED ADS

As Betty was seeking information about salespeople whose earnings depend on commissions, she discovered the following classified ads in the employment section of her Sunday newspaper. The jobs seemed very attractive. But Betty thought that there was more to the matter than she could see from the ads.

Sales

ONE-CALL CLOSERS

If your are willing to learn and want to earn

$650–$1200
(weekly commission)

- CASH PAID DAILY
- NO INVESTMENT
- DAYTIME SALES
- COMPLETE TRAINING
- AUTO REQUIRED
- MANAGEMENT OPPTY.

CALL MONDAY ONLY

SALES/GOLF

Earn more than you require. $40-$120,000 per year commissions. Self-starting closer.

TIRED OF FALSE PROMISES?

(I was)

- Join one of the fastest growing industries in the country.
- Enjoy success as hundreds of our representatives do.
- Earn over $100,000 per year.
- FRESH FREE LEADS
- NO SATURDAYS
- 10% OF BACK-END PAID/MO.

ADVENTURE INN BIG MONEY

Direct sales experience required. Seminar sales. Sell vacation club. Prospects furnished. Work evenings and weekends. 5-day work week required. Straight commission. Earn up to $1600 a week. Closed Monday.

CAREER SALES OPPORTUNITY

Would You Believe $144/day?

General Research is expanding its headquarters to the suburbs. We are hiring 2-3 people of legal age for marketing and management training program. Minimum $1500/month guaranteed commissions and bonuses to start. All fringe benefits including hospitalization, medical and life insurance. Must have dependable auto. Call today.

Betty remembered talking with a friend who sells cars. She found out that if a rebate or discount is given, that fact affects how much commission is earned. Suppose that a car is priced at $6000 and a $420 discount is offered to make the customer want to buy the car. The salesperson receives a commission on only $5580 (that is, on $6000 - 420$). Because of details such as this, Betty realizes that finding a job requires a lot of research. She cannot just read the ads.

Ask Yourself

1. How can people earn the high pay offered in ads such as those shown?
   By making a lot of sales.
2. How does a graduated commission work?
   The percent of commission increases as the sales increase.
3. What is push money?
   A cash incentive paid by the manufacturer for selling a product.
4. How is push money different from a commission?
   See Additional Answers.

ADDITIONAL ANSWERS

4. Push money is from the manufacturer while commission is from the store. Push money is for a certain product, commission is for all products.

SHARPEN YOUR SKILLS

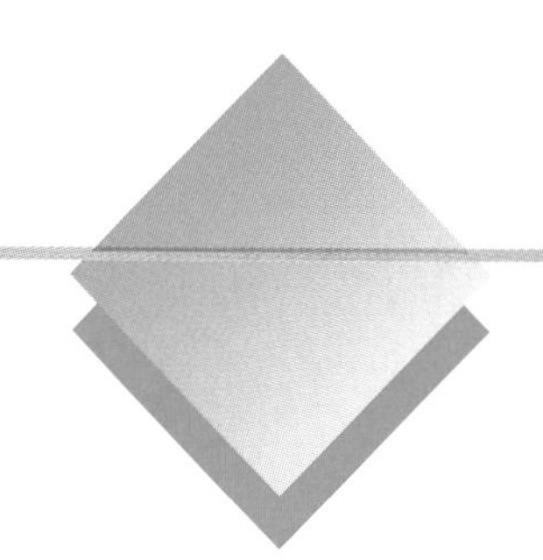

SKILL 1

You can use a formula to find the commission C when you are given the rate of commission r, the price p, and the number of items n sold. Remember to round commission to the nearest cent.

Commission = price • commission rate • number sold

Commission Formula

$C = prn$ where p = price of one item
r = commission rate
n = number of items sold

INSTRUCTION

TEACHING THE LESSON

This lesson presents commission selling from the viewpoint of both a consumer and a salesperson. However, since the consumer viewpoint is adequately represented in other lessons of the book, the emphasis here is on the problems faced by the responsible salesperson. Point out how direct rewards follow not only from hard work, but from understanding the mathematical basis that underlies those rewards. For example, in Example 5 the use of algebra reveals the exact point at which a certain payment option offered by an employer will become profitable, allowing the employee to focus on his or her ability to reach that point.

EXAMPLE 1 Betty's friend Rosalie is a salesperson at a department store. She receives a 7% commission for selling small appliances and housewares. Among the many items that she sells is a fancy ceramic mug. Rosalie would like to know how many mugs she needs to sell to earn at least $50.

COMPUTER MUG of glazed earthenware with terminal and keyboard as handle. Dishwasher safe. For hot or cold beverages, pencils, and more. Holds 14 oz.
#45678 mug $12.95

QUESTION How can Rosalie use her graphing calculator to find out how many mugs she needs to sell to earn $50?

SOLUTION

Rosalie uses the commission formula.

$C = prn$ $p = 12.95,\ r = 7\% = 0.07$
$C = 12.95 \cdot 0.07 \cdot n$
$C = 0.9065n$

To use her graphing calculator to graph this equation, Rosalie has to change the variable C to y and the variable n to x:

$y = 0.9065x$

Rosalie enters the equation into her graphing calculator. She sets the range as shown and graphs the equation.

Xmin:	0	Ymin:	−10
Xmax:	60	Ymax:	60
Xscl:	10	Yscl:	10

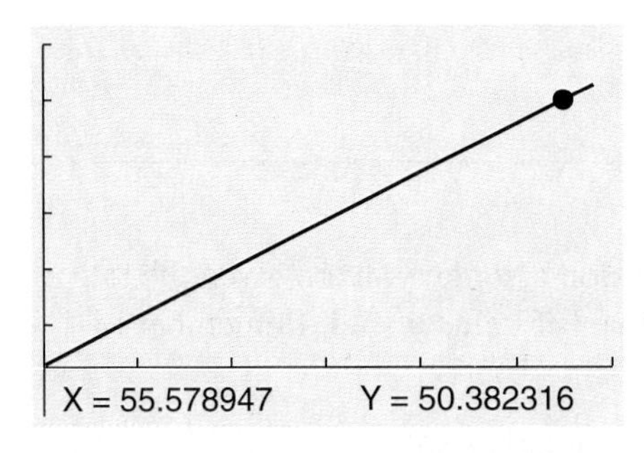

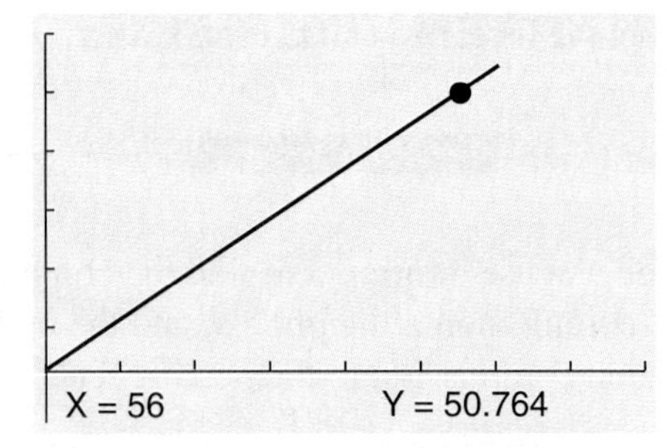

FOCUS ON ALGEBRA

This lesson continues the work with formulas and the solving of equations that was presented in earlier lessons. In addition, there is some preliminary application of linear systems (simultaneous equations) in Example 5. *(NCTM Standard 1, p. 137; Standard 5, p.150; Standard 6, p. 154)*

The graph is a line as shown at the left above. Rosalie uses the TRACE function to move the cursor along the line until the y value is close to \$50. She concludes that she can earn over \$50 in commissions by selling 56 mugs.

She would like x values to be integers, so she uses the INTEGER function of her graphing calculator or sets the range to a "friendly window" and redraws the graph as shown at the right above.

To confirm the calculator results, Rosalie substitutes $n = 56$ into the commission equation.

$C = 0.9065n$
$C = 0.9065 \bullet 56$ $\quad$ $n = 56$
$C = 50.764$

The equation method confirms the results of the graphing calculator. Rosalie's commission is \$50.76 for selling 56 mugs.

SKILL 2

COMMON ERRORS

Some students will need to be reminded of certain algebraic conventions such as the fact that the missing coefficient in the simple expression "x" is 1, not 0. Thus, $x + 8x - 2x$ is $7x$, not $6x$. To forestall such errors, review the multiplication properties of 0 and 1 followed by a discussion of *Algebra Review* Exercises 10-12.

You can use a formula to find the earnings E when you are given the monthly salary s, the commission rate r, and the total sales t.

> **Earnings with Commission Formula**
>
> $E = s + rt$ $\quad$ where s = monthly salary
> r = commission rate
> t = total sales for the month

EXAMPLE 2 William earns a salary of \$1000 per month selling cars plus a commission of 6% on each car that he sells. In his first month on the job, he sold 3 cars for a total of \$29,500 in car sales.

QUESTION How much did William earn in all?

SOLUTION

$E = s + rt$ $\quad$ Use the earnings formula.
$E = 1000 + 0.06(29{,}500)$ $\quad$ $s = 1000,\ r = 6\% = 0.06,\ t = 29{,}500$
$E = 2770$

You can enter the calculation in your graphing calculator as written, since the calculator will multiply before adding.

William earned $2770 in salary and commissions during the month.

EXAMPLE 3 Yvette is a real-estate salesperson. Her earnings are based on a *graduated commission.* She receives 4% of any sale up to $100,000. For any amount over $100,000 she receives a 6% commission. Last Sunday she sold the house described below.

> **MINT CONDITION!!** $109,000
> You'll appreciate the love and care that went into this stunning 3-bdrm, 2-bath home. Vaulted ceilings, custom drapes, mst. bath w/whirlpool/double vanity.

COOPERATIVE LEARNING

The graduated-commission problems of *Exercise Your Skills* Exercises 9-11 and Example 3 of Skill 2 can be done by a computer spreadsheet program. Have students work in groups to create the appropriate formulas and test them against the results obtained using other methods.

QUESTION How much did she make selling the home?

SOLUTION

Commission = price • rate

Find the commission C_1 (read "C sub one") on $100,000. Then find the commission C_2 (read "C sub two") on the amount over $100,000.

$C_1 = 100{,}000 \cdot 0.04 = 4000$

$C_2 = 9000 \cdot 0.06 = 540$ $109{,}000 - 100{,}000 = 9000$

Total commission $= C_1 + C_2$
Total commission $= 4000 + 540 = 4540$

Yvette's commission is $4540.

SKILL 3

To solve problems involving piece rate, use the following formula.

Piece Rate Formula

$E = rn$ where r = the piece or item rate
n = the number of items

INTERDISCIPLINARY INVESTIGATION

Interested students may wish to investigate ways in which products are marketed other than by over-the-counter selling. For example, direct-mail selling is an increasingly popular practice among computer makers. Shopping by TV is also increasingly popular among the American public.

EXAMPLE 4 Carole decorates wedding cakes at Sloane's Bakery. Mrs. Sloane pays her $16.00 for each cake that she decorates.

QUESTION How much money did Carole earn in a week in which she decorated 5 wedding cakes?

SOLUTION

Substitute in the formula for piece work.

$E = rn$
$E = 16.00 \cdot 5$
$E = 80$

Carole earned $80 for 5 wedding cakes.

RETEACHING

Some parts of the lesson can be developed through the solving of inequalities rather than equations. This is illustrated by Example 5. Since $12 < 14.50$, Carole already knows that if she works hard enough, Plan 2 will be more profitable. Therefore, she can ask herself for what values of n it is the case that

$105 + 14.50n > 120 + 12n$

When solved, this inequality leads directly to $n > 6$.

EXAMPLE 5 Mrs. Sloane needs Carole to help her with customers when she is not busy decorating cakes. She has offered Carole a choice of one of two payment plans.

Plan 1: Salary of $120 for a 20-hour week and $12 for each cake
Plan 2: Salary of $105 for a 20-hour week and $14.50 for each cake

QUESTION Which plan should Carole choose?

SOLUTION

Carole is not sure which plan is better for her. She thinks that she can decorate at least 7 cakes a week. Carole writes two equations, one for each plan.

Plan 1: $E = 120 + 12n$ Earnings = salary + piece-rate pay
Plan 2: $E = 105 + 14.50n$

Carole wants to know the least value of n that will make the same money with either plan. She solves the *system of equations* by setting the two earnings equal to each other.

$120 + 12n = 105 + 14.5n$

ENRICHMENT

Some students may benefit from examples of linear systems of equations that are not as simple to solve as that of Example 5. Have them solve the system

$x + y = 17.5$
$x - y = 3.5$

by adding the left and right sides of the two equations and then finding the value of x and of y that makes both equations true. **($x = 10.5$, $y = 7$)**

When solving an equation, you may subtract the same quantity from each side of the equation. Carole decides to subtract $12n$ to obtain a simpler equation.

$$\begin{aligned}
120 + 12n &= 105 + 14.5n \\
120 + 12n - 12n &= 105 + 14.5n - 12n && \text{Subtract } 12n \text{ from both sides.} \\
120 &= 105 + 2.5n && \text{Combine similar terms: } 12n - 12n = 0;\ 14.5n - 12n = 2.5n \\
120 - 105 &= 105 + 2.5n - 105 && \text{Subtract 105 from both sides.} \\
15 &= 2.5n && \text{Simplify.} \\
\frac{15}{2.5} &= \frac{2.5n}{2.5} && \text{Divide both sides by 2.5.} \\
6 &= n && \text{Simplify.}
\end{aligned}$$

Carole will earn the same money under both plans if she decorates 6 cakes.

Plan 1: $E = 120 + 12n = 120 + 12(6) = 192$ $\quad n = 6$
Plan 2: $E = 105 + 14.50n = 105 + 14.50(6) = 192$ $\quad n = 6$

She also calculates her earnings under both plans if she decorates 7 cakes.

Plan 1: $E = 120 + 12(7) = 204$ $\quad n = 7$
Plan 2: $E = 105 + 14.5(7) = 206.50$ $\quad n = 7$

Plan 2 is better for her when $n = 7$. Carole concludes that she earns more money under Plan 2 if $n > 6$ (*n is greater than 6*) and more money under Plan 1 if $n < 6$ (*n is less than 6*).

TRY YOUR SKILLS

The commission rate, the price of each item, and the number of items sold are given for each salesperson.

a. Calculate the commission earned by selling the first number of items sold (for example, 10 items in Exercise 1).

b. Write an equation to show the commission y earned by selling x number of items.

c. Graph the equation on your graphing calculator. Then use the TRACE function to find the commissions for all four amounts. Remember to use an integer function or a "friendly window."

	Person	Price	Rate	Number Sold
1.	Carole	$16.75	9%	10; 18; 24; 32
2.	José	21.90	8.5%	8; 16; 22; 28
3.	Gabe	21.95	7.5%	5; 12; 15; 18

1. **a.** $15.08 **b.** $y = 1.5075x$ **c.** $15.08; $27.14; $36.18; $48.24
2. **a.** $14.89 **b.** $y = 1.8615x$ **c.** $14.89; $29.78; $40.95; $52.12
3. **a.** $ 8.23 **b.** $y = 1.64625x$ **c.** $8.23; $19.76; $24.69; $29.63

GUIDED PRACTICE

Students are asked to calculate the commission for 10 items in part a of Exercise 1 (and to perform similar calculations in Exercises 2 and 3) in order to test the correctness of the equation that they devise in part b.

Tracey and several of her friends have jobs working for a monthly salary plus commission. Find the month's earnings for each person. Monthly salary, commission rate, and the amount of sales are shown. Use a computer spreadsheet.

	Name	Monthly Salary	Commission Rate	Total Sales	Commission Earned	Total Earnings
4.	Tracey	$ 350	12%	$6125	$735.00	$1085.00
5.	Isabel	475	11.5%	6250	718.75	1193.75
6.	Damien	1000	8.5%	7100	603.50	1603.50

7. Tom assembles radios and tape recorders in a factory. He receives $9 for each radio and $12 for each tape recorder. How much did he earn in a week in which he assembled 26 radios and 38 tape recorders? $690
8. Alice has her own barber shop. She charges $9 for a regular haircut and $16 for a personalized haircut. Last week, 26 of her customers had a regular haircut, and 8 customers had the personalized haircut. What were her earnings for the week? $362
9. Patrick sells annual memberships in a health club. His boss has offered him a choice between Plan 1, no salary but a straight $30 for each membership sold, and Plan 2, a weekly base salary of $300 plus $15 for each membership he sells. How many memberships would Patrick have to sell to make the same amount of money under either plan? 20 memberships

ADDITIONAL ANSWERS

2. Compute your average monthly expenses. Then keep that amount in a savings account.

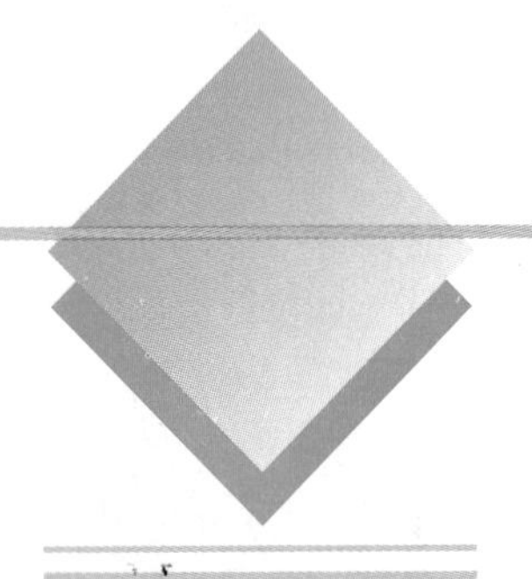

EXERCISE YOUR SKILLS

1. If you are earning a commission, why is it to your advantage to sell a relatively expensive item? Since commission is a percent of sales, you will earn more.
2. A person working on commission cannot predict exactly what next month's income is going to be. What is a good way to handle money so that there is enough to pay the bills? See Additional Answers.
3. Imagine that you work on commission in a large department store and that during the month before Christmas and Hanukkah, your department has a high volume of sales. As a result, your salary is three times what it was the previous month. Would you assume that next month's salary will be equally high and so be tempted to spend most of your earnings? Why or why not? No. Next month's sales will probably be back to average or below.
4. Reread each ad that Betty read in this lesson. Decide what is positive and what is negative about each position being advertised. Make a list of questions that you would ask if you went for an interview for each job. Which jobs appeal to you the most? Which jobs appeal the least? Answers may vary.

KEY TERMS

base salary
commission
graduated commission
item rate
piece rate
push money

The commission rate, the price of each item, and the number of items sold are given for each salesperson.

a. Calculate the commission earned by selling the first number of items sold (for example, 18 items in exercise 5).

b. Write an equation to show the commission y earned by selling x number of items.

c. Graph the equation on your graphing calculator. Then use the TRACE function to find the commissions for all four amounts. Remember to use an integer function or a "friendly window."

	Person	Price	Rate	Number Sold
5.	Ellen	$ 3.75	10%	18; 28; 38; 48
6.	Ivy	22.80	8.2%	6; 10; 14; 18
7.	Peter	7.99	9.5%	15; 25; 35; 45

8. Jenny earns a base monthly salary of $2150 plus a commission of $7\frac{1}{2}$% on her sales. In July her sales were $7350. How much did she earn in July? $2701.25

9. Samantha sells used cars. Her earnings are based on a graduated commission. She receives 3% of her total sales up to $20,000 and $4\frac{1}{4}$% of all sales over $20,000. Find Samantha's total earnings for a week in which she sold cars priced at $6,399, $2,495, $7,450, and $10,188. $877.61

Eduardo is a real-estate salesperson who specializes in selling houses. His earnings are based on a graduated commission. He receives 2% on each sale up to $100,000 and 3% of the amount above $100,000. How much would Eduardo earn for selling each of these houses? **10.** $4997 **11.** $3497

10.

BEAUTY & SPACE
$199,900 5 bedrooms, 3 1/2 baths, formal living and dining. Study. Master bedroom. Lots of storage. Jacuzzi.

11.

LOCATION PLUS $149,900
4 bedrooms, 2 1/2 baths, formal living and dining. Family room, game room.

The salespeople below are offered a choice of two monthly payment plans.

Plan 1: Base salary + commission
Plan 2: A higher rate of commission and no base salary

Let t represent the amount of sales and E represent the earnings. Write an equation for each payment plan. By solving the equations together, find the amount of sales needed to make the same earnings with either plan.

12. Eric
Plan 1: 9% of sales + $900
Plan 2: 12% of sales
$E = 0.09t + 900$; $E = 0.12t$;
$t = \$30,000$

13. Wanda
Plan 1: 7% of sales + $2400
Plan 2: 10% of sales
$E = 0.07t + 2400$; $E = 0.10t$,
$t = \$80,000$

ASSIGNMENT GUIDE

The exercises all deal with actual problems that could be faced by salespeople working on commission or by the piece. The first four exercises ask students to think about their potential roles in the selling world. Exercises 5-13 help the future salesperson to calculate the reward offered by various commission formulas. Exercises 14-16 show the somewhat different problems of the piece worker. Unlike a lucky or industrious salesperson on commission, the piece worker is limited by the time that is available to reach a monetary goal. The kind of frustration that a piece worker may feel is illustrated in Exercise 16 by Boris's inability to make $700.

ADDITIONAL ANSWERS

5. a. $6.75
b. $y = 0.375x$
c. $6.75; $10.50; $14.25; $18.00

6. a. $11.22
b. $y = 1.8696x$
c. $11.22; $18.70; $26.17; $33.65.

7. a. $11.39
b. $y = 0.75905x$
c. $11.39; $18.98 $26.57; $34.16

Boris repairs TV and VCR remote-control units at $7.50 an hour in an electronic equipment repair store. He also gets $4.00 for each unit that he repairs. He works a 40-hour week.

14. How many units does he have to repair to earn at least $500 per week? 50 units

CLOSURE

Ask students which method of earning money they find more appealing, working on commission or piece work. Point out the advantages and disadvantages of each method. Also mention the importance of one's own personality. Salespeople, unlike piece workers, often need to be outgoing and openly enthusiastic.

LESSON QUIZ

Ivan's boss offers him a choice of either making or selling widgets. In each case, how much will he earn on 200 widgets that sell for $45 each?

1. selling: 3% commission on the first $2000 of sales, 4% of sales over $2000 **($340)**
2. making: $1 for each of the first 100 widgets, $2 for each widget over 100 **($300)**

15. It takes Boris an average of $\frac{1}{2}$ hour to repair a unit. What is the most money that he can earn in one week? $620

16. How many units does Boris have to repair to earn $700 in one week? Explain your answer. 100 units. Boris would need to work faster and repair a unit in 24 minutes (40 h • 60 min/h ÷ 100).

MIXED REVIEW

Determine the total weekly earnings for each employee described in each classified ad below. Use an equation of the form $E = rh + t$, where r represents the regular hourly rate, h represents the number of hours worked, and t represents the tips or bonus, if any. **1.** $450 **2.** $317.50

1. 1-1

Health Care
LAB TECHNICIAN $10/h
45 hours/week

2. 1-1

Manicurist
English and Spanish speaking. Guaranteed pay $6.50/h to start. Tips can be up to $90 for a 35-hour week!

For each company fringe benefits policy described below, find

1-2 **a.** the amount that the company must pay for extra taxes, retirement, health and life insurance, and other fringe benefits.

b. how much salary is paid for the number of nonworking days in a total of 260 operating days.

	EMPLOYEE	SALARY	FRINGE BENEFITS POLICY
3.	Bookkeeper	$40,000	7.65% of gross for FICA taxes
	a. $5460.00		3% of gross for retirement
	b. $4000.00		$100/month for life and health insurance
			26 nonworking days
4.	Cashier	$18,500	7.65% of gross for FICA taxes
	a. $2015.25		$50/month for health insurance
	b. $1565.38		8 holidays, 4 sick days, 10 vacation days
5.	Bank manager	$55,000	7.65% of gross for FICA taxes
	a. $8047.50		$120/month for life and health insurance
	b. $3807.69		$200/month expense allowance
			18 nonworking days

1-4 DEDUCTIONS: WHO GETS WHAT?

QUICK REFERENCE

TEACHER'S RESOURCES AND ANSWER KEY

Lesson Quiz Answers 1-4, p. 40
Reteaching Activity Answers 1-4, p. 91
Enrichment Activity Answers 1-4, p. 92

EXTENSION ACTIVITIES

Reteaching Activity 1-4, p. 4
Enrichment Activity 1-4, p. 8

Luis has known Alex since they were in the seventh grade together. He and Alex have been comparing notes about the increasing costs of dating, keeping a car running, and getting any kind of schooling after high school. Luis realized three years ago that the only way to come up with the money he needed was to find a job. He started out at the supermarket, first stocking the shelves, then bagging the groceries. Now he runs the cash register and sometimes gets to work behind the deli counter.

What a shock it was to Luis the first time he received a weekly paycheck. He knew that he had worked 20 hours and that he was supposed to get $4.75 an hour. He figured that the $95.00 was just enough to buy one new tire for his car. When he saw that the check was for only $86.73, he immediately wanted to know what had happened to the rest of the money.

Luis knew that all the deductions from the check were somehow related to the mass of papers that got spread around the family dining room every April. His mother would perform numerous calculations with her calculator and transcribe some of them onto sheets of paper about taxes. It was all very mysterious, so Luis didn't ask questions. After April 15, all the papers vanished for another year.

Luis has heard his parents mention Medicare frequently in the last year. His grandfather has been hospitalized for a chronic illness and has had numerous medical tests and treatments. Some of these costs are covered by Medicare.

OBJECTIVES: *In this lesson, we will help Luis to:*

- *Understand why income taxes are deducted from a paycheck and what they are used for.*
- *Learn what FICA taxes are for.*
- *Realize why a new employee fills out a W-4 form.*
- *Calculate take-home pay by subtracting deductions from gross pay.*

TAXES: A TOOL OF GOVERNMENT

When Luis began to investigate the matter, he learned that the taxes collected by the federal government every year are used for many purposes, the largest of which are the nation's defense, the funding of the Social Security system and Medicare, and interest on the national debt. Federal taxes also support a variety of other activities including national parks, airport construction, agriculture, nutritional programs, school lunch programs, and libraries.

Luis's taxes—and yours, too, of course—are used to support not only federal activities, but also activities of state and local governments. In our society, government performs several broad functions, many of which significantly affect the general economy. These include the upkeep and funding of national parks and public highways, the redistribution of income to care for the needy, regulation of certain industries such as communications, and the establishment and preservation of the legal framework for the protection of the rights of citizens.

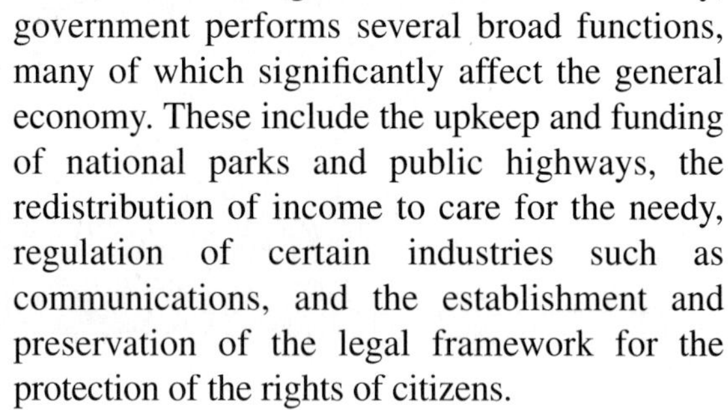

Luis knows that parks, highways, schools, and the police department are supported by taxes. He does not know how the government redistributes income so that some citizens pay for government benefits to other citizens. Taxes are the government's major means of making these transfer payments to support federal programs such as Social Security, programs for disabled veterans, disaster relief, and various welfare services.

Luis knows that, as a citizen, he is entitled to suggest changes in the way his taxes are spent. The legislators representing him are the people who can actually make the changes in the tax laws. In doing so, they must take into account the effect that new tax legislation will have on workers, consumers, business and industry, and the economy in general. This may sound easy, but it is not. Legislators realize that no one likes tax increases, and so they avoid raising taxes if they possibly can.

WHY SOCIAL SECURITY?

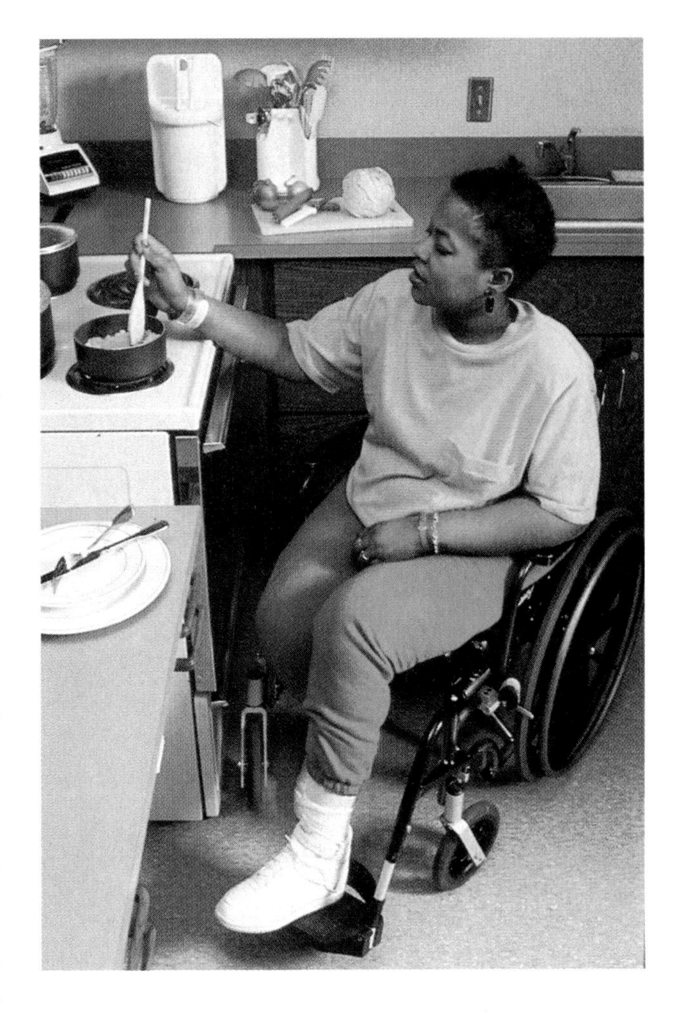

So government operations are supported by income taxes. Are Social Security taxes just another form of income tax?

From the moment we are born until we die, people of all ages are affected if **income security** is absent. Children are affected if their parents lose their ability to earn an income. Young adults are concerned about how they will survive if they are disabled by illness or injury. Middle-aged people are concerned about how their children would be supported and educated if they were to die or become disabled. Older retired people often worry that their retirement income will not cover their living expenses.

In 1935 the U.S. government established **Social Security** to help protect citizens from economic insecurity. Simply put, the Social Security system is a combination of disability insurance and a pension plan that is compulsory for most workers. It provides a base on which individuals may build protection for themselves and their dependents. The cost of the system is paid for by both workers and employers.

Most employed people have **FICA** (Federal Insurance Contributions Act) taxes deducted from their paychecks. Employers also pay into the program by matching the amounts paid by the employees. A portion of the FICA taxes is paid into the Social Security trust fund, which entitles the employee eventually to receive Social Security benefits. A portion is used to pay for **Medicare.**

A few of the benefits paid for by FICA taxes include the following:

SOCIAL SECURITY

- **Disability insurance** Income to individuals who become disabled and thus cannot work. Disability-income benefits are paid after a five-month waiting period during which it is determined whether the physical or mental disability is severe enough to prevent the person from working for at least a year.
- **Survivors insurance** Money for monthly living expenses to survivors of deceased workers who were eligible for Social Security benefits. These benefits go to children and in some cases to a widow or widower.
- **Retirement benefit** Monthly Social Security payments made to a retired worker and to the worker's spouse.

MEDICARE

- **Health insurance** A two-part health insurance program for people over age 65. Medicare benefits consist of hospital insurance and medical (doctor) insurance.

CRITICAL THINKING

Ask students whether they think that most other countries have a Social Security pension similar to our own. Which other highly developed countries have no such system at all? Which countries (highly developed or not) induce citizens to save for their retirement by sponsoring savings plans with attractive tax advantages? In addition to IRA plans in the United States (see Lesson 7-3) suggest Japan and Chile as two countries to investigate.

FORMS

When Luis began his job at the supermarket, he was asked to fill out a **Form W-4,** *Employee's Withholding Allowance Certificate.* The information that he put on this form determined the amount that his employer deducts from Luis's paycheck for income taxes. The amount depends on Luis's income level, his marital status, and the number of withholding allowances that he claims.

Form **W-4** Department of the Treasury Internal Revenue Service

Employee's Withholding Allowance Certificate

▶ **For Privacy Act and Paperwork Reduction Act Notice, see reverse.**

OMB No. 1545-0010 19--

1 Type or print your first name and middle initial — Last name: Luis Estevez

2 Your social security number: 123-45-6789

Home address (number and street or rural route): 456 Cedar Street

3 ☒ Single ☐ Married ☐ Married, but withhold at higher Single rate.
Note: *If married, but legally separated, or spouse is a nonresident alien, check the Single box.*

City or town, state, and ZIP code: Inland, In. 12345

4 If your last name differs from that on your social security card, check here and call 1-800-772-1213 for more information . . . . ▶ ☐

5 Total number of allowances you are claiming (from line G above or from the worksheets on page 2 if they apply) . — 5: 1

6 Additional amount, if any, you want withheld from each paycheck . . . . — 6: $

7 I claim exemption from withholding for 1993 and I certify that I meet **ALL** of the following conditions for exemption:
- Last year I had a right to a refund of **ALL** Federal income tax withheld because I had **NO** tax liability; **AND**
- This year I expect a refund of **ALL** Federal income tax withheld because I expect to have **NO** tax liability; **AND**
- This year if my income exceeds $600 and includes nonwage income, another person cannot claim me as a dependent.

If you meet all of the above conditions, enter "EXEMPT" here . . . . ▶ 7

Under penalties of perjury, I certify that I am entitled to the number of withholding allowances claimed on this certificate or entitled to claim exempt status.

Employee's signature ▶ — **Date** ▶ , 19

8 Employer's name and address (Employer: Complete 8 and 10 only if sending to the IRS) — 9 Office code (optional) — 10 Employer identification number

Cat. No. 10220Q

FOCUS

ALGEBRA CONNECTIONS

Suppose that g stands for gross pay, w for income-tax withholding, and f for FICA taxes. Which expression could not represent a person's take-home pay?

$g - (w + f)$
$g - w - f$
$g - w + f$
$-w - f + g$
$(g - w + f)$

His employer does not have to withhold any tax from Luis's income if he did not have any income-tax liability (that is, did not owe any taxes) last year and if he does not expect to have any tax liability this year. Students often qualify for this no-withholding status because they have only part-time or summer jobs.

READING YOUR PAYCHECK

When he got his first job, Luis really looked forward to his first paycheck. He knew that his income from this job would be the first step toward financial

Evans Foods
1000 Center Street
Inland, IN 12345

Pay Statement

Check No.	Name	Social Security Number
3705	Luis Estevez	123-45-6789
Pay Date	**Pay Period**	**Type**
07/10/--	07/03/-- - 07/09/--	Salary

	Gross Pay	Federal Inc. Tax	FICA	State Inc. Tax	Other	Net Pay
	$95.00	$1.00	$7.27			$86.73

Evans Foods
1000 Center Street
Inland, IN 12345

3705

July 10 19 *--*

PAY
TO THE
ORDER OF *Luis Estevez* $ *86.73*

Eighty-six and $\frac{73}{100}$ DOLLARS

Commercial Bank
Inland, Indiana

FOR ______ *Amelia Evans, manager*

independence. That check is shown above. A pay statement summarizing his deductions is shown at the bottom of page 38.

Young workers are often surprised by the difference between total pay, or **gross pay,** and net pay, or take-home pay. **Take-home pay** is the amount remaining after federal and state taxes and other deductions have been taken out of the check. A **deduction** is an amount of money subtracted from a person's gross pay for such items as FICA taxes, federal and state income taxes, insurance premiums, and union dues. After deductions your take-home pay is likely to be about 20% to 30% less than your gross pay.

Ask Yourself

1. What are five broad economic functions of our government?
   See Additional Answers.
2. What are four examples of the redistribution of income that the government accomplishes through taxation and spending?
   See Additional Answers.
3. Who pays for the federal Social Security system?
   Employed people and employers
4. How does your employer determine how much of your income to withhold for taxes?
   From your Form W-4
5. When you first get a job, which federal tax form are you asked to complete and submit to your employer?
   Form W-4
6. How is take-home pay computed?
   Gross pay minus all federal and state taxes and other deductions.

ALGEBRA *REVIEW*

Percent means "per hundred."

Example $7\% = 0.07 = \frac{7}{100}$

Express each as a decimal.

1. 5% — 0.05
2. 20% — 0.20
3. 7.5% — 0.075
4. 0.5% — 0.005
5. 7.65% — 0.0765
6. 125% — 1.25

Express each as a percent.

7. 0.30 — 30%
8. 0.02 — 2%
9. 1.5 — 150%
10. 0.004 — 0.4%

Find each amount.

11. 5% of 120 — 6
12. 7% of $4.56 — 0.3192
13. 120% of $808 — 969.6
14. 7.5% of $9.10 — 0.6825
15. 7.65% of $4920 — 376.38

SHARPEN YOUR SKILLS

SKILL 1

INSTRUCTION

TEACHING THE LESSON

This lesson is an introduction to some of the practical problems of dealing with federal tax deductions. (These issues are covered in more depth in Chapters 4 and 9.) Students are shown how the manner in which they fill out a W-4 form directly affects their tax obligation with federal government. They should realize that if they choose "0 deductions," then they are likely to have a refund if the number of actual deductions turns out to be greater than 0. Emphasize to students that the FICA taxes are collected to fund not only their Social Security pension beginning at age 62 or later, but also the Medicare benefits that they will be eligible for at age 65.

To find the amount to withhold for federal income tax, use a table like the one shown for a single person with weekly wages.

SINGLE Persons—WEEKLY Payroll Period

(For Wages Paid in 19--)

If the wages are—		And the number of withholding allowances claimed is—					
At least	But less than	0	1	2	3	4	5
		The amount of income tax to be withheld is—					
$0	$50	$0	$0	$0	$0	$0	$0
50	55	1	0	0	0	0	0
55	60	1	0	0	0	0	0
60	65	2	0	0	0	0	0
65	70	3	0	0	0	0	0
70	75	4	0	0	0	0	0
75	80	4	0	0	0	0	0
80	85	5	0	0	0	0	0
85	90	6	0	0	0	0	0
90	95	7	0	0	0	0	0
95	100	7	1	0	0	0	0
100	105	8	1	0	0	0	0
105	110	9	2	0	0	0	0

To find FICA taxes (Social Security and Medicare), use the following.

> If annual income is less than $57,600, then the deduction for FICA taxes is 7.65% of gross pay.

EXAMPLE 1 Luis's salary for this week is $95. Luis entered one withholding allowance on his W-4 form.

QUESTIONS
1. How much will be withheld for income tax?
2. How much will be deducted for Social Security and Medicare?
3. How much is Luis's take-home pay?

SOLUTIONS

1. Find the amount to withhold on the table shown above. Notice that Luis's wages, $95.00, are on the eleventh line, "at least 95, but less than 100." Read across to the withholding allowance column headed 1. The number in that column is the dollar amount to be withheld. The amount withheld for income tax from Luis's check will be $1.

2. Luis's salary is less than \$57,600, so the deduction for FICA taxes is 7.65% of the gross pay.

   $0.0765 \cdot 95 = 7.2675$ 7.65% = 0.0765

   The deduction for FICA taxes is \$7.27.

3. Luis can find his take-home pay by writing and solving an equation for his take-home pay T, where g = gross pay, w = amount withheld for income taxes, and f = deduction for FICA taxes.

   Take-home pay = gross pay − all deductions

   $$T = g - (w + f)$$
   $$T = 95 - (1 + 7.27)$$
   $$T = 95 - 8.27$$
   $$T = 86.73$$

   Luis's take-home pay is \$86.73.

FOCUS ON ALGEBRA

The formula $T = g - (w + f)$ is introduced as an aid in finding the take-home pay that remains after deductions for income and FICA taxes have been subtracted from gross income. *(NCTM Standard 1, p.137; Standard 5, p.150)*

ESL STUDENTS

Students who are unfamiliar with technical terms in English may need special attention. The word "security" may mislead recently arrived residents into overestimating the size of those benefits.

SKILL 2

EXAMPLE 2 Luis wanted to see how he could use a spreadsheet to represent his take-home pay.

SOLUTION

He set up a spreadsheet in the following way.

	A	B	C	D	E
1	Name	Luis Estevez			
2					
3	Gross	Withholding	Income Tax	FICA	Take-Home
4	Pay	Allowances	Withholding	Withholding	Pay
5					
6					

He used row 1 for the Name line and rows 3 and 4 for the five column headings. He completed row 6 by typing the following into cells A6–E6. He formatted columns D and E for two decimal places.

6	95	1	1	0.0765*A6	+A6−(C6+D6)

On the computer screen, row 6 actually appears as follows.

6	95	1	1	7.27	86.73

By using formulas in columns D and E, Luis was able to find the take-home pay for any value of the gross pay that he wishes by filling in columns A and C with the correct numbers and letting the computer calculate the values in cells D6 and E6.

TECHNOLOGY HINT

In the computer spreadsheet solution for Skill 2, remind students about the spreadsheet program's COPY command that allows them to generate several rows of calculations in columns D and E using the basic formulas of row 6.

COMMON ERRORS

When typing formulas in a computer spreadsheet program, students may forget some of the special rules for expressing such formulas. Remind them, for example, that a plus sign (+) needs to precede the first character of a formula if that character is a letter. Thus, in the spreadsheet of Skill 2, 0.0765*A6 can be entered without the plus sign but A6−(C6+D6) must be entered as +A6−(C6+D6).

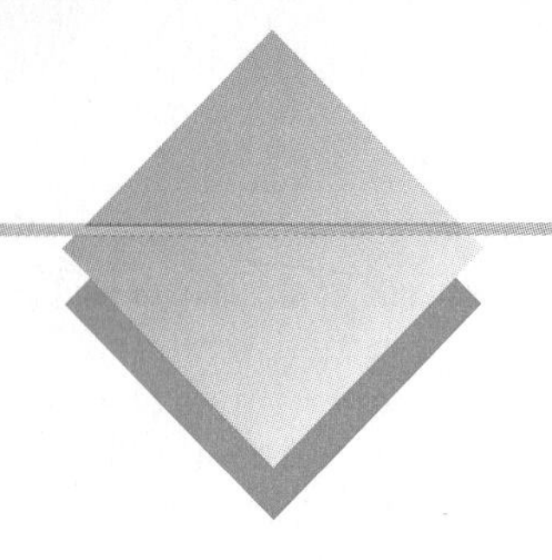

TRY YOUR SKILLS

Use the income tax withholding table in the Reference Section to find the take-home pay for the given wages and withholding allowances. Assume that each person is married. For the FICA deduction, find 7.65% of gross pay. Use the formula for take-home pay

$$T = g - (w + f)$$

where g is the gross pay, w is the amount withheld for taxes, and f is the deduction for FICA.

GUIDED PRACTICE

Students are asked to calculate their take-home pay using the take-home pay formula. They do this directly in Exercises 1-10 and indirectly through a spreadsheet formula in Exercises 11-20.

	Monthly Salary	Withholding Allowances			Monthly Salary	Withholding Allowances	
1.	$2400	0	$1930.40	**2.**	$ 700.00	3	$ 646.45
3.	3400	5	2850.90	**4.**	4380.00	1	3391.93
5.	1280	5	1182.08	**6.**	4600.00	1	3528.10
7.	2000	3	1710.00	**8.**	2208.33	5	1930.39
9.	2550	4	2168.92	**10.**	1666.67	0	1367.17

11.–20. Create a computer spreadsheet to calculate the take-home pay for the people in Exercises 1–10. Use formulas for the columns for the FICA deduction and take-home pay. See Additional Answers.

EXERCISE YOUR SKILLS

1. Why is it difficult to make changes in tax policies? See Additional Answers.
2. Why did the federal government start the Social Security system?
3. The Social Security taxes that you are paying now are being transferred to current beneficiaries. Who will pay for the benefits that you will receive in the future? The workers in the future.
4. Why do you suppose that the Social Security system, unlike private insurance, is not voluntary for the majority of workers? Answers may vary.
5. The provisions for retirement benefits under the Social Security system eliminate or alleviate certain risks for its recipients. What are some of these risks? Retirement income will not cover living expenses.

KEY TERMS

deduction
disability insurance
FICA deduction
form W-4
gross pay
health insurance
income security
Medicare
retirement benefits
Social Security
survivors insurance
take-home pay

Use the income tax withholding table in the Reference Section to find the take-home pay for the given wages and withholding allowances. Assume that each person is married. For the FICA deduction, take 7.65% of gross pay. Use the formula for take-home pay

$$T = g - (w + f)$$

where g is the gross pay, w is the amount withheld for taxes, and f is the deduction for FICA (Social Security and Medicare).

	Monthly Salary	Withholding Allowances			Monthly Salary	Withholding Allowances	
6.	$2640	1	$2146.04	**7.**	$2500.00	1	$2040.75
8.	4720	2	3659.92	**9.**	1500.00	5	1384.25
10.	4000	5	3315.00	**11.**	1500.00	4	1355.25
12.	3000	3	2483.50	**13.**	4166.67	2	3305.92
14.	4700	1	3598.45	**15.**	3500.00	3	2873.25

16.–25. Create a computer spreadsheet to calculate the take-home pay for the people in Exercises 6–15. Use formulas for the columns for Social Security/Medicare and take-home pay. See Additional Answers.

MIXED REVIEW

1. Marsha's job at the library pays her $12.50/h plus $1\frac{1}{2}$ times her regular wage rate for each hour over 35 hours. How much did she earn in a week during which she worked 42 hours? $568.75
1-1

2. A law firm is offering a new employee the following fringe benefits: $1100 in annual health insurance premiums, 6% of gross pay to match the employee's retirement contributions, and a monthly travel allowance of $100. The annual salary is $53,000, and the FICA taxes are 7.65% of gross pay. Find the total annual cost that the law firm has to budget for the new employee. $62,534.50
1-2

3. Alvin receives a 4% commission on his total automobile sales in a given month plus an additional 1.5% on all sales over $60,000 during the same month. In April, Alvin sold $87,000 worth of automobiles. What were his total earnings for April? $3885
1-3

4. Inez receives a weekly salary of $280 at the real-estate office where she works and also receives a 3.6% commission for each house that she sells. In a recent year she worked for 45 weeks and was credited with sales of $600,000. What was her annual income? $34,200
1-3

5. Ricardo owns a tour-bus company. He offers his tour guides a choice between receiving $1.50 for each tourist who rides on the guide's tour bus and receiving a salary of $150 for a five-day week plus $1.00 for each tourist. Each guide conducts two tours a day, and each bus holds 40 tourists. On the average, what is the least percent of the bus that should be full for the straight-commission option (no salary) to be the more attractive choice? 75%; $1.50n \cdot 5 = 150 + 5n$; $n = 60$; 60 is 75% of 80.
1-3

6. Katrina receives a salary of $46,000 and takes vacation days and holidays for 24 of the company's 260 operating days. How much of her salary is paid for the days during which she does not work? $4246.15
1-2

7. Diane assembles electronic beepers in a factory. She receives $15 for each deluxe model and $11 for each standard model. How much did she earn in a week when she assembled 17 deluxe models and 13 standard models? $398
1-3

ASSIGNMENT GUIDE

Exercises 1-5 are intended to encourage students to reflect upon and discuss governmental tax policies, especially those that relate to pension benefits. Exercises 6-25 provide further familiarization with the effects of deductions on one's paycheck.

RETEACHING

Show students how the way in which they fill out their W-4 form affects their tax payment or refund. Have them calculate their deductions using the following numbers of withholding allowances: 0, 1, and 2.

ENRICHMENT

Have students contact local small business owners to see how they process income tax and FICA tax deductions for employee payrolls.

CLOSURE

Reemphasize that the choice of the number of withholding allowances on a W-4 form affects the amount withheld. Ask students whether they prefer to over-withhold or under-withhold.

LESSON QUIZ

Find the take-home pay for a single person with the indicated gross income and number of allowances.

1. $260 a week; 1 **($214.11)**

2. $3,200 a month; 2 **($2449.20)**

CHAPTER 1 REVIEW

1. If your company requires a lot of overtime, would you prefer to work for an hourly rate or for a monthly salary. Why? Hourly. You would be paid for overtime.
2. Why would you not trust the advice of a salesperson whom you knew was getting push money? See Additional Answers.
3. Why is your take-home pay less than your gross earnings?
4. Why would you expect a person with little education, job training, or both to have more difficulty getting a job than a person with a college degree and work experience? Many jobs require more education or experience.

ADDITIONAL ANSWERS

2. The salesperson may say the item being pushed is better when it isn't.
3. Deductions are subtracted.

Find the total weekly pay based on an overtime rate of $1\frac{1}{2}$ times the regular hourly rate for all working time over 20 hours.

	Number of Hours	Wages/Hour	Bonuses	
5.	42.0	$ 6.50	$10.00	$ 354.50
6.	35.5	18.75	20.00	830.94
7.	40.0	21.50	41.00	1116.00

Find the total number of hours you would work during the week if you worked on the following time schedule.

8. Saturday 3–8 P.M. Sunday 3–8 P.M. Monday 4–11 P.M.
   Tuesday off Wednesday 4–8 P.M. Thursday 6–10 P.M. 25 h
9. Assume that you work on the schedule in Exercise 8. Find your total weekly salary if you make $5.25 per hour and an overtime rate of $1\frac{1}{2}$ times your regular hourly rate for any time over 20 hours. $144.38
10. Find the total annual cost to an employer for a new employee hired at $30,000 per year. The annual insurance premiums are $900 for health insurance and $120 for life insurance. The FICA taxes are 7.65% of gross pay. $33,315
11. Find the total annual cost to an employer to hire a lab technician at $44,000 per year. Health insurance is $260 per month. Retirement is 3% of gross pay and the FICA taxes are 7.65% of gross pay. $51,806

Find the amount of total sales and total commissions made on the basis of the given information.

	Number of Items	Unit Price	Commission Rate	Total Sales	Commission
12.	25	$ 108.92	15%	$ 2,723	$408.45
13.	3	3405.00	8.2%	10,215	837.63
14.	42	23.50	6.3%	987	62.18
15.	150	19.50	7%	2,925	204.75

Solve each problem.

16. Your salary is $300 per week plus 15% commission. What is your week's pay if your total sales for the week are $4175? $926.25

17. Katy assembles craft kits. She receives $8 for each regular kit and $12 for each advanced kit. How much did she earn in a week in which she assembled 35 regular kits and 22 advanced kits? $544

18. Shawn sells annual memberships in an exercise club. His boss has offered him a choice between Plan 1, no salary but a straight $45 for each membership sold, and Plan 2, a weekly base salary of $250 plus $20 for each membership sold. How many memberships would Shawn have to sell to make the same amount of money under either plan? 10 memberships

19. Your graduated commission pays you 3% on all sales up to $7000 and 5% on all sales over $7000. What is your pay if your total sales are $10,800? $400

20. Your piece rate at a local toy factory is $2.75 for each assembled toy, and your weekly salary is $170. What are your earnings in a week during which you assembled 110 toys? $472.50

Use the income tax withholding tables in the Reference Section to find the income tax and FICA withholding for each single person making the monthly salaries given below. What will be the take-home pay for each person?

	Monthly Salary	Withholding Allowances	Income Tax Withholding	FICA Withholding	Take-Home Pay
21.	$1500.00	0	$193	$114.75	$1192.25
22.	2728.00	2	372	208.69	2147.31
23.	1731.20	5	83	132.44	1515.76

For each company fringe benefits policy described below, find

a. the amount that the company must pay for extra taxes, retirement, health and life insurance, and other fringe benefits.

b. how much salary is paid for the indicated number of nonworking days in a total of 260 operating days.

	Employee	Salary	Fringe Benefits Option
24.	Auto mechanic	$32,000	7.65% of gross for FICA taxes
	a. $5208.00		$150/month for health insurance
	b. $3446.15		3% of gross for retirement
			28 nonworking days
25.	Bank teller	$23,500	7.65% of gross for FICA taxes
	a. $3237.75		$120/month for health insurance
	b. $1988.46		22 nonworking days

CHAPTER 1 TEST

1. If you work 60 hours each week, would you make more money working for $6 an hour or for $1200 a month? $6/h
2. How does your take-home pay differ from your gross earnings?
3. Which of the two following employment positions provides more money in salary and fringe benefits? Atwood

	Atwood Electronics	***Technology, Inc.***
Salary	$38,500	$36,000
Medical insurance	$85/month	$115/month
Retirement benefits	3% of gross pay	3.5% of gross pay
Expense account	$450/month	$130/month

Atwood: $46,075; Technology, Inc.: $40,200

4. If your pay is $500 per week plus a 12% commission, what is your week's pay during a week in which your total sales are $2723? $826.76
5. Your graduated commission pays 3% for sales up to $8500 and 5% on sales over $8500. Find your pay for total sales of $12,000. $430

Find the total weekly pay based on an overtime rate of $1\frac{1}{2}$ times the regular hourly rate for all working time over 20 hours.

6. You work $43\frac{1}{2}$ hours at $4.75 per hour and receive $19.50 in tips. $281.94
7. You work 37 hours at $4.60 per hour and receive $28.00 in tips. $237.30

Assume that 40 hours = 1 workweek, 52 weeks = 1 year, and 12 months = 1 year. Find the hourly, weekly, monthly and yearly salaries for the following pay rates. See Additional Answers.

8. $6392 per month **9.** $1276.80 per week **10.** $31.70 per hour

Find the total sales and total earnings based on the information shown.

11. You sell 17 sewing kits at $35 each. Your piece rate is $4 per kit. $595; $68
12. You sell 78 VCRs at $383.60 each. Your commission rate is 10%.
13. You sell 5 virtual-reality consoles at $2050 each. Your commission rate is 10%. $10,250; $1025

Use the tax tables in the Reference Section to complete the following chart for each single person making the monthly salaries given.

	Monthly Salary	**Withholding Allowances**	**Income Tax Withholding**	**FICA Withholding**	**Take-Home Pay**
14.	$3456	3	$519	$264.38	$2672.62
15.	2880	4	307	220.32	2352.68
16.	4000	1	785	306.00	2909.00

ALTERNATIVE ASSESSMENT

Ask students to investigate the form of payment young working people in their community actually receive. What percent of young workers are paid hourly, weekly, by the piece, or by commission? Have the student researchers report their findings to the class. The United States government periodically revises its rules about the income maximums for Social Security taxes (currently at $57,600) and for Medicare taxes (currently at $135,000). Have students find out the most recent changes in these maximums and also determine whether there is any plan to revise the 7.65% rate upward or downward. As a hint for the students, suggest that they ask the IRS for the most recent version of Publication 15 (Circular E, The Employer's Tax Guide).

ADDITIONAL ANSWERS

2. Take-home pay = gross pay − deductions

8. Hourly: $36.88
Weekly: $1,475.08
Yearly: $76,704

9. Hourly: $31.92
Monthly: $5,532.80
Yearly: $66,393.60

10. Weekly: $1,268
Monthly: $5,494.67
Yearly: $65,936

12. $29,920.80; $2992.08

CUMULATIVE REVIEW

1-1 Determine the total weekly earnings for each employee described in each classified ad below. Use an equation of the form $E = rh + t$, where r represents the regular hourly rate, h represents the number of hours worked, and t represents the tips or bonus, if any.

1.

Dental Care
LAB TECHNICIAN $13/h
40 hours/week $520

2.

Hair Stylist
English and Spanish speaking. $15.50/h to start. Tips can be up to $150 for a 35-hour week! $692.50

1-1 **3.** Find the weekly earnings of a server who earned $4.80 an hour for 25 hours at a restaurant and who received $130 in tips during that time. $250

1-1 **4.** Keith earns $7.50 an hour and $1\frac{1}{2}$ times that amount for time over 20 hours in 1 week. What is his pay for 23 hours of work? $183.75

1-2 **5.** Does the size of a fringe benefit necessarily increase when your salary is raised? Explain your answer. See Additional Answers.

1-2 **6.** An auto parts store is offering a new manager the following fringe benefits: $180/month in health insurance premiums and 5% of gross pay to match the employee's retirement contributions. The annual salary is $47,500 and the FICA taxes are 7.65% of gross pay. Find the annual cost that the auto parts store has to budget for the manager. $55,668.75

1-3 **7.** Carmen sells real estate. She receives 5% of the first $100,000 of the selling price of a house and 6% of the amount over $100,000. How much commission does Carmen receive on a house that sells for $225,000?

1-3 **8.** Sylvia has a part-time business videotaping weddings. Her rate is $75 for the ceremony and $150 if both the ceremony and the reception are covered. Last year she was able to cover 25 weddings, 10 of which included the reception. Her expenses for the year were $200. What were her earnings for the year? $2425

1-3 **9.** James has a monthly base salary of $900 and a commission rate of 9% on his sales. Last month he sold 25 stereo systems at $1450 each and 18 VCRs at $425 each. What were his earnings last month? $4851.00

1-4 Use the income tax withholding tables in the Reference Section to find the income tax and FICA withholding for each single person making the monthly salaries given below. What will be the take-home pay for each person?

	Monthly Salary	Withholding Allowances	Income Tax Withholding	FICA Withholding	Take-Home Pay
10.	$4440	2	$854	$339.66	$3246.34
11.	2990	1	494	228.74	2267.26
12.	1250	0	157	95.63	997.37
13.	2200	3	213	168.30	1818.70

ADDITIONAL ANSWERS

5. No. Some benefits, such as health insurance, are the same for all employees.
7. $12,500

PROJECT 1–1: **College Information**

Obtain information from colleges you might wish to attend.

1. Call or write three colleges for a copy of their college catalog.
2. Make a chart comparing costs of tuition; room and board; transportation to school; personal expenses; and so on. **Note:** Transportation to school may or may not be important depending upon how many times during each year you return home.
3. List the field(s) of study you are considering.
4. List the degrees or majors offered by each school.
5. Bring in three college catalogs. Turn in your cost comparison chart. Turn in a list of majors and degrees offered.
6. Write about which college you think you would like to attend and explain why.

Extension

Interview people who have attended college. Write a one-page report on their experiences.

PROJECT 1–2: **Career Interview**

1. Make arrangements for an interview with someone who has a job in a career field or occupational area that interests you.
2. Prepare for the interview by making a list of questions that you would like answered. The following list of job topics should help you prepare questions for the interview.

Place of work and daily tasks	Hours worked
Working environment	Education required
Appearance needed for job	Experience needed
Special talents needed	Strength needed
Hazards at work	Time pressures on job
Entry-level earnings	Top-level earnings
Type of work—easy/hard	Future of job potential
Fringe benefits of job	Security of job
Work as part of a team	Work individually

3. Complete the interview. Write a report that summarizes what you have learned. Include specific questions and answers from your interview.

Solve each system of equations for x and y using substitution.

Example

$$x + y = 10 \quad (1)$$
$$2x - y = 8 \quad (2)$$

$y = 10 - x$ — Solve (1) for y.

$2x - (10 - x) = 8$ — Substitute $10 - x$ for y in (2).

$2x - 10 + x = 8$ — Simplify.

$3x = 18$ — Collect terms.

$x = 6$

$6 + y = 10$ — Substitute 6 for x in (1).

$y = 4$

The solution is $x = 6$ and $y = 4$. Check the solution in both original equations.

1. $x - 2y = 11$
$2x + y = 7$
(5, −3)

2. $2x + 3y = -1$
$-5x - y = 9$
(−2, 1)

3. $x + y = 4$
$x - y = 1$
(2.5, 1.5)

4. $y = -2.5$
$2x + 3y = 2.5$
(5, −2.5)

A system of linear equations can also be solved using the addition and subtraction method. First multiply or divide one or both equations by numbers so that one pair of like terms has opposite coefficients. Then add or subtract the equations to eliminate one variable.

Solve each system of equations using the addition and subtraction method.

Example

$$x + y = 1 \quad (1)$$
$$2x + 5y = -4 \quad (2)$$

$-2x - 2y = -2$ — Multiply equation (1) by −2.

$2x + 5y = -4$ — Equation (2)

$3y = -6$ — Add equations.

$y = -2$ — Divide by 3.

$x + (-2) = 1$ — Substitute −2 for y in equation (1).

$x = 3$ — Add 2 to both sides.

The solution is (3, −2). Check the solution in both original equations.

5. $x + y = 7$
$2x - 3y = -1$
(4, 3)

6. $2x + y = 3$
$x - 2y = 9$
(3, −3)

7. $2x + 3y = 7$
$3x - 2y = 4$
(2, 1)

8. $x - y = 1$
$-3x + 2y = 0$
(−2, −3)

9. $x + 5y = 3$
$2x - y = 6$
(3, 0)

10. $x - y = 0$
$5x + 4y = -9$
(−1, −1)

CHAPTER 2

Check Out the Account

IN THIS CHAPTER THE young people in our story will learn about checking accounts. Jeff will examine the differences among services offered with checking accounts at different banks and will compare charges. He will calculate new checking account balances based on the cost of checks, service charges, extra fees, and interest earned.

Latoya will show us how to write a check, endorse it correctly, and keep a proper check register. She imagines herself the treasurer of the Eldorado Open Golf Tournament, writing checks for thousands of dollars to the winners. If she pursues her interest in business and finance, she may some day be responsible for large amounts of money.

Larry will discover the necessity of keeping a close watch on how much money he has in his checking account and will reconcile his check register with a bank statement. He is fascinated by automated teller machines, or ATMs, which give him money from his checking account at convenient locations all over town–just for entering the correct codes. But he is still a little uncertain about exactly what happens at the other end–in his checking account at the bank. He needs to learn more.

Many young people open their own checking accounts when they begin working at regular jobs and handling money that is truly their own. A checking account can help you use money safely and efficiently. But it needs to be managed carefully so that you always know how much money you have!

Eldorado Golf Tournament — 001

Eldorado Golf Club
Inland, IN 47304
Telephone 555-7902

March 4, 19 *– –*

Pay to the order of *Cynthia Alvarez* $ *352,000.00*

Three hundred fifty-two thousand and 00/100 Dollars

Inland Bank
15 Commercial Street
Inland, IN 47304

For: ______

Signature *Latoya S. Marshall*

2-1 CHOOSING A BANK: IS THERE A DIFFERENCE?

QUICK REFERENCE

TEACHER'S RESOURCES AND ANSWER KEY

Lesson Quiz Answers 2-1, p. 42
Reteaching Activity Answers 2-1, p. 93
Enrichment Activity Answers 2-1, p. 93

EXTENSION ACTIVITIES

Reteaching Activity 2-1, p. 9
Enrichment Activity 2-1, p. 12

Jeff has been keeping track of his own money ever since he started receiving a dollar a week in allowance when he was seven. Jeff's older brother, Jeremy, never seems to know where all his money goes, but Jeff keeps very careful records of what he has earned, spent, and saved. Jeff earned $305 last summer tutoring some of his classmates as they took algebra in summer school. He spent only $15 of his summer earnings and saved the rest.

Jeremy has asked to borrow money from Jeff several times. The last time, he borrowed $25 for a gift he wanted to buy for his girlfriend, Ellen. Jeremy did not pay the money back when he said he would. Jeff has decided to open a checking account so that his money will be a little safer. He will be doing some more tutoring now that semester exams are coming up, and he knows that a bank account will help him keep track of his income in preparation for filing a tax return next spring.

Jeff also has bills to pay now, too, since he applied for a gasoline credit card in his own name. Jeff does quite a lot of driving when he does the tutoring. He also pays the automobile insurance premium.

Jeff knows that different banks offer many of the same services but that service charges, interest rates, and minimum balance requirements can vary from bank to bank. Before he chooses which bank to go to, he will investigate the differences.

OBJECTIVES: *In this lesson, we will help Jeff to:*

- *Learn what different types of checking accounts are available.*
- *Calculate the interest and bank charges for a checking account.*
- *Compare the costs of banking services offered by various banks.*
- *Use a spreadsheet to compare costs of various checking account options.*

FOCUS

ALGEBRA CONNECTIONS

In the formula, $y = bx$, y could be used to represent distance traveled, amount earned, interest paid, or area. What would b and x stand for in each case? Write numerical examples.

CHECKING ACCOUNTS

The check is the most widely used means of transferring money. Nearly 90% of all monetary transactions, such as buying goods and paying rent, are made with checks. **Checks** are orders written by a depositor directing a bank to pay out money. Checks are safe and convenient. They provide proof of payment and records for tax and budget purposes.

Most young people open their first checking account shortly after they get a job or go to college. When a person opens a checking account, the bank and the customer establish a contractual agreement that allows the customer to deposit money in the bank and to write checks on the account. The bank agrees to maintain the account, provide records, and honor checks when they are presented for collection. When a customer writes a check, it is a demand on the customer's deposits so checks are called **demand deposits.**

Following is an explanation of the different types of checking accounts that are available.

CRITICAL THINKING

As you discuss the questions, ask students how a minimum balance might be calculated. Average daily balance is covered later in the text, not in this lesson. But as students suggest different ways for computing balance, you can point out the possibility of using an average balance.

Cost-Per-Check Account Cost-per-check accounts are considered thrift accounts or minimum use accounts. Generally, one has to maintain a minimum balance and pay a fee for each check. The fee may range from \$0.02 to \$0.25 per check. Some banks also charge a service charge or maintenance fee that varies from \$0.50 to \$2.50 or more per month even if there is no activity in the account.

Minimum-Balance Account Minimum-balance accounts require the customer to maintain a certain balance. The minimum balance may be a low balance or an average balance. With a **low-balance account** the customer is charged a service fee even if the account falls below the minimum only one day in a month. An **average-balance account** can drop to zero as long as the customer deposits enough money during the month to bring the account average for the month up to the minimum required. Some banks advertise minimum-balance accounts as "free checking accounts." They may be low-cost accounts, but they are not free.

Free Checking Account Some banks provide totally free checking; that is, there are no minimum balance requirements or service charges. Some banks provide free checking in the belief that checking account customers will also use the bank's other services such as savings accounts, consumer loans, and safe-deposit boxes.

NOW (Negotiable Order of Withdrawal) Account Another type of checking service is an interest-bearing account known as a **NOW account.** One drawback to NOW accounts is the minimum balance requirement. If the account balance falls below the required balance, the bank charges a monthly service fee or a check-handling fee.

ALGEBRA *REVIEW*

Copy and complete the table for each equation by substituting the given values. Describe in words how the values for *y* in Exercises 2 through 5 relate to the values in Exercise 1.

x	1	2.5	7
y			

1. $y = 3x$
   3, 7.5, 21
2. $y = 3x + 1$
   4, 8.5, 22; They are one greater.
3. $y = -3x$
   −3, −7.5, −21; They are opposites.
4. $y = 3x - 1$
   2, 6.5, 20; They are one less.
5. $y = -3x - 1$ −4, −8.5, −22; They are one less than the opposites.

Complete the table for each equation.

x	−2	0	2
y			

6. $y = 4x - 2.5$
   −10.5, −2.5, 5.5
7. $y = x$
   −2, 0, 2
8. $y = 1.5x + 1.05$
   −1.95, 1.05, 4.05
9. $y = -2x + 2.7$
   6.7, 2.7, −1.3
10. $y = -(3x + 4.1)$
    1.9, −4.1, −10.1

Ask Yourself

1. Why is a check referred to as a demand deposit?
   A check is a demand on the customer's deposits.
2. Are minimum-balance accounts free?
   See Additional Answers.
3. Does a NOW account require a minimum balance?
   Yes
4. Which is the least expensive account?
   Cost-per-check account

ADDITIONAL ANSWERS

2. No. Minimum-balance accounts require the customer to maintain a certain balance.

SHARPEN YOUR SKILLS

SKILL 1

In Chapter 1 you worked with percentages to find commissions. In this lesson you will work with percents less than 1%.

1% means 1 part out of 100.
0.5% means $\frac{1}{2}$ of 1%.
As a decimal, 0.5% = 0.005.

Interest is an amount of money paid for the use of money. Interest is based on percent, which means parts out of one hundred.

To find 0.5% of $246, convert 0.5% to a decimal and multiply.

$0.005 \cdot 246 = 1.23$ 0.5% = 0.005
0.5% of $246 is $1.23

At the first bank he visits, Jeff reads a brochure about a Presto account. The brochure states that a checking account earns interest of 0.45% per month and that there is a charge of 5 cents per check.

Jeff remembers that interest is an amount paid based on percentage. Jeff does a quick mental calculation and sees that if he left $100 in the bank for a month, it would earn $0.0045 \cdot 100 = \$0.45$ in interest. Not much, he thinks, but it is something.

Each month the bank pays any interest due and charges any fees due. The interest that is paid is determined by multiplying the monthly average balance by the monthly interest rate. The charge for checks is determined by multiplying the number of checks by the charge for each check. The amount earned or charged can be expressed by the following formula.

Amount earned or charged = interest − charge for checks

$A = br - cn$ where A = amount earned or charged
b = monthly average balance
r = monthly interest rate
n = number of checks
c = charge for each check

EXAMPLE 1 Presto checking accounts pay 0.45% interest per month and charge $0.05 per check.

QUESTION What is the amount earned or charged if 17 checks are written during one month and the average balance is $620?

TEACHING THE LESSON

Explain the checking account vocabulary and why checks are used. Ask questions to be sure that students understand the meaning of balance and interest.

Use the *Algebra Connections* to explain that a formula is a short way of writing a sentence about a mathematical relationship. Skills 1 and 2 present information about checking account charges and balances in words and in algebraic formulas. Have students explain the meaning of each letter and the operations connecting the letters in each formula.

Review the meaning of percents less than 1% in both fraction and decimal form. Have students calculate 50%, 5%, and 0.5% of $300. **($150, $15, $1.5)**

FOCUS ON ALGEBRA

Point out that, in this lesson, information about checking account balances is given using different representations: words, equations, tables, and graphs. (*NCTM Standard 4, p. 146*)

COMMON ERRORS

The fact that many students are frightened of formulas is not so much an error as a mental block. Continue making the point that the letters in a formula stand for words that students know and understand very well.

AT-RISK STUDENTS

The fact that the mathematical ideas in this lesson are presented in different ways—equation, table, graph—may bother some students who are accustomed to receiving information in discrete bits. Taking time to explain the connections will strengthen the understanding and confidence of such students.

SOLUTION

Remember to add the interest paid and subtract the charges for checks.

$A = br - cn$	Use the formula.
$A = 620(0.0045) - 0.05(17)$	$b = 620$, $r = 0.0045$, $c = 0.05$, $n = 17$
$A = 1.94$	Use your calculator.

The amount earned is $1.94.

SKILL 2

Jeff visits another bank to compare costs. This bank charges a service charge in addition to the charge for checks. So we need to adjust the formula.

> Amount earned or charged = interest − charge for checks − service charge
>
> $A = br - cn - s$ where
> A = amount earned or charged
> b = monthly average balance
> r = monthly interest rate
> n = number of checks
> c = charge for each check
> s = service charge

EXAMPLE 2 Super checking accounts earn interest of 0.4% per month and have the following service charges.

For balances of $500 or more, there is no service charge and no charge for checks.
For balances under $500, there is a charge of $0.025 per check.
For balances of $200 to $499.99, there is a service charge of $4.00 per month.
For balances under $200, the service charge is $6.00 per month.

QUESTION What amount will be earned or charged for the following average balances?

a. $650 with 20 checks

b. $400 with 18 checks

c. $150 with 16 checks

SOLUTION
Jeff used the formula above.

$A = br - cn - s$

a. $A = 650(0.004) - 0(20) - 0 = 2.6$
The account earned \$2.60.

b. $A = 400(0.004) - 0.025(18) - 4 = -2.85$
The account was charged \$2.85.

c. $A = 150(0.004) - 0.025(16) - 6 = -5.8$
The account was charged \$5.80.

TECHNOLOGY HINT

Use the spreadsheet in Skill 3 to reinforce the algebraic equations, pointing out where each term of the equation appears in the spreadsheet. In this way, the technology will not be seen as something added on, but rather will be organically connected with the money applications and the algebra used.

You may wish to show students how to use the percent key on the calculator. That is, to find 0.4% of \$150, enter 150 X 0.4% and read the answer. This will help them avoid errors in converting from percent to decimal.

SKILL 3

EXAMPLE 3 Jeff learned that there are a number of differences in the interest rates and charges that banks use for checking accounts. He decided to make a spreadsheet to compare various interest rates and charges.

QUESTION What column headings and formulas should be used in a spreadsheet to calculate the interest and charges on Super checking accounts?

SOLUTION
For Super checking accounts there are three different ways of charging fees as shown in the spreadsheet.

	A	B	C	D	E	F	G	H
1		Interest	Interest	Number	Cost per	Cost of	Service	Amount Earned
2	Balance	Rate	Earned	of Checks	Check	Checks	Charge	or Charged
3	500 or more							+C4−F4−G4
4	650	0.004	2.6	20	0	0	0	2.60
5	200 to 499.99		+A4∗B4					
6	400	0.004	1.6	18	0.025	0.45	4	(2.85)
7	under 200					+D6∗E6		
8	150	0.004	0.6	16	0.025	0.4	6	(5.80)
9								
10								
11								

The cost of checks and the service charge are different for all three cases. The formulas for the case of a \$500 balance are shown below.

Interest earned: +A4*B4	Interest rate times the average balance.
Cost of checks: +D4*E4	Number of checks times the cost per check.
Amount earned or charged: +C4−F4−G4	Interest earned minus the cost of checks minus the service charge.

COOPERATIVE LEARNING

Until students become accustomed to using the graphing calculator, they will need direction in setting the range and entering each equation. This can be accomplished in cooperative pairs. Have one student at a time use the calculator while the other watches and checks on the accuracy of each step. If possible, show the graphs on the chalkboard or use the overhead projector.

RETEACHING

If students do not fully grasp the concepts and procedures, do not go on to the next lesson. Go back to the beginning and use several numerical examples to show how a checking account balance might increase due to interest or decrease due to charges. Gradually draw out from students the relationships and formulas for each situation.

ENRICHMENT

More advanced students might be asked to devise a method for calculating average daily balance.

You can use the COPY function of your spreadsheet to create the formulas for rows 6 and 8 of the spreadsheet. For example, if you copy cell C4 (interest earned) to cell C6, then the program will create the formula +A6*B6 for that cell. You will see the value 1.6 in the cell.

You may wish to format the column for "Amount Earned or Charged" for currency as shown. Usually a negative amount will be shown in parentheses as in the spreadsheet on page 57.

SKILL 4

EXAMPLE 4 Jeremy's father owns a small business and has to decide between two checking accounts. Bank A does not pay interest and does not charge for checks but has a \$4 per month service charge. Bank B pays 0.4% interest per month and charges \$0.075 per check. Jeremy's father expects to maintain a balance of \$1000 but is not sure how many checks he will write.

QUESTION How does the monthly number of written checks affect the interest and charges for the two accounts?

SOLUTION

Write equations like the ones used before for the interest and charges. Let y be the total of the interest and charges. Let x be the number of checks. The interest for Bank B is $0.004(1000) = 4$.

For Bank A: $y = -4$
For Bank B: $y = 4 - 0.075x$

Graph the equations on a graphing calculator using these range values.

Xmin:	−5	Ymin:	−6
Xmax:	125	Ymax:	6
Xscl:	10	Yscl:	1

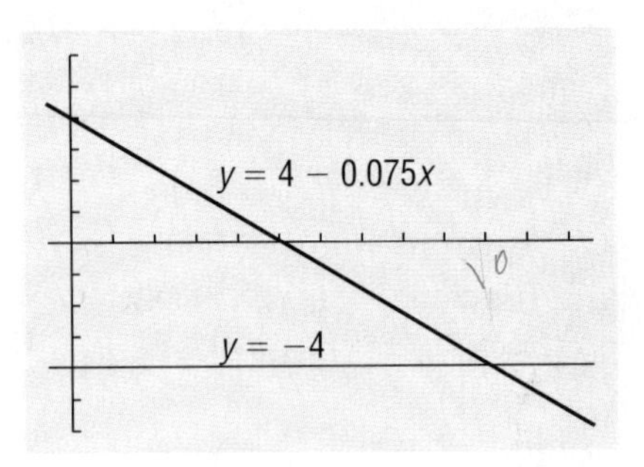

The trace function shows that the graphs intersect at approximately $x = 107,\ y = -4$.

To confirm the calculator results, solve the system of equations by substituting the value for y in the second equation.

$-4 = 4 - 0.075x$
$-8 = -0.075x$
$x = 106.67$ To the nearest hundredth

Both methods show that the interest-bearing account in Bank B will be less expensive if fewer than 106 checks are written per month, and more expensive if 107 or more checks are written per month.

TRY YOUR SKILLS

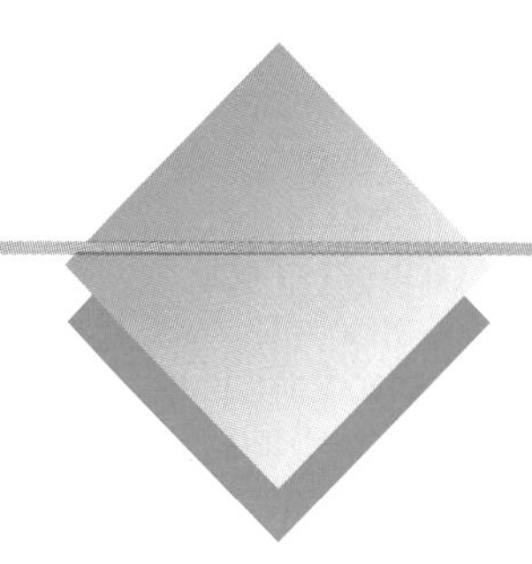

Three banks offer the following arrangements for checking accounts.

National: Interest 0.5% per month, a charge of \$0.09 per check for balances under \$500.
Central: Interest 0.54% per month, a charge of \$0.02 per check, \$4 service charge for balances under \$200.
Western: Interest 0.35% per month, a charge of \$0.06 per check for balances under \$1000.

The *Try Your Skills* exercises offer an opportunity for you to move around the classroom observing students' work. Some will be uncertain about when to add and when to subtract amounts from the bank balance.

Find the amount earned or charged for each of the following average balances for one month.

1. \$400 at National, with 22 checks written \$0.02 earned
2. \$500 at Western, with 9 checks written \$1.21 earned
3. \$175 at Central, with 31 checks written \$3.68 charged
4. \$760 at National, with 40 checks written \$3.80 earned
5. \$340 at Central, with 28 checks written \$1.28 earned
6. \$1280 at Western, with 60 checks written \$4.48 earned
7. Copy and complete the table comparing the interest earned, charges, and changes in balance at each of the three banks for a balance of \$400 with 20 checks written.

Balance	Interest Earned	Cost of Checks	Service Charge	Amount Earned or Charged
National	\$2.00	\$1.80	\$0	+ \$0.20
Central	2.16	0.40	0	+ \$1.76
Western	1.40	1.20	0	+ \$0.20

8. Complete a table like that in Exercise 7 for an average balance of \$1220 with 30 checks written.

National:	\$6.10	\$0	\$0	+ \$6.10
Central:	\$6.59	\$0.60	\$0	+ \$5.99
Western:	\$4.27	\$0	\$0	+ \$4.27

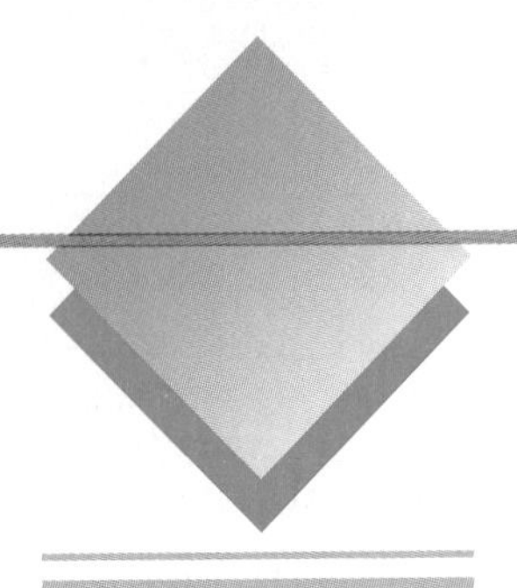

EXERCISE YOUR SKILLS

KEY TERMS

average-balance account
checks
cost-per-check account
demand deposits
free checking account
interest
low-balance account
minimum-balance account
NOW account

INDEPENDENT PRACTICE

ASSIGNMENT GUIDE

Exercises 1-6 reinforce an understanding of checking accounts and review the mathematical tools used in the lesson.

Exercises 7-13 take students step by step through the check register spreadsheet.

Exercises 14-16 cover the lesson's skills in spreadsheet format.

Exercises 17-22 explore the material in verbal form. These are slightly more advanced but all students should be able to do them.

1. Write a general algebraic equation that can be used to find the amount earned or charged in a checking account in a bank that pays interest, charges a monthly fee, and charges for each check written. $A = br - cn - s$
2. What are two things that a bank agrees to do as its part of the agreement when a person opens a checking account? See Additional Answers.
3. Explain how a spreadsheet will be useful to Jeff in organizing data and explaining checking accounts to his friends.
4. Why might a person choose a bank even though it does not have high interest rates or low costs per check?
5. What might lead a person to pay special attention to the cost per check that a bank charges? Writing large numbers of checks
6. What is the difference between a low-balance account and an average-balance account? See Additional Answers.
7. Set up a spreadsheet with headings and formulas as explained in Skill 3 and shown.

	A	B	C	D	E	F	G
1		Interest	Interest	Number of	Cost per	Cost of	Amount
2	Balance	Rate	Earned	Checks	Check	Checks	Earned
3			[+A3*B3]			[+D3*E3]	[+C3−F3]

8. To use the spreadsheet correctly, in which cells must you enter numbers?
9. Which cells do you leave blank, letting the spreadsheet fill them in?
10. Write an algebraic expression to calculate the amount that will appear in cell C3. br
11. Write an algebraic expression to calculate the amount that will appear in cell F3. cn
12. Write an algebraic equation to calculate the amount that will appear in cell G3. $A = br - cn$
13. Show how the equation that you wrote is related to the formulas used in the spreadsheet. $b = A3$, $r = B3$, $c = E3$, $n = D3$, $A = G3$

For Exercises 14–16, use a spreadsheet with headings like the ones shown in the tables to calculate the interest, costs, and new balance for the banks as described. To find the new balance, it is necessary to add monthly interest and subtract monthly charges.

14. At First Bank:

    No interest is paid on the balance.

    There are no service charges.

    The cost per check is \$0.0225. If the beginning balance is below \$1000, First Bank charges an additional \$0.23 for each check written.

	Beginning Balance	Number of Checks	Interest Earned	Service Charges	Cost of Checks	Extra Charges	New Balance
a.	$650	18	0	0	$0.41	$4.14	$645.45
b.	625	26	0	0	0.59	5.98	618.43
c.	480	20	0	0	0.45	4.60	474.95
d.	190	31	0	0	0.70	7.13	182.17

15. At Second Bank:

Interest is paid on the beginning balance at the rate of 0.5% per month.

The service charge is $3.00 per month.

The cost per check is $0.0225. There are no additional charges whether or not a minimum balance is maintained.

	Beginning Balance	Number of Checks	Interest Earned	Service Charges	Cost of Checks	Extra Charges	New Balance
a.	$650	18	$3.25	$3.00	$0.41	0	$649.84
b.	625	26	3.13	3.00	0.59	0	624.54
c.	480	20	2.40	3.00	0.45	0	478.95
d.	190	31	0.95	3.00	0.70	0	187.25

16. At Third Bank:

Interest is paid at the rate of 0.52% per month on the beginning balance.

There are no service charges.

The cost per check is $0.0225. If the beginning balance is below $500, Third Bank charges an additional $0.25 for every check written.

	Beginning Balance	Number of Checks	Interest Earned	Service Charges	Cost of Checks	Extra Charges	New Balance
a.	$650	18	$3.38	0	$0.41	0	$652.97
b.	625	26	3.25	0	0.59	0	627.66
c.	480	20	2.50	0	0.45	$5.00	477.05
d.	190	31	0.99	0	0.70	7.75	182.54

17. Write a formula for the amount earned or charged in a checking account for which interest is paid but for which there is no service charge or charge per check. $A = br$

18. Sarah's checking account has an average balance of $120 for one month, and it earns 42 cents in interest. Find the monthly interest rate at that bank. 0.35%

19. Jason's bank has a monthly service fee of $3. He finds that an average balance of $180 causes a decrease in his account of $2.28. Find the monthly interest rate at the bank. 0.4%

To be sure that students understand the underlying concepts, ask what would cause a checking account balance to increase. These would include deposits and interest earned. Then ask what would cause the balance to decrease. These would be checks written, per check charges, and bank fees.

LESSON QUIZ

1. A person has a monthly account balance of $525. Find her balance after she earns 0.35% interest and pays fees of $4.50. **($522.34)**
2. With a balance of $200 and 25 checks written, is it better to pay 5 cents a check and earn 0.4% monthly interest, or to pay nothing for checks and earn no interest? **(No check fee, no interest)**

ADDITIONAL ANSWERS (EXERCISE YOUR SKILLS)

21. Bank S: $y = 4.4 - 0.07x$
Bank T: $y = 2 - 0.025x$
$x = 53.3$
The account at Bank S will be more expensive if fewer than 53 checks are written and less expensive if more than 53 checks are written.

22. Bank K: $y = 1.75 - 0.08x$
Bank M: $y = -0.025x$
$x = 31.8$
The account at Bank M will be more expensive if 31 or fewer checks are written and less expensive if more than 31 checks are written.

(MIXED REVIEW)

1. $275.50

20. Marcia's bank has no service charge for a checking account but pays interest and charges $0.04 for each check written. With a balance of $160 and an interest rate of 0.3%, Marcia finds that her account shows a bank charge of $0.04 for the month. How many checks did she write? 18 11

21. Bank S pays interest of 0.55% and charges $0.07 per check. Bank T pays interest of 0.25% and charges $0.025 per check. Susan expects to maintain a balance of $800 but is not sure how many checks she will write. Write and graph equations as in Example 4 to determine which account is less expensive. Explain your results. See Additional Answers.

22. Henry expects to maintain a balance of $500 in his checking account. Write and graph equations to determine which account is less expensive. Bank K pays interest of 0.35% and charges $0.08 per check. Bank M pays no interest and charges $0.025 per check. Explain your results.

MIXED REVIEW

1-1 **1.** At $7.25 an hour, how much does Jessica earn for 38 hours of work?

1-1 **2.** Richard earns $6.80 an hour but gets $1\frac{1}{2}$ times his regular pay for every hour that he works over 40 hours. What is his pay for 46 hours of work? $333.20

1-2 **3.** Jorge receives a salary of $39,000 and takes vacation days and holidays for 24 of the company's 260 operating days. How much of his salary is paid for days during which he does not work? $3600

1-3 **4.** What is the amount of a 7.5% commission on sales of $1245? $93.38

1-3 **5.** Karen is paid a commission of $40 on a sale of $200. What is her rate of commission? 20%

1-4 In a spreadsheet, cell B2 contains an hourly rate of pay, cell B3 contains the number of hours, cell B5 contains FICA deductions at 7.65%, and cell B6 contains all other deductions. Use this information to find each of the following.

6. a spreadsheet formula for the gross wages B4 = B2*B3

7. a spreadsheet formula for the FICA withholding B5 = 0.0765*B4

8. a formula for the total deductions B7 = B5+B6

9. a formula for the take-home pay B8 = B4−B7

2-2 USING CHECKS: WRITE IT DOWN

QUICK REFERENCE

TEACHER'S RESOURCES AND ANSWER KEY

Lesson Quiz Answers 2-2, p. 42
Reteaching Activity Answers 2-2, p. 93
Enrichment Activity Answers 2-2, p. 94

EXTENSION ACTIVITIES

Reteaching Activity 2-2, p. 10
Enrichment Activity 2-2, p. 13

Latoya receives a weekly check from the Hometown Diner, the restaurant where she works after school. The weekly paychecks are printed on a computer, but they are all signed by the manager. Latoya imagines how good it would feel to sign thousands of dollars' worth of checks every month. Latoya's parents write checks to pay the bills that come to their house every month. They do not seem to enjoy the task all that much, but maybe that is because there are usually more bills than there is money to pay them with.

Some of Latoya's earnings are used to help pay those bills. Latoya would like to help write the checks, too, and keep track of how much money her parents have in their checking account. What Latoya really wants to do is to write checks for a large corporation for thousands of dollars of someone else's money! But for the moment she will settle for learning how to write a proper check and maintain an accurate check register. She plans to open her own checking account when she feels comfortable with these procedures.

OBJECTIVES: *In this lesson, we will help Latoya to:*

- *Write checks and maintain a check register.*
- *Endorse checks properly.*
- *Make a bank deposit.*
- *Use a spreadsheet to maintain a check register.*

FOCUS

ALGEBRA CONNECTIONS

If a, b, and c are all positive numbers, which is larger: $a - (b + c)$ or $a - b - c$? **(They are equal.)** When will $a - (b + c)$ be a negative number? **(When $b + c$ is greater than a.)**

WRITING AND ENDORSING CHECKS

Most people make money transactions using checks. When checks are properly written, as shown on page 65, they provide proof of payment and are an excellent record of transactions. However, it is important to prevent someone from altering your checks. You must carefully fill in the **payee** (the person to whom the check is written), sign it as the **drawer** (the person from whose account the funds are to be withdrawn), and keep a careful record in your check register. A **check register** is a separate form on which a checking account holder keeps a record of deposits and checks written. The small, portable check register is called a checkbook.

Endorsements It is also important to sign and endorse checks properly. Banks and other businesses that cash or accept checks for payment are careful that the signature on the front of the check or the endorsement on the back is that of the person presenting the check for payment. That is why businesses often ask for identification before accepting a check from a person they do not know.

An **endorsement** is a signature and a message to the bank telling it to cash a check, deposit it, or transfer your right to the check to someone else. It is written on the back of the check, at the left-hand end. Endorsements are made on the back of the check exactly as your name appears on the face of the check. Even if your name is misspelled, you should sign it as it is written on the face and then add the correct spelling. There are three common types of endorsements: blank, restrictive, and full.

A **blank endorsement** is your signature only. A check with a blank endorsement is like cash. Anyone who has the check can present it for payment at a bank. For this reason you should not blank endorse a check unless you are at the bank.

A **restrictive endorsement** is your signature and a message that limits the use of the check. A restrictive endorsement usually reads "For deposit only." This type of endorsement allows you to send the check by mail for deposit without fear of loss. If a check endorsed "for deposit only" is lost or stolen, it cannot be cashed.

A **full endorsement** is your signature and a message that directs the transfer of the check to someone else whom you designate. A full endorsement is written: "Pay to the order of . . . " followed by the name of the recipient and your signature. This endorsement transfers the right of payment to the new payee.

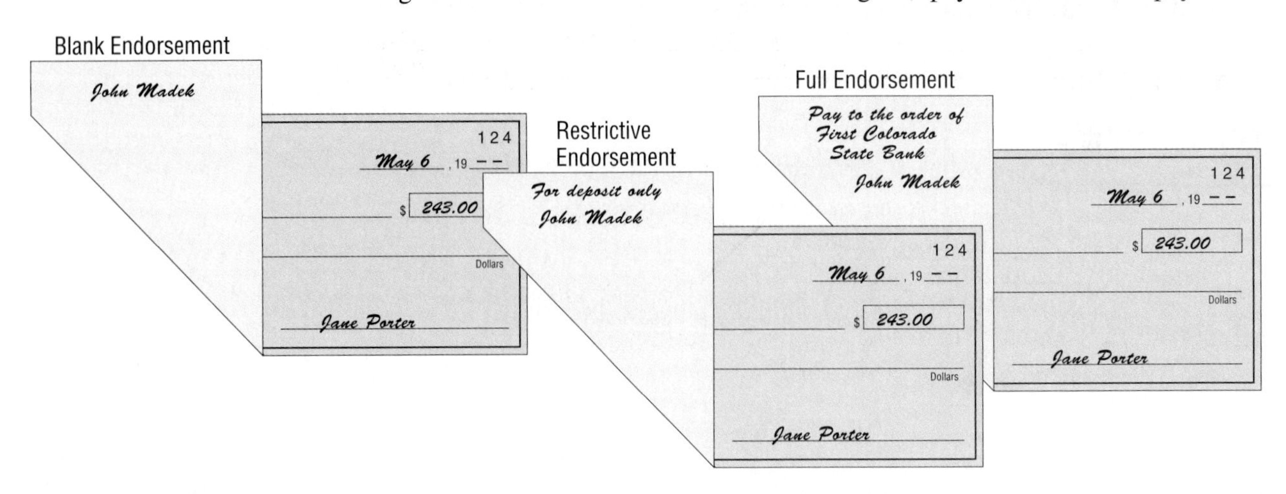

Latoya follows the procedures listed below when writing a check and keeping a record in her check register. The numbers of the steps correspond to the circled numbers in the illustration. It is important to use permanent ink to write checks, never pencil or erasable ink.

These are the steps to be followed when writing a check.

1. Before writing the check, record in the check register: the check number, the date, the payee, the purpose of the check, and the amount of the check.
2. On the check itself, write the date. This is important for your own records and in case the check is ever needed as proof of payment.

PLEASE BE SURE TO DEDUCT ANY PER CHECK CHARGES OR MAINTENANCE CHARGES THAT AFFECT YOUR ACCOUNT

CHECK NO.	DATE	CHECKS DRAWN OR DEPOSITS MADE	BALANCE FORWARD ➡	✓		
D	4/10	TO *Deposited paycheck*	DEDUCT CHECK – ADD DEPOSIT +		258	49
		FOR *for 3/16 – 3/31*	BALANCE ➡		258	49
001 (1)	4/11	TO *The Slax and Jax Shop*	DEDUCT CHECK – ADD DEPOSIT +		15	95
		FOR *pullover*	BALANCE ➡		242	54
002	4/13	TO *Cash*	DEDUCT CHECK – ADD DEPOSIT +		40	00
		FOR *current expenses*	BALANCE ➡		202	54
D	4/16	TO *Deposit*	DEDUCT CHECK – ADD DEPOSIT +		20	50
		FOR	BALANCE ➡		223	04
003	4/22	TO *Spangler Gifts, Inc.*	DEDUCT CHECK – ADD DEPOSIT +			88
		FOR *postage due on 4/8 order*	BALANCE ➡		222	16
AT	4/29	TO *Cash from automatic teller*	DEDUCT CHECK – ADD DEPOSIT +		30	00
		FOR *game tickets and dinner*	BALANCE ➡		192	16

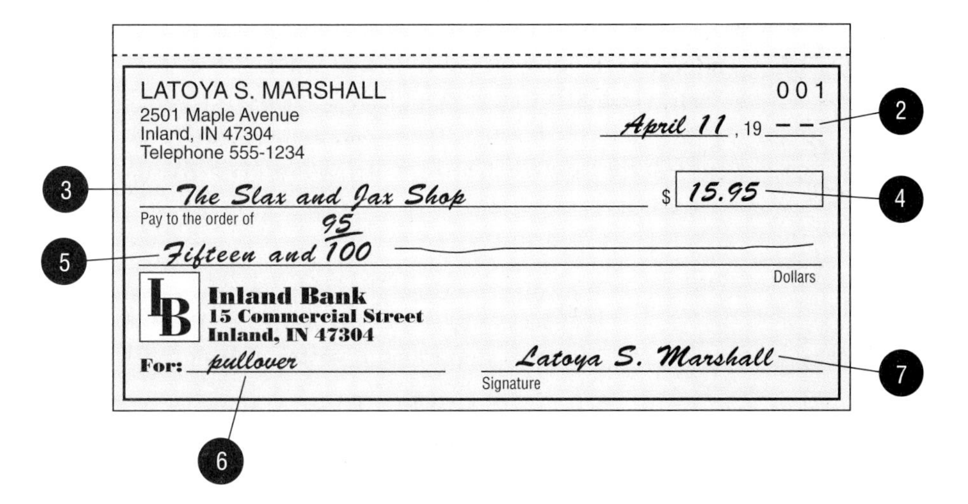
LATOYA S. MARSHALL
2501 Maple Avenue
Inland, IN 47304
Telephone 555-1234

001

(2) *April 11*, 19 – –

(3) *The Slax and Jax Shop*
Pay to the order of

$ (4) *15.95*

(5) *Fifteen and 95/100* —————— Dollars

Inland Bank
15 Commercial Street
Inland, IN 47304

For: (6) *pullover*

(7) *Latoya S. Marshall*
Signature

3. Designate the payee; that is, write the name of the person or organization to whom you are paying the money.
4. Write the amount of the check, in numbers, next to the dollar sign.
5. Write the amount of the check, in words, on the middle line and draw a line to fill in the space to the word "Dollars."

CRITICAL THINKING

After discussing the questions, tell students that in some countries people receive their weekly or monthly pay in cash. They don't use checking accounts. Ask about some of the disadvantages of this arrangement. Ask about the different ways in which people without checking accounts pay their bills.

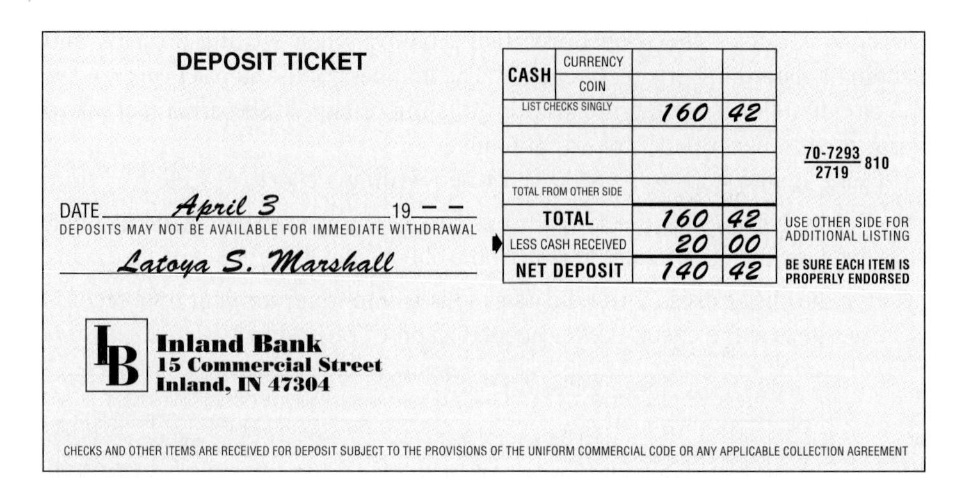

DEPOSIT TICKET

DATE April 3 19 – –

DEPOSITS MAY NOT BE AVAILABLE FOR IMMEDIATE WITHDRAWAL

Latoya S. Marshall

CASH	CURRENCY		
	COIN		
LIST CHECKS SINGLY		160	42
TOTAL FROM OTHER SIDE			
TOTAL		160	42
LESS CASH RECEIVED		20	00
NET DEPOSIT		140	42

70-7293/2719 810

USE OTHER SIDE FOR ADDITIONAL LISTING

BE SURE EACH ITEM IS PROPERLY ENDORSED

Inland Bank
15 Commercial Street
Inland, IN 47304

CHECKS AND OTHER ITEMS ARE RECEIVED FOR DEPOSIT SUBJECT TO THE PROVISIONS OF THE UNIFORM COMMERCIAL CODE OR ANY APPLICABLE COLLECTION AGREEMENT

ALGEBRA *REVIEW*

Write an equation to express the relationship shown. Ask yourself how the *y* values can be derived from the *x* values.

1.

x	1	3	5
y	2	6	10

$y = 2x$

2.

x	−1	0	1
y	1	0	−1

$y = -x$

3.

x	2	4	6
y	5	9	13

$y = 2x + 1$

4.

x	1	2	3
y	2	5	8

$y = 3x - 1$

5.

x	−2	0	2
y	5	1	−3

$y = -2x + 1$

Write the equation of a line through each set of points.

6. (2, 3), (3, 4), (4, 5)
$y = x + 1$
7. (2, 4), (3, 6), (4, 8)
$y = 2x$
8. (1, −3), (2, −6), (3, −9)
$y = -3x$
9. (1, 2.5), (2, 5.5), (3, 8.5)
$y = 3x - 0.5$

6. Write the purpose for which each check is written on the line at the bottom of the check.
7. Sign your name as the drawer on the signature line.

DEPOSIT TICKETS

When you deposit money in the bank, you must fill out a deposit ticket as shown above. Notice above how a deposit ticket is completed. Generally, if you withdraw cash, you must sign the deposit ticket as shown. The bank will stamp the deposit slip and return it to you for your records.

In this transaction, Latoya gives the bank a check for $160.42 made out to her. She also asks for $20 in cash from the check, so her deposit is $140.42.

Ask Yourself

1. Why do people use checks for money transactions?
See Additional Answers.
2. Why is it important to write checks properly?
3. Why does the bank require an endorsement when cashing a check?

ADDITIONAL ANSWERS

1. Answers may vary. (Ex.: Checks are safe, convenient, and provide proof of payment for tax and budget purposes.)
2. Answers may vary. (Ex.: When properly written, checks provide proof of payment and are excellent records of transactions.)
3. To make sure that the signature on the front of the check or the endorsement on the back is the same as the signature of the person presenting the check for payment.

SHARPEN YOUR SKILLS

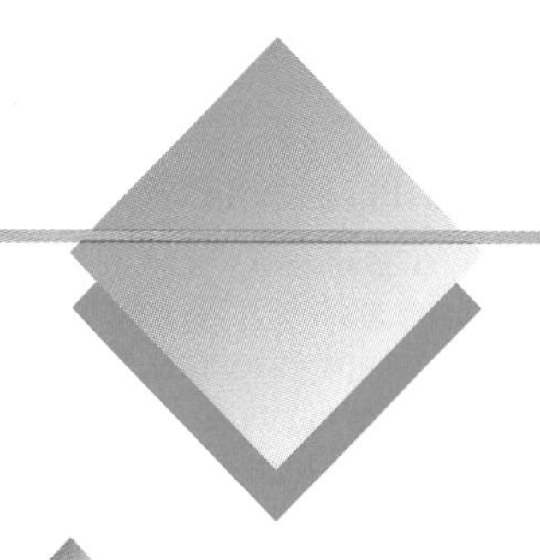

SKILL 1

EXAMPLE 1 You are about to mail out checks to pay the following bills: Central Telephone Co., $43.78; Harvey's Computer Store, $102.05; Downtown Music, $18.75. The checks you are using begin with number 101. The date is October 6. Your initial balance is $201.68.

QUESTION How would you complete the check register, entering each transaction and finding the balance after each transaction?

SOLUTION

Following is the check register with the entries. Payments are subtracted to find each new balance; deposits are added.

CHECK NUMBER	DATE	CHECKS/DEPOSITS	AMOUNT	BALANCE
101	10/6	TO: Central Telephone		201.68
		FOR: Phone bill	43.78	-43.78
102	10/6	TO: Harvey's Computer		157.90
		FOR: Software	102.05	-102.05
103	10/6	TO: Downtown Music		55.85
		FOR: CD	18.75	-18.75
		TO:		37.10
		FOR:		

SKILL 2

EXAMPLE 2 Latoya read in the paper about the Eldorado Golf Tournament played in her city on March 4. She imagined what it would be like to write checks to the prize winners. The paper listed the top five as follows:

Cynthia Alvarez	$352,000
Mark Louis	$240,000
Peggy Race	$150,000
Pablo Chosa	$ 75,000
Tina Marin	$ 50,000

QUESTION How would Latoya make out the largest check?

INSTRUCTION

TEACHING THE LESSON

If needed, review operations with integers on the number line. Students should be confident in working with negative numbers.

Attention to detail in maintaining a checkbook can easily be used to point out the need for care in general, when completing forms or doing the basic calculations connected with personal money management.

Students generally will enjoy completing simulated checks, but some may complain about the requirement that each part be done in a precise way. If convenient, use an overhead projector to show a check and point out how each part is to be written. Point out that the standard format is necessary since the check can be worth a large amount of money and there should not be any guessing about what is intended.

FOCUS ON ALGEBRA

Each number used in this lesson corresponds to a concrete expenditure or deposit. Further, the use of written numbers and the expression of mathematical ideas in writing can help clarify concepts and operations. (*NCTM Standard 2, p 140.*)

SOLUTION

The following is a check correctly made out to the winner.

Eldorado Golf Tournament
Eldorado Golf Club
Inland, IN 47304
Telephone 555-7902
001
March 4, 19 – –
Cynthia Alvarez
Pay to the order of
$ 352,000.00
Three hundred fifty-two thousand and 00/100
Dollars
Inland Bank
15 Commercial Street
Inland, IN 47304
For:
Latoya S. Marshall
Signature

COMMON ERRORS

You can point out to students that most adults make more mathematical errors in their checkbooks than in any other area of applied math. Because each answer depends on the previous answer, errors are carried forward. Point out that it can take a long time to correct a careless mistake.

SKILL 3

EXAMPLE 3 Latoya has learned that she should endorse a check on the back exactly as her name appears on the face of the check. Even if her name is misspelled, she should sign it as it is written on the face and then write her full name correctly. Her grandfather recently gave her a check using "Toya," his pet name for her. Latoya is mailing the check to the bank for deposit.

QUESTION How should Latoya endorse the check?

SOLUTION

Latoya should write her name as it appears on the face of the check and then write her full name. She should add the restriction, as shown, so that no one will be able to cash the check.

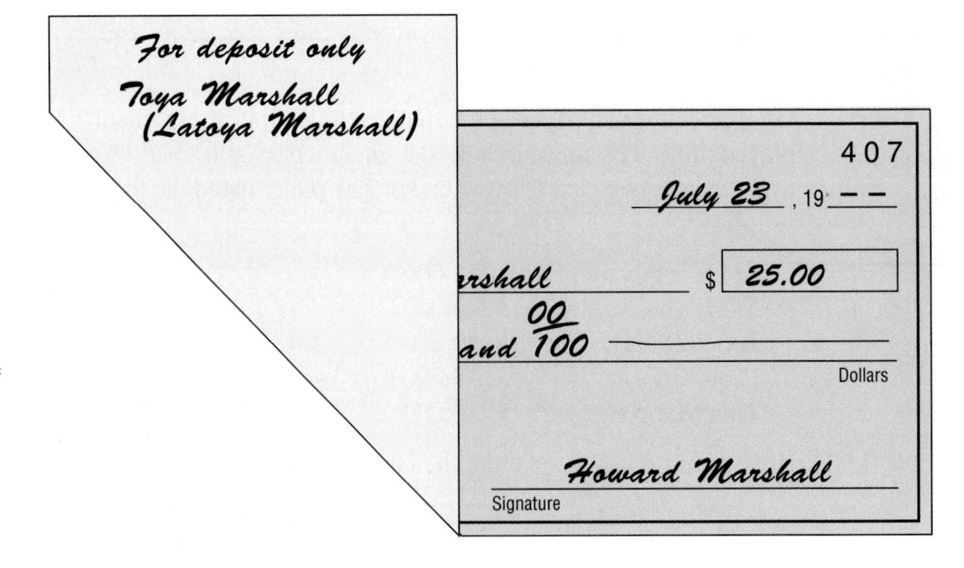

INTERDISCIPLINARY INVESTIGATION

Computer Science: Have students research how banks have changed their record keeping methods since the introduction of computers. They might interview someone who works in a bank.

SKILL 4

Some people use a spreadsheet as a check register. A spreadsheet is useful for this purpose because it is set up almost like a checkbook and because of the computations that can be done by using a spreadsheet.

EXAMPLE 4 Latoya has the following transactions written down and will enter them as she starts her spreadsheet. All of the transactions were done on May 3. Balance: $178.85; Check 301, Home Hardware, $37.12; Check 302, Campus Bookstore, $45.60; Check from Mark Li, $40.00 for deposit.

QUESTION How do you set up a spreadsheet as a check register?

SOLUTION

To set up a check register spreadsheet, you must set the columns needed and write the formulas for automatic computation. A model is shown below.

	A	B	C	D	E
1	Number	Date	Checks and Deposits	Amount	Balance
2					178.85
3	301	5/3	Home Hardware	−37.12	141.73
4	302	5/3	Campus Bookstore	−45.60	96.13
5		5/3	gift from Mark Li, deposit	40.00	136.13
6					

In the "Amount" column, D, it is important to enter deposits as positive amounts and checks as negative amounts. Then, when the formulas are entered for addition, the spreadsheet will compute the new balance correctly for each transaction. Each new balance is found by adding the amount (negative for a check; positive for a deposit) to the previous balance.

When you set up your check register spreadsheet, you may insert a formula in the "Balance" column, as shown below. On your computer screen, cell E3 will show a true balance, not the formula.

	A	B	C	D	E
1	Number	Date	Checks and Deposits	Amount	Balance
2					178.85
3					+E2+D3
4					

You insert the information for columns A to D yourself. The formula will calculate the new balance, which will appear in cell E3. When you proceed to row 4, use the COPY command to copy the formula from cell E3 to cell E4. A new formula, +E3+D4, will be automatically created for that cell.

TECHNOLOGY HINT

Students should be gaining familiarity with the use of the spreadsheet. Have a student explain the formula used in the spreadsheet for Skill 4. Ask how the formula for the balance differs from row to row. Ask why the addition formula will work for each balance. **(It works because the subtraction sign is included with the amount. This is an application of the principle that subtracting is the same as adding the opposite.)**

RETEACHING

If reteaching seems needed, bring in an assortment of completed checks and the corresponding check register. Have students study them, looking for mistakes in the way they are completed or in the records kept in the check register. Some students will respond to this sort of "detective work."

ENRICHMENT

Challenge interested students to design checks using clip art and whatever computer programs you have available for formatting.

TRY YOUR SKILLS

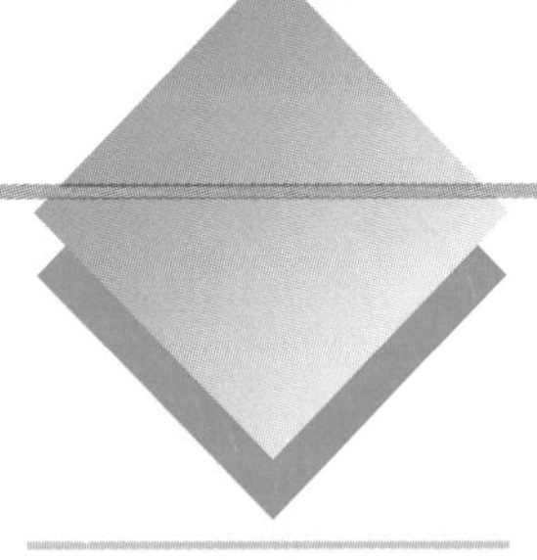

GUIDED PRACTICE

Point out the importance of each piece of information. Be sure that students include the date, check number, purpose of the check, and calculate the balance.

Enter the following transactions in a check register, finding the balance after each transaction. Each exercise will be a separate line on your simulated check register. Use a check register form supplied by your teacher, or draw up your own check register form. See Additional Answers.

		Balance
1.	Starting balance $400	$400.00
2.	Check 101, Feb 2, to Josh Harmon, $175.60 for electrical work	$224.40
3.	Check 102, Feb 3, to Sunrise Shop, $27.90 for a jacket	$196.50
4.	Check 103, Feb 5, to Bright Spot, $46.23 for a lamp	$150.27
5.	Feb 6, deposit, $45.60	$195.87
6.	Check 104, Feb 6, to Garden Center, $19.25 for plants	$176.62

7. Write the check for the Garden Center (Exercise 6). See Additional Answers.

EXERCISE YOUR SKILLS

KEY TERMS

blank endorsement
check register
drawer
endorsement
full endorsement
payee
restrictive endorsement

1. Why should you write the information in the check register before you make out a check? So you will not forget to enter it later.

2. What is meant by "writing your name the same way" each time you sign a document? Answers may vary. (Ex.: for consistency of endorsements)

3. If you have done all your computations correctly, what does the checkbook balance tell you? How much money remains in the account.

4–7. For the golf tournament of Example 2, write checks to the second through fifth place finishers listed and for the amounts listed. See Additional Answers.

8. Write an endorsement as you would on the back of a check that you intend to mail to the bank for deposit. For deposit only Student's Name

9–39. Create a check register spreadsheet, and enter each of the following transactions, finding the balance after each transaction. Record an initial balance of $660. See Additional Answers.

COOPERATIVE LEARNING

This lesson provides a good opportunity for students to check each other's work in groups of three. Have them go over each transaction. When their answers do not match, they should re-do the exercise together to find the correct answer.

	CHECK NUMBER	DATE	CHECKS/DEPOSITS		AMOUNT
9.	301	1/15	To: Panasonic Bike Shop	For: Bicycle	$ 181.50
10.	302	1/18	To: Brian's Sports	For: Helmet	25.00
11.	303	1/20	To: Friendly Feet	For: Shoes	25.00
12.	304	1/21	To: 50K Ride-A-Thon	For: Entry Fee	5.00
13.	305	1/23	To: Gargantua's Dept. Store	For: Socks	14.26
14.	306	1/23	To: Pandora's Hi Fashion	For: Shorts	15.04
15.	307	1/25	To: Amalgam Dept. Store	For: Bike Rack	115.00
16.		1/25	Deposit—Paycheck		311.19
17.	308	1/25	To: Sweatshirts Outrageous	For: Jacket	71.81
18.	309	1/25	To: Bike World	For: Tires	111.26
19.	310	1/25	To: Bruce's Sports	For: Racing Pedals	40.17
20.		1/25	Deposit—Savings		200.00
21.	311	1/25	To: Camper's Delight	For: Back Pack	98.50
22.	312	1/25	To: Light The Night	For: Lantern	24.00
23.	313	1/25	To: Army Surplus	For: Mess Kit	25.00
24.	314	1/25	To: Cash		50.00
25.	315	1/25	To: Sounds Galore	For: Tapes	63.00
26.		1/25	Deposit—Loan		120.00
27.	316	1/25	To: Rent-A-Tent	For: Tent	125.00
28.	317	1/26	To: Cash		25.00
29.	318	1/28	To: The Grocer	For: Groceries	55.30
30.	319	1/30	To: Energy Source	For: Batteries	9.50
31.		2/1	Deposit—Tutoring		100.00
32.	320	2/1	To: County Vehicles	For: Registration	5.00
33.	321	2/1	To: Biking News	For: Magazine	18.28
34.	322	2/3	To: Ozurka	For: Water	4.50
35.	323	2/4	To: Familiar Pharmacy	For: First Aid	7.55
36.	324	2/5	To: Emergency	For: Clinic	15.00
37.	325	2/5	To: Familiar Pharmacy	For: Medicine	9.11
38.	326	2/6	To: Sticks and Stones	For: Crutches	95.00
39.	327	2/6	To: Forever Photo	For: Album	11.71

40. You have received a check from your part-time job. Your boss used your full name, including your middle initial. You are at the bank. Show how you would endorse the check. Answers may vary. (Ex.: Latoya S. Marshall)

41. You have received a check from your part-time job. Your boss used your full name, including your middle initial. This time you are going to mail the check to the bank. Show how you would endorse the check. Answers may vary. (Ex.: For deposit only. Latoya S. Marshall.)

42. Your aunt mailed you a check. She used your nickname, "Rusty," instead of your real name. You are at the bank. Show how you would endorse the check. Answers may vary. (Ex.: Rusty Marshall (Latoya Marshall))

43. Your father gave you a check. He used your first initial and last name only. You want to give the check to your friend, Judy Bruns, to repay a loan she made you. Show how you would endorse the check. Answers may vary. (Ex.: Pay to the order of Judy Bruns. L. Marshall.)

INDEPENDENT PRACTICE

ASSIGNMENT GUIDE

Exercises 1-3 help determine students' understanding of the reasons for certain procedures.

Exercises 4-8 have students write checks and endorsements.

Exercises 9-39 provide practice in keeping the checkbook.

Exercises 40-44 offer brief problem scenarios for students to solve.

CLOSURE

Discuss how the calculator has helped people save time in keeping track of their checking records. Explain the importance of looking closely at amounts that are added or subtracted in order to estimate results and recognize how keying mistakes lead to wrong answers. Have students use rounding in order to estimate answers, then review with the calculator.

LESSON QUIZ

1. A bank balance is $450. A check is written for $178.96 and a deposit of $67.43 is made. Find the new balance. **($338.47)**
2. A bank balance is $167.98. One check is written and then the balance is $89.30. What was the amount of the check? **($78.68)**

44. Your friend Tracy has a checking account but says that she doesn't bother with the check register. She says that the bank keeps records of her deposits and expenditures. What do you say to convince Tracy to keep an accurate record of transactions in her check register?
Answers may vary. (Ex.: You may avoid bouncing a check.)

MIXED REVIEW

1-1 **1.** What is the rate for time-and-a-half based on an hourly rate of \$9.40? \$14.10

1-1 **2.** Write an equation for total pay at x hours for regular time and y hours of overtime at \$9.40 an hour and time-and-a-half for overtime. $T = 9.40x + 14.10y$

1-1 **3.** If Sharon works 45 hours at the rates given in Exercise 2, what will be her total pay if 40 hours are regular hours? \$446.50

1-1 **4.** Lou gets paid \$16 an hour plus time-and-a-half for any hours over 40. What does he earn if he works for 49 hours? \$856

1-3 **5.** Jack receives a base pay of \$240 a week and 9% commission on a sale of \$200. What are his earnings for the week? \$258

1-3 **6.** Leslie receives 4.5% commission on her real estate sales. This month she sold two houses, one for \$113,000 and one for \$79,500. What was her commission this month? \$8662.50

1-3 **7.** Jeffrey sells real estate. This month he sold a house for \$225,000. The commission was 5% but he only received half of the commission because he had to split the commission with another agent. What was his commission? \$5625

1-4 **8.** Alfredo's weekly take-home pay is \$103 and his deductions are \$27. What percent of his gross pay is his take-home pay? 79.2%

2-1 Northern Bank offers a checking account that pays interest of 0.3% per month for all accounts and charges \$0.08 per check for balances of less than \$500. Find the amount earned or charged for each of the following average balances for one month.

9. \$350 with 18 checks written \$0.39 charged

10. \$550 with 20 checks written \$1.65 earned

2-3 RECONCILIATION: DO THE BANK AND I AGREE?

QUICK REFERENCE

TEACHER'S RESOURCES AND ANSWER KEY

Lesson Quiz Answers 2-3, p. 44
Reteaching Activity Answers 2-3, p. 93
Enrichment Activity Answers 2-3, p. 94

EXTENSION ACTIVITIES

Reteaching Activity 2-3, p. 11
Enrichment Activity 2-3, p. 14

Larry still watches in fascination as the $10 and $20 bills come tumbling down out of the automated teller machine at the convenience store down the street. The first time Larry used the machine, he wasn't sure he would know what buttons to push, but the instructions on the computer screen were very simple; after one visit, Larry knew what to do. In fact, one weekend Larry spent Saturday afternoon driving around town locating other machines that would accept his card, and he came home with $240 in his pocket in tens and twenties. Driving around so much had made Larry a little hungry; he spent some money for pizza, some for a beef sandwich, some more for popcorn, and finally some for frozen yogurt.

Next to the yogurt store was one of Larry's favorite haunts—the music store. Larry spotted a new CD he wanted and bought two copies—the extra one for his girlfriend, Lorrie Anne. Larry considered making some notes on the amounts of money he had spent, but he forgot about it while he was thinking how much Lorrie Anne would like the present.

As Larry was gathering up his keys and change and getting ready to drive to Lorrie Anne's house, one of the slips of paper from the ATM, which Larry had stuck in his pocket, fell to the floor and caught Larry's attention. Printed near the end of the receipt was "Available Balance: $14.78." "Wait a minute, now," Larry thought. "I wrote two checks yesterday that haven't reached the bank yet. They totaled more than $150. How can I only have $14.78 left? Why, I deposited $300 two days ago! What became of the $300?"

Larry had a sinking feeling that he knew what had become of the money. What would happen at the bank tomorrow if those checks he wrote went through his account? Larry wondered whether there was any way to stuff some of that cash back into the machine.

FOCUS

ALGEBRA CONNECTIONS

After Jack uses a check to pay $60 and Sue uses a check to pay $50, they both have the same amount in their bank accounts. Who had more money in the bank account before they used the checks? Write an equation for each situation. **(Jack; Jack: $B_1 - 60 = N$; Sue: $B_2 - 50 = N$; B_1 and B_2 represent original amount, N represents the new amount in each account.)**

OBJECTIVES: *In this lesson, we will help Larry to:*

- *Assess the advantages and disadvantages of using electronic fund transfers.*
- *Reconcile a bank statement with the corresponding check register balance.*

AUTOMATED TELLER MACHINES (ATMS)

No longer do customers have to rush to the bank before it closes. **Automated teller machines (ATMs)** allow customers easy access to their accounts during banking or nonbanking hours. By using a special ATM card and punching in a personal identification number, a customer can deposit or withdraw money and even obtain a loan at the site of an ATM. Automated teller machines are placed in convenient locations such as airports, shopping malls, and street corners. They may also be located on the premises of the bank.

But what if your ATM card is lost, stolen, or used without permission? If you lose or misplace your ATM card, you must notify the bank immediately. Usually, if you notify the bank within two business days, the most you will have to pay is $50.

Follow these rules of bookkeeping when using electronic fund services:

1. Always keep the record of your transaction. Check the date, amount, location, and type of transaction. You will later use this information to verify your monthly statement.
2. If a mistake is made at the time of the transaction, call your bank for direct customer service. If you cannot get service at the time, contact the bank as soon as possible.
3. Enter your debit transactions (that is, money taken out) in your check register just as you would a check transaction. This practice allows you to maintain an accurate running total of how much money you have.

The Checkless Society A few years ago, people thought that the checkless society, in which all transactions are made electronically, would soon be upon us. But this was not to be. Many bank customers are still reluctant to use ATMs and computers. They fear breakdowns of the electronic equipment or computer errors that could tie up their funds for days or weeks. Nonetheless, electronic banking is widely used and will continue to grow.

KEEPING A RECORD OF YOUR MONEY

Customers receive monthly **statements** from their banks that reflect all checking account transactions: deposits, checks cleared, service charges, and the ending balance of the account at the close of the statement period. Several days elapse from the time the statement is prepared to the time the customer receives the statement. During this time, additional checks may be written and deposits made and recorded in the register. Also, if the owner of the checking account has written a check but the recipient has not yet cashed it, the bank will not have subtracted the amount of that check from the account. Such checks are called **outstanding checks.** Each of these reasons will cause the customer's check register and statement to differ. The process of finding the correct balance is called **reconciliation.** (The procedure to follow for reconciling your bank statement and check register is described in detail in Sharpen Your Skills.)

After the reconciliation procedure is completed, the adjusted balances should agree. They represent the correct amount of money that remains in the checking account.

Each month you should reconcile the bank's statement with the balance in your register. Then make a brief notation in the register to record the fact that your account is in balance.

If the check register and statement do not agree, you should first check your own arithmetic and then—even though the statement was probably created on a computer and is therefore likely to be accurate—check the bank's arithmetic as well. You will have received your checks back from the bank with your statement. These are called **canceled checks.** Compare these checks with your register to see whether you missed any outstanding checks. Compare the amount you recorded for each check with the register amount and the statement amount for that check. When you find an error, recalculate the balance affected, and compare it with the other balance.

If there is still no agreement, report the matter to the bank. You should do this within ten working days of receiving the statement. Bank bookkeeping departments work with customers to determine whether there has been an error by the bank and will help customers find the reason for the differences. Some banks charge a fee for assistance with reconciliation.

Ask Yourself

1. What should you do if you lose your ATM card?
   Notify the bank immediately.
2. What is a monthly statement?
   See Additional Answers.
3. What are outstanding checks?
   Written checks that the recipients have not yet cashed.

CRITICAL THINKING

Some students may think that the need for reconciliation is due to a mistake in the checkbook or the bank records. Explain that a check paid goes to several places before it reaches the bank and is recorded. Ask students to write out a sequence of events, paying special attention to dates, to illustrate how and when a check would not be included on the bank statement.

ADDITIONAL ANSWERS

2. A monthly statement reflects all checking account transactions for the statement period.

ALGEBRA *REVIEW*

Solve each equation for x.

1. $x - 2 = 0$
   $x = 2$
2. $x + 5 = 2$
   $x = -3$
3. $2x - 8 = 0$
   $x = 4$
4. $2x - 1 = 5$
   $x = 3$
5. $3x = 2x + 1$
   $x = 1$
6. $5x - 1 = 2x + 8$
   $x = 3$
7. $x - 1 = 8 - 2x$
   $x = 3$
8. $1.2x = 7 + 1.4x$
   $x = -35$
9. $2(x - 1) = 3(x - 2)$
   $x = 4$
10. $2.5x - 1.75 = 3.25 - 1.25x$
    $x = 1.3$

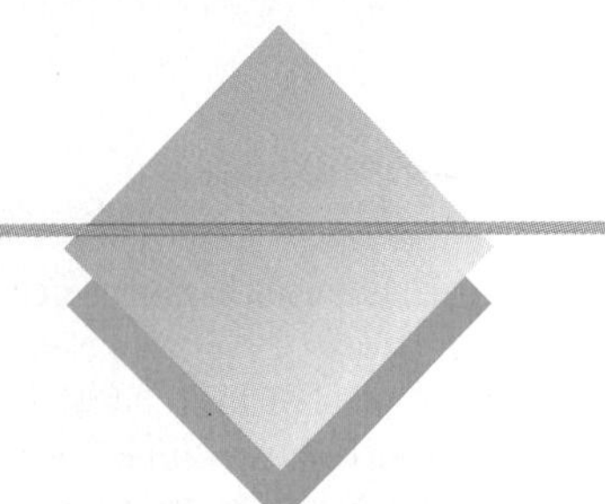

SHARPEN YOUR SKILLS

SKILL 1

EXAMPLE 1 The bank statement below shows the monthly transactions for Larry Lender. Notice the following entries on the statement.

1. Account balance at the beginning of the period, $186.43.
2. Service charge, $1.90. It is important to record this in your checkbook.
3. The record of checks written, received, and processed by the bank.
4. Deposits made by Larry Lender during the month.
5. Balance at the close of the statement period.

INLAND BANK

BANK STATEMENT

TO:

LARR Y LENDER
4115 N. LINCOLN
INLAND, IN 47304

ACCOUNT NUMBER:
75-197-66

DATE	YOUR BALANCE WAS	WE SUBTRACTED		SERVICE CHARGE	WE ADDED		MAKING YOUR PRESENT BALANCE
		NO.	CHECKS		NO.	DEPOSITS	
8/31/-	(1) 186.43	14	586.65	1.90	2	706.09	(5) 303.97

DATE	CHECKS	DEPOSITS	BALANCE
			186.43
8/2		286.75 (4)	473.18
8/6	125.00 (3)		348.18
8/9	23.46 40.00		284.72
8/10	9.45 15.00		260.27
8/12	15.74 139.00		105.53
8/15	34.10		71.43
8/16		419.34	490.77
8/17	21.19		469.58
8/19	8.00		461.58
8/22	14.86 10.00 45.00		391.72
8/27	85.85		305.87
8/31	1.90SC (2)		303.97

Please notify bank immediately of any change of address. The account will be considered correct if errors are not reported immediately.

Symbol code:
SC Service Charge
OD Overdrawn Account
SP Stop Payment
EC Error Correction

The balance in Larry's check register is $405.87. There are no outstanding checks, but he has made one deposit for $100 that does not appear on the statement.

TEACHING THE LESSON

Adults know that reconciling the checkbook with the bank statement can be challenging. But if the method is introduced step-by-step young people can avoid most of the anxiety associated with the process.

To begin, use round numbers and explain what happens if only one check or ATM withdrawal is missing from a bank statement. (The amount must be subtracted from the statement balance.) Then ask what happens if one deposit is missing. (The amount must be added to the statement balance.) Finally, ask about a bank fee that is on the bank statement but not included in the checkbook. (The amount is subtracted from the checkbook balance.)

All the procedures for reconciliation build on these three situations. Gradually use these quantities in different combinations until students can sort and use them with confidence.

ESL STUDENTS

Be sure to go over each heading on the bank statement and reconciliation form with any non-native speakers. The meaning of these words is critical to the successful understanding of the lesson and the completion of the exercises.

QUESTION What might be the reasons why Larry's balance is not the same as the balance on the bank statement?

SOLUTION

The first reason is that the deposit is not on the bank statement. Add the deposit to the bank balance:

$$303.97 + 100.00 = 403.97$$

Compare the adjusted bank balance with Larry's balance:

$$405.87 - 403.97 = 1.90$$

Look on the bank statement for an amount of 1.90. There is a service charge of 1.90. Larry discovers that he has not subtracted this charge from his balance. When he does, the adjusted balances are the same.

FOCUS ON ALGEBRA

Reading mathematics is now recognized as an important component in learning math. This text, being thoroughly practical, lends itself to student reading. Students should be encouraged to read the opening scenarios and to continue with the skill sections. The combination of verbal and mathematical explanations will enhance understanding. (*NCTM, Standard 2, p 140*)

SKILL 2

In the process of reconciling the balance on the bank statement with the balance in your checkbook, it generally happens that you must adjust both balances. This adjustment or correction does not mean that one or the other of the balances is wrong; it simply means that with new information the balances must be brought up to date.

How is the process of adjusting the bank statement balance similar to the process of adjusting the checkbook balance and how is it different?

The procedures for making the adjustments are similar. In both cases, amounts that were not available earlier must be added to or subtracted from the initial balance to find the adjusted balance. Adjustments of the bank statement and checkbook follow the equation

$$\text{Balance} + \text{deposits not recorded} - \text{checks and fees not recorded} = \text{new balance}$$

Differences are found in the numbers used. The beginning balance and the amounts of the unrecorded deposits, checks, and fees might all be different. But if all of the information is correct and the computations are done correctly, then the adjusted balances will be the same.

With each bank statement the bank also sends the customer a reconciliation form. This form is set up to help you compare the bank's records with your records and reconcile the account. That is, you adjust your balance and the bank's balance to arrive at the correct current balance.

COMMON ERRORS

Most errors in doing bank reconciliation result from mixing up addition and subtraction. If possible, use play money to show that a deposit increases the bank balance and a check or ATM withdrawal decreases the balance. This is true whether the balance is in the checkbook or on the bank statement.

EXAMPLE 2 In another month Larry received a bank statement showing a balance of $254.78. Larry's checkbook balance is $211.23. Larry compares the statement with his checkbook and finds the following, not shown on the statement: deposit, $60; checks, No. 226, $23.65, and No. 231, $82.30. On the statement, but not in his checkbook, is a bank charge of $2.40.

TECHNOLOGY HINT

Sometimes we can take a lesson from the way machines work. Explain to students the series of checkpoints contained in an ATM transaction. After each step, you are asked to OK your number, the account involved, the amount requested, and the denominations of bills you would like.

RETEACHING

A lack of comprehension will more often be due to the procedures than the arithmetic in the lesson. Reteaching can be done using round numbers so that the calculations do not obscure the method. Start with a checkbook balance of $90 and bank statement balance of $100 and an outstanding check of $10. When this is understood include an outstanding deposit of $20, and then a bank fee of $2.

ENRICHMENT

Challenge students to develop a script or flowchart of instructions and responses that would guide an ATM cash withdrawal transaction. Tell them that the machine must know the customer's code number and the amount to be withdrawn, and that the machine will also check the code number and whether the account balance is sufficient for the transaction. Each entry of the customer must have a response from the machine and vice-versa.

QUESTION How does Larry use the reconciliation form to reconcile his checkbook with the bank and to find new adjusted balances?

SOLUTION

The transactions in question are listed on the reconciliation form below. Notice where each transaction appears and how it is used.

1. The balance on the bank statement and in the checkbook.
2. Deposits not on the statement.
3. Checks not on the statement.
4. Service charges not in the checkbook.
5. Adjusted balances.

YOU CAN EASILY
BALANCE YOUR CHECKBOOK
BY FOLLOWING THIS PROCEDURE

FILL IN BELOW AMOUNTS FROM YOUR CHECKBOOK AND BANK STATEMENT

BALANCE SHOWN ON BANK STATEMENT (1) $ 254.78

ADD DEPOSITS NOT ON STATEMENT (2) $ 60.00

TOTAL $ 314.78

SUBTRACT CHECKS ISSUED BUT NOT ON STATEMENT (3)

#	$
226	23.65
231	82.30

TOTAL $ 105.95

BALANCE (5) $ 208.83

BALANCE SHOWN IN YOUR CHECKBOOK (1) $ 211.23

ADD ANY DEPOSITS AND OTHER CREDITS NOT ALREADY ENTERED IN CHECKBOOK $

TOTAL

SUBTRACT SERVICE CHARGES AND OTHER BANK CHARGES NOT IN CHECKBOOK

$ 2.40 (4)

TOTAL $ 2.40

BALANCE (5) $ 208.83

THESE TOTALS REPRESENT THE CORRECT AMOUNT OF MONEY YOU HAVE IN THE BANK AND SHOULD AGREE. DIFFERENCES, IF ANY, SHOULD BE REPORTED TO THE BANK WITHIN TEN DAYS AFTER THE RECEIPT OF YOUR STATEMENT

TRY YOUR SKILLS

1. What amounts must be added to the bank statement balance or to the check register balance in preparing the adjusted balance? See Additional Answers.
2. What amounts must be subtracted from the bank statement or from the check register balance in preparing the adjusted balance?
3. Your bank statement shows a closing balance of $75.65. There are no outstanding checks or deposits. Your checkbook shows a balance of $77.95. What might account for the different balances?

Your bank statement shows a closing balance of $102.60. The following are not on the statement: deposit, $50.00; check, $23.88; ATM withdrawal, $40.00. There is a service charge of $1.50 on the statement but not in your checkbook.

4. What amount(s) must be added to the bank statement balance? $50.00
5. What amount(s) must be subtracted from the bank statement balance? $23.88; $40.00
6. What should be the balance in your checkbook before you do a reconciliation? $90.22
7. What should be the balance after you do a reconciliation? $88.72
8. Roger has a balance of $218.52 in his check register. The balance on his bank statement is $487.35. Not reported on his bank statement are a deposit for $335.56 and checks for $572.00, $39.83, and $77.11. The service charge was $7.43. Roger noticed checks for $52.12 and $25.00 were not in his check register. He had not recorded the service charge in his check register. What should the balance be in his check register after he reconciles his account? $133.97

ADDITIONAL ANSWERS

1. deposits not recorded
2. checks and fees not recorded
3. check fees, service charges, ATM withdrawal

GUIDED PRACTICE

The first five *Try Your Skills* exercises can be done orally with a brief discussion after each. You can keep a close watch as students do Exercises 6–8, making sure that they correctly follow the procedures.

EXERCISE YOUR SKILLS

1. What are two advantages of ATMs? See Additional Answers.
2. What is a disadvantage of ATMs?
3. Why is the current balance on the bank's statement often different from the balance in your check register?
4. What can you do to balance your account if the initial use of the reconciliation form does not yield matching balances?

Larry received the following monthly statement from his bank.

INLAND BANK

BANK STATEMENT

TO:

LARR Y LENDER
4115 N. LINCOLN
INLAND, IN 47304

ACCOUNT NUMBER:
75-197-66

DATE	YOUR BALANCE WAS	WE SUBTRACTED		SERVICE CHARGE	WE ADDED		MAKING YOUR PRESENT BALANCE
		NO.	CHECKS		NO.	DEPOSITS	
9/30/- -	303.97	11	561.45	2.30	2	400.00	140.22

DATE	CHECKS			DEPOSITS	BALANCE
					303.97
9/1	100.25				203.72
9/3	75.00				128.72
9/9				200.00	328.72
9/11	214.00				114.72
9/16	12.95	15.45			86.32
9/19				200.00	286.32
9/21	11.15	20.00	34.50		220.67
9/24	5.40	7.75			207.52
9/28	65.00				142.52
9/30	2.30SC				140.22

Please notify bank immediately of any change of address. The account will be considered correct if errors are not reported immediately.

Symbol code:
SC Service Charge
OD Overdrawn Account
SP Stop Payment
EC Error Correction

KEY TERMS

automated teller machine (ATM)
canceled checks
outstanding checks
reconciliation
statement

INDEPENDENT PRACTICE

ASSIGNMENT GUIDE

Exercises 1-9 look into students' understanding of the banking procedures covered in the lesson.

Exercises 10-13 require the reconciliation of checkbooks and bank statements.

Exercise 14 asks students to compare and contrast the thinking of a banker with that of a customer, regarding the maintenance of records.

Exercises 10-13 present problem situations for students to solve.

ADDITIONAL ANSWERS

1. Answers may vary. (Ex.: convenient, open 24 hours)
2. Answers may vary. (Ex.: cards may be lost or stolen, transactions not recorded)
3. Additional deposits made after statement prepared, outstanding checks
4. Recheck register, check the bank's arithmetic, then call bank's customer service department.

5. How many check amounts are reported on the statement? 11
6. What is the initial balance for the account? $303.97
7. What is the closing balance for the account? $140.22
8. What is the total dollar amount of the deposits reported on the statement? $400
9. What service charge is reported on the statement? $2.30
10. Not reported on Larry's bank statement are two checks: No. 238, $25.00, and No. 241, $38.05. Also not reported is a deposit for $50.00. The service charge listed on the statement is not recorded in Larry's check register. The balance in Larry's check register is $129.47. Use this information to reconcile Larry's account using a reconciliation form provided by your teacher, or copy the one shown earlier in the lesson. See Additional Answers.
11. Cheryl has a balance of $478.83 in her check register. The balance on her bank statement is $345.75. Not reported on her bank statement are a deposit for $250 and three checks for $85.00, $54.32, and $129.75. She forgot to record two ATM withdrawals of $50 and $100 in her check register. She also needs to record the service charge of $2.15 in her check register. What should the balance be in her check register after she reconciles her account? $326.68
12. Following are a portion of Rachel Rosen's check register and the bank statement that she received. Reconcile her bank statement with her check register. Obtain a reconciliation form from your teacher, or make a copy of the one shown earlier in the lesson. See Additional Answers.

Check Register

CHECK NUMBER	DATE		CHECKS/DEPOSITS	AMOUNT	BALANCE $1840 63
201	3/7	TO:	FOREVER PHOTOS	183 70	
		FOR:	CAMERA		1656 93
202	3/8	TO:	LIGHTNING DEVELOPING	63 50	
		FOR:	PICTURES		1593 43
203	3/8	TO:	WE-HAVE-IT DEPT. STORE	23 80	
		FOR:	ALBUMS		1569 63
204	3/9	TO:	FLASH BACK	12 95	
		FOR:	CAMERA CASE		1556 68
	3/10	TO:	DEPOSIT	363 90	
		FOR:	PAYCHECK		1920 58
205	3/12	TO:	M-MART	12 50	
		FOR:	FILM		1908 08
ATM	3/15	TO:	ATM	50 00	
		FOR:	CASH		1858 08
206	3/17	TO:	CAMERA CASE	85 60	
		FOR:	LENSES		1772 48
	3/17	TO:	DEPOSIT	250 00	
		FOR:			2022 48
ATM	3/17	TO:	ATM	50 00	
		FOR:	CASH		1972 48
207	3/19	TO:	LIGHTNING DEVELOPING	26 85	
		FOR:	CHEMICALS		1945 63
208	3/21	TO:	ELECTRIC COMPANY	95 75	
		FOR:	ELECTRIC BILL		1849 88

CLOSURE

You might conclude the lesson by having a student explain the reconciliation process to the class. If possible let him or her use an overhead transparency. The other students should be permitted to ask questions.

LESSON QUIZ

1. A bank statement balance is $356.00. There are two outstanding checks: $45.60 and $176.20; one deposit of $80 not shown; no fees. What should be the checkbook balance? **($214.20)**
2. You deposit a check for $100 and withdraw $90 in cash. Should you record both transactions or just list a deposit of $10? **(Record both.)**

ADDITIONAL ANSWERS

10. 140.22 + 50 − 25 − 38.05 = 127.17; 129.47 − 2.30 = 127.17
12. 1870.23 − 26.85 = 1843.38; 1849.88 − 6.50 = 1843.38

Checking Account Statement

ACCT. 190-12566
DATE 3/31/- -
PAGE 1

National City Bank

RACHEL ROSEN
717 NOR TH WILSON PLACE
INLAND, IN 47304

BALANCE FORWARD	NO. OF CHECKS	TOTAL CHECK AMOUNT	NO. OF DEP.	TOTAL DEPOSIT AMOUNT	SERVICE CHARGE	BALANCE THIS STATEMENT
1840.63	7	577.80	2	613.90	6.50	1870.23

CHECKS AND OTHER DEBITS		DEPOSITS AND OTHER CREDITS	DATE	BALANCE
201	183.70		3/10	1656.93
202	63.50		3/10	1593.43
203	23.80		3/11	1569.63
204	12.95	363.90	3/11	1920.58
205	12.50		3/15	1908.08
206	85.60		3/20	1822.48
208	95.75	250.00	3/22	1976.73
	50.00ATM		3/23	1926.73
	50.00ATM		3/24	1876.73
	6.50SC		3/25	1870.23

PLEASE EXAMINE AT ONCE.
IF NO ERRORS ARE REPORTED WITHIN 10 DAYS, THE ACCOUNT WILL BE CONSIDERED CORRECT.

PLEASE ADVISE US
IN WRITING OF ANY CHANGE IN YOUR ADDRESS

KEY TO SYMBOLS

AD AUTOMATIC DEPOSIT
AP AUTOMATIC PAYMENT
AR AUTOMATIC REVERSAL
CB CHARGE BACK
CC CERTIFIED CHECK
CM CREDIT MEMO
CO CHARGE OFF
DM DEBIT MEMO
EC ERROR CORRECTED
IE INTEREST EARNED
OD OVERDRAWN
RC RETURN CHECK CHG
RT RETURN ITEM
SC SERVICE CHARGE

13. Jose has a balance of $899.10 in his check register. The balance on his bank statement is $569.93. Not reported on his bank statement are two checks for $120.60 and $53.77 and a deposit for $175. He had forgotten to record in his check register a transfer payment to his credit card for $125 and two ATM withdrawals of $100 each. He had also not recorded the service charge of $3.54. What should the balance be in his check register after he reconciles his account? $570.56

14. Both the bank and the customer who has an account at the bank want records to be accurate. But their reasons are not exactly the same. Think of yourself as a banker, and write a paragraph on why you want all accounts to be accurate and up to date. Then think of yourself as a person with a bank account, and write a paragraph about why you try to keep your banking records accurate and up to date. Answers may vary. (Ex.: As a banker, I would want all accounts to be as accurate as possible. This would enable me to know how much money is available to invest as well as how much is available to lend. With accurate records, the bank's debits and credits would balance, thus the bank's funds would be secure for the customer and the business.

As a bank customer, I would want all accounts to be as accurate and up-to-date as possible so I could be sure that the checks I write would be paid by the funds in my account. I would be assured that my money was both secure and accessible.)

MIXED REVIEW

1-1 **1.** At $8.50 per hour, how much does Nancy earn for 26.5 hours of work?

1-3 **2.** If you earn a commission of 3% on sales up to $80,000 and 5% on sales over $80,000, what is your commission on sales of $96,000?

1-3 **3.** Norma Torres is paid a base salary of $375 a month and a commission of 9%. Find her monthly earnings if her total sales are $8700. $1158

1-1 **4.** Jeremy works from 3:00 P.M. until 6:30 P.M., five days a week. How many hours does he work a week? 17.5 hours

2-1 **5.** Sarah's checking account pays interest of 0.35% a month and charges 6 cents per check. What amount will be earned or charged if her average balance is $235 and she writes 17 checks? $0.20 charged

2-2 **6.** Find the balance for a checking account after the following transactions are made: initial balance, $260.50; check written for $76.45; check written for $19.87; deposit, $35.00. $199.18

1-4 **7.** For most employees, the FICA tax is 7.65%, which includes a 6.2% tax for Social Security. The remainder of the tax is for Medicare. What is the Medicare tax on Basilio's gross weekly income of $300? $4.35

1-4 **8.** For most employees, the FICA tax is 7.65%, which includes a 6.2% tax for Social Security. The remainder of the tax is for Medicare. What is the Medicare tax on Tricia's gross monthly income of $2900? $42.05

2-2 **9.** Find the balance for a checking account after the following transactions are made: initial balance $549.13; check written for $137.58; check written for $89.60; check written for $111.25; deposit, $100.00; ATM withdrawal, $60.00. $250.70

2-3 **10.** Enid has a balance of $1235.54 in her check register. The balance on her bank statement is $689.66. Not reported on her bank statement are a deposit for $575, a transfer from her savings account for $250, and two checks for $195.00 and $437.85. She forgot to record two ATM withdrawals of $150 and $200 in her check register. She also needs to record the service charge of $3.73 in her check register. What should the balance be in her check register after she reconciles her account? $881.81

2-3 **11.** Pete has a balance of $99.10 in his check register. The balance on his bank statement is $432.51. Not reported on his bank statement are two checks for $325.50 and $22.25, an ATM withdrawal for $200.00 and a deposit for $155. He had forgotten to record in his check register a transfer payment to his credit card for $60. He had also not recorded the interest earned of $0.66. What should the balance be in his check register after he reconciles his account? $39.76

1-3 **12.** Ernestine sells real estate. This month she sold a house for $375,000. The commission was 5% but she only received one-third of the commission because she had to split the commission with two other agents. What was her commission? $6250

ADDITIONAL ANSWERS

1. $225.25
2. $3200

CHAPTER 2 REVIEW

1. Find the interest for one month on $120 at 0.35% interest per month. $0.42

2. A bank pays interest of 0.45% per month and charges a finance fee of $4 on accounts with balances of less than $500. Find the total payment or charge for an account with a balance of $467. $1.90 charged

3. If you had $487.51 in the bank and paid someone $63.94 by check, what would be your new bank balance? $423.57

4. Give three reasons why the balance on a bank statement might be different from the balance in your check register. See Additional Answers.

5. The Ridgewood Savings Bank pays 0.65% monthly interest and charges $0.10 per check. The Second National Bank pays 0.35% monthly interest and charges $0.04 per check. Write equations for the monthly charge and payment at each bank for a balance of $1000 but an unknown number of checks. Graph the equations using a graphing calculator. Use the graph to find the number of checks for which each bank is more economical.

6. The cost of Presto+ checks is $0.0225 per check. There is an additional charge of $0.15 for each check over 20. No service charge is charged for beginning balances of $500 and above; $5.00 is charged for beginning balances below $500. Interest of 0.45% per month is paid on the beginning balance. Copy and complete the chart.

	Beginning Balance	Number of Checks	Cost of Checks	Extra Charges	Service Charge	Interest Earned	New Balance
a.	$ 500	20	$0.45	0	0	$2.25	$ 501.80
b.	1000	10	0.23	0	0	4.50	1004.27
c.	600	16	0.36	0	0	2.70	602.34
d.	400	24	0.54	$0.60	$5.00	1.80	395.66
e.	300	30	0.68	1.50	5.00	1.35	294.17
f.	100	32	0.72	1.80	5.00	0.45	92.93

7. Find the new balance at the end of each transaction. The beginning balance is $0.00.

Date	Check No.	Check/Deposit	Amount	Balance
2–17		Deposit	$455.78	$455.78
2–18	405	Ravine Variety	85.95	369.83
2–18	406	Power & Light	38.53	331.30
2–18	407	Douglas Water	7.45	323.85
2–19	408	Folly Gasoline	14.76	309.09
2–20	409	Folly Service Station	53.90	255.19
2–20		Deposit	104.88	360.07
2–21	410	Bay Oil Company	16.28	343.79

ADDITIONAL ANSWERS

4. Answers may vary. (Ex.: outstanding checks, arithmetic errors, deposits made after statement prepared)

5. Ridgewood Savings Bank:
$y = 6.5 - 0.10x$
Second National Bank:
$y = 3.5 - 0.04x$
$x = 50$
The account at Ridgewood Savings Bank will be more expensive if fewer than 50 checks are written and less expensive if more than 50 checks are written.

8. Reconcile the bank statement and check register summarized below.

BANK STATEMENT			
Date	**Transaction**	**Amount**	**New Balance**
6–11	Balance forward	$400.00	$400.00
6–12	ATM cash withdrawal	50.00	350.00
6–12	Check #435	19.37	330.63
6–14	Check #437	35.00	295.63
6–17	Check #436	40.55	255.08
6–19	Deposit	50.00	305.08
6–19	ATM cash withdrawal	25.00	280.08
6–21	Check #439	11.16	268.92
6–22	Check #438	1.00	267.92
6–25	Service charge	5.00	262.92
6–26	Check #440	85.25	177.67
6–26	ATM cash withdrawal	80.00	97.67
6–27	ATM transfer payment	50.00	47.67

CHECK REGISTER			
Date	**Transaction**	**Amount**	**New Balance**
6–05	Balance forward	$400.00	$400.00
6–11	Check #435	19.37	380.63
6–12	ATM cash	50.00	330.63
6–12	Check #436	40.55	290.08
6–13	Check #437	35.00	255.08
6–14	Check #438	1.00	254.08
6–15	Check #439	11.16	242.92
6–19	ATM cash	25.00	217.92
6–19	Deposit	50.00	267.92
6–29	Deposit	40.00	307.92

9. Set up a spreadsheet to calculate charges and new balances for the checking account described in Exercise 6. Find the interest earned and new balance for the following: \$155, 11 checks; \$278, 17 checks; \$505, 21 checks; \$789, 27 checks. See Additional Answers.

10. Write a restrictive endorsement as you would to deposit a check in the bank. For deposit only. Signature

ADDITIONAL ANSWERS

8. $47.67 + 40 = 87.67$; $307.92 - 85.25 - 80 - 50 - 5 = 87.67$

CHAPTER 2 TEST

ALTERNATIVE ASSESSMENT

Have students interview two or three adults about what they look for when choosing a bank in which to open a checking account. Help them draft five or six questions about financial benefits, customer service, and convenience.

Ask students to write to several large banking chains requesting information about the benefits and costs of their accounts. Students in small groups can consolidate their data and make written or oral reports.

Ask students to start with a simulated bank balance of $250, and then to purchase five items based on newspaper prices. They should write out the checks and keep the checkbook for these items. One or two students should give an oral report and the others hand in their checks and checkbook pages for review.

Ask students to write a set of instructions to be given to a bank officer telling him or her what to do when a customer asks for help reconciling the bank statement with the checkbook.

1. The cost of Presto+ checks is $0.0225 per check. There is an additional charge of $0.15 for each check over 20. No service charge is charged for beginning balances of $500 and above; $5.00 is charged for beginning balances below $500. Interest of 0.45% per month is paid on the beginning balance. Copy the chart, and complete it using the information given.

	Beginning Balance	Number of Checks	Cost of Checks	Extra Charges	Service Charge	Interest Earned	New Balance
a.	$ 400	20	$0.45	0	$5.00	$1.80	$ 396.35
b.	150	20	0.45	0	5.00	0.68	145.23
c.	1000	32	0.72	$1.80	0	4.50	1001.98

2. Find the new balance at the end of each transaction.

Date	Check No.	Check/Deposit	Amount	Balance
4–15		Deposit	$1214.85	$1214.85
4–15	415	VISA	101.97	1112.88
4–18	416	Craft Supply Co.	34.85	1078.03
4–19	417	Village Variety	83.66	994.37
4–20		Deposit	320.00	1314.37

3. Reconcile the bank statement and check register shown below.

BANK STATEMENT

Date	Transaction	Amount	New Balance
4–2	Balance forward	$200.00	$200.00
4–4	Check #426	35.00	165.00
4–6	Check #425	13.19	151.81
4–15	Deposit	230.00	381.81
4–20	Service charges	8.30	373.51
4–22	ATM cash withdrawal	150.00	223.51

CHECK REGISTER

Date	Transaction	Amount	New Balance
4–1	Balance forward	$200.00	$200.00
4–1	Check #425	13.19	186.81
4–2	Check #426	35.00	151.81
4–22	Deposit	500.00	651.81
4–22	Check #427	413.00	238.81
4–25	Check #428	187.18	51.63

$223.51 + 500 - 413 - 187.18 = 123.33$; $51.63 + 230 - 150 - 8.30 = 123.33$

CUMULATIVE REVIEW

1-1 **1.** You worked 10 hours at $5.25 per hour. You received $11.50 in tips. Find your total earnings. $64.00

1-1 **2.** You worked 24 hours at $4.90 per hour. You received $1\frac{1}{2}$ times your hourly rate for any hours over 20. Find your total wages. $127.40

1-1 **3.** You worked the schedule shown below. You earned the same as in Exercise 2. Find your total wages for the week. $112.70

Monday	Tuesday	Wednesday	Thursday	Friday	Saturday
4–7 P.M.	4–7 P.M.	5–10 P.M.	4–7 P.M.	4–7 P.M.	9 A.M.–2 P.M.

1-3 **4.** You earn a base salary of $400 per month plus commission of 8% of sales. If your sales for one month are $4000, how much will you earn that month? $720

1-3 **5.** As a realtor, you sold a house for $210,000. You earn 4% commission on the first $100,000 and 5% commission on any amount over $100,000. How much commission did you earn on the house? $9500

1-4 **6.** You earned $125.75. Your income tax withholding is $5, and your FICA withholding is 7.65% of your gross pay. Find your take-home pay. $111.13

1-4 **7.** You earned $181.25. Your income tax withholding is $13, and your FICA withholding is 7.65% of your gross pay. Find your take-home pay. $154.38

1-2 **8.** You earn $7.10 per hour. You work 40 hours per week for 50 weeks. You get two weeks of paid vacation plus $2500 worth of fringe benefits. Find your yearly earnings, including benefits. $17,268

2-1 **9.** You wrote 25 checks one month. The bank charges $0.025 per check for the first 20 checks and $0.10 for each check over 20. How much are the charges for that month? $1.00

1-2 **10.** You earn $46,000 per year plus 16% of your salary in fringe benefits. Find your yearly earnings, including benefits. $53,360

2-2 **11.** Your uncle wrote you a check and used your full legal name. Write your name as you would to endorse the check. Answer may vary.

2-2 **12.** What is a blank endorsement? When is the only time you should use it?

2-2 **13.** You wrote a check for $24.65. You had $95.16 in your checking account. Show how you would enter this check in your check register.

2-1 **14.** Use a graphing calculator to compare the following kinds of accounts. For each one, assume a balance of $1000 but an unknown number of checks. Bank A pays interest of 0.4% per month and charges $0.085 per check. Bank B pays interest of 0.2% per month and charges $0.055 per check. Use the trace function to find the approximate point of intersection for the two graphs. Under what conditions is each bank better?

ADDITIONAL ANSWERS

12. Your signature only; at the bank.

13.

Amount	Balance
	95.16
24.65	−24.65
	70.51

14. Bank A: $y = 4 - 0.085x$; Bank B: $y = 2 - 0.055x$; $x = 66.6$
The account at Bank A will be less expensive if 66 or fewer checks are written and more expensive if more than 66 checks are written.

PROJECT 2–1: **Banking Institutions**

Gather information about a commercial bank and a savings and loan institution in your community. Compare your information with that of your classmates. Below are some possible questions.

1. Are checking accounts available?
2. What are the service charges?
3. Are savings accounts available?
4. What method is used to determine interest paid?
5. Are special checks provided?
6. Are ATM cards available?
7. Does the institution have deposit insurance?
8. For what amount are the accounts insured?
9. Can electronic fund transfers be made?
10. What types of loans are available?
11. Are safe-deposit boxes available?
12. Is estate planning guidance available?
13. Is tax assistance or counseling provided?

PROJECT 2–2: **Automated Teller Machines**

Automated teller machines allow customers 24-hour access to banking services. Because anyone can walk or drive up to a machine and use it, banking institutions take security precautions to protect against theft. To use your card, you must insert it into the machine and also enter your personal identification number (PIN) into the machine. The bank sometimes allows you to choose your own four- or five-digit PIN. If anyone should have your card and know your PIN, he or she could take money from your account.

1. Visit local banks, and find how you obtain a PIN. Some people use their birthdate as their PIN. Is this a good idea? Why or why not?
2. Visit local banks and find what security precautions they take with regard to ATM cards.

ALGEBRA *REFRESHER*

The *commutative* and *associative* properties of addition and multiplication can help you simplify a numerical or algebraic expression.

	Addition	Multiplication
Commutative property	$a + b = b + a$	$a \cdot b = b \cdot a$
Associative property	$(a + b) + c = a + (b + c)$	$(a \cdot b) \cdot c = a \cdot (b \cdot c)$

The associative property allows you to express $3 + (0.06 + 8)$ and $(5.4 \cdot 2) \cdot 10$ without parentheses.

$$3 + (0.06 + 8) = 3 + 0.06 + 8 = 3.06 + 8 = 11.06$$
$$(5.4 \cdot 2) \cdot 10 = 5.4 \cdot 2 \cdot 10 = 10.8 \cdot 10 = 108$$

Simplify.

EXAMPLE $(11.23 + x) + 1.77$

$= (x + 11.23) + 1.77$ Commutative property for addition
$= x + (11.23 + 1.77)$ Associative property for addition
$= x + 13$ Simplify.

1. $8a + (12 + 3a)$ $11a + 12$

2. $(200)(50y)(5)(2x)$ $100{,}000xy$

3. $2x + 3y - x + 4y$ $x + 7y$

4. $0.5(8x)(2a)$ $8xa$

Two other properties that can help you simplify expressions are:

Distributive property	$a(b + c) = ab + ac$
Definition of subtraction	$a - b = a + (-b)$

Remember to subtract a number is to add its opposite.

EXAMPLE $11x - 6(x - 1)$

$= 11x + (-6)[x + (-1)]$ Definition of subtraction
$= 11x + [(-6)x + (-6)(-1)]$ Distributive property
$= [11x + (-6x)] + 6)$ Associative property of addition
$= 5x + 6$ Add like terms.

5. $4.03x + 5.97 - 0.03x$
$4x + 5.97$

6. $28 - (8.5 - 10c)$
$19.5 + 10c$

7. $40a - 20b - 9a$
$31a - 20b$

8. $(4.2 - 3p) - 2p$
$4.2 - 5p$

9. $-1(x^2 - 1.06) + x$
$-x^2 + x + 1.06$

10. $y^3 - (7y^3 + 7)$
$-6y^3 - 7$

CHAPTER 3

A DOLLAR Saved . . .

MARIA IS IN THE SAME HIGH SCHOOL as Alex, Jeff, and their friends. She and her friends face some of the same tough choices as their other classmates. Most of them have earned some money of their own and are enjoying the sense of independence it has brought them. They are eager to be out of school and completely on their own. They also realize the importance of choosing a career that they will find satisfying. For many this choice means attending college for further training, thus delaying total independence for a while. Many of these students' parents have already been saving money toward this extra education and have been encouraging their sons and daughters to save as well.

In this chapter, Maria and her friend Nelson will help us examine how people go about saving money for purchases that require more cash than can be obtained easily from a regular salary. Maria dreams of buying her own car and saves money toward that goal each month. As a result, she learns about various savings institutions and how they pay interest on her money. Nelson looks into interest in greater detail, discovering how often interest may be paid on a savings account and what is meant by an annual interest rate.

Their friend Olivia will help explain to us the impact of the Federal Reserve System on both savers and consumers. She discovers that transferring money over great distances takes only minutes. She also learns that the Federal Reserve System plays a key role in managing the country's money supply.

3-1 SAVINGS: SAVE NOW—BUY LATER

QUICK REFERENCE

TEACHER'S RESOURCES AND ANSWER KEY

Lesson Quiz Answers 3-1, p. 44

Reteaching Activity Answers 3-1, p. 94

Enrichment Activity Answers 3-1, p. 95

EXTENSION ACTIVITIES

Reteaching Activity 3-1, p. 15

Enrichment Activity 3-1, p. 18

Maria does not think of herself as extravagant but ever since she began earning money on her own, she has noticed several items that she would like to buy. She was even able to buy a small car just three months ago after she persuaded her father that she would be able to pay for the monthly installment charges, insurance, gasoline, and regular maintenance. He helped her pick out the car, which had only 14,000 miles on it and had been owned and maintained by a nationwide car-rental chain. So far she has not had to buy so much as a new windshield wiper blade.

It was not easy for Maria to convince Dad that she could keep up the car payments. For six months she put half of her weekly paycheck from a fast-food restaurant into her savings account at the bank. Dad had told her to keep the money for her college expenses next year, but what he really wanted to find out was whether Maria would be able to save up money for the car. After six long months, Dad was convinced that she could.

Maria will share with us some of what she has learned about saving money. She has decided that saving isn't so hard after all and that maybe she will save even more so that she can buy some of the things she did without during the six months. Perhaps she can even take her friend Melody out to lunch to celebrate Melody's birthday.

OBJECTIVES *In this lesson, we will help Maria to:*

- *Recognize why people save money.*
- *Identify the places where people commonly deposit savings.*
- *Explain factors such as interest rates and liquidity that influence the return that your money can earn at financial institutions.*
- *Explain the differences between regular savings accounts, money market accounts, and certificates of deposit.*

ALGEBRA CONNECTIONS

A newborn infant was given one billion dollars in cash at birth. How much would she need to spend on the average each day to use it all up in eighty years, assuming none was invested to earn interest?

WHY PEOPLE SAVE

People have many reasons for saving—to pay for a vacation, stereo or sports equipment, or a major appliance, for example. Many young people also save to help pay for college.

Maria's father had two reasons for saving. First, soon after Maria began working, he realized that, apart from Maria's own wishes, the family actually did need another car, since arranging transportation for everyone too often required skillful coordination. Second, he realized that by helping Maria follow through with her plan to pay for the car, he could not only ease the family's transportation problem but, perhaps more important, also help his daughter to develop the self-discipline she needed to save regularly.

As most people do, Maria's family knows that unforeseen events such as an accident, illness, or the loss of a job can cause a sudden loss of income. For this reason, financial advisors recommend a savings reserve of at least three to six months' salary.

It takes self-discipline and consistency of purpose to maintain a habit of saving each month, so developing these admirable qualities is a strong reason for acquiring that habit. In addition to its positive effect on your character, the saving habit also is directly rewarding. Whenever you let a financial institution such as a bank use your money, it will pay you interest on that money.

WHERE PEOPLE SAVE

Once Maria had made a firm decision to open a savings account, she had to decide what kind of institution she would choose. She knew that she had several options such as commercial banks, savings banks, savings and loan associations, and credit unions. Saving in these institutions is usually safe and offers a relatively fixed rate of return on money.

Commercial banks offer a wide variety of services that persuade many families to select this kind of bank as the center of their financial affairs.

These services include offering checking and savings accounts, making loans, issuing credit cards, and renting safe-deposit boxes. Commercial banks usually also sell traveler's checks and money orders. They also offer financial counseling and trust and investment services.

Savings banks operate in much the same way as commercial banks; their services include savings accounts, check cashing, safe-deposit boxes, and savings bank life insurance. They also specialize in real-estate loans. Savings banks are common in the northeastern part of the United States.

Savings and loan associations, sometimes also called *thrift institutions,* generally lend money for home purchases and home construction.

Credit unions are not-for-profit financial institutions that pay interest on members' savings and use these savings to make loans to their members. Credit unions have membership requirements. The members are people who belong to a professional organization, church, company, or other group that has a common social, economic, political, or religious basis. Maria's father, along with 7500 other federal employees, is a member of their county's Federal Employees Credit Union.

HOW TO SELECT A SAVINGS ACCOUNT

After Maria discovered the many types of savings institutions that are eager to use her money, she began to wonder why she should choose one kind of savings institution over another. One reason is that some institutions pay more **interest** than others.

When Maria's money is put to work, she should expect it to work as hard as possible to earn interest with a high degree of safety. *Interest* is the return savers get from letting someone else use their money. The *interest rate* is the most important factor in determining the interest a savings account will earn. By calling around to various institutions, Maria found out that interest rates vary from place to place. She also learned that the *liquidity* of a savings instrument influences the interest rate that she can get on her savings. **Liquidity** means the ease and speed with which savings can be converted to cash. The longer the savers are willing to tie up their money, the higher the interest rate they will receive.

Regular Savings Accounts The most liquid or flexible choice for Maria is a **passbook** or **regular savings account,** an account that allows her to make deposits and withdrawals at any time. A regular savings account accepts deposits of any size and allows you to withdraw any amount that does not exceed your total balance. Maria's family already has its emergency-fund savings in a regular savings account.

Money Market Accounts If you can do without the convenience of a regular savings account, you may be able to obtain a higher rate of interest than would otherwise be the case. One way you can do this is to open a **money market account**. In exchange for a higher rate of interest, you may have to accept some restrictions on your access to the account. For more details on money market accounts see Lesson 8–1.

✓ History

Certificates of Deposit Another way to obtain a higher rate of interest is to purchase a **certificate of deposit (CD)**. A certificate of deposit is less liquid than a regular savings account. It is purchased at a bank by filling out a special form that records the amount of savings a person chooses to place in a special account. The CD specifies a fixed amount of money that must be deposited and a period of time during which the saver promises not to withdraw money from the account. The interest rates for six-month certificates are tied to the weekly U.S. Treasury bill interest rate.

Since the bank has the use of your CD money for a guaranteed period of time, it can afford to pay a higher rate of interest than on a regular savings account. The federal government requires the bank to impose stringent penalties if you make an early withdrawal from a certificate of deposit account. For example, you can be penalized three months' interest on certificates of less than one year if you withdraw the money before the date that you originally agreed to.

Maria's family has placed her school tuition money in certificates of deposit because it is willing to accept the restrictions on its ability to withdraw the money in exchange for the relatively high interest that the certificates pay.

Ask Yourself

1. What emergencies do people usually save for?
   accidents, illness, unemployment
2. How much money do financial advisors recommend that people save to cover emergencies such as sickness or the loss of a job?
   3 to 6 months' salary
3. What are three types of institutions in which people usually keep their savings? commercial banks, savings banks, savings and loan associations, also credit unions
4. What are four of the services offered by commercial banks? checking and savings accounts, loans, credit card issuing, safe-deposit boxes
5. What examples can you give of the kind of people who might belong to the same credit union?
   See Additional Answers.
6. What is interest?

ADDITIONAL ANSWERS

5. Answers may vary. (Ex.: professional organizations, employee groups)
6. Interest is the return savers get for letting someone else use their money.

ALGEBRA *REVIEW*

In each formula, substitute the given values. Then solve to find the value of the remaining variable.

Example

$p = 2l + 2w; p = 80, l = 15$

$80 = 2(15) + 2w$

$80 = 30 + 2w$

$50 = 2w$ Subtract 30 from each side.

$25 = w$ Divide each side by 2.

1. $A = lw; l = 2.6, w = 7$
   $A = 18.2$
2. $i = 250rt; r = 0.07, t = 3$
   $i = 52.5$
3. $s = l + w; l = 16, w = 4.6$
   $s = 20.6$
4. $B = p + i, B = 302.5, p = 250$
   $i = 52.5$
5. $i = p(0.045)t; p = 5000, t = 15$
   $i = 3375$
6. $i = prt; p = 1500, r = 0.05, t = 10$
   $i = 750$
7. $B = p + i; p = 750, i = 30.63$
   $B = 780.63$
8. $B = p + prt; p = 100, r = 3\%, t = 4$
   $B = 112$

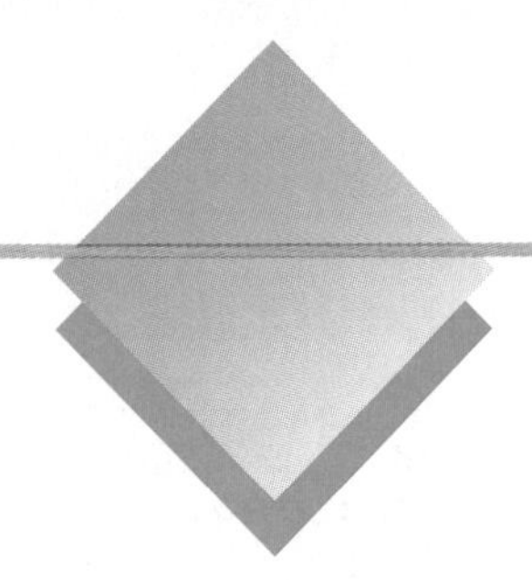

SHARPEN YOUR SKILLS

SKILL 1

INSTRUCTION

TEACHING THE LESSON

Explain simple interest formula. Emphasize that interest can be earned only on money already deposited; that is, if you save $100 per week for 52 weeks, the first deposit earns interest for 51 weeks, the second deposit for 50 weeks, and the 52nd deposit earns no interest in the first year. After completing Skills 2 and 3, have students compute the minimum amount that Dwight must save weekly to buy the golf bag on sale. Suggest that they find the number of weeks Sarita must save if she saves at a rate of $20 per week or $30 per week.

CRITICAL THINKING

After the students have completed Skill 1, have them consider the following: $1000 is invested at 6% compounded annually. For each of the next four years an additional $1000 is added to this account where it too earns interest. Compute the total interest earned after five years. Then compare that result to the interest earned on $5000 compounded annually at 6% for five years. (Hint: Find the total in the account after one year, then compute interest on that amount for the second year, and so on.)

The amount of money invested is called the **principal** p. **Simple interest** i can be determined using the following formula. The **interest rate** r is a percent of the principal.

Simple Interest Formula

$i = prt$ where i = interest
p = principal
r = interest rate
t = time

EXAMPLE 1 Maria can save $85 per week from her paycheck. After saving for a year, she decides to buy a one-year CD that pays 4% simple interest.

QUESTIONS

1. How much will Maria save in 52 weeks?
2. How much interest will the CD earn in 1 year?
3. How much will Maria's CD be worth after 1 year?

SOLUTIONS

1. Total savings = weekly savings • number of weeks
 = 85 • 52
 = 4420

 In 52 weeks, Maria will save $4420.

2. To find the interest after 1 year, Maria uses the formula for simple interest.

 $i = prt$
 $i = 4420 \cdot 0.04 \cdot 1$
 $i = 176.80$

 The CD earns $176.80 in interest.

3. The balance in the CD will be the principal plus the interest.

 Balance = principal + interest
 $B = p + i$
 $B = 4420 + 176.80$
 $B = 4596.80$

 The total amount in the CD at the end of one year is $4596.80.

If Maria leaves the money in the CD, then in the second year the bank will pay interest on both the original principal and on the previous year's interest. In other words, instead of being simple interest, as in year 1, the interest for year 2 and all later years will be *compound interest.* Compound interest is discussed in Lesson 3–2.

SKILL 2

EXAMPLE 2 Maria's friend Dwight wants to buy a golf bag. He finds one on sale for $59.84. The sale price will be in effect for 1 month.

> **Golf Bag . . . . . . SALE $59.84**
> Less than half original price! Genuine leather, sturdy construction. Just like the ones the pros use!

QUESTION If Dwight can save $22 per week, will he have enough money to buy the bag before the sale is over?

SOLUTION

Weeks needed = cost of golf bag ÷ amount saved each week
= 59.84 ÷ 22
= 2.72

Dwight will have enough money in three weeks, so Dwight will have enough money to buy the golf bag before the sale is over.

FOCUS ON ALGEBRA

In this lesson, formulas for computing savings and simple interest give students practice with decimal computation and with substitution of numerical values for variables. (*NCTM Standard 5, p. 150.*)

COMMON ERRORS

Be sure that students understand that they must always round up in a situation such as that illustrated in Example 2. Even if "weeks needed" were equal to 2.1, one would still round to three weeks.

SKILL 3

EXAMPLE 3 Maria's sister Sarita would like to buy one or more of the items shown below. She can save $25 each week from her salary.

1. **Audio Box** 3-bank EQ.
   Local-distance switch.
   Time/frequency display switch.
   Sale $59.87 Reg. $75.00

2. **Quartz Watch**
   Yellow case with black band.
   Measures elapsed and lap time.
   Sale $89.99 Reg. $135.00

3. **26" 10-speed Bicycle**
   Men's or women's mountain bike.
   Front and rear brakes.
   Sale $109.99 Reg. $119.95

4. **35mm Camera** $139.99
   Completely automatic. No focusing. Auto flash. Uses 2 AA batteries (not included).

QUESTION How many weeks will Sarita have to save to buy each item?

COOPERATIVE LEARNING

Have students work in small groups to compute interest earned on the CD in *Try Your Skills* Exercises 1-6 using interest rates of 3% and 4%. Tell them to use a graphing calculator or spreadsheet program to display the different interest earnings at 3%, 3.5%, and 4% for each exercise.

AT-RISK STUDENTS

Have students figure out how much money they can save each week. Tell them to choose an item to enhance a student hangout (such as a compact disc player or a couch) and to compute the number of weeks they will need to save enough money to buy this item.

ESL STUDENTS

Explain to non-native speakers the meaning of traveler's checks, money orders, safe-deposit boxes, credit cards and real estate loans.

TECHNOLOGY HINT

To determine the number of weeks Sarita needs to save for each item, she must count the grid lines on the bar graph. As a follow-up activity, have students prepare an XY graph using the spreadsheet program. To generate x-values, students should divide each price by 25 and use the @ROUND(x,n) function to round up to the nearest whole number.

SOLUTION

To find the number of weeks of savings required, Sarita prepares a bar graph using a spreadsheet program. She wants the graph to show equally spaced $25 levels so that she can easily see the number of weeks of savings required for each of the four items. Sarita begins by creating the following spreadsheet:

	A	B
1	stereo	59.87
2	watch	89.99
3	bicycle	109.99
4	camera	139.99

Most spreadsheet programs allow you to graph by selecting an option such as GRAPH or CHART. Choose the type of graph, in this case, a **bar graph.** Then use the option that allows you to choose minimum and maximum values on each axis.

Sarita chooses 0 for the minimum and 150 for the maximum on the y-axis.

You can also select an option to show horizontal gridlines. You may be able to select the spacing between gridlines by using a command called SCALE, INCREMENT, or something similar.

Sarita selects 25 for the scale.

The spreadsheet program will automatically label the y-axis (the left vertical axis) with the labels 0, 25, 50, 75, 100, 125, and 150 with or without dollar signs, as you choose. The spreadsheet menu will also let you choose an option with a name such as INTERIOR LABELS that lets you position each of the four sales prices above its graph. You will probably have to type the text labels yourself. Fortunately, you can instruct most spreadsheet programs to place certain kinds of text inserts (headings, for example) in their proper position.

Sarita chooses the INTERIOR LABELS option. Then she types in the labels for the top, bottom, and right side of the graph. The graph is shown below.

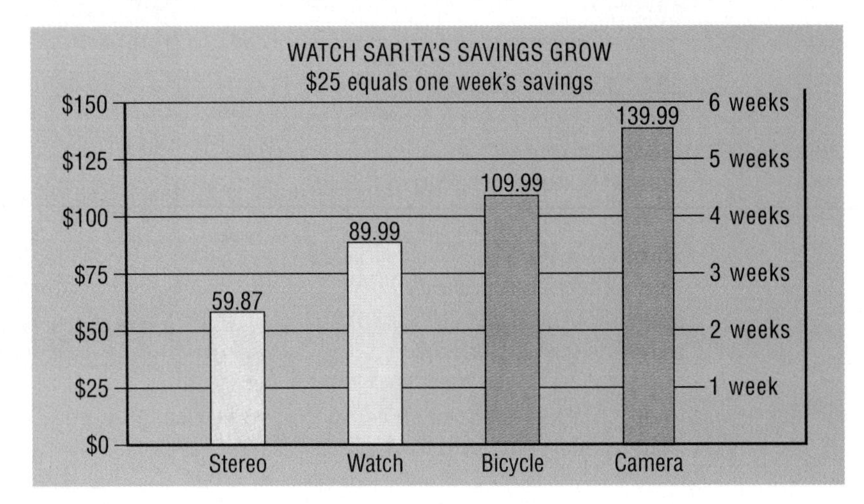

From the graph, it will take 3 weeks to save enough for the stereo, 4 weeks for the watch, 5 weeks for the bicycle, and 6 weeks for the camera.

TRY YOUR SKILLS

A bank pays 3.5% simple interest on a one-year CD. The table below shows the amounts that Maria and her friends have saved for a year to buy a CD. Find out how much is saved, how much interest is earned by the CD in one year, and the total amount of money in each account after the bank has credited the interest to the accounts.

	Amount Saved per Week	Number of Weeks	Total Saved	Interest Earned	Total in Account
1.	$10	52	$ 520	$18.20	$ 538.20
2.	15	48	720	25.20	745.20
3.	20	50	1000	35.00	1035.00
4.	25	52	1300	45.50	1345.50
5.	30	50	1500	52.50	1552.50
6.	40	48	1920	67.20	1987.20

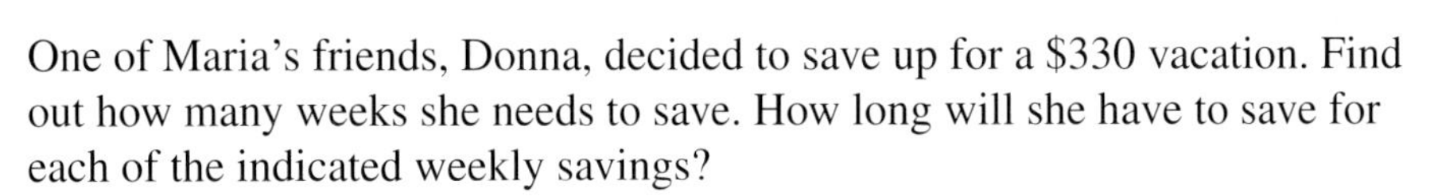

One of Maria's friends, Donna, decided to save up for a $330 vacation. Find out how many weeks she needs to save. How long will she have to save for each of the indicated weekly savings?

	Amount Saved per Week	Number of Weeks Needed	Total Saved
7.	$20	17	$340
8.	32	11	352
9.	18	19	342
10.	30	11	330
11.	35	10	350
12.	40	9	360

Draw a bar graph to show the number of weeks it will take to save enough money to buy the items listed in Exercises 13–16. Assume that you save $25 per week. See Additional Answers.

13. Car stereo system, $139.99 6 wk

14. Portable AM/FM stereo, $219.99 9 wk

15. Computer, $499 20 wk

16. 10-speed bicycle, $339.99 14 wk

RETEACHING

Some students might need to review conversion between percents and decimals, particularly where the percent is expressed as a mixed number (such as $6\frac{1}{2}$%). Emphasize that "percent" means "per 100" to help students relate percent to decimals.

ENRICHMENT

Challenge students to find a formula for computing interest compounded semiannually and quarterly.

GUIDED PRACTICE

Have students work in groups on the *Try Your Skills* exercises. Tell them to provide graphs for 1-6 and 7-12.

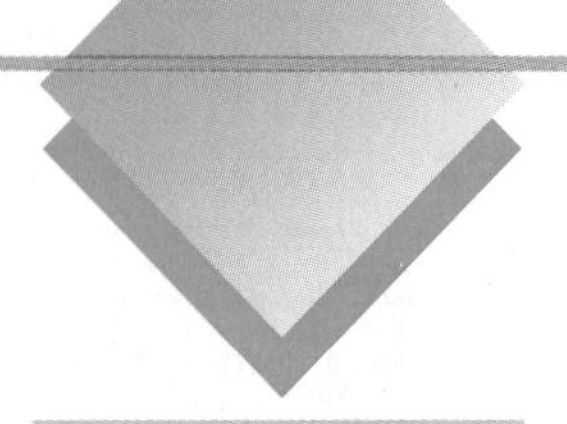

EXERCISE YOUR SKILLS

1. Why would a less liquid savings account earn higher interest than a more liquid account? See Additional Answers.
2. Why do you think that the federal government imposes heavy penalties for liquidating a certificate of deposit before its maturity date?
3. Why should savers deposit emergency savings in a regular savings account rather than in a certificate of deposit?

A bank pays 4% simple interest on a one-year CD. The following table shows amounts saved for a year to buy a CD. Find out how much is saved, how much interest is earned by the CD in 1 year, and the total amount of money in each account after the bank has credited the interest to the accounts.

	Amount Saved per Week	Number of Weeks	Total Saved	Interest Earned	Total in Account
4.	$ 30	52	$1560	$ 62.40	$1622.40
5.	50	48	2400	96.00	2496.00
6.	100	50	5000	200.00	5200.00

Dwight is considering buying some items. Find out how much he can save and how long he will have to save to be able to afford each item below when he saves at the stated rate.

	Amount Saved per Week	Number of Weeks Needed	Total Saved
7.	$15	10	$150
8.	20	7	140
9.	25	6	150
10.	50	3	150

Sale: $139.99 . . Reg. $159.95 Murray's "Pro-Master" 3–9 irons, pitching wedge, 1–3–5 woods. Lightweight steel shafts. Stainless-steel woods.

	Amount Saved per Week	Number of Weeks Needed	Total Saved
11.	$15	14	$210
12.	20	10	200
13.	25	8	200
14.	50	4	200

Diamond/Sapphire Ring 14K gold. Elegant setting.
Regular $275.00
Special Price $199.99

	Amount Saved per Week	Number of Weeks Needed	Total Saved
15.	$15	10	$150
16.	20	8	160
17.	25	6	150
18.	50	3	150

Sale: $149.99 . . . . Reg. $200 Teak Portable CD Player with 3-beam laser pick-up. Music shuffle for random playback. Uses 4 AA batteries (not included).

KEY TERMS

bar graph
certificate of deposit (CD)
commercial banks
credit unions
interest
interest rate
liquidity
money market account
principal
passbook or regular savings account
savings and loan associations
savings banks
simple interest

INDEPENDENT PRACTICE

ASSIGNMENT GUIDE

Exercises 1-3 ask students to display their understanding of different forms of savings programs.

Exercises 4-6 concentrate on computing total savings and interest earned on that savings.

Exercises 7-18 focus on the length of time required to save for specific purchases.

Exercises 19-24 require graphing manually or using a computer spreadsheet to display time required to save for purchases.

Draw a bar graph to show the number of weeks it will take to save enough money to buy the items listed in Exercises 19–21. Assume that you save $25 per week. See Additional Answers

19. Auto-focus camera, $189.99 8 wk

20. VCR, $199.99 8 wk

21. Keyboard, $89.95 4 wk

22.–24. Draw a bar graph like the one for Exercises 19–21 assuming you save $40 per week. 5 wk; 5 wk; 3 wk. See Additional Answers.

CLOSURE

Have students write a paragraph to explain how to calculate simple interest earned on savings after one year and how to determine the number of weeks needed to save for a particular purchase. Tell them to explain how different interest rates affect the amount of interest earned.

LESSON QUIZ

1. Compute the simple interest on a savings of $2500 invested at 7% for one year. Repeat for a savings of $3000 at $6\frac{1}{2}$% for one year. Which is greater? (**$175; $195; $3000 at $6\frac{1}{2}$% is greater.**)
2. Find the number of weeks needed to save for a $300 purchase saving $22 per week and $23 per week. (**14,14**)

MIXED REVIEW

1-3 **1.** A real-estate salesperson receives 4% of the first $100,000 of the selling price of a home and 6% of any amount over $100,000. How much commission does the salesperson receive on a home that sells for $250,000? $13,000

1-3 **2.** Jared has a part-time business videotaping weddings. His rates are $60 for the ceremony and $110 if both the ceremony and the reception are covered. Last year he was able to videotape 29 weddings, 4 of which included the reception. His expenses for the year were $900. What were his net earnings for the year? $1040

2-1 Fill in the table below using the following information about Essex National Bank checking accounts.

Interest is paid on the beginning balance at the rate of 0.4% per month.
The service charge is $3.50 per month.
The cost per check is $0.03.
There are no additional charges regardless of the amount of the balance.

	Beginning Balance	Number of Checks	Interest Earned	Service Charges	Cost of Checks	New Balance
3.	$ 800	14	$ 3.20	$3.50	$0.42	$ 799.28
4.	16,000	27	64.00	3.50	0.81	16,059.69
5.	166	8	0.66	3.50	0.24	162.92

Suppose that on October 7 your checking account balance is $426.67.

2-2 **6.** What will be your balance after depositing a $75.81 check from another state on October 8? $502.48

2-2 **7.** What will be your balance after you write a check of $350 against your account on October 9? $152.48

2-2 **8.** Why might it not be a good idea to write a check for $80 on October 10? The out-of-state check may not have cleared. Then the balance would be only $76.67.

1-1 **9.** Keith earns $7.50 an hour and $1\frac{1}{2}$ times that amount for time over 20 hours in 1 week. What is his pay for 23 hours of work? $183.75

3-2 COMPOUND INTEREST: MONEY THAT GROWS

QUICK REFERENCE

TEACHER'S RESOURCES AND ANSWER KEY

Lesson Quiz Answers 3-2, p. 46

Reteaching Activity Answers 3-2, p. 94

Enrichment Activity Answers 3-2, p. 95

EXTENSION ACTIVITIES

Reteaching Activity 3-2, p. 16

Enrichment Activity 3-2, p. 19

Maria's friend Nelson was impressed by the discipline that she imposed on herself to save money to cover the operating expenses of a car. He knew that there were times when Maria was tempted to spend some of her savings, but he also saw how fast those savings were growing. In fact, he discovered that her total account balance was larger than the money she had put in the bank. Maria explained that her savings were earning interest. Nelson saw this as the bank's way of rewarding her for keeping her money in the bank.

Nelson spends his extra time writing programs in the computer lab. He received recognition at a science competition for a simulation program that projects the changing growth patterns of the great forests in the Western Hemisphere and the impact those changes could have on human lives. He became interested in the topic when a lumber company wanted to cut some of the trees from a national forest where he and his family have gone camping.

As a result of his study, Nelson learned how long it takes to grow new trees. He began to wonder if growing new money from old money was anything like that.

OBJECTIVES: *In this lesson, we will help Nelson to:*

- *Compute the total interest for a savings account when the interest is compounded annually, semiannually, or quarterly.*
- *Compute interest in a savings account using the compound interest formula.*

COMPOUND INTEREST

Just as Maria discovered that interest rates vary in different institutions, Nelson found that the way in which interest is calculated can vary, too.

The interest rate is always expressed as an annual percent. This means that if Maria kept $1000 in a CD for one year with an interest rate of 5%, she would earn $50 in simple interest. If p represents the principal, r represents the annual rate of interest, and t represents the time in years, then the interest i is found from the simple interest formula.

$$i = prt$$
$$i = 1000 \bullet 0.05 \bullet 1$$
$$i = 50$$

Her simple interest is $50.

If Maria were to earn simple interest on her investment of $1000 for a second year, then she would earn 1000 • 0.05 • 1, or $50 in the second year also. However, in practice, interest is not calculated that way. All savings institutions pay **compound interest,** that is, interest on the principal and on *the previously paid interest,* assuming that the interest is left in the account. In the second year, Maria will receive

$$1050 \bullet 0.05 \bullet 1, \qquad \text{or } \$52.50$$

in interest if the compounding is done only once a year.

However, most banks compound at least *semiannually.* This means that at least twice a year the interest earned is added to the previous balance, so the principal balance on which interest is paid grows a little faster than if the compounding were done only once a year. Some banks pay compound interest on whole dollar amounts only. But most large banks now compute interest on *exact amounts, compounded daily.* In this book, compound interest will be figured on exact amounts.

FOCUS

ALGEBRA CONNECTIONS

Which would you prefer, to be paid $1000 per day for one month's work, or a penny the first day that is doubled every day for 30 days?

If Maria received 5% **compounded semiannually** on her original $1000, then her interest for the first year would be calculated as follows. To find half of the yearly interest, divide the yearly interest rate by 2. If the yearly rate is 5%, divide by 2 to obtain a semiannual rate of $2\frac{1}{2}\%$.

$1000 \bullet 0.025 = 25.00$	First 6 months: interest on initial principal
$1025 \bullet 0.025 = \underline{25.63}$	Second 6 months: interest on principal and interest
50.63	Total interest for 1 year

Of course, the interest has been rounded to the nearest cent. Maria earned 63 cents more than she would have if the interest had not been compounded.

You can calculate the interest that is **compounded quarterly** in a similar manner. Since the interest is calculated and paid four times a year (every three months), the quarterly interest rate is found by dividing the yearly rate by 4. If the yearly rate is 5%, divide by 4 to obtain a quarterly rate of 1.25%.

Increasing the frequency of compounding does not greatly increase the amount of interest that you actually get. The table below shows the total interest that you would receive over ten years on $1000 at a rate of 5% under several compounding methods.

CRITICAL THINKING

Have students develop a formula for compounding interest on a monthly and a daily basis (assume 365 days per year). Have them compare simple and daily compounded interest after one year on a $10,000 investment that pays 5% annual interest. Tell them to determine the amount of principal necessary to obtain a difference of at least $100 between simple and daily compounded interest at 5%.

	Annually	Semiannually	Quarterly	Continuously
Interest received	$628.89	$638.62	$643.62	$648.72

Even if your original $1000 were compounded *continuously* (every split second), in ten years you would gain only about $20 more than by compounding annually. So it does not pay to put a lot of emphasis on the frequency of compounding when you choose a bank. However, *all* of the compounding methods give you much more money than you would receive if the interest were calculated just as simple interest, as shown below:

$i = prt$	
$i = 1000 \bullet 0.05 \bullet 10$	5% interest paid on only the original $1000
$i = 500$	10 years of simple interest

You would get only $500 in simple interest at 5% on your original $1000 at the end of ten years. Compound interest would pay you at least $128 more than that.

Sometimes you will want to have an approximate idea of the long-range effect that compounding will have on your savings or investments. Fortunately, there is an easy calculation that you can perform to help you do this. It is explained at the top of the next page.

RULE OF 72

If you invest $1000 at an annually compounded rate of 3% and make no withdrawals, your balance will double to $2000 in about $23\frac{1}{2}$ years; the time for the $2000 to redouble to $4000 at a 3% rate is also about $23\frac{1}{2}$ years. The **Rule of 72** offers a fast and reasonably accurate way to determine how long it will take for a sum of money to double when it is compounded over several years. To find the time for an investment to double, divide 72 by the annual interest rate times 100. Since 10% • 100 = 10, just ignore the percent symbol. For example, if your $5000 investment has an annual return of 10%, then in roughly 7 years (72 ÷ 10 = 7.2) your money will have doubled to $10,000.

Another use for the Rule of 72 is to determine how long it will take for a currency to lose half its purchasing power. If the annual inflation rate stays at 6% for 12 years (72 ÷ 6 = 12), then the currency's purchasing power will drop by 50%.

The Rule of 72 is useful in calculating how long it will take you to save for a special long-term purpose, such as college expenses or retirement.

Ask Yourself

1. What is simple interest?
   interest earned on principal
2. What is compound interest?
   interest earned on principal and previously paid interest
3. Do you think you will earn more interest on a savings account if the compounding is done quarterly rather than semiannually? Will the difference in the interest be very large?
   yes; no
4. What advantages do you give up to earn interest in a savings account rather than in some other kind of investment?
   Answers may vary. (Ex.: other investments earn higher interest rates)

ALGEBRA *REVIEW*

The definition of an exponent is

$$a^m = \underbrace{a \cdot a \cdot \ldots \cdot a}_{m \text{ factors}}$$

When multiplying like bases, add exponents.

$$a^m \cdot a^n = a^{m+n}$$

Simplify.

Example

$$(x^2 \cdot x^3) \cdot x \cdot x \cdot x$$
$$= (x^2 \cdot x^3) \cdot x^3$$
$$= x^5 \cdot x^3$$
$$= x^8$$

1. $x^1 \cdot x^3$
   x^4
2. $2 \cdot 2^3$
   2^4
3. $x^6 \cdot x^2$
   x^8
4. $b^6 \cdot b^2$
   b^8
5. $3y^4 \cdot 5y^4$
   $15y^8$
6. $(3y)^4 (5)^4 y$
   $3^4 5^4 y^5$
7. $(a \cdot a) \cdot a^4$
   a^6
8. $p(1 + r)^n \cdot (1 + r)$
   $p(1 + r)^{n+1}$

Simplify. Use the exponent key x^y of a calculator. Give answers to the nearest hundredth.

Example.

$$[(1.02)^4]\,[(1.02)(1.02)(1.02)]$$
$$= (1.02)^4\,(1.02)^3$$
$$= 1.02^7$$
$$= 1.15$$

9. $15{,}000(1.01)^3(1.01)$
   15,609.06
10. $(1.035)^2(1.035)(1.035)$
   1.15

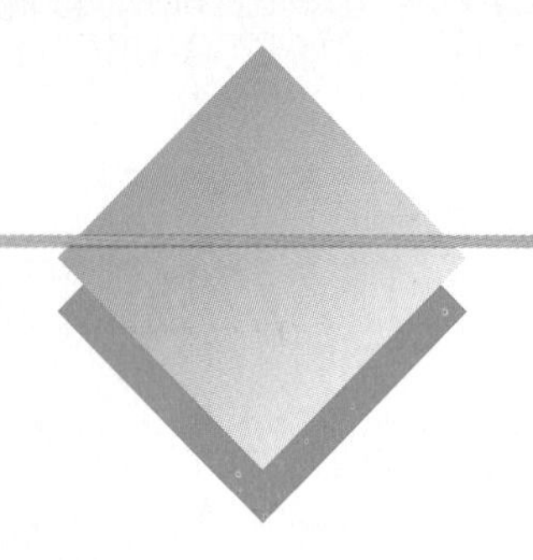

SHARPEN YOUR SKILLS

SKILL 1

INSTRUCTION

TEACHING THE LESSON

Have students practice obtaining semiannual and quarterly rates by suggesting that they compute 6% ÷ 2, 6% ÷ 4, 7% ÷ 2, 7% ÷ 4, and so on. Tell them to compare the interest earned at the end of the year on $10,000 invested at 5% for one year on an annual, semiannual and quarterly basis. Repeat this comparison for two years and three years. Elicit from students that the difference between annual and quarterly yields becomes greater as the period of investment is increased.

FOCUS ON ALGEBRA

In this lesson the compound interest formula necessitates a review of order of operations before multiplication, laws of exponents ($x^a x^b = x^{a+b}$), and decimal division (to obtain semiannual and quarterly rates). (*NCTM Standard 5, p. 150.*)

To find semiannual interest, multiply the principal p by the semiannual rate $r \div 2$ since $i = prt$ and $t = \frac{1}{2}$.

EXAMPLE 1 Nelson's parents have a certificate of deposit in the amount of $10,000. It is held by a bank that pays 5% interest, compounded semiannually.

QUESTION How much will Nelson's parents have in this account after 2 years?

SOLUTION

Remember that the semiannual rate for 5% annual interest is 5% ÷ 2, or $2\frac{1}{2}$%.

i = principal • semiannual rate
$i = 10{,}000 \cdot 0.025$
$i = 250$

This interest is added to the principal of $10,000. The new balance is $10,250, so the next time that interest is paid, it will be computed on a balance of $10,250. Each time, the interest is added to the balance before the next interest is calculated.

At the end of 2 years the family's balance sheet will look like this.

Time Period	Interest Earned	Principal Balance
Beginning	$ 0.00	$10,000.00
End of first half-year	250.00	10,250.00
End of first year	256.25	10,506.25
End of third half-year	262.66	10,768.91
End of second year	269.22	11,038.13

To obtain the interest earned in two years, use the formula

$$p + i = B$$

where p is the original principal, i is the interest and B is the balance at the end of two years.

$p + i = B$
$i = B - p$ — Subtract p from both sides.
$i = 11{,}038.13 - 10{,}000$ — $B = 11{,}038.13$ and $p = 10{,}000$
$i = 1038.13$

The interest earned over two years is $1038.13.

SKILL 2

Nelson wanted to know whether he could predict how much interest the family would have obtained if the CD had been held for more than two years. He decided to see whether there was a pattern to the numbers in the table. First he wrote the bank balance at the end of six months in the following form.

$$
\begin{aligned}
p + i &= 10{,}000 + 10{,}000 \cdot 0.025 \\
&= 10{,}000 \cdot 1 + 10{,}000 \cdot 0.025 && x = x \cdot 1 \\
&= 10{,}000 \cdot (1 + 0.025) && ab + ac = a(b + c)
\end{aligned}
$$

The balance at the end of six months was 10,000 • 1.025. Nelson had found the following pattern:

$$
\begin{aligned}
\text{New balance} &= \text{old balance} \cdot (1 + 0.025) \\
B_1 &= p(1.025)
\end{aligned}
$$

where p is the original principal and B_1 (read as "B sub one") is the total balance at the end of one six-month period.

Next Nelson noticed that at the end of two interest periods, the previous new balance became the new old balance. Thus at the end of two periods (1 year),

$$
\begin{aligned}
\text{New balance} &= \text{old balance} \cdot 1.025 && 1 + 0.025 = 1.025 \\
B_2 &= p(1.025) \cdot 1.025 \\
B_2 &= p(1.025)(1.025) \\
B_2 &= p(1.025)^2 && a \cdot a = a^2
\end{aligned}
$$

where B_2 is the account balance at the end of two six-month periods.

Nelson continued this pattern and found that at the end of three six-month periods the account balance would be

$$
\begin{aligned}
B_3 &= p(1.025)^2 \cdot (1.025) \\
&= p(1.025)^2 \cdot (1.025)^1 \\
&= p(1.025)^3 && a^2 \cdot a^1 = a^{2+1} = a^3
\end{aligned}
$$

From this pattern, Nelson was able to discover the following **compound interest formula** for the balance B in the account at any future date.

Compound Interest Formula

$B = p(1 + r)^n$ where B = balance
p = original principal
r = interest rate for the time period
n = total number of time periods

Nelson wondered how he might use the new formula that he had discovered.

COMMON ERRORS

Be sure that students understand order of operations with exponents and know the proper use of the exponent key (x^y, ^, or a^b) on the calculator.

AT-RISK STUDENTS

Generate interest by having students simulate investing $3000 at three banks in their community. Tell them to ask for rates at these banks. Have them work in groups to compare interest earned over five years at each institution.

ESL STUDENTS

Explain to non-native speakers the meaning of key words such as principal, interest, accumulate, and new balance.

RETEACHING

Some students might require a review lesson on substitution and laws of exponents as well as additional practice in computing on a calculator.

EXAMPLE 2 Nelson decided to try the formula to find how much the CD would be worth at the end of five years.

QUESTION How much will a $10,000 CD at 5% interest, compounded semiannually, be worth in five years?

SOLUTION

$B = p(1 + r)^n$ Compound interest formula

$B = 10{,}000\,(1 + 0.025)^{10}$ $r = 5\% \div 2$, or 2.5%; $n = 10$ six-month periods

$B = 12{,}800.85$ Use your calculator.

The CD would be worth about $12,800 at the end of five years.

To enter the calculation in your graphing calculator use the exponent key [xʸ], sometimes labeled [^] or [aᵇ]. The Enter key is sometimes labeled [EXE].

10000 [(] 1 [+] 0.025 [)] [xʸ] 10 [ENTER]

SKILL 3

According to the Rule of 72, Nelson can find the speed at which his invested money will double by dividing 72 by 100 times the annual interest rate. Nelson would like to know whether he will be a millionaire by investing in CDs.

ENRICHMENT

Challenge students to figure out why the Rule of 72 works. Have them show examples using 3%, 4%, 8% and 12%.

Rule of 72

$$\frac{72}{\text{Annual interest rate} \bullet 100} = \text{years to double}$$

EXAMPLE 3 Nelson invests $10,000 in a CD that pays 6% compounded quarterly.

QUESTION How long will it take his investment to double?

SOLUTION Using the Rule of 72,

$72 \div 6 = 12$ $6\% \bullet 100 = 6$

It will take 12 years to double the investment.

Nelson decided to check this result using the compound interest formula. If the interest is compounded quarterly, then the rate for each period is

$r = 0.06 \div 4 = 0.015$

If he invests the money for 12 years, then the number of periods is

$n = 12 \bullet 4 = 48$

Substitute in the compound interest formula.

$B = p(1 + r)^n$ Compound interest formula

$B = 10{,}000(1 + 0.015)^{48}$ Use a calculator.

$B = 20{,}435$ To the nearest dollar

Since 20,435 is approximately 2 • 10,000, the result is confirmed; $10,000 has approximately doubled to $20,435.

EXAMPLE 4 Next, Nelson decided to continue the doubling to see how many 12-year periods he needed to reach $1,000,000.

QUESTION How long will it take his investment to be worth more than 1 million dollars?

SOLUTION

He noticed that in 12 years, the value of the investment had grown to about 2.04 times its earlier value (20,435 ÷ 10,000 = 2.0435). This was not exactly a double, but since he did not need an exact result, he believed that he could get a very good estimate by using 2 instead of 2.04. He made a table to keep track of the doublings.

Doublings	Years Passed	Balance
1	12	$ 20,435
2	24	40,870
3	36	81,740
4	48	163,480
5	60	326,960
6	72	653,920
7	84	1,307,840

It will take 7 doublings or 84 years to reach one million dollars. If Nelson is to become a millionaire through investments, he should either invest more money to begin with or find an investment that will pay a higher rate of return.

SKILL 4

EXAMPLE 5 Nelson has given up on the idea of becoming a millionaire quickly. Instead, he will invest $20,000 in an investment plan that he hopes will provide 10% interest, compounded quarterly.

QUESTION Nelson now wants to have $100,000 in his investment plan by the time he retires. How long will that take?

SOLUTION

Nelson substitutes all the values he can into the compound interest formula. The current value of his principal p is $20,000, the balance B that he hopes to have in the future is $100,000, and the quarterly rate of interest r that he hopes the investment will pay is 0.10 ÷ 4 = 0.025.

$B = p(1 + r)^n$ Use the compound interest formula.

$100{,}000 = 20{,}000(1 + 0.025)^n$ $B = 100{,}000$; $p = 20{,}000$; $r = 0.025$

$5 = 1.025^n$ Divide each side by 20,000.

TECHNOLOGY HINT

If you are using a graphing calculator that does not have a TRACE feature, place the cursor at the intersection point and read the coordinates from the display.

Nelson is not sure how to solve this equation for n. He decides to use his graphing calculator. For each side of the above equation he writes a new equation, using the variable x for n.

$y = 1.025^x$
$y = 5$

He graphs them using these range values.

Xmin: 0
Xmax: 95
Xscl: 10
Ymin: 0
Ymax: 10
Yscl: 1

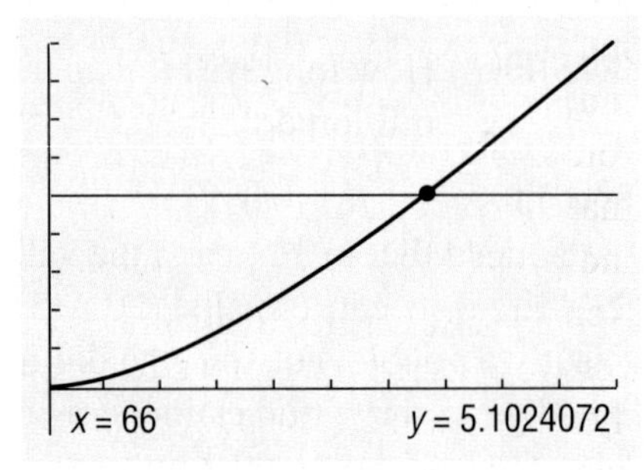

The horizontal line is the graph of $y = 5$ and the steep, rising curve (called an **exponential curve**) is the graph of $y = 1.025^x$. Nelson moves the trace cursor along the curve until he gets as close as he can to $y = 5$ without being less than 5. He finds that when $x = 66$, $y = 5.1024072$, so he substitutes 66 for n in the compound interest formula.

$B = 20{,}000(1.025)^{66}$
$B = 102{,}048$ To the nearest dollar

COOPERATIVE LEARNING

Have students work in small groups to graph the information from the *Try Your Skills* section using a spreadsheet.

Nelson's \$20,000 investment will grow to over \$100,000 when $n = 66$. Since n represents the number of *quarters* (not years), Nelson needs to divide by 4 to find the number of years that it will take for him to retire.

Number of years $= 66 \div 4 = 16\frac{1}{2}$

If the annual interest rate remains at 10%, then Nelson can retire with \$100,000 in $16\frac{1}{2}$ years.

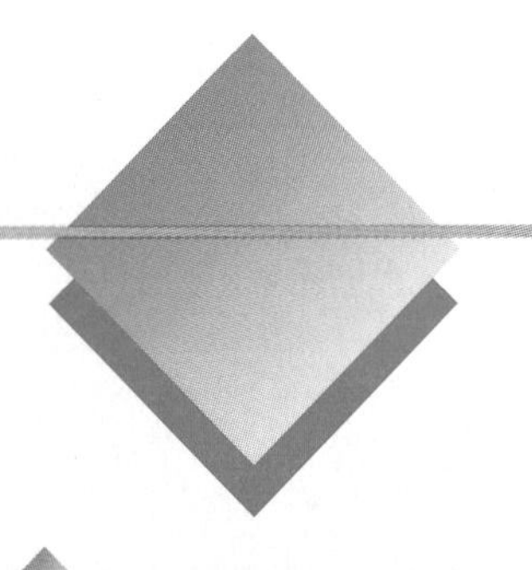

TRY YOUR SKILLS

GUIDED PRACTICE

Have students work in groups of three or four on *Try Your Skills* exercises. Tell them to write a paragraph explaining the difference between the final results of Exercises 1 and 2 and have them read their explanations to each other.

Use a calculator to find the amount of interest and the new balance that will accumulate over two years on the following principal amounts at the given interest rate compounded as shown.

	Principal	Interest Rate	How Often Compounded	Interest Earned		New Balance
1.	\$1000	6%	Annually	First period:	\$60.00	\$1060.00
				Second period:	63.60	1123.60
2.	\$1000	6%	Semiannually	First period:	30.00	1030.00
				Second period:	30.90	1060.90
				Third period:	31.83	1092.73
				Fourth period:	32.78	1125.51

Suppose that in 1776, Benjamin Franklin and other founders of our nation each decided to invest a small amount of money for their descendants. Assume that they were able to find a bank willing to pay 4% interest, compounded semiannually for the indefinite future. Use the compound interest formula $B = p(1 + r)^n$ to find the amount that each investment will be worth after the indicated number of years have passed. Remember that for twice-a-year compounding the number of periods is twice the number of years and the interest rate for each 6-month period is $0.04 \div 2$, or 0.02.

3. Thomas Jefferson invested $100.00 for 30 years. $328.10
4. Paul Revere invested $50.00 for 90 years. $1766.04 semiannual. 4%
5. John Hancock invested $20.00 for 200 years. $55,093.29
6. Use the Rule of 72 and your calculator to find the year in which $1 that Benjamin Franklin invested in the year 1776 at 8% compounded quarterly would have become worth over $2000. Organize your results in a table like the one below. 99 years later in 1875

Doublings	Years Passed	Balance
1	9	$ 2
2	18	4
3	27	8
4	36	16
5	45	32
6	54	64

See Additional Answers.

ADDITIONAL ANSWERS

6.	7	63	128
	8	72	256
	9	81	512
	10	90	1024
	11	99	2048

7. Assume that Nelson has $25,000 to invest and that he expects to earn 9% per year on his investment. The interest will be compounded quarterly. Use the Rule of 72 to approximate how long it will take for Nelson's money to double repeatedly until it has grown to more than 1 million dollars. 48 years
8. Use the compound interest formula and a graphing calculator as in Skill 4 to check your answer to Exercise 7. Which method do you think is easier? Which is more accurate? Tell which method you prefer, and give a reason for your answer. $40 = 1.0225^n$; 41.5 years; Rule of 72 is easier. Graphing calculator is more accurate.

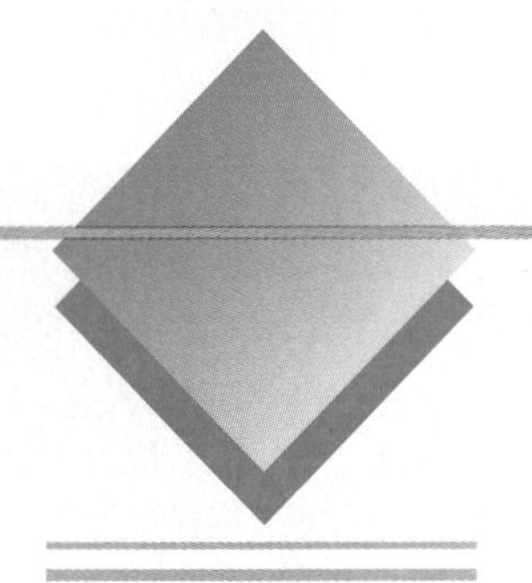

EXERCISE YOUR SKILLS

KEY TERMS

compound interest
compounded quarterly
compounded semiannually
Rule of 72

INDEPENDENT PRACTICE

ASSIGNMENT GUIDE

Exercises 1-5 ask students to display basic understanding of compound interest.

Exercises 6-13 ask students to record the interest earned on a given principal through eight quarters of earnings.

Exercises 14-17 require the use of the compound interest formula for large numbers of interest periods.

Exercises 18-21 give students practice with the Rule of 72.

Exercises 22-26 ask students to graph the cumulative interest on a given principal.

Exercise 27 challenges students to set up complex expressions for interest-earning investments.

Fill in the missing numbers. (Do not write in your book.)

1. Interest that is compounded annually is paid every ? 12 months.
2. Interest that is compounded semiannually is paid every ? 6 months.
3. Interest that is compounded quarterly is paid every ? 3 months.
4. 5% interest compounded annually means ? 5 % paid every ? 12 months.
5. 6% interest compounded quarterly means ? 1.5 % paid every ? 3 months.

Use a calculator to find the amount of interest and the new balance that will accumulate over two years on the following principal amount at the given interest rate compounded as shown. The first quarter is done for you as an example.

	Principal	Interest Rate	How Often Compounded	Interest Earned		New Balance
6.	$1000	6%	Quarterly	First period:	$15	$1015.00
7.				Second period:	$15.23	$1030.23
8.				Third period:	$15.45	$1045.68
9.				Fourth period:	$15.69	$1061.37
10.				Fifth period:	$15.92	$1077.29
11.				Sixth period:	$16.16	$1093.45
12.				Seventh period:	$16.40	$1109.85
13.				Eighth period:	$16.65	$1126.50

Suppose that in 1789, John Adams invested $7.50 in a bank at 4.5% compounded semiannually and stipulated in his will that the money was to be allowed to accumulate in the account after his death. Use the compound interest formula to find the value of the account in each case.

14. After 50 years $69.41 **15.** After 70 years $169.02 **16.** After 100 years $642.28

17. In Exercise 16, suppose that the interest rate is 6% instead of 4.5%. Find the value of the account after 100 years. Compare the two results. What conclusion do you draw from them? See Additional Answers.

Nelson has invested $5000 in a CD. The interest is compounded annually. Use the Rule of 72 to find how many years it will take for the investment to grow to $40,000 at the given rates of interest.

18. 2% 108 y **19.** 6% 36 y **20.** 8% 27 y **21.** 10% 21.6 y

In the year that Nelson was born, Nelson's father invested $15,000 in an investment fund that has been paying 8% interest compounded quarterly. Use the graphing method outlined in Skill 4 to determine how long it will take for

ADDITIONAL ANSWERS

17. $2770.17; slightly higher interest rates yield much higher interest over time

the $15,000 to grow enough to pay for Nelson's costs of attending college. Nelson estimates the costs of several colleges to be as given below.

22. $75,000 20.5 y **23.** $45,000 14 y **24.** $37,500 11.75 y **25.** $52,500 16 y

26. Nelson plans to begin college at age 18. Which of the above amounts will be impossible to acquire by then? $75,000

27. At the age of 19, Robert placed $2000 into an investment at 6% compounded annually and added no money to the investment for the rest of his life. His friend Amos, who was the same age, waited until he was 30 to begin his investment program by also investing $2000 at 6% compounded annually. Later, at age 40, Amos invested another $2000 at 6% compounded annually, and still later, at age 50, he invested the final $2000 of his $6000 investment on the same interest terms. Use the compound interest formula to show that at age 60, when they both retire next year, Robert's investment will have grown to a larger amount than Amos's. How much will each investment be worth? See Additional Answers.

MIXED REVIEW

1-3 Find the commission earned and the total monthly earnings for each salesperson.

1. Paul, with a monthly base salary of $900 and a commission rate of 9% on sales of $6975 $627.75; $1527.75

2. Gabe, with a monthly base salary of $2500 and a commission rate of 7% on sales of $7475 $523.25; $3023.25

2-3 Suppose your bank statement shows a closing balance of $236.50 and the following are not on the statement: deposit: $80.00; check: $18.37; ATM withdrawal: $110.00. There is a service charge of $1.50 on the statement but not in your checkbook.

3. What amounts must be added to the bank statement balance? $80 deposit

4. What amounts must be subtracted from the bank statement balance? $18.37 check; $110.00 ATM withdrawal

5. What should be the balance in your checkbook before you do a reconciliation? $189.63

6. What should be the checkbook balance after you do a reconciliation? $188.13

2-1 **7.** Pauline thinks she will write about 120 checks a month in her new business. She expects to maintain an average balance of $1500. Which of the following banks offers her the better deal? Explain your answer.

Bank A: interest of 0.35% per month paid on the average balance and a charge of $0.06 for each check.

Bank B: no interest paid on the checking account and no charge for checks but a monthly service charge of $6. See Additional Answers.

CLOSURE

Have students write a paragraph explaining the similarities and differences between simple and compound interest. Tell them to explain how to determine the number of interest periods in a given situation as well as the effective rate of interest for each period. Elicit an informal recap of the Rule of 72.

LESSON QUIZ

1. Compute the interest on $7000 at 6% annually. Then compute the interest, this time compounded quarterly. (**$420,$429.54**)

2. How many times is interest credited to your account in five years if the interest is compounded semiannually? (**10**)

ADDITIONAL ANSWERS (EXERCISE YOUR SKILLS)

27. Robert:
$2000(1.06)^{41} = \$21,805.72$
Amos:
$2000(1.06)^{30} = 11,486.98$
$2000(1.06)^{20} = 6,414.27$
$2000(1.06)^{10} = 3,581.70$
21,482.95

(MIXED REVIEW)

7. Bank A; Bank B is a better deal only if the number of checks reaches 188 checks a month.

3-3 THE FEDERAL RESERVE: THE BANK'S BANK

QUICK REFERENCE

TEACHER'S RESOURCES AND ANSWER KEY

Lesson Quiz Answers 3-3, p. 46

Reteaching Activity Answers 3-3, p. 95

Enrichment Activity Answers 3-3, p. 95

EXTENSION ACTIVITIES

Reteaching Activity 3–3, p. 17

Enrichment Activity 3–3, p. 20

Every year for as long as she can remember, Olivia has been receiving a birthday gift from her Aunt Millicent. Every November, a $20 check travels from California to Olivia's Indiana home. Last year Olivia's mother flew to California and stayed three weeks when Aunt Millicent had surgery.

Olivia knew that while in California, Mom needed more money than she had taken with her and that she was able to cash a check on an Indiana bank at Aunt Millicent's bank. Olivia thought that it was amazing that the California bank could somehow take the cash that Mom had in the bank in Indiana and give it to Mom in California in a matter of minutes. Dad said that it had something to do with the "Federal Reserve" and "computers" and declared that "computers are taking over the world." Dad has a tendency to complain. Many times, he has told Olivia how much more you could buy with $20 when he was a kid.

Olivia herself has noticed that Aunt Millicent's birthday check doesn't buy as much as it used to. Olivia took that as a sign that she was growing up, since the larger-size clothes that she now needs to buy cost more than the clothes she wore when she was little. The way her father talks about inflation, taxes, government spending, and cost-of-living increases reminds her of something she heard about the Federal Reserve System in economics class. Olivia decided to find out more about how the Federal Reserve System works.

OBJECTIVES: *In this lesson, we will help Olivia to:*

- *Describe the organization of the Federal Reserve System.*
- *List four important functions of the Federal Reserve System.*
- *Describe how the Federal Reserve System controls monetary policy.*
- *Observe and calculate the multiplier effect on the nation's money supply.*

FUNCTIONS OF THE FEDERAL RESERVE SYSTEM

The **Federal Reserve System** consists of 12 Federal Reserve banks, 25 branch banks, a board of governors, and the Federal Open Market Committee, which directs the purchases and sales by the reserve banks of Federal government securities and other obligations in the open market. The following map of the Federal Reserve System shows the locations of the district banks and branches. Olivia has seen this map in her economics class. She noticed that she lives in District 7 and Aunt Millicent lives in District 12.

ORGANIZATION OF THE FEDERAL RESERVE SYSTEM

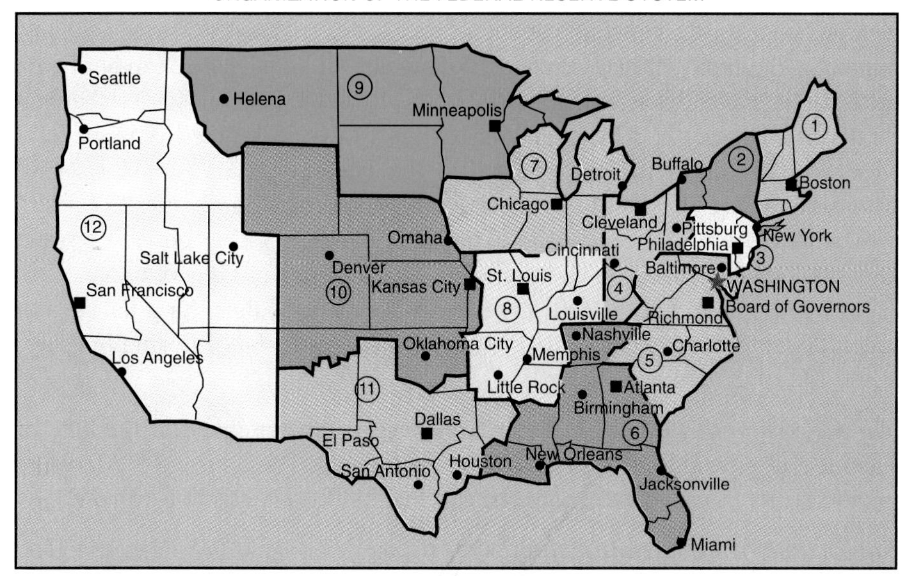

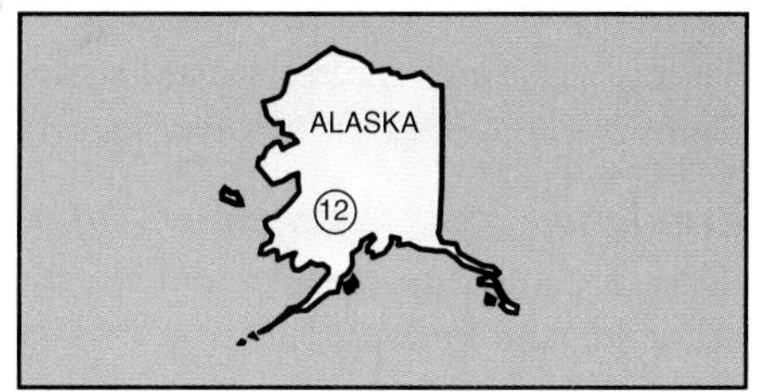

LEGEND

— BOUNDARIES OF FEDERAL RESERVE DISTRICTS

— BOUNDARIES OF FEDERAL RESERVE BRANCH TERRITORIES

★ BOARD OF GOVERNORS OF THE FEDERAL RESERVE SYSTEM

■ CITY WHERE FEDERAL RESERVE BANK IS LOCATED

● CITY WHERE A BRANCH OF FEDERAL RESERVE BANK IS LOCATED

(3) FEDERAL RESERVE DISTRICT NUMBER. THIS NUMBER APPEARS ON THE CURRENCY ISSUED BY THE FEDERAL RESERVE BANK IN THE DISTRICT

FOCUS

ALGEBRA CONNECTIONS

Dana received $1000 in cash gifts in honor of her high school graduation. She plans to put $400 in a CD earning 8% semiannually and the rest in a savings account. What semiannual interest rate must the savings account pay for Dana's total gift money to earn at least $70 next year?

The important functions of the Federal Reserve System are the following:

1. To act as a central bank.
2. To serve as a bank for the U.S. government.
3. To supervise financial institutions.
4. To regulate and manage the nation's money supply.

The Nation's Central Bank The Federal Reserve System issues the country's main type of currency, the **Federal Reserve note.** Such notes are **legal tender;** that is, they are money that by law must be accepted for payments of debts and taxes. This currency is printed by the Bureau of Printing and Engraving.

In addition to issuing currency, the Federal Reserve banks act as a major clearinghouse for the collection and return of checks so that depositors' accounts are credited quickly. When Olivia deposits Aunt Millicent's check in her bank, the bank transmits this information electronically to the Federal Reserve bank in its district, District 7, which includes Olivia's town in Indiana. This Federal Reserve bank sends the information to the Federal Reserve bank in Aunt Millicent's district, District 12, which includes California. The District 12 bank in turn contacts Aunt Millicent's bank. The use of computers has made this process very fast, and the banks can now complete such transfers in a day or two.

In its role as a central bank the Federal Reserve System supplies cash to meet the temporary needs of its member banks, for instance during holidays when consumer demand for cash is high.

The Government's Bank The Federal Reserve System handles the flow of income and expenditures for the federal government. The issuance of Medicare payments, for instance, may come from a Federal Reserve bank near you.

Supervision of Financial Institutions All member banks are supervised and regulated by the Federal Reserve System. Periodically, the Federal Reserve System examines the records of member banks to determine whether they are conforming to Federal Reserve standards.

Regulation and Management of the Nation's Money Supply Olivia's father's comments about the decrease in the purchasing power of money are related to the Federal Reserve's most important function: the regulation and management of the nation's **money supply**—the total amount of coins, paper currency, and demand deposits (checking-account money) in circulation in the economy. The Federal Reserve System tries to maintain the right amount of money in circulation. If there is too much money, **inflation** results; that is, the overall purchasing power of money declines. If the money supply is too low, economic activity will diminish and may even lead to a recession and an increase in unemployment.

If business activity is slowing down and unemployment does begin to rise, the Federal Reserve may try to expand the supply of money and credit. To do this, it will encourage financial institutions to borrow short-term money from the Federal Reserve banks at its relatively low interest rates. With such low rates, businesses will be encouraged to borrow money to expand their operations and wage earners will be encouraged to borrow money to build more homes and buy more goods. On the other hand, suppose that business activity expands too rapidly. This will be characterized by some positive things such as full employment but also by rising prices and too much borrowing by both businesses and consumers. Then the Federal Reserve banks will try to cool off the economy by reducing the money and credit supply. If the money supply is limited in this way, then interest rates will rise, and businesses and individuals will be discouraged from borrowing.

By maintaining a balance in the amount of money in circulation, the Federal Reserve tries to ensure that credit is plentiful enough to allow expansion of the economy but not so plentiful that rapid inflation occurs. A rough indication of the rate of inflation is provided by the **Consumer Price Index (CPI),** a statistic that is calculated monthly by the Bureau of Labor Statistics. This index reflects the price of a specific group of goods and services used by the average household. This "market basket" consists of about 400 goods and services in the areas of food, housing, transportation, clothing, entertainment, and medical and personal care. (You will learn more about the CPI in Chapter 9.)

In summary, by controlling the money supply the Federal Reserve System affects the prices that consumers pay for goods and services. When interest rates are low and the money supply increases, people spend more freely. Suppliers can more easily raise their prices because demand for goods and services is greater than their supply. Thus higher inflation may occur. On the other hand, when interest rates are high and money is tight, people buy fewer goods and services. Therefore merchants have an oversupply of inventory and must lower their prices to attract buyers.

THE CREATION OF DEPOSITS

By law, depository institutions (banks) must deposit a specified percent of their customers' deposits with the Federal Reserve System. Such a deposit is known as a **required reserve.** It serves as a safeguard against a financial institution's placing too much of its customers' deposits into various investments. These investments might range from simple loans to the purchase of government securities.

The Federal Reserve can exert several minor controls on deposits in financial institutions; these controls are quite powerful tools. However, the real strength that the Federal Reserve exerts is due to its control over an institution's required reserves. This is due to the **multiplier effect** that these reserves create throughout the banking system.

For any given *demand deposit* (money in a checking account) a financial institution has the right to invest or extend loans from the **excess reserves.** As an example, assume that the reserve requirement is 20%; that leaves 80% in excess reserves. This is illustrated below by the simple balance sheet for a $1000 demand deposit. The financial institution actually creates an extra $800 of money or credit. This is done by lending the $800 rather than holding it in the vault as a reserve.

INITIAL DEPOSIT			
Assets		**Liabilities**	
Required reserves (20%)	$200	Demand deposit	$1000
Loans and investments (80%)	800		
Total	$1000	Total	$1000

If the $800 that is used in making loans is deposited by the borrower in the same or other institution, still more money or credit can be created. The second deposit creates an extra $640 of money or credit, as shown below.

SECOND DEPOSIT			
Assets		**Liabilities**	
Required reserves (20%)	$160	Demand deposit	$800
Loans and investments (80%)	640		
Total	$800		$800

We can also expect this $640 to be deposited by the new borrower. The multiplier effect continues as shown below.

THIRD DEPOSIT			
Assets		**Liabilities**	
Required reserves (20%)	$128	Demand deposit	$640
Loans and investments (80%)	512		
Total	$640	Total	$640

The cycle continues until there is not a penny left to deposit. By this time the original $1000 deposit will have enabled close to $4000 in new money to be created.

CRITICAL THINKING

After the students have answered the questions in *Ask Yourself*, have them consider the impact of $500,000 in pennies that are not deposited in banks on the multiplier effect. Have them estimate lost interest assuming 5% compounded quarterly over a five year period and lost new money assuming a 20% required reserve.

MANAGING THE MONEY SUPPLY

The Federal Reserve System uses the reserve requirements to manage the money supply. If the reserve requirement is increased, then the multiplier effect for creating new money or credit is reduced. This is known as a **tight-money policy.** If the reserve requirement is reduced, then the multiplier effect is increased. This is known as an **easy-money policy.**

Economic conditions can prevent the multiplier effect from reaching its full force. For example, if the economy is in a period of high inflation and interest rates are also quite high, businesses and consumers may not wish to borrow money. In this case the lack of demand for money means that not all the excess reserves available will be loaned out, and the multiplier effect will be reduced. If the economy is in a period of recession, businesses and consumers may be afraid to borrow even though interest rates are low. This, too, will tend to keep the multiplier effect from reaching its full potential, since financial institutions will not be able to make the loans that they would like. The multiplier effect can reach its full potential in creating new money and credit only when the demand for loans equals the supply of money available for those loans.

Another condition that can keep the multiplier effect from reaching its full potential is personal behavior regarding money and financial institutions. Some people do not trust depository institutions but keep their money at home or in safe-deposit boxes. If this money does not get into circulation (through spending or depositing), then the multiplier effect will be diminished.

Olivia and her brother Orson will certainly do their part to keep their money in circulation. They aren't very big savers; they spend whatever comes their way and wherever they can, for example, at the local music store. Perhaps the owner of the store will deposit its cash receipts in his checking account at the local bank, thus allowing the bank to multiply that money four or five times.

Ask Yourself

1. What states make up District 12 of the Federal Reserve System?
   See Additional Answers.
2. What are four important functions of the Federal Reserve System?
3. What is legal tender?
4. How does the Federal Reserve System try to expand the money supply?
5. What is a reserve requirement?
6. What is the difference between a tight-money policy and an easy-money policy?
7. What is the multiplier effect?
8. Describe the Federal Reserve System.

9 Explain Consumer price index

ALGEBRA *REVIEW*

For each series (sum of two or more terms), tell whether there is a common ratio between successive terms. If there is, give that ratio.

Examples

a. $1 + 2 + 4 + 8 + 16$
$2 \div 1 = 2$
$4 \div 2 = 2$
$8 \div 4 = 2$
$16 \div 8 = 2$
Yes; 2

b. $1 + 2 + 3 + 4$
$2 \div 1 = 2$
$3 \div 2 = 1.5$
No

1. $8 + 4 + 2 + 1 + \frac{1}{2}$
   yes; 0.5
2. $x + 4x + 8x + 16x$
   no
3. $0.04 + 0.08 + 0.12 + 0.16$
   no
4. $0.01 + (0.01)(0.01) + (0.01)(0.01)(0.01)$
   yes; 0.01
5. $0.04 + (0.04)^2 + (0.04)^3$
   yes; 0.04
6. $0.3 + 2(0.3) + 3(0.3) + 4(0.3)$
   no
7. $100(0.07)^2 + 100(0.07)^3 + 100(0.07)^4 + 100(0.07)^5$
   yes; 0.07

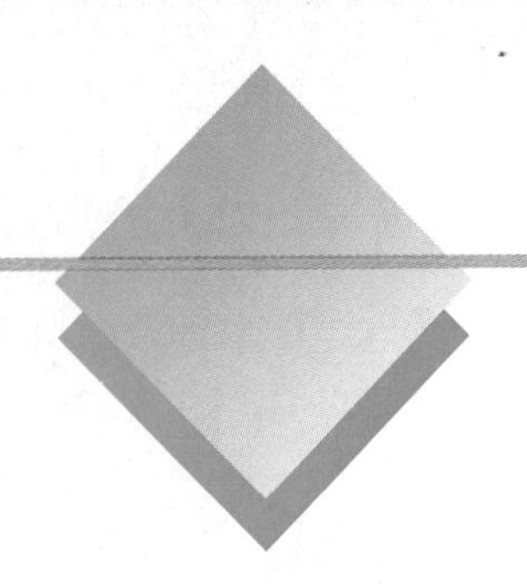

SHARPEN YOUR SKILLS

SKILL 1

INSTRUCTION

TEACHING THE LESSON

Make a chart showing how the Federal Reserve banks expedite check cashing. Have students read about the multiplier effect. Set up a chart such as that shown in Example 1 to create more examples using different demand deposits and reserve requirements. Then have students display this data through ten levels using a spreadsheet program. Do some examples through 20 levels and point out the decreasing amount of money available for loans and investments in later levels. This observation will help establish that an infinite number of levels generates a finite sum of new money.

Make sure students can define the Key Terms listed in *Exercise Your Skills.*

FOCUS ON ALGEBRA

This lesson requires a review of partial and infinite geometric series and uses order of operations to perform complicated calculations. (*NCTM Standard 4, p. 146.*)

EXAMPLE 1 Olivia wants to find out the multiplier effect on an initial deposit of $1000 through five levels.

QUESTION What is the total amount of extra money or credit created through five levels of the multiplier effect?

SOLUTION

Olivia made the following table to show the first five levels and the total additional money available for loans and investments after five levels. The required reserve is 20% of the demand deposit.

$$0.20(1000.00) = 200.00$$

The amount of loans and investments is the demand deposit minus the required reserves.

$$1000.00 - 200.00 = 800.00$$

At each level beginning with Level 2, the demand deposit is the loan amount from the previous level.

Level	Demand Deposit	Required Reserve (20%)	Loans and Investments
1	1000.00	200.00	800.00
2	800.00	160.00	640.00
3	640.00	128.00	512.00
4	512.00	102.40	409.60
5	409.60	81.92	327.68
		Total after 5 levels	2689.28

This shows how the money supply will multiply if the initial $800 loan is all redeposited in the banking system and if the resulting $640 loan also is all redeposited, and so on through five levels of loans and redeposits. The result is the creation of $2689.28 that did not exist before the inital $1000 deposit.

SKILL 2

EXAMPLE 2 Olivia wants to use a spreadsheet program to find out the multiplier effect on an initial deposit of $1000 through ten levels.

QUESTION What is the total amount of extra money or credit created through ten levels of the multiplier effect?

SOLUTION

Olivia used formulas in the spreadsheet program to give the numerical values shown in the table in Example 1. The spreadsheet shows the actual values (not the formulas). Thus in cell D5 you will see "512.00," not "+B5−C5."

	A	B	C	D
1	Level	Demand	Required	Loans and
2		Deposit	Reserve (20%)	Investments
3	1	1000.00	@round(0.2*B3,2)	+B3−C3
4	2	+D3	@round(0.2*B4,2)	+B4−C4
5	3	+D4	@round(0.2*B5,2)	+B5−C5

In Column C, Olivia used a "rounding" function (often expressed as @ROUND) so that the spreadsheet would use the rounded value in the remaining calculations. If she did not do this, a calculated value would not actually be rounded; it would only be displayed as rounded and small errors would be created. For cell C3, Olivia typed @ROUND (0.2*B3,2) to obtain a value rounded to two decimal places.

Olivia used the spreadsheet's COPY command to extend the formulas created in Rows 3 and 4 to the other rows. The spreadsheet creates the formulas for the new rows but displays the values for those formulas.

TECHNOLOGY HINT

If students use the COPY command to transfer formulas to additional rows or columns, have them verify cell references to ensure that they are accurate.

To find the total amount of money created, Olivia used the spreadsheet's "sum" function (often expressed as @SUM) to add the entries of Column D from Row 3 to Row 12 and display that total in Column D, Row 13. For cell D3, Olivia typed @SUM(D3 . . D12).

	A	B	C	D	
1	Level	Demand	Required	Loans and	
2		Deposit	Reserve (20%)	Investments	
3	1	1000.00	200.00	800.00	
4	2	800.00	160.00	640.00	
5	3	640.00	128.00	512.00	
6	4	512.00	102.40	409.60	
7	5	409.60	81.92	327.68	
8	6	327.68	65.54	262.14	
9	7	262.14	52.43	209.71	
10	8	209.71	41.94	167.77	
11	9	167.77	33.55	134.22	
12	10	134.22	26.84	107.38	
13			**Total after 10 levels**	3570.50	@SUM(D3..D12)

The total amount of extra money or credit created through ten levels of the multiplier effect is $3570.50.

If you extend the spreadsheet even further, to 20 levels, 30 levels, and so on, you will create more and more money as summarized below.

Level	Amount of New Money Created
5	2689.28
10	3570.50
20	3953.88
30	3995.05
40	3999.47

Even though the deposits and redeposits continue to ever higher levels without end, the total amount of new money created by the initial \$1000 deposit seems to have a limiting value of \$4000. In other words, it seems that the sum

$$800 + 640 + 512 + 409 + \ldots$$

with an infinite number of terms (indicated by the three dots at the end) has a finite sum, \$4000. This is actually the case.

The sum can also be written as

$$800 + 800(0.8) + 800(0.8)^2 + 800(0.8)^3 + 800(0.8)^4 + \ldots$$

You can see that the ratio of each pair of successive terms is always the same, namely, 0.8. Any series that has a **common ratio** (such as 0.8) is called a **geometric series.** If the series has an infinite number of terms and if it has a sum (many series do not), then that sum is called the **sum of an infinite geometric series.** The series above *is* infinite and *does* have a sum. The formula for the sum of an infinite geometric series is

$$S = \frac{a}{1 - r}$$

where a is the first term 800 and r is the common ratio 0.8. Since $1 - 0.8 = 0.2$, the expression $1 - r$ is the reserve requirement 0.2.

EXAMPLE 3 Olivia wants to find out the maximum multiplier effect on an initial deposit of \$1000.

QUESTION What is the maximum amount of money a \$1000 deposit can create and what is the multiplier?

SOLUTION

For a \$1000 deposit, initial loan amount of \$800, and 20% reserve requirement,

COMMON ERRORS

Students may be confused by the idea that an infinite geometric series can have a finite sum. Do several simple examples such as $1 + \frac{1}{2} + \frac{1}{4} + \ldots$; $6 + 2 + \frac{2}{3} + \ldots$; $4 + 1 + \frac{1}{4} + \ldots$; $9 + .9 + .09 + \ldots$ to give them more practice.

RETEACHING

Have students simulate the multiplier effect by using play money that they have created. Assign students to keep a record of required reserves, excess reserves and cumulative available loan money. Tell them to compute the multiplier for each of their examples.

ENRICHMENT

Have students research the Consumer Price Index and its relation to the Federal Reserve's money policy.

$$S = \frac{a}{1 - r}$$ $a = 800; 1 - r = 0.2$

$$S = \frac{800}{0.2}$$ Reserve requirement = 0.2

$$S = 4000$$

The original $1000 deposit can create as much as $4000 in new money. To find the multiplier add the amount created to the original amount and divide by the original amount.

$$\frac{1000 + 4000}{1000} = 5$$

The multiplier is 5.

NTERDISCIPLINARY INVESTIGATION

Economics: Have students research the founding of the Federal Reserve system. Direct them to report on how the Federal Reserve Bank's policies affect the economy.

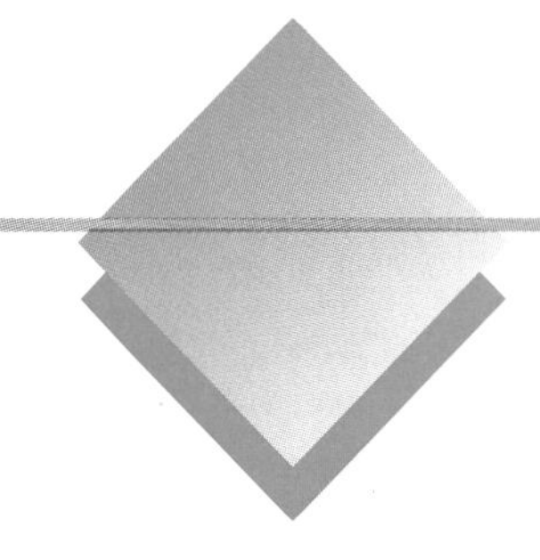

TRY YOUR SKILLS

Make a table to show the multiplier effect on the amounts below if the Federal Reserve requirement is 20%. Show the first ten levels and find the total extra money that is generated.

1. Initial deposit: $500
2. Initial deposit: $2500

See Additional Answers.

GUIDED PRACTICE

Have students work in groups to calculate the levels of the multiplier effect and to compute the total extra money generated. Instruct them to write a paragraph explaining the different results using 20% and 25%.

Make a table or use a spreadsheet to show the multiplier effect on the amounts below if the Federal Reserve requirement is 25%. Show the first ten levels and find the total extra money that is generated. See Additional Answers.

3. Initial deposit: $500

4. Initial deposit: $2500

5.–8. For the initial deposits of Exercises 1–4, find the maximum amount of new money created by the multiplier effect.

9.–12. What is the multiplier for Exercises 5–8?

13. Consider your answers to Exercises 1–12. How does changing the reserve requirement affect the extra money generated? The greater the reserve requirement, the less extra money generated.

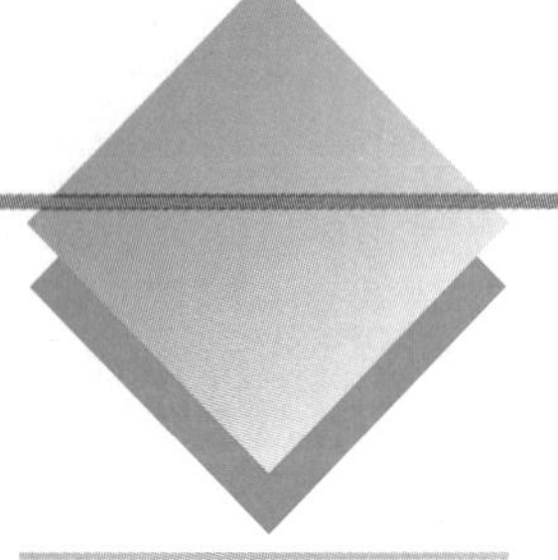

EXERCISE YOUR SKILLS

1. Does one bank actually send cash to another bank across the country to cash a check drawn on an account there? Why or why not? See Additional Answers.

2. What happens if there is too much money in circulation? too little money in circulation?

3. How does the multiplier effect work?

4. How can the Federal Reserve reduce or increase the multiplier effect?

Make a table or use a spreadsheet to show the multiplier effect on the amounts below if the Federal Reserve requirement is as given. Show the first ten levels and find the total extra money that is generated.

5. Reserve requirement: 15%; initial deposit: $500

6. Reserve requirement: 15%; initial deposit: $2500

7. Reserve requirement: 15%; initial deposit: $12,500

8. Reserve requirement: 18%; initial deposit: $12,500

9. Reserve requirement: 20%; initial deposit: $12,500

10. Reserve requirement: 25%; initial deposit: $12,500

11.–16. For the initial deposits of Exercises 5–10, find the maximum amount of new money created by the multiplier effect.

17.–22. What is the multiplier for Exercises 11–16?

23. Consider your answers to Exercises 1–22. How does changing the reserve requirement affect the extra money generated?

24. Find the maximum amount of extra money that can be generated by a savings account deposit of $4000 if the Federal Reserve System's reserve requirement is 18%. $18,222.22

25. Find the maximum amount of extra money that can be generated by a savings account deposit of $1000 if the Federal Reserve System's reserve requirement is 16%. $5250

KEY TERMS

- common ratio
- Consumer Price Index (CPI)
- easy-money policy
- excess reserves
- Federal Reserve note
- Federal Reserve System
- geometric series
- inflation
- legal tender
- money supply
- multiplier effect
- required reserve
- sum of an infinite geometric series
- tight-money policy

MIXED REVIEW

1-3 An employer has offered two of his salespeople the two options shown. What is the least amount of sales that each person would need for Option 2 to be the better choice?

1. Ellen: Option 1 is a base salary of $1224 and an 8% commission on all sales; Option 2 is no base salary and 12% commission on all sales. $30,600
2. Peter: Option 1 is a base salary of $724 and a 10% commission on all sales; Option 2 is no base salary and 12% commission on all sales. $36,200

3-1 For the given weekly savings, find out how much Dwain can save and how long he will have to save to be able to afford a business calculator that usually sells for $209.00 but that is on sale for $179.99. The sale ends nine weeks from now.

3. $50 per week $200; 4 wk
4. $25 per week $200; 8 wk
5. $20 per week $180; 9 wk
6. $15 per week $210; 14 wk; regular price

3-2 7. How much will $15,500 be worth at the end of three years if it earns 7% interest, compounded quarterly? $19,087.31

3-2 8. How much will $4900 be worth at the end of five years if it earns 4% interest, compounded semiannually? $5973.07

2-3 9. Suppose that your checkbook shows a balance of $290.56. Your bank deducts a service charge of $2.50 each month but does not pay interest and does not charge for each written check. The only check that you wrote during the month was for $80.60 and was mailed the day before the bank's monthly statement arrived. What will the bank show as the closing balance? $368.66

3-2 Use the Rule of 72 to determine how long it will take your $3300 savings account to double in value if it is growing at a rate of

10. 4% per year 18 y
11. 6% per year 12 y
12. 10% per year 7.2 y
13. 12% per year 6 y

1-4 Susan's paycheck is for $998.43. There was an amount of $276 withheld for income taxes as well as 7.65% of the gross salary withheld for FICA taxes.

14. Let x represent the gross salary and write an equation for the gross salary. $998.43 + 276 + 0.0765x = x$
15. What is Susan's gross salary? $1380
16. How much money is withheld in all? $381.57

INDEPENDENT PRACTICE

ASSIGNMENT GUIDE

Exercises 1-4 ask students to display their understanding of the Federal Reserve system.

Exercises 5-10 require tables to show levels of the multiplier effect.

Exercises 11-16 elicit maximum amounts of new money created.

Exercises 17-22 involve using a simple formula to compute the multiplier.

Exercise 23 asks for a general explanation of the reserve requirement's importance.

Exercises 24 and 25 combine the work of Exercises 5-22.

CLOSURE

Have students discuss why a greater or lesser supply of money can result in inflation or recession, and the resultant problems in these outcomes. Direct them to write an explanation of how a required reserve affects the amount of new money generated by excess reserves. Have them define each of the key terms listed in *Exercise Your Skills*.

LESSON QUIZ

1. Compute the amount of new money generated from an initial deposit of $450 with a reserve requirement of 15%. Repeat for 10%.(**$2550, $4050**)
2. Find the multiplier for both parts of Question 1. (**6.7, 10**)

CHAPTER 3 REVIEW

How much can be saved over the following time periods if $45 can be saved each week?

1. 4 weeks $180
2. 12 weeks $540
3. 26 weeks $1170
4. 1 year (52 weeks) $2340

How long will it take to save $165 if you can save the following amounts each week? What is the total amount saved in that time period?

	Amount Per Week	Number of Weeks	Total Amount Saved
5.	$15	11	$165
6.	25	7	175
7.	50	4	200

8. Ivan can save $30 a week to buy one or more of the following items.

   A watch for $111.99 4 wk
   A camera for $199.99 7 wk
   A VCR for $260.00 9 wk
   A computer monitor for $309.00 11 wk

   Draw a graph or use a spreadsheet graphing program to show the number of weeks required to save up for each item. See Additional Answers.

A bank pays 3% simple interest on a one-year CD. Find out how much is saved, how much interest is earned by the CD in one year, and the total amount of money in each account after the bank has credited the interest to the accounts.

	Amount Saved per Week	Number of Weeks	Total Saved	Interest Earned	Total in Account
9.	$ 25.00	52	$1300.00	$ 39.00	$1339.00
10.	32.50	48	1560.00	46.80	1606.80
11.	68.00	50	3400.00	102.00	3502.00
12.	110.00	40	4400.00	132.00	4532.00

Use a calculator to find the amount of interest and the new balance that will accumulate over two years on the following principal amount at the given annual interest rate compounded semiannually.

	Principal	Interest Rate	Interest Earned		New Balance
13.	$2000	5%	First period:	$50.00	$2050.00
14.			Second period:	51.25	2101.25
15.			Third period:	52.53	2153.78
16.			Fourth period:	53.84	2207.62

Use a calculator to find the amount of interest and the new balance that will accumulate over two years on the following principal amount at the given annual interest rate compounded quarterly.

	Principal	Interest Rate	Interest Earned		New Balance
17.	$2000	5%	First period:	$25.00	$2025.00
18.			Second period:	25.31	2050.31
19.			Third period:	25.63	2075.94
20.			Fourth period:	25.95	2101.89
21.			Fifth period:	26.27	2128.16
22.			Sixth period:	26.60	2154.76
23.			Seventh period:	26.93	2181.69
24.			Eighth period:	27.27	2208.96

25. Use the compound interest formula to find out how much $2000 will be worth at the end of two years, at an annual interest rate of 5% compounded semiannually and compounded quarterly. Do your results agree with Exercises 16 and 24? $2207.63; $2208.97; They only differ by one cent.

Suppose that in 1750, when he was 44 years old, Benjamin Franklin invested 50 pounds (£50) in a bank at 3.5% compounded semiannually and stipulated in his will that the money was to be allowed to accumulate in the account after his death. Use the compound interest formula to find the value of the account in each case.

26. After 5 years £59.47 **27.** After 50 years £283.41 **28.** After 75 years £674.73

Peggy's grandmother opened a savings account for her that pays interest compounded semiannually. Use the Rule of 72 to find how many years it will take for the original principal of $500 to double at the given rates of interest.

29. 3% 24 y **30.** 4% 18 y **31.** 8% 9 y **32.** 10% 7.2 y

Make a table to show the multiplier effect on the amounts below if the Federal Reserve requirement is as given. Show the first five levels, and find the total extra money that is generated. See Additional Answers.

33. Reserve requirement: 20%; initial deposit: $2000

34. Reserve requirement: 25%; initial deposit: $1000

35. Reserve requirement: 17%; initial deposit: $11,500

36. Reserve requirement: 15%; initial deposit: $25,000

37.–40. For the initial deposits of Exercises 33–36, find the maximum amount of new money created by the multiplier effect.

41.–44. What is the multiplier for Exercises 37–40?

45. Find the maximum amount of extra money that can be generated by a checking account deposit of $800 if the Federal Reserve System's reserve requirement is 18%. $3644.44

CHAPTER 3 TEST

How much can be saved over the following time periods if $75 can be saved each week?

1. 4 weeks $300

2. 12 weeks $900

3. 26 weeks $1950

4. 1 year (52 weeks) $3900

How long will it take to save $355 if you can save the following amounts each week? What is the total amount saved in that time period?

	Amount Per Week	Number of Weeks	Total Amount Saved
5.	$20	18	$360
6.	30	12	360
7.	60	6	360

ALTERNATIVE ASSESSMENT

Ask students to write a report on interest rates and their effect on both individual savings and behavior in the economy as a whole. Tell them to explain the Federal Reserve's four main functions and to elaborate on the effects of different required reserves. Have them give examples and display data using graphs and/or computer spreadsheets.

A bank pays 4% simple interest on a one-year CD. Find out how much is saved, how much interest is earned by the CD in one year, and the total amount of money in each account after the bank has credited the interest to the accounts.

	Amount Saved per Week	Number of Weeks	Total Saved	Interest Earned	Total in Account
8.	$ 35.00	52	$1820.00	$ 72.80	$1892.80
9.	42.50	52	2210.00	88.40	2298.40
10.	96.80	48	4646.40	185.86	4832.26
11.	125.00	46	5750.00	230.00	5980.00

Suppose that you have a savings account with $8500 in it. It pays 7% interest compounded as shown below. Find the value at the end of each year for the next four years. See Additional Answers.

12. Annually

13. Semiannually

14. Quarterly

15. Monthly

Use the Rule of 72 to find how many years it will take for $7000 to double at the given rates of interest.

16. 9% 8 years

17. 3% 24 years

A $3000 deposit is made in a checking account when the Federal Reserve System's reserve requirement is 16%.

18. Make a table to show the multiplier effect on the deposit. Show the first five levels, and find the total extra money that is generated. See Additional Answers.

19. Find the maximum amount of extra money that can be generated. $15,750

20. What is the multiplier? 6.25

CUMULATIVE REVIEW

1-2 For each company fringe benefits policy described below, find

a. the amount that the company must pay for extra taxes, retirement, health and life insurance, and other indicated fringe benefits.

b. how much salary is paid for the indicated number of nonworking days in a total of 240 operating days.

	EMPLOYEE	SALARY	FRINGE BENEFITS POLICY
1.	Office manager	$45,000	7.65% of gross for FICA taxes;
	a. $6742.50		4% of gross for retirement;
	b. $4500.00		$125/month for life and health insurance;
			24 nonworking days
2.	Restaurant cashier	$12,000	7.65% of gross for FICA taxes;
	a. $1878		$80/month for health insurance;
	b. $ 700		10 holidays, 4 sick days, no vacation days

1-3 Basil's earnings as a real-estate salesperson are based on a graduated commission. He receives 4% on each sale up to $125,000 and 6% of the amount above $125,000. How much would he earn for selling each house below?

3.

PRICED TO SELL
$108,500 Imagine yourself in this 2000 square foot home.

$4340

4.

LOW-MAINTENANCE YARD
$139,950 Large living room, formal dining and breakfast area, 4 bedrooms, 2 baths.

$5897

2-1 **5.** Suppose you wrote 35 checks in July. Your bank charges $0.035 per check for the first 25 checks and $0.15 for each check over 25. How much were your July bank charges? $2.38

3-3 **6.** Find the maximum amount of extra money that can be generated by a checking account deposit of $2200 if the Federal Reserve System's reserve requirement is 18%. $10,022.22

3-2 **7.** How much will $9500 be worth at the end of five years if it earns 8% interest, compounded quarterly? $14,116.50

3-2 **8.** How much will $22,500 be worth at the end of five years if it earns 4% interest, compounded semiannually? $27,427.37

2-2 **9.** Suppose that you wrote a check for $243.50. Before writing the check, you had $982.61 in your checking account. Show how you would enter this check in your check register. See Additional Answers.

3-2 Peggy's grandmother opened a savings account for her that pays interest compounded semiannually. Use the Rule of 72 to find about how many years it will take for the original principal of $500 to double at the given rates of interest.

10. 3% 24 y　　**11.** 4% 18 y　　**12.** 8% 9 y　　**13.** 10.5% 6.9 y

PROJECT 3–1: **Savings Accounts**

As you know, there are many different kinds of savings accounts. These accounts may vary greatly. Visit local banks and savings and loan associations to find information on various types of savings accounts.

1. Name of institution
2. Types of accounts available
3. Rate of interest for each type of account
4. Minimum balance requirements
5. How interest is compounded
6. When interest is paid

PROJECT 3–2: **Federal Reserve System**

As you learned in this chapter, the Federal Reserve System is a very important agency. It acts as a fiscal agent for the U.S. government. The Federal Reserve banks regulate our economy by regulating the money supply.

Gather information about the Federal Reserve System by doing research at your local library, by interviewing bankers, and/or by reading news magazines and newspapers. Answer the following questions.

1. What Federal Reserve District do you live in?
2. How does the Federal Reserve System actually process your checks?
3. What is the current condition of the money supply?
4. Is the Federal Reserve currently trying to expand or reduce the money and credit supply?
5. What are the current implications of the above activity?
6. What is the current required reserves for banks?

Remember that the *slope m* of a line is found by taking any two points on the line and dividing the change in y by the change in x.

$$m = \frac{y_2 - y_1}{y_2 - x_1}$$

For the graph and points shown the slope m is

$$m = \frac{5 - (-1)}{-2 - 3} = \frac{6}{-5} = -\frac{6}{5}$$

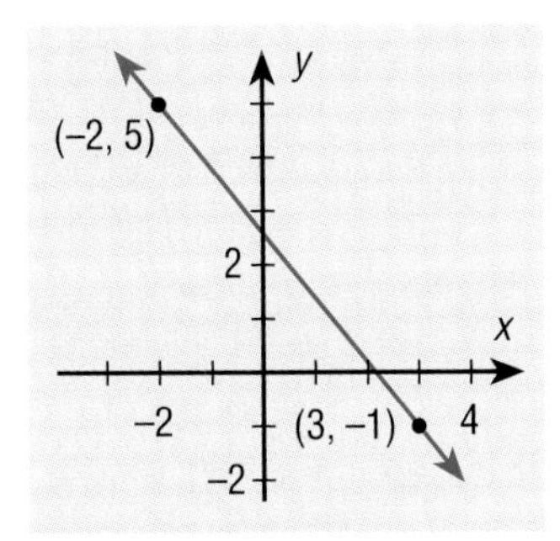

In an equation of the form $y = mx + b$, m is the slope and b is the y-intercept. Use the slope-intercept method to graph each equation.

Example $2y + 3x = 12$

$2y = -3x + 12$ Solve for y.

$y = -\frac{3}{2}x + 6$

When $x = 0$, $y = 6$. The y-intercept is 6.

The slope is the coefficient of x, $-\frac{3}{2}$.

1. $y = 2x - 3$
   See Additional Answers.
2. $3y = 2x - 6$
3. $2x - y = 10$
4. $x = 10 - 5y$
5. $3y + 2x = -12$
6. $y - x = 0$

To graph a *linear inequality,* first graph the related equation, then test points to see which side of the line satisfies the inequality. To graph several inequalities, shade the regions for each one and see where they overlap. Remember to include the line of the equation in the solution for $\leq$ or $\geq$.

Find the solution of each system.

Example $y \leq 2x + 3$ (1)

$y > -x + 1$ (2)

Equation (1) has slope 2 and y-intercept 3. Equation (2) has slope -1 and y-intercept 1. The solution is the points in the green region, including all points on or below $y = 2x + 3$ and all points above but not on $y = x + 1$.

7. $y \leq 3x - 4$
   $2 \geq -x$
   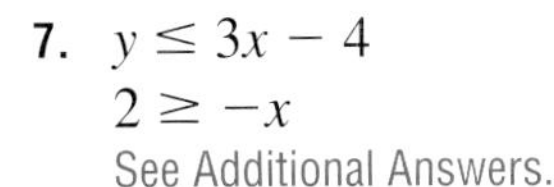
   See Additional Answers.
8. $2x + 3y \leq 12$
   $x - y \geq 5$
9. $2y < x - 4$
   $y > 2x - 2$
10. $x + y > 0$
    $x + y < 0$

A VENTURE INTO BUSINE$$

EVELYN AND THREE OF HER FRIENDS have decided to operate a small business to earn money for college. They have read books with suggestions about everything from house– and pet–sitting to selling popcorn and soft drinks at ball games to singing telegrams and printing mailing labels.

They now realize that there is a lot to think about in starting a business. Will there be an interest in and demand for the product or service that they offer? Can they do the work efficiently enough to make money? What will be the costs of producing and marketing their service or product? How will they keep track of their sales revenue, costs, and profit? This chapter is about how the group of students answers these and other business questions.

In the process of starting their business, the students learn about payroll records and deductions for different kinds of taxes. They also discover that there are many costs and expenses that they had not previously considered. All that they learn will be useful to them in the future, whether they work for a small or large company, whether they work for a private business such as a store or a public agency such as the post office. Careful planning and care with expenses are always important.

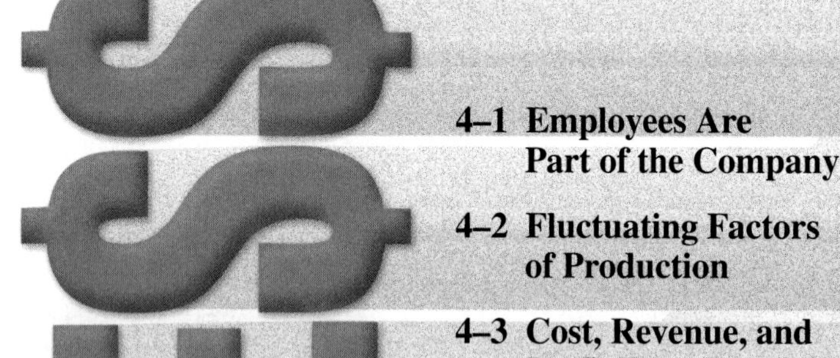

Evelyn is a natural take-charge person and entrepreneur. She first thinks about starting a word-processing business. She has identified a market among fellow students who do not have access to computers or who do not type. She has already done some typing for people.

Greg will help with the typing. He is a design student who brings other talents to the business. He has already painted designs on sweatshirts and sold them. As it turns out, this leads to a more successful business than word processing.

Freda, a friend of Evelyn and Greg, is brought into the business because of her marketing ability. She can sell anything. Her record-keeping and financial skills are not as impressive.

They all need Hari, who can use a spreadsheet program to calculate the effect of different costs and sales on revenue and profit. He can find the break-even point and use linear programming to find how different factors affect the business. This will help the group decide how much to produce and what price to charge.

4-1 EMPLOYEES ARE PART OF THE COMPANY

QUICK REFERENCE

TEACHER'S RESOURCES AND ANSWER KEY

Lesson Quiz Answers 4-1, p. 48

Reteaching Activity

Answers 4-1, p. 96

Enrichment Activity

Answers 4-1, p. 97

EXTENSION ACTIVITIES

Reteaching Activity 4-1, p. 21

Enrichment Activity 4-1, p. 26

Evelyn is a junior in high school. She and her friends have been studying business principles in their economics class. They have several ideas for services or products that they could sell without having to invest large sums of money.

One of these services is word processing. Evelyn's family bought a computer several years ago, and Evelyn has used it as a word processor to prepare reports and essays for school. She has also learned how to prepare a spreadsheet and has experimented with some functions that the spreadsheet program enables her to do. She realizes that the computer should help her in several ways as she considers starting a business. It could be used to provide a service, and it could also be used to organize her company's records.

Some of her classmates do not have access to a word processor. Or if they do have one, they do not know how to type. Perhaps they would be willing to pay to have research papers professionally typed and printed on a word processor. What is more, there is a college nearby that Evelyn now sees as being full of potential customers. There is no end to the term papers and research papers college students must write. Word processing is clearly a business worth considering.

If Evelyn and her friends decide to offer this service, they will have to charge a fair and reasonable price so that it will be worth the time and effort they will have to spend. Managing any business requires making many decisions. Evelyn will investigate some of these decisions.

OBJECTIVES: In this lesson, we will help Evelyn to:

- *Examine some of the functions of management in a business.*
- *Prepare a payroll showing wages, salaries, deductions, and take-home pay for employees in a small company.*
- *Use a spreadsheet to display payroll information.*

FOCUS

ALGEBRA CONNECTIONS

The formula for pay p based on hourly rate r and hours worked h is $p = rh$. The formula for the area of a rectangle A based on width w and length l is $A = wl$. Draw a rectangle in which the width represents rate of pay and the length represents hours worked. Use numbers to show the "area" of a 40-hour work week and also to show overtime at a different rate.

LABOR

If Evelyn and her friends are going to invest their physical effort, mental effort, and skills in a new venture, they want to see a return on their investment. The efforts and skills that they invest are their **labor.** The word *labor* is also used for those workers who do not share in the ownership or the executive decision making of the company they work for. In a small company like Evelyn's, all the employees perform the functions of labor, management, and ownership. Even in very large firms, labor is sometimes represented on management boards. This is because workers have unique experience and insight into the production process.

MANAGEMENT

Management is one key to a successful business operation. Suppose Evelyn comes up with the best idea there is and acquires the most intelligent and industrious labor force. She cannot have a successful business without good organization and management. In all businesses, plans must be made, workers hired, materials obtained, and equipment purchased or built. Putting all these factors together is called **entrepreneurship.** This is the role Evelyn is playing now as she envisions a business and seeks to make it real.

MARKET ECONOMY

As manager, Evelyn must be aware that it is her responsibility to sell the best product at a fair price and still make the highest profit possible. In a **market economy,** profit is the incentive, or reason, for the **producers** to satisfy the wants of **consumers,** or users of the service or product. The owners of property are free to determine how they can most efficiently use available resources. Evelyn's competition—that is, other businesses that offer the same product or service—may be more efficient. She must find out what competition she has.

In a market economy, consumers influence what will be produced by the way they spend their income. Of course, businesses try to influence consumer demand through promotion and advertising, but the final decision is made by the consumers. Consumers decide whether or not to buy, in what quantities, and at what price. If consumers cannot obtain what they want from one producer, there is usually another producer that is happy to satisfy them.

If a product or service is of high quality, it will usually sell for a higher price than one of low quality. In other words, in a competitive market we pay for our goods and services according to how well they satisfy our needs and wants.

Evelyn and her friends are thinking about starting a word-processing business. Evelyn and her friend Greg would do the typing. Both are good typists, and Greg has learned how to use Evelyn's computer. Their friend Freda would be responsible for advertising. She would deliver flyers to neighborhood homes and students at school. She would also contact people at the community college and ask permission to put up posters in the student recreation center and some of the classroom buildings. Another friend, Hari, would be in charge of the finances. He has a special skill with figures and can show Evelyn how to set up their records.

They hope to provide a high-quality service at a competitive price and still be able to pay reasonable wages and manage their other resources as well.

ALGEBRA *REVIEW*

Graph the line through each set of points. Give the equation of the line and its slope.

See Additional Answers.

1. $(-2, -5), (0, -1), (2, 3)$
   $y = 2x - 1; m = 2$
2. $(-1, 0), (2, 3), (5, 6)$
   $y = x + 1; m = 1$
3. $(3, -2), (1, 0), (-2, 3)$
   $y = -x + 1; m = -1$
4.

x	3	7	9
y	−3	−7	−9

   $y = -x; m = -1$

5.

x	−1	0	1
y	2	3	4

   $y = x + 3; m = 1$

6.

x	−3	−1	1
y	7	3	−1

   $y = -2x + 1; m = -2$

For Exercises 7–10, let *x* represent hours worked and *y* represent wages. Write an equation for wages received at the given rate.

7. $6.50 per hour
   $y = 6.50x$
8. $5.20 per hour
   $y = 5.20x$
9. Find the slope of the graph of the equation representing wages received at a rate of $7.25 per hour.
   $m = 7.25$
10. Find the hourly rate and equation for the wages received using the following values.

x	18	25	40
y	$89.10	$123.75	$198.00

   $4.95; $y = 4.95x$

Ask Yourself

1. What incentives do businesses have to satisfy the wants of consumers?
   See Additional Answers.
2. How do consumers influence what will be produced?
3. How do businesses attempt to influence consumers in their choices?

ADDITIONAL ANSWERS

1. Answers may vary. (Ex.: If consumers cannot buy what they want or need from one producer, they can find another one to satisfy them.)
2. Answers may vary. (Ex.: Consumers influence what will be produced by the way they spend their income.)
3. Answers may vary. (Ex.: Businesses attempt to influence consumers' choices through promotion and advertising.)

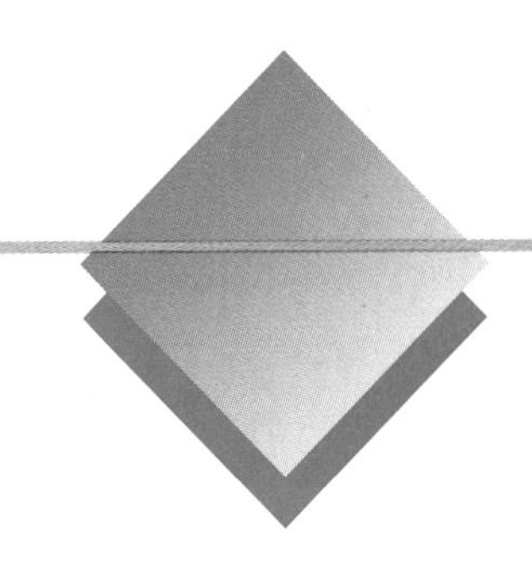

SHARPEN YOUR SKILLS

SKILL 1

EXAMPLE 1 Evelyn plans to keep accurate records of the wages earned by each person in the business. The first record will be a weekly listing of the gross pay for the employees. This list is shown in the spreadsheet developed by Evelyn and Hari. Each heading on the spreadsheet contains important information. Greg and Hari are supporting themselves, so that they each can claim one exemption and pay at lower tax rates than Evelyn and Freda, who are supported by their parents and do not claim themselves as exemptions. Taxes also depend on marital status.

PAYROLL REGISTER FOR WEEKLY EARNINGS

	A	B	C	D	E	F
1	Employee	Number of	Marital	Hourly	Hours	Gross
2		Exemptions	Status	Rate	Worked	Pay
3	Evelyn	0	S	6.50	14	91.00
4	Greg	1	S	6.50	12	78.00
5	Freda	0	S	5.50	16	88.00
6	Hari	1	S	5.50	4	22.00
7	Total					279.00

+D3∗E3

@SUM(F3..F6)

QUESTIONS
a. What algebraic formula can be used to calculate gross pay?
b. What formula should be used to calculate gross pay for the spreadsheet?

SOLUTION

To calculate gross pay, multiply hourly rate by the number of hours worked.

a. The algebraic formula is

$p = r \cdot h$ where p = gross pay
r = hourly rate
h = number of hours worked

For Evelyn, $p = 6.5 \cdot 14 = 91$. Her gross pay is \$91.00.

b. For the spreadsheet calculation, use the formula +D3∗E3, in cell F3. Then copy this formula in the F column for the gross pay of the other employees. To find the total gross pay in Row 7, use the following sum function.

@SUM(F3..F6)

TECHNOLOGY HINT

This lesson depends on the use of the spreadsheet. This will give you the opportunity to reinforce spreadsheet concepts and procedures, such as the @SUM function. Remind students that the function @SUM(F3..F6) is equivalent to writing the formula +F3+F4+F5+F6.

INSTRUCTION

TEACHING THE LESSON

Reading the opening scenario will give students a context for the payroll figures and spreadsheets. The *Algebra Connections* can help some students gain a visual insight into the meaning of hourly and weekly pay. Each skill builds on the preceding skill. It will be useful for you to refer back and point out the connections. Skill 1 covers the calculation of gross pay based on hourly rate and the number of hours worked. It moves from the formula to the spreadsheet presentation. Skill 2 calculates deductions for taxes based on gross pay. Skill 3 considers an individual's cumulative gross and net pay. Skill 4 shows how to compute a monthly payroll summary.

FOCUS ON ALGEBRA

The lesson emphasizes equivalent representations of the same concept. Wages based on hourly pay are described in words, algebraic formulas, and spreadsheet tables. These representations should be used both to reinforce the basic concepts and to show that mathematics has many interconnected branches. (*NCTM Standard 4, p. 146*)

CRITICAL THINKING

Ask students if they would rather be paid a salary of $400 a week or a wage of $9 an hour and time and a half for overtime. This is open ended. Encourage students to consider other factors and the kinds of questions they might ask a potential employer.

SKILL 2

EXAMPLE 2 The next record that Evelyn and Hari prepare is for payroll deductions and take-home pay.

QUESTION How would you set up a spreadsheet for payroll deductions?

SOLUTION

Evelyn and Hari must consult tax tables as shown in Lesson 1–4 to determine the federal income tax withholding. Use the table for single persons with a weekly payroll period in the Reference Section. For example, Evelyn earned $91 and claims 0 exemptions. The amount of withholding for her is $7.00, found in the 90–95 row and the zero column.

FICA withholding is found by taking 7.65% of gross pay. For Evelyn, 91 • 0.0765 = 6.96. Her FICA withholding is $6.96. A formula is entered into the spreadsheet to calculate the FICA withholding.

PAYROLL REGISTER FOR WEEKLY DEDUCTIONS

	A	B	C	D	E	F
1	Employee	Gross	Income Tax	FICA	Total	Take-Home
2		Pay	Withholding	Withholding	Deductions	Pay
3	Evelyn	91.00	7.00	6.96	13.96	77.04
4	Greg	78.00	0.00	5.97	5.97	72.03
5	Freda	88.00	6.00	6.73	12.73	75.27
6	Hari	22.00	0.00	1.68	1.68	20.32
7	Total	279.00	13.00	21.34	34.34	244.66

For Evelyn, the formulas to be entered in the spreadsheet are as follows.

FICA taxes in cell D3: +B3*0.0765
Total deductions in cell E3: +C3+D3
Take-home pay in cell F3: +B3−E3

For the other employees these formulas are copied into the appropriate cells. For row 7 the sum formula is used to add each column.

SKILL 3

There is a limit to the amount of your income that is subject to FICA tax. The current rule is as follows:

> Social Security: 6.2% of all income at or under $57,600
> Medicare: 1.45% of all income at or under $135,000

For most people, the percent of income collected is 6.2% + 1.45%, or 7.65%. From time to time, the above rule is modified by the government, usually by raising one or both of the income maximums.

EXAMPLE 3 Evelyn's father earns $68,000 a year and Hari's mother earns $150,000 a year.

QUESTION How much do Evelyn's father and Hari's mother pay in FICA taxes?

SOLUTION

The amount paid by Evelyn's father is

0.062 • 57,600 =	3571.20	6.2% of $57,600
0.0145 • 68,000 =	986.00	1.45% of $68,000
Total	4557.20	

The amount paid by Hari's mother is

0.062 • 57,600 =	3571.20	6.2% of $57,600
0.0145 • 135,000 =	1957.50	1.45% of $135,000
Total	5528.70	

Evelyn's father pays $4557.20 in FICA taxes and Hari's mother pays $5528.70 in FICA taxes.

SKILL 4

EXAMPLE 4 The third record that Evelyn and Hari must prepare is a monthly payroll summary for each employee. This form contains the payroll information for an individual, week by week.

QUESTION How should Evelyn and Hari use a spreadsheet to prepare a monthly payroll summary?

SOLUTION

The summary should show gross pay, deductions, and take-home pay, week by week. A form is shown below.

MONTHLY PAYROLL SUMMARY

Employee: Evelyn

	A	B	C	D	E
1	Week	Gross	Income Tax	FICA	Take-Home
2	Ending	Pay	Withholding	Withholding	Pay
3	3/7	91.00	7.00	6.96	77.04
4	3/14	65.00	3.00	4.97	57.03
5	3/21	104.00	8.00	7.96	88.04
6	3/28	78.00	4.00	5.97	68.03
7	Totals	338.00	22.00	25.86	290.14

+B3−C3−D3

+B3*0.0765

In this spreadsheet, take-home pay is calculated by subtracting both deductions from gross pay. Remember to use the sum function for the totals.

COMMON ERRORS

The errors in this lesson will be related to the use of the computer. If computers are readily available, you can observe and correct student work in spreadsheet formatting, data entry, and the use of formulas.

RETEACHING

The vocabulary of the lesson may require more reinforcement and reteaching than the math skills. Bring in magazine or newspaper articles using these words.

ENRICHMENT

Challenge interested students to prepare a class presentation on payroll deductions: what taxes or benefits they are for, and how they differ from one employer to another.

TRY YOUR SKILLS

GUIDED PRACTICE

The *Try Your Skills* exercises review the calculations one by one. Students should do them individually. Then go over each one, explaining the process and checking for understanding.

One week, Freda works 25 hours. Freda is single and claims no exemptions.

1. Find her gross pay if she makes $5.50 per hour. $137.50
2. Find the amount of her federal income tax withholding. $13.00
3. Find the amount of her FICA withholding. $10.52
4. Find her take-home pay. $113.98

Jackie is hired to deliver work and run errands. One week, she works 15 hours at $4.80 an hour. She is single and does not claim herself as an exemption.

5. Find her gross pay. $72.00
6. Find the amount of her federal income tax withholding. $4.00
7. What will be her FICA withholding? $5.51
8. Find her take-home pay. $62.49

EXERCISE YOUR SKILLS

KEY TERMS

consumers
entrepreneurship
labor
management
market economy
producers

See Additional Answers.

1. What are several qualities or factors that are needed to make a small business successful?
2. What are some of the responsibilities of a good manager?
3. Give two different meanings of the word *labor.*
4. What investment do workers make in a business?
5. Construct a payroll register for weekly deductions to record the amount for federal income tax withholding based on the tax table in the Reference Section, for FICA withholding based on 7.65% of gross pay, and for take-home pay. Kendra worked 20 hours at a rate of $5.25 an hour, with no exemptions. Paulo worked 9 hours at $8.20 an hour, with one exemption. Michael worked 29 hours at $6.10 an hour, with no exemptions. Also record the totals for each category. All of the employees are single. See Additional Answers.

ADDITIONAL ANSWERS

1. Answers may vary. (Ex.: good management, good market, quality product)
2. Answers may vary. (Ex.: making plans, empowering workers)
3. (1) efforts and skills (2) workers
4. time and labor

6. Use a computer spreadsheet to prepare a payroll register for weekly earnings like the one shown in Example 1. You will create four spreadsheets, one for each week in the month. See Additional Answers.

Employee	Number of Exemptions	Marital Status	Hourly Rate	Hours Worked Week Ending			
				4/4	4/11	4/18	4/25
Catlyn	0	S	$3.75	16	14	21	32
Sara	1	S	4.50	20	28	25	18
Joleen	0	S	8.00	26	16	16	24
Hernando	1	S	6.24	14	10	30	20

7. Use a computer spreadsheet with the headings shown below to prepare four weekly payroll registers for each of the employees in Exercise 6. Refer to the tables for Federal Income Tax Withholding in the Reference Section to find the federal income tax to be withheld. For FICA withholding, use 7.65% of gross pay.

PAYROLL REGISTER FOR WEEKLY DEDUCTIONS					
Employee	Gross Pay	Income Tax Withholding	FICA Withholding	Total Deductions	Take-Home Pay

8. Use a spreadsheet program to prepare a monthly payroll summary with the headings shown below for each employee from Exercises 6 and 7. Show the weekly totals for payroll information to prepare this summary.

MONTHLY PAYROLL SUMMARY				
Employee: ______________				
Week	Gross Pay	Income Tax Withholding	FICA Withholding	Take-Home Pay

9. Construct a computer spreadsheet that will be a payroll register combining the information and calculations for earnings with those of deductions. Use the spreadsheet to compute gross pay, deductions, and take-home pay for the following employees. All of them are single.

Employee	Exemptions	Hourly Rate	Hours
Adikes	1	$7.85	10
Carney	1	5.30	18
Inez	0	6.78	27
Sun-Li	0	9.36	17

How would you need to adjust the FICA tax for a person who earns the following in a year.

10. More than $57,000, but less than $135,000.

11. More than $135,000.

ASSIGNMENT GUIDE

Exercises 1-4 ask about the context of the lesson. A short discussion can again remind students that what they are learning is an important part of an adult's working life.

Exercises 5-9 assess students ability to produce the spreadsheets or tables and do the calculations needed for payroll.

Exercises 10-17 assess students ability to find FICA withholding on salaries greater than $57,600.

Tell students to draw a paper and pencil payroll table with books closed. Tell them to include the necessary categories and to make their own figures. See how well they recall the different categories and procedures.

LESSON QUIZ

1. A person earns $7 an hour for 40 hours and time and a half for 5 hours of overtime. What is her gross pay? (**$332.50**)
2. At 7.65%, what will be the FICA withholding amount for the person in Exercise 1 above? (**$25.44**)

ADDITIONAL ANSWERS

10. When the total gross salary has reached $57,600, find only 1.45% of the salary.
11. When the total gross salary has reached $135,000, there are no FICA taxes for the remainder of the year.

ADDITIONAL ANSWERS

13. $4824.00

15. $5528.70

Find the amount of FICA taxes that are to be paid by the following employees.

12. Ms. Dupont, a salesperson who earned $75,000 last year $4658.70

13. Mr. DeCecco, a master carpenter who earned $7200 a month last year

14. Mr. Andersen, a tugboat captain who earned $143,000 last year $5528.70

15. Mrs. Ramirez, a magazine publisher who earned $210,000 last year

16. What is the maximum amount of FICA taxes that a person could pay in one year? $5528.70

17. A person who pays the maximum FICA tax in one year must make a salary greater than what amount? $135,000

MIXED REVIEW

2-2 Complete the missing portions of the check register shown.

	Number	Date	Checks/Deposits	Amount	Balance
					$335.60
1.	435	4/17	to Frank Brown	$110.00	225.60
2.	—	4/20	from Sue Dunn	45.80	271.40

3-1 Find the simple interest on each CD.

3. $1500 at 5% for one year $75 **4.** $2800 at 4.5% for one year $126

3-1 **5.** Liz is saving $15 a week. How long will it take her to save enough money for a blender that costs $149.50? 10 wk

3-2 **6.** Write the formula for finding the value of an investment that is earning compound interest. $B = p(1 + r)^n$

3-2 Find the interest on each investment.

7. $900 invested for 3 years at 6% compounded annually $1071.91

8. $900 invested for 3 years at 6% compounded quarterly $1076.06

1-3 **9.** Katy prepares flower arrangements for Rodger's Flower Store. She receives $8 for each budget arrangement, $12 for each regular arrangement, and $20 for each deluxe arrangement. How much did she earn in a week in which she assembled 40 budget arrangements, 20 regular arrangements, and 10 deluxe arrangements? $760

3-3 **10.** Find the maximum amount of extra money that can be generated by a checking account deposit of $4350 if the Federal Reserve System's reserve requirement is 20%. $17,400

4-2 FLUCTUATING FACTORS OF PRODUCTION

QUICK REFERENCE

TEACHER'S RESOURCES AND ANSWER KEY

Lesson Quiz Answers 4-2, p. 48

Reteaching Activity Answers 4-2, p. 96

Enrichment Activity Answers 4-2, p. 97

EXTENSION ACTIVITIES

Reteaching Activity 4-2, p. 22

Enrichment Activity 4-2, p. 27

After a few weeks, Evelyn and Greg discover that with only Evelyn's computer to use, the word-processing business cannot be run efficiently. Evelyn recalls how she typed a 15-page English paper for Greg last year. In return, Greg painted a beautiful eagle on a sweatshirt for Evelyn. Many of their friends admired the sweatshirt. That suggested another possible business idea. Greg is wondering what costs would be involved in painting designs on sweatshirts or T-shirts to sell to his friends. He also would need to know the prices at which they would have to sell sweatshirts to be competitive and still make a profit.

OBJECTIVES: *In this lesson, we will help Greg to:*

- *Examine the many costs of producing items for sale in a small business.*
- *Establish prices that cover the costs of production.*

FOCUS

ALGEBRA CONNECTIONS

Find the total cost of 20 T-shirts at $5.50 each and 25 sweatshirts at $7.50 each. Select variables and write an equation for the combined cost of a number of T-shirts and a different number of sweatshirts.

ADD-ON COSTS

Greg has contacted a wholesale supplier who is willing to provide plain T-shirts and sweatshirts for a reasonable price. Transforming a plain T-shirt into a painted one is a matter of having the proper paints, a place to work, design ideas, and enough time to do the painting.

Greg is aware that it takes a long chain of events to turn a **commodity** into a consumer good. In this chain are costs—referred to as **add-on costs**—that contribute to the price of the end product. Add-on costs are those that add to the price of a raw commodity as it goes through the processing and marketing steps in the channel of distribution. Add-on costs cover such things as labor, advertising, energy, and transportation.

Labor At every stage of the production process, labor costs are incurred. The business partners set wages for themselves that are then figured as costs of the business. These contribute to the add-on cost of any final product. Labor costs include not only wages but also employee benefits such as pension plans and medical, dental, and life insurance. Evelyn and Hari have already designed a spreadsheet to keep track of the basic labor costs they would incur. Greg will also have to take this into consideration when marketing T-shirts and sweatshirts.

Packaging Packaging costs reflect both the material and the design of the package. Packaging plays an important role in marketing a product. Frequently, a product's package is responsible for a buyer's initial reaction. Therefore, packaging is often designed to project a particular image or attract a specific audience. Greg thinks the artistic design he will paint on his shirts, while not actually a package, will project an image that he can sell.

Advertising This category includes all kinds of materials and activities used to promote sales of a particular product. In addition to media promotion (television, magazine, and radio), promotional materials might include educational pamphlets, booklets, posters, and other publications. Advertising costs vary greatly from one product to another. Evelyn and Greg have discussed wearing some of the shirts as a kind of "walking advertisement" for their products. When Evelyn wears the shirt Greg made for her last year, friends often ask where they can get one like it.

Energy Energy expenses include electricity, fuel oil, natural gas, or any other energy source used in the manufacturing, distribution, and sales of a product. Energy is necessary to complete each step in the marketing and distribution chain. For example, storage may require energy for special heat and humidity controls. Evelyn and Greg have discussed with their parents the idea of paying a portion of the **utility** (energy) bills for the home they work in.

Transportation The cost of transportation is added to the cost of a product at practically every step in the marketing and distribution channel. Transportation costs include fuel, maintenance and upkeep, depreciation, and even labor costs. Transportation for this enterprise mainly involves Freda's driving to deliver flyers and posters and to talk with people, plus transporting their finished products to their customers.

Ask Yourself

1. What do labor costs include, besides wages?
2. What are several ways of advertising a product?
   television, radio, magazines, pamphlets, booklets, posters
3. What do energy expenditures cover?

ADDITIONAL ANSWERS

1. Employee benefits such as pension plans and medical, dental, and life insurance.
3. Electricity, fuel oil, natural gas or any energy source used in manufacturing, distribution and sales of products.

ALGEBRA *REVIEW*

Solve for *x*.

1. $3x = 30$
   $x = 10$
2. $2x + 50 = 280$
   $x = 115$
3. $2.5x - 3 = 12$
   $x = 6$
4. $x + 2.7 = 2x - 5.6$
   $x = 8.3$
5. $15.6x + 188 = 500$
   $x = 20$
6. $9x + 120 = 660$
   $x = 60$
7. $-3.8x - 12.2 = -6.5$
   $x = -1.5$
8. $8.1x + 11.7 = 7.2x - 9.9$
   $x = -24$
9. $55 - 37x = 75 + 23x$
   $x = -\frac{1}{3}$
10. $2(x - 4) = 13$
    $x = 10.5$
11. $25 = 5(3x - 7)$
    $x = 4$
12. $23x = 19 - 3(x + 2)$
    $x = 0.5$

Graph each equation. Determine the slope of each graph. See Additional Answers.

13. $y = 5x$
    $m = 5$
14. $2y + 4x = 8$
    $m = -2$
15. $y - 3x - 5 = 0$
    $m = 3$
16. $4x = 3y - 6$
    $m = \frac{4}{3}$

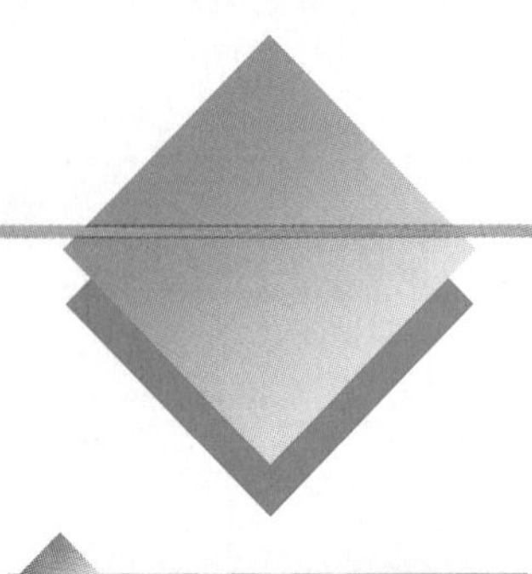

SHARPEN YOUR SKILLS

SKILL 1

EXAMPLE 1 Evelyn, Greg, Freda, and Hari want to find the costs involved in producing and selling hand-painted T-shirts and sweatshirts. They have the following information.

Materials
  12 T-shirts, \$5.50 per shirt; 12 Sweatshirts, \$7.50 per shirt; Paints, \$1.25 per shirt
Labor, \$5.00 per hour
  Evelyn and Greg, 15 hours each; Freda and Hari, 10 hours each
Packaging
  Small plastic bags, \$2.00 per 100
Advertising
  Flyers, \$5.00 per 100
Energy
  Part of utility bill at Evelyn's house, \$2.00 per week
Transportation
  Freda's car, \$25 per week

QUESTION What is the total cost to make 24 shirts (12 T-shirts and 12 sweatshirts) in one week?

SOLUTION

They use the information to determine the total cost.

Materials		
T-shirts	12 • 5.50	66.00
Sweatshirts	12 • 7.50	90.00
Paints	24 • 1.25	30.00
Labor at \$5.00 per hour		
Evelyn and Greg	2 • 15 • 5.00	150.00
Freda and Hari	2 • 10 • 5.00	100.00
Packaging, 2.00 ÷ 100	= 0.02 per bag	
24 bags	0.02 • 24	0.48
Advertising, flyers		5.00
Energy, part of utilities		2.00
Transportation, Freda's car		25.00
	Total	468.48

The total cost to make 24 shirts in one week is \$468.48.

INSTRUCTION

TEACHING THE LESSON

This chapter leans towards business mathematics and can be used to stimulate interest in this area. Point out to students that starting a small business can be simple or complicated, inexpensive or costly—depending on the kind of work involved and the way in which it is undertaken.

Ask students to suggest several kinds of independent businesses that they themselves might attempt. Discuss the tasks related to the work, the costs, and potential problems.

You can then use the interest that has been created to bring in the financial and mathematical aspects of a small business. This lesson is short and the skills are not difficult. As a change of pace, with some classes, you might assign the entire lesson as a self-study unit.

FOCUS ON ALGEBRA

The Standards look for student growth in the ability "to judge the validity of an argument." This skill is developed in this text both in mathematical and consumer situations. (*NCTM Standard 3, p. 143*)

SKILL 2

One of a businessperson's most important tasks is establishing prices. Prices must be high enough to pay all the costs and earn money for the business but low enough to sell as many of the product as possible.

EXAMPLE 2 Evelyn, Greg, Freda, and Hari want to price the hand-painted T-shirts and sweatshirts in Example 1 so that they cover all of their costs.

QUESTION What should be the price of the T-shirts and sweatshirts to pay all costs?

SOLUTION

Evelyn, Greg, Freda, and Hari use the total cost and the number of items to be sold to determine the average cost of a shirt to the nearest cent.

$468.48 \div 24 = 19.52$ The total cost is $468.48 for 24 shirts.

To the nearest dollar, a price of $20 will pay all the costs.

Since the T-shirts cost less to produce than the sweatshirts, the students will put a lower selling price on the T-shirts. But the average of the two prices should be $20. Thus they might sell the T-shirts for $17 and the sweatshirts for $23. There is no absolute formula or rule for establishing prices. Businesses try different prices to find out what is successful.

TECHNOLOGY HINT

The expense categories in Skill 1 include unit costs multiplied by numbers, and then a running subtotal. This is an excellent opportunity to use the calculator memory.

CRITICAL THINKING

In connection with Skill 2, ask students to consider and name some factors, other than covering costs, that might be considered when establishing prices. These might include: prices of competing products; the possibility that not all the products will be sold; possible future increases in production costs. How does a business use these factors in its plans?

TRY YOUR SKILLS

Three students are planning to make a video about their school. Find each of the following costs.

1. Materials: 25 video tapes at $2.40 each $60.00
2. Labor: 18 hours at $5.60 an hour $100.80
3. Advertising: 200 flyers at $0.035 each $7.00
4. Find the total costs for the items in Exercises 1–3.
5. To cover costs, what price to the nearest dollar should the students place on the tapes in Exercises 1–4? $7.00
6. Two students are planning to make kites for resale. Find the total of the following costs. $458

Materials:	40 kite kits at $5.20 $208
Labor:	55 hours at $4.00 an hour
Promotion:	4 ads at $7.50 each $30

7. To cover costs, what price to the nearest dollar should the students place on the kites in Exercise 6? $12.00

ADDITIONAL ANSWERS

4. $167.80
6. Labor: $220

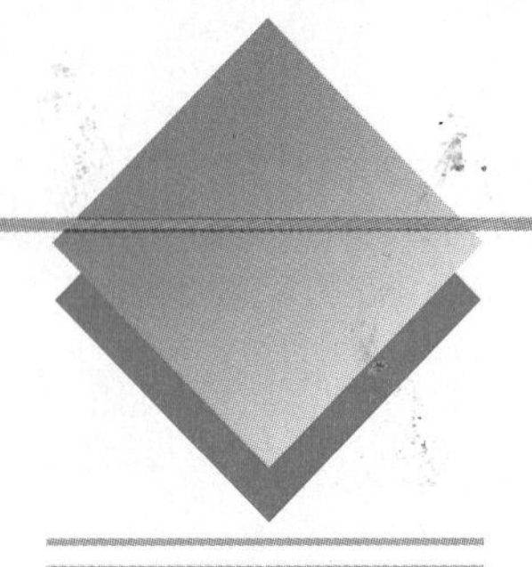

EXERCISE YOUR SKILLS

1. What add-on costs increase the price of a product? See Additional Answers.
2. How does packaging play a role in marketing a product or service?
3. When you purchase a nationally advertised item, are you paying for any advertising costs? Yes, probably a lot
4. What are some of the things a business must consider when establishing the price of a product?

In Exercises 5–16, assume that the advertising, energy, and transportation costs are the following.

Advertising:	$ 5.00
Energy:	2.50
Transportation:	25.00
Total:	$32.50

5. Three students have decided to sell bumper stickers advertising school spirit slogans. Find the cost to produce 200 stickers each week if the costs include $443.00

Labor:	45 hours at $4.50 per hour	$202.50
Materials:	$1.04 per sticker	$208.00
Advertising, energy, and transportation costs		$32.50

6. What selling price to the nearest dollar should be placed on the bumper stickers of Exercise 5 to cover costs? $3.00
7. Four students have decided to applique letters on towels to personalize them. They think they can prepare 100 towels per week. Find the cost to make 100 towels if the costs include $502.50

Labor:	60 hours at $4.00 per hour	$240.00
Materials:	Towels at $2.00 each	$200.00
	Letters at $0.20 per towel	$20.00
Packaging:	Cardboard boxes at $0.10 each	$10.00
Advertising, energy, and transportation costs		$32.50

8. What selling price to the nearest dollar should be placed on the towels in Exercise 7 to cover costs? $6.00
9. Two students plan to sell pennants with the school logo on them. Find the cost to produce 200 pennants in a week if the costs include $452.50

Labor:	40 hours at $4.50 per hour	$180.00
Materials:	Pennants at $1.00 each	$200.00
	Paints at $0.20 per pennant	$40.00
Advertising, energy, and transportation costs		$32.50

4. Prices must be high enough to pay for costs and earn a profit but low enough to sell as many units as possible.

KEY TERMS

add-on costs
commodity
utility

COMMON ERRORS

In this lesson the most common errors will result from the careless use of the calculator. In particular, the placement of the decimal point in addition and multiplication should be watched and checked.

AT-RISK STUDENTS

This lesson can be a confidence builder for at-risk students. The concepts are real-world and "adult," but the math skills involve no more than the basic operations on decimals. Try to be sure that all your at-risk students experience success with the exercises.

ESL STUDENTS

Some frequently used terms in the chapter include: packaging, advertising, energy, and transportation. These are common and yet are abstract and should be reviewed with nonnative speakers.

ADDITIONAL ANSWERS

1. labor, advertising, energy and transportation, packaging
2. The more enticing the packaging, the more likely it is that a customer will want to buy the product.

10. To cover costs, what selling price to the nearest dollar should the students place on the pennants of Exercise 9? $3.00

11. Four students want to make hand-painted coffee mugs for use in cars. Find the costs to produce 50 mugs if the costs include $460.00

Labor:	50 hours at $5.00 per hour	$250.00
Materials:	Mugs at $3.00 each	$150.00
	Paints at $0.40 per mug	$20.00
Packaging:	Boxes at $0.15 each	$7.50
Advertising, energy, and transportation costs		$32.50

12. What selling price to the nearest dollar should be placed on the mugs of Exercise 11 to cover costs? $10.00

13. Two students plan to sell tennis balls at the tennis courts. They will purchase the tennis balls through a discount store and resell them. Find the cost of 100 cans of tennis balls if the costs include $327.50

Labor:	20 hours at $4.00 per hour	$80.00
Materials:	Balls at $2.15 per can	$215.00
Advertising, energy, and transportation costs		$32.50

14. What selling price to the nearest dollar should be placed on the cans of tennis balls of Exercise 13 to cover costs? $4.00

15. Two students plan to order school book bags from a manufacturer and sell them in their school. They will paint the school's insignia on the bags. Find the cost to prepare and sell 50 book bags if the costs include $390.00

Labor:	35 hours at $4.50 per hour	$157.50
Materials:	Bags at $3.50 each	$175.00
	Paints at $0.50 per bag	$25.00
Advertising, energy, and transportation costs		$32.50

16. What selling price to the nearest dollar should be put on the book bags in Exercise 15 to cover the costs and make $50 in profit? $9.00

The following are strategies from which the students might choose in different situations: raise prices, lower prices, increase the number of products for sale, find ways to decrease costs, enhance the quality of the product. Tell which of these strategies you would choose in each of the following situations, and give reasons for your choice. You may choose more than one strategy for a given situation.

17. Very few of the products sell in the first few days. See Additional Answers.
18. All of the products sell out on the first day.
19. People tell you that they like the product but it costs too much.

ADDITIONAL ANSWERS

17. lower price, enhance quality of product
18. raise price, increase number of products for sale
19. decrease price, find ways to decrease costs

RETEACHING

If you find mistakes in basic computation, it would be useful to review estimating strategies so that students can be aware of errors in using the calculator.

ENRICHMENT

Have interested students interview a local merchant and report back on how prices are established.

GUIDED PRACTICE

Have students create a table for the *Try Your Skills* exercises. This approach will organize their work, help them see the connectedness among the different parts, and help you check their work.

INDEPENDENT PRACTICE

ASSIGNMENT GUIDE

Exercises 1-4 review the packaging and marketing ideas of the lesson.

Exercises 5-16 reinforce the mathematical skills used to compute basic expenses and establish prices.

Exercises 17-19 investigate marketing and selling circumstances. They require students to think about problem-solving alternatives and could be used in a cooperative arrangement.

CLOSURE

To prevent students from getting lost in details, summarize the main concept: the cost of operating a business. This cost involves more than first meets the eye. But it is important that students remember that everything discussed and all of the math in the lesson relate to this single fact.

LESSON QUIZ

1. Find the total expenses of a business that pays $5.75 for 50 employee-hours worked, $44.20 for advertising, and $21.80 for utilities. (**$353.50**)
2. A restaurant spends $450 dollars to prepare 50 meals. What should it charge per meal to make a profit of $100? (**$11**)

MIXED REVIEW

1-3 **1.** Alison earns a commission of 6.5% on each of her sales. What is her commission on a sale of $858? $55.77

1-4 **2.** Carol's take-home pay is $416.75. There was an amount of $45.00 withheld for income taxes as well as 7.65% of the gross salary withheld for FICA taxes (Social Security and Medicare). Let x represent the gross salary and write an equation for the gross salary. Then find how much money is withheld in all. $x = 416.75 + 45 + 0.0765x$; $x = 500$; $83.25

2-1 **3.** Tom has a balance of $413 in his checking account. What will be the new balance after monthly interest of 0.5% is credited to the account? $415.07

2-2 **4.** Sean has a bank balance of $235.67. What is his balance after he deposits $56.78 and writes a check in the amount of $15.49? $276.96

3-2 **5.** How much will $7200 be worth at the end of 3 years if it earns 5% interest, compounded semiannually? $8349.79

3-3 Make a table to show the first two levels of new money that is created for the Federal Reserve requirement that is given. See Additional Answers.

6. Reserve requirement: 20%; Initial deposit: $2,400

7. Reserve requirement: 15%; Initial deposit: $1,800

8. Reserve requirement: 25%; Initial deposit: $1,800

2-3 **9.** Suppose that your checkbook shows a balance of $375.21. Your bank deducts a service charge of $5.50 each month but does not pay interest and does not charge for each written check. The only check that you wrote during the month was for $160.32 and was mailed the day before the bank's monthly statement arrived. What will the bank show as the closing balance? $530.03

4-1 **10.** Carmen's annual salary is $78,000. How much does Carmen pay for FICA taxes each year? $4702.20

4-1 **11.** Michael's annual salary is $150,000. How much does Michael pay for FICA taxes each year? $5528.70

4-1 **12.** Julie's annual salary is $55,000. How much does Julie pay for FICA taxes each year? $4207.50

3-2 Use the Rule of 72 to find how many years it will take for $3000 to grow to $24,000 at the given rates of interest.

13. 4.5% 48 y **14.** 6% 36 y **15.** 8% 27 y

3-3 A $1600 deposit is made in a checking account when the Federal Reserve System's reserve requirement is 17%.

16. Find the maximum amount of extra money that can be generated. $7811.76

17. What is the multiplier? 5.9

4-3 COST, REVENUE, AND PROFIT FUNCTIONS

QUICK REFERENCE

TEACHER'S RESOURCES AND ANSWER KEY

Lesson Quiz Answers 4-3, p. 50

Reteaching Activity Answers 4-3, p. 96

Enrichment Activity Answers 4-3, p. 97

EXTENSION ACTIVITIES

Reteaching Activity 4-3, p. 23

Enrichment Activity 4-3, p. 28

Freda has always been a good salesperson. She has sold everything from sunglasses to video tapes, from book bags to personalized towels. She has a knack for understanding people and exactly what motivates them to buy something they might not know they want or need.

The desire to earn money keeps Freda selling, even though she never seems to be able to hold onto her profits. In fact, she doesn't always know whether she has made a profit or not. A lot of cash passes through her hands every month, but her savings account has not grown much.

When Freda joined Evelyn and Greg's business, she hoped that they could help her find out where all her money goes. In return, Freda could help Evelyn and Greg understand the effects of competition on sales. As a result they could all make a profit.

OBJECTIVES: *In this lesson, we learn how to:*

- *Explain how competition works.*
- *Find the profit or loss from the sales of a company's products.*
- *Use the cost function, revenue function, and profit function.*

FOCUS

ALGEBRA CONNECTIONS

Write (or say) as much as you can about the equation $y = 5x$. This is a good open-ended way of reviewing some basic algebraic connections. If students have trouble getting started, ask more specific questions: What would a table look like? What happens as x increases? What will the graph look like?

COMPETITION

Competition is a rivalry among sellers for consumers' dollars or a rivalry among producers for the factors of production. Competition keeps producers from charging unreasonably high prices and making excessive profits. Certain conditions have to exist for competition to work as it should.

For Competition to work:

- There should be a large number of buyers and sellers.
- Buyers should be informed about the quality and price of products.
- Buyers should have choices among products.
- New products should have easy entry into the marketplace.

COST, REVENUE, AND PROFIT

Revenue is the money that a business receives from customers for its products or services. **Profit** is the difference between revenues and costs. If revenues are greater than costs, then a business makes a profit. If costs are greater than revenues, then the business suffers a loss.

Usually, if a company's profits are high, other companies will present similar products at lower prices and take away some of the business and the profits. Evelyn, Greg, and Freda hope to offer a quality product at a low price. To offer the lowest price possible, they analyze their costs. They separate costs into *fixed costs* and *variable costs.* **Fixed costs** are costs that remain the same each week. Based on the assumption that the employees work the same number of hours each week, these costs include labor, transportation, advertising, and energy. **Variable costs** vary according to how many items are produced; these include material and packaging. The cost per item is called the **unit cost.** Similarly the price per unit is called the **unit price.**

Ask Yourself

1. What conditions must be present for competition to work?
2. Without competition, what might happen to prices?

ADDITIONAL ANSWERS (ASK YOURSELF)

1. large number of buyers and sellers; buyers informed about quality and price; buyers have choices among prices; new products have easy entry into market
2. Prices could become unreasonably high since one business could have a monopoly on the product.

ADDITIONAL ANSWERS (ALGEBRA REVIEW)

Answers may vary. Sample answers are given.

7. r = sales revenue, p = unit price, n = number sold; $r = pn$
8. c = unit cost, t = total cost, n = number of units; $c = t \div n$
9. x = unit price of first item, y = unit price of second item, a = average unit price; $a = (x + y) \div 2$
10. P = profit, s = sales, c = cost; $P = s - c$

ALGEBRA *REVIEW*

Write an algebraic expression for each word phrase.

1. The cost of 12 shirts at g dollars each
   $c = 12g$
2. The cost of n books at \$5.75 each
   $c = 5.75n$
3. The cost of n books at h dollars each
   $c = hn$
4. The cost of one book if n books cost \$70
   $c = 70 \div n$
5. The cost of one shirt if 15 shirts cost e dollars
   $c = e \div 15$
6. The cost of one disk if n disks cost d dollars
   $c = d \div n$

For each word sentence, select two or more variables and write an equation. See Additional Answers.

7. Sales revenue is equal to the unit price multiplied by the number sold.
8. The unit cost is equal to the total cost divided by the number of units.
9. The average unit price of two items is equal to the sum of the two individual prices divided by two.
10. Profits equal sales minus costs.

SHARPEN YOUR SKILLS

SKILL 1

INSTRUCTION

TEACHING THE LESSON

After students have read the opening scenario and vocabulary section, discuss the concepts. The ideas of cost, price, revenue, and profit will be important in the students' further education and in their general understanding of how business works. The term "function" is introduced informally in the terms "cost function," and "profit function." Although the equations are true functions, the definition of function is not stressed at this time. It is enough that students gain familiarity with the word. The formal meaning will be covered in a later chapter.

FOCUS ON ALGEBRA

The function is an important unifying idea in mathematics. The increased use of technology has only added to the strength of the function concept. Many input-output operations on the computer and calculator are functions. (*NCTM, Standard 6, p. 154*)

The total cost incurred by a business includes both fixed and variable costs. Total cost can be expressed algebraically.

Cost Function

$c = un + f$ where c = total cost
u = unit cost
n = number of units
f = fixed cost

In the above equation for the cost function, the expression un represents the variable cost. As an example of the cost function, if the students purchase and paint n shirts at a cost of \$6.00 each, then \$6.00 is the unit cost and an equation for the cost function is

$c = 6n + 0$ The fixed cost is assumed to be 0.
$c = 6n$

Revenue, the amount of money a company receives for its products or services, can also be expressed algebraically.

Revenue Function

$r = sn$ where r = revenue
s = selling price per unit, or unit price
n = number of units sold

For example, if n of the above shirts are sold at \$22 each, then an equation for the revenue function is $r = 22n$.

The profit is the amount left when a company subtracts costs from revenue.

Profit Function

$p = r - c$ where p = profit
r = revenue
c = total cost

In the case of the painted shirts, an equation for the profit function is $p = 22n - 6n$, or $p = 16n$.

EXAMPLE 1 Evelyn and her friends have decided to work a fixed number of hours and to start their business by doing only T-shirts. They have analyzed their costs and separated them into fixed and variable categories as shown.

FIXED COSTS: LABOR, ADVERTISING, ENERGY, TRANSPORTATION	
Fixed Costs	
Labor, 50 hours at $5 per hour	$250.00
Advertising	5.00
Energy	2.00
Transportation	25.00
Total fixed cost	$282.00

VARIABLE COSTS: MATERIALS FOR PRODUCTION AND PACKAGING		
Unit Cost per T-Shirt		
Materials:	T-shirts	$5.50
	Paint	1.25
	Package	0.02
Total unit cost		$6.77

QUESTION

a. What is the cost function for the business?

b. With a selling price of $19, what is the revenue function?

c. What will be the profit in selling 30 T-shirts?

SOLUTION

a. The cost function includes the unit cost, the number purchased, and the fixed costs. For the purchase of n T-shirts the cost function is

$c = un + f$

$c = 6.77n + 282$ $\quad u = 6.77;\ f = 282$

b. The revenue function is the product of the selling price and the number sold. For the sale of n T-shirts at $19 each,

$r = sn$

$r = 19n$ $\quad s = 19$

c. The profit function states that profits are found by subtracting the total cost from revenue,

$p = r - c$

For 30 T-shirts, the revenue is

$r = 19n$

$= 30(19) = 570$

COMMON ERRORS

Be sure that students grasp the importance of negative numbers when discussing profit and loss. Some students may think that negative numbers simply mean that the formula did not work out. Explain that a business can find itself "in the red," which means that it lost money. Mathematically, the equation or spreadsheet shows a negative number. This number is just as real and as important as a positive number.

COOPERATIVE LEARNING

If your school has a computer lab, have students work in groups of twos or threes to prepare the spreadsheet formulas. After working at the computer, compare printouts to show differences and alternate approaches.

INTERDISCIPLINARY INVESTIGATION

Business: Have students collect from newspapers and magazines the graphs and tables that track business revenues, profits, and losses. Some might write a short paper on the fortunes or misfortunes of a company they have studied.

CRITICAL THINKING

Ask students to derive a single equation, in six variables, that expresses the spreadsheet relationships of Skill 2. The equation is: $p = mk - f - nc$, where p is profit; m is the number sold; k is the selling price; f is fixed costs; n is the number produced; c is the unit cost of production. Clearly, different letters can be used. If students get stuck, you can give hints.

The cost to make 30 shirts is

$$\begin{aligned} c &= 6.77n + 282 \\ &= 6.77(30) + 282 \\ &= 203.10 + 282 \\ &= 485.10 \end{aligned}$$

The profit is the revenue minus the cost.

$$\begin{aligned} p &= r - c \\ &= 570 - 485.10 \\ &= 84.90 \end{aligned}$$

The weekly profit for selling 30 T-shirts is $84.90. This is not a lot, but the students are paying themselves hourly wages that are included in the fixed costs so they believe that they do not need a large profit.

RETEACHING

An approach to reteaching would be to start with a table, simpler than the spreadsheet, showing costs, revenue from selling, and profit or loss. Have students look carefully at the entries in the table and the trends represented. From this they might move more easily to the abstract level of cost and profit functions.

SKILL 2

Profits in the T-shirt business depend on costs, selling price, and sales quantities. The spreadsheet can be a powerful tool for analyzing the effect of these factors on the business.

EXAMPLE 2 Hari wants to compare profits for different quantities of T-shirts.

QUESTION How can he use a spreadsheet to analyze the profits for various quantities?

SOLUTION

To compare profits for different sales quantities, Hari develops the following spreadsheet.

+A3∗B3+C3

+B3∗E3

+F3−D3

	A	B	C	D	E	F	G
1	Unit	Number	Fixed	Total	Unit		Profit
2	Cost	Produced	Cost	Cost	Price	Revenue	(Loss)
3	6.77	15	282.00	383.55	19.00	285.00	($98.55)
4	6.77	20	282.00	417.40	19.00	380.00	($37.40)
5	6.77	25	282.00	451.25	19.00	475.00	$23.75
6	6.77	30	282.00	485.10	19.00	570.00	$84.90
7	6.77	35	282.00	518.95	19.00	665.00	$146.05

TECHNOLOGY HINT

The spreadsheet in Skill 2 can be used to extend the lesson by creating an XY graph with the X input as B3..B7, and the Y input as G3..G6.

Hari is pleased that the enterprise seems to be becoming more profitable as the number produced rises to 30.

Notice in column G that a loss is indicated by a number in parentheses. Spreadsheet programs allow you to represent a loss of $98.55 either as −98.55 or as ($98.55). Select the style you prefer using the menu choice for formatting.

SKILL 3

We have seen how costs, sales, and profits are related. We have also learned that different tools are available to analyze business finances. These include algebraic equations and the computer spreadsheet that we used in the previous two skills. Graphing is another method that is useful for financial analysis.

EXAMPLE 3 Hari wants to graph the cost function.

QUESTION How can you use the graphing feature of a graphing calculator to graph the cost function from Example 1?

SOLUTION

The equation for the cost function from Example 1 is $c = 6.77n + 282$. To enter this equation in a graphing calculator you need to use the variables x and y instead of n and c. When the equation is rewritten using x and y, the result is

$$y = 6.77x + 282$$

Choose the range values.

Xmin:	−2	Ymin:	−75
Xmax:	45.5	Ymax:	600
Xscl:	2	Yscl:	50

The graph of the cost function is shown.

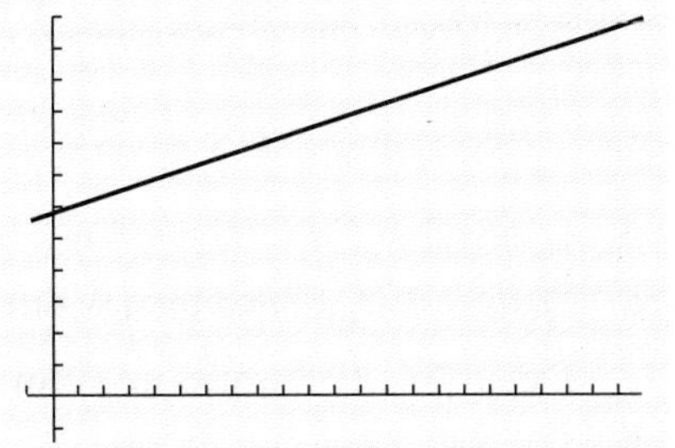

TECHNOLOGY HINT

The example in Skill 3 is elementary and can be used to reteach the use of the graphing calculator. For reinforcement, have students draw an accurate graph based on what they obtain using the calculator. In particular, remind them to include a title and scales on the axes.

TRY YOUR SKILLS

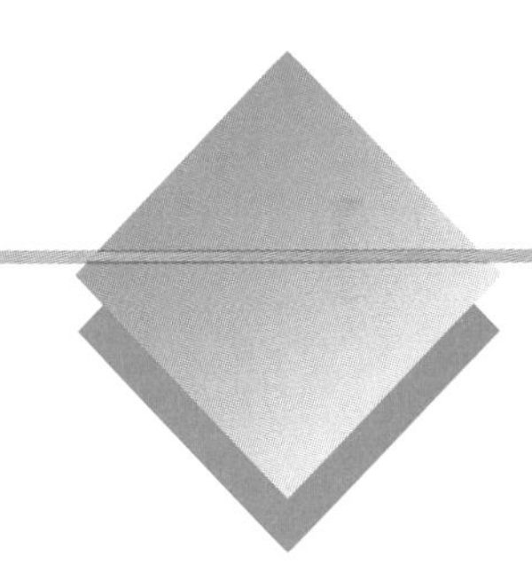

Use the costs in Example 1.

1. Determine the cost of producing 23 T-shirts. $c = 6.77(23) + 282 = \$437.71$
2. Determine the profit or loss in selling 23 T-shirts at $19 each. loss: $0.71

A mug is purchased for $3.75, painted for $0.37, and packaged for $0.08. The fixed costs are $75. The mugs sell for $8 each.

3. Find the unit cost. $4.20
4. Determine the cost function. $c = 4.20n + 75$
5. Find the cost of preparing 32 mugs. $209.40
6. What is the revenue function? $r = 8n$
7. What is the profit function? $p = 3.8n - 75$
8. What is the profit if all 32 mugs are sold? $46.60

GUIDED PRACTICE

Go over the *Try Your Skills* exercises orally, asking for the method to be used in doing each one. Then ask students to do them individually. Finally, check answers by calling on individuals.

ENRICHMENT

Have several interested students or groups of students do research and prepare brief presentations on how companies establish prices. In particular, they should pay attention to consumer reaction, competing products, and profits. Some industries to study would be: airline, automotive, and electronics.

Students selling bumper stickers at $2.00 per sticker have the following costs.

Fixed costs: Labor, 45 hours at $4.50 per hour; Advertising, energy, and transportation, $32.50

Variable costs: $1.04 per sticker

A partial spreadsheet is shown.

	A	B	C	D	E	F	G
1	Unit	Number	Fixed	Total	Unit	Revenue	Profit
2	Cost	Produced	Cost	Cost	Price	Per Unit	(Loss)
3	1.04	200	235.00	443.00	2.00	400.00	($43.00)
4	1.04	300	235.00	547.00	2.00	600.00	$53.00
5	1.04	400	235.00	651.00	2.00	800.00	$149.00

9. Determine the total cost formula for cell D4. +A4*B4+C4
10. Determine the revenue formula for cell F4. +B4*E4
11. Determine the profit (loss) formula for cell G4. +F4−D4
12. Complete the spreadsheet. See above.

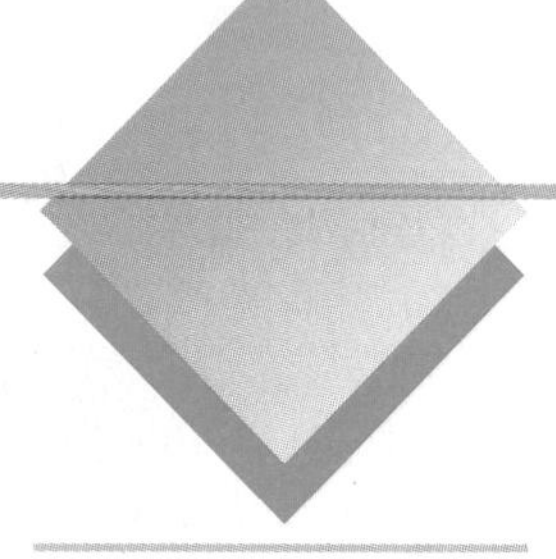

EXERCISE YOUR SKILLS

1. What are some of the factors that limit the amount that a business can charge for its products or services? competition, demand
2. Why do businesses have sales in which they reduce prices?
3. How are the cost and revenue functions related when a business does not make a profit? The cost is greater than the revenue.
4. Why might a business reduce prices so much that it does not make a profit?
5. Price fixing occurs when different companies agree on what they will charge for particular products. Why is there a law against price fixing?

KEY TERMS

competition
fixed costs
profit
revenue
unit cost
unit price
variable costs

ADDITIONAL ANSWERS

2. In order to create more business.
4. to introduce a product, to take away business from a competitor
5. It can drive some companies out of business. It can create unfair high prices.

Students selling personalized towels at \$5.00 per towel have the following costs.

Fixed costs: Labor, 60 hours at \$4.00 per hour
Advertising, energy, and transportation, \$32.50
Variable costs: Towels, \$2.00 each
Letters, \$0.20 per towel
Cardboard boxes at \$0.10 each

6. Determine the cost function for n towels. $c = 2.30n + 272.50$

7. Find the cost of 100, 200, and 300 towels. \$502.50; \$732.50; \$962.50

8. Determine the revenue function for n towels. $r = 5n$

9. Find the revenue for selling 100, 200, and 300 towels. See Additional Answers.

10. Use the cost and revenue functions to determine a profit function.

11. Find the profit in selling 100, 200, and 300 towels. (\$2.50); \$267.50; \$537.50

Students selling pennants at \$2.00 each have the following costs.

Fixed costs: Labor, 40 hours at \$4.50 per hour
Advertising, energy, and transportation, \$32.50
Variable costs: Pennants, \$1.00 each
Paints, \$0.20 per pennant

12. Determine the cost, revenue, and profit functions. See Additional Answers.

13. Set up a spreadsheet to show the cost, revenue, and profit for quantities of 200, 300, and 400.

14. Graph the cost function. See Additional Answers.

Students selling hand-painted mugs at \$10.00 each have the following costs.

Fixed costs: Labor, 50 hours at \$5.00 per hour
Advertising, energy, and transportation, \$32.50
Variable costs: Mugs, \$3.00 each
Paints, \$0.40 per mug
Packaging, \$0.15 each

15. Determine the cost, revenue, and profit functions. See Additional Answers.

16. Set up a spreadsheet to show costs, sales, and profits for quantities of 25, 50, and 75.

17. Graph the cost function.

Students selling tennis balls at \$4.00 per can have the following costs.

Fixed costs: Labor, 20 hours at \$4.00 per hour
Advertising, energy, and transportation, \$32.50
Variable costs: Balls, \$2.75 per can

18. Determine the cost, revenue, and profit functions. See Additional Answers.

19. Set up a spreadsheet to show costs, revenues, and profits for quantities of 100, 200, and 300 cans.

20. Graph the cost function.

ASSIGNMENT GUIDE

Exercises 1-5 are about some of the strategies employed when businesses are setting prices. Not all of the answers are covered explicitly in the lesson. They encourage thinking and imagination. A discussion of these questions will help students understand how mathematics serves the needs of business.

In Exercises 6-23 students are taken though five different situations and asked questions that can be answered based on the skill development in the lesson. In the first situations, questions review each necessary subskill. As the exercises proceed, the questions become more general, requiring students to do the intervening steps.

ADDITIONAL ANSWERS

9. \$500.00; \$1000.00; \$1500.00

10. $p = 2.7n - 272.50$

12. $c = 1.20n + 212.50$; $r = 2n$; $p = 0.80n - 212.50$

15. $c = 3.55n + 282.50$; $r = 10n$; $p = 6.45n - 282.50$

18. $c = 2.75n + 112.50$; $r = 4n$; $p = 1.25n - 112.50$

After covering the three skills, explain how equations, spreadsheets, and graphs can describe the same situation. Have students discuss the benefits and uses of each. When is a spreadsheet most useful? When is a graph the best way to present data?

LESSON QUIZ

1. A small business has variable costs of $200, a unit cost of $3.50, and produces 40 of its product. The business sells 30 of the product at a selling price of $6.00. Find the profit or loss. (**Loss, $160; Explain that a business does not always sell all of its product. Ask what it might do in trying to sell the remainder.**)
2. Graph both equations on the same axes: the cost of x items at $3.50; the sales revenue from x items at $6.00 each.

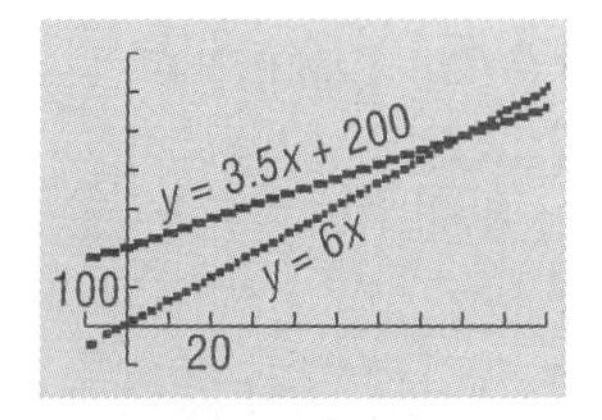

ADDITIONAL ANSWERS (EXERCISE YOUR SKILLS)

21. $c = 4n + 190$; $r = 7.5n$; $p = 3.5n - 190$

(MIXED REVIEW)

5. To: Farley Smith; For: VCR repair; Balance: $363.50

7. To: Abe's Supermarket; For: groceries; Balance: $668.73

8. No; there is a $1.40 discrepancy.

Students selling book bags at $7.50 per bag have the following costs.

Fixed costs: Labor, 35 hours at $4.50 per hour
Advertising, energy, and transportation, $32.50
Variable costs: Bags, $4.00 each

21. Determine the cost, revenue, and profit functions. See Additional Answers.

22. Set up a spreadsheet to show costs, revenues, and profits for quantities of 25, 50, and 75.

23. Graph the cost function.

MIXED REVIEW

1-1 **1.** Tracy earns $130 for working 24 hours. Of that time, 20 hours are at her regular rate and the remainder is at $1\frac{1}{2}$ times the regular rate. What is her regular hourly rate? $5.00 per hour

1-2 **2.** Janice earned $29,670 last year. Her fringe benefits were 14.5% of her salary. What was the value of the fringe benefits? $4302.15

1-3 **3.** Louis earns a salary of $145 a week and also a commission of 4.5% on his sales. What are his total weekly earnings in a week in which his sales are $1,324? $204.58

1-3 **4.** A house has been sold for $190,000. The real-estate agent who sold the house earned a commission of 3.5%. The broker for whom she worked earned 2.5%. How much did each receive? agent: $6650; broker: $4750

2-2 Enter the following transactions in a check-register form. Make up your own form with the following headings: Check Number, Date, Checks/Deposits, Amount, and Balance. Under "Checks/Deposits" have two lines for each entry, the top one labeled "To:" and the bottom one labeled "For:". Find the new balance after each transaction. The starting balance is $620. See Additional Answers.

5. Check 701, February 17, to Farley Smith, $256.50 for VCR repair

6. Deposit on February 19 of $400 For: Deposit; Balance: $763.50

7. Check 702, February 19, to Abe's Supermarket, $94.77 for groceries

2-3 **8.** Earl's check register shows a balance of $104.56. He knows that he has a $20 check outstanding. His bank statement shows a balance of $125.96. Does his account balance? Explain your answer.

1-3 **9.** Tabitha sells annual memberships in an exercise club. Her boss has offered her a choice between Plan 1, no salary but a straight $50 for each membership sold, and Plan 2, a weekly base salary of $225 plus $25 for each membership sold. How many memberships would Tabitha have to sell to make the same amount of money under either plan? 9 memberships

3-2 **10.** Use the Rule of 72 to find how long it will take your $4550 savings account to double in value if it is growing at a rate of 6.6% per year. 11 years

4-4 BREAK-EVEN POINT

QUICK REFERENCE

TEACHER'S RESOURCES AND ANSWER KEY

Lesson Quiz Answers 4-4, p. 50

Reteaching Activity Answers 4-4, p. 96

Enrichment Activity Answers 4-4, p. 98

EXTENSION ACTIVITIES

Reteaching Activity 4-4, p. 24

Enrichment Activity 4-4, p. 29

Evelyn and her friends, like all businesspeople, have no guarantees. When they began their business venture, they did not know whether they would have many, few, or even no customers. These students were determined but also a little anxious. They wanted to succeed. In business the primary indicator of success is making a profit. As we have seen, this will occur when revenues are greater than costs.

Usually, the more efficient a business is, the greater its profits will be. A business can increase its profits by offering consumers better products, lower prices, and better services than competitors. Thus consumers benefit from the efforts of businesses to make a profit.

Profits are important to a business, not because businesspeople are greedy, but because extra money is needed to investigate new business opportunities, to set aside reserves for times when the company does not make a profit, and to provide for emergencies.

Profit, then, is the incentive or reward for improved production and services. It was the desire for profit that led Evelyn and the others to decide on the T-shirt business rather than word processing, and it was the profit motive that helped them determine the appropriate quantities to buy and sell. Finally, it was their interest in profits that led these students to ask Hari to help them understand the finances of the businesses.

The first thing Hari taught the other students was that they should be aware of the break-even point that separates losses from profits.

OBJECTIVES: In this lesson, we will help Hari to:

- *Graph the cost and revenue functions.*
- *Determine the break-even point algebraically.*
- *Determine the break-even point using the graphing calculator.*

ALGEBRA *REVIEW*

Write in slope-intercept form, and find the slope.

1. $x + y = 10$
   $y = -x + 10$; $m = -1$
2. $y = 2x - 3$
   $y = 2x - 3$; $m = 2$
3. $3y = 2x + 6$
   $y = \frac{2}{3}x + 2$; $m = \frac{2}{3}$
4. $2x - 4y = 20$
   $y = \frac{1}{2}x - 5$; $m = \frac{1}{2}$

Graph each equation.

5. $y = 2x - 5$
   See Additional Answers.
6. $2x - 3y = 12$

Find the point of intersection of the graph of each pair of equations.

7. $x + y = 8$
   $x - y = 12$
   (10, −2)
8. $y = 2x$
   $2x + 2y = 18$
   (3, 6)

Use a graphing calculator to approximate the points of intersection to the nearest hundredth.

9. $2x + 5y = 17$
   $3x - y = 21$
   (7.18, 0.53)
10. $4y - 3.7x = 16$
    $3.4x + 2y = 34.6$
    (5.07, 8.69)

BREAK-EVEN POINT

We have seen that establishing prices is a most important step for any business. Prices and the number of products sold determine a company's revenue. If a business's total cost is greater than its revenue, then the business incurs a *loss.* If the revenue is greater than the total cost, then the business makes a *profit.* The point at which cost and revenue are equal is called the **break-even point.** This point is given numerically by the number of products sold at a given price. It is also useful to illustrate the break-even point by graphing the line for cost and the line for revenue. The break-even point is where these two lines intersect.

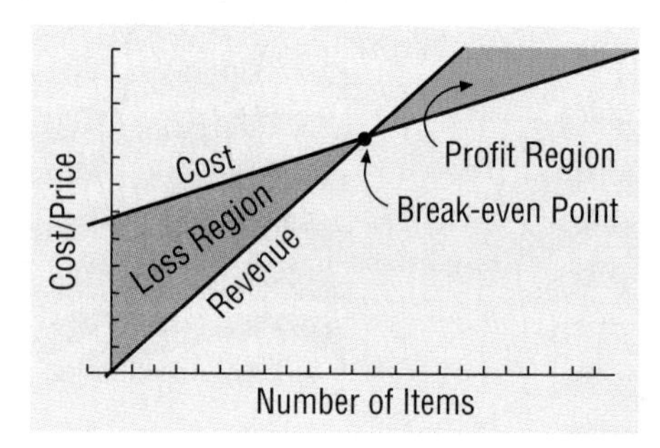

The region between the lines and above the break-even point is the **profit region.** The region between the lines and below the break-even point is the **loss region.**

Ask Yourself

1. Why are profits important to a business?
   See Additional Answers.
2. What is the break-even point?
   The point at which cost and revenue are equal.
3. What factors determine the break-even point?
   Cost, revenue, number of products sold, selling price

ADDITIONAL ANSWERS

1. Profits are needed to investigate new business opportunities and to provide reserves for emergencies.

SHARPEN YOUR SKILLS

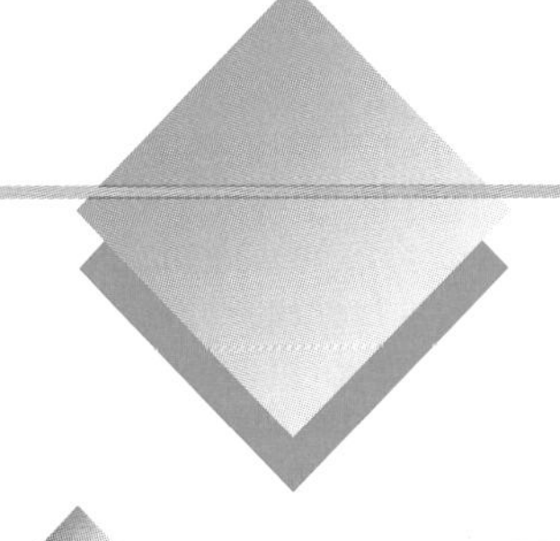

SKILL 1

EXAMPLE 1 Hari wants to find the break-even point for the T-shirt business when the fixed cost is \$282, the unit cost is \$6.77, and the selling price is \$19.

QUESTION How can a graph be used to show the break-even point for Hari's T-shirt business?

SOLUTION

From Lesson 4-3, Example 1, the equation for the cost function is

$c = 6.77n + 282$ n = the number of T-shirts made

and the equation for the revenue function is

$r = 19n$ n = the number of T-shirts sold

Since c and r both refer to amounts of money, you can use the same axis to graph them. Since the amount of money depends on the number of T-shirts, use y for c and r; and x for n, the number of T-shirts. The equations are

$y = 6.77x + 282$
$y = 19x$

Graph these equations on a graphing calculator with the following range values.

Xmin:	−2	Ymin:	−75
Xmax:	45.5	Ymax:	600
Xscl:	2	Yscl:	50

The break-even point is the point where the graphs intersect. The region between the lines and above the break-even point is the *profit region.* In this region the revenue is greater than the cost. The region between the lines and below the break-even point is the *loss region.* In this region the cost is greater than the revenue. The break-even point is the point where the cost and the revenue are *equal.* Using the trace function the break-even point appears to be (23, 437.71).

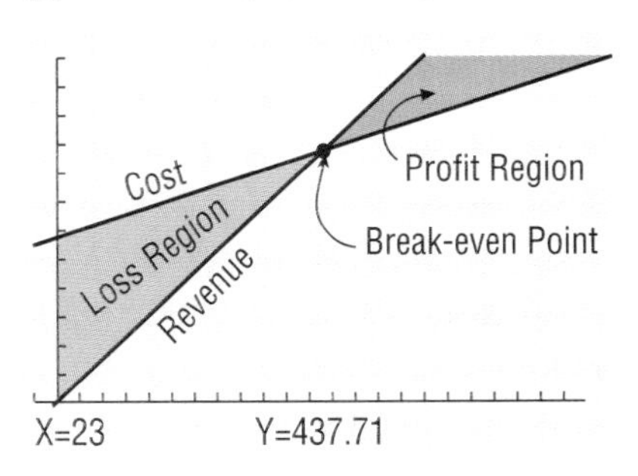

FOCUS

ALGEBRA CONNECTIONS

Write an equation such that each y value is 10 more than twice the x value. Write another equation such that each y value is three times the x value. Find the pair of numbers (x, y) that is a solution for both equations. **(10, 30)**

INSTRUCTION

TEACHING THE LESSON

The introductory scenario and description of break-even give a useful review of what businesses try to do. They could be read aloud in a few minutes, with follow-up questions to insure that students get the break-even idea and how it fits naturally with the concepts of business costs and profits. The skills are sequenced to use the graphing calculator while reviewing standard methods of solving simultaneous equations. In Skill 1, we find the solution graphically; in Skill 2, we find the solution algebraically.

ESL STUDENTS

Instruction and discussion of the break-even point can help you review key vocabulary and the concepts of revenue, costs, profit, and loss.

RETEACHING

For students who have trouble with systems of equations, it is helpful to return to familiar ground. Take the equations one at a time, form a table, plot points, and graph. Look at the point of intersection, and estimate the coordinates of that point. Then return to the algebraic solution of systems of equations.

ENRICHMENT

Show the following equations $y = x$, $y = 2x$, $y = x + 1$, $y = -x + 1$. Ask students for ideas about how the graphs of these equations will look. Have them graph the equations. Tell them to make statements about the graphs and the equations.

FOCUS ON ALGEBRA

Current thinking on mathematics teaching encourages increased use of the calculator and computer as tools for learning. In this lesson the graphing calculator (or computer) is used to review and reinforce the solutions of systems of simultaneous linear equations. (*NCTM Standard 1, p. 137*)

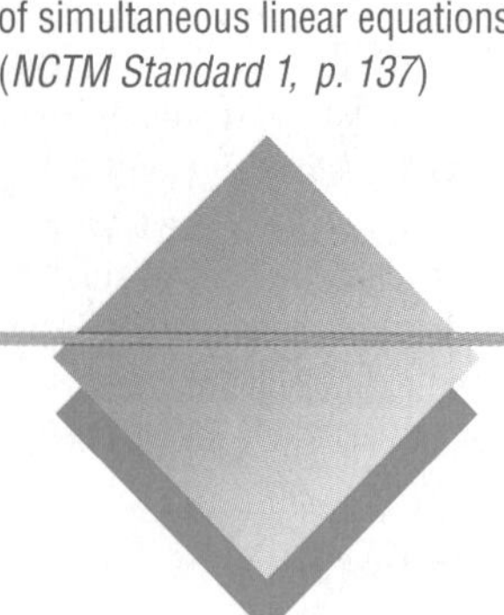

GUIDED PRACTICE

Have students do the *Try Your Skills* exercises individually. Have students do the solutions on the chalkboard. Be sure that everyone understands the solutions. Those who do not will not be able to do the following exercises.

SKILL 2

EXAMPLE 2 Hari wonders if there is a way to determine the break-even point algebraically.

QUESTION How can Hari solve the equations of Example 1 algebraically to find the break-even point?

SOLUTION

A system of two linear equations can be solved by the addition and subtraction method or by substitution. In this case, substitution appears to be the more direct method.

$y = 6.77x + 282$	(1) Cost equation
$y = 19x$	(2) Revenue equation
$19x = 6.77x + 282$	Substituting for y in (1)
$12.23x = 282$	Subtract $6.77x$ from both sides.
$x = 23.06$	To the nearest hundredth
$y = 6.77(23.06) + 282$	Substitute for x in (1).
$y = 438.1$	To the nearest tenth.
$y = 19(23.06)$	Substitute for x in (2).
$y = 438.1$	To the nearest tenth.

We need to interpret the meaning of the point (23.06, 438.1). Since only whole T-shirts can be made, x values need to be whole numbers. Since making 23 T-shirts would result in a loss, although small, the students need to make 24 T-shirts to break even. The break-even point is 24 T-shirts.

If $x = 23$
Cost: $6.77(23) + 282 = 437.71$
Revenue: $19(23) = 437$
Loss: $437 - 437.71 = -0.71$

If $x = 24$
Cost: $6.77(24) + 282 = 444.48$
Revenue: $19(24) = 456$
Profit: $456 - 444.48 = 11.52$

TRY YOUR SKILLS

Students selling bumper stickers at $2.00 each have the following costs.

Fixed costs: Labor, 50 hours at $4.20 per hour
Advertising, energy, and transportation, $32.50
Variable costs: $1.04 per sticker

1. What is the cost function? $c = 1.04n + 242.5$
2. What is the revenue function? $r = 2n$
3. Using algebraic methods, solve the cost function and revenue function simultaneously to find the break-even point. 253 stickers
4. Graph the equations to show the regions of profit and loss. See Additional Answers.

EXERCISE YOUR SKILLS

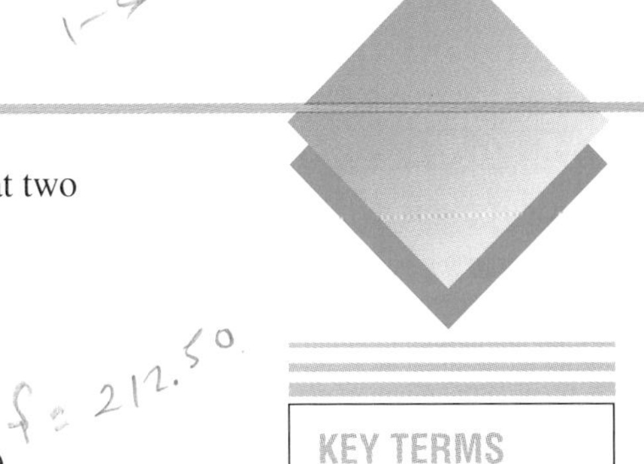

1. The break-even point is the intersection of the graphs of what two functions? cost function and revenue function
2. Why is the break-even point important to a businessperson?

Students selling pennants at $2.00 each have the following costs.

Fixed costs: Labor, 40 hours at $4.50 per hour
Advertising, energy, and transportation, $32.50
Variable costs: Pennants, $1.00 each
Paints, $0.20 per pennant

3. What is the cost function? $c = 1.20n + 212.50$
4. What is the revenue function? $r = 2n$
5. Find the break-even point algebraically. 266 pennants

KEY TERMS

break-even point
loss region
profit region

Students selling tennis balls at $4.00 per can have the following costs.

Fixed costs: Labor, 20 hours at $4.00 per hour
Advertising, energy, and transportation, $32.50
Variable costs: Balls, $2.75 per can

6. What is the cost function? $c = 2.75n + 112.50$
7. What is the revenue function? $r = 4n$
8. Graph the cost function and revenue function. See Additional Answers.
9. Find the break-even point using calculator or computer features. 90 cans

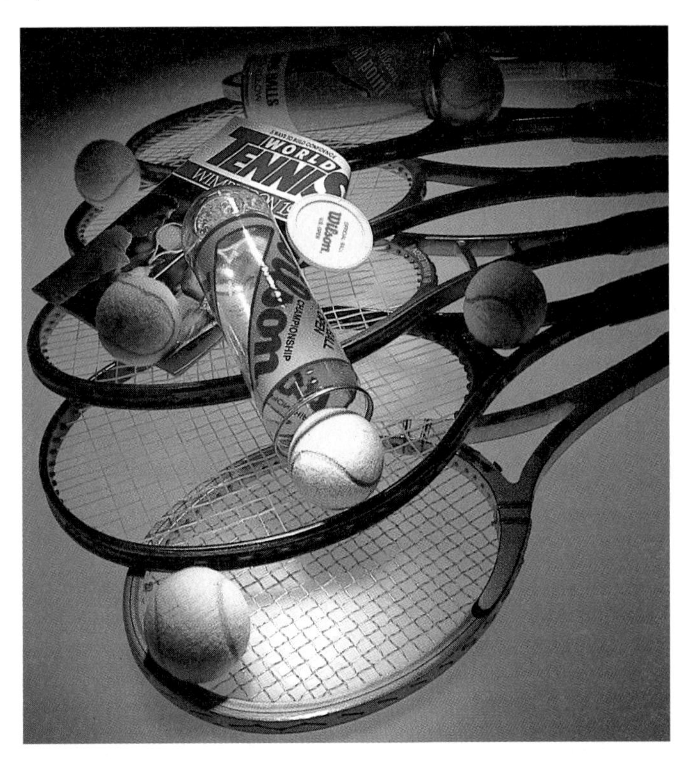

Students selling book bags at $9.00 each have the following costs.

Fixed costs: Labor, 35 hours at $4.50 per hour
Advertising, energy, and transportation, $32.50
Variable costs: Bags, $4.00 each

10. What is the cost function? $c = 4n + 190$
11. What is the revenue function? $r = 9n$
12. Find the break-even point using any method. 38 book bags
13. Graph the cost function and revenue function. See Additional Answers.
14. What is the profit function? $p = 5n - 190$
15. Find the point at which the business suffers a loss of $65. 25 book bags
16. Find the point at which the business gains a profit of $75. 53 book bags

AT-RISK STUDENTS

Solving systems of linear equations requires a level of abstraction that can cause difficulty. Make continued reference to the graphs of the equations and the point of intersection that is found in the algebraic solution. The graphing calculator helps make these connections.

INDEPENDENT PRACTICE

ASSIGNMENT GUIDE

Exercises 1 and 2 review the main underlying concepts of the lesson. Exercises 3-23 apply the break-even skills in four similar situations, but the activities are different. The first asks for the break-even point algebraically. The second asks for the break-even point using the graphing calculator. The third and fourth ask for additional information about points where a particular profit or loss is found.

CLOSURE

Have students explain each part of the graph from which break-even is obtained: the cost and sales function lines, the area that shows a loss, the area that shows a profit. Have them point out similarities and differences between the graphs of the two functions.

LESSON QUIZ

The cost function for product A is $y = 25x + 2000$. The sales function is $y = 40x$.

1. Graph the two equations on the same axes.

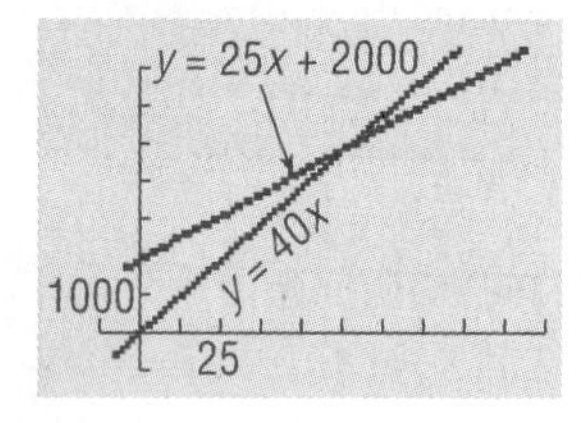

2. Find the break-even point. (**Approx. 1330**)
3. Describe the scale (or range) that you selected. (**x min = 0, x max = 200; y min = 0, y max = 10,000**)

Students selling personalized towels at \$5.00 each have the following costs.

Fixed costs: Labor, 60 hours at \$4.00 per hour
Advertising, energy, and transportation, \$32.50
Variable costs: Towels, \$2.00 each
Letters, \$0.20 per towel
Cardboard boxes, \$0.10 each

17. What is the cost function? $c = 2.30n + 272.50$

18. What is the revenue function? $r = 5n$

19. Graph the cost and revenue functions. See Additional Answers.

20. Find the break-even point. (101, 505)

21. What is the profit function? $p = 2.7n - 272.50$

22. Find the point at which the business suffers a loss of \$78.10. 72 towels

23. Find the point at which the business gains a profit of \$194.60. 173 towels

MIXED REVIEW

1-3 **1.** Veronica earns a commission of \$72 on a sale of \$600. What is her percent of commission? 12%

3-2 **2.** Find the value to which \$500 grows when it is invested for 5 years at 6.5% interest compounded annually. \$685.04

3-3 In Exercises 3–6, suppose that the Federal Reserve requires that 20% of an initial deposit of \$5,000 be held in reserve but that the balance may be lent out by the bank.

3. Show the first two levels of extra money that is generated.

4. What is the total amount of new money that is potentially created by the multiplier effect? \$20,000

5. What is the multiplier for the case in which the reserve requirement is 20%? 5

6. If the reserve requirement were raised to 25%, what would be the value of the multiplier? 4

4-2 Find the cost of producing the given amounts of the following items. In both cases, the cost of advertising, energy, and transportation is \$47.50.

7. Five students are making personalized school calendars. They expect to prepare 25 of the calendars per week. What will be the cost of production if the time required is 20 hours at \$4.75 per hour and the cost of materials is \$1.20 per calendar? \$172.50

8. Two students plan to produce coin dispensers for highway tolls. How much will it cost to produce 30 dispensers if the time for the labor is 25 hours at \$5.50 per hour and the cost of materials is \$0.83 per dispenser? \$209.90

4-5 LINEAR PROGRAMMING

QUICK REFERENCE

TEACHER'S RESOURCES AND ANSWER KEY

Lesson Quiz Answers 4-5, p. 52

Reteaching Activity Answers 4-5, p. 97

Enrichment Activity Answers 4-5, p. 98

EXTENSION ACTIVITIES

Reteaching Activity 4-5, p. 25

Enrichment Activity 4-5, p. 30

As the T-shirt company grew, the students came to appreciate Hari more and more. He knows, both from experience and from studying, how mathematics helps in planning and evaluating business situations.

Hari was especially helpful when his friends learned that there are limits to what they can and cannot do in business. For example, they may have only between 10 and 15 hours a week available to work at their business, and during those hours they can do only so much work.

These limits, or constraints, are found in every business. The students buying coffee mugs or school supplies for resale discovered that they could purchase only as many of these products as their money would allow. Sometimes they could not buy as many as they wanted because their suppliers had a limited number on hand to sell to them. They also discovered that there were limits to the demand for the products that they prepared. They might be able to sell 50 personalized lunch boxes, but they soon found that they could not expect to sell 500.

Smart businesspeople recognize the constraints under which their businesses must operate. They identify these constraints and make their plans within the boundaries of the constraints.

OBJECTIVES: *In this lesson, we will help Hari to:*

- *Write inequalities for some of the business's constraints.*
- *Graph the inequalities and find the points of intersection.*
- *Test the points to find the numbers that minimize costs or maximize profits.*

PROFIT

You have seen that, like all businesspeople, Evelyn and her friends want to make a profit. If they sell too few T-shirts, they will not cover their expenses. If they purchase too many, they will be left with unsold **inventory,** or stock on hand.

A more **efficient** business generally achieves greater profits at a lower cost. Thus consumers also benefit from efficiency.

CONSTRAINTS

Constraints are the conditions that limit business activities. Some examples of constraints are the amount of money available for investment, the time or materials available for production, and the demand for the product or service. There are many other constraints. Each business must discover its own limits and plan accordingly.

Linear programming is a mathematical method for planning within given constraints. The constraints are generally expressed as linear inequalities. A linear inequality is an inequality that has as its graph a half-plane bounded by a straight line.

Maximum and Minimum We use linear programming models to maximize or minimize certain factors in a business situation. To **maximize** means to find the greatest value within the constraints; to **minimize** means to find the least value within the constraints. For example, a business wishes to maximize profits and minimize costs.

Ask Yourself

1. What are three of the constraints that a person starting a business must consider?
   See Additional Answers.
2. What does a business try to minimize?
   costs
3. What does a business try to maximize?
   profits
4. How does a business become more efficient?
   Answers may vary. (Ex.: cutting costs)

ADDITIONAL ANSWERS

1. amount of money available for investment, time or materials available, demand for the product or service

ALGEBRA *REVIEW*

Graph each equation.

1. $y = 2x - 3$
   See Additional Answers.
2. $3x + 2y = 24$

Find the coordinates of the point of intersection.

3. $x + y = 5$
   $2x - y = 4$
   (3, 2)
4. $3y = 5x - 1$
   $y - x = 1$
   (2, 3)

Solve by graphing.

5. $x + 2y \geq 2$
   $2x - y \leq 3$
   See Additional Answers.
6. $x \leq 3y$
   $2y - 1 \leq x$

7. Solve the system by graphing. Find the coordinates of the corners of the region bounded by the inequalities.
   $x \leq 4$
   $y \leq 6$
   $x + 2y \geq 12$
   (0, 6), (4, 4), (4, 6)

SHARPEN YOUR SKILLS

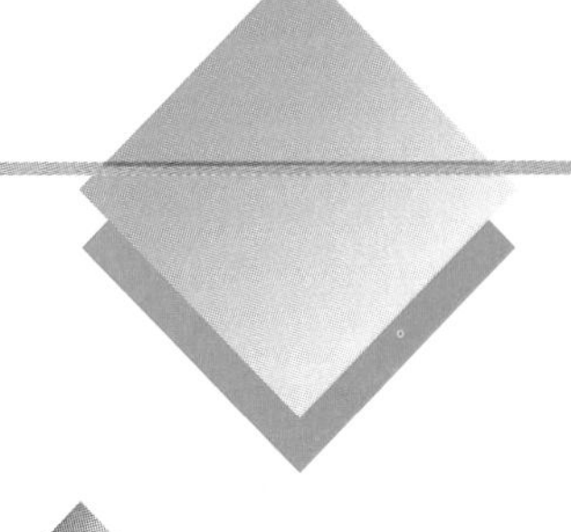

SKILL 1

EXAMPLE 1 Evelyn and her family are planning a car trip that will take several days. They must consider the following constraints.

a. On the first day they are leaving at about noon and don't want to drive after dark. They have only 7 hours in which to drive.

b. They want to travel at least 300 miles that day.

c. They want to stay within the speed limit of 55 miles per hour.

QUESTION How can you use linear programming to graph the constraints as inequalities and show different possible ways to make the trip?

SOLUTION

Let x represent hours and y represent miles. Write inequalities for each constraint.

a. The trip must be no more than 7 hours. $x \leq 7; x \geq 0$

b. The distance must be at least 300 miles. $y \geq 300$

c. Obey the speed limit of 55 miles per hour. Distance = rate • time
$y \leq 55x$

Note that x must be greater than zero, since time is positive.

FOCUS

ALGEBRA CONNECTIONS

To be a member of a certain precision drill team, you must be at least 5 ft 10 in. tall but not more than 6 ft 2 in. tall. You must also weigh at least 150 lb but not more than 210 lb. Draw a graph with heights on one axis and weights on the other. Use a rectangle to show the possible heights and weights that would qualify.

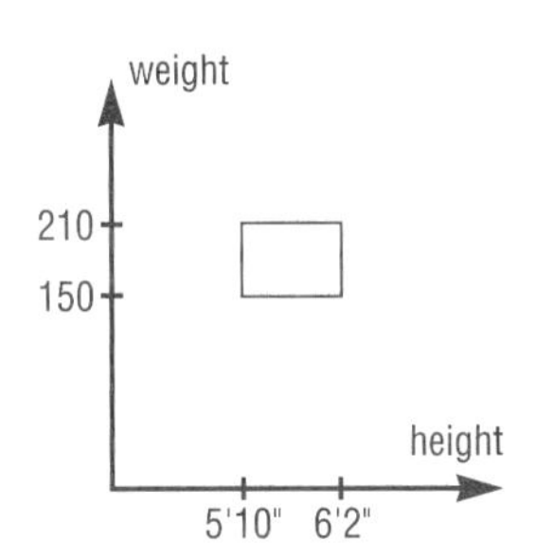

TEACHING THE LESSON

This lesson contains a number of new concepts and skills in which students might be rusty. It may take as much as three or four days to cover the material adequately. After reading the introductory scenario, explain the meaning of maximum and minimum with examples and discussion. Speed limits indicate maximums (and sometimes minimums). Dietary suggestions often indicate recommended minimum daily amounts of vitamins and maximum calories. Ask for other examples. Skill 1 uses linear programming to explore a situation within everyone's experience. Skill 2 uses linear programming to find a minimum, and Skill 3 considers a maximum. As you do each example, place special emphasis on the objective function, how it is determined, what it means, and how it is used.

FOCUS ON ALGEBRA

Linear programming is recommended as an area of discrete mathematics. Although the linear equations that are used to set up a linear programming situation are continuous, the relevant regions are made up of a finite number of line segments. (*NCTM Standard 12, p. 176*)

CRITICAL THINKING

Ask students to think of several real-life situations that are best represented by inequalities rather than equations. You might suggest areas such as shopping, performance in sports, and test scores.

Graph the system of inequalities as shown. The triangular shaded region satisfies all three inequalities. You may not be able to graph $x \leq 7$ on your graphing calculator since it is not a function.

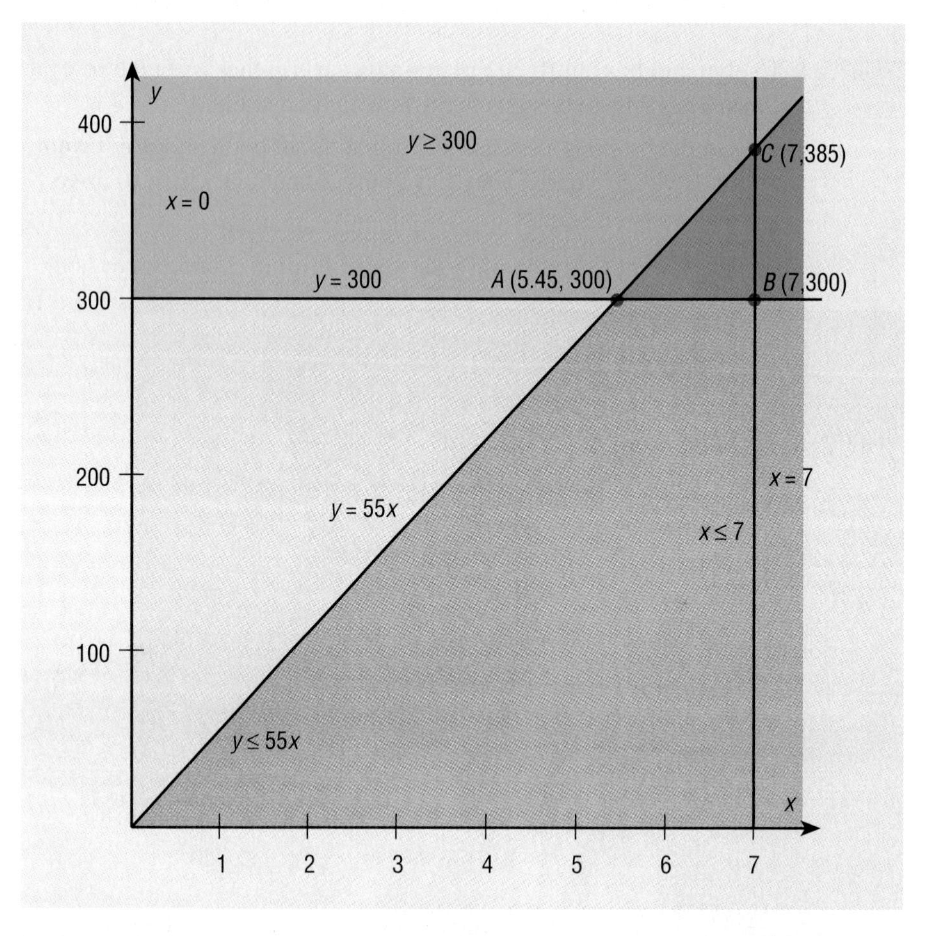

We can find the corners of the triangular region by solving each pair of equations that intersect at the given point.

Point A	*Point B*	*Point C*
$y = 300$	$y = 300$	$x = 7$
$y = 55x$	$x = 7$	$y = 55x$
$55x = 300$		$y = 55x$
$x = 5.45$		$y = 55(7)$
$y = 300$		$y = 385$
$A(5.45, 300)$	$B(7, 300)$	$C(7, 385)$

Although all the points in the shaded region satisfy the three inequalities, the vertices have special meaning.

$A(5.45, 300)$ represents the shortest driving time, 5 h 27 min. It would give the least amount of time in the car.
$B(7, 300)$ represents the lowest speed possible, 43 miles per hour. It would offer the most leisurely drive.
$C(7, 385)$ represents the greatest distance, 385 miles. You would use this to travel as far as possible before dark.

SKILL 2

EXAMPLE 2 Evelyn, Freda, Greg, and Hari have begun selling both T-shirts and sweatshirts. They purchase the T-shirts for $5.50 and the sweatshirts for $7.50. They have become successful but have also discovered constraints that affect their business.

a. To satisfy the demand and not disappoint customers, they must produce a total of at least 52 shirts a week.

b. Their supplier can supply them with no more than 45 blank sweatshirts per week.

c. Because of the coming cool weather, they must be prepared to sell at least as many sweatshirts as T-shirts, and possibly more.

d. Because of the time available, they cannot prepare more than a total of 70 shirts of both kinds per week.

QUESTION How can they use linear programming to find the lowest costs within the given constraints?

SOLUTION

To solve the problem using linear programming, you must first do two things:

1. Choose a quantity that you want to maximize or minimize and write an equation for that quantity. This quantity is called the **objective function.**
2. Identify the constraints.

In this case we want to minimize the total cost of the T-shirts and sweatshirts. This cost is expressed by the following equation.

$c = 5.50x + 7.50y$ where $c =$ the cost
$x =$ the number of T-shirts
$y =$ the number of sweatshirts

COMMON ERRORS

The most common student error in this lesson is writing inequalities in the wrong direction. Tell students that after setting up each inequality, they should test it with numbers and then translate the numerical inequality back into words to be sure that it expresses the original relationship correctly.

COOPERATIVE LEARNING

Linear programming naturally lends itself to cooperative learning. Assign three students to a group with the tasks rotating as follows:

(1) write the inequalities;
(2) check the inequalities;
(3) graph the inequalities;
(1) check the graphs; (2) find the intersection points;
(3) check the points of intersection. Have each student try the relevant coordinates and then check each other's work.

INTERDISCIPLINARY INVESTIGATION

History: Have interested students research the origins of linear programming. To stimulate their curiosity tell them that it began in World War II to set up bombing raids and supply lines within various constraints.

Expressing the constraints as inequalities, we have the following.

a. The students must sell a total of more than 52 shirts.

$x + y \geq 52$

In slope-intercept form, this is written as

$y \geq -x + 52$

b. They can obtain no more than 45 blank sweatshirts.

$y \leq 45$

c. They will sell at least as many sweatshirts as T-shirts.

$y \geq x$

d. They cannot prepare more than a total of 70 shirts.

$y + x \leq 70$ or $y \leq -x + 70$

We now graph the lines for each of these inequalities. This can be done by using a graphing calculator.

TECHNOLOGY HINT

Depending on the graphing calculator or computer utility in use, it may be necessary to point out to students that they may or may not be able to show regions determined by inequalities. If they cannot, it is best to redraw the lines and indicate the inequality regions with paper and pencil. Remind students that they cannot graph equations such as $x = 4$ using most graphing calculators because such an equation is not a function. Again, they will have to move from the graphing calculator to pencil and paper graphs.

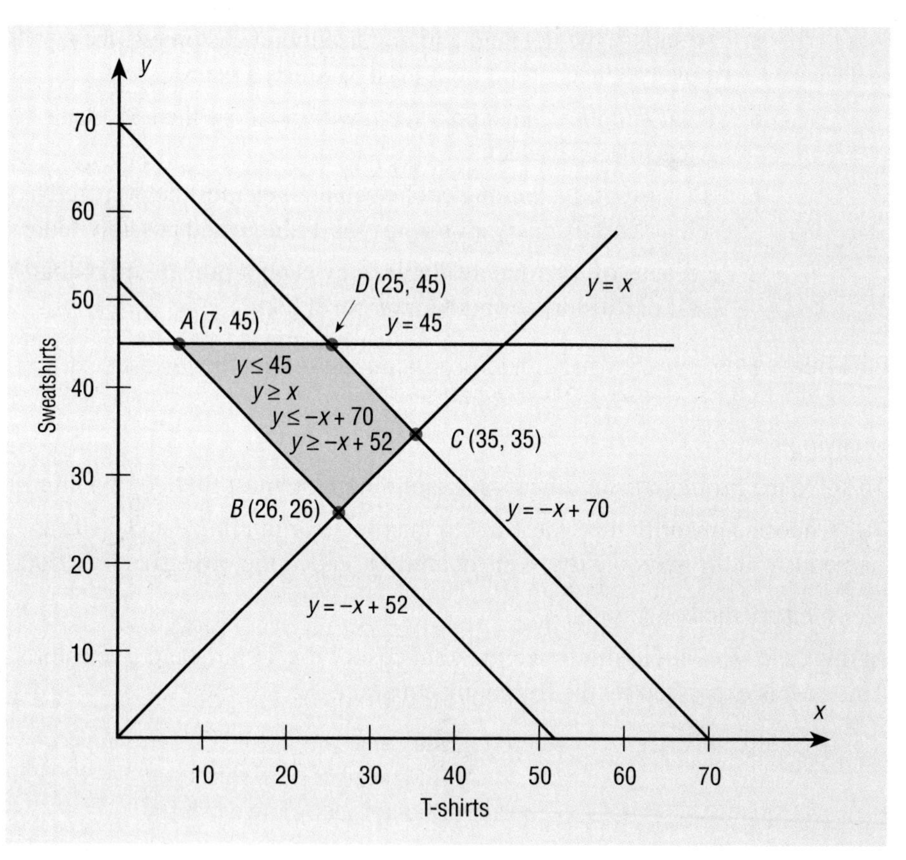

The shaded region satisfies all four inequalities.

Solve each pair of equations that intersect at the given point.

Point A	*Point B*	*Point C*	*Point D*
$y = 45$	$y = x$	$y = x$	$y = 45$
$x + y = 52$	$x + y = 52$	$x + y = 70$	$x + y = 70$
$x + 45 = 52$	$2x = 52$	$2x = 70$	$x + 45 = 70$
$x = 7, y = 45$	$x = 26, y = 26$	$x = 35, y = 35$	$x = 25, y = 45$
$A(7, 45)$	$B(26, 26)$	$C(35, 35)$	$D(25, 45)$

Substitute these coordinates in the objective function, $c = 5.50x + 7.50y$.

For point A, $5.50(7) + 7.50(45) = 376$
For point B, $5.50(26) + 7.50(26) = 338$
For point C, $5.50(35) + 7.50(35) = 455$
For point D, $5.50(25) + 7.50(45) = 475$

The lowest cost will be achieved by using the coordinates at point B; that is, by making and selling 26 T-shirts and 26 sweatshirts.

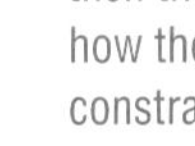

RETEACHING

A number of students may have difficulty with the material in this lesson. If this is the case, it is best to go back to Skill 1. Explain each inequality and graph each one by itself. Then put two of the inequalities on the same axes; then three, explaining gradually how the region satisfies all three constraints.

SKILL 3

ENRICHMENT

Challenge students to find a real-life example of how linear programming was used to make a business decision.

The constraints listed and graphed in Example 2 will be the same whether we are considering cost or revenue.

EXAMPLE 3 Evelyn's group can raise some additional money to invest, so they want to maximize profit. They sell the T-shirts for \$17 each and the sweatshirts for \$23 each.

QUESTION How can they find the sales quantities that will give the maximum profit within the given constraints?

SOLUTION

We use the same linear programming model that was used in Example 2. However, we use a different objective function for revenue. This is the revenue function, the number of items sold multiplied by unit price.

$r = 17x + 23y$ where r = the total revenue
17 = the selling price of the T-shirts
x = the number of T-shirts sold
23 = the selling price of the sweatshirts
y = the number of sweatshirts sold

For intersection points on the graph the revenues are

$A(7, 45)$: $17(7) + 23(45) = 1154$
$B(26, 26)$: $17(26) + 23(26) = 1040$
$C(35, 35)$: $17(35) + 23(35) = 1400$
$D(25, 45)$: $17(25) + 23(45) = 1460$

The maximum revenue is at point D. But this is not the maximum profit.

Profit equals revenue minus cost: $p = r - c$.

We now use the costs from Example 2 to find profit at each point.

	revenue	−	*cost*	=	*profit*
$A(7, 45)$	1154	−	376	=	778
$B(26, 26)$	1040	−	338	=	702
$C(35, 35)$	1400	−	455	=	945
$D(25, 45)$	1460	−	475	=	985

Within the given constraints the greatest profit occurs at point $D(25, 45)$, which represents 25 T-shirts and 45 sweatshirts. We can also find the maximum profit by writing an objective function for profit.

$$p = 17x + 23y - (5.50x + 7.50y)$$
$$p = 11.5x + 15.5y$$

TRY YOUR SKILLS

GUIDED PRACTICE

The *Try Your Skills* exercises first review necessary skills. Then students are taken through a problem with constraints somewhat similar to those in Skill 1. That is, the situation does not ask for a maximum or minimum, but reinforces the use of inequalities.

1. Find the coordinates of the point of intersection of the graphs $y = 8$ and $3x + y = 20$. (4, 8)
2. Draw the triangle bounded by the inequalities $y \leq 8$, $x \leq 12$, $2y + x \geq 20$. See Additional Answers.
3. Find the corners of the triangle drawn in Exercise 2. (4, 8), (12, 8), (12, 4)
4. Write inequalities for the following constraints.
   a. You want to work no more than 10 hours per week. $x \leq 10$
   b. You want to earn at least $45 per week. $y \geq 45$
   c. You won't be able to get a job earning more than $6.00 per hour. $y \leq 6x$
5. Draw the triangle for the constraints listed in Exercise 4, and explain what each corner of the triangle means. See Additional Answers.

EXERCISE YOUR SKILLS

Students selling pennants have joined those selling bumper stickers. They can purchase pennants for $1.00 each and bumper stickers for $1.25 each. They want to earn as much money as possible but have discovered that they must work under the following constraints. Let x represent pennants, and y represent bumper stickers.

a. They can obtain no more than 200 pennants. $x \leq 200$
b. They can obtain no more than 400 bumper stickers. $y \leq 400$
c. They must sell a combined total of at least 300. $x + y \geq 300$

1. Write inequalities to express the constraints. Let x represent the number of pennants and y represent the number of bumper stickers. See above.

2. Graph the constraint inequalities, and shade the region that satisfies all the inequalities. See Additional Answers.
3. Find the coordinates of the corners of the region in Exercise 2.
4. Write the objective function to minimize costs. $c = 1.00x + 1.25y$
5. Write the objective function to maximize revenue if the pennants and bumper stickers sell for \$2 each. $r = 2x + 2y$
6. Write the objective function to maximize profit. $p = x + 0.75y$
7. Substitute the corner coordinates in the objective function to find the number of pennants and stickers that they should sell to maximize profit.

A friend of Hari's goes to work for a computer company that manufactures computer chips. To meet demands, the company must manufacture at least 4000 chips of type A per day and 6000 chips of type B per day. The testing department cannot process more than a total of 15,000 per day. The company earns a profit of \$0.50 for each chip of type A sold and \$0.40 for each chip of type B sold. Let x stand for the number of A chips and y stand for the number of B chips.

8. Write the objective function for the profit. $p = 0.5x + 0.4y$
9. Write three inequalities to express the constraints. See Additional Answers.
10. Graph the inequalities.
11. Find the coordinates of the corners of the region defined by the inequalities.
12. Test the coordinates of the corners in the objective function to find the pair that gives the greatest profit. (9000, 6000); \$6900 profit

KEY TERMS

constraints
efficient
inventory
linear programming
maximize
minimize
objective function

ADDITIONAL ANSWERS

3. A(0, 300); B(200, 100); C(200, 400); D(0, 400)
7. 200 pennants; 400 stickers
9. $x \geq 4000$; $y \geq 6000$; $x + y \leq 15{,}000$
11. A(4000, 6000); B(9000, 6000); C(4000, 11,000)

INDEPENDENT PRACTICE

ASSIGNMENT GUIDE

In Exercises 1-31 students are led through the linear programming steps in five different situations. Students sort out the information, set up objective functions, write and graph inequalities, and determine solutions.

CLOSURE

Have students discuss the meaning of the objective function. Ask for examples of a function representing an amount of money that a business would want to maximize; that it would want to minimize. In a similar way discuss some of the constraints that affect the way businesses operate.

LESSON QUIZ

1. Graph the following inequalities on the same axes. $y \leq 2x$, $x \leq 10$, $y \geq 5$

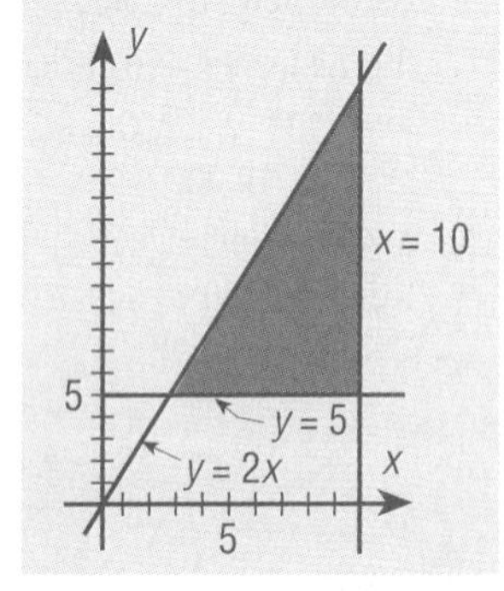

2. Find the corners of the region bounded by the inequalities. {**(2.5, 5)**, **(10, 5)**, **(10, 20)**}
3. Which point identified in Exercise 2 will maximize the function $p = 2x - y$? **(10, 5)**

ADDITIONAL ANSWERS

19. $x \geq 40$; $y \geq 50$; $x \leq 80$; $y \leq 2x$; $x + y = 210$

21. A(40, 50); B(80, 50); C(80, 130); D(70, 140); E(40, 80)

22. 80 acres of crop x and 130 acres of crop y

28. A(5, 10); B(30, 10); C(30, 15); D(15, 30)

Students selling coffee mugs have joined with those selling desk organizers. Costs are \$3.55 for each coffee mug and \$6.40 for each desk organizer. To make a profit, they must sell a combined total of at least 50 coffee mugs and desk organizers. The total spent on the supplies for both items must be less than \$350. Because of arrangements with suppliers, the number of mugs must be less than four times the number of organizers. The number of organizers must be less than 40. Let x represent the coffee mugs and let y represent the organizers.

13. Write the objective function to minimize costs. $c = 3.55x + 6.40y$

14. Write inequalities to express the constraints. $x + y \geq 50$; $x < 4y$; $y < 40$

15. Graph the constraint inequalities and shade the region that satisfies all the inequalities. See Additional Answers.

16. Find the corners of the region in Exercise 15. A(10, 40); B(40, 10); C(160, 40)

17. What is the number of coffee mugs and desk organizers they should sell to minimize their cost? 40 mugs, 10 organizers

Freda's father is a farmer with 210 acres on which to plant. From crop x he will earn \$400 an acre; from crop y he will earn \$350 an acre. To meet demands, he must plant at least 40 acres of crop x and 50 acres of crop y. Because of the conditions of the soil, he may not plant more than 80 acres of crop x. According to state regulations, the acres for crop y must be no more than twice the number of acres for crop x.

18. Write the objective function to maximize revenue. $r = 400x + 350y$

19. Write inequalities to express the constraints. See Additional Answers.

20. Graph the constraint inequalities and shade the region that satisfies all the inequalities.

21. Find the corners of the region in Exercise 20.

22. How many acres of each crop should he plant to maximize his earnings?

Some students who are starting a business plan to sell earrings for \$4 a pair and necklaces for \$10 each. Their unit costs are \$2.50 for the earrings and \$9 for the necklaces. They must work within the following constraints. The store through which they are selling will take only up to a total of 45 pairs of earrings and necklaces combined. Their source of supplies is such that the number of necklaces that they can make will be less than twice the number of pairs of earrings. They must sell at least 10 necklaces to maintain their agreement with the store. Demand will not permit them to sell more than 30 pairs of earrings. Let x represent the earrings and y represent the necklaces.

23. Write the objective function to minimize cost. $c = 2.5x + 9y$

24. Write the objective function to maximize revenue. $r = 4x + 10y$

25. Write the objective function to maximize profit. $p = 1.5x + y$

26. Write inequalities to express the constraints. $x + y \leq 45$; $y < 2x$; $y \geq 10$; $x \leq 30$

27. Graph the constraint inequalities and shade the region that satisfies all the inequalities. See Additional Answers.

28. Find the corners of the region in Exercise 27.

29. How many of each should they sell to minimize cost? See Additional Answers.

30. How many of each should they sell to maximize revenue?

31. How many of each should they sell in order to maximize profit?

ADDITIONAL ANSWERS

29. 5 pairs of earrrings; 10 necklaces

30. 15 pairs of earrrings; 30 necklaces

31. 30 pairs of earrrings; 15 necklaces

MIXED REVIEW

1-3 Joseph repairs watches at $10 an hour and receives $2 for each watch that he repairs. He works 30 hours a week.

1. How many watches does he have to repair in order to earn at least $350 per week? 25

2. It takes Joseph an average of 20 minutes to repair a watch. What is the most money that he can reasonably expect to make in one week? $480

3-2 Suppose that $100 is invested at 5% annually. Find the value of the investment after the given number of years and at the given frequency of compounding. Use the calculator's exponent key x^y.

3. 50 years, compounded annually $1146.74

4. 50 years, compounded quarterly $1199.52

5. 100 years, compounded annually $13,150.13

4-1 A partially completed monthly payroll summary is shown below. Complete the table. Use a spreadsheet program, if available.

	Week Ending	Gross Pay	Income Tax Withholding	FICA Withholding	Take-Home Pay
6.	9/3	$ 87.00	$10	$ 6.66	$ 70.34
7.	9/10	123.00	19	9.41	94.59
8.	9/17	167.50	23	12.81	131.69
9.	9/24	201.00	28	15.38	157.62
10.	**Totals**	578.50	80	44.26	454.24

2-3 11. What is the limit of your liability if you immediately report the loss or theft of your ATM card? $50

4-3 12. The fixed costs in Amanda's business are $325 per week. The variable costs are $3.50 per item. What are her total costs if she produces 150 items in one week? $850

CHAPTER 4 REVIEW

Find the following earnings:

	Number of Hours	Hourly Rate	Weekly Earnings		Number of Hours	Hourly Rate	Weekly Earnings
1.	38	$ 3.35	$127.30	**2.**	26	$ 7.50	$195.00
3.	41	22.70	930.70	**4.**	22	14.80	325.60

Find the weekly take-home pay for each of these single people if each claims one exemption. Use the tables for federal income tax withholding in the Reference Section.

	Gross Pay	Income Tax Withholding	FICA Withholding	Take-Home Pay
5.	$127.30	$ 5.00	$ 9.74	$112.56
6.	195.00	16.00	14.92	164.08
7.	930.70	183.00	71.20	676.50
8.	325.60	35.00	24.91	265.69

Duanita and Alisha want to make and sell stuffed kittens. Use the information given in the box to help you find the total production costs for manufacturing the amounts of toys in Exercises 9, 10, and 11.

Fixed Costs		Variable Costs	
Labor	$3.50 per hour	Materials	$2.25 each
	for 30 hours	Packaging	$0.02 each
Advertising	$3.00		
Energy	$3.58		
Transportation	$5.92		

9. 20 kittens $162.90 **10.** 40 kittens $208.30 **11.** 60 kittens $253.70

Duanita and Alisha will sell stuffed kittens for $7.00 each. Use the costs in Exercises 9-11 to find the profit or loss from selling each amount.

12. 20 kittens **13.** 40 kittens **14.** 60 kittens

ADDITIONAL ANSWERS

12. loss: $22.90

13. profit: $71.70

14. profit: $166.30

Use your results from Exercises 9–14 to make a spreadsheet that shows the following information for producing and selling 20, 40, and 60 kittens.

	Fixed Cost	Unit Cost	Number Produced	Total Cost	Unit Price	Revenue	Profit (Loss)
15.	$117.50	$2.27	20	$162.90	$7.00	$140.00	($22.90)
16.	117.50	2.27	40	208.30	7.00	280.00	71.70
17.	117.50	2.27	60	253.70	7.00	420.00	166.30

For Exercises 18–23, use your results from Exercises 9–17.

18. Write the cost function. $c = 2.27n + 117.50$

19. Write the revenue function. $r = 7n$

20. Write the profit function. $p = 4.73n - 117.50$

21. Graph the cost function. See Additional Answers.

22. Graph the revenue function on the same set of axes as the cost function.

23. Find the break-even point algebraically and using graphing techniques.

24. Why is it important to know how much profit (or loss) a business makes?

25. If a company does not make enough money to break even, what kinds of changes should be considered? Write a paragraph explaining the changes you would consider if your company were losing money and what you would do next if each strategy did not work.

26. After one year, the employees in Oscar's company expected to receive a raise in salary. Explain how an employee's thinking about a raise in salary might differ from the owner's thinking. What should Oscar consider before deciding whether to give raises? Why might it be better for the company and for the employees themselves if salary raises are small? When might salary raises be large?

Use the following information to do Exercises 27–30. When Duanita and Alisha were successful in the stuffed kitten business, they expanded and began to make stuffed puppies as well. Their cost for materials for the kittens continued to be \$2.27, but the materials for the puppies cost \$3.00. They sold the kittens for \$7.00 and the puppies for \$7.50. To meet the demand, they had to sell more kittens than puppies. They could not obtain materials to make more than 40 kittens per week and more than 20 puppies per week. To pay their expenses, they had to sell a total of more than 50 per week. Let x represent the stuffed kittens and y represent the stuffed puppies.

27. Write the objective function to minimize cost. $c = 2.27x + 3y$

28. Write the objective function to maximize revenue. $r = 7x + 7.5y$

29. Write the objective function to maximize profit. $p = 4.73x + 4.5y$

30. Write inequalities for the constraints. $x > y$; $x \leq 40$; $y \leq 20$; $x + y > 50$

31. Graph the constraints and shade the region that satisfies all the inequalities. See Additional Answers.

32. Find the corners of the region in Exercise 31. A(30, 20); B(40, 10); C(40, 20)

33. Find the quantities that will minimize cost within the constraints.

34. Find the quantities that will maximize revenue within the constraints.

35. Find the quantities that will maximize profit within the constraints.

ADDITIONAL ANSWERS

23. 25 kittens

24. It helps the company to analyze its costs and to plan expansion or condense production.

25. Answers may vary. (Ex.: higher prices, higher production, lower costs, and so on)

26. Answers may vary. (Ex.: The employees may not realize that getting raises affects costs, and therefore profit or loss. When profits are high, raises could be large. Small raises may insure the existence of the company.)

33. 40 kittens and 10 puppies

34. 40 kittens and 20 puppies

35. 40 kittens and 20 puppies

CHAPTER 4 TEST

Find the weekly earnings for each. See Additional Answers.

1. 42 hours at $4.80 per hour **2.** 33 hours at $17.25 per hour

Find the weekly take-home pay for each gross pay. Each person is single and claims one withholding allowance (exemption). Tables for federal income tax withholding are in the Reference Section.

3. $201.60 **4.** $569.25 **5.** $202.50 **6.** $844.20

Akira sells stadium cushions. Find the total production cost to manufacture the numbers of cushions indicated in Exercises 7–9.

Fixed Costs		Variable Costs	
Labor	$4,25 per hour for 30 hours	Materials	$1.40 each
		Packaging	$0.50 each
Advertising	$2.00		
Energy	$2.58		
Transportation	$4.42		

7. 30 cushions **8.** 60 cushions **9.** 90 cushions

Find the profit or loss for selling the following amounts at $5.50 each.

10. 30 cushions **11.** 60 cushions **12.** 90 cushions

Use your results from the previous exercises to complete the spreadsheet.

	Fixed Cost	Unit Cost	Number Produced	Total Cost	Unit Price	Revenue	Profit (Loss)
13.	$136.50	$1.90	30	$193.50	$5.50	$165.00	($28.50)
14.	136.50	1.90	60	250.50	5.50	330.00	79.50
15.	136.50	1.90	90	307.50	5.50	495.00	187.50

16. Refer to Exercises 7–15. Graph the cost and revenue functions. See Additional Answers.

17. Refer to Exercises 7–15. Find the break-even point algebraically. 38 cushions

Jod and Jessie sell stadium cushions and caps. The cushions cost $1.90; the caps cost $2.25. They sell the cushions for $5.00 and the caps for $6.00. They can obtain no more than 100 cushions and 75 caps per week. To meet demands, they have to sell a total of at least 120 of the two together. They cannot package more than 150 per week. Let x represent the cushions and y represent the caps.

18. Use linear programming to graph the given constraints, and find the points that will give maximum and minimum quantities. See Additional Answers.

19. What quantities will give the minimum cost? 100 cushions and 20 caps

20. What quantities will give the maximum revenue? 75 cushions and 75 caps

21. What quantities will give the maximum profit? 75 cushions and 75 caps

ALTERNATIVE ASSESSMENT

Have students prepare a 5- to 10-minute presentation on a small business of their choice. The presentation should include: the fixed costs; the factors that contribute to the variable costs of production; and what considerations enter into the establishment of prices. A student should also explain whether the company makes a profit and the outlook for the future.

Have students write a brief scenario about a production business of their choice. Ask them to develop cost and revenue functions and to draw graphs.

Provide a graph with a polygonal region of the type covered in Lesson 4-5. Include the inequalities for each line. Ask students to write a minimum or maximum problem for which the given graph would be the solution.

ADDITIONAL ANSWERS

1. $201.60 **2.** $569.25
3. $169.18 **4.** $445.70
5. $170.01 **6.** $621.62
7. $193.50 **8.** $250.50
9. $307.50
10. Loss: $28.50
11. Profit: $79.50
12. Profit: $187.50

CUMULATIVE REVIEW

1-1 **1.** Your regular rate of pay is \$6.22 per hour. You also receive $1\frac{1}{2}$ times your hourly rate for any hours over 20 in one week. Last week you worked 26 hours. You also received \$12.50 in tips. Find your total earnings for the week. \$192.88

1-1 **2.** You work according to the time schedule shown. You earn \$5 per hour. Find your total earnings for the week if you receive \$22 in tips that week.

Monday	Tuesday	Wednesday	Thursday	Friday	Saturday
3–6 P.M.	3–7 P.M.	3–6 P.M.	3–6 P.M.	4–9 P.M.	8 A.M.–12 P.M.

1-2 **3.** You earn \$9.30 per hour. You work 35 hours a week for 50 weeks and get 2 weeks of paid vacation as well as 18% of your base salary in fringe benefits. Find your yearly earnings including benefits. \$19,972.68

1-3 **4.** You are a real-estate agent and sold a house for \$110,000. You earn a 4% commission on the first \$100,000 of each sale and 6% commission on the amount over \$100,000. How much commission did you earn on the sale?

1-4 **5.** You earned \$94.00. Your income tax withholding is \$7 and your FICA withholding is 7.65% of your earnings. Find your take-home pay. \$79.81

2-1 **6.** You wrote 33 checks one month. The bank charges \$0.0225 for the first 20 checks that you write and \$0.08 for each written check over 20. How much are the charges for that month? \$1.49

2-2 **7.** Your employer made out your paycheck with your first and middle initials and full last name. Write your name as you would to endorse the check.

3-1 **8.** You save \$28 per week. You want to buy a stereo system that costs \$589. For how many weeks must you save before you can have the stereo system? 22 wk

3-1 **9.** Felix withdraws the interest from his savings account as soon as the interest is posted twice a year. What is the simple interest earned in $2\frac{1}{2}$ years on his \$500 balance? The annual interest rate is 3.5%. \$43.75

3-2 **10.** Find the value of a \$1000 investment after interest has been compounded quarterly for 1 year at an 8.5% annual rate. \$1087.75

3-3 **11.** Find the maximum amount of extra money generated by a deposit of \$3000 if the reserve requirement is 12%. \$22,000

4-3 **12.** The fixed cost of producing hand-painted sweatshirts is \$285 per week. The variable costs are \$8 per shirt and \$1 per shirt for paints. What is the total cost of producing 25 shirts in one week? \$510

4-3 **13.** Use the information in Exercise 12 to find the profit or loss, given that each sweatshirt is sold for \$20. Loss: \$10

4-4 **14.** Write the equations for the cost function and revenue function of Exercises 12 and 13. Solve these equations to find the break-even point for the sweatshirt-selling operation. $c = 9n + 285$; $r = 20n$; 26 sweatshirts

ADDITIONAL ANSWERS

2. \$132.00
4. \$4600
7. Answers may vary.

PROJECT 4–1: **Retail Sales**

Work in small groups to set up imaginary companies. Select items that your company would like to sell or services that it would provide. You may use information from the chapter to plan costs or do your own research to find the costs of producing various services or products. Tell how you arrived at these costs for each product or service that your company produces.

1. Decide on the product or service for your company. Describe that product or service. Be sure to include all elements including advertising, energy use, transportation, and packaging.
2. Decide on the number of hours per week each employee (member of the group) will work. Decide the hourly wages for each person. Decide the marital status and number of exemptions for each employee.
3. Complete a payroll register showing each employee's earnings for four consecutive weeks.
4. Furnish a four-week payroll summary for each employee.
5. List your fixed and variable costs. Use a spreadsheet to show fixed costs, variable costs, number produced, total costs, average sale (or sale per unit), total sales, and profit or loss. Your spreadsheet should have entries that result in a loss and others that result in a profit.
6. Write the cost, revenue, and profit functions for your product.
7. Construct a line graph showing the break-even point for the business.
8. List several realistic constraints related to quantities and costs. Use linear programming to minimize costs or maximize sales revenue within the given constraints.

Extensions

1. After you have completed a plan for your company, get together with other groups in your class and try to sell them your product.
2. On the basis of feedback from other groups, redesign your product or service to be more useful or attractive to potential customers.
3. Listen in turn to other groups' presentations, and give them feedback about their product or service. Try to make comments that will help them in redesigning their product or rethinking their service.

Evaluate each expression. Recall that for all real numbers *a*, $a \neq 0$, and for all integers *n*:

$$a^{-n} = \frac{1}{a^n}$$

Examples: $3^{-2} = \frac{1}{3^2} = \frac{1}{9}$ $\quad (2^{-4})^2 = 2^{-8} = \frac{1}{2^8} = \frac{1}{256}$

1. 4^{-3} $\frac{1}{64}$
2. 5^{-5} $\frac{1}{3125}$
3. $(3^{-4})^3$ $\frac{1}{531{,}441}$
4. $(2^{-3})^2$ $\frac{1}{64}$
5. $(4^{-2})^{-3}$ 4096
6. $(3^{-1})^{-3}$ 27

Express with positive exponents. Recall that for all nonzero real numbers *a* and all integers *m* and *n*:

$$\frac{a^m}{a^n} = a^{m-n} \text{ if } m \geq n \qquad \frac{a^m}{a^n} = \frac{1}{a^{n-m}} \text{ if } m < n \qquad a^0 = 1$$

Example: Express $\frac{x^{-3}y^2z^{-5}}{x^2y^{-4}z^{-5}}$ with positive exponents: $\frac{x^{-3}y^2z^{-5}}{x^2y^{-4}z^{-5}} = \frac{y^6}{x^5}$

7. $\frac{x^{-3}y^{-4}z^{-1}}{x^{-5}y^3z^{-12}}$ $\frac{x^2z^{11}}{y^7}$
8. $\frac{x^0y^3z^{-4}}{x^{-8}y^2z^0}$ $\frac{x^8y}{z^4}$
9. $\frac{x^9y^{-8}z^{-7}}{x^{-3}y^{-4}z^7}$ $\frac{x^{12}}{y^4z^{14}}$

Express each percent as a decimal. Recall that *percent* means *per hundred.*

Examples: 7.5% = 0.075 $\quad$ 0.02% = 0.0002 $\quad$ 135% = 1.35

10. 8.9% 0.089
11. 13.6% 0.136
12. 0.095% 0.00095
13. 0.78% 0.0078
14. 128% 1.28
15. 1600% 16

Evaluate to the nearest hundredth. Use your calculator.

Example: $\frac{1500(0.07)(1 + 0.07)^{24}}{(1 + 0.07)^{24} - 1}$

Use the exponent key [x^y] or [^] and parentheses as necessary.

1500 [×] 0.07 [×] [(] 1 [+] 0.07 [)] [x^y] 24 [÷]
[(] [(] 1 [+] 0.07 [)] [x^y] 24 [−] 1 [)] [ENTER]
130.78 To the nearest hundredth

16. $\frac{1275(0.005)(1 + 0.005)^{48}}{(1 + 0.005)^{48} - 1}$ 29.94
17. $\frac{1225(0.0025)(1 + 0.0025)^{60}}{(1 + 0.0025)^{60} - 1}$ 22.01
18. $\frac{2400[1 - (1 + 0.015)^{-13}]}{0.015}$ 28,155.68
19. $\frac{325[1 - (1 + 0.0125)^{-24}]}{0.0125}$ 6702.88

CONSUMER CREDIT

USING SOMEONE ELSE'S MONEY WHILE we wait to acquire our own gives us an inflated sense of wealth. We can now buy almost everything that is for sale without using money that we currently possess.

With an installment loan, we can have a house or a car before we can afford to pay for it. Banks and other lending institutions are willing to finance these purchases if we enter into an installment loan agreement that guarantees repayment of the borrowed money. Banks generate profits in the process because we are willing to repay two or three times as much as we borrow if it enables us to purchase a house.

As Americans, we think of ourselves as honest and hardworking. We have been convinced that owing money and being in debt are fine as long as we can keep up with the payments. Skillful advertisers understand our basic desires for more than we have. Once we buy one item on credit, we find it easier and easier to buy more items on credit. Lending institutions capitalize on our false sense of wealth by offering more and more credit to those of us who can make the installment payments.

Owning something now that we cannot afford to pay for now, except in small monthly installments, is a very attractive and luxurious way to live. However, we can overextend ourselves easily without even realizing it. We may continue to enjoy the use of credit, but as intelligent consumers we should be aware of the hazards. We should know the high price of spending what we do not have and who receives the profits from our use of credit.

In this chapter our high-school students will examine installment buying. They will become aware of the risks of using credit and learn ways to reduce its cost.

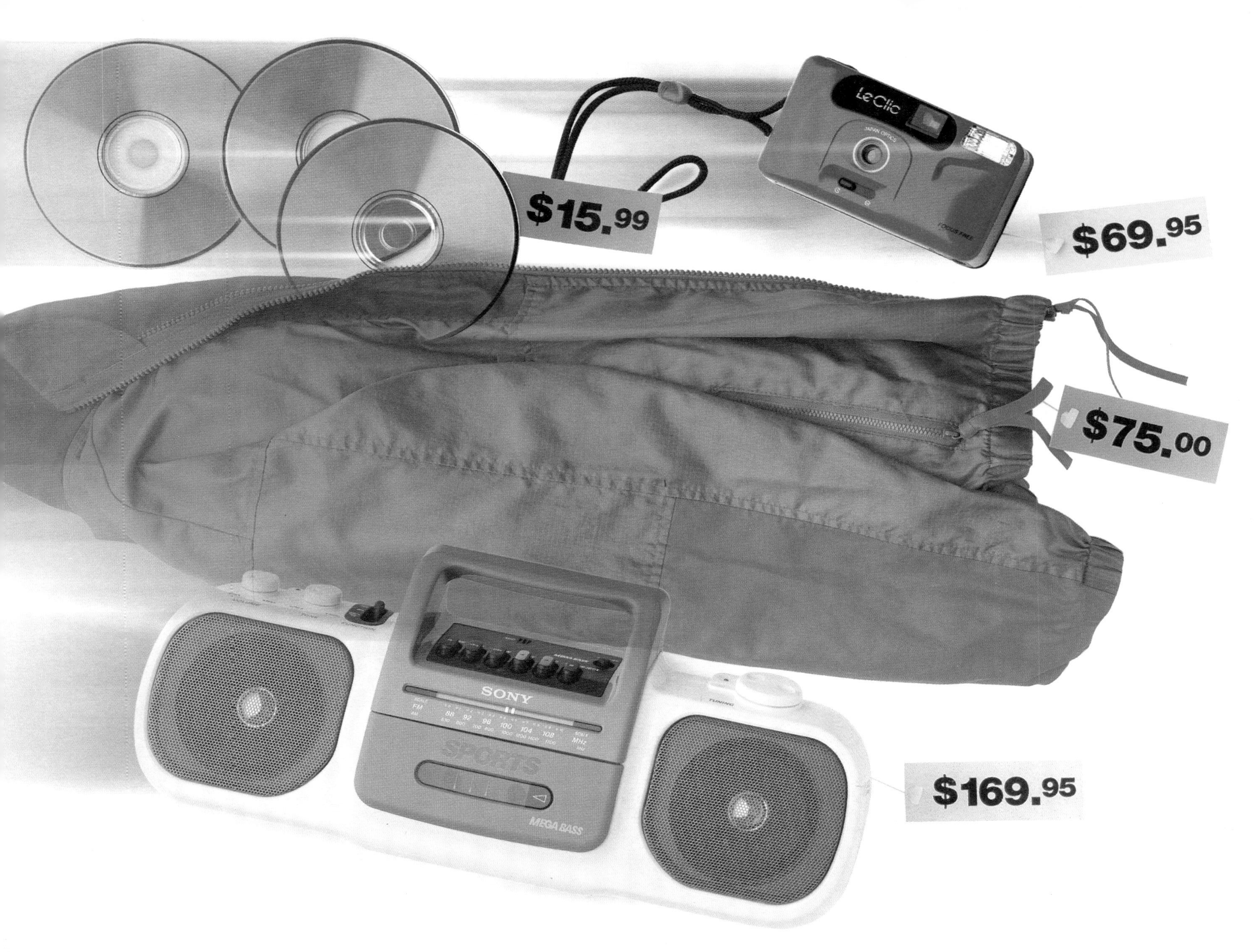
$15.99
Le Clic
JAPAN OPTICS
FOCUS FREE
$69.95
$75.00
SONY
TUNING
SPORTS
MEGA BASS
$169.95

5-1 MONTHLY PAYMENTS: HOW MUCH CAN YOU AFFORD TO BORROW?

QUICK REFERENCE

TEACHER'S RESOURCES AND ANSWER KEY

Lesson Quiz Answers 5-1, p. 52

Reteaching Activity Answers 5-1, p. 98

Enrichment Activity Answers 5-1, p. 99

EXTENSION ACTIVITIES

Reteaching Activity 5-1, p. 31

Enrichment Activity 5-1, p. 36

Patrick is impressed by Maria's ability to save so much money from her weekly paycheck. He has a job at Paradise Department Store, where he is a cashier in the camera/luggage department. He will graduate from high school this year and would like to attend a community college for at least a year or two.

Patrick and his father have been trying to save money for the state university tuition, but it has been very difficult since Patrick's mother died three years ago. They lost her income, and her illness used up a large portion of their savings. Therefore if Patrick wants to go to college, he will have to help pay the expenses.

Before making a loan, a lending institution must have faith in the borrower's ability to repay it at a future date. Patrick knows that Maria's father is helping her to borrow money to buy a car. Maria will be making monthly payments to the bank to pay back the loan.

Patrick hopes that he too can borrow from the bank for school tuition. He can afford only a certain amount in monthly payments, and he would like to know how much he will be able to borrow.

Maria told Patrick that he will have to pay back more than he borrows over the course of his loan. He is not sure that this is fair, but he wants to go to college, and he knows that the bank will need to make a fair profit as a condition of granting him a loan.

OBJECTIVES: *In this lesson, we will help Patrick to:*

- *Identify the major functions of credit.*
- *Recognize the features of installment credit.*
- *Calculate monthly payments on a loan.*
- *Compute total payments on a loan.*
- *Determine how much he can afford to borrow.*

NATURE OF CREDIT

Patrick has heard the phrase "Buy now, pay later" in commercial advertising. The use of credit has become a way of life for many consumers, and Patrick thinks that he can handle it too. His economics teacher mentioned that the word *credit* actually means *debt.* People do not like the sound of being in debt, so they use the word *credit* instead.

Credit is a form of debt that occurs whenever cash, goods, or services are provided in exchange for a promise to pay at a future date. Many people cannot afford to pay cash for a large purchase such as a house or an automobile. Therefore they borrow the amount they need for the purchase and pay back that amount plus interest. The total amount paid to the bank is known as the **deferred payment price.** The deferred payment price is considerably more than the original loan amount. The difference between the two amounts is the *interest* or the **finance charge.** It is the profit made by the bank or lending institution. It is also the price paid by the borrower for using someone else's money.

FOCUS

ALGEBRA CONNECTIONS

Would you prefer to make a car payment of $250 per month for 3 years or $300 per month for 2 years? (**$300 per month for 2 years costs $1800 less.**)

FUNCTIONS OF CREDIT

Patrick has discovered from doing research that the use of credit has the following positive effects:

1. *Credit stabilizes the economy.* It steadies economic activity because it enables individuals and businesses to purchase goods and services even when their incomes are temporarily limited.
2. *Credit promotes business growth.* Many people borrow money to start new businesses or to maintain established ones.

3. *Credit expands productivity and production.* By borrowing money to attend college, Patrick may increase his earning potential. Similarly, a business that must sell what it produces will have no income until its products are sold. The initial costs of production must be financed by funds that are already in the business or by borrowed funds.
4. *Credit raises the standard of living.* Individuals or couples do not have to wait to build up savings before buying things that make life comfortable, such as a house, a car, furniture, and appliances.

USING INSTALLMENT CREDIT

Using installment credit means making purchases and then making regular payments over a period of months or years. Maria will be getting an **installment loan** when she buys a car. Purchases made on an installment plan often have the following features:

1. A **down payment** is required at the time of the purchase. A down payment is the portion of the purchase that must be paid up front in cash.
2. A substantial finance charge is added to the price.
3. Payments of equal amounts are spread over a specified period of time.
4. To insure that payments on a purchase loan are made as scheduled, protection may be provided to the lender in the form of a security agreement.

Ask Yourself

1. What are the four major functions of credit?
   See Additional Answers.
2. What are three features of installment credit?
   downpayment, finance charge, and security agreement
3. In what must a bank have faith when it approves a loan?

ADDITIONAL ANSWERS

1. Stabilizes the economy; promotes business growth; expands productivity and production; and raises the standard of living
3. A bank must have faith in the borrower's ability to repay the loan at a future date.

ALGEBRA REVIEW

Evaluate. Round answers to the nearest hundredth.

1. $x = \frac{2^3}{3^2}$
   0.89
2. $x = \frac{6^4}{3^3}$
   48
3. $x = 4(3 + 2)^3$
   500
4. $x = \frac{5}{(5 + 3)^2}$
   0.08
5. $x = (1 + 0.01)^3$
   1.03
6. $x = \frac{(1 + 0.02)^6}{0.02}$
   56.31
7. $x = \frac{1200(2)^5}{(5 - 3)^3}$
   4800
8. $x = \frac{600(1 + 5)^3}{(6 - 3)^4}$
   1600
9. $x = \frac{98(0.005)(1 + 0.005)^{24}}{(1 + 0.005)^{24} - 1}$
   4.34
10. $x = \frac{900(0.0025)(1 + 0.0025)^{36}}{(1 + 0.0025)^{36} - 1}$
    26.17

Solve for x.

11. $y = x(1 - r)^2$ $x = \frac{y}{(1 - r)^2}$
12. $y = \frac{2x(1 + r)}{r^2 z}$ $x = \frac{r^2 yz}{2(1 + r)}$
13. $z = \frac{3rq^{24}}{x(q + 2)^{24}}$ $x = \frac{3rq^{24}}{z(q + 2)^{24}}$

SHARPEN YOUR SKILLS

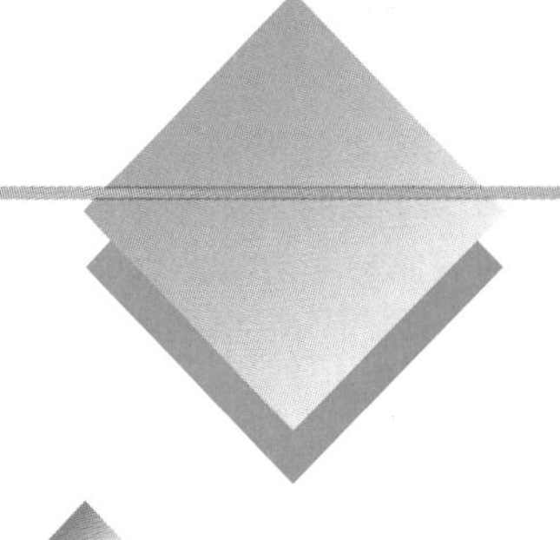

SKILL 1

Maria can use the monthly payment formula to determine the monthly cost of a car loan. The monthly interest rate is the annual interest rate divided by 12.

Monthly Payment Formula

$$M = \left(\frac{Pr(1+r)^n}{(1+r)^n - 1}\right)$$

where M = monthly payment
P = amount of loan
r = monthly interest rate
n = number of payment periods

EXAMPLE 1 Maria finds a used car that she might want to buy. To pay for the car, she must borrow \$6500.

QUESTION What is her monthly payment if she borrows \$6500 for a 3-year period at an annual interest rate of 9%? How much will she pay for the car in the course of 3 years?

SOLUTION

Use the monthly payment formula and your calculator.

$$M = \frac{Pr(1+r)^n}{(1+r)^n - 1}$$

$$M = \frac{6500(0.0075)(1+0.0075)^{36}}{(1+0.0075)^{36} - 1}$$ $P = 6500$; $r = 0.09 \div 12 = 0.0075$; $n = 3 \bullet 12 = 36$

$M = 206.70$ To the nearest cent

To input this calculation in your calculator, remember to use the exponent key [x^y] (sometimes labeled ^ or a^b) and to use parentheses around the denominator as follows:

6500 [×] 0.0075 [×] [(] 1 [+] 0.0075 [)] [x^y] 36 [÷]
[(] [(] 1 [+] 0.0075 [)] [x^y] 36 [−] 1 [)] [ENTER]

Maria's monthly payment is \$206.70.

The answer to the above calculation is given to many decimal places in your calculator. Use that answer to calculate the deferred payment price by using the answer key [ANS] on your calculator.

[ANS] [×] 36 = 7441.14 Deferred payment price

Notice that $206.70 \bullet 36 = 7441.20$. The bank adjusts your last payment so that the deferred payment price is \$7441.14.

INSTRUCTION

TEACHING THE LESSON

Have students read the opening scenario. Use the *Algebra Connections* on p. 187 to demonstrate that selecting a lower monthly payment is not necessarily the best option. Have students read the sections on the nature, functions, and use of installment credit. Make sure that students understand that the word credit actually means debt.

After completing Skills 1 and 2, have students examine the role of *r* in the monthly payment formula.

Have students compare the monthly payments and deferred payment prices for a \$10,000 loan for 3 years at 8%, 9%, and 10%.

(**8%: \$313.36, \$11,281.09; 9%: \$318.00, \$11,447.90; 10%: \$322.67, \$11,616.19**)

CRITICAL THINKING

After the students have answered the questions, ask them to consider the impact that the use of credit has had on their individual families. Ask students to research the advantages and disadvantages of buying a used car.

FOCUS ON ALGEBRA

In this lesson, the monthly payment formula allows a student to determine whether or not he/she can afford a car or other major purchase that will influence his/her daily life. (*NCTM Standard 3, p. 143*)

EXAMPLE 2 If Maria buys a new car instead, she will have to borrow $12,825.

QUESTION What is her monthly payment if she borrows $12,825 for a 4-year period at an annual interest rate of 6%? How much will she pay for the car in the course of 4 years?

SOLUTION

Use the monthly payment formula and your calculator.

$$M = \frac{Pr(1 + r)^n}{(1 + r)^n - 1}$$

$$M = \frac{12{,}825(0.005)(1 + 0.005)^{48}}{(1 + 0.005)^{48} - 1}$$ $P = 12{,}825$; $r = 0.06 \div 12 = 0.005$; $n = 4 \bullet 12 = 48$

$$M = 301.20$$ To the nearest cent

Maria's monthly payment is $301.20. Over four years she will pay [ANS] [×] 48 = $14,457.38, which is the deferred payment price or total payment.

COMMON ERRORS

Students may erroneously insert the yearly interest rate instead of the monthly interest rate in the monthly payment formula.

Be sure that students understand the placement of parentheses around the denominator when using a calculator to determine the monthly payment.

ESL STUDENTS

Non-native speakers may have difficulty understanding that the word credit actually means debt.

EXAMPLE 3 Maria is also considering five other cars on which she can obtain 8% financing for 3 years, 4 years, or 5 years.

QUESTION What are her monthly and total payments (deferred payment price) on loans of $4200, $5500, $6325, $8275, and $9750?

SOLUTION

Use a spreadsheet program to find the monthly and total payments for the loan amounts at 8% over the periods of 3 years, 4 years, and 5 years. When entering the formula for the monthly payment, use 0.08/12 for the rate. For 3 years at 8% the monthly payment formula in cell C3 is

```
+A3*(0.08/12)*(1+(0.08/12))^36/((1+(0.08/12))^36-1)
```

For the monthly payment formula for 4 years, use 48 for the number of payment periods. You can use the COPY command of your spreadsheet program to copy the formula of cell C3 to C4. This will automatically change A3 to A4. Then you can use the EDIT command to change 36 to 48.

+A4*(0.08/12)*(1+(0.08/12))^48/((1+(0.08/12))^48−1)

The monthly payment formula for 5 years is

+A5*(0.08/12)*(1+(0.08/12))^60/((1+(0.08/12))^60−1)

For the total payment, type 12*B3*C3 in cell D3. Then use the COPY command to copy the formula in the total payment column.

	A	B	C	D
1	Loan	Number	Monthly	Total
2	Amount	of Years	Payment	Payment
3	4200	3	131.61	4,738.06 ← 12*B3*C3
4	4200	4	102.53	4,921.65 ← 12*B4*C4
5	4200	5	85.16	5,109.65
6	5500	3	172.35	6,204.60
7	5500	4	134.27	6,445.01
8	5500	5	111.52	6,691.21
9	6325	3	198.20	7,135.29
10	6325	4	154.41	7,411.76
11	6325	5	128.25	7,694.89
12	8275	3	259.31	9,335.10
13	8275	4	202.02	9,696.81
14	8275	5	167.79	10,067.23
15	9750	3	305.53	10,999.06
16	9750	4	238.03	11,425.25
17	9750	5	197.69	11,861.69

If you use (12*B3) instead of 36 in cell C3, you can use the COPY command to complete the entire monthly payment column. The formula is

+A3*(0.08/12)*(1+(0.08/12))^(12*B3)/((1+(0.08/12))^(12*B3)−1)

SKILL 2

EXAMPLE 4 Patrick is an excellent student and will probably receive a scholarship for college. However, his father explains that Patrick will still have to borrow money.

QUESTION If Patrick can afford monthly payments of $225 and would like to borrow an even multiple of $1000, what is the largest amount of money that he can borrow at 8% for 3 years, 4 years, and 5 years?

TECHNOLOGY HINT

Explain that students must substitute A6-A8, A9-A11, A12-A14, and A15-A17 into the formulas presented in the solution to Example 3 to obtain the monthly payments for $5500, $6325, $8275, and $9750, respectively. An easy way to repeat the calculations is to copy Rows 3, 4, and 5 to Rows 6, 7, and 8 and then substitute $5500 for $4200 in A6-A8. The spreadsheet program will then automatically compute the values in columns C and D for each principal amount and number of years. The process can be repeated for $6325, $8275, and $9750.

The spreadsheet formula in the text sets forth in detail the actual monthly payment formula. Most spreadsheet programs have a function which computes monthly payments. Students should not use this function until they understand the monthly payment formula itself.

Save this spreadsheet to use in Lesson 5-2.

AT-RISK STUDENTS

Generate interest by having students gather information on several cars that they would like to buy. Have them work in groups to compute the monthly payments if they borrow the purchase prices of the cars at 8% for 3 years.

SOLUTION

Use a spreadsheet program to determine the monthly payments for $7000 to $12,000 in increments of $1000. The monthly payment formulas are

```
+A3*(0.08/12)*(1+(0.08/12))^36/((1+(0.08/12))^36−1)    For 3 years
+A3*(0.08/12)*(1+(0.08/12))^48/((1+(0.08/12))^48−1)    For 4 years
+A3*(0.08/12)*(1+(0.08/12))^60/((1+(0.08/12))^60−1)    For 5 years
```

	A	B	C	D
1	Amount	3 years	4 years	5 years
2	Borrowed			
3	7,000	219.35	170.89	141.93
4	8,000	250.69	195.30	162.21
5	9,000	282.03	219.72	182.49
6	10,000	313.36	244.13	202.76
7	11,000	344.70	268.54	223.04
8	12,000	376.04	292.96	243.32

Patrick can afford to borrow $7000 for 3 years, $9000 for 4 years, or $11,000 for 5 years.

Reviewing order of operations and converting annual rates of interest to monthly rates of interest may be beneficial to students. To reteach the lesson, concentrate on these skills.

Challenge students to program their graphing calculators or computers to calculate the monthly payment for a given loan amount, interest rate, and payment period.

SKILL 3

Patrick solves the monthly payment formula for P, the amount of the loan.

$$M = \frac{Pr(1+r)^n}{(1+r)^n - 1}$$

$$M[(1+r)^n - 1] = Pr(1+r)^n \quad \text{Multiply both sides by } [(1+r)^n - 1].$$

$$\frac{M[(1+r)^n - 1]}{r(1+r)^n} = P \quad \text{Divide both sides by } r(1+r)^n.$$

He now has a formula to find the exact amount he can borrow.

Amount Formula

$$P = \frac{M[(1+r)^n - 1]}{r(1+r)^n}$$

where P = amount of loan
r = monthly interest rate
n = number of payment periods
M = monthly payment

EXAMPLE 5 If Patrick does not restrict himself to borrowing multiples of $1000, he can borrow more money over each time interval.

QUESTION What is the exact amount of money Patrick can borrow at 8% for 3 years if he can afford a monthly payment of $225?

SOLUTION

Use the amount formula. Since the monthly interest rate r is $0.08 \div 12$, which is a repeating decimal, calculate this value, and then use the [ANS] key to enter it in the formula.

$$P = \frac{225((1 + \text{ANS})^{36} - 1)}{\text{ANS}(1 + \text{ANS})^{36}}$$ $M = 225$; $n = 3 \cdot 12 = 36$

$P = 7180.16$ To the nearest cent

To input this in your calculator use the following keystrokes.

0.08 [÷] 12 [ENTER]
225 [(] [(] 1 [+] [ANS] [)] [x^y] 36 [−] 1 [)] [÷] [ANS] [(] 1 [+] [ANS] [)] [x^y] 36 [ENTER]

Patrick can afford to borrow $7180.16 at 8% for 3 years.

The [ANS] key is a temporary memory that stores the latest calculation. You can also store $0.08 \div 12$ in one of the regular memories of your calculator.

0.08 ÷ 12 [STO →] [ALPHA] [A] [ENTER]

To recall the number stored in memory A, you would enter [ALPHA] [A] in place of [ANS] in the previous calculation.

TRY YOUR SKILLS

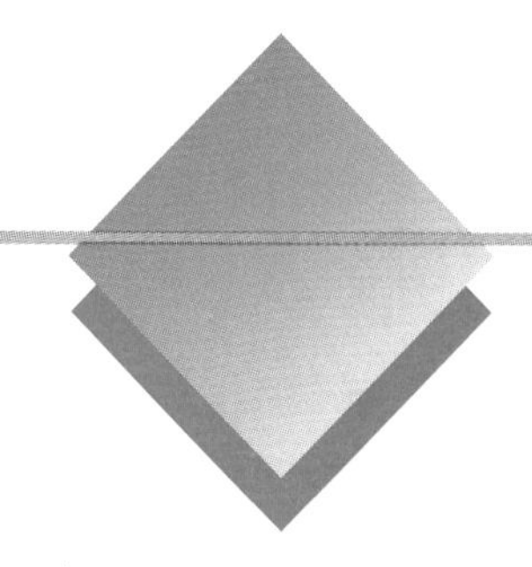

Determine the monthly payment and total payments (deferred payment price) for each car loan. Remember that the deferred payment price is [ANS] times the number of payments.

1. $10,000 at 7% for 5 years $198.01; $11,880.72
2. $12,000 at 8% for 4 years $292.96; $14,061.84

Paola can afford a monthly payment of $150 and would like to borrow an even multiple of $1000. Complete the table below to determine the largest amount of money that she can borrow at 10% for 3 years, 4 years, and 5 years.

	Amount Borrowed	3 years	4 years	5 years
3.	$3000	$ 96.80	$ 76.09	$ 63.74
4.	4000	129.07	101.45	84.99
5.	5000	161.34	126.81	106.24
6.	6000	193.60	152.18	127.48
7.	7000	225.87	177.54	148.73
8.	8000	258.14	202.90	169.98

9. Use the amount formula to determine the exact amount of money that Paola can borrow for each time period in Exercise 3. Remember that Paola can borrow (or afford) a payment of $150 a month. $4648.69; $5914.22; $7059.81

GUIDED PRACTICE

You may wish to have students work in groups to calculate the monthly payments and the deferred payment prices in Exercises 1-9. Observe each group as they decide which numbers to use in the monthly payment formula and provide guidance on how to calculate r and n.

ADDITIONAL ANSWERS

3.-8. Paola can borrow $4000 for 3 years; $5000 for 4 years; or $7000 for 5 years.

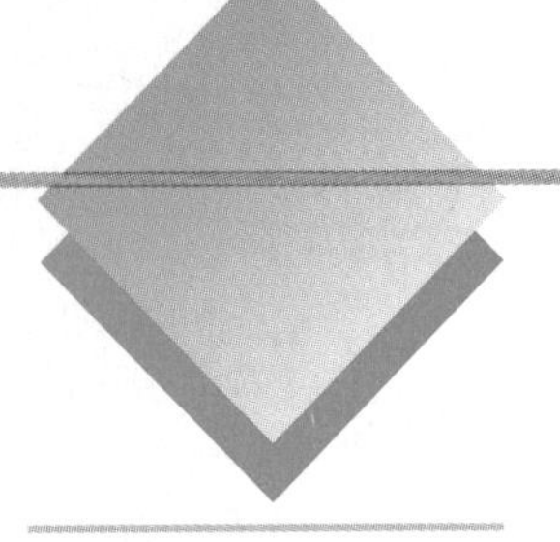

EXERCISE YOUR SKILLS

1. Why is the term *credit* used instead of *debt*? See Additional Answers.
2. How does the use of credit raise the standard of living?
3. Why do you think consumers are willing to pay substantial finance charges as a condition of an installment loan?
4. Who receives the revenue from finance charges paid by consumers?
5. What do you think the bank or other lending institution would do if a borrower did not make the payments on an installment loan?
6. What is the deferred payment price?

Determine the monthly payment and total payment (deferred payment price) for each car loan.

7. $13,500 at 7% for 4 years
8. $10,500 at 9% for 5 years

Patrick's friend Ricky is also considering various loans for college tuition. Use a spreadsheet program to find the monthly payment and total payment for the loan amounts listed below at 12.5% over the periods of 3 years, 4 years, and 5 years. (Save your results to use in Lesson 5–2.)

	Loan Amount	Number of Years	Monthly Payment	Total Payment
9.	$ 6,500	3	$217.45	$ 7,828.15
10.	6,500	4	172.77	8,292.96
11.	6,500	5	146.24	8,774.20
12.	9,500	3	317.81	11,441.14
13.	9,500	4	252.51	12,120.48
14.	9,500	5	213.73	12,823.82
15.	6,000	3	200.72	7,225.98
16.	6,000	4	159.48	7,655.04
17.	6,000	5	134.99	8,099.26
18.	12,000	3	401.44	14,451.97
19.	12,000	4	318.96	15,310.08
20.	12,000	5	269.98	16,198.52
21.	17,500	3	585.44	21,075.78
22.	17,500	4	465.15	22,327.20
23.	17,500	5	393.71	23,622.84
24.	11,650	3	389.73	14,030.45
25.	11,650	4	309.66	14,863.54
26.	11,650	5	262.10	15,726.06
27.	8,775	3	293.56	10,568.00
28.	8,775	4	233.24	11,195.50
29.	8,775	5	197.42	11,845.16

Cost =

KEY TERMS

credit
deferred payment price
down payment
finance charge
installment loan

INDEPENDENT PRACTICE

ASSIGNMENT GUIDE

Exercises 1-6 ask students questions about various aspects of credit. Exercises 7-37 assess students' ability to apply the formulas taught in the lesson.

ADDITIONAL ANSWERS

1. People do not like the sound of being "in debt."
2. People do not have to wait to build up savings before buying things that make life comfortable.
3. The loan enables someone to acquire a home or a car long before he or she could save the full amount needed to purchase it.
4. the lending institution
5. repossess the item(s) bought with the credit
6. It is the total amount paid to the bank; the loan amount plus the interest.
7. $323.27; $15,517.17
8. $217.96; $13,077.76

Kareem can afford a monthly payment of $300 and would like to borrow a multiple of $1000. Complete the table below to determine the largest amount of money that he can borrow at 9% for 3 years, 4 years, and 5 years.

	Amount Borrowed	3 years	4 years	5 years
30.	$ 9,000	$286.20	$223.97	$186.83
31.	10,000	318.00	248.85	207.58
32.	11,000	349.80	273.74	228.34
33.	12,000	381.60	298.62	249.10
34.	13,000	413.40	323.51	269.86
35.	14,000	445.20	348.39	290.62
36.	15,000	477.00	373.28	311.38

37. Determine the exact amount of money that Kareem can borrow for each time period in Exercises 30–36.

MIXED REVIEW

2-1 **1.** Matthew's monthly check costs are $0.025 per check for the first 15 checks and $0.12 for each check over 15. If he writes 33 checks this month, how much will the checks cost him? $2.54

3-1 **2.** Crystal saves $12.50 per week. At the end of a year, she deposits her savings into a CD that earns 2% for each six-month period. When the CD matures, she withdraws the interest and reinvests the principal. How much interest will she have earned at the end of two years? $52

3-2 **3.** How much will $6200 be worth at the end of 4 years if it earns 7% interest, compounded semiannually? $8164.22

1-2 **4.** Olivia earns $9.50 per hour. She works $37\frac{1}{2}$ hours per week for 49 weeks and receives 3 weeks of paid vacation. Her other benefits cost her employer $4000 per year. What is the annual cost to the company of her salary and benefits? $22,525

3-1 **5.** Arnold saves $33 per week. He wants to buy a desk that costs $294. How many weeks must he save before he can buy the desk? 9 wk

4-2 **6.** Alan and Meg are producing and selling sports pennants. The raw materials for each pennant cost $2.13 and the labor is $5.50 an hour. Advertising and other fixed costs are $45. What is the cost of production for 250 pennants, 20 of which can be produced in one hour? $646.25

4-4 Suppose that the fixed costs of a business are $350 per week and the variable costs are $6.50 per item.

7. What is the cost of manufacturing 92 items in one week? $948

8. Each item is sold for $10.99. What is the weekly profit? $63.08

9. What is the break-even point? 78 items

ADDITIONAL ANSWERS

30.-36. Kareem can borrow $9000 for 3 years; $12,000 for 4 years or $14,000 for 5 years.

37. $9434.04 for 3 years; $12,055.43 for 4 years; $14,452.01 for 5 years

CLOSURE

Ask students to define each of the variables in the monthly payment formula, and to give you the value of *r* for various annual rates of interest. Ask students to explain how to calculate the deferred payment price if they know the monthly payment. Have students discuss whether Patrick should restrict himself to borrowing an integral multiple of $1000.

LESSON QUIZ

1. Compute the monthly payment and deferred payment price on a loan of $8700 for 4 years at 9%. Then repeat the exercise for a loan of the same amount of money for 5 years at 8%. (**$216.50, $10,391.99; $176.40, $10,584.28**)

2. How much money can you afford to borrow at 9% for 3 years if you can afford a monthly payment of $235? (**$7390.00**)

5-2 PROBLEMS WITH CREDIT: CREDIT OVERLOAD

QUICK REFERENCE

TEACHER'S RESOURCES AND ANSWER KEY

Lesson Quiz Answers 5-2, p. 54

Reteaching Activity Answers 5-2, p. 98

Enrichment Activity Answers 5-2, p. 100

EXTENSION ACTIVITIES

Reteaching Activity 5-2, p. 32

Enrichment Activity 5-2, p. 37

Joan feels that she must warn Patrick and Maria about credit. The first installment loan that her family accepted seemed quite reasonable because it was a mortgage for their house. Then they bought a brand new car on credit.

After a series of violent thunderstorms, the roof in the den started to leak, so the family took out a second mortgage to pay for a new roof and to replace the damaged carpeting. They will be paying on the second mortgage for 7 years.

When Joan's mom and dad were both working, they were easily making all of the installment payments. Four months ago, Joan's mother lost her job. The electronics company at which she was working had been forced to eliminate some jobs in her department because the company lost money last year.

More bills arrive every day. Joan's parents even borrowed money from the credit union to pay some of the bills, but now they are having trouble making the payments to the credit union. The family members are a little stunned by recent events and are wondering how they managed to get into such a mess so quickly.

Joan's mother feels that one of their problems is that they borrowed the money for short time periods and have high monthly payments. She thinks that they should have taken a second mortgage with a payback period of 15 years instead of 7 years. Her father gets upset because he claims that the total payments and total cost are higher when the term of the loan is longer. Joan will be glad when Mom finds another job. Then Mom and Dad will stop being upset, and the family can get back to normal.

OBJECTIVES: *In this lesson, we will help Joan to:*

- *Cite several problems with using credit.*
- *Recognize indications that a family has used too much credit.*
- *Compare the effects of the payback period of a loan on the total payments and interest charges.*

CREDIT: TOOL OR TRAP?

Many Americans borrow money from a bank or other lending institution to obtain the use of merchandise before paying for it. Americans are heavy credit users. Almost everyone uses credit, especially for expensive purchases such as a new car or home repairs. What went wrong for Joan's parents, and how can credit create a problem for millions of others like them? You should always keep the following points about credit in mind.

1. *Credit is rented money.* Borrowers pay heavily for the privilege of using credit. Banks and other lending institutions encourage borrowers to indulge in this privilege because the credit that is granted is very profitable for the lenders.
2. *Credit ties up future income.* You are actually spending future income or earnings when you make credit purchases. If you tie up too much future income in making credit payments, you are putting a burden on yourself. If your income happens to disappear as Joan's mother's did, making the payments may become impossible.
3. *Credit makes it easy to overspend.* Businesses with products or services to sell are cashing in on consumers' willingness to use credit for a purchase that they would not consider buying if they had to pay cash. A furniture store owner can sell more bedroom suites by offering "Convenient credit terms—No payments for 6 months."

ALGEBRA CONNECTIONS

If the interest rate on your loan is 9% and you make monthly payments of approximately $250 for 3 years, you can purchase an $8000 car. If you make approximately the same monthly payments for 5 years, you can purchase a $12,000 car. Which car would you select? (**Answers may vary.**)

MANAGING CREDIT

People, businesses, and even governments can be forced into bankruptcy through the unwise use of credit. Excessive debt can undermine job performance, marriage, and health. Installment buying can be helpful if you use it wisely. However, credit can be harmful if you abuse it.

One convenient yardstick for measuring your ability to handle debt is the following: No matter what your income, 10% of take-home pay, excluding a home mortgage, is a comfortable amount to spend making payments on credit

ALGEBRA *REVIEW*

Evaluate. Store the results in your calculator's memory. Give the results to the nearest hundredth.

1. $x = \dfrac{6700(0.002)(1 + 0.002)^{40}}{(1 + 0.002)^{40} - 1}$
   174.46
2. $y = \dfrac{3300(0.0033)(1 + 0.0033)^{30}}{(1 + 0.0033)^{30} - 1}$
   115.72

Use the results of Exercises 1 and 2 to determine z and q.

3. $z = 40x$
   $z = 6978.40$
4. $q = 30y$
   $q = 3471.60$

Use the results of Exercises 3 and 4 to determine s and t.

5. $s = z - 6700$
   $s = 278.40$
6. $t = q - 3300$
   $t = 171.60$

Evaluate. Store the results in your calculator's memory. Give the results to the nearest hundredth.

7. $x = \dfrac{1800(0.004)(1 + 0.004)^{44}}{(1 + 0.004)^{44} - 1}$
   $x = 44.70$
8. $y = \dfrac{6600(0.0066)(1 + 0.0066)^{44}}{(1 + 0.0066)^{44} - 1}$
   $y = 173.32$
9. Use the stored results of Exercises 7 and 8 to solve for z if $z = 44x - 44y$.
   $z = -5659.61$

Determine which expression is smaller.

10. 36(424.70) or 60(247.64)
    60(247.64) is smaller.
11. 48(359.28) or 36(482.12)
    48(359.28) is smaller.

accounts or installment loans, 15% is a manageable amount, and more than 20% is a dangerous credit overload. Indications that you are suffering from credit overload include the following:

- You must miss some installment payments to make the monthly mortgage payment.
- You seek a new loan before repaying an old one.
- You continue to use your credit cards, even though you can pay only the minimum amount due on your credit card accounts.
- You take out loans to combine debts or ask for extensions on existing loans.
- You receive repeated overdue notices from creditors.
- You have little or no savings or are drawing on savings to pay regular bills that you used to pay out of monthly income.
- You receive telephone calls or letters from creditors demanding payment on overdue bills.

Ask Yourself

1. What are three problems with using credit?
   See Additional Answers.
2. What are four indicators that a borrower is experiencing credit overload?
3. Why is it easy to borrow too much money?
   It is a deceptively easy way to get what you want.

ADDITIONAL ANSWERS

1. Borrowers pay heavily for the privilege of using credit; credit ties up future income; credit makes it easy to overspend.
2. missed installment payments; can only make minimum payments; receive repeated overdue notices; little or no savings

SHARPEN YOUR SKILLS

SKILL 1

Maria is learning more about borrowing money and paying it back in monthly amounts. She has discovered that bankers and loan agencies give a name to this "payback" procedure. To make payments on a loan is to **amortize** the loan. The word amortize is derived from the French *à mort* which means "to death." When Maria amortizes her loan, she is gradually "bringing her loan to death."

The cost of a loan is equal to the amount of interest paid over the term of the loan. It is found by subtracting the original loan amount from the total payment.

Total cost = total payment − original loan amount

EXAMPLE 1 In Example 3 of Lesson 5–1, Maria considered loans of $4200, $5500, $6325, $8275, and $9750 at 8%. As the term of a loan gets longer, her monthly payments get smaller, and her total payment gets larger. She wonders how long she should amortize her loan.

QUESTION Does the cost of financing a loan increase or decrease as the term of the loan increases?

SOLUTION
Using a spreadsheet program, add a column to the table in Example 3 of Lesson 5–1 to show the total cost (interest payments) of each loan.

	A	B	C	D	E
1	Loan	Number	Monthly	Total	Total
2	Amount	of Years	Payment	Payment	Cost
3	4200	3	131.61	4,738.06	538.06
4	4200	4	102.53	4,921.65	721.65
5	4200	5	85.16	5,109.65	909.65
6	5500	3	172.35	6,204.60	704.60
7	5500	4	134.27	6,445.01	945.01
8	5500	5	111.52	6,691.21	1191.21
9	6325	3	198.20	7,135.29	810.29
10	6325	4	154.41	7,411.76	1086.76
11	6325	5	128.25	7,694.89	1369.89
12	8275	3	259.31	9,335.10	1060.10
13	8275	4	202.02	9,696.81	1421.81
14	8275	5	167.79	10,067.23	1792.23
15	9750	3	305.53	10,999.06	1249.06
16	9750	4	238.03	11,425.25	1675.25
17	9750	5	197.69	11,861.69	2111.69

+D3−A3

INSTRUCTION

TEACHING THE LESSON

Have students read the opening scenario. Use the *Algebra Connections* on p. 197 to demonstrate that the interest cost of the 5-year loan is greater than the interest cost of the 3-year loan, even though the monthly payments are approximately the same. Have students read the sections on credit and managing credit. Make sure that students understand that buying on credit can be very helpful if it is done wisely, but it can lead to disaster if done unwisely.

After completing Skill 1, have students compare the total cost of borrowing $10,000 at 12% for 3 years, 4 years, and 5 years. (**$1957.15, $2640.24, $3346.67**)

CRITICAL THINKING

After students have answered the questions, ask them to consider how a store can make up the interest revenue that it loses when it allows its customers to make "no payments for 6 months." Ask students to research the interest rates that are currently available on car loans.

FOCUS ON ALGEBRA

In this lesson, students learn to compute the cost of borrowing money. They learn that it costs more to finance a given amount of money at a given interest rate as the amortization period increases. They can then make decisions that affect their daily lives. (*NCTM Standard 1, p. 137*)

TECHNOLOGY HINT

The most appropriate graph to show the relationship between amortization periods and interest payments is a bar graph, with the principal amount of each loan and the amortization periods on the *x*-axis and the total cost on the *y*-axis. If you block the total cost column only, the spreadsheet program will automatically group the three bar graphs (representing amortization periods of 3, 4, and 5 years) for each loan amount.

COMMON ERRORS

Be sure that students understand the difference between total payment and total cost. Explain that the total payments are equal to the deferred payment price, and that total cost is equal to the amount of interest paid over the term of the loan.

Use the graph function of your spreadsheet program to graph the relationship between amortization periods of 3 years, 4 years, and 5 years and interest payments.

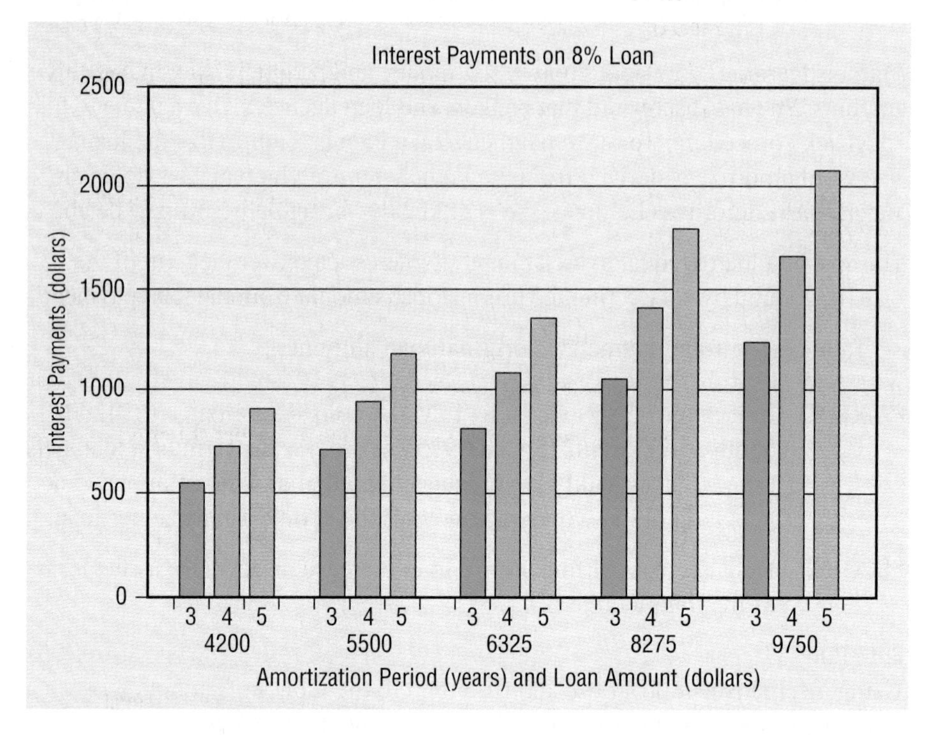

It costs more to finance a specified loan amount at a given interest rate as the amortization period increases.

It is possible to borrow a larger sum if you make the same monthly payments over a longer period of time. Maria must decide whether she prefers a lower monthly payment or a lower total cost for a given loan amount.

SKILL 2

EXAMPLE 2 Patrick points out that Maria can buy a car priced at $9750 or a car priced at $6325 for a monthly payment of approximately $198.

QUESTION Should Maria automatically choose the more expensive car without further investigation, since it appears to cost the same amount of money on a monthly basis? What other factor(s) should she take into account?

SOLUTION

No, she must also consider the total cost of each loan. Since she will be paying for the more expensive car for 5 years instead of 3 years, the total cost of the 5-year loan will be greater—$2111.69 instead of $810.29 for the smaller loan.

EXAMPLE 3 Joan wants to settle the dispute between her parents. Their second mortgage on the house was a loan for \$40,000 at 12% for 7 years.

QUESTION Would an amortization period of 15 years, rather than 7 years, have been better for Joan's parents?

SOLUTION

Use your spreadsheet program to create a table showing monthly payments, total payments, and total cost for amortization periods of 7 through 15 years. The monthly rate is 0.12/12 = 0.01, and the loan amount is \$40,000. The number of periods for 7 years is 12 • 7 = 84, so the spreadsheet formula for the monthly payment formula for 7 years for cell B6 is

40000*(0.01)*(1+0.01)^84/((1+0.01)^84−1)

For the monthly payment formulas for the other years, adjust the number of payments.

	A	B	C	D
1		Amortization Schedule—7 to 15 years		
2		\$40,000 loan at 12% interest		
3				
4	Amortization	Monthly	Total	Total
5	Period (years)	Payment	Payment	Cost
6	7	706.11	59,313.18	19,313.18
7	8	650.11	62,410.91	22,410.91
8	9	607.37	65,595.88	25,595.88
9	10	573.88	68,866.06	28,866.06
10	11	547.12	72,219.20	32,219.20
11	12	525.37	75,652.94	35,652.94
12	13	507.47	79,164.77	39,164.77
13	14	492.57	82,752.06	42,752.06
14	15	480.07	86,412.10	46,412.10

+C6−40000

+B6*(12*A6)

In Lesson 5–1, Example 3, you were shown how to modify a spreadsheet formula by replacing a number by a general formula, thus reducing the amount of work required to complete a long table. The same technique can be used for the spreadsheet amortization table above. Instead of 84 in the formula for cell B6, you can write the formula (12*A6). Then you can use the spreadsheet's COPY command to complete all of the Monthly Payment column, column B. For the total payment, use the formula +B6*(12*A6) in cell C6 instead of +B6*84.

ESL STUDENTS

Non-native speakers may have difficulty understanding the difference between the terms "total payment" and "total cost."

Use the graph function of your computer spreadsheet program to graph the relationship between amortization periods of 7–15 years and monthly payments. This graph is shown at the top of the next page.

INTERDISCIPLINARY INVESTIGATION

Political Science: Have students research the city of Bridgeport, Connecticut. How did the city end up in bankruptcy? How does the city continue to function?

COOPERATIVE LEARNING

Have students work together in small groups to use the graph function of their spreadsheet program to graph the relationship between amortization periods of 3 years, 4 years, and 5 years and total cost in Exercises 1-12 in the *Try Your Skills* section of this lesson.

RETEACHING

Some students may benefit from a review of creating and interpreting bar graphs. To reteach the lesson, concentrate on this skill.

ENRICHMENT

Challenge students to program their graphing calculators or computers to calculate the total cost of a loan for a given amount, interest rate, and amortization period.

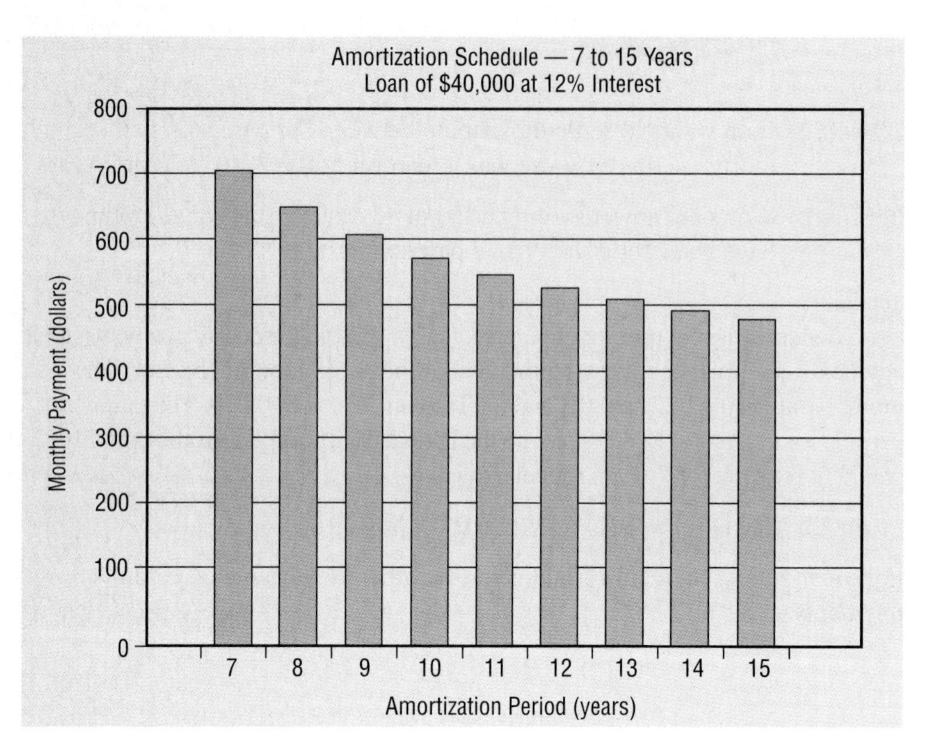

Note that the monthly payments decrease as the amortization period increases. As Joan's mother said, they would have approximately $226 more each month if they had amortized the loan over 15 years instead of 7.

Create another graph that shows the relationship between amortization periods of 7–15 years and total cost.

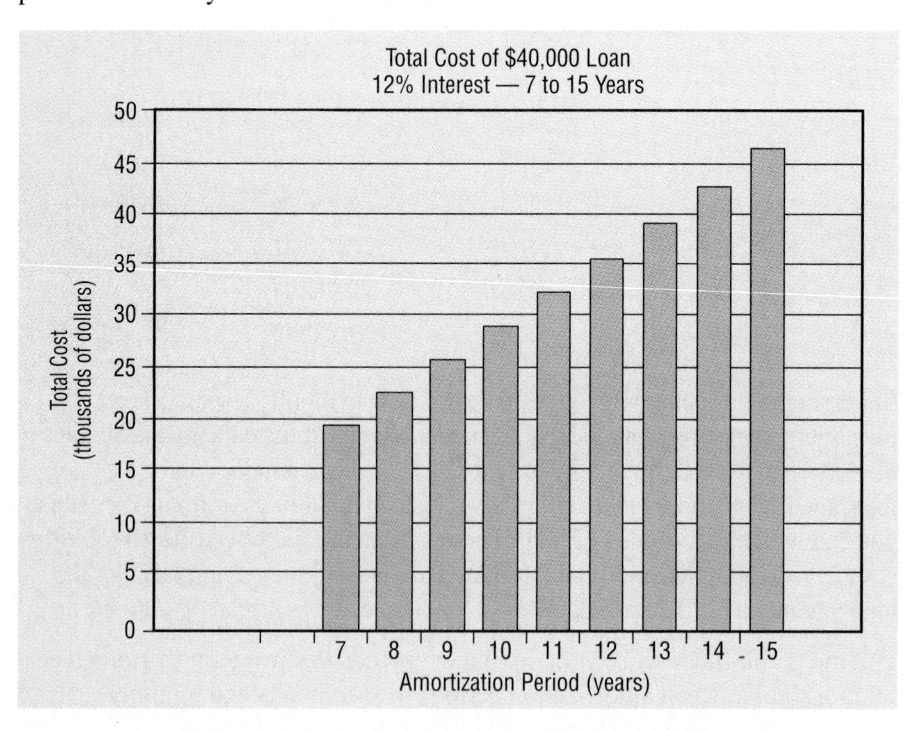

Note that the total cost increases as the amortization period increases. As Joan's father said, the cost of the loan would be approximately $27,000 more over the course of the loan if they had amortized it over 15 years instead of 7 years.

Joan notes that both her parents have strong support for their arguments. However, she feels that they should have discussed the situation before applying for the second mortgage. Perhaps they should have considered a greater total cost since it would have made their monthly cash flow lower. Maybe they should have compromised on an amortization period of 11 years.

TRY YOUR SKILLS

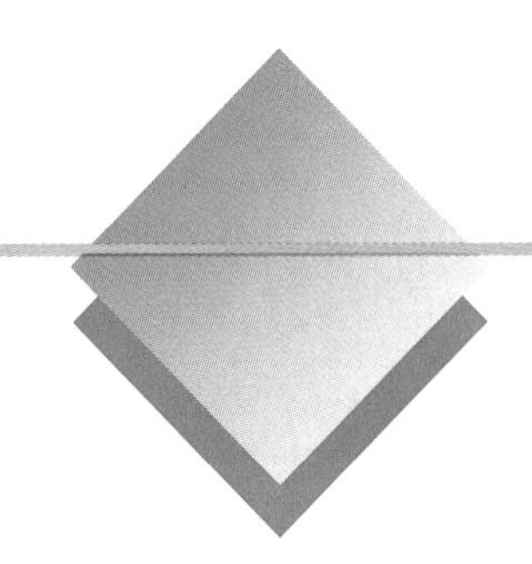

Use the monthly payment formula and a calculator to complete the table below to determine the total cost of borrowing each sum of money at 10% per year over 3 years, 4 years, and 5 years.

	Loan Amount	Number of Years	Monthly Payment	Total Payment	Total Cost
1.	$ 2,500	3	$80.67	$2904.05	$404.05
2.	2,500	4	63.41	3,043.51	543.51
3.	2,500	5	53.12	3,187.06	687.06
4.	5,000	3	161.34	5,808.09	808.09
5.	5,000	4	126.81	6,087.02	1087.02
6.	5,000	5	106.24	6,374.11	1374.11
7.	7,500	3	242.00	8,712.14	1212.14
8.	7,500	4	190.22	9,130.53	1630.53
9.	7,500	5	159.35	9,561.17	2061.17
10.	10,000	3	322.67	11,616.19	1616.19
11.	10,000	4	253.63	12,174.04	2174.04
12.	10,000	5	212.47	12,748.23	2748.23

GUIDED PRACTICE

You may wish to have students work in groups to determine the monthly payment, the total payment, and the total cost in Exercises 1-12. Observe each group to ensure that they are using the correct values of *r* and *n* in the monthly payment formula. Make sure that students are using the answer key on their calculators appropriately.

13. In Exercises 1–12, you can see that the payments on a $5000 loan for 3 years and a $7500 loan for 5 years are both approximately $160. Should you automatically select the $7500 loan? Why or why not? Total cost is $2061.17 versus $808.09. You might consider total cost as a factor in your choice.

KEY TERM

amortize

ADDITIONAL ANSWERS

1. You could find yourself with excessive debt.
2. Anything below 20% is manageable; anything above is a dangerous credit overload.
3. The stress and worry can undermine one's health.

INDEPENDENT PRACTICE

ASSIGNMENT GUIDE

Exercises 1-3 ask students questions about credit management. Exercises 4-24 ask students to add a column to the spreadsheet they created for Exercises 9-29 in Lesson 5.1 to determine the total costs of each loan. Exercises 25-27 assess students' ability to use a spreadsheet program to generate payments and costs, and to display this information in spreadsheet graphs. Exercise 28 asks students to discuss the information they gathered in Exercises 25-27.

EXERCISE YOUR SKILLS

1. Why should you refrain from taking out a new loan to help make payments on a previous loan if you are having difficulty making these payments? See Additional Answers.
2. What percent of a family's income can be spent making installment payments, excluding a home mortgage, before the family begins to experience credit overload?
3. How can excessive debt damage an individual's health?

Ricky would like to go backpacking with his friends in Colorado before going off to college. If he did this every summer, he would not be able to earn as much money working at summer jobs. Therefore it would take him longer to repay his loan. He wonders how much it will cost him to amortize his loan over 4 or 5 years instead of 3 years. Using your results from Exercises 9–29 of Lesson 5–1 and a spreadsheet program, add a column to the table to show Ricky the total cost (interest payments) of each loan.

	Loan Amount	Number of Years	Monthly Payment	Total Payment	Total Cost
4.	$ 6,500	3	$217.45	$7828.15	$1328.15
5.	6,500	4	172.77	8,292.96	1792.96
6.	6,500	5	146.24	8,774.20	2274.20
7.	9,500	3	317.81	11,441.14	1941.14
8.	9,500	4	252.51	12,120.48	2620.48
9.	9,500	5	213.73	12,823.82	3323.82
10.	6,000	3	200.72	7,225.98	1225.98
11.	6,000	4	159.48	7,655.04	1655.04
12.	6,000	5	134.99	8,099.26	2099.26
13.	12,000	3	401.44	14,451.97	2451.97
14.	12,000	4	318.96	15,310.08	3310.08
15.	12,000	5	269.98	16,198.52	4198.52
16.	17,500	3	585.44	21,075.78	3575.78
17.	17,500	4	465.15	22,327.20	4827.20
18.	17,500	5	393.71	23,622.84	6122.84
19.	11,650	3	389.73	14,030.45	2380.45
20.	11,650	4	309.66	14,863.54	3213.54
21.	11,650	5	262.10	15,726.06	4076.06
22.	8,775	3	293.56	10,568.00	1793.00
23.	8,775	4	233.24	11,195.50	2420.50
24.	8,775	5	197.42	11,845.16	3070.16

CLOSURE

Ask students to define total payment and total cost. Ask them to explain how to calculate the total cost if they know the total payment. Have students explain why it costs more to finance a specified loan amount at a given interest rate as the amortization period increases. Have students debate whether they agree with Joan's mother or father.

Mary Kim wants to borrow $20,000 at 7.5%. She wants to decide whether to borrow the money for 5, 6, 7, 8, 9, or 10 years. See Additional Answers.

25. Create a spreadsheet that shows the monthly payments, total payments, and total cost for amortization periods of 5–10 years for the loan.

26. Use the graph function of your computer spreadsheet program to graph the relationship between amortization periods and monthly payments.

27. Create a graph that shows the relationship between the amortization periods and the total costs.

28. Discuss the advantages and disadvantages of borrowing the money for the different time periods. Answers may vary.

LESSON QUIZ

1. How much does it cost to borrow $11,250 at 7% for 4 years? (**$1680.97**)
2. What are the total costs of borrowing $15,000 at 8% for 3 years, 4 years, and 5 years? (**$1921.64; $2577.30; $3248.75**)
3. Discuss the advantages and disadvantages of borrowing the money for the different time periods in question 2. (**3-year loan: lowest total cost and highest monthly payment; 5-year loan: lowest monthly cost and highest total cost.**)

MIXED REVIEW

1-4 **1.** Lloyd earns $405.98 in gross income each week. How much will be deducted for Social Security and Medicare? $31.06

5-1 **2.** If Debbie can afford a monthly payment of $250, how much can she afford to borrow at a yearly interest rate of 8.5% for 4 years? $10,142.69

4-2 **3.** Irma and Louise have decided to crochet ponytail holders for hair. They can prepare 75 hair holders per week. How much will these 75 hair holders cost to make if the costs include 20 hours of labor at $5.00 per hour and the following materials costs:

Twine at $0.95 per holder
Beads at $0.75 per holder
Hair elastics at $0.15 per holder $238.75

3-1 **4.** Julio saves $35 from his salary each week. For how many weeks will he have to save to be able to buy a computer word-processing system that costs $500? 15 wk

3-2 **5.** Use the Rule of 72 to determine how long it will take your $2000 savings account to double in value if it is growing at a rate of 8%. 9 y

5-1 **6.** Jen can afford a monthly payment of $250 on her credit card account. How much can she afford to borrow for 4 years at an APR of 8.5%?

ADDITIONAL ANSWERS

6. $10,142.69

5-3 CREDIT MANAGEMENT: KEEPING CREDIT COSTS DOWN

QUICK REFERENCE

TEACHER'S RESOURCES AND ANSWER KEY

Lesson Quiz Answers 5-3, p. 54

Reteaching Activity Answers 5-3, p. 99

Enrichment Activity Answers 5-3, p. 100

EXTENSION ACTIVITIES

Reteaching Activity 5-3, p. 33

Enrichment Activity 5-3, p. 38

Raul wants to take Joan to the prom this spring, but she is unwilling to discuss the matter. They have been dating for almost two years, so Raul feels that he knows Joan quite well. He also knows that her mother has lost her job, and he imagines that the financial situation at her house is difficult. He has concluded that Joan is reluctant to go to the prom because she cannot afford a new dress.

Raul is thinking about borrowing money to cover his prom expenses, a dress for Joan, and a portable keyboard. However, he wants to avoid making the same mistakes her family made. Therefore he will explore ways to reduce the cost of credit and then determine whether he can afford it.

OBJECTIVES: *In this lesson, we help Raul to:*

- *Study financial advisers' suggestions regarding the wise use of credit.*
- *Determine how much can be saved by understanding the various terms that an installment loan may carry.*

GUIDELINES FOR USING CREDIT WISELY

Credit can be an important tool in money management if it is used wisely. By doing some research, Raul has discovered the following suggestions on the proper use of credit made by financial advisers and credit counselors.

1. *Limit installment debt to 15–20% of take-home pay.* Monthly installment payments, excluding a home mortgage, are likely to be manageable if they are limited to 15–20% of take-home pay. Any amount over 20% may reduce an individual's or family's ability to pay for food, shelter, transportation, clothing, and other essentials. An authority on money matters has recommended that you not owe more than one-third of your discretionary income for the year. **Discretionary income** is the income you have left after paying for basic needs such as food, clothing, and shelter.
2. *Purchase durable products that will outlast the payment period.* If you are going to use credit for items that do not appreciate (increase in value), use it only on the purchase of durable products such as household appliances or cars. Be careful to choose products carefully. For instance, you do not want to make payments on a used car for 3 years if it only lasts for 15 months.

If you are thinking of borrowing money or opening a charge account, determine the cost, and then decide whether you can afford it. Then shop around for the best credit terms.

WAYS TO REDUCE THE COST OF INSTALLMENT LOANS

Raul has also discovered the following ways to minimize credit costs:

1. *Select a short payment period.* As you discovered in Lesson 5–2, the cost of financing an item is greater with a longer payment period and smaller payments.
2. *Compare interest rates.* The interest rate on a loan can vary dramatically from lender to lender, so it is wise to shop for interest rates just as you would shop for the product itself.
3. *Make a large down payment.* The larger your down payment, the less you will have to borrow.

You can save money in interest charges by paying for a larger portion and financing a smaller portion.

Ask Yourself

1. What are two guidelines for using credit wisely?
   See Additional Answers.
2. What should you consider first if you are thinking about borrowing money?
   determine cost and decide if it is affordable
3. What are three ways to reduce the cost of installment loans?

ALGEBRA *REVIEW*

Determine each of the following.

1. 20% of \$5500
   \$1110
2. 10% of \$7500
   \$750
3. 30% of \$4300
   \$1290

Evaluate. Give the results to the nearest hundredth.

4. $y = 5750 - (0.2)(5750)$
   4600
5. $z = 3600 - (0.4)(3600)$
   2160
6. Express y as the product of two numbers by factoring Exercise 4.
   $y = 5750(0.8)$
7. Express z as the product of two numbers by factoring Exercise 5.
   $z = 3600(0.6)$

ADDITIONAL ANSWERS (ASK YOURSELF)

1. Limit debt to 15-20% of take-home pay; use credit only for durable goods.
3. Select a short payment period; compare interest rates; make a large down payment.

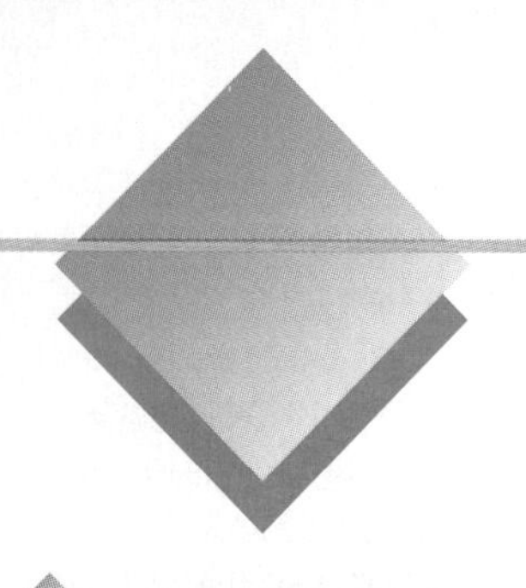

SHARPEN YOUR SKILLS

SKILL 1

FOCUS

ALGEBRA CONNECTIONS

Your local department store is offering two payment plans on the television set that you want to buy. Plan A requires a $100 down payment followed by 12 monthly payments of $40. Plan B requires $25 down followed by 12 monthly payments of $55. Which plan provides the lower price? (**Plan A**)

CRITICAL THINKING

After students have answered the questions, ask them to consider the impact that fluctuating interest rates have on the cost of homes. Ask them to research the refinancing of mortgage loans.

FOCUS ON ALGEBRA

In this lesson, students learn that obtaining a lower interest rate and making a larger down payment can reduce the cost of a loan. (*NCTM Standard 1, p. 137*)

EXAMPLE 1 Raul is thinking about buying a portable keyboard. He has compared prices and financing costs at various stores. He can purchase it at one store for $650 at an interest rate of 12% and at another store for $650 at an interest rate of 9%. Both loans must be repaid over a period of 2 years.

QUESTION How much money will he save if he purchases the keyboard at the store that offers 9% interest?

SOLUTION

Use the monthly payment formula and your calculator to determine the monthly and total payments at each rate of interest.

$$M = \frac{Pr(1 + r)^n}{(1 + r)^n - 1} \qquad P = 650;\ n = 2 \cdot 12 = 24$$

Monthly Payment at 12%	Total Payment at 12%	Monthly Payment at 9%	Total Payment at 9%	Total Savings
$30.60	$734.35	$29.70	$712.68	$21.67

Remember to use the answer key to find the total payments. However, to find the total savings, use the exact amounts. Why?

EXAMPLE 2 In Example 3 of Lesson 5-1, Maria considered loans of $4200, $5500, $6325, $8275, and $9750 at 8% over periods of 3 years, 4 years, and 5 years. The car dealership is offering 5% financing for one day only.

QUESTION How much can she save on each loan if she finances at a rate of 5% instead of 8%?

SOLUTION

Use a spreadsheet program to find the total payments on each loan at 8% and 5%, and then find the difference (savings). To subtract the exact amount of the total payments, use the function of your spreadsheet that formats a column in exact amounts, such as @round. Remember when you use a display function that a computer or calculator still uses the unrounded amounts in calculations unless you use special formatting functions.

For 5 years at 8%:

```
@ROUND(+60*A5*(0.08/12)*(1+(0.08/12))^60/((1+(0.08/12))^60-1),2)
```

For 5 years at 5%:

```
@ROUND(+60*A5*(0.05/12)*(1+(0.05/12))^60/((1+(0.05/12))^60-1),2)
```

	A	B	C	D	E
1	Loan	Number	Total	Total	Total
2	Amount	of Years	Payment at 8%	Payment at 5%	Savings
3	4200	3	4,738.06	4,531.60	206.46
4	4200	4	4,921.65	4,642.71	278.94
5	4200	5	5,109.65	4,755.55	354.10
6	5500	3	6,204.60	5,934.24	270.36
7	5500	4	6,445.01	6,079.73	365.28
8	5500	5	6,691.21	6,227.51	463.70
9	6325	3	7,135.29	6,824.37	310.92
10	6325	4	7,411.76	6,991.69	420.07
11	6325	5	7,694.89	7,161.63	533.26
12	8275	3	9,335.10	8,928.33	406.77
13	8275	4	9,696.81	9,147.24	549.57
14	8275	5	10,067.23	9,369.57	697.66
15	9750	3	10,999.06	10,519.78	479.28
16	9750	4	11,425.25	10,777.71	647.54
17	9750	5	11,861.69	11,039.67	822.02

Use the graph function of your computer spreadsheet program to graph the relationship between amortization periods of 3 years, 4 years, and 5 years and total payments for each amount at 8% and 5%.

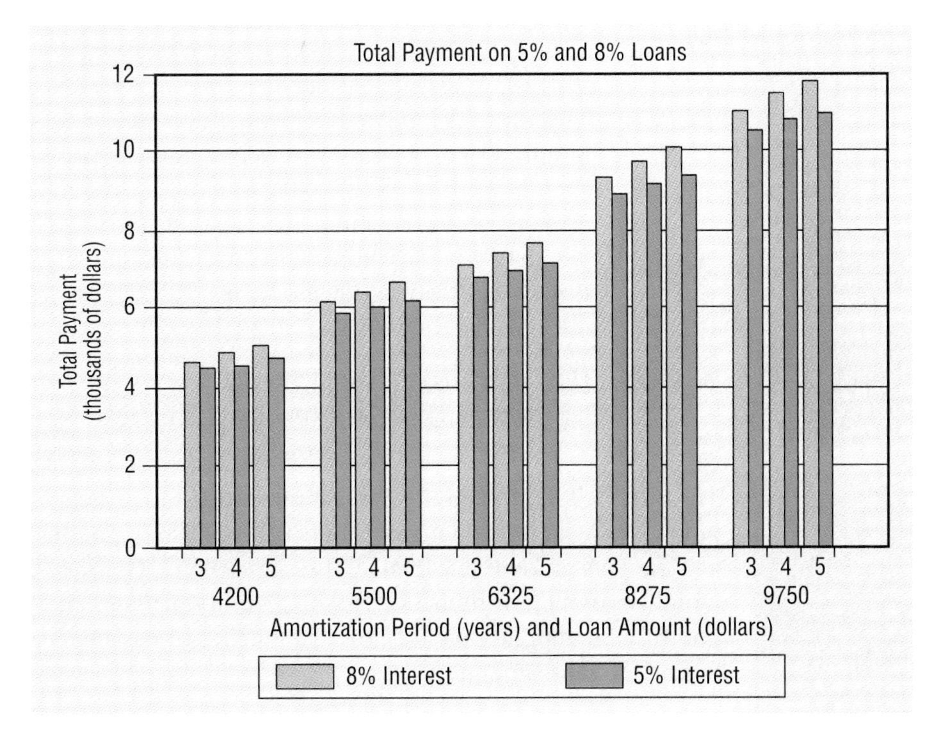

INSTRUCTION

TEACHING THE LESSON

Have students read the opening scenario. Use the *Algebra Connections* on p. 208 to illustrate that the size of the down payment may influence the total cost of a purchase. Have students read the sections on using credit wisely and reducing the costs of loans. Make sure that students understand that there are ways to minimize credit costs.

After completing Skills 1 and 2, have students explain why lower interest rates and larger down payments reduce credit costs. Have students compare the total costs of a $15,000 loan at 9% for 3 years with down payments of 10%, 20%, and 30%. (**$16,954.67, $16,737.48, $16, 520.30**)

TECHNOLOGY HINT

The total savings on the 5% loan versus the 8% loan can be seen in the different sizes of the two bars for each loan amount and term. Because of the low relative value of total savings compared to total payments, it would be beneficial to have the students graph total savings on separate axes.

The total payments, and therefore the total cost, on a loan for a given amount of money over a specified amortization period are always lower when the interest rate is lower.

SKILL 2

Many loans require that you make a *down payment,* which is a percentage of the cost or price. To find the down payment D multiply the cost C by the percent for the down payment r. To find the loan amount P, subtract the down payment D from the cost C. The total amount T is the sum of the down payment D and all the monthly payments nM.

Loans with Down Payment

$D = rC$ where D = down payment
r = percent for down payment
C = cost

$P = C - D$ where P = loan amount
C = cost
D = down payment

$T = nM + D$ where T = total amount
n = number of payments
M = monthly payment
D = down payment

EXAMPLE 3 Joan's aunt and uncle just bought a boat for \$14,950 at an interest rate of 12% for 5 years. They are very excited because they were required to make only a 10% down payment. Joan is worried that her aunt and uncle will end up in the same position as her parents did. She wonders whether they could have saved money if they had made a down payment of 20% or 30% instead of 10%.

QUESTION How much money could they have saved if they had made a down payment of 20% or 30%?

SOLUTION

The cost is \$14,950 and the interest rate is 12% for 5 years. For 10% down, the down payment is $0.10(14{,}950) = 1{,}495$ and the loan amount is $14{,}950 - 1495 = 13{,}455$. Use the monthly payment formula to find the monthly payment and multiply the monthly payment by the number of payments to find the total amount.

Percent Down, r	Down Payment, D	Loan Amount, P	Monthly Payment, M	Total Amount, T	Savings over 10% Down
10%	$1495	$13,455	$299.30	$19,452.94	
20%	2990	11,960	266.04	18,952.62	$ 500.32
30%	4485	10,465	232.79	18,452.29	1000.65

They could have saved $500.32 with 20% down and $1000.65 with 30% down. How much do you think they would have saved with 40% down?

EXAMPLE 4 Maria knows that there is another way to lower the total cost of each loan if she cannot reach a decision on a car in time for the one-day 5% financing offer. She wants to know how much money she can save if she makes a down payment of 20% and therefore borrows only 80% of the cost of the car.

QUESTION How much will Maria save on loans of $4200, $5500, $6325, $8275, and $9750 at 8% over periods of 3 years, 4 years, and 5 years if she makes a down payment of 20% instead of borrowing the entire amount?

SOLUTION

Use a spreadsheet program to find the total amount for the 8% loan without a down payment and with a 20% down payment. Then find the difference (savings).

	A	B	C	D	E	F	G
1	Loan	Loan Amount	Number	Total Payment	Total Payment	Total Amount	Total
2	Amount	20% Down	of Years	0% Down	20% Down	20% Down	Savings
3	4200	3360	3	4,738.06	3790.45	4,630.45	107.61
4	4200	3360	4	4,921.65	3937.32	4,777.32	144.33
5	4200	3360	5	5,109.65	4087.72	4,927.72	181.93
6	5500	4400	3	6,204.60	4963.68	6,063.68	140.92
7	5500	4400	4	6,445.01	5156.01	6,256.01	189.00
8	5500	4400	5	6,691.21	5352.97	6,452.97	238.24
9	6325	5060	3	7,135.29	5708.23	6,973.23	162.06
10	6325	5060	4	7,411.76	5929.41	7,194.41	217.35
11	6325	5060	5	7,694.89	6155.91	7,420.91	273.98
12	8275	6620	3	9,335.10	7468.08	9,123.08	212.02
13	8275	6620	4	9,696.81	7757.45	9,412.45	284.36
14	8275	6620	5	10,067.23	8053.78	9,708.78	358.45
15	9750	7800	3	10,999.06	8799.25	10,749.25	249.81
16	9750	7800	4	11,425.25	9140.20	11,090.20	335.05
17	9750	7800	5	11,861.69	9489.35	11,439.35	422.34

+A3−0.2∗A3

0.2∗A17+E17

RETEACHING

Some students may benefit from a review of how and when decimals should be rounded. To reteach the lesson, concentrate on this skill.

ENRICHMENT

Challenge students to write a computer program that calculates the total savings between one loan for a given amount at a given interest rate, amortization period, and percent down payment and another loan for the same amount with a different interest rate, amortization period, and/or percent down payment.

Use the graph function of your program to show the savings for each amortization period and loan amount.

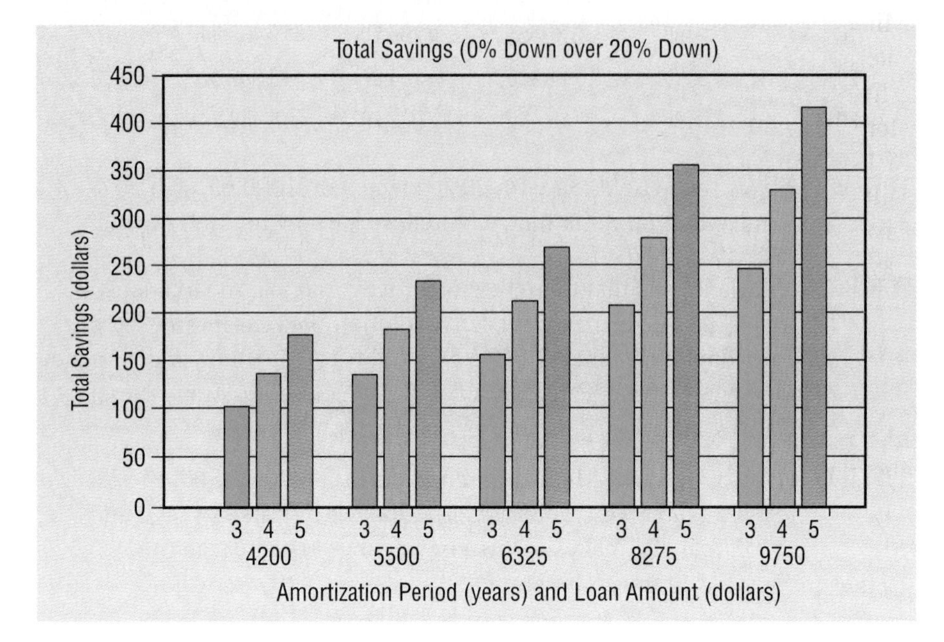

EXAMPLE 5 Bernard is going to buy a stereo for $1200 on which he will make monthly payments. He has collected information about the terms of various loans. These loans have different amortization periods, rates of interest, and down payment requirements. He can amortize the loan over 1 year, 2 years, or 3 years. He can finance it at 9% or 12%. He can make a down payment of 10% or 20%.

QUESTION Which combination of amortization period, interest rate, and down payment yields the lowest total payments? Which yields the highest?

SOLUTION

Use your calculator and the monthly payment formula.

Number of Years	Percent Down	Down Payment	Loan Amount	Total Amount at 9%	Total Amount at 12%
1	10%	$120.00	$1080.00	$1253.37	$1271.48
1	20%	240.00	960.00	**1247.44**	1263.54
2	10%	120.00	1080.00	1304.15	1340.14
2	20%	240.00	960.00	1292.58	1324.57
3	10%	120.00	1080.00	1356.37	**1411.37**
3	20%	240.00	960.00	1339.00	1387.89

COMMON ERRORS

Make sure that students know the proper use of the answer key on the calculator and why they should use the exact amount to find the total savings.

Students may forget to add the down payment amount when they calculate the total payments.

The lowest total payment is $1247.44, which occurs with the shorter amortization period, the higher down payment, and the lower interest rate. The highest total payment is $1411.37, which occurs with the longer amortization period, the lower down payment, and the higher interest rate.

TRY YOUR SKILLS

Mindy is going to purchase new furniture. She has compared prices and interest charges at two stores. She can purchase a dining room set for $3300, a living room set for $4200, and a bedroom set for $1800 at both stores. One store offers an interest rate of 15%, and the other store offers an interest rate of 10%. Both stores require that the loan be repaid in 3 years. At neither store will she be required to make a down payment. How much money can she save on each item by financing at 10% instead of 15%? Use the monthly payment formula and your calculator to complete the following chart.

GUIDED PRACTICE

You may wish to have students work in groups to find the total savings for each exercise. Remind students to add the down payment when they calculate the total payments.

	Cost of Item	Monthly Payment at 15%	Total Payment at 15%	Monthly Payment at 10%	Total Payment at 10%	Total Savings
1.	$3300	$114.40	$4118.24	$106.48	$3833.34	$284.90
2.	4200	145.59	5241.40	135.52	4878.80	362.60
3.	1800	62.40	2246.31	58.08	2090.91	155.40

Determine the total savings on a loan of $12,450 at 8.5% for 25 years if you make a down payment of 20% or 30% instead of 10%.

	Percent Down	Down Payment, *D*	Loan Amount, *A*	Monthly Payment, *M*	Total Amount, *T*	Total Savings Over 10%
4.	10%	$1245	$11,205	$90.23	$28,312.71	
5.	20%	2490	9,960	80.20	26,550.19	$1762.52
6.	30%	3735	8,715	70.18	24,787.66	3525.05

EXERCISE YOUR SKILLS

1. What is discretionary income? the income left after paying for basic needs
2. Why does increasing the size of your down payment for a purchase save you money in the long run? You borrow less, so you save money in interest charges.
3. Why would a financial institution encourage you to make a smaller down payment and increase the amortization period for an installment loan? The financial institution makes higher profits.

Steve bought the items listed below and amortized the payments at 12% over 3 years with no down payment. Use the monthly payment formula and your calculator to determine how much he could have saved if the interest rate had been 9% instead of 12%.

CD player for $500
Computer for $2500
Stereo cassette deck for $250
TV/VCR for $1000

INDEPENDENT PRACTICE

ASSIGNMENT GUIDE

Exercise 1 asks students to define a credit term. Exercises 2-3 asks students to discuss the size of a down payment from two points of view. Exercises 4-11 require students to apply the formulas taught in the lesson using a calculator. Exercises 12-17 asks students to use the formulas taught in the lesson in a spreadsheet program. Exercises 18-19 asks students to interpret the information in their spreadsheet. Exercise 20 asks students to translate numerical information into a general statement.

	Item Cost	Monthly Payment at 12%	Total Payment at 12%	Monthly Payment at 9%	Total Payment at 9%	Total Savings
4.	$ 500	$16.61	$ 597.86	$15.90	$ 572.40	$ 25.46
5.	2500	83.04	2989.29	79.50	2861.98	127.31
6.	250	8.30	298.93	7.95	286.20	12.73
7.	1000	33.21	1195.72	31.80	1144.79	50.93

Use your results from Exercises 4–7 to determine how much Steve would have saved if he had made a down payment of 30% instead of making no down payment. The interest rate remains at 12%.

	Item Cost	Monthly Payment 0% Down	Total Payment 0% Down	Loan Amount 30% Down	Monthly Payment 30% Down	Total Payment 30% Down	Total Savings
8.	$ 500	$16.61	$ 597.86	$ 350.00	$11.63	$ 568.50	$ 29.36
9.	2500	83.04	2989.29	1750.00	58.13	2842.50	146.79
10.	250	8.30	298.93	175.00	5.81	284.25	14.68
11.	1000	33.21	1195.72	700.00	23.25	1137.00	58.72

Daniel is going to buy a new all-terrain vehicle for $42,000 on which he will make monthly payments. He has collected information about the terms of various loans. These loans have different amortization periods, rates of interest, and down payment requirements. He can amortize the loan over 3 years, 4 years, or 5 years. He can finance it at 6% or 8%. He can make a down payment of 10% or 20%. Create a spreadsheet program to find the total payments for each situation.

	Number of Years	Percent Down	Down Payment	Loan Amount	Total Payment at 6%	Total Payment at 8%
12.	3	10%	$4200.00	$37,800	$45,598.17	$46,842.53
13.	3	20%	8400.00	33,600	45,198.38	46,304.47
14.	4	10%	4200.00	37,800	46,811.24	48,494.81
15.	4	20%	8400.00	33,600	46,276.65	47,773.16
16.	5	10%	4200.00	37,800	48,046.79	50,186.86
17.	5	20%	8400.00	33,600	47,374.93	49,277.21

18. Which combination of amortization period, interest rate, and down payment yields the lowest total payments? 3 y; 6%; 20% down

19. Which combination of amortization period, interest rate, and down payment yields the highest total payments? 5 y; 8%; 10% down

20. Use the results of Exercises 12–19 to write a general statement about the effect of amortization period, interest rate, and down payments on total payments. For the lowest total payments, it is best to put down as much money as you can, have a low interest rate and a short amortization period.

MIXED REVIEW

1-3 **1.** You are a real-estate agent who has just sold a house for $325,000. Determine your total commission if you earn 5% on the first $150,000 and 8% on any amount over $150,000. $21,500

2-2 **2.** You just received a check with your full first name, middle initial, and full last name. Write your name as you would to endorse the check. Answers may vary.

2-2 **3.** You have $446.77 in your checking account, and you write a check for $57.49. Show how you would enter this check in your check register.

3-2 **4.** Determine the interest for 6 months if the principal is $5525, the rate is 5%, and the interest is compounded quarterly. $138.99

1-2 **5.** You earn $26,000 per year plus 15% of your salary in fringe benefits. Find your yearly earnings, including benefits. $29,900

5-2 **6.** What percent of your take-home pay spent on credit payments (excluding mortgage payments) is regarded as a "manageable" amount (safer than "dangerous" but not as safe as "comfortable")? 15%

2-1 **7.** Given a choice, which method would you prefer your bank to use when determining whether a service charge will be imposed; the minimum-balance method or the average-balance method? Explain your answer.

4-4 The cost function for a small business is $c = 2.8n + 40$. The revenue function for the business is $r = 6n$.

8. Graph the cost and revenue functions. See Additional Answers.

9. Find the break-even point. 13 items

CLOSURE

Ask students to state the percent of take-home pay to which they should limit monthly installment payments. Have them explain how to calculate the total payment if they know the down payment, monthly payment and number of payment periods. Have students discuss whether Bernard should automatically select the 1-year loan at 9% with 20% down.

LESSON QUIZ

1. Calculate the savings on an $8500 loan at 7% for 3 years compared to an $8500 loan at 10% for 3 years. (**$425.37**)

2. Calculate the savings on a $9995 loan at 12% for 4 years with 30% down compared to a $9995 loan at 12% for 4 years with 10% down. (**$527.79**)

3. Greg is going to buy a new mini-van for $25,000 on which he will make monthly payments. He can amortize the loan over 3 years, 4 years, or 5 years. He can finance the loan at 6% or 7%. He can make a down payment of 5% or 15%. Create a spreadsheet program to find the total payments for each situation. Which combination of amortization period, interest rate, and down payment yields the lowest total payments? The highest? (**A 3-year amortization period at 6% with 15% down yields the lowest. A 5-year amortization period at 7% with 5% down yields the highest.**)

5-4 AMORTIZATION SCHEDULES: SHRINKING INTEREST PAYMENTS

QUICK REFERENCE

TEACHER'S RESOURCES AND ANSWER KEY

Lesson Quiz Answers 5-4, p. 56

Reteaching Activity Answers 5-4, p. 99

Enrichment Activity Answers 5-4, p. 100

EXTENSION ACTIVITIES

Reteaching Activity 5-4, p. 34

Enrichment Activity 5-4, p. 39

Maria knows that she must be careful to choose a loan for which she can afford the monthly payment. Patrick recently told her that he received a raise from Paradise Department Store. Maria hopes that she will receive a pay increase in the next six months too. Her boss also indicated that she might receive a holiday bonus. She plans to save any additional money that she earns.

Maria wonders whether she can pay off her loan early if her financial situation continues to improve. Patrick has read that there can be penalties associated with prepaying a loan. However, Maria's uncle has just told the family that he saved over $500 by paying the remainder of his car loan 15 months early. She also knows that her mother prepaid her contract on her microwave oven. Therefore Maria is certain that there must be substantial advantages to prepaying a loan.

Maria has a feeling that the money she can save if she prepays her loan is related to the portion of her monthly payment that is interest. Since her monthly loan payment is fixed, she wonders whether her monthly interest payment is constant also.

OBJECTIVES: *In this lesson, we will help Maria to:*

- *Compute the interest due, note reduction, and unpaid balance on her loan on a monthly basis.*
- *Create an amortization schedule for a loan.*
- *Determine prepayment amounts paid on a loan.*
- *Identify the amount saved by prepaying a loan.*

AMORTIZATION SCHEDULES

When Maria receives her car loan, her monthly payment will be identical each month. However, the portion of that payment that is applied as interest changes each month. Banks or other lending institutions provide borrowers with schedules, called **amortization schedules,** that list the interest portion of monthly loan payments. These amortization schedules also list the payment number, the monthly payment, the reduction in unpaid balance (note reduction), and the remaining unpaid balance for each month.

The **payment number** corresponds to the number of months that have elapsed since the money was borrowed. The **interest due** is determined by multiplying the monthly interest rate by the previous month's unpaid balance. The **note reduction** is found by subtracting the interest due from the monthly payment and shows how much of the monthly payment actually goes to paying off the loan. The **unpaid balance** is calculated by subtracting the note reduction from the previous month's unpaid balance and represents how much of the original loan is still unpaid.

PREPAYMENT

Most financial institutions will allow a borrower to prepay the balance of an installment loan. This right of prepayment (paying off the contract before it comes due) saves the borrower interest charges. However, the amount saved by prepaying a loan varies according to each loan agreement. Most loan contracts require the borrower to pay more interest in the early months of the loan period when the lender's risks are greater and less interest near the maturity date when the lender's risks are less.

Many financial institutions use a standard formula to determine the remaining portion of a loan at the time the borrower chooses to prepay it. The formula is based upon the monthly payment the borrower has been making, the monthly interest rate, and the number of remaining payment periods.

FOCUS

ALGEBRA CONNECTIONS

Hillary owes her mother $300, which she must repay in 6 months. Her mother offers her the 2 following payment plans:

Plan A:

pay $\frac{1}{6}$ of $300 per month

Plan B:

pay $\frac{6}{21}$ of $300 in Month 1

pay $\frac{5}{21}$ of $300 in Month 2

pay $\frac{4}{21}$ of $300 in Month 3

pay $\frac{3}{21}$ of $300 in Month 4

pay $\frac{2}{21}$ of $300 in Month 5

pay $\frac{1}{21}$ of $300 in Month 6

If she expects money to be tight in months 4 and 5, which payment plan should she select? (**Plan B**)

CRITICAL THINKING

After students have answered the questions, ask them to discuss whether they would like to have a loan contract in which each payment contained a fixed amount of interest. Have students research whether or not the banks in their town charge a prepayment penalty on mortgage loans.

RULE OF 78

Some stores apply the **Rule of 78,** instead of the prepayment formula used by financial institutions, to buyers who purchase merchandise on credit for 1 year and wish to prepay. Since the sum of the numbers from 1 to 12 is 78, the buyer pays a portion of the yearly interest equal to $\frac{12}{78}$ in month 1, $\frac{11}{78}$ in month 2, $\frac{10}{78}$ in month 3, and so on, until $\frac{1}{78}$ is paid in month 12. The largest amount of interest is paid in month 1, and the smallest amount of interest is paid in month 12. By the end of the year, all the interest is paid.

In general, prepaying a loan saves the borrower money. However, some lending institutions charge a **prepayment penalty** to offset a portion of the lost revenues and additional clerical costs that they incur when a borrower prepays a loan. A borrower who is considering prepaying a loan must check the prepayment stipulations stated in the loan contract to determine the costs, if any, of prepayment. This is one of the many reasons that it is very important that the borrower understand the terms of a credit agreement before signing it.

ALGEBRA *REVIEW*

1. If the annual interest rate is 15%, what is the monthly interest rate?
   1.25%
2. Express your answer to Exercise 1 as a decimal.
   0.0125

Evaluate. Round to the nearest hundredth.

3. $x = 9 + 8 + 7 + \ldots + 2$
   44
4. $x = 15 + 14 + 13 + \ldots + 5$
   110
5. $x = \frac{3}{78} + \frac{2}{78}$
   0.06
6. $x = \frac{7}{29} + \frac{6}{29} + \frac{5}{29}$
   0.62
7. $x = 3 + 2^4$
   19
8. $x = 3 + 2^{-4}$
   3.0625
9. $x = 1 + (3 + 2)^{-4}$
   1.0016
10. $x = 1 + (3 + 2)^4$
    626
11. $x = 4.5[100 - (6 + 1)^{-12}]$
    450
12. $x = 3.9[98 - (2.30122)^{-3}]$
    381.88
13. $x = \frac{400[1 - (1 + 0.015)^{-10}]}{0.015}$
    3688.87
14. $x = \frac{325[1 - (1 + 0.0125)^{-24}]}{0.0125}$
    6702.88

Ask Yourself

1. What does an amortization schedule show?
   the interest portion of monthly loan payments
2. Why do most loan contracts require the borrower to pay more interest in the early months of the loan period? The early months are when the lender's risks are greater.
3. Do you think that you will save money if you prepay a loan? Perhaps; it depends upon the amount of any prepayment penalty.
4. Do you think that it is fair for a bank to charge a prepayment penalty?
   Answers may vary.

SHARPEN YOUR SKILLS

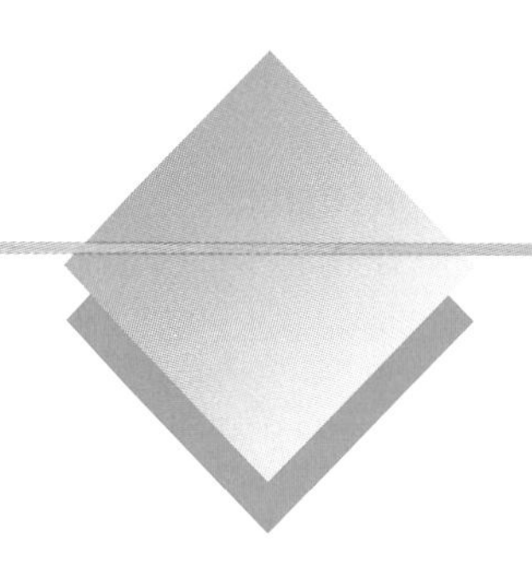

SKILL 1

EXAMPLE 1 Maria is considering a $4500 loan for 2 years at 9%. Her monthly payment is $205.58.

QUESTION What are the interest due, note reduction, and unpaid balance for the first 2 months?

SOLUTION

To determine I_1, the interest due at the end of month 1, Maria multiplied the loan amount L by the monthly interest rate r.

$I_1 = rL$ $\quad 0.0075 \cdot 4500 = \$33.75$ $\quad r = 0.09 \div 12 = 0.0075$; $L = 4500$

To find R_1, the note reduction in month 1, she subtracted I_1, the interest due at the end of month 1, from the payment P.

$R_1 = P - I_1$ $\quad 205.58 - 33.75 = \$171.83$ $\quad P = 205.58$; $I_1 = 33.75$

To calculate B_1, the unpaid balance at the end of month 1, she subtracted the note reduction R_1 from the original loan amount L.

$B_1 = L - R_1$ $\quad 4500 - 171.83 = \$4328.17$ $\quad L = 4500$; $R_1 = 171.83$

To determine I_2, the interest due at the end of month 2, she multiplied B_1, the unpaid balance from month 1, by the monthly interest rate r.

$I_2 = rB_1$ $\quad 0.0075 \cdot 4328.17 = \$32.46$ $\quad r = 0.09 \div 12 = 0.0075$; $B_1 = 4328.17$

To find the note reduction R_2 in month 2, she subtracted the interest I_2 due from the payment P.

$R_2 = P - I_2$ $\quad 205.58 - 32.46 = \$173.12$ $\quad P = 205.58$; $I_2 = 32.46$

To calculate the new unpaid balance B_2, she subtracted the note reduction R_2 from the unpaid balance from month 1.

$B_2 = B_1 - R_2$ $\quad 4328.17 - 173.12 = 4155.05$ $\quad B_1 = 4328.17$; $R_2 = 173.12$

Maria was able to see a pattern in her calculations and to use the pattern to predict the following formula for the unpaid balance after 3 months.

$B_3 = B_2 - R_3$ $\quad$ where $R_3 = P - I_3$ and $I_3 = rB_2$

Maria realized that the pattern she observed could be extended to cover more than just the first 3 months of the loan. The pattern could be used to create an amortization schedule as long as she wanted.

INSTRUCTION

TEACHING THE LESSON

Have students read the opening scenario. Use the *Algebra Connections* on p. 217 to demonstrate different payment plans. Have students read the sections on amortization schedules, prepayment, and the Rule of 78. Make sure they understand that they are paying a greater portion of the interest charges in the earlier months when they make a purchase at a store that applies the Rule of 78.

After completing Skill 1, make sure students understand that the interest portion of an installment loan decreases each month even though the monthly payment remains constant. After completing Skills 2 and 3, ask students to calculate how much they would save by prepaying an $18,000 loan at 9% for 5 years. They have been making monthly payments of $373.65 and have 28 months remaining. (**$1057.16**)

COMMON ERRORS

Be sure students understand that the formulas presented in the solution to Example 1 can be extended beyond two months. Remind them to use the unpaid balance from the previous month when calculating the interest due in any given month beyond the first. Make sure they use the sign change key instead of the subtraction key in the prepayment formula.

AMORTIZATION SCHEDULE FORMULAS

$I_1 = rL \quad I_2 = rB_1$

$R_1 = P - I_1 \quad R_2 = P - I_2$

$B_1 = L - R_1 \quad B_2 = B_1 - R_2$

where L = loan amount
r = monthly interest rate
P = payment amount
I_1 = interest due at end of month 1
R_1 = loan reduction at end of month 1
B_1 = balance at end of month 1

EXAMPLE 2 Maria wants to know what her loan amortization schedule will look like for the first year of her loan for $4500 for 2 years at 9%.

QUESTION What are her interest due, note reduction, and unpaid balance for the first 12 months?

SOLUTION

Maria used her spreadsheet program to create an amortization schedule that shows the payment number, payment amount, interest due, note reduction, and unpaid balance for months 1–12. The rate $r = 0.09 \div 12 = 0.0075$.

FOCUS ON ALGEBRA

In this lesson, formulas for prepaying a loan give students the opportunity to explore mathematical functions that apply to their daily lives. (*NCTM Standard 6, p. 154*)

	A	B	C	D	E
1	Payment	Payment	Interest	Note	Unpaid
2	Number	Amount	Due	Reduction	Balance
3		@ROUND(E3*0.0075,2)	+B4−C4	+E3−D4	4500.00
4	1	205.58	33.75	171.83	4328.17
5	2	205.58	32.46	173.12	4155.05
6	3	205.58	31.16	174.42	3980.63
7	4	205.58	29.85	175.73	3804.90
8	5	205.58	28.54	177.04	3627.86
9	6	205.58	27.21	178.37	3449.49
10	7	205.58	25.87	179.71	3269.78
11	8	205.58	24.52	181.06	3088.72
12	9	205.58	23.17	182.41	2906.31
13	10	205.58	21.80	183.78	2722.53
14	11	205.58	20.42	185.16	2537.37
15	12	205.58	19.03	186.55	2350.82

TECHNOLOGY HINT

Note that the @ROUND function must be used to hold the calculations in column C to two decimal places.

Next she used the graph function on her computer spreadsheet program to graph the portion of each monthly payment that is interest.

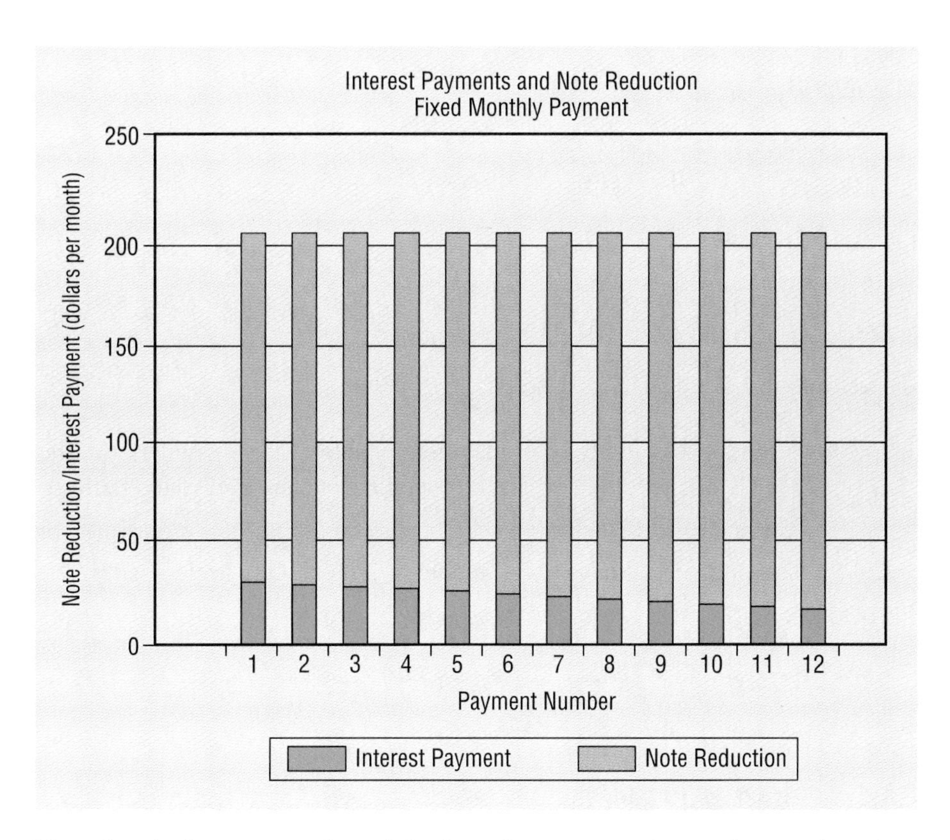

Note that the interest portion of the loan decreases each month even though the monthly payment remains constant.

SKILL 2

Maria would like to know how much she will save if she prepays her loan. She can use the prepayment formula to determine the remaining payment if she prepays it at any given time. She can then compare this amount with the amount that she would have paid if she did not prepay the loan.

PREPAYMENT FORMULA

$$A = \frac{M[1 - (1 + r)^{-q}]}{r}$$

where M = monthly payment
r = monthly interest rate
q = number of remaining payment periods
A = prepayment amount

principal + interest

EXAMPLE 3 Maria's uncle decides to prepay his car loan. He has been making payments of \$444.89 per month on a 5-year loan of \$20,000 at 12%.

QUESTION How much does he owe if he prepays the loan 15 months early? How much money will he save?

SOLUTION

Use the prepayment formula.

$$A = \frac{444.89[1 - (1 + 0.01)^{-15}]}{0.01}$$

$M = 444.89$
$r = 0.12 \div 12 = 0.01$
$q = 15$

$A = \$6168.42$

If he does not prepay the loan, he will pay 15(444.89) = 6673.35.

6673.35 − 6168.42 = 504.93   Subtract.

Therefore, he saves $504.93 by prepaying the loan.

RETEACHING

Some students may benefit from a review of math skills such as raising an expression to a negative exponent and multiplying fractions. To reteach the lesson, concentrate on these skills.

EXAMPLE 4 Some banks charge a prepayment penalty. Maria wonders whether her uncle would have saved money if his bank had charged a prepayment penalty of $250.

QUESTION Would he still have saved money if he had prepaid the loan?

SOLUTION

504.93 − 250.00 = 254.93   Subtract the penalty from the original savings.

Yes, he would have saved $254.93.

ENRICHMENT

Challenge students to program their graphing calculators or computers to calculate the savings that result from prepaying a loan at any given time.

SKILL 3

RULE OF 78

The buyer pays a portion of the yearly interest equal to

$\frac{12}{78}$ in month 1

$\frac{11}{78}$ in month 2

$\frac{10}{78}$ in month 3

⋮

$\frac{1}{78}$ in month 12

EXAMPLE 5 Maria's mother purchased a microwave oven for $360, putting $60 down. She signed a contract stating that the balance of $300 was to be paid in 12 equal installments with an interest rate of 21% per year.

QUESTION If she decides to prepay her contract at the time of her ninth payment and the store applies the Rule of 78, how much interest will she save?

SOLUTION

Find the yearly interest.

$0.21 \cdot 300 = 63$ The yearly interest rate is 21%.

After 9 months, Maria's mother has three remaining payments. By prepaying the loan she will not have to pay the portion of yearly interest due in the last three months. Apply the Rule of 78.

$$\frac{3}{78} + \frac{2}{78} + \frac{1}{78} = \frac{6}{78}$$

$$\frac{6}{78} \cdot 63 = 4.85$$ The yearly interest is $63.

Therefore she saves $4.85.

COOPERATIVE LEARNING

Have students work in small groups to write a spreadsheet program that will display the first 12 months of the amortization schedule for the loan described in Exercises 1-3 in the *Try Your Skills* section. Ask them to graph the shrinking interest payments. Make sure that they use an appropriate scale.

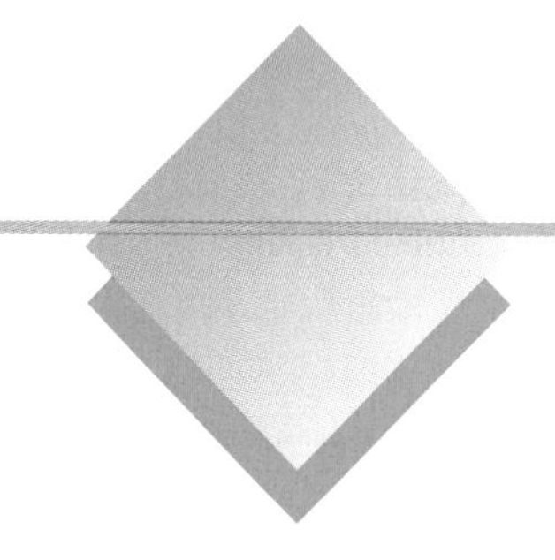

TRY YOUR SKILLS

Create an amortization schedule for the first 3 months of a $2000 loan at 6% for 4 years for which the monthly payment is $46.97.

	Payment Number	Payment Amount	Interest Due	Note Reduction	Unpaid Balance
1.	1	$46.97	$10.00	$36.97	$1963.03
2.	2	46.97	9.82	37.15	1925.88
3.	3	46.97	9.63	37.34	1888.54

GUIDED PRACTICE

You may wish to have students work in groups to create the amortization schedule in Exercises 1-3 and to calculate the prepayment savings in Exercises 4-7. Observe each group as they assemble information, and provide guidance when necessary.

Determine the prepayment savings on a bank loan of $25,000 at 12% for 30 years for which you have been paying $257.15 each month under each condition. Assume that there are no prepayment penalties.

4. It is prepaid 20 months early. Use the prepayment formula.

$$A = \frac{M[1 - (1 + r)^{-q}]}{r}$$ Substitute $M = 257.15$, $r = 0.12 \div 12 = 0.01$, and $q = 20$.

$$A = \frac{257.15[1 - (1 + 0.01)^{-20}]}{0.01}$$

$502.59

5. It is prepaid 3 years early.

$1515.26

A $600.00 television purchase is financed at 18% per year with no down payment at a store that applies the Rule of 78. It must be paid in 12 equal installments. Determine the prepayment savings under each condition.

6. It is prepaid after 4 months. $49.85

7. It is prepaid after 7 months. $20.77

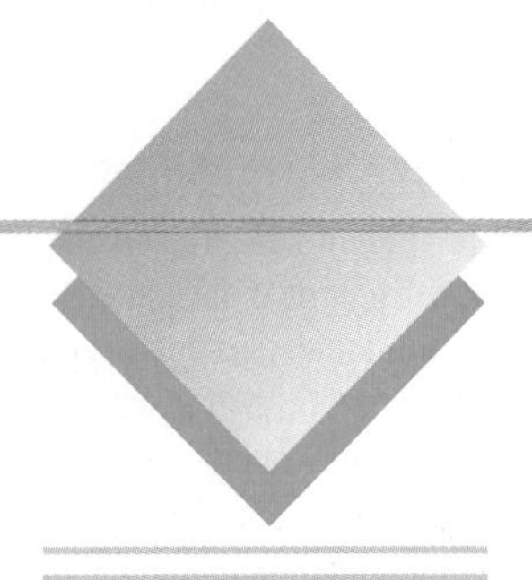

EXERCISE YOUR SKILLS

1. If your bank charges a $200 prepayment penalty, under what condition will you consider prepaying your loan? See Additional Answers.
2. Why does the interest portion of a monthly payment decrease each month?
3. What is an amortization schedule?

Gregory borrows $9500 from a bank to purchase a car. He finances it at 8% for 4 years, and his monthly payment is $231.92. Use a spreadsheet program to create an amortization schedule that shows the payment number, payment amount, interest due, note reduction, and unpaid balance for the first year.

	Payment Number	Payment Amount	Interest Due	Note Reduction	Unpaid Balance
4.	1	$231.92	$63.33	$168.59	$9331.41
5.	2	231.92	62.21	169.71	9161.70
6.	3	231.92	61.08	170.84	8990.86
7.	4	231.92	59.94	171.98	8818.88
8.	5	231.92	58.79	173.13	8645.75
9.	6	231.92	57.64	174.28	8471.47
10.	7	231.92	56.48	175.44	8296.03
11.	8	231.92	55.31	176.61	8119.42
12.	9	231.92	54.13	177.79	7941.63
13.	10	231.92	52.94	178.98	7762.65
14.	11	231.92	51.75	180.17	7582.48
15.	12	231.92	50.55	181.37	7401.11

Determine the prepayment savings on a bank loan of $21,400 at 11% for 25 years for which you have been paying $209.74 each month under each condition. Assume that there are no prepayment penalties. See Additional Answers.

16. It is prepaid 37 months early.
17. It is prepaid 5 years early.

Determine the prepayment savings on a bank loan of $100,000 at 12.5% for 15 years for which you have been paying $1232.52 each month under each condition. Assume that there is a prepayment penalty of $500.

18. It is prepaid 42 months early.
19. It is prepaid 6 years early.

Use the Rule of 78 to determine how much interest will be saved by prepaying a one-year loan of $1500 at 10% after each time period.

20. 4 months $69.23
21. 6 months $40.38
22. 10 months $5.77

KEY TERMS

amortization schedule
interest due
note reduction
payment number
prepayment penalty
Rule of 78
unpaid balance

INDEPENDENT PRACTICE

ASSIGNMENT GUIDE

Exercises 1-3 ask students to demonstrate their understanding of the concepts of the lesson. Exercises 4-15 ask students to use the amortization formulas in a spreadsheet program. Exercises 16-22 assess students abilities to apply the prepayment formulas taught in the lesson.

CLOSURE

Ask students to define interest due, note reduction, and unpaid balance. Ask students to explain how each is calculated. Ask students to define each of the variables in the prepayment formula and to explain why an individual would choose to prepay a loan. Have students explain when a prepayment penalty would discourage them from prepaying a loan.

MIXED REVIEW

5-1 **1.** Determine which of the following loans has a greater total cost:

a. $10,000 at 12% for 5 years **b.** $11,000 at 11% for 3 years

1-1 **2.** During the first week of April, Brielle worked 34 hours and received $44.95 in tips. If her hourly rate of pay is $5.75, how much did she earn that week? $240.45

3-2 **3.** Carmen has $8500 in a CD that pays 6% interest, compounded semiannually. How much is her CD worth at the end of 2 years?

1-4 **4.** Angela earns $125 in gross income each week and is single and claims one exemption. How much will her employer withhold for income tax? $5

1-3 **5.** Rachel receives a commission of 4% on sales up to $10,000 and 6% on all sales over $10,000. How much has she earned in commission if her sales are $18,250? $895

4-5 James Wilcox is a farmer with 160 acres on which to plant. From crop x he can earn $200 per acre; from crop y he can earn $150 per acre. To meet demands, he must plant at least 40 acres of crop x and 60 acres of crop y. Soil conditions will permit him to plant no more than 100 acres of crop x.

6. Write the objective function for his total income. $I = 200x + 150y$

7. Write inequalities that express each constraint. See Additional Answers.

8. Graph the inequalities.

9. Find the corners of the region in Exercise 8. (40, 60), (40, 120), (100, 60)

10. What is the number of acres of each crop that James Wilcox should plant to maximize his income? Crop x: 100; Crop y: 60

LESSON QUIZ

1. Create an amortization schedule for the first 3 months of a $3500 loan at 8% for 3 years for which the monthly payment is $109.68.

Create a 5-column schedule with the following column heads: PAYMENT NUMBER, PAYMENT AMOUNT, INTEREST DUE, NOTE REDUCTION, and UNPAID BALANCE.

The three rows in the schedule will include the following entries:

Row 1: 1; 109.68; 23.33; 86.35; 3413.65

Row 2: 2; 109.68; 22.76; 86.92; 3326.73

Row 3: 3; 109.68; 22.18; 87.50; 3239.23

2. You have a bank loan of $22,500 at 10% for 15 years for which you have been paying $241.79 each month. Assuming there is no prepayment penalty, determine the amount you could save if you prepaid the loan 4 years early. (**$2072.59**)

3. A $1600 appliance is financed at 21% per year with no down payment at a store that applies the Rule of 78. The loan must be paid in 12 equal installments. Determine the savings if it is prepaid after 6 months. (**$90.46**)

5-5 THE LURE OF CREDIT TERMS: THE MERCHANT PROFITS

QUICK REFERENCE

TEACHER'S RESOURCES AND ANSWER KEY

Lesson Quiz Answers 5-5, p. 56

Reteaching Activity Answers 5-5, p. 99

Enrichment Activity Answers 5-5, p. 100

EXTENSION ACTIVITIES

Reteaching Activity 5-5, p. 35

Enrichment Activity 5-5, p. 40

Maria discovered that there are methods to reduce the cost of credit. Selecting a shorter amortization period, a lower interest rate, and a larger down payment enables the borrower to reduce the total payments on a loan. In addition, prepaying a loan can often save the borrower money. Many car ads mention terms that include cash-back opportunities and significantly lower interest rates than financial institutions are offering.

Maria wonders whether she can save money on the purchase of her car by taking advantage of some of these terms. Patrick warns that merchants can increase their profits by enticing customers to choose credit terms that are favorable to the merchants.

OBJECTIVES: *In this lesson, we will help Maria to:*

- *Learn how merchants generate credit income and attract customers with special credit plans.*
- *Calculate the total financed price with lower-than-market-rate interest and rebate plans.*
- *Determine the merchants' profits with lower-than-market-rate interest and rebate plans.*
- *Observe the drawbacks associated with rent-to-own plans.*

SPECIAL CREDIT PLANS

1. Since stores must pay credit card companies 4–7% of the purchase price of each item they sell on credit, many customers feel that they are entitled to discounts on their cash purchases. By law, retailers have the right to offer **cash discounts** to customers who pay for their purchases in cash as long as the discounts are offered to all potential customers. However, the law prohibits retailers from adding extra charges to credit card purchases to cover their credit card expenses.
2. The **rent-to-own plan** enables you to purchase expensive items now and pay for them with small payments over a specified period of time. These payments often add up to a larger amount than they would have if you had financed the purchase outright. If you decide to buy the items, then your rental fees are applied to the purchase price. However, if you decide that you cannot afford the items or do not want them, you take them back to the store, which then keeps all the money you have already paid. This is the price you pay for renting. You are spared the embarrassment of breaking an installment loan contract and having your credit rating blemished, but you have lost the money you paid to rent the items.
3. One particularly attractive incentive that is used to promote sales is the extension of credit terms at an interest rate that is lower than the **market rate**, which is the rate currently available on the open market. Automobile companies use this device extensively. Since the dealers can offer to finance your loan through the company's subsidiary finance corporation, they can offer you a lower interest rate than one that might be available through your bank.
4. Some automobile dealers offer an alternative to a low-interest-rate loan called a **rebate** or cash-back plan. When you purchase the car, you receive cash back to be applied to the purchase price, thereby lowering the purchase price.

Ask Yourself

1. Why would a merchant be willing to offer a discount for a cash purchase?
   See Additional Answers.
2. What is a rent-to-own plan?
   rental fees are applied to the purchase price
3. What companies frequently offer credit at interest rates that are lower than those offered by banks?
   car dealerships

CRITICAL THINKING

After students have answered the questions, have them consider why a car company would create a financing subsidiary. Ask them to research the interest rates currently being offered by these financing subsidiaries.

ALGEBRA *REVIEW*

Evaluate to the nearest thousandth.

1. $\dfrac{1850(4)^6}{(7+3)^6}$
   7.578
2. $\dfrac{7500(2+0.62)^3}{(9.7-2.1)^3}$
   307.273
3. $\dfrac{79(0.006)(1+0.006)^{60}}{(1+0.006)^{60}-1}$
   1.572
4. $\dfrac{900(0.0025)(1+0.0025)^{72}}{(1+0.0025)^{72}-1}$
   13.674

Determine which quantity is smaller.

5. $x = \dfrac{10{,}000(0.0083)(1+0.0083)^{48}}{(1+0.0083)^{48}-1}$

   $y = \dfrac{10{,}500(0.005)(1+0.005)^{48}}{(1+0.005)^{48}-1}$

   y
6. $q = \dfrac{12{,}500(0.006)(1+0.006)^{12}}{(1+0.006)^{12}-1}$

   $t = \dfrac{14{,}500(0.0055)(1+0.0055)^{12}}{(1+0.0055)^{12}-1}$

   q
7. $48x$ or $48y$
   $48y$
8. $12q$ or $12t$
   $12q$

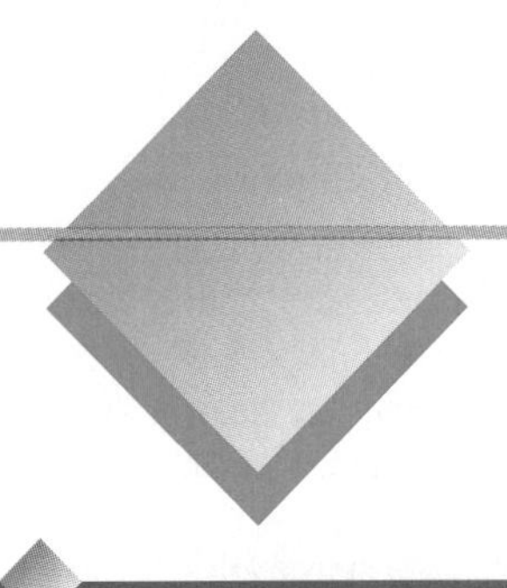

SHARPEN YOUR SKILLS

FOCUS

ALGEBRA CONNECTIONS

Would you prefer to buy a car for $12,000 financed at 15.9% for 4 years or to buy the same car for $13,000 financed at 5% for 4 years? (**$13,000 at 5% for 4 years yields the lower total financial price.**)

INSTRUCTION

TEACHING THE LESSON

Have students read the opening scenario. Use the *Algebra Connections* above to illustrate that there are often different payment plans for purchasing an automobile. Have students read the section on special credit plans. Have them discuss the benefits to the merchant as well as the buyer under each plan. After completing Skills 1-3, have students decide which of the following plans yields the lowest total financed price to the consumer and which plan yields the highest profit to the company on the purchase of a new $25,395 car.

Plan 1: Rebate = 3000.00

Plan 2: Rebate = 1500
Rate = 8.5%
Time = 48 mos.

Plan 3: Rebate = 0
Rate = 5.9%
Time = 48 mos.

Assume that the consumer can borrow the money at a market rate of 11.5% for 4 years in Plan 1. (**Plan 1 yields the lowest total financed price to the consumer. Plan 3 yields the highest profit to the company.**)

SKILL 1

Maria has just discovered the ad shown for a new luxury model car. She observes that the $1000 cash-back plan offers no financing. Therefore she assumes that she will have to apply for her own bank financing at 13.75% over 48 months if she selects this plan.

New Luxury Model

$1000 CASH BACK

— OR —

$750 CASH BACK Plus
6.9% APR

— OR —

2.9% APR
Available

FACTORY AIR

- Wide Vinyl Bodyside Mold
- AM/FM 4 Speaker Stereo
- Tinted Glass
- Power Steering
- Interval Wipers
- Rear Window Defroster

MSRP	$10,154
FACTORY DISC.	–744
FACTORY REBATE	–1000
OUR DISC.	–1035
YOUR PRICE	**$ 7375**

- Instrumentation Group
- Digital Clock w/over Console Light/Security Group
- Dual Elect. Mirrors
- Luxury Wheel Covers
- 1.9L EFI Engine

EXAMPLE 1 Plan 1 offers a purchase price of $7375 and a $1000 rebate with no financing. (13.75% bank financing is available.)

Plan 2 offers a purchase price of $7625 and a $750 rebate at 6.9% over 48 months.

Plan 3 offers a purchase price of $8375 and no rebate at 2.9% over 48 months.

QUESTION Which plan yields the lowest total financed price to the consumer?

SOLUTION

Use the monthly payment formula and your calculator to find the monthly payment and total financed price for the car under each plan.

	Plan 1	Plan 2	Plan 3
Rebate	$1000	$750	0
Rate	13.75%	6.9%	2.9%
Time	48 months	48 months	48 months
Loan amount	$7375	$7625	$8375
Monthly payment	$200.61	$182.24	$185.01
Total financed price	$9629.24	$8747.35	$8880.24

The lowest total financed price of $8747.35 results from Plan 2 with a rebate of $750 and financing at 6.9%. Maria must carefully consider the details of each financing deal. Looking at the lowest purchase price, the lowest interest rate, or the highest rebate is not enough!

FOCUS ON ALGEBRA

In this lesson students explore different payment plans and how these plans impact the cost of major purchases. (*NCTM Standard 1, p. 137*)

COMMON ERRORS

Students may have difficulty looking at various payment plans from the point of view of the seller. Point out that the automobile company and its finance subsidiary save money when the consumer does not take the rebate and that the interest income goes to the company if the consumer accepts financing from it instead of getting a bank loan.

SKILL 2

EXAMPLE 2 Maria realizes that the automobile company and its financing subsidiary would save money if she did not take either rebate. She wonders which plan the company would prefer her to select.

QUESTION Which plan yields the highest profit to the company?

SOLUTION

Calculate how much the company would save if the customer did not take the rebate and how much interest income it would make for each plan. This represents profit *over the profit built into the price of the car.* The company earns interest income only in Plans 2 and 3, and the amount of that interest is equal to the difference between the total financed price and the amount borrowed.

	Plan 1	Plan 2	Plan 3
Interest	0	$1122.35	$ 505.24
Rebate not paid	0	250.00	1000.00
Profit	0	1372.35	1505.24

In Plan 1, the company gets to keep none of its factory rebate. In Plan 2 it keeps $250 of the rebate, paying out $750. In Plan 3 it keeps all of the $1000 rebate. The most advantageous plan for the company is Plan 3. An automobile company may give its dealers incentives to encourage you to select the plan that is most profitable for the company.

INTERDISCIPLINARY INVESTIGATION

Finance: Ask students to research the founding of GMAC, General Motors' financing subsidiary. Have them research how money is exchanged between the two companies and whether one company can have a profitable year while the other loses money.

RETEACHING

Some students may benefit from a review of translating information presented in words in an advertisement or article into numbers that can be analyzed mathematically. To reteach the lesson, concentrate on this skill.

ENRICHMENT

Challenge students to compare and contrast the effects on total payments of rebates and down payments.

SKILL 3

Rent-to-own plans are especially attractive to young people who have not yet established a credit rating and would not qualify for an installment loan. In addition, people who have already overextended their credit and are not eligible for any more may decide to extend themselves even further by using this plan. This plan has the downside risk that all rental money is forfeited if the customer decides not to keep the item.

EXAMPLE 3 Charro wants to buy a TV/VCR, but she cannot afford it at this time. The Houston Home Company offers a rent-to-own credit plan that will allow her to rent it now at \$30 per month for 24 months. If she decides to buy it, her rental fees will be applied to the purchase price. If she does not choose the rent-to-own plan, she can purchase the TV/VCR for \$600 with an installment loan at 12% interest over 2 years.

QUESTION How much money does she save by obtaining the installment loan rather than using the rent-to-own plan? How much will she lose if she can no longer afford the TV/VCR after 9 months?

SOLUTION

With the rent-to-own plan, she will pay 24(30) = \$720.

To determine her total payments with the installment loan, Charro used her calculator and the monthly payment formula. She will pay \$677.86.

This is a savings of 720 − 677.86 = \$42.14 over the rent-to-own plan.

She will lose 9(30) = \$270 if she can no longer afford the rental payments after 9 months.

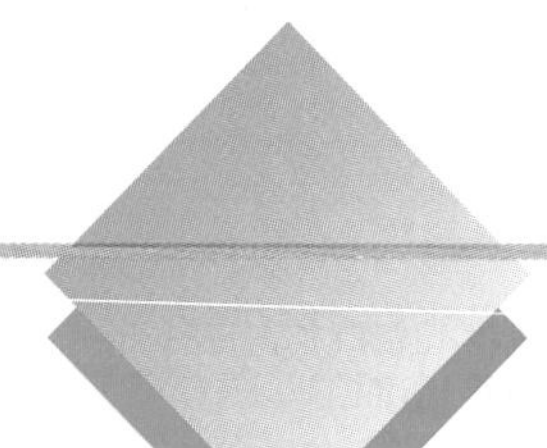

TRY YOUR SKILLS

A car company is offering a new car priced at \$34,600. The following financing plans are available. Assume the loans in Plan 1 are from a bank or other lending institution and that the loans in Plans 2 and 3 are from the car manufacturer's financing subsidiary.

	Plan 1	Plan 2	Plan 3
Rebate	\$2000	\$1000	0
Rate	10.5%	7.5%	3.0%
Time	48 months	48 months	48 months

22. Create an amortization schedule to show the interest due, note reduction, and unpaid balance for the first six months on an $8500 loan for 2 years at 7.5% for which the monthly payments are $382.50. See Additional Answers.

23. How much money will you save by prepaying 13 months early a loan of $37,000 at 9% for 3 years for which the monthly payment is $1176.59? Assume that there are no prepayment penalties. $773.79

A $900.00 stereo system purchase is financed at 14% per year with no down payment at a store that applies the Rule of 78. It must be paid in 12 equal installments. Determine the prepayment savings under each condition.

24. It is prepaid after 5 months. **25.** It is prepaid after 8 months.

26. Determine the plan that would provide the lowest total financed price for a car priced at $19,500 by completing the table below. Assume that the loan in Plan 1 is from a bank or other lending institution and that the loans in Plans 2 and 3 are from the car manufacturer's financing subsidiary. Plan 3

	Plan 1	Plan 2	Plan 3
Rebate	$1750.00	$1250.00	0
Rate	10.75%	9.9%	3%
Time	36 months	36 months	36 months
Loan amount	$17,750.00	$18,250.00	$19,500.00
Monthly payment	$579.01	$588.02	$574.85
Total financed price	$20,844.47	$21,168.71	$20,694.62

27. Determine which plan yields the highest profit to the company in Exercise 26 by completing the table below. Plan 2

	Plan 1	Plan 2	Plan 3
Interest	0	$2918.71	$1194.62
Rebate not paid	0	500.00	1750.00
Profit	0	3418.71	2944.62

Use the information that you learned in this chapter to answer each of the following. See Additional Answers.

28. Name three indicators that a borrower is experiencing credit overload.

29. Is it always better to choose a shorter amortization period? Explain your answer.

30. What are three ways to reduce the cost of installment loans?

31. Why does the interest portion of a monthly payment decrease each month?

32. Will the financing plan with the largest rebate always yield the lowest total financed price? Why or why not?

ADDITIONAL ANSWERS

24. $45.23

25. $16.15

28. missed installment payments; repeated overdue notices; little or no savings

29. No; sometimes you cannot afford the higher monthly payments.

30. longer amortization period; lower interest rates; higher down payment

31. The interest due is the product of the monthly interest rate and the previous month's unpaid balance. Since the unpaid balance decreases each month, the interest due also decreases.

32. No; other factors will affect the outcome (interest rates, amortization period).

CHAPTER 5 TEST

Find the monthly payment and total payment on each loan. See Additional Answers.

1. $13,333 at 5% for 4 years
2. 87,500 at 7.5% for 25 years

3. How much money can you borrow at 5% for 5 years if you can afford a monthly payment of $340? $18,016.84
4. Make a chart with the headings below to determine how much money you can save by borrowing $10,000 at 12% for 3 years or 4 years instead of 5 years. See Additional Answers.

Number of Years	Monthly Payment	Total Payment	Savings from 5 Years

5. Determine the total cost of a loan for $12,222 at 9.5% for 6 years.

Complete the chart below to find the total payments at 9% and 6% and the savings associated with the 6% interest rate for each loan.

	Loan Amount	Number of Years	Total Payment 9%	Total Payment 6%	Total Savings
6.	$4500	3	$5151.56	$4928.35	$223.21
7.	6750	4	8062.75	7609.15	453.60

8. Make a chart with the headings below to determine how much money you can save on a loan of $6550 at 7.5% for 3 years if you make a down payment of 20% or 30% instead of 10%. See Additional Answers.

Percent Down	Down Payment	Loan Amount	Total Amount	Savings Over 10% Down

9. Determine the interest due, note reduction, and unpaid balance for the first two months on a $3600 loan for 3 years at 6% for which the monthly payments are $109.50. See Additional Answers.
10. How much money will you save by prepaying a loan of $25,000 at 6% for 5 years for which the monthly payment is $483.32 if you prepay it 20 months early? Assume that there are no prepayment penalties. $9177.00
11. Determine the plan that would provide the lowest total financed price for a car priced at $27,500. Assume the loan in Plan 1 is from a bank and the loans in Plans 2 and 3 are from the car manufacturer's financing subsidiary.

	Plan 1	Plan 2	Plan 3
Rebate	$1950.00	$1000.00	0
Rate	12.75%	9.9%	4.9%
Time	48 months	48 months	48 months

12. Determine which plan yields the highest profit to the company in Exercise 11. Plan 2

ADDITIONAL ANSWERS

1. $307.05; $14,738.38
2. $646.62; $193,985.18
5. $3859.44
11. Plan 3

ALTERNATIVE ASSESSMENT

Ask students to write a report on stores that offer rent-to-own plans. Have them gather information on the percent of buyers that choose to return a purchase made on this plan before they have made all the payments. Are more of these items returned in the earlier months of the rental period? Remind students that some information in their reports should be represented graphically.

Ask students to research the interest rates that are currently being offered by automobile financing companies. Does one company offer better rates than another? Are there any companies that offer 0% financing? Does the same company offer different rates for different makes or models in their line? Why or why not? Does each dealer of a certain car offer the same financing options? Remind students that some information in their reports should be represented graphically.

CUMULATIVE REVIEW

2-1 **1.** Tyrone's monthly check costs are $0.035 per check for the first 20 checks and $0.15 for each check over 20. If he writes 42 checks this month, how much will the checks cost him? $4.00

3-1 **2.** Joanie buys a CD for $450 that matures in 1 year, earning 8% simple interest. Every time the CD matures, she withdraws the interest and buys another one-year $450 CD at the same interest rate. If she does this for 3 years in a row, how much interest will she earn in that time? $108

1-2 **3.** Olga earns $7.75 per hour. She works 35 hours per week for 50 weeks and receives two weeks of paid vacation. Her other benefits cost her employer $3500 per year. What is the annual cost to the company of her salary and benefits? $17,605

5-1 **4.** If David can afford a monthly payment of $485, how much can he afford to borrow at a yearly interest rate of 9% for 10 years? $38,286.72

3-2 **5.** Use the Rule of 72 to determine how long it will take your $2300 savings account to double in value if it is growing at an annual rate of 4.5%. 16 years

1-3 **6.** You are a real-estate agent who has sold a house for $275,000. Determine your total commission if you earn 6% on the first $100,000 and 7.5% on any amount over $100,000. $19,125

2-2 **7.** You have $1546.68 in your checking account, and then you write a check for $597.99. Show how you would enter this check in your check register. 1546.68 − 597.99 = 948.69

3-2 **8.** Determine the interest for 9 months if the principal is $1775, the rate is 6%, and the interest is compounded quarterly. $81.08

5-4 **9.** Determine the prepayment savings on a bank loan of $75,000 at 7.5% for 25 years for which you have been paying $554.24 per month if it is prepaid 5 years early. Assume that there is no prepayment penalty. $5594.88

5-2 **10.** Determine which of the following loans has a greater total cost:

a. $20,000 at 10% for 5 years **b.** $25,000 at 12% for 3 years

3-2 **11.** Drucilla has $25,500 in a CD that pays 8% interest, compounded quarterly. How much is her CD worth at the end of 2 years? $29,877.31

4-1 **12.** Jacquie earns $1025 in gross income each week. How much will her employer withhold for FICA taxes? $78.41

5-4 **13.** Use the Rule of 78 to determine how much interest can be saved by prepaying a one-year loan of $1375 at 8% after 7 months. $21.15

5-2 **14.** How much do you save by financing a $65,000 loan at 7% for 5 years instead of 10 years? $13,339.94

5-3 **15.** Explain why making a large down payment decreases the total financed price of a loan. The amount of the loan is lower and therefore the interest payments are less.

ADDITIONAL ANSWERS

10. b is $4396.43 more costly

PROJECT 5–1: **Buying a Car**

You are about to leave for college, and you need a car. You will have to buy a used car because you cannot afford a new one. Start by looking at newspaper advertisements to find the prices of some of the cars in which you might be interested.

1. Cut out six classified ads for cars that are for sale in your area. Attach the ads in a column along the left-hand side of a large sheet of paper.
2. Divide the remainder of your sheet of paper into columns to record the following information: loan amount, monthly payment, total payments, and total cost.
3. Find the loan amount by subtracting a 10% down payment from the price of each of the six cars you chose.
4. Use the monthly payment formula to find the monthly payment for a loan of 5 years for each of the six cars.
5. Find the total payments made over 5 years for each of the six cars.
6. Find the total cost for each of the six cars you chose.
7. Estimate your possible savings and income by the time you will be ready for college. Will you be able to afford to buy one of these cars?
8. Visit the car dealers that are offering the cars in which you are interested. Compare the values of each of the cars to the sale price. Write a report explaining which of the cars is the best deal.

PROJECT 5–2: **Selecting a Loan**

Suppose you buy a car for $5000 and have $1000 for a down payment. You must finance $4000. To find the best terms for a loan, you investigate possible sources of credit in your neighborhood. Possible sources of credit are a bank, a finance company, the car dealer, and a credit union. Also, remember that not all banks have the same terms for financing loans, so it is wise to check out several.

1. Select four sources of credit.
2. Complete a chart with the following headings for each of the sources.

Source of Credit	Monthly Payment	Number of Payments	Amount Paid	Finance Charges	Total Cost

3. Compare the total cost of the car for each source.
4. Which source of credit saves you the most money?
5. How much money can you save by using this source of credit?
6. Compare your findings with those of your classmates.

ALGEBRA *REFRESHER*

When simplifying expressions, use the following rules for *order of operations:*

- First, perform operations within parentheses or other grouping symbols.
- Next, simplify any exponents.
- Then, multiply and divide in order from left to right.
- Last, add and subtract in order from left to right.

Simplify each expression. Round your answer to the nearest hundredth.

1. $7 + (4 \cdot 2)$
   15
2. $(7 + 4) \cdot 2$
   22
3. $3 \cdot \left(8 + \frac{4}{3}\right)^4$
   22,765.04
4. $3 \cdot 8^4 + \frac{4}{3}$
   12,289.33
5. $13^2[(2 \cdot 9)^3(7 \div 9)]$
   766,584
6. $[13^2(2 \cdot 9^3)][7 \div 9]$
   191,646
7. $2.78(9876 - 500 \div 25)$
   27,399.68
8. $16[4 - 3.1(42 - 3^4 \div 9) + 7]$
   −1460.80

A *relation* is a set of ordered pairs of numbers. The set of ordered pairs $\{(0, 2), (1, 3), (2, 5), (3, 10), (5, 14)\}$ is a relation. The set of replacements for the first variable x is called the *domain* of the relation. The set of replacements for the second variable y is called the *range* of the relation. A *function* is a relation in which each member of the domain is paired with exactly one member of the range. The above relation *is* a function because for each x there is only one y. The relation $\{(2, 3), (3, 4), (2, 6)\}$ is *not* a function because the x value of 2 is paired with y values of 3 *and* 6.

Find the domain and range of each relation.

9. $\{(7, 3), (3, 7), (2, 4)\}$
   d: {7, 3, 2}; r: {3, 7, 4}
10. $\{(8, 3), (3, 8), (8, 8)\}$
    d: {3, 8}; r: {3, 8}
11. Is the relation in Exercise 9 a function? Why or why not?
    Yes; each member of the domain is paired with a different member of the range.
12. Is the relation in Exercise 10 a function? Why or why not?
    No; the *x* value of 8 is paired with *y* values of 3 *and* 8.

Solve for y if $x = 16$.

13. $y = x^2 + x$
    272
14. $y = x^3$
    4096
15. $y = x^{\frac{1}{4}}$
    2
16. $y = x^4 - x$
    65,520
17. $y = \sqrt{x}$
    4
18. $y = \sqrt[3]{x + 11}$
    3
19. The Berger family spent the following amount of money on food in 1993:

    January: $500 6.89% February: $610 8.41% March: $715 9.85%
    April: $650 8.96% May: $509 7.01% June: $636 8.76%
    July: $702 9.67% August: $653 9.00% September: $619 8.53%
    October: $515 7.10% November: $598 8.24% December: $550 7.58%

    Determine the percent of their 1993 food costs that occurred in each month to the nearest hundredth of a percent.

CREDIT CARDS

THE AMERICAN ECONOMY HAS GRADUALLY shifted away from a system in which individuals and families produced or grew most of what they needed, and offered these products in exchange for goods and services provided by others. In America today, the production and distribution of goods and services is on a much larger scale. Our economy is so specialized that companies exist not only to create products and services, but also to convince us that we should acquire more of these products. The banking industry processes the money exchanged among businesses and offers many services that keep our economy flowing.

One major service offered to consumers by banks, retailers, and other companies is the credit card (or charge card). Bankers and merchants have realized that if they make credit convenient and painless for us, merchants will sell more. The banks can then charge both the consumer and the merchant for the use of their credit cards. We seem to be more willing to hand over a plastic card than money, so we have a tendency to spend more with credit cards.

If we use our credit cards for a period of time and pay small finance charges each month, we may be tempted to use them over and over again until the monthly finance charges become quite large. Our wants become needs. We can no longer afford to buy items that we really need because all of our extra money is spent keeping up the minimum payments for the items that we previously bought on credit.

Fortunately, we do not have to go completely under and declare bankruptcy.

If we watch our spending and begin to repay our debts, we can slowly diminish our credit load. Congress has passed laws to help protect us from ourselves. However, part of the price we pay for living in a free society is the opportunity to make bad choices and the responsibility for coping with the consequences of those choices.

In this chapter our high school students will take a close look at the costs and conveniences of using credit cards. They will learn how individuals qualify for credit cards. They will observe some of the warning signs of credit overload, as well as some of the difficulties we face once we are in over our heads. They will also examine some steps to take to regain financial stability.

6-1 CREDIT CARDS: PLASTIC MONEY

QUICK REFERENCE

TEACHER'S RESOURCES AND ANSWER KEY
Lesson Quiz Answers 6-1, p. 58
Reteaching Activity Answers 6-1, p. 101
Enrichment Activity Answers 6-1, p. 102

EXTENSION ACTIVITIES
Reteaching Activity 6-1, p. 41
Enrichment Activity 6-1, p. 48

Sylvia Shawn enjoys spending money. She has a bumper sticker on her car that reads "When the Going Gets Tough, the Tough Go Shopping." Her parents are divorced, and she lives with her mother. Since her mother rarely takes her shopping, Sylvia looks forward to shopping with her father during the two weekends that she sees him every month. Paying for everything seems simple for him, but he rarely carries any cash.

Their first stop is usually at a gas station to fill up the tank of his yellow sports car. While there, Sylvia picks up a snack and a soft drink, and her father grabs several packs of baseball cards. He pays for these purchases with his handy oil credit card. Then they are off to Sylvia's favorite shopping mall, where he waits patiently while she tries on eight or ten outfits before selecting two or three. He also lets her buy jewelry and accessories to match her outfits. He always pays with a credit card. It seems as though he has a credit card for every department store!

Since Sylvia will be going away to college next year, her father has been helping her select an elaborate nine-piece set of luggage. For this purchase, he uses a card called a VISA card. He has several VISA cards from different banks across the country. He calls them his "plastic money" cards.

When they finish shopping, Sylvia's dad often takes her to a fancy restaurant for dinner, where he uses his American Express card. He calls it his "entertainment expense account" card, and he uses it for restaurant meals, airline tickets, car rentals, and other travel expenses.

Sylvia does not understand how he got so many cards and why he does not carry real money instead. Her mother has several credit cards, but she rarely uses them. When her parents were still married, they frequently argued about the use of credit cards. Her mother is particularly against using credit cards for cash advances. Sylvia would like to know more about credit cards.

OBJECTIVES: In this lesson, we will help Sylvia to:

- *Describe how credit cards are used.*
- *Identify three categories of credit cards and several companies that offer the cards in each category.*
- *Determine the new balance that will be shown on a credit card statement after a purchase is made.*
- *Calculate the effective rate of interest on a credit card purchase.*

ALGEBRA CONNECTIONS

Katie has some pennies, nickels and dimes. She has less than 50 cents. What is the most money she can have and not be able to give change for a quarter? (**44 cents**)

WHAT IS A CREDIT CARD AND HOW DOES IT WORK?

A **credit card** is a small plastic card, approximately 2" by $3\frac{1}{2}$", that identifies the holder and extends to the cardholder an unsecured line of credit. Most cards entitle the holder to charge purchases up to a maximum **credit limit.** By signing the application for the card, the cardholder agrees to pay for purchases according to the credit card terms. The **Truth in Lending Act** requires that the credit card company disclose all terms associated with its card, including the annual percentage rate, finance charges, **late payment penalties,** grace period, membership fee, and so on. Examples of companies that issue these cards are local department stores, national retail chains, car rental companies, travel and entertainment companies, airlines, telephone companies, oil companies, hotel/motel chains, and commercial banks.

When a customer uses a credit card in a store, the merchant turns in the charge to the store's bank. The store's bank notifies the bank that issued the card that a purchase has been made. The issuing bank then bills the cardholder for all purchases made that month. The cardholder writes one check for all purchases payable to the issuing bank. Through the bank card system, the card issuer then pays the merchant's bank. The merchant's bank has already paid the merchant.

TYPES OF CARDS

There are four categories of cards: (1) single-purpose credit cards; (2) multipurpose travel, food, and entertainment cards; (3) all-purpose bank credit cards; and (4) debit cards.

Single-purpose credit cards include those from oil companies such as Exxon, Mobil, and Shell and those from stores such as Sears, Montgomery Ward, and Neiman Marcus. Sylvia's father uses an oil company card and cards from the stores in the mall, which are all single-purpose cards. There is no fee for this kind of card, and the card allows him to make purchases or obtain services at the store, station, or company that issued it. The purpose of such a card is to encourage you to buy exclusively from a particular company. If you have a credit card from one specific oil company, you are likely to buy much of your gasoline at that company's stations.

Multipurpose travel and entertainment cards include American Express, Diners Club, and Carte Blanche. There is an annual membership fee for these cards. For this membership fee, these cards

CRITICAL THINKING

Suppose you have $6000 in a savings account earning 5% compounded semiannually. Your VISA bill is $1300 and accrues finance charges at a rate of 1.5% monthly. Determine your total net worth after 6 months if you pay in monthly installments of $300. Compare that to your net worth after 6 months if you pay the entire VISA bill immediately. What can you conclude from this?

ADDITIONAL ANSWERS (ASK YOURSELF)

1. 1) annual percentage rates, 2) finance charges, 3) payment penalties

provide a variety of services that some credit cards do not provide, such as guaranteed check cashing at hotels and airline counters, free travel insurance, and emergency card replacement. Usually no credit limit is imposed on the cardholder. However, these are *charge cards,* not credit cards, so the cardholder is expected to pay the entire unpaid balance upon receipt of the monthly statement.

All-purpose bank cards include VISA, MasterCard, and Discover cards. These cards usually carry an annual membership fee. They are widely accepted by retailers as well as by restaurants, hotels, airlines, and other companies that wish to cash in on the boost in sales that accompanies the acceptance of credit cards. Some banks offer cardholders special purchase insurance and worldwide travel assistance.

An advantage to using credit cards is that cardholders are covered by Regulation Z, which is the regulation that enforces the Truth in Lending Act passed by Congress. This regulation protects consumers in cases of error, if merchandise is not received, or if services are not rendered. Card users may receive more protection than they would if they paid by check or cash, since it may be easier to get a credit card credit than to get a cash refund.

You can often use these cards to get instant cash up to a limited amount at the bank, credit union, or 24-hour teller facility. Some banks provide special checks with your credit card that you can use to pay for a purchase or to make a deposit into your regular checking or savings account. Using these checks is similar to getting a cash advance.

Debit cards are essentially electronic checks. They enable cardholders to use their automated teller machine (ATM) cards to pay for purchases at local stores. The purchase amount is deducted directly from the cardholder's checking account. Debit cards enable merchants to use credit card companies' existing authorization, draft transmission, and settlement systems.

ALGEBRA *REVIEW*

Express each percent as a decimal.

1. 12.5%
   0.125
2. 3.6%
   0.036
3. 119%
   1.19
4. 125.6%
   1.256

Express each decimal as a percent.

5. 33.3
   3330%
6. 122.3
   12,230%
7. 0.76
   76%
8. 1.234
   123.4%

Evaluate. Give answer to the nearest hundredth.

9. $\left(1 + \frac{0.125}{12}\right)^{12} - 1$
   0.13
10. $\left(1 + \frac{0.065}{12}\right)^{11} - 1$
   0.06

ASK YOURSELF

1. What are three terms that must be disclosed by a bank according to the Truth in Lending Act?
   See Additional Answers.
2. What are two companies that offer credit or charge cards in each of the following categories?
   a. Single-purpose cards
      Answers may vary.
   b. Multipurpose travel and entertainment cards
      Answers may vary.
   c. All-purpose bank cards
      Answers may vary.

SHARPEN YOUR SKILLS

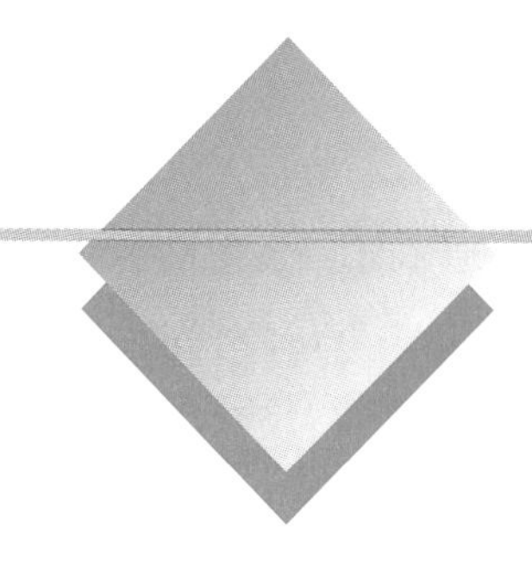

SKILL 1

One major difference between an *installment credit* purchase and a *credit card* purchase is the amount of time that elapses before you must pay interest on the purchase. Remember that interest is charged in the first month on an installment loan. However, for a credit card, when you receive the monthly statement, it tells you how much time you have to pay the bill before a finance charge is added. This amount of time is called the **grace period.** Thus you will have the amount of time that elapses between the time of your purchase and the time you receive your statement *plus* the grace period to pay for the purchase before receiving any finance charges. (Usually there is no grace period if you use your credit card to get a cash advance.) Assume here and throughout the lesson that payments on credit card accounts are received on the last day of the month.

FOCUS ON ALGEBRA

In this lesson, students must convert between decimals and percents and use the effective interest rate formula which requires order of operations and correct calculator manipulation. (*NCTM Standard 5, p. 150.*)

EXAMPLE 1 Sylvia wants her father to buy her a sewing machine that costs \$550. He charges it on a new VISA card with which he has made no other purchases. He will make monthly payments of \$60.

QUESTION What new balance will be shown on his first three monthly statements if his bank applies a 1.5% monthly finance charge and he refrains from using the credit card during this time?

SOLUTION

His first monthly statement will show a previous balance of \$0 and a purchase of \$550. No interest is charged this month, so the new balance for the first month is \$550.

His second monthly statement will show a previous balance of \$550, an interest charge of $0.015(550) = \$8.25$, and a payment of \$60, which was received on the last day of the month. Therefore

$$\begin{aligned}\text{New balance} &= \text{old balance} + \text{interest} - \text{payment}\\ &= 550.00 + 8.25 - 60.00\\ &= 498.25\end{aligned}$$

The new balance for the second month is \$498.25.

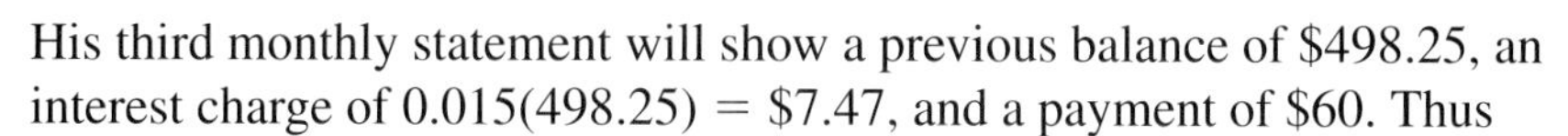

His third monthly statement will show a previous balance of \$498.25, an interest charge of $0.015(498.25) = \$7.47$, and a payment of \$60. Thus

$$\text{New balance} = 498.25 + 7.47 - 60.00 = 445.72$$

The new balance for the third month is \$445.72.

TEACHING THE LESSON

Spend 2 days on this lesson. On the first day, make a chart summarizing the four categories of cards and their basic functions. Pose situations to students and ask them which type of card would be appropriate to the given situation. Have students consider which card they would prefer to have if they were able to qualify for exactly one card and tell them to give reasons.

After reading the Skill 1 section have students practice determining monthly balances with varying initial balances such as $800, $1200, and $1500. Have them repeat their calculations, changing the monthly finance charge to 1.8% and tell them to compare results.

On the second day, introduce APR and effective interest rate. Have students compute effective interest rates based on 16%, 19%, and 20%, both with and without the 1-month grace period.

EXAMPLE 2 Sylvia noticed that there is a section on the credit card statement that summarizes information each month. For the second month, it appeared as follows:

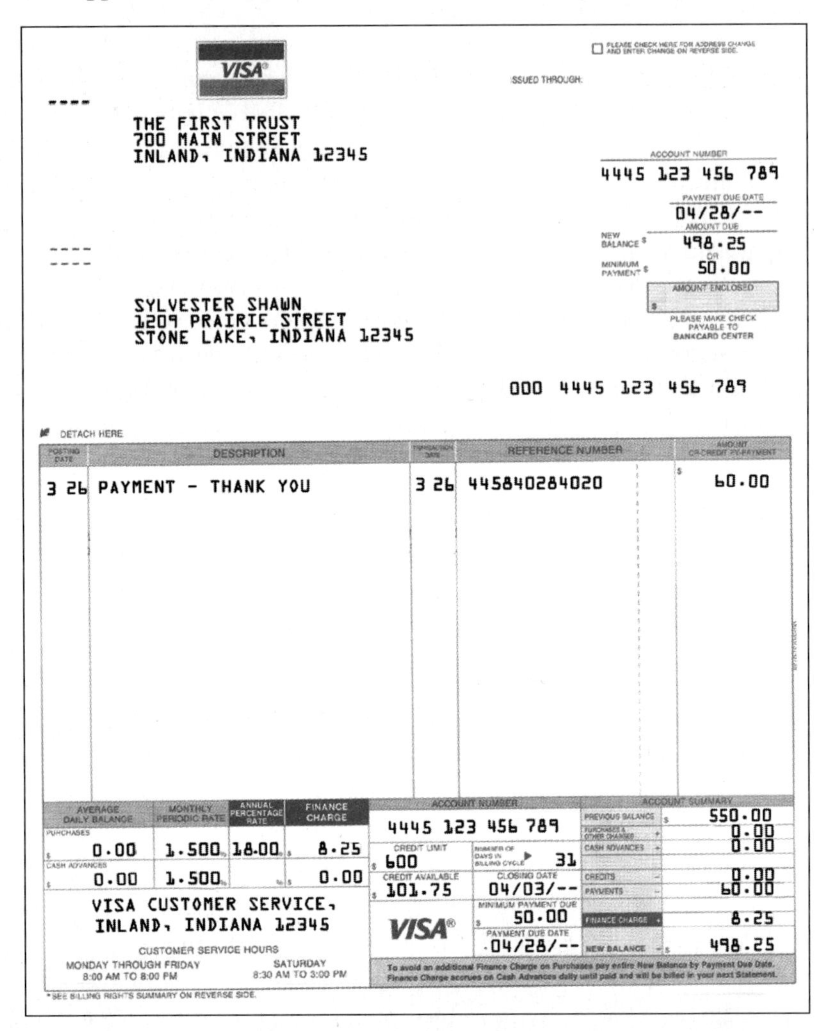

VISA

ISSUED THROUGH:

THE FIRST TRUST
700 MAIN STREET
INLAND, INDIANA 12345

ACCOUNT NUMBER
4445 123 456 789

PAYMENT DUE DATE
04/28/--

AMOUNT DUE
NEW BALANCE $ 498.25
OR
MINIMUM PAYMENT $ 50.00

AMOUNT ENCLOSED
$

PLEASE MAKE CHECK PAYABLE TO BANKCARD CENTER

SYLVESTER SHAWN
1209 PRAIRIE STREET
STONE LAKE, INDIANA 12345

000 4445 123 456 789

DETACH HERE

POSTING DATE	DESCRIPTION	TRANSACTION DATE	REFERENCE NUMBER	AMOUNT CR-CREDIT PY-PAYMENT
3 26	PAYMENT - THANK YOU	3 26	445840284020	$ 60.00

	AVERAGE DAILY BALANCE	MONTHLY PERIODIC RATE	ANNUAL PERCENTAGE RATE	FINANCE CHARGE
PURCHASES	$ 0.00	1.500%	18.00%	$ 8.25
CASH ADVANCES	$ 0.00	1.500%	%	$ 0.00

VISA CUSTOMER SERVICE,
INLAND, INDIANA 12345

CUSTOMER SERVICE HOURS
MONDAY THROUGH FRIDAY 8:00 AM TO 8:00 PM
SATURDAY 8:30 AM TO 3:00 PM

ACCOUNT NUMBER
4445 123 456 789

CREDIT LIMIT	NUMBER OF DAYS IN BILLING CYCLE
$ 600	31
CREDIT AVAILABLE	**CLOSING DATE**
$ 101.75	04/03/--
VISA®	MINIMUM PAYMENT DUE $ 50.00
	PAYMENT DUE DATE 04/28/--

ACCOUNT SUMMARY	
PREVIOUS BALANCE	$ 550.00
PURCHASES & OTHER CHARGES +	0.00
CASH ADVANCES +	0.00
CREDITS −	0.00
PAYMENTS −	60.00
FINANCE CHARGE +	8.25
NEW BALANCE =	$ 498.25

To avoid an additional Finance Charge on Purchases pay entire New Balance by Payment Due Date. Finance Charge accrues on Cash Advances daily until paid and will be billed in your next Statement.

*SEE BILLING RIGHTS SUMMARY ON REVERSE SIDE.

QUESTION What numbers would be included in this section of the statement for the first 8 months of the loan?

SOLUTION

Use a spreadsheet program to calculate the previous balance, new charges, payment received, finance charges, and new balance for each month. Remember that there is a 1.5% monthly finance charge. Be sure to use a rounding function when calculating the finance charges.

	A	B	C	D	E	F
1	Month	Previous	New	Finance	Payment	New
2		Balance	Charges	Charges	Received	Balance
3	1	0.00	550.00	0.00	0.00	550.00
4	2	550.00	0.00	8.25	60.00	498.25
5	3	498.25	0.00	7.47	60.00	445.72
6	4	445.72	0.00	6.69	60.00	392.41
7	5	392.41	0.00	5.89	60.00	338.30
8	6	338.30	0.00	5.07	60.00	283.37
9	7	283.37	0.00	4.25	60.00	227.62
10	8	227.62	0.00	3.41	60.00	171.03

@ROUND(0.015*B4,2)

+B4+C4+D4−E4

TECHNOLOGY HINT

The formula @ROUND(0.015*B4, 2) ensures that the finance charge will be rounded to two decimal places. If @ROUND was not used, the computation 0.015 • 498.25 would yield $7.47375 (a number that includes fractions of a cent).

COOPERATIVE LEARNING

Have students work in groups of three or four to complete a table such as that shown in Example 2 assuming a previous balance of $600, monthly payments of $60 and an APR of 18%. Tell them to continue the table until the new balance is 0. Have them compare total interest paid to total payment received.

COMMON ERRORS

Students may assume that the APR is in fact the effective interest rate. Have them do several examples in converting from the APR to the effective interest rate to impress upon them the difference.

SKILL 2

As interest rates in the economy fluctuate, banks may change the interest rates they charge their credit card customers. When interest rates are high, banks often advertise their *monthly* rates. When interest rates are low, they often advertise their *yearly* rates.

When an **annual percentage rate (APR)** of 21% is advertised, the 21% is simple interest. However, credit card interest is charged on a monthly basis. The **monthly interest rate** for an APR of 21% is $0.21 \div 12 = 0.0175$, or 1.75%. Because the interest charge is compounded, you must determine the **effective interest rate** being charged. The following formula can be used to determine the effective interest rate i_{eff} when the APR i is known.

Effective Interest Rate Formula

$$i_{eff} = \left(1 + \frac{i}{12}\right)^{12} - 1$$

where i_{eff} = effective interest rate
i = APR (annual percentage rate)

The effective interest rate on an APR of 21% is

$$i_{eff} = \left(1 + \frac{0.21}{12}\right)^{12} - 1 \qquad i = 21\%$$

$$= 0.2314, \quad \text{or } 23.14\%$$

EXAMPLE 3 Sylvia's mother has a credit card with an APR of 15%. Sylvia claims that the effective interest rate is actually more than 15%. Her mother claims that she pays less than 15% during the first 12 months following a purchase because of the 1-month grace period.

QUESTIONS

a. What is the effective interest rate on the credit card?

b. What is the effective interest rate for the first year following a purchase?

c. Who is correct, Sylvia or her mother?

d. What is the monthly interest rate?

SOLUTIONS

a. To determine the effective rate on an APR of 15%, substitute $i = 0.15$ into the effective interest rate formula.

$$i_{\text{eff}} = \left(1 + \frac{0.15}{12}\right)^{12} - 1 \quad = 0.1608$$

The effective interest rate is 16.08%.

b. To determine the effective rate for the first year, note that Sylvia's mother pays interest for only 11 of the 12 months following a purchase because of the grace period. Her effective rate for the first 12 months is

$$i_{\text{eff}} = \left(1 + \frac{0.15}{12}\right)^{11} - 1 \quad = 0.1464$$

The effective interest rate is 14.64%.

c. Both are correct. An APR of 15% on a credit card produces an effective rate of 16.08%. However, the rate is only 14.64% in the 12 months following a purchase because of the grace period.

d. The monthly interest rate is $0.15 \div 12 = 0.0125$, or 1.25%.

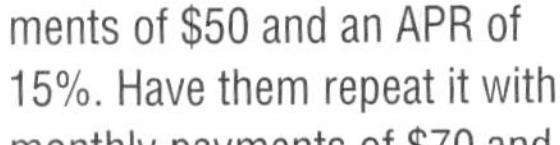

RETEACHING

Have students use spreadsheets to organize data about a purchase of $700 with monthly payments of $50 and an APR of 15%. Have them repeat it with monthly payments of $70 and $100. Tell them to determine the number of months needed to pay off the bill and the total interest payments in each case.

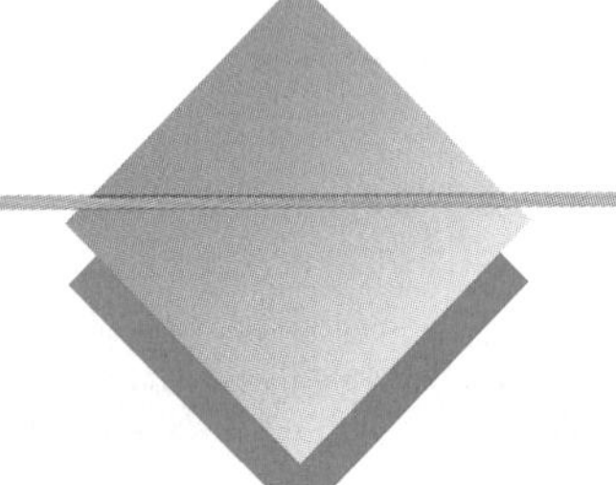

TRY YOUR SKILLS

Using the chart below, determine the new balance shown on your monthly statement for the first 4 months after making a $2300 purchase with your new MasterCard. The card has an APR of 12% and a 1-month grace period. Your monthly payment is $275 and you make no additional purchases.

	Month	Previous Balance	New Charges	Finance Charges	Payment Received	New Balance
1.	1	0.00	$2300.00	0.00	0.00	$2300.00
2.	2	$2300.00	0.00	$23.00	$275.00	2048.00
3.	3	2048.00	0.00	20.48	275.00	1793.48
4.	4	1793.48	0.00	17.93	275.00	1536.41

5. Determine the effective interest rate on a credit card with an APR of 12%.

6. Determine the effective interest rate for the credit card in Exercise 5 during the first 12 months following a purchase if the card has a 1-month grace period. 11.57%

GUIDED PRACTICE

Have students work in groups on the *Try Your Skills* exercises. Tell them to show the results for 1-4 on a spreadsheet.

ADDITIONAL ANSWERS

5. 12.68%

EXERCISE YOUR SKILLS

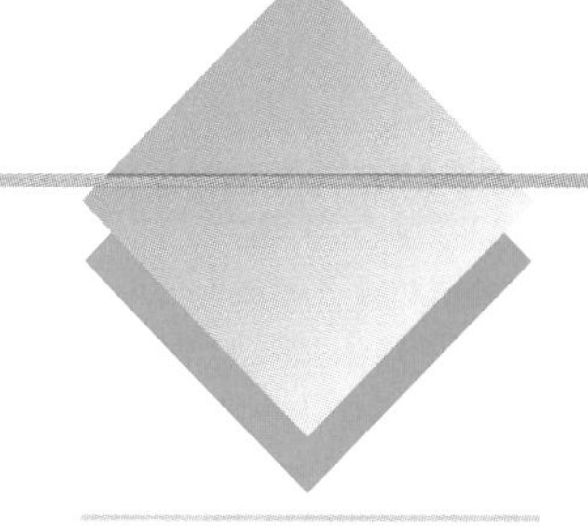

1. Name one major difference between an installment credit purchase and a credit card purchase. See Additional Answers.
2. Describe the difference between an annual percentage rate and an effective interest rate. See Additional Answers.
3. What is the major difference between what you must pay VISA and American Express each month? See Additional Answers.

Complete the chart below to determine the new balance shown on your monthly statement for the first 3 months after you make a $3000 purchase on your new VISA card that has an APR of 16%. The card carries a 1-month grace period. Assume that you make monthly payments of $325 and that you make no additional purchases.

	Month	Previous Balance	New Charges	Finance Charges	Payment Received	New Balance
4.	1	0.00	$3000.00	0.00	0.00	$3000.00
5.	2	$3000.00	0.00	$40.00	$325.00	2715.00
6.	3	2715.00	0.00	36.20	325.00	2426.20

Use a spreadsheet program to determine the previous balance, new charges, payment received, finance charges, and new balance that would be shown on Ming's first 12 MasterCard statements after a purchase of $3215 if he makes monthly payments of $345. The card has a 1-month grace period and an APR of 13.9%. Assume that he has never used the card and that he makes no further charges on it during this time period.

	Month	Previous Balance	New Charges	Finance Charges	Payment Received	New Balance
7.	1	0.00	$3215.00	0.00	0.00	$3215.00
8.	2	$3215.00	0.00	$37.24	$345.00	2907.24
9.	3	2907.24	0.00	33.68	345.00	2595.92
10.	4	2595.92	0.00	30.07	345.00	2280.99
11.	5	2280.99	0.00	26.42	345.00	1962.41
12.	6	1962.41	0.00	22.73	345.00	1640.14
13.	7	1640.14	0.00	19.00	345.00	1314.14
14.	8	1314.14	0.00	15.22	345.00	984.36
15.	9	984.36	0.00	11.40	345.00	650.76
16.	10	650.76	0.00	7.54	345.00	313.30
17.	11	313.30	0.00	3.63	316.93	0.00
18.	12	0.00	0.00	0.00	0.00	0.00

KEY TERMS

- all-purpose bank cards
- annual percentage rate (APR)
- credit card
- credit limit
- debit cards
- effective interest rate
- grace period
- late payment penalties
- monthly interest rate
- multipurpose travel and entertainment cards
- single-purpose credit cards
- Truth in Lending Act

INDEPENDENT PRACTICE

ASSIGNMENT GUIDE

Exercises 1-3 explore the reading at the beginning of the lesson.

Exercises 4-6 ask students to compute balances on a VISA card when making monthly payments on a major purchase.

Exercises 7-18 ask students to use a spreadsheet program to compute balances and finance charges assuming a large installment credit purchase.

Exercises 19-26 require students to find effective and monthly interest rates given the APR.

ENRICHMENT

Have students go to a local bank to find out when the APR on its credit cards and when interest rates on savings accounts have changed over the past 10 years. Tell them to display the information in a table and see if there appears to be any relationship between these two events.

ADDITIONAL ANSWERS (EXERCISE YOUR SKILLS)

19. 14.37%; 1.13%

20. 12.57%; 0.99%

21. 17.23%; 1.33%

22. 21.82%; 1.66%

(MIXED REVIEW)

7. $150

9. $195.89

CLOSURE

Have students write a paragraph discussing the four categories of cards and their uses. Tell them to demonstrate how finance charges are calculated by using a spreadsheet to show monthly payments of $100 on a $1000 charge assuming an APR of 18%, first with and then without a 1-month grace period.

LESSON QUIZ

1. Compute the effective interest rate on an APR of 24% assuming first a 1-month grace period, then no grace period. (**26.8%, 24.3%**)
2. How many months will it take to pay off a balance of $900 assuming monthly payments of $100, an APR of 21%, and no grace period? (**10**)

Determine the effective interest rate and monthly interest rate for each APR. Give answers to the nearest hundredth of a percent. See Additional Answers.

19. 13.5% **20.** 11.9% **21.** 16% **22.** 19.9%

Determine the effective interest rate on each credit card during the first 12 months following a purchase if each card has a 1-month grace period. Give answers to the nearest hundredth of a percent.

23. 13.5% 13.1% **24.** 11.9% 11.47% **25.** 16% 15.68% **26.** 19.9% 19.83%

MIXED REVIEW

5-1 **1.** Determine the monthly payment on a loan of $68,500 at 7.9% for 15 years. $650.67

5-4 **2.** How much will Sanchez save by prepaying a loan of $5000 at 9% for 5 years for which his monthly payments are $103.79 if he repays it 3 years early? $472.58

3-2 **3.** How much will $7500 be worth at the end of 2 years if it earns 4% interest, compounded quarterly? $8121.43

1-2 **4.** Alberto earns $12.50 per hour. He works 35 hours per week for 48 weeks and receives 4 weeks of paid vacation. His other benefits cost his employer $5200 per year. What is the annual cost to the company of his salary and benefits? $27,950

3-1 **5.** Lynn saves $52 per week. She wants to buy a sofa that costs $798. How many weeks must she save before she can buy the sofa? 16 wk

5-4 **6.** Peter has a loan of $12,000 at an annual rate of 5%. The term of the loan is 7 years and the monthly payments are $169.61. How much will he save by repaying the loan 2 years early? $204.57

2-3 Suppose that your bank statement shows a closing balance of $325.89. The following transactions occurred but were not shown in the statement: a deposit of $150, an ATM withdrawal of $80, a check in the amount of $200.

7. What amount or amounts must be added to the balance in the statement?

8. What amount or amounts must be subtracted from the statement balance? $80, $200

9. If no error has occurred, what should your current checkbook balance be?

4-5 A company manufactures widgets in two models, plain and fancy. The unit cost for the plain model is $4 and for the fancy model is $6. The selling price is $10 for the plain model and $20 for the fancy model. Let x represent the plain model and y represent the fancy model.

10. What is the objective function for the cost? $c = 4x + 6y$

11. What is the objective function for the revenue? $r = 10x + 20y$

12. What is the objective function for the profit? $p = 6x + 14y$

13. The company can produce 50 widgets a week. Find the weekly profit if it decides to manufacture 30 of the fancy model. $540

6-2 CREDIT CARD BALANCES: HOW LONG DO THEY LAST?

QUICK REFERENCE

TEACHER'S RESOURCES AND ANSWER KEY

Lesson Quiz Answers 6-2, p. 58
Reteaching Activity Answers 6-2, p. 101
Enrichment Activity Answers 6-2, p. 103

EXTENSION ACTIVITIES

Reteaching Activity 6-2, p. 42
Enrichment Activity 6-2, p. 49

Trevor grew up next door to Sylvia and her mother. As kids, he and his younger sister Tracey spent a lot of time playing with Sylvia. Sylvia always wanted to pretend that they were a family so that she could take Tracey shopping for new clothes and toys. Trevor's mother gave them some outdated credit cards with which to play. (She cut out a small portion of each one so that they could not be used to make real purchases.) Sylvia would pull out credit cards to pay for all the imaginary purchases that she made for Tracey. Since Sylvia had always seen her father with his credit cards, she assumed that everyone made all their purchases with credit cards.

Trevor realized even then how much Sylvia enjoyed shopping on credit. His parents have always had their share of credit cards too, but they do not use them as much as Sylvia's father does. Trevor would like to identify the advantages and disadvantages of credit cards, since Sylvia's father and his parents have different views on their use.

OBJECTIVES: *In this lesson, we will help Trevor to:*

- *Identify advantages and disadvantages of credit cards.*
- *Determine how long it takes to pay off a credit card balance if a fixed payment is made each month.*
- *Determine the size of the monthly payment that must be made in order to pay off a credit card balance in a fixed amount of time.*

ADVANTAGES OF CREDIT CARDS

1. Credit cards are very convenient and simple to use when making purchases.
2. You can delay payment for purchases until a predetermined date.
3. A record of your purchases is made automatically.
4. The danger of losing money while shopping is minimized, since you do not need cash at the time of your purchase.
5. It is easy to order merchandise by mail or telephone with a credit card.
6. Salespeople and business owners may come to recognize you if you are a frequent charge customer and may thus provide you with better service.
7. You can make several purchases at one time.
8. Credit cards help you to take advantage of sales.
9. When you have a credit card, you can use it to obtain an immediate cash advance (up to a specified amount) at an automated teller machine (ATM).

DISADVANTAGES OF CREDIT CARDS

1. A credit card encourages people who have a tendency to overspend to charge beyond their income and ability to pay.
2. Using a credit card does not have the same impact as paying cash. You may forget how much you have charged until the bill comes at the end of the month.
3. Since you can make so many charges on the telephone and through the mail, anyone who knows your credit card number, your name, and other information that is easily obtainable can make unauthorized purchases to your account without having your card in their possession.
4. Credit cards can be stolen. If yours are stolen, you should notify the issuing company immediately, or you could be held liable for unauthorized charges. In general, you will not be liable for unauthorized charges once you have notified the company.

ASK YOURSELF

1. What are four advantages of credit cards?
   See Additional Answers.
2. What are three disadvantages of credit cards?
3. If you lose your credit card, why should you immediately notify the company that issued it?

ADDITIONAL ANSWERS (ASK YOURSELF)

1. 1) convenient; 2) can delay payment; 3) creates record of purchase; 4) easy to order by mail or phone
2. 1) can encourage over-spending; 2) unauthorized purchases; 3) can be stolen
3. so you will not be liable for unauthorized purchases

(ALGEBRA REVIEW)

3. No; the domain value of 9 is paired with range values of 2 and 8.
4. Yes; each member of the domain is paired with only one member of the range.

ALGEBRA *REVIEW*

Find the domain and range of each relation.

1. $\{(9, 2), (3, 8), (9, 8)\}$
   d: {9, 3}; r: {2, 8}
2. $\{(2, 5), (3, 5), (6,6)\}$
   d: {2, 3, 6}; r: {5, 6}
3. Is the relation in Exercise 1 a function? Why or why not?
   See Additional Answers.
4. Is the relation in Exercise 2 a function? Why or why not?
   See Additional Answers.

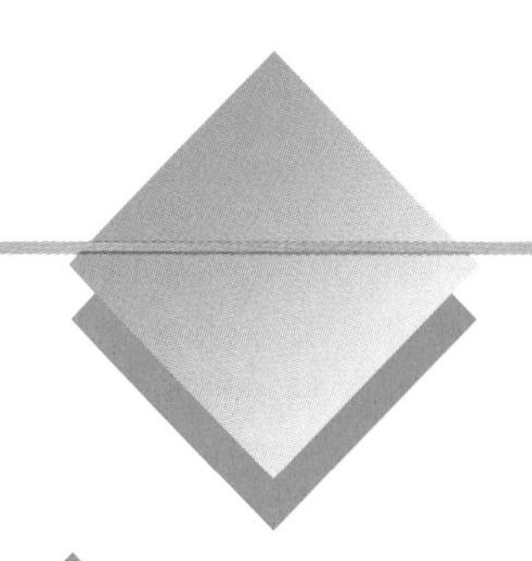

SHARPEN YOUR SKILLS

SKILL 1

In this lesson you will use the **common logarithm function,** $y = \log_{10} x$. For this function, y represents the exponent to which the number 10 must be raised to obtain x. For example, $\log_{10} 100 = 2$ because $10^2 = 100$, and $\log_{10} 100{,}000 = 5$ because $10^5 = 100{,}000$. When writing the common logarithm $y = \log_{10} x$, it is customary to omit the base 10. Thus $\log x = \log_{10} x$.

Your graphing calculator has a special common logarithm key. Evaluate $y = \log 1{,}000{,}000$ by pressing the log key on your calculator followed by 1,000,000. Then press the enter key to obtain 6.

A special formula allows you to calculate the length of time that it will take to pay the remaining balance on a credit card if a specified monthly payment M is made. The formula uses the common logarithm function.

Time-to-Pay-Off Formula

$$n = \frac{\log\left(\frac{M}{M - Pr}\right)}{\log(1 + r)}$$

where P = amount of loan
r = monthly interest rate
M = monthly payment
n = number of payment periods

EXAMPLE 1 Trevor's father has a balance of \$1200 on one of his MasterCard cards that has an APR of 15%. He can afford monthly payments of \$150.

QUESTION If he makes no further purchases with this card, how long will it take him to pay off the balance?

SOLUTION

$$n = \frac{\log\left(\frac{M}{M - Pr}\right)}{\log(1 + r)}$$ Use the time-to-pay-off formula.

$$n = \frac{\log\left(\frac{150}{150 - (1200)(0.0125)}\right)}{\log(1 + 0.0125)}$$ $P = 1200$; $r = 0.15 \div 12 = 0.0125$; $M = 150$

$n = 8.48$ To the nearest hundredth

Use the following keystrokes: [(] [LOG] [(] 150 [÷] [(] 150 [−] 1200 [×] 0.0125 [)] [)] [)] [÷] [(] [LOG] [(] 1.0125 [)] [)] [ENTER]

It takes 9 months to pay off the balance. In the last month he owes less than \$150.

FOCUS

ALGEBRA CONNECTIONS

Becca is making a long distance call from a pay phone. She has \$4.00 in change. The phone company charges \$0.85 for the first three minutes and 23 cents per minute for each additional minute. How many minutes can she talk? **(16 minutes)**

INSTRUCTION

TEACHING THE LESSON

Spend 2 days on this lesson.

On the first day tell students to read the opening text about advantages and disadvantages of credit cards. Review relations and functions and have them try exercises in the *Algebra Review* section. Then practice using the log key by finding log 10, log 100, log 1000, and log 10,000.

Introduce the *Time-to-Pay-Off Formula*, emphasizing the importance of using parentheses often:

$$\frac{\log\left(\frac{M}{m - Pr}\right)}{\log(1 + r)}$$

On the second day practice using the Skill 1 formula. Then introduce the *Monthly Payment Formula for Paying Off Balance*. Point out the need for multiple nested parentheses in this formula as well.

EXAMPLE 2 Trevor's father just received an unexpected raise. As a result, he may be able to make a larger monthly payment to MasterCard.

QUESTION How long will it take him to pay off his bill if he makes monthly payments of \$200 instead of \$150?

SOLUTION

$$n = \frac{\log\left(\frac{M}{M - Pr}\right)}{\log(1 + r)}$$

$$n = \frac{\log\left(\frac{200}{200 - (1200)(0.0125)}\right)}{\log(1 + 0.0125)}$$

$P = 1200$; $r = 0.15 \div 12 = 0.0125$; $M = 200$

$$n = 6.28$$

It takes 7 months to pay off the balance if he makes a monthly payment of \$200 instead of \$150.

TECHNOLOGY HINT

To ensure accuracy, students using a graphing calculator should check for proper use of parentheses before entering the calculation.

FOCUS ON ALGEBRA

In this lesson students review the definition and uses of functions as they apply to specific formulas. Students also employ the log key on their calculators in computing. (*NCTM Standard 6, p. 154.*)

COMMON ERRORS

Students will be tempted to "cancel" the log in both numerator and denominator of the *Time-to-Pay-Off Formula* in Skill 1. Do several simple examples such as $\frac{\log 1000}{\log 100}$ and $\frac{\log 10{,}000}{\log 10}$ to demonstrate that canceling here alters the value of the expression and is thus incorrect.

CRITICAL THINKING

Have students compute the total interest payment in Example 3 by creating a spreadsheet for the 2-year period that the uncle plans to use to pay off his VISA card. Ask students if they can think of another way to compute total interest in this case. (**\$215.82*24 − \$4200 = interest payment**)

SKILL 2

You may wish to pay off a credit card balance before a certain event occurs or before you need to make another purchase. For example, you may want to pay off a balance before you begin college. In this case you will want to know how much you have to pay each month to bring the balance to zero in a specified period of time.

Since the process of paying off a credit card balance is similar to making monthly payments on an installment bank loan, you can apply the monthly payment formula that you studied in Chapter 5. To use the formula for this purpose, let P represent the current balance.

Monthly Payment Formula for Paying Off Balance

$$M = \frac{Pr(1 + r)^n}{(1 + r)^n - 1}$$

where P = current balance
r = monthly interest rate
n = number of payment periods
M = monthly payment

EXAMPLE 3 Trevor's cousin Adam will be going to college in 2 years. Trevor's uncle would like to pay off his VISA card balance before he begins making tuition payments.

QUESTION If his current balance is \$4200 and his card carries an annual interest rate of 21%, how large must his monthly payment be if the balance is to be paid off in 2 years?

ENRICHMENT

Challenge students to use laws of logarithms and general algebraic technique to convert the *Time-to-Pay-Off Formula* into *Monthly Payment Formula for Paying Off Balance.*

SOLUTION

$$M = \frac{Pr(1 + r)^n}{(1 + r)^n - 1}$$

$$M = \frac{4200(0.0175)(1 + 0.0175)^{24}}{(1 + 0.0175)^{24} - 1} \qquad P = 4200;\ r = (0.21 \div 12) = 0.0175;\ n = 2 \cdot 12 = 24$$

$$M = 215.82$$

He must make monthly payments of $215.82 to pay off the balance in 2 years.

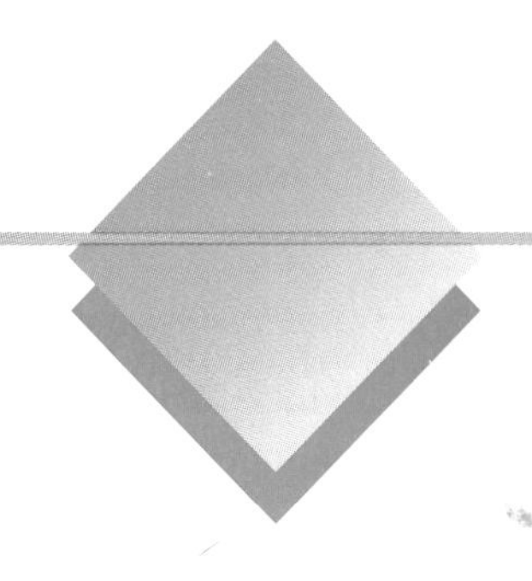

TRY YOUR SKILLS

Evaluate to the nearest hundredth.

1. log 1000 ÷ log 17 2.44
2. log 129 ÷ log 1.02 245.41
3. log (1200 ÷ 1100) ÷ log 1.75 0.16
4. log (180 ÷ 165) ÷ log 1.01 8.74

How long will it take to pay off each credit card balance with the given monthly payment at the given APR? Express your answer as a whole number of months.

5. $1500; $170; 12% 10 mo
6. $3500; $200; 14% 20 mo
7. $1850; $225; 16% 9 mo
8. $2000; $210; 15% 11 mo

Determine the monthly payment that must be made to reduce each MasterCard balance to zero in the specified period of time. Assume that there is an annual finance charge of 15.9%.

9. $28,000 in $2\frac{1}{2}$ years $1137.18
10. $16,000 in 25 months $756.03

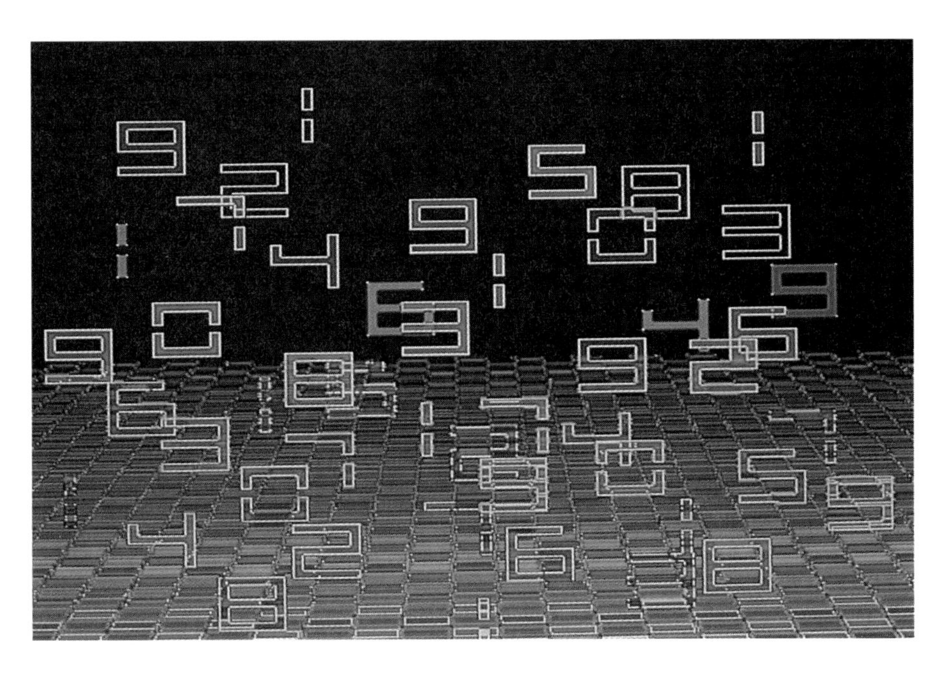

GUIDED PRACTICE

Have students work in groups of three or four on the *Try Your Skills* exercises. Suggest that group members compare results to assure accuracy.

RETEACHING

Students may need more intensive review of functions and more detailed practice in using the log key when computing. Give them additional examples with simple functions and use a basic form of the *Time-to-Pay-Off Formula* such as

$$z = \frac{\log x}{\log y}$$

Build up to more complicated formulas incrementally:

$$z = \frac{\log(x + 1)}{\log(y + 1)}$$

$$z = \frac{\log(3x + 1)}{\log(y + 1)}$$

$$z = \frac{\log\left(\frac{x}{3x + 1}\right)}{\log(y + 1)}$$

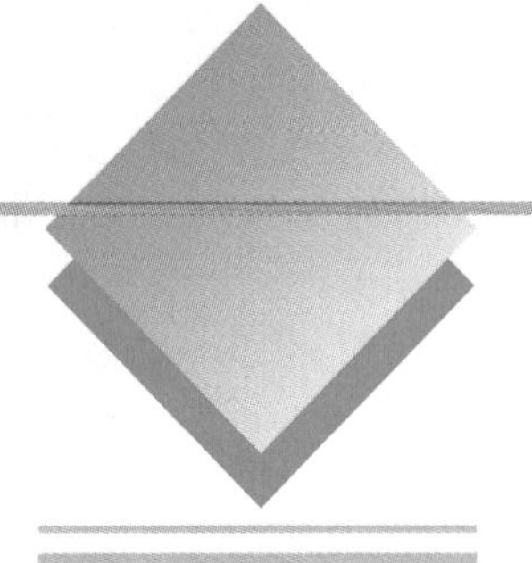

EXERCISE YOUR SKILLS

KEY TERM

common logarithm function

INDEPENDENT PRACTICE

ASSIGNMENT GUIDE

Exercises 1-2 check students' understanding of the process of paying off a credit card balance.

Exercises 3-6 give students practice computing with logarithms on calculators.

Exercises 7-18 ask students to determine the number of months needed to pay off various balances at a given monthly payment.

Exercises 19-24 ask students to determine the size of payment needed to pay off various balances in a given time.

CLOSURE

Have students write a paragraph discussing the advantages and disadvantages of credit cards.

LESSON QUIZ

1. How many months will it take to pay off a balance of $1500 assuming monthly payments of $175 and an APR of 15%? (**10**)
2. Determine the monthly payment necessary to pay off a VISA balance of $1500 in 2 years assuming an APR of 15%. (**$72.73**).

1. If you determine that it will take 7.2 months to pay off a credit balance, why would you round your answer to 8 months instead of 7 months?
2. Why is the process of paying off a credit card balance similar to making monthly payments on an installment bank loan? Answers may vary.

Evaluate to the nearest hundredth.

3. $\log 1280 \div \log 25$ 2.22
4. $\log 277 \div \log 1.15$ 40.24
5. $\log (1600 \div 1450) \div \log 1.22$ 0.50
6. $\log (210 \div 195) \div \log 1.0167$ 4.47

How long will it take to pay off each credit card balance with the given monthly payment at the given APR? Express your answer as a whole number of months. See Additional Answers.

7. $1500; $110; 12%
8. $170; $15; 12%
9. $1350; $85; 10.9%
10. $180; $36; 10.9%
11. $2750; $180; 16.9%
12. $325; $15; 16.9%
13. $3000; $225; 18%
14. $400; $28; 18%
15. $4500; $220; 12%
16. $525; $42; 12%
17. $5100; $325; 21%
18. $610; $45; 21%

Determine the monthly payment that must be made to reduce each VISA balance to zero in the specified period of time at the given APR.

19. $1900; $1\frac{1}{2}$ years; 15% $118.53
20. $2350; 22 months; 16.9% $124.96
21. $775; 16 months; 17.5% $54.66
22. $895; 15 months; 18% $67.08
23. $1675; $3\frac{1}{2}$ years; 9% $46.64
24. $1958; $2\frac{1}{2}$ years; 16% $79.62

MIXED REVIEW

5-4 1. How much money will you save by prepaying a loan of $34,000 at 12.5% for 10 years for which you have been making monthly payments of $497.68 if you prepay it at the end of 7 years? $3039.77

6-1 2. Determine the effective rate of interest to the nearest hundredth of a percent if the APR is 14%. 14.93%

4-1 3. Bevilaqua earns $566.66 in gross income each week. How much will be deducted for Social Security and Medicare? $43.35 per wk

5-1 4. If Marci can afford a monthly payment of $360, how much can she afford to borrow at a yearly interest rate of 9.5% for 5 years? $17,141.34

3-2 5. Use the Rule of 72 to determine how long it will take your $5500 savings account to double in value if it is growing at a rate of 3%. 24 y

5-3 6. If you buy living room furniture for $4100 and finance it at 16% over 3 years, how much will you save by making a down payment of 20% instead of 10%? $108.92

4-4 7. Suppose that the equation for the cost function of a small business is $c = 1.9n + 118.50$ and the equation for the revenue function is $r = 5n$. Use a graphing calculator to find the break-even point. 39 items

The answer is *yes* if *all* of the following are true:

- You have handled credit responsibly in the past.
- You use the card as a budgeting convenience.
- You recognize the dangers as well as the attraction of using credit cards.

To use credit cards to your best advantage, adhere to the following rules:

1. Keep only the cards that you will use fairly often. For most people in the middle-income bracket, a couple of oil company cards for gasoline and one bank card are enough. The businessperson may wish to use a travel and entertainment card also.
2. Consider every charged purchase as though you were paying cash. Ask yourself: Can I repay the charge promptly and easily?
3. Do not spend more than 20% of your take-home pay on credit payments. As described in earlier chapters, if a family uses more than this amount on credit payments (not including the mortgage payment), it may suffer from credit overload.
4. At the beginning of each month, set a limit on the total amount of charges you will be able to repay easily. Stay within that limit and repay the charges promptly to avoid additional interest charges.
5. Keep your receipts until you receive your statement to verify your spending with the statement.

ASK YOURSELF

1. What are two ways in which consumers use credit cards?
   Answers may vary.
2. Why is it a good idea to charge only what you can pay for each month? so you can pay off the balance each month without paying interest
3. What are three reasons not to use a credit card?
   1) overspending; 2) impulse buying; 3) no steady income

ALGEBRA *REVIEW*

Express each percent as a decimal.

1. 13.9%
   0.139
2. 1.85%
   0.0185
3. 245%
   2.45
4. 111.1%
   1.111

Express each decimal as a percent.

5. 456
   45,600%
6. 233.4
   23,340%
7. 0.09
   9%
8. 9.876
   987.6%

Find.

9. What is 1.6% of $1600?
   $25.60
10. What is 0.85% of $2100?
    $17.85
11. What is 18% of $4433.22?
    $797.98
12. $1500 is what percent of $7500?
    20%
13. $12,750 is what percent of $85,000?
    15%
14. $41.25 is what percent of $750?
    5.5%
15. 25 is 100% of what number?
    25
16. $13.75 is 20% of what number?
    $68.75
17. $7800 is 15% of what number?
    $52,000

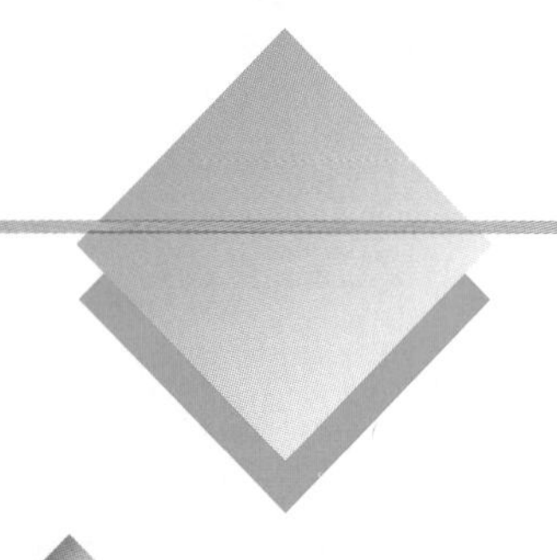

SHARPEN YOUR SKILLS

SKILL 1

INSTRUCTION

TEACHING THE LESSON

Spend one day on this lesson. Have the students read the opening text and review characteristics of a good credit card user versus a bad credit card user. Tell them to do the *Algebra Review* section. Then work through Examples 1 and 2 together. Repeat, assuming monthly payments of 15%. Elicit from students that larger monthly payments reduce total interest payments. Discuss Example 3, repeating for APRs of 15% and 24%.

FOCUS ON ALGEBRA

In this lesson students must make extensive use of spreadsheet programs and must round accurately. (*NCTM Standard 1, p. 137.*)

Banks and other lending institutions set forth a **minimum monthly payment** that must be made by their credit card holders. However, you can set your own monthly payment as long as it is above the required amount. The amount of your payment affects the length of time it will take you to pay off your balance. It also affects the amount of interest that you will pay.

EXAMPLE 1 The current balance on Vernon's mother's VISA card is \$785.00, and her monthly finance charge is 1.5% of the amount owed. Her bank requires a minimum monthly payment of 10% of her unpaid balance rounded to the nearest dollar or \$20, whichever is larger. If the amount she owes drops below \$20, then she must make a payment equal to the total amount owed.

QUESTION How much interest will she pay in the next year, assuming that she makes a payment of 10% of the amount owed on the last day of the month?

SOLUTION

In the first month she has a balance of \$785.00.

$$\begin{aligned}\text{Interest charge} &= \text{balance} \cdot 0.015\\ &= 785.00 \cdot 0.015\\ &= 11.78 \qquad \text{To the nearest cent}\end{aligned}$$

$$\begin{aligned}\text{Amount owed} &= \text{balance} + \text{interest}\\ &= 785.00 + 11.78\\ &= 796.78\end{aligned}$$

$$\begin{aligned}\text{Payment} &= 10\% \text{ of amount owed}\\ &= 0.10 \cdot 796.78\\ &= 80 \qquad \text{To the nearest dollar}\end{aligned}$$

In the second month she has a balance of 796.78 − 80 = \$716.78.

Use a spreadsheet program to work through the first 12 payments, and then total the interest column. To total the interest column use a sum function, such as @SUM(C2..C13) which adds the numbers in cells C2 through C13. In the columns for interest and 10% payment, you will need to use a rounding function. However, the interest is rounded to the nearest cent but the payment is rounded to the nearest dollar. Therefore the formulas are

For cell C2: @ROUND(0.015*B2,2)

For cell E2: @ROUND(0.1*D2,0)

	A	B	C	D	E
1	Month	Balance	Interest	Amount Owed	10% Payment
2	1	785.00	11.78	796.78	80.00
3	2	716.78	10.75	727.53	73.00
4	3	654.53	9.82	664.35	66.00
5	4	598.35	8.98	607.33	61.00
6	5	546.33	8.19	554.52	55.00
7	6	499.52	7.49	507.01	51.00
8	7	456.01	6.84	462.85	46.00
9	8	416.85	6.25	423.10	42.00
10	9	381.10	5.72	386.82	39.00
11	10	347.82	5.22	353.04	35.00
12	11	318.04	4.77	322.81	32.00
13	12	290.81	4.36	295.17	30.00
14		Total interest	90.17		

@ROUND(0.1*D2,0)

@ROUND(0.015*B2,2)

@SUM(C2..C13)

She pays \$90.17 in interest during the first year.

EXAMPLE 2 Vernon wonders whether his mother's interest payments for the first year would increase if her bank allowed her to make smaller monthly payments.

QUESTION How much interest would she pay in the first year if her bank required only that she make a minimum monthly payment of 5% of her unpaid balance (rounded to the nearest dollar) or \$20, whichever is larger? How much more interest would she pay in this case?

SOLUTION

In the first month her balance, interest, and amount owed would be the same as in Example 1. However, her payment would be $0.05 \cdot 796.78 = \$40$, rounded to the nearest dollar.

Copy the spreadsheet program from Example 1 and change the payment to 5% of the amount owed as shown on page 264.

Vernon's mother would pay \$116.62 in interest during the first year. In the first year, she would pay $116.62 - 90.17 = \$26.45$ more in interest if she made monthly payments of 5% of the amount owed rather than 10% of the amount owed.

@ROUND(0.05*D19,0)

	A	B	C	D	E
18	Month	Balance	Interest	Amount Owed	5% Payment
19	1	785.00	11.78	796.78	40.00
20	2	756.78	11.35	768.13	38.00
21	3	730.13	10.95	741.08	37.00
22	4	704.08	10.56	714.64	36.00
23	5	678.64	10.18	688.82	34.00
24	6	654.82	9.82	664.64	33.00
25	7	631.64	9.47	641.11	32.00
26	8	609.11	9.14	618.25	31.00
27	9	587.25	8.81	596.06	30.00
28	10	566.06	8.49	574.55	29.00
29	11	545.55	8.18	553.73	28.00
30	12	525.73	7.89	533.62	27.00
31		Total interest	116.62		

COMMON ERRORS

When computing a percent payment (such as 10% of unpaid balance) students may forget to include the interest payment as part of that balance.

TECHNOLOGY HINT

As an additional activity, have students create graphs of "Amount Owed," "Balance," and "Payment" as the Y input. Remind students to adjust the Y-Range as needed.

To compare the interest paid for monthly payments of 5% and 10%, Vernon created a bar graph as shown. He used the months for x values, the interest for 10% payments as the first data range, and the interest for 5% payments as the second data range.

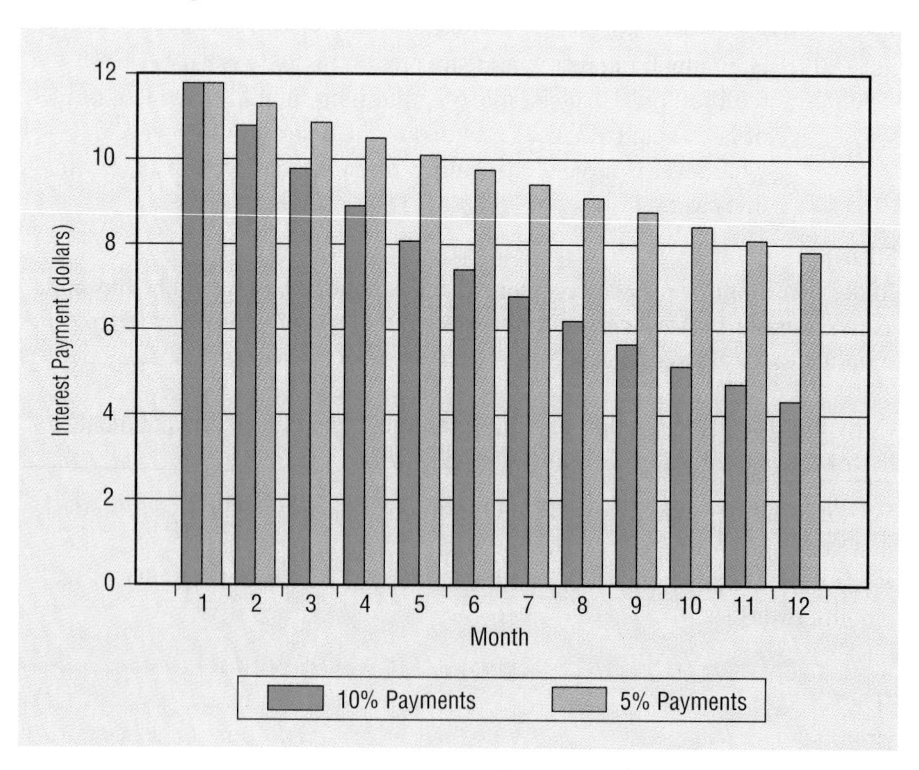

SKILL 2

Since Christmas and Hanukkah shopping entices many consumers to buy more than they do at other times of the year, many banks open new MasterCard and VISA accounts for potential shoppers in preparation for the holiday season. The APRs that are offered vary from bank to bank.

EXAMPLE 3 Karl's and Aubrey's parents have two MasterCard accounts. They currently have balances of $450.00 on each account. One account has an APR of 21%, and the other has an APR of 14%.

QUESTION They make monthly payments of 10% of the amount owed (rounded to the nearest dollar) or $20, whichever is larger, on the last day of each month on both accounts. How much will they pay in interest charges during the next year on the account with an APR of 21%? On the account with an APR of 14%? How much will be saved at the lower interest rate?

SOLUTION

Use a computer spreadsheet program to determine the total interest paid on each account during the next year. Remember the monthly interest is the APR divided by 12.

When 10% of the amount owed is less than $20, the payment is $20. Notice that in cell E14 this happens:

$0.10(188.94) = 19$ To the nearest dollar

Therefore in cell E14 and also in cell E15 you must enter 20. The spreadsheet automatically recalculates the remaining cells when you do this.

In working an exercise you may get a negative number in your spreadsheet. This may be because the amount owed is less than $20. Remember that if the amount owed is less than $20, you will only pay the amount owed, so you will need to adjust your spreadsheet.

COOPERATIVE LEARNING

Suggest that students work in groups of three or four on problems they have designed themselves, choosing a current balance, minimum percent payments and APR. Have group members check each other's calculations.

CRITICAL THINKING

Ask students why, in Example 3, Karl's and Aubrey's parents might choose to use the card with the higher APR given that the other card allows lower interest payments.

	A	B	C	D	E
1	MASTERCARD ACOUNT WITH 21% APR				
2	21% APR = 0.0175 per month				
3	Month	Balance	Interest	Amount Owed	Payment
4	1	450.00	7.88	457.88	46.00
5	2	411.88	7.21	419.09	42.00
6	3	377.09	6.60	383.69	38.00
7	4	345.69	6.05	351.74	35.00
8	5	316.74	5.54	322.28	32.00
9	6	290.28	5.08	295.36	30.00
10	7	265.36	4.64	270.00	27.00
11	8	243.00	4.25	247.25	25.00
12	9	222.25	3.89	226.14	23.00
13	10	203.14	3.55	206.69	21.00
14	11	185.69	3.25	188.94	20.00
15	12	168.94	2.96	171.90	20.00
16		Total interest	60.90		
17					
18					
19	MASTERCARD ACOUNT WITH 14% APR				
20	14% APR = 0.0116666 per month				
21	Month	Balance	Interest	Amount Owed	Payment
22	1	450.00	5.25	455.25	46.00
23	2	409.25	4.77	414.02	41.00
24	3	373.02	4.35	377.37	38.00
25	4	339.37	3.96	343.33	34.00
26	5	309.33	3.61	312.94	31.00
27	6	281.94	3.29	285.23	29.00
28	7	256.23	2.99	259.22	26.00
29	8	233.22	2.72	235.94	24.00
30	9	211.94	2.47	214.41	21.00
31	10	193.41	2.26	195.67	20.00
32	11	175.67	2.05	177.72	20.00
33	12	157.72	1.84	159.56	20.00
34		Total interest	39.56		

@ROUND(0.1*D4,0) (callout to cell E4)

change 19.00 to 20.00 (callout to cell E14)

change 18.00 to 20.00 (callout to cell E32)

RETEACHING

Review the procedure shown in Example 1 for calculating interest charges and payments. Rework Example 1 assuming a current balance of $1012. Then have students compute total interest payments for 1 year using a spreadsheet program.

ENRICHMENT

Suggest that students visit a local bank to find out how banks keep records of their credit card customers' payments.

TECHNOLOGY HINT

Some of your students may want to know what formula could be used to determine if the payment is less than $20. This requires a conditional formula of the form @IF(conditional, true, false).

For the spreadsheet of Example 3, the formula for cell E4 would be @IF(@ROUND(0.1*D4,0)>20, @ROUND(0.1*D4,20), 20)

Karl's and Aubrey's parents pay $60.90 in interest during the year with an APR of 21%. They pay $39.56 in interest during the year with an APR of

14%. This represents a savings of $60.90 - 39.56 = \$21.34$ over the interest charges on the card with an APR of 21%.

To compare the interest payments, create a bar graph as shown.

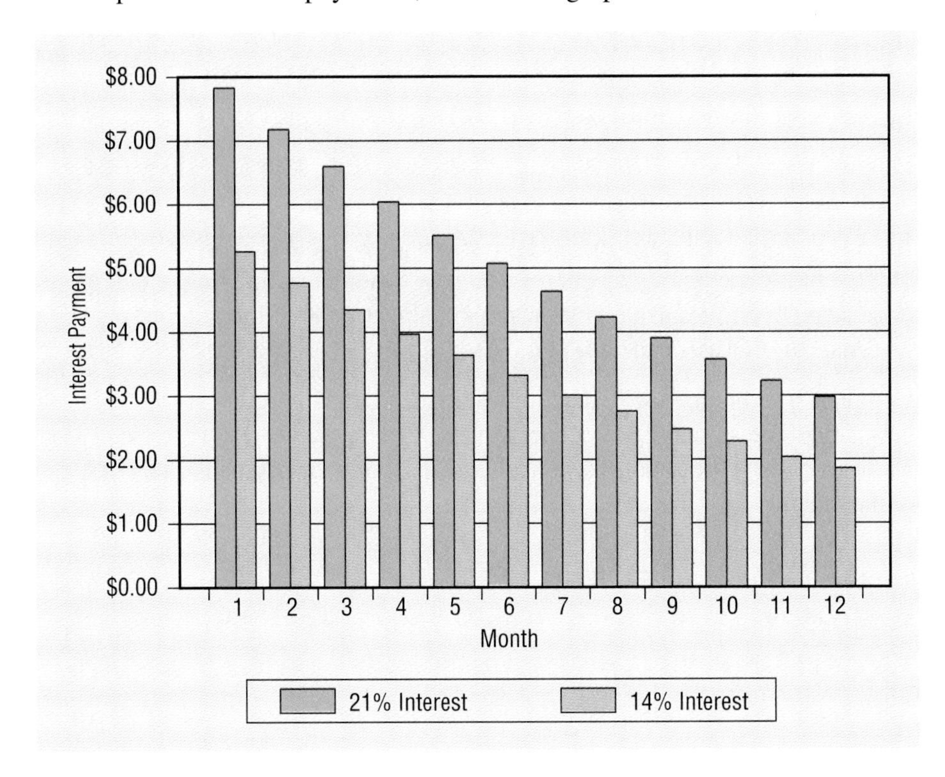

The interest charge is greater when the APR is larger if all other variables are the same.

TRY YOUR SKILLS

Complete the following chart to determine the interest charges over the next four months on a balance of $995.00. Assume that your card has an APR of 15% and you make monthly payments of 5% of the amount due, rounded to the nearest dollar. Assume that you make your payment on the last day of the month.

	Month	Balance	Interest	Amount Owed	Payment
1.	1	$995.00	$12.44	$1007.44	$50.00
2.	2	957.44	11.97	969.41	48.00
3.	3	921.41	11.52	932.93	47.00
4.	4	885.93	11.07	897.00	45.00

5. What is the total interest for the credit card account of Exercises 1–4? $47.00

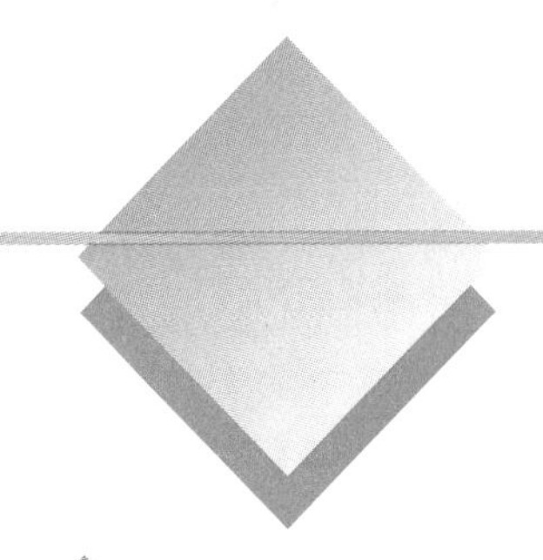

GUIDED PRACTICE

Have students work in small groups on the *Try Your Skills* exercises using a spreadsheet program.

For the credit card account of Exercises 1–5, find the interest cost for the first 4 months if you pay 10% of the amount due instead of 5% of the amount due.

	Month	Balance	Interest	Amount Owed	Payment
6.	1	$995.00	$12.44	$1007.44	$101.00
7.	2	906.44	11.33	917.77	92.00
8.	3	825.77	10.32	836.09	84.00
9.	4	752.09	9.40	761.49	76.00

10. What is the total interest for the credit card account of Exercises 6–9? $43.49

11. Compare the results of Exercises 5 and 10 to determine how much more interest you pay when you make payments of 5% of the amount owed instead of 10% of the amount owed. $3.51

Complete the chart below to determine how much the interest cost will be if you pay 10% of the amount due each month, rounded to the nearest dollar, and your card carries an APR of 9%. Assume that you make your payment on the last day of the month.

	Month	Balance	Interest	Amount Owed	Payment
12.	1	$995.00	$7.46	$1002.46	$100.00
13.	2	902.46	6.77	909.23	91.00
14.	3	818.23	6.14	824.37	82.00
15.	4	742.37	5.57	747.94	75.00

16. What is the total interest for the credit card account of Exercises 12–15? $25.94

17. Compare the results of Exercises 10 and 16 to determine how much more interest you pay if the APR on your card is 15% instead of 9%. $17.55

EXERCISE YOUR SKILLS

1. How can you use credit cards to your best advantage? Answers may vary.
2. Why is it easier to overspend when you use a credit card? Answers may vary.
3. Why do the interest payments decrease when you make larger monthly payments? the balance due decreases, thus the interest on the balance also decreases

KEY TERM
minimum monthly payment

The current balance on Pedro's VISA card is $400.00, and his card carries a monthly finance charge of 1.75% of the amount owed. His bank requires a minimum monthly payment of 10% of his unpaid balance rounded to the nearest dollar or $20, whichever is larger. If the amount he owes drops below $20, then he must make a payment equal to the total amount owed. Use a spreadsheet program to determine how much interest he will pay in the next year, assuming that he makes his payment on the last day of the month.

	Month	Balance	Interest	Amount Owed	Payment
4.	1	$400.00	$7.00	$407.00	$41.00
5.	2	366.00	6.41	372.41	37.00
6.	3	335.41	5.87	341.28	34.00
7.	4	307.28	5.38	312.66	31.00
8.	5	281.66	4.93	286.59	29.00
9.	6	257.59	4.51	262.10	26.00
10.	7	236.10	4.13	240.23	24.00
11.	8	216.23	3.78	220.01	22.00
12.	9	198.01	3.47	201.48	20.00
13.	10	181.48	3.18	184.66	20.00
14.	11	164.66	2.88	167.54	20.00
15.	12	147.54	2.58	150.12	20.00

16. What is the total interest that Pedro must pay? $54.12

Complete the chart below to determine his interest cost for the year if Pedro pays 5% of the amount due instead of 10% of the amount due.

	Month	Balance	Interest	Amount Owed	Payment
17.	1	$400.00	$7.00	$407.00	$20.00
18.	2	387.00	6.77	393.77	20.00
19.	3	373.77	6.54	380.31	20.00
20.	4	360.31	6.31	366.62	20.00
21.	5	346.62	6.07	352.69	20.00
22.	6	332.69	5.82	338.51	20.00
23.	7	318.51	5.57	324.08	20.00
24.	8	304.08	5.32	309.40	20.00
25.	9	289.40	5.06	294.46	20.00
26.	10	274.46	4.80	279.26	20.00
27.	11	259.26	4.54	263.80	20.00
28.	12	243.80	4.27	248.07	20.00

29. What is the total interest that Pedro must pay? $68.07

30. Compare the results of Exercises 16 and 29 to determine how much more interest he pays if he makes payments of 5% of the amount owed instead of 10% of the amount owed. $13.95

31. Use the graph function of your computer spreadsheet program to graph Pedro's monthly payments at 10% and 5% of the amount owed.
See Additional Answers.

INDEPENDENT PRACTICE

ASSIGNMENT GUIDE

Exercises 1-3 ask students to display their understanding of the opening text.

Exercises 4-30 and 32-45 require students to determine monthly payments and interest charged using a spreadsheet program and to compare total interest paid assuming differing minimum monthly payments.

Exercises 31 and 46 direct students to display information using a graph function.

CLOSURE

Suggest that students write an explanation of how initial balances and accrued interest help determine monthly payments and how total interest paid over a period of months can be computed.

LESSON QUIZ

1. Determine your interest costs at the end of 6 months on a current balance of $560 with a minimum payment of 10% and a monthly finance charge of 1.25%. (**$33.68**)
2. Repeat Exercise 1 for a minimum payment of 20%. (**$26.49**)

ADDITIONAL ANSWERS (EXERCISE YOUR SKILLS)

45. $24.25

(MIXED REVIEW)

1. **a.** $4753.14
   b. $3608.78
   Loan **b** has the greater total cost.
6. $860 or less

Complete the chart below to determine how much Pedro's interest cost will be if he pays 10% of the amount due each month and his card carries a monthly finance charge of 1% of the amount owed.

	Month	Balance	Interest	Amount Owed	Payment
32.	1	$400.00	$4.00	$404.00	$40.00
33.	2	364.00	3.64	367.64	37.00
34.	3	330.64	3.31	333.95	33.00
35.	4	300.95	3.01	303.96	30.00
36.	5	273.96	2.74	276.70	28.00
37.	6	248.70	2.49	251.19	25.00
38.	7	226.19	2.26	228.45	23.00
39.	8	205.45	2.05	207.50	21.00
40.	9	186.50	1.87	188.37	20.00
41.	10	168.37	1.68	170.05	20.00
42.	11	150.05	1.50	151.55	20.00
43.	12	131.55	1.32	132.87	20.00

44. What is Pedro's total interest? $29.87

45. Compare the results of Exercises 16 and 44 to determine how much more interest Pedro pays if the monthly finance charge is 1.75% instead of 1%.

46. Use the graph function of your computer spreadsheet program to graph Pedro's monthly payments with a monthly finance charge of 1.75% and 1%. See Additional Answers.

MIXED REVIEW

5-2 **1.** Determine which of the following loans has a greater total cost.
a. $12,000 at 14% for 5 years **b.** $15,000 at 11% for 4 years

1-1 **2.** During the first week of September, Samantha worked $37\frac{1}{2}$ hours and received $75.75 in tips. If her hourly rate of pay is $6.95, how much did she earn that week? $336.38

6-2 **3.** Determine the monthly payment that must be made to reduce a $3333.33 MasterCard balance to zero in 28 months if the card carries an APR of 17.5% $145.86

1-3 **4.** Zachary receives a commission of 5.5% on sales up to $12,000 and 7.5% on all sales over $12,000. How much has he earned in commission if his sales are $19,950? $1256.25

6-2 **5.** How long will it take to pay off a credit card balance of $1800.75 with monthly payments of $200 if the APR is 11%? Express your answer as a whole number of months. 10 mo

5-2 **6.** Suppose that your family's take-home pay is $4,300 per month. How much can you afford to spend for credit-card payments each month?

6-4 CREDIT CARDS: HOW TO PROTECT THEM

QUICK REFERENCE

TEACHER'S RESOURCES AND ANSWER KEY

Lesson Quiz Answers 6-4, p. 60
Reteaching Activity Answers 6-4, p. 101
Enrichment Activity Answers 6-4, p. 103

EXTENSION ACTIVITIES

Reteaching Activity 6-4, p. 44
Enrichment Activity 6-4, p. 51

Sylvia Shawn's mother has several credit cards of her own that she obtained after she and Mr. Shawn were divorced. She was able to get credit in her own name after the divorce even though all of the accounts were in Sylvia's father's name while they were still married. The reason that this was possible is that she was protected by the Equal Credit Opportunity Act.

Sylvia remembers how difficult it was for her father when his credit cards were stolen. He had left them in the glove compartment of his car one day at the golf course. Sylvia was with him at the time. As they left the golf course and stopped to buy gasoline, he realized that all his credit cards had been taken. He spent the next three hours making phone calls to all the banks and charge account offices. The thief had spent over $10,000 before the information could be entered into the various computer networks! The thief charged $2650 worth of ski equipment, $3240 worth of video equipment, and a $4500 computer system to Mr. Shawn's now-closed accounts.

For the next three months, Mr. Shawn dealt with the credit card companies. He discovered that under the Fair Credit Reporting Act and the Fair Credit Billing Act, his credit rating would be preserved and he would have to pay only $50 of the unauthorized charges on each card. (This was still a substantial amount because he had so many cards.) He also learned what an inconvenience it is to have all your credit accounts changed and to have to wait for new cards to arrive from every creditor.

From her mother's and father's experiences, Sylvia learned that credit cards are just as valuable as cash.

FOCUS

ALGEBRA CONNECTIONS

A friend is thinking of an integer between 1 and 100. You must guess the number by asking questions with yes-or-no answers. How many questions do you need to ask in order to be sure of guessing the number? (**7**)

OBJECTIVES: In this lesson, we will help Sylvia to:

- *Discover how consumers are protected by*
  - *the Fair Credit Reporting Act.*
  - *the Fair Credit Billing Act.*
  - *the Equal Credit Opportunity Act.*
- *Recognize ways to protect credit cards.*
- *Calculate credit account interest, payments, and balances.*
- *Calculate the increase in monthly interest caused by new purchases.*

CONSUMER RIGHTS

The **Fair Credit Reporting Act** forbids credit agencies from giving out incorrect credit information about consumers. An individual who has been denied credit, employment, or insurance because of an inaccurate credit agency report may ask to know the source of the report. The individual can then have both incorrect information and any information that cannot be proven removed from the file. The act also allows consumers to examine information in their credit files. If consumers feel that they have been misrepresented, they can add short statements to the files that give their side of the story.

Under the **Fair Credit Billing Act,** consumers can preserve their credit ratings while settling disputes with stores and credit card companies. During this time a **creditor** cannot report you as delinquent to any credit agency for your failure to pay the portion of your bill that is under dispute.

The **Equal Credit Opportunity Act** prevents discrimination on the basis of sex, marital status, race, color, religion, age, or national origin. For instance, women applying for credit must be judged by the same standards as men. If they have a steady income and can qualify in other respects as good credit risks, they are equally entitled to credit.

The law specifically states that creditors cannot:

- Deny credit on the basis of sex, marital status, race, color, religion, national origin, or age (if an individual is old enough to enter into a binding contract).
- Deny credit because an applicant receives any income from a public program.
- Ask questions concerning birth control practices and plans for children, or assume that a female applicant is likely to become pregnant and have an interruption of income.
- Ask about an applicant's marital status (unless a spouse will be contractually liable for the loan, a spouse's income is counted on to repay a loan, or a spouse plans to use the loan).

- Refuse to consider part-time income of a working spouse, alimony, child support, or Social Security payments.
- Cancel a divorced or widowed person's credit when a marriage ends unless the income has dropped so much that the person may not be able to pay.

This act does not entitle you to credit whenever you want it. You must still pass the creditor's tests that indicate your financial ability and willingness to pay.

HOW TO PROTECT YOUR CREDIT CARDS

If your credit cards are stolen, the maximum amount of money that you will have to pay is $50 per card. However, if you carry many cards, like Sylvia's father, you will still have to pay a large sum of money if the thief works quickly before you notice that your cards are gone.

You can take the following precautions to protect yourself:

- Destroy any cards that you do not use. Cut unwanted cards in half and throw them away. Then notify the credit card company to close your account.
- Make a list of all the credit cards that you have (with the account numbers and the name and address of each issuer), and keep this list at home.
- Be sure that you get your card back every time you use it. Dishonest employees of legitimate establishments may retain your card to give to fraudulent users.
- Do not leave your credit cards in the glove compartment of your car. This is one of the first places a credit card thief will look.
- Do not underestimate the value of a credit card. Carrying a credit card is similar to carrying cash. The thief does not underestimate its value!

ASK YOURSELF

1. How does the Fair Credit Reporting Act protect consumers? prevents credit agencies from giving out incorrect credit information
2. How does the Fair Credit Billing Act protect consumers? preserves consumers' credit ratings while settling disputes with stores and credit card companies
3. What does the Equal Credit Opportunity Act prevent? prevents discrimination on the basis of sex, marital status, race, color, religion, age, or national origin
4. What are three ways in which you can protect your credit cards? 1) destroy credit cards you do not use; 2) make a list of all the credit cards you have; 3) be sure to get your credit card back every time you use it

ALGEBRA *REVIEW*

Simplify each expression. Round your answer to the nearest hundredth.

1. $2 + (3 \cdot 4)$
   14.00
2. $(2 + 3) \cdot 4$
   20.00
3. $2 \cdot (5 + \frac{5}{4})^4$
   3051.76
4. $2 \cdot 5^4 + \frac{5}{4}$
   1251.25
5. $18^2[(7 \cdot 8)^3(1 \div 8)]$
   7,112,448.00
6. $[18^2(7 \cdot 8^3)][1 \div 8]$
   145,152.00
7. $0.22(1200 + 25 - 175)$
   231.00
8. $1.51(6666 - 444 + 22)$
   9428.44
9. $\frac{(6 + 3 \cdot 4)^5}{(2 + 0.75 \cdot 12)^2}$
   15,616.26
10. $\frac{[(6 + 3) \cdot 4]^5}{[(2 + 0.75) \cdot 12]^2}$
    55,524.50
11. $\frac{25(500) + 450 + 2(450)}{28}$
    494.64
12. $\frac{5(980) + 875 + 24(875)}{30}$
    892.50

SHARPEN YOUR SKILLS

SKILL 1

In Lesson 6–3 you learned that the amount of your monthly payment and the monthly rate of interest influence your credit card balance and the amount of interest that you pay each month. Another factor that influences your balance and interest charges is additional purchases that you make before paying for earlier ones.

FOCUS

TEACHING THE LESSON

Spend one day on this lesson. Direct students to read the text on *Consumer Rights* and *How to Protect Your Credit Cards*. Have them write answers in their notebooks to the questions in *Ask Yourself*. Go through Examples 1 and 2 carefully, offering variations for extra practice if necessary.

FOCUS ON ALGEBRA

In this lesson students apply formulas for computing interest and use a spreadsheet program to determine account balances. (*NCTM Standard 2, p. 140.*)

CRITICAL THINKING

Have students determine their year-end balance if they start with a balance of \$650, make a 10% payment each month, accrue a 1.5% finance charge, and add \$60 in new charges each month.

EXAMPLE 1 Saul's mother is trying to restrain herself from making any more purchases with her MasterCard until she pays off her current balance. She is quite successful at making no further purchases during some months, but she is unable to resist making more purchases during others. Saul is worried that her balance may be larger at the end of a 12-month period than it was at the beginning of the period. They have therefore decided to follow the account closely for one year.

QUESTION Will she owe more or less at the end of the year if the following hold true?

In month 1 she makes no purchases, and her balance is \$1265.80.
The monthly interest rate on the card is 1.5%.
She makes monthly payments of 10% of the amount owed, rounded to the nearest dollar.
Purchases are shown on her statements as follows:
\$250.75 in month 3,
\$75.00 in month 5,
\$380.50 in month 8, and
\$45.90 in month 11.

SOLUTION

Use a spreadsheet program to work through the first 12 months. Purchases are added to the amount owed from the previous month and payments are subtracted, so

Balance = amount owed + purchases − payments

Remember to use the ROUND function for interest in column D and the payment in column F. Interest is rounded to the nearest cent so for cell D2 the formula is

@ROUND(C2*0.015,2)

The payment is rounded to the nearest dollar so for cell F2 the formula is

@ROUND(E2*0.1,0)

	A	B	C	D	E	F
1	Month	Purchases	Balance	Interest	Amount Owed	Payment
2	1	0.00	1265.80	18.99	1284.79	128.00
3	2	0.00	1156.79	17.35	1174.14	117.00
4	3	250.75	1307.89	19.62	1327.51	133.00
5	4	0.00	1194.51	17.92	1212.43	121.00
6	5	75.00	1166.43	17.50	1183.93	118.00
7	6	0.00	1065.93	15.99	1081.92	108.00
8	7	0.00	973.92	14.61	988.53	99.00
9	8	380.50	1270.03	19.05	1289.08	129.00
10	9	0.00	1160.08	17.40	1177.48	118.00
11	10	0.00	1059.48	15.89	1075.37	108.00
12	11	45.90	1013.27	15.20	1028.47	103.00
13	12	0.00	925.47	13.88	939.35	94.00
14			Total interest	203.40		

+E3−F3+B4

Despite her purchases, Saul's mother reduced her balance from $1265.80 in month 1 to $925.47 in month 12. She paid a total of $203.40 in interest.

SKILL 2

As you have learned, many bank cards carry an annual fee. If you are late making your scheduled payment, your bank may also charge a late payment penalty.

EXAMPLE 2 In month 9, Saul's mother had to pay her annual fee of $50.00. She was also late making her month 11 payment. Therefore, she received a late penalty fee of $15.00 on her month 12 statement.

QUESTION How do these charges affect her month 12 balance and the interest that she pays for the year?

SOLUTION

Copy the spreadsheet program from Example 1. Insert the annual fee and late penalty payments in the "Purchases" column to recalculate using the new information.

TECHNOLOGY HINT

The formulas needed to complete row 4 of the spreadsheet in Example 1 are: cell D4: @ROUND(0.015*C4, 2); cell E4: +C4+D4; cell F4: @ROUND(0.1*E4, 0).

INTERDISCIPLINARY INVESTIGATION

Consumer Affairs: Many credit card companies offer incentives such as frequent flier miles, a percent of purchases returned to the customer, or delayed payment schedules. Suggest that students contact companies to learn about these policies.

In addition to the terms listed in the *Key Terms* box near *Exercise Your Skills*, help non-native speakers with terminology such as *credit agency, credit file, delinquent,* and *fraudulent.*

COMMON ERRORS

Students may inadvertently list the additional purchases a month too early or too late, thus incorrectly computing the associated interest charges.

RETEACHING

Have students write out the answers to the questions in the *Ask Yourself* section. Work though additional examples based on Examples 1 and 2 without a spreadsheet program so that students see on a step-by-step basis how balances go up and down.

ENRICHMENT

Challenge students to write a formula that calculates the increase in monthly interest caused by new purchases.

	A	B	C	D	E	F
1	Month	Purchases	Balance	Interest	Amount Owed	Payment
2	1	0.00	1265.80	18.99	1284.79	128.00
3	2	0.00	1156.79	17.35	1174.14	117.00
4	3	250.75	1307.89	19.62	1327.51	133.00
5	4	0.00	1194.51	17.92	1212.43	121.00
6	5	75.00	1166.43	17.50	1183.93	118.00
7	6	0.00	1065.93	15.99	1081.92	108.00
8	7	0.00	973.92	14.61	988.53	99.00
9	8	380.50	1270.03	19.05	1289.08	129.00
10	9	**50.00**	1210.08	18.15	1228.23	123.00
11	10	0.00	1105.23	16.58	1121.81	112.00
12	11	45.90	1055.71	15.84	1071.55	107.00
13	12	**15.00**	979.55	14.69	994.24	99.00
14			Total interest	206.29		

Her month 12 balance is 979.55 − 925.47 = $54.08 larger with the late fee and annual fee. She pays 206.29 − 203.40 = $2.89 more interest.

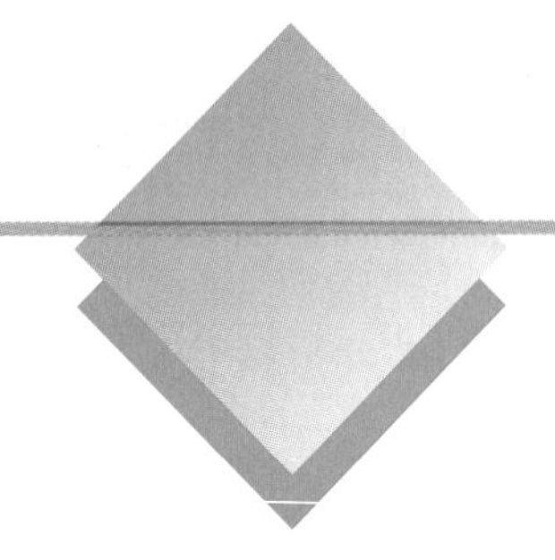

TRY YOUR SKILLS

Use a spreadsheet program to determine whether Yolanda owes more or less on her VISA account in month 12 than at the beginning of the year. Find her yearly interest payment. Her monthly interest charge is 1.5%, and she makes monthly payments of 10% of the amount owed, rounded to the nearest dollar.

	Month	Purchases	Balance	Interest	Amount Owed	Payment
1.	1	0.00	$554.75	$8.32	$563.07	$56.00
2.	2	$76.35	583.42	8.75	592.17	59.00
3.	3	0.00	533.17	8.00	541.17	54.00
4.	4	0.00	487.17	7.31	494.48	49.00
5.	5	75.99	521.47	7.82	529.29	53.00
6.	6	22.80	499.09	7.49	506.58	51.00
7.	7	0.00	455.58	6.83	462.41	46.00
8.	8	97.88	514.29	7.71	522.00	52.00
9.	9	75.00	545.00	8.18	553.18	55.00
10.	10	31.00	529.18	7.94	537.12	54.00
11.	11	66.00	549.12	8.24	557.36	56.00
12.	12	67.00	568.36	8.53	576.89	58.00

GUIDED PRACTICE

Have students work in groups of three or four on the exercises in the *Try Your Skills* section. Suggest that they check the program's calculations using a calculator.

ADDITIONAL ANSWERS

1.-12. Yolanda owes more after 12 months.

13. What is the total interest that Yolanda pays? $95.12

14.–26. Recalculate Exercises 1–13 to reflect late fees of $10.00 in months 2 and 7 and an annual fee of $40.00 in month 4. See Additional Answers.

INDEPENDENT PRACTICE

ASSIGNMENT GUIDE

Exercises 1-3 assess students' understanding of the opening text.

Exercises 4-55 use spreadsheet programs to compute monthly interest and running balances as they are affected by additional purchases, late fees, and annual fees.

EXERCISE YOUR SKILLS

1. Why is it important for people to be able to examine information in their credit files? Answers may vary.
2. What might happen if consumers were not protected by the Fair Credit Billing Act? Answers may vary.
3. How does the Equal Credit Opportunity Act help a woman who becomes divorced or widowed? Creditors cannot cancel a divorced or widowed person's credit unless the income drops dramatically.

KEY TERMS

creditor
Equal Credit Opportunity Act
Fair Credit Billing Act
Fair Credit Reporting Act

CLOSURE

Have students write a paragraph discussing the Fair Credit Reporting Act, the Fair Credit Billing Act, and the Equal Credit Opportunity Act. Tell them to explain the difference between computing monthly interest with and without additional purchases.

LESSON QUIZ

1. Compute the total interest owed at the end of 3 months assuming an initial balance of $580, monthly payments of 10% of the amount owed, a monthly interest charge of 1.5% and a purchase in the second month of $120. (**$27.35**)
2. Repeat Exercise 1, assuming an annual fee of $30 charged in the second month. (**$28.21**)

Use a spreadsheet program to determine whether Roger owes more or less on his VISA account in month 12 than at the beginning of the year. Find the interest that he pays for the year. His account carries a monthly interest charge of 1.5%, and he makes monthly payments of 10% of the amount owed, rounded to the nearest dollar. Roger owes less after 12 months.

	Month	Purchases	Balance	Interest	Amount Owed	Payment
4.	1	0.00	$875.66	$13.13	$888.79	$89.00
5.	2	0.00	799.79	12.00	811.79	81.00
6.	3	0.00	730.79	10.96	741.75	74.00
7.	4	0.00	667.75	10.02	677.77	68.00
8.	5	0.00	609.77	9.15	618.92	62.00
9.	6	$158.80	715.72	10.74	726.46	73.00
10.	7	0.00	653.46	9.80	663.26	66.00
11.	8	85.00	682.26	10.23	692.49	69.00
12.	9	92.00	715.49	10.73	726.22	73.00
13.	10	0.00	653.22	9.80	663.02	66.00
14.	11	0.00	597.02	8.96	605.98	61.00
15.	12	123.00	667.98	10.02	678.00	68.00

16. What is the total interest that Roger pays? $125.54

Use a spreadsheet program to determine whether Yory owes more or less on her VISA account in month 12 than at the beginning of the year. Find the interest that she pays for the year. Her account carries a monthly interest charge of 1.5%, and she makes monthly payments of 10% of the amount owed, rounded to the nearest dollar. Yory owes less after 12 months.

	Month	Purchases	Balance	Interest	Amount Owed	Payment
17.	1	0.00	$1450.63	$21.76	$1472.39	$147.00
18.	2	0.00	1325.39	19.88	1345.27	135.00
19.	3	0.00	1210.27	18.15	1228.42	123.00
20.	4	$234.00	1339.42	20.09	1359.51	136.00
21.	5	0.00	1223.51	18.35	1241.86	124.00
22.	6	0.00	1117.86	16.77	1134.63	113.00
23.	7	0.00	1021.63	15.32	1036.95	104.00
24.	8	0.00	932.95	13.99	946.94	95.00
25.	9	0.00	851.94	12.78	864.72	86.00
26.	10	314.00	1092.72	16.39	1109.11	111.00
27.	11	45.00	1043.11	15.65	1058.76	106.00
28.	12	68.00	1020.76	15.31	1036.07	104.00

29. What is the total interest that Yory pays? $204.44

30.–42. Recalculate Exercises 4–16 to reflect late fees of $20.00 in months 3 and 9 and an annual fee of $40.00 in month 8. See Additional Answers.

43.–55. Recalculate Exercises 17–29 to reflect late fees of $15.00 in months 2, 4 and 9 and an annual fee of $30.00 in month 10. See Additional Answers.

MIXED REVIEW

4-1 **1.** If you earn $25,900.00 and your FICA withholding is 7.65% of your earnings, how much do you pay for Social Security and Medicare?

2. If you can afford a monthly payment of $285, how much money can you borrow at 8.75% for 5 years? $13,809.99

6-2 **3.** Determine the monthly payment that must be made to reduce a $2599.90 VISA balance to zero in 42 months if the card carries an APR of 14.5%. $79.30

6-1 **4.** Determine the effective rate of interest on a credit card purchase if the APR is 17.9%. 19.44%

5-2 **5.** How much money can you save by financing a $215,000 mortgage at 7.5% over 15 years rather than 30 years? See Additional Answers.

5-2 **6.** How much money can you save by financing a $10,500 car at 9% over 3 years rather than 5 years? $1057.46

4-5 A cookie factory makes at least 2000 plain cookies per day and at least 3000 chocolate chip cookies per day. However, the packaging machinery cannot process more than 7500 cookies per day. The profit on each plain cookie is $0.20 and $0.25 on each chocolate chip cookie. Let x and y represent the number of plain and chocolate chip cookies, respectively, that are produced each day.

7. Write the objective function for the profit on the day's production of cookies. $p = 0.2x + 0.25y$

8. Write the inequalities to express the constraints. See Additional Answers.

9. Graph the inequalities and use the graph to find the number of plain and chocolate chip cookies that gives the greatest profit.

5-5 Matilda wants to have a stereo system that costs $700. The store offers to let her purchase the set with an installment loan at 12.5% over 2 years. They also offer her a rent-to-own credit plan that allows her to rent the set for a fixed number of months and to apply all of the rent payments to the purchase price. Find the cost of each of the following rental options.

10. $35 per month for 24 months with no other payment required $840

11. $24 per month for 36 months with no other payment required $864

12. $30 per month for 24 months with an additional payment of $50 due at the end of the 24 months $770

13. Which, if any, of the rental options in Exercises 10–12 is a better deal than the installment loan? See Additional Answers.

ADDITIONAL ANSWERS

1. $1981.35

5. $182,438.25
Total cost for 15 years = $358,753.78
Total cost for 30 years = $541,192.03

8. $x + y \leq 7500$; $x \geq 2000$; $y \geq 3000$

9. $1775 when $x = 2000$ and $y = 5500$

13. The option of Exercise 12 is better than the cost of the installment option, $794.76.

6-5 AVERAGE DAILY BALANCE: WHEN SHOULD YOU PAY YOUR BILLS?

QUICK REFERENCE

TEACHER'S RESOURCES AND ANSWER KEY

Lesson Quiz Answers 6-5, p. 62
Reteaching Activity Answers 6-5, p. 102
Enrichment Activity Answers 6-5, p. 103

EXTENSION ACTIVITIES

Reteaching Activity 6-5, p. 45
Enrichment Activity 6-5, p. 52

Ursula likes her job as a sales clerk at Joyful Toys. There is only one part of her job that she does not enjoy doing. Each time a customer presents a credit card for a purchase, Ursula must verify that the customer's bank will approve the charge. Sometimes the credit is denied. Ursula does not like to inform a customer of this fact.

One time a customer became angry with Ursula when she told him that his credit was denied. He acted as though she was at fault. He yelled at her and demanded to see the store manager. As he left, he hurled a handful of action figures and a model rocket toward the display case, knocking over a tower of blocks and almost breaking the glass. The store manager did not blame Ursula for the mess or for the fact that several customers left while all this was happening. Nevertheless the incident made her feel very uncomfortable.

The bank that issues a customer's credit card may refuse to pay Joyful Toys if Ursula or another clerk neglects to verify the customer's credit. If that happens, Joyful Toys will lose the entire cost of the items that are purchased. The store already pays a service fee to the bank each time a purchase with a bank card is made.

When Ursula handles a customer's credit card, she must be careful to process the information correctly. After verifying the charge with the customer's bank, she makes an imprint of the card on the sales slip. She circles the expiration date on the sales slip. Ursula then asks the customer to sign the sales slip and compares the customer's signature on the sales slip with the signature on the card. She can then be satisfied that she has done everything possible to verify the customer's right to use the card.

OBJECTIVES: *In this lesson, we will help Ursula to:*

- *Identify costs paid by retailers for selling on credit.*
- *Find the average daily balance of a credit card account for the month.*
- *Explore how the timing of a monthly payment affects the interest charge.*

FOCUS

ALGEBRA CONNECTIONS

Hannah opened a savings account. She deposited $85 the first week. The second week and each following week she deposited $9 more than she deposited the previous week. How much money has she put in her account after seven weeks? (**$784**)

THE COST TO RETAILERS OF ISSUING THEIR OWN CARDS

Retailers who issue their own cards to sell on credit incur additional costs on every sale. These costs are a result of the following factors:

- The clerical work necessary to record sales and collect payments
- Losses from customers who fail to make payments
- An increased tendency of charge customers to return goods for exchange

Retailers who neglect to investigate a customer's ability to pay are likely to have high losses from unpaid debts. Stores that recklessly advertise generous credit terms to everyone may have to raise their prices to offset their potential losses. Stores that have sound credit policies are likely to have few losses from customers who fail to make payments. Some retailers elect to accept MasterCard, VISA, or American Express instead of (or in addition to) extending their own credit.

Banks that issue all-purpose cards like VISA and MasterCard collect a percentage of every sale made on their cards from retailers. This charge provides some protection for the banks against the money they stand to lose if

ADDITIONAL ANSWERS

1. The charge covers them against the money they stand to lose if the cardholders fail to pay for the purchases charged.
2. by adding the unpaid balances for each day of the billing cycle then dividng by the number of days

the cardholders fail to pay for the purchases they charged. Even though the salesperson takes every precaution, a few illegitimate uses of cards are bound to slip through.

Banks also pass some of their costs on to their credit card customers by charging interest on unpaid balances. The maximum amount of interest that a bank can charge a customer is set by state laws. Differences in state interest rate limitations caused at least one large bank to move its national credit card operation from New York to South Dakota. In South Dakota the interest rate that a bank can charge is higher than it is in New York.

As was stated in Lesson 6-1, banks often allow customers a grace period before they must pay the bill. If they pay the bill on time, they can avoid interest charges. If cardholders do not wish to pay the entire balance due, they can make a minimum payment (on some types of credit card accounts) and pay interest charges on the remainder.

The amount of interest that a customer must pay is based upon the **average daily balance,** which is found by adding the unpaid balances for each day of the billing cycle and then dividing this number by the number of days in the billing cycle. In this chapter a billing cycle is assumed to be a calendar month. Customers can save money by making their payments as early in the month as is possible.

ALGEBRA *REVIEW*

When an *average* is to be determined from a number of scores in which some scores are repeated, the number of times a score is repeated is called the *frequency.*

Find the average of the following scores with the indicated frequencies.

Example:

Score	82	75	91
Frequency	3	2	4

$$\text{Average} = \frac{3(82) + 2(75) + 4(91)}{9} = 84.4$$

1.

Score	35	61	78
Frequency	5	3	2

51.4

2.

Score	468	721	325
Frequency	1	20	3

660.96

3.

Score	10.2	15.5	25.3
Frequency	18	5	2

12.47

4.

Score	804	246	760
Frequency	2	28	1

298.58

5.

Score	65	945	607
Frequency	20	4	6

290.73

ASK YOURSELF

1. Why do banks that issue cards such as MasterCard and VISA charge the retailers who accept these cards?
   See Additional Answers.
2. How is the average daily balance computed?
3. How is the maximum amount of interest that a bank can charge determined?
   set by state laws

SHARPEN YOUR SKILLS

SKILL 1

The average daily balance b is determined by dividing the sum of the daily balances s by the total number of days d in the billing cycle.

Average Daily Balance Formula

$b = \frac{s}{d}$ where b = average daily balance
s = sum of the daily balances
d = total number of days in the billing cycle

EXAMPLE 1 Scott's MasterCard account has a monthly interest rate of 1.5%. He had the following daily balances and payments for the month of May:

From 5/1 through 5/16 his daily balance was \$385.00.
On 5/17 he made a payment of \$80.
From 5/18 through 5/31 his daily balance remained at \$305.00.

QUESTION What are the average daily balance, the monthly finance charge, and the ending balance during the month of May?

SOLUTION
Complete the following chart to find the sum of the daily balances.

Dates	Payment	Balance at End of Day	Number of Days	Sum of Daily Balances
5/1–5/16	0.00	\$385.00	16	\$ 6160.00
5/17	\$80.00	305.00	1	305.00
5/18–5/31	0.00	305.00	14	4270.00
			Total 31	\$10,735.00

$b = \frac{s}{d}$ Use the average daily balance formula.

$b = \frac{10,735}{31}$

$b = 346.29$ To the nearest cent

Average daily balance: $10,735 \div 31 = 346.29$
Finance charge (May 17): $0.015(346.29) = 5.19$
Ending balance for May: $305.00 + 5.19 = 310.19$

INSTRUCTION

TEACHING THE LESSON

Spend 2 days on this lesson.

On the first day, have students read the opening text and discuss the advantages and disadvantages of retailers' accepting credit card payments or issuing their own cards. Spend time on the *Algebra Review* exercises. Go over Examples 1 and 2 in detail. Redo the examples assuming an initial balance of \$800 to help students see the larger differentials in monthly interest that follow from larger balances.

On the second day discuss Example 3, creating variations by changing the date of the \$50 purchase to see how an earlier or later purchase date affects the average daily balance.

FOCUS ON ALGEBRA

In this lesson students use order of operations to compute average daily balances and use decimal multiplication and rounding to compute interest. (*NCTM Standard 1, p. 137.*)

COMMON ERRORS

Students may forget to enclose numerators of fractions in parentheses before dividing by denominators. Remind them to do so to avoid inaccurate results.

CRITICAL THINKING

Have students use Example 2 to compute the difference in monthly interest if Scott pays his entire bill on May 1, then on May 31.

RETEACHING

Have students rework Example 1 in Lesson 6-4 assuming that Saul's mother makes her payment on the first day of each month (assume Month 1 is January, Month 2 is February and so on). Direct them to compute her total interest and to compare it to the amount shown in the original example.

ENRICHMENT

Have students collect credit card advertisements from the business section of their local newspapers. Tell them to read the details carefully and make sure that they understand the lender's terms concerning finance charges, APR, average daily balance and so on.

EXAMPLE 2 Scott always makes his MasterCard payments in the middle of the month when he receives his second bimonthly paycheck. He has often wondered whether the timing of his payment influences the amount he pays in monthly interest.

QUESTION What would his interest payment have been if he paid his May MasterCard bill on the first day of the month instead of on the seventeenth day? On the last day instead of the seventeenth day?

SOLUTION

Set up two charts similar to the chart in Example 1, one with a payment on May 1 and the other with a payment on May 31.

Dates	Payment	Balance at End of Day	Number of Days	Sum of Daily Balances
5/1	$80.00	$305.00	1	$ 305.00
5/2–5/31	0.00	305.00	30	9150.00
			Total 31	$9455.00

Average daily balance: $9455.00 \div 31 = 305.00$
Finance charge (May 1): $0.015(305.00) = 4.58$
Ending balance for May: $305.00 + 4.58 = 309.58$

Dates	Payment	Balance at End of Day	Number of Days	Sum of Daily Balances
5/1–5/30	0.00	$385.00	30	$11,550.00
5/31	$80.00	305.00	1	305.00
			Total 31	$11,855.00

Average daily balance: $11{,}855.00 \div 31 = 382.42$
Finance charge (May 31): $0.015(382.42) = 5.74$
Ending balance for May: $305.00 + 5.74 = 310.74$

Paying his bill on May 1 instead of May 17 saves: $5.19 - 4.58 = 0.61$
Paying his bill on May 17 instead of May 31 saves: $5.74 - 5.19 = 0.55$
Paying his bill on May 1 instead of May 31 saves: $5.74 - 4.58 = 1.16$

Although the savings in one month do not seem like a lot of money, they will be greater if your balance is larger. The savings accumulate over the months and years. Also, you may have many credit cards with balances. Therefore you can save a substantial amount of money on all of these cards by making credit card payments as early in the month as possible.

When finding the average daily balance be sure to use the correct number of days. Remember that April, June, September and November only have 30 days and February has 28 days unless it is a leap year.

SKILL 2

EXAMPLE 3 Consumers often make additional charges on their accounts each month. These additional charges affect the average daily balance in the account.

QUESTION If Scott had purchased a new jacket for $50.00 on 5/11 in Example 1, what would have been his average daily balance, finance charge, and ending balance?

SOLUTION

Use a spreadsheet program.

Dates	Payment	Purchase	Balance at End of Day	Number of Days	Sum of Daily Balances
5/1–5/10	0.00	0.00	$385.00	10	$ 3850.00
5/11	0.00	$50.00	435.00	1	435.00
5/12–5/16	0.00	0.00	435.00	5	2175.00
5/17	$80.00	0.00	355.00	1	355.00
5/18–5/31	0.00	0.00	355.00	14	4970.00
			Total	31	$11,785.00

Average daily balance: $11{,}785 \div 31 = 380.16$

Finance charge (May 31): $0.015(380.16) = 5.70$

Ending balance for May: $355.00 + 5.70 = 360.70$

TECHNOLOGY HINT

In the spreadsheet for Example 3, formulas for "Average Daily Balance," "Finance Charge," and "Ending Balance" are: @ROUND(F7/E7, 2); @ROUND(0.015∗ "Average Daily Balance", 2); +D6+ "Finance Charge", respectively. Note that cell locations for "Average Daily Balance," "Finance Charge," and "Ending Balance" have not been assigned.

TRY YOUR SKILLS

1. Complete the following chart to determine the average daily balance, the finance charge, and the ending balance shown on Jerry's VISA statement for the month of September. Assume that his VISA card carries a monthly interest rate of 1.25%.

Dates	Payment	Balance at End of Day	Number of Days	Sum of Daily Balances
9/1–9/14	0.00	$642.00	14	$8988.00
9/15	$75.00	567.00	1	567.00
9/16–9/30	0.00	567.00	15	8,505.00
		Total	30	$18,060.00

a. Average daily balance $602.00 **b.** Finance charge $7.53

c. Ending balance $574.53

GUIDED PRACTICE

Have students work in small groups on the *Try Your Skills* exercises. Direct group members to check each other's computations.

2. Complete the chart below to find how much Jerry can save if he makes his September payment on 9/1 instead of 9/15.

Dates	Payment	Balance at End of Day	Number of Days	Sum of Daily Balances
9/1	$75.00	$567.00	1	$ 567.00
9/2–9/30	0.00	567.00	29	16,433.00
			Total 30	$17,010.00

a. Average daily balance $567.00 b. Finance charge $7.09

c. Ending balance $574.09

d. How much can he save if he makes his payment on 9/1 instead of 9/15?

3. Complete the chart below to determine the average daily balance, the finance charge, and the ending balance shown on Savannah's VISA statement for the month of August. Assume that her VISA card carries a monthly interest rate of 1.75%.

Dates	Payment	Purchase	Balance at End of Day	Number of Days	Sum of Daily Balances
8/1–8/12	0.00	0.00	$675.00	12	$8100.00
8/13	$95.00	0.00	580.00	1	580.00
8/14–8/18	0.00	0.00	580.00	5	2900.00
8/19	0.00	$60.00	640.00	1	640.00
8/20–8/31	0.00	0.00	640.00	12	7680.00

a. Average daily balance $641.94 b. Finance charge $11.23

c. Ending balance $651.23

ADDITIONAL ANSWERS (TRY YOUR SKILLS)

2. d. $0.44

(EXERCISE YOUR SKILLS)

1. to verify the customer's right to use the card
2. Yes; interest is charged on the average daily balance; thus the earlier you pay, the lower the average daily balance and interest.

EXERCISE YOUR SKILLS

KEY TERM

average daily balance

1. Why does a clerk use a computer scanner to verify a customer's credit card before allowing the customer to use it for a purchase?
2. Does the day of the month on which you make a credit card payment influence the interest charge? Why or why not?

Complete each chart. Then determine the average daily balance, the finance charge, and the ending balance that will be shown on each monthly MasterCard statement. Assume that the card carries a monthly finance charge of 1.5%.

ASSIGNMENT GUIDE

Exercises 1-2 check students' understanding of the opening text.

Exercises 3-22 ask students to determine average daily balance, finance charges and ending balance assuming payments made on specified days of the month.

Exercises 23-25 require average daily balance, finance charges and ending balance assuming additional purchases during the month.

3.

Dates	Payment	Balance at End of Day	Number of Days	Sum of Daily Balances
1/1–1/7	0.00	$1948.50	7	$13,639.50
1/8	$98.00	1850.50	1	1,850.50
1/9–1/31	0.00	1850.50	23	42,561.50

a. Average daily balance $1872.63 **b.** Finance charge $28.09

c. Ending balance $1878.59

4.

Dates	Payment	Balance at End of Day	Number of Days	Sum of Daily Balances
2/1–2/10	0.00	$1878.59	10	$18,785.90
2/11	$93.00	1785.59	1	1,785.59
2/12–2/28	0.00	1785.59	17	30,355.03

a. Average daily balance $1818.80 **b.** Finance charge $27.28

c. Ending balance $1812.87

5.

Dates	Payment	Balance at End of Day	Number of Days	Sum of Daily Balances
3/1–3/15	0.00	$1812.87	15	$27,193.05
3/16	$88.00	1724.87	1	1,724.87
3/17–3/31	0.00	1724.87	15	25,873.05

a. Average daily balance $1767.45 **b.** Finance charge $26.51

c. Ending balance $1751.38

6.

Dates	Payment	Balance at End of Day	Number of Days	Sum of Daily Balances
4/1–4/12	0.00	$1751.38	12	$21,016.56
4/13	$83.00	1668.38	1	1,668.38
4/14–4/30	0.00	1668.38	17	28,362.46

a. Average daily balance $1701.58 **b.** Finance charge $25.52

c. Ending balance $1693.90

CLOSURE

Have students write a paragraph about issues faced by retailers who sell on credit. Tell them to explain how average daily balance is computed, taking into account additional purchases and timing of monthly payments.

LESSON QUIZ

Use information for Questions 1 and 2 by setting up a 5-column table with the following headings: Dates, Payment, Balance at End of Day, Number of Days, and Sum of Daily Balances. The first three rows in your table will include the following information. Note that there are no entries in columns 4 and 5.

Row 1: Dates = 8/1-8/13
Payment = 0.00
Balance at End of Day = 350.00

Row 2: Dates = 8/14
Payment = 90.00

Row 3: Dates = 8/15-8/31
Payment = 0.00

1. Find the average daily balance. (**$297.74**)
2. Compute the finance charges, assuming a monthly interest rate of 1.5%. (**$4.47**)
3. Find the ending balance. (**$264.47**)

7.

Dates	Payment	Balance at End of Day	Number of Days	Sum of Daily Balances
5/1–5/20	0.00	$1693.90	20	$33,878.00
5/21	$79.00	1614.90	1	1,614.90
5/22–5/31	0.00	1614.90	10	16,149.00

a. Average daily balance $1665.87 **b.** Finance charge $24.99
c. Ending balance $1639.89

8.

Dates	Payment	Balance at End of Day	Number of Days	Sum of Daily Balances
6/1–6/13	0.00	$1639.89	13	$21,318.57
6/14	$75.00	1564.89	1	1,564.89
6/15–6/30	0.00	1564.89	16	25,038.24

a. Average daily balance $1597.39 **b.** Finance charge $23.96
c. Ending balance $1588.85

9.

Dates	Payment	Balance at End of Day	Number of Days	Sum of Daily Balances
7/1–7/18	0.00	$1588.85	18	$28,599.30
7/19	$71.00	1517.85	1	1,517.85
7/20–7/31	0.00	1517.85	12	18,214.20

a. Average daily balance $1559.08 **b.** Finance charge $23.39
c. Ending balance $1541.24

10.

Dates	Payment	Balance at End of Day	Number of Days	Sum of Daily Balances
8/1–8/9	0.00	$1541.24	9	$13,871.16
8/10	$68.00	1473.24	1	1,473.24
8/11–8/31	0.00	1473.24	21	30,938.04

a. Average daily balance $1492.98 **b.** Finance charge $22.39
c. Ending balance $1495.63

11.

Dates	Payment	Balance at End of Day	Number of Days	Sum of Daily Balances
9/1–9/13	0.00	$1495.63	13	$19,443.19
9/14	$64.00	1431.63	1	1,431.63
9/15–9/30	0.00	1431.63	16	22,908.08

a. Average daily balance $1459.36 **b.** Finance charge $21.89
c. Ending balance $1453.52

12.

Dates	Payment	Balance at End of Day	Number of Days	Sum of Daily Balances
10/1–10/21	0.00	$1453.52	21	$30,523.92
10/22	$61.00	1392.52	1	1392.52
10/23–10/31	0.00	1392.52	9	12,532.68

a. Average daily balance $1433.84 **b.** Finance charge $21.51

c. Ending balance $1414.03

13. If you pay the bill in Exercise 3 on the first day of the month, how much will you save? $0.33

14. If you pay the bill in Exercise 4 on the first day of the month, how much will you save? $0.50

15. If you pay the bill in Exercise 5 on the first day of the month, how much will you save? $0.64

16. If you pay the bill in Exercise 6 on the first day of the month, how much will you save? $0.49

17. If you pay the bill in Exercise 7 on the first day of the month, how much will you save? $0.77

18. If you pay the bill in Exercise 8 on the first day of the month, how much will you save? $0.49

19. If you pay the bill in Exercise 9 on the first day of the month, how much will you save? $0.62

20. If you pay the bill in Exercise 10 on the first day of the month, how much will you save? $0.29

21. If you pay the bill in Exercise 11 on the first day of the month, how much will you save? $0.42

22. If you pay the bill in Exercise 12 on the first day of the month, how much will you save? $0.62

Complete each chart to determine the average daily balance, the finance charge, and the ending balance shown on your VISA statement for the month. Assume that your VISA card carries a monthly interest rate of 1.25%

23.

Dates	Payment	Purchase	Balance at End of Day	Number of Days	Sum of Daily Balances
8/1–8/18	0.00	0.00	$875.00	18	$15,750.00
8/19	$90.00	0.00	785.00	1	785.00
8/20–8/25	0.00	0.00	785.00	6	4,710.00
8/26	0.00	$135.00	920.00	1	920.00
8/27–8/31	0.00	0.00	920.00	5	4,600.00

a. Average daily balance $863.39 **b.** Finance charge $10.79

c. Ending balance $930.79

24.

Dates	Payment	Purchase	Balance at End of Day	Number of Days	Sum of Daily Balances
2/1–2/11	0.00	0.00	$500.00	11	$5500.00
2/12	0.00	$90.00	590.00	1	590.00
2/13–2/20	0.00	0.00	590.00	8	4720.00
2/21	$65.00	0.00	525.00	1	525.00
2/22–2/28	0.00	0.00	525.00	7	3675.00

a. Average daily balance $536.07 **b.** Finance charge $6.70

c. Ending balance $531.70

25.

Dates	Payment	Purchase	Balance at End of Day	Number of Days	Sum of Daily Balances
4/1–4/8	0.00	0.00	$1275.00	8	$10,200.00
4/9	0.00	$225.00	1500.00	1	1,500.00
4/10–4/17	0.00	0.00	1500.00	8	12,000.00
4/18	$175.00	0.00	1325.00	1	1,325.00
4/19–4/30	0.00	0.00	1325.00	12	15,900.00

a. Average daily balance $1364.17 **b.** Finance charge $17.05

c. Ending balance $1342.05

MIXED REVIEW

5-2 **1.** How much money can you save by financing an $8800 car at 10.5% over 3 years rather than 5 years? $1052.01

5-4 **2.** A $760.00 television purchase is financed at 15% per year with no down payment at a store that applies the Rule of 78. It must be paid in 12 equal installments. Determine the prepayment savings if it is prepaid after 5 months. $40.92

5-3 **3.** If you buy a computer system for $3500 and finance it at 18% over 2 years, how much will you save by making a down payment of 25% instead of 10%? $104.04

1-1 **4.** What are your weekly earnings if you work 30.5 hours at $14.75 per hour? $449.88

3-1 **5.** Calculate the simple interest for one year on a $1200 deposit earning 3.75%. $45.00

3-1 **6.** Nancy always withdraws the interest on her savings account as soon as it is posted. Calculate the simple interest for three years on her original $1200 deposit that earns 5% per year. $180

5-3 **7.** If you buy a used automobile for $7500 and finance it at 11% over 3 years, how much will you save by making a down payment of 25% instead of 10%? $200.92

6-6 CREDIT RATINGS: HOW TO DETERMINE YOUR SCORE

QUICK REFERENCE

TEACHER'S RESOURCES AND ANSWER KEY

Lesson Quiz Answers 6-6, p. 62
Reteaching Activity Answers 6-6, p. 102
Enrichment Activity Answers 6-6, p. 104

EXTENSION ACTIVITIES

Reteaching Activity 6-6, p. 46
Enrichment Activity 6-6, p. 53

Daryl has watched in fascination as his father unfolded his packet of credit cards. He has five oil company cards, six charge accounts at department stores, an American Express card, an ATM cash card, four MasterCards, and six VISA cards. Daryl thinks that carrying that many cards around must be a chore! How in the world can anyone manage to acquire so many cards, much less need them all? There must be a lot of checks to write each month when the bills come. Daryl carries only four cards: his driver's license, his school library card, a credit card for gasoline, and his ID card for the video rental store.

Once when Daryl asked his father about all the credit cards, he explained that his credit rating is excellent. He has lived in the same house for 8 years, he has been a high-salaried employee at the same business for 12 years, he has bought three different cars with installment loans, making prompt payments every month, and he is 41 years old. Creditors are more than willing to give him credit because he has shown he is capable and consistent in making payments. Over a 4-year period, 12 different banks from around the country offered him credit cards. He took most of them up on their offers, realizing how easy it is to use the cards to buy much more than he could afford to pay for with cash. Now he makes the minimum payment each month to each bank and continues to use almost all the credit available to him on each card.

When Daryl watches his father writing all the checks each month, he is not sure carrying 23 credit cards is truly necessary. However, he would like to know how to get at least one general purpose card for himself. Since he has never had a bank card and probably does not even have a credit rating, he is wondering where to begin.

ALGEBRA CONNECTIONS

The price of a $9000 car was reduced by $600 and an additional amount of $50 for each day it was still on the market. If the car sold for $7200, how many days was it on the market? (**24 days**)

OBJECTIVES: *In this lesson, we will help Daryl to:*

- *Determine an individual's credit rating, using the judgment of the creditor or a credit-scoring table.*
- *Learn how to get credit for the first time.*

HOW CREDIT IS GRANTED

A **credit rating** is an indication of a person's ability to secure goods, services, and money in return for the promise to pay. Daryl knows a favorable credit rating does not come automatically. His father's good rating has developed slowly over time. It has been nurtured, fostered, strengthened, and improved. A good credit rating is an asset of tremendous value to those who develop it over a long period, but it can easily be destroyed. A good credit rating is sensitive to abuse and usually continues only as long as it is justified. Daryl's father has so much credit now that were he to apply for more, he would probably be told he already has sufficient credit.

The decision to grant credit lies with the *judgment of the creditor*. The creditor considers the "three C's" of credit—capacity, character, and collateral.

Capacity Do you have the capacity to repay the loan? How long have you worked, and how much do you earn? Creditors will also ask about your expenses such as rent, mortgage, car payments, and so on.

Character Have you been responsible about meeting financial obligations in the past?

Collateral Are the creditors protected from loss if you fail to pay? Creditors want to know if you have a savings account or other assets (car, home, or other valuable assets) to offer as security for a loan.

Because of laws that prohibit discrimination, large credit-lending institutions have developed **credit-scoring systems** that can be considered more objective than personal judgment. Credit-scoring systems award points for various factors to determine credit worthiness. A scoring system is shown on page 293. The principles underlying a scoring system are as follows:

- In general, you are a better *credit risk* if you are older. People in their mid-30's often face unforeseen expenses due to divorce or other causes of financial stress. Consequently their credit rating points may be lower.
- You are considered more stable the longer you have lived in the same place, and your score will increase if you own your home rather than rent it. People with cars get more points than those without, and owners of newer cars generally earn more points than owners of older cars.

- The length of time you have held the same job is another factor that is considered. The kind of job you have, your income, and your amount of current debt also make a difference.
- It is favorable to have either a savings account or a checking account. If you have both, you score more points.
- If you have one or more credit cards in good standing, you will score higher than those who have none. Department store charge cards, travel and entertainment cards, and oil company cards usually rate somewhat lower than all-purpose bank cards.
- If you have borrowed recently at high interest from a finance company, you will probably lose several points. If you have applied and been turned down for credit more than once in the last 6 months, you may have difficulty getting new credit at all.

CRITICAL THINKING

Why do you think the credit-scoring table awards more points for auto payments of $151-$199 than for those of $126-$150? Why is 5-9 years at a given address worth more points than 3-5 years?

Hypothetical Credit-Scoring Table

Fill out your credit profile by answering the nine questions below in Table 1. Circle the one response that applies to you, and then find your total score by adding up the points you got for each response. The points are found in the lower right-hand corner of each box. (For example: if you are 25 years old, you get 5 points.) Once you've totaled your score, look at Table 2 to find out how good a credit "bet" you may be.

1.

1.	age?	under 25 12	25–29 5	30–34 0	35–39 1	40–44 18	45–49 22	50 or over 31
2.	time at address?	less than 1 yr. 9	1yr. 0	2–3 yrs. 5	4–5 yrs. 0	6–9 yrs. 5	10 yrs. or more 21	
3.	age of auto?	none 0	0–1 yrs. 12	2 yrs. 16	3–4 yrs. 13	5–7 yrs. 3	8 yrs. or more 0	
4.	monthly auto payment?	none 18	less than $125 6	$126–$150 1	$151–$199 4	$200 or more 0		
5.	housing cost?	less than $274 0	$275–$399 10	$400 or more 12	owns clear 12	lives with relatives 24		
6.	checking and savings accounts	both 15	checking only 2	savings only 2	neither 0			
7.	finance company reference	yes 0	no 15					
8.	major credit cards?	none 0	1 5	2 or more 15				
9.	ratio of debt to income?	no debts 41	1%–5% 16	6%–15% 20	16% or over 0			

2.

A lender using this scoring table selects a cutoff point from a table like this, which gauges how likely applicants are to repay loans.

Total Score	Probability of Repayment
90	89 in 100
95	91 in 100
100	92 in 100
105	93 in 100
110	94 in 100
115	95 in 100
120	95.5 in 100
125	96 in 100
130	96.25 in 100

Source: Federal Reserve Board. Developed by Fair, Isaac, and Co., Inc. Modified to update.

Daryl would like to know how to obtain *first-time credit.* Credit is easy to get if you have had it before. But what about getting credit for the first time? Below are some suggestions for Daryl to help him establish a good credit reputation. He can do the following:

- Open a charge account at a retail store where his parents have an account.
- Establish checking and savings accounts at the local bank.
- Get one of his parents to **cosign** a small loan. By cosigning, Daryl's father promises to pay if Daryl does not. When Daryl repays on time, he establishes a good credit record.
- Join a credit union.
- Be a responsible employee. Creditors generally ask for length of employment at a job and for personal references.

As Daryl begins to establish his own credit rating, he will be careful not to abuse it. He will treat it as the valuable asset that it is, and continue to strengthen it as time goes on.

ASK YOURSELF

1. What are the "3 Cs" that creditors use to judge a person's acceptability for credit?
   capacity, character, collateral
2. How does a lending institution determine the credit worthiness of an applicant?
   using a credit-scoring system
3. How might you establish a good credit reputation?
   Answers may vary.

ALGEBRA *REVIEW*

When you take a test in school, you often receive a score. Mrs. Esposito assigned the following point values to each question on her mathematics test:

Questions 1–10: 5 points each
Questions 11–14: 7 points each
Questions 15–16: 11 points each
Bonus Question: 10 points

Determine each test score.

1. Sally missed Questions 2 and 12, but answered the others correctly. She also answered the bonus question correctly.
   98
2. Tom missed Questions 13 and 14, but answered the others correctly. He received no bonus points.
   86
3. Janet missed Questions 5, 6 and 11, but answered the others correctly. She received no bonus points.
   83

4.–6. If Mrs. Esposito assigns grades as follows, determine each student's grade on the test.

90+ → A
80–89 → B
70–79 → C
65–69 → D
0–64 → F

Sally: A; Tom: B; Janet: B

7. Write a general formula for calculating a test score.

ADDITIONAL ANSWERS (ALGEBRA REVIEW)

7. Let S represent the test score, x represent the number of questions 1-10 answered correctly, y represent the number of questions 11-14 answered correctly, and z represent the number of questions 15-16 answered correctly. $S = 5x + 7y + 11z + 10$, if the bonus question is answered correctly and $S = 5x + 7y + 11z$, if the bonus question is not answered correctly.

SHARPEN YOUR SKILLS

SKILL 1

EXAMPLE 1 Daryl wonders how his father qualifies for so many credit cards. He knows the following facts about his father.

1. He is 41 years old.
2. He has lived at his current address for 8 years.
3. His current car is 2 years old.
4. He has no monthly car payment.
5. His housing cost is $482 per month.
6. He has a checking account and a savings account.
7. He has not been referred to a finance company.
8. He has 23 major credit cards.
9. His ratio of debt to income is 20%.

QUESTION How would a potential creditor rate his credit worthiness? According to the credit-scoring system, what is the probability that he will repay a loan?

SOLUTION

	Points	
1.	18	Use the credit-scoring table to find his credit score.
2.	5	
3.	16	
4.	18	
5.	12	
6.	15	
7.	15	
8.	15	
9.	0	
Score:	114	

Look in part 2 of the credit-scoring table to find the probability that Daryl's father will repay his loan. Since 114 is not listed, use the lower score of 110. The probability that he will repay a loan is 94 in 100, or 94%.

INSTRUCTION

TEACHING THE LESSON

Spend one day on this lesson. After students have read the opening text, ask them to make a list of characteristics that creditors consider advantageous and disadvantageous for credit seekers. Tell students to do the *Algebra Review* section, then practice using the credit-scoring table by going over the information in Example 1.

FOCUS ON ALGEBRA

In this lesson students compute based on information from a table and relate their findings to another chart. (*NCTM Standard 1, p. 137.*)

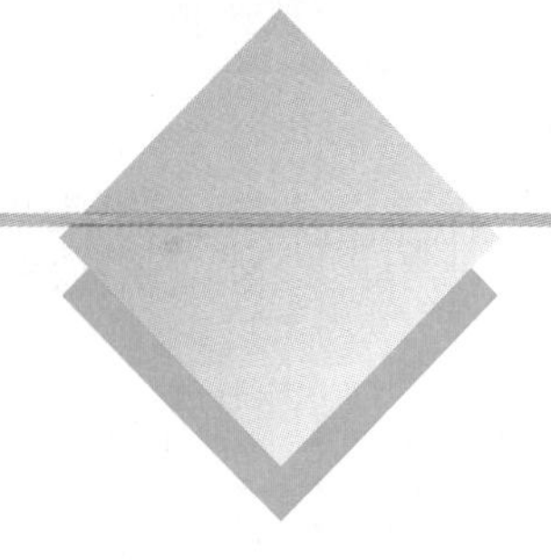

TRY YOUR SKILLS

Use the credit-scoring table to find Brooke's credit score for each item.

1. She is 28 years old. 5
2. She has lived at her current address for 1 year. 0
3. Her current car is 4 years old. 13
4. She has no car payment. 18
5. She lives with relatives. 24
6. She has a checking account and a savings account. 15
7. She has not been referred to a finance company. 15
8. She has no major credit cards. 0
9. She has no debts. 41
10. What is Brooke's total score? 131
11. What is the probability that Brooke will repay a loan? 96.25%

GUIDED PRACTICE

Have students work in small groups on the situation in *Try Your Skills*. Tell them to vary the data (current address for 5 years, owning two major credit cards, current car is 9 years old) to see the effect on an overall credit score.

EXERCISE YOUR SKILLS

1. Why is it important to establish a favorable credit rating? See Additional Answers.
2. If you were a lender, what is the minimum score on the credit-scoring table that you would accept on an applicant? Why? See Additional Answers.
3. It has been said that a person's credit rating is one of his or her most valuable assets. Do you agree or disagree? Why? Answers may vary.

Use the credit-scoring table on page 293 to determine the credit score for each head of household described below. Then determine the probability that the person will repay a loan.

4. **Family 1**
   Age of the head of the household—35 1
   Lived at the current address—4 years 0
   Age of current car—3 years 13
   Monthly car payment—$165 4
   Housing cost—$650 12
   Has a checking account and a savings account. 15
   Has not been referred to a finance company. 15
   Has 7 major credit cards. 15
   The ratio of debt to income is 12%. 20; Score: 95; 91%

KEY TERMS

collateral
cosign
credit rating
credit-scoring systems

COMMON ERRORS

Students may be unsure of how to compute ratio of debt to income. Advise them that debt does not include mortgage payments but does include unpaid credit card balances.

AT-RISK STUDENTS

Tell students to use the credit-scoring table to rate their own home situation in consultation with a parent. Have them discuss their findings in small groups.

ADDITIONAL ANSWERS

1. to secure credit for a large loan
2. Answers may vary.

5. **Family 2**
   Age of the head of the household—48 22
   Lived at the current address—12 years 21
   Age of current car—5 years 3
   Monthly car payment—$0.00 18
   Housing cost—$385 10
   Has a checking account and a savings account. 15
   Has not been referred to a finance company. 15
   Has 6 major credit cards. 15
   The ratio of debt to income is 19%. 0; Score: 119; 95%
6. **Family 3**
   Age of the head of the household—26 5
   Lived at the current address—2 years 5
   Age of current car—1 year 12
   Monthly car payment—$364 0
   Housing cost—$741 12
   Has a checking account but not a savings account. 2
   Has not been referred to a finance company. 15
   Has 12 major credit cards. 15
   The ratio of debt to income is 24%. 0; Score: 66; less than 89%
7. **Family 4**
   Age of the head of the household—43 18
   Lived at the current address—16 years 21
   Age of current car—1 year 12
   Monthly car payment—$0.00 18
   Housing cost—$640 12
   Has a checking account and a savings account. 15
   Has not been referred to a finance company. 15
   Has 3 major credit cards. 15
   The ratio of debt to income is 8%. 20; Score: 146; greater than 96.25%
8. **Family 5**
   Age of the head of the household—32 0
   Lived at the current address—3 years 5
   Age of current car—1 year 12
   Monthly car payment—$237 0
   Housing cost—$853 12
   Has a checking account but not a savings account. 2
   Has been referred to a finance company. 0
   Has 12 major credit cards. 15
   The ratio of debt to income is 32%. 0; Score: 46; less than 89%

INDEPENDENT PRACTICE

ASSIGNMENT GUIDE

Exercises 1-3 check students' understanding of credit issues presented in the opening text. Exercises 4-11 ask students to determine credit scores using a credit-scoring table.

RETEACHING

Review the "three Cs" of credit with the class. Give students additional practice with the credit-scoring table by creating new family profiles to rate.

ENRICHMENT

Have students write to several oil companies to determine their criteria for issuing credit cards to high school and college students. Suggest that they ask the oil companies about using a credit-scoring table.

CLOSURE

Have students write a strategy for qualifying for their first credit card. Tell them to rate themselves using the credit-scoring table to see how a prospective lender might view their potential for repaying loans.

LESSON QUIZ

Determine the credit score for each person using the credit-scoring table on p. 293. Then determine for each person the probability of repayment of a loan.

1. Age—40
Lived at current address—8 years
Age of current car—1 year
No monthly car payment.
Housing cost–$868
Has only a checking account.
Has been referred to a finance company.
Has 8 major credit cards.
Ratio of debt to income is 15%.
(**102; 92%**)

2. Age—25
Lived at current address—2 years
Age of current car—1 year
Monthly car payment—$244
Housing cost–$359
Has only a savings account.
Has not been referred to a finance company.
Has 12 major credit cards.
Ratio of debt to income is 24%.
(**64; less than 89%**)

ADDITIONAL ANSWERS

3. 778.22 − 105.89 = 672.33

9. Family 6
Age of the head of the household—52 31
Lived at the current address—22 years 21
Age of current car—3 year 13
Monthly car payment—$0.00 18
Housing cost—$332 10
Has a checking account and a savings account. 15
Has not been referred to a finance company. 15
Has 8 major credit cards. 15
The ratio of debt to income is 4%. 16; Score: 154; greater than 96.25%

10. Family 7
Age of the head of the household—34 0
Lived at the current address—6 months 9
Age of current car—8 year 0
Monthly car payment—$0.00 18
Housing cost—$213 0
Has a checking account but not a savings account. 2
Has not been referred to a finance company. 15
Has 2 major credit cards. 15
The ratio of debt to income is 8%. 20; Score: 79; less than 89%

11. Family 8
Age of the head of the household—23 12
Lived at the current address—17 years 21
Age of current car—1 year 12
Monthly car payment—$233 0
Housing cost—$150 0
Has a checking account and a savings account. 15
Has not been referred to a finance company. 15
Has 1 major credit card. 5
The ratio of debt to income is 3%. 16; Score: 96; 91%

MIXED REVIEW

5-1 **1.** If you borrow $1350.00 for 4 years at 12% to buy furniture, what is the deferred payment price of the furniture? $1706.43

4-3 **2.** The fixed costs in Nicolette's business are $300 per week. The variable costs are $4.50 per item. What is the total cost if she produces 120 items in one week? $840

2-2 **3.** You have $778.22 in your checking account and you write a check for $105.89. Show how you would enter this check in your check register.

3-2 **4.** Determine the interest for 9 months if the starting principal is $5525, the annual rate is 5%, and the interest is compounded quarterly. $209.79

1-2 **5.** You earn $18,600 per year plus 12% of your salary in fringe benefits. Find your yearly earnings, including benefits. $20,832

6-7 REGAINING FINANCIAL STABILITY: SOLUTIONS TO DEBT PROBLEMS

QUICK REFERENCE

TEACHER'S RESOURCES AND ANSWER KEY

Lesson Quiz Answers 6-7, p. 64
Reteaching Activity Answers 6-7, p. 102
Enrichment Activity Answers 6-7, p. 104

EXTENSION ACTIVITIES

Reteaching Activity 6-7, p. 47
Enrichment Activity 6-7, p. 54

Sylvia and Daryl were talking and began to discuss credit card use. They compared facts about their fathers and their use of credit cards. One credit story led to another and, by the end of the conversation, Sylvia and Daryl were both relieved to know they were not alone in having serious doubts about some of the choices made by their respective fathers. Daryl was especially relieved to know he did not have the only father in town whose stack of credit cards was so thick he appeared to be shuffling a deck of playing cards every time he used one. Once when his father dropped them all on the floor in a restaurant, Daryl was so embarrassed that he rushed out to the parking lot to wait, pretending he did not know the man crawling around on the floor looking for his cards.

Daryl's father was embarrassed, too, but not as much as when he was so far behind in paying all the credit bills that he actually considered trying to pawn some of the new video equipment he had just purchased. At times he has even considered filing for bankruptcy, but he knows that is a drastic step. A person who declares bankruptcy cannot ask for any credit for at least seven years and may be turned down even then.

Sylvia and Daryl are discussing some less radical methods to help their fathers with their finances.

ALGEBRA CONNECTIONS

A group of students went to a restaurant and spent a total of $80 for dinner. They agreed to split the bill equally. When two students discovered that they had forgotten their wallets, the others decided to pay an extra $2 each to cover the bill. How many people were in this group? (**10 students**)

OBJECTIVES: *In this lesson, we will help Sylvia and Daryl to:*

- *Discover what happens to debtors who cannot pay their debts.*
- *Recognize the signs of carrying too much debt.*
- *Learn methods to safely get out of debt.*
- *Calculate percent of take-home pay that is used for credit payments.*

DEBT COLLECTION PRACTICES

Creditors often employ professional **collection agencies** to collect overdue accounts and repossess articles on which money is due. The collection process has been known to include everything from collection letters, late-night telephone calls, abusive language, and threats of having consumers fired from their jobs because of nonpayment of debt. In order to stop these practices, Congress passed the **Fair Debt Collection Practices Act,** which declares that the following collection actions are illegal:

- Threatening violence, using obscene language, publishing shame lists of debtors, and making harassing phone calls at night.
- Calling a debtor at work or contacting a debtor's employer.
- Claiming the collector is from a state or federal agency or is a government official.
- Revealing the existence of a bad debt to third parties such as neighbors or employers.
- Using false or deceptive means to obtain information about a debtor.

The act also provides that debtors have the right to notify a collection agency in writing that they do not wish to hear from the agency again, except for legal notices, and notices of possible further action. It is important to note that the law applies only to debt collection agencies—not to banks and other financial institutions or stores.

Creditors can use a legal procedure, know as **garnishment,** to withhold a part of a debtor's earnings for the payment of a debt. However, the law also prohibits an employer from firing an employee because of garnishment of wages for indebtedness.

DEBT PROBLEMS

What kinds of families face debt problems? Following is a profile of the typical family that has problems with debt:

- The family is young, with more than the average number of children but only an average income.
- The parent or parents are carefree. They are impulsive shoppers who cave in easily to high-pressure salesmen, and do not postpone buying things when they want them.

- The family does not read much, not even the daily newspaper. Television is their main form of entertainment and their primary source of news and information. Television influences the family's buying decisions.
- The parent or parents tend to blame their situation on "unavoidable circumstances" (such as pregnancy, temporary unemployment, the purchase of a new car) and do not take responsibility for their problems.
- The family moves more often than the average family.
- No one assumes clear responsibility for managing the family finances.
- The family has a single adult at its head (either by choice or as a result of divorce, separation or the death of a partner).

The following are *clear danger signals* of too much debt:

- You continually lengthen the repayment periods on your installment loans and make smaller and smaller down payments.
- The balances on your revolving charge accounts continue to go up.
- The bills for each month begin piling up before you have finished paying the previous month's bills.
- You slowly, but steadily, use a larger part of your income to pay your debts each month.
- You are taking cash advances on your credit card to pay regular monthly bills, such as utilities, rent and food.

SOLUTIONS TO DEBT PROBLEMS

Is there a safe way out? Sylvia and Daryl are convinced there are ways to stop the debt spiral. They have found the following information:

Wage Earner Plan—Chapter XIII Chapter XIII of the United States Bankruptcy Code is an alternative to declaring bankruptcy. Chapter XIII allows the debtor, creditors, and a judge, acting together, to set a monthly amount for the debtor to pay over an extended repayment period. This is called a *wage earner plan.* The debts are not wiped out, but the court takes a portion of each paycheck and distributes it to the creditors. Chapter XIII does not require the debtor to give up assets such as personal property. Usually, as part of this arrangement, the debtor cannot make additional credit purchases from a creditor until the original debt has been paid.

Bankruptcy Straight bankruptcy, also called *liquidation bankruptcy,* requires debtors to sell most of their assets at public sale through a trustee in return for a discharge from most, if not all, of their outstanding debts. The concept behind bankruptcy is to wipe out all debts and give the debtor a new start.

The wage earner plan and bankruptcy plan are drastic moves, but not as drastic as the next two that Sylvia and Daryl found.

Pawnbrokers When individuals or families are not able to get loans from financial institutions such as banks or small finance companies, they sometimes turn to pawnbrokers. This form of borrowing money can be very expensive and should usually be avoided.

For a cash loan, pawnbrokers accept collateral such as jewelry, art objects, watches, and clothing. The pawnbroker actually lends approximately 40% of the resale value of the article at a high interest rate of 36 to 50%. The pawnbroker keeps the collateral for the period of the loan and returns it when the loan has been repaid. Problems occur when people do not save the money to repay the loan and then lose their collateral, which is invariably worth more than the loan.

Loan Sharks Desperate or unknowing borrowers may turn to loan sharks. They are illegal credit lenders who charge very high rates for money. For example, they will give you $50 now in exchange for your giving them $75 when you get paid. Usually such loans are for one week or one month. Rates are not quoted by these illegal or unethical lenders because they are extremely high—ranging from 100 to 500%. Collection practices often involve threats of harm and violence.

Sylvia and Daryl hope that their fathers will not turn to pawnbrokers or loan sharks. Most states have laws governing small loan practices, but some do not. Some unscrupulous lenders operate without a license in states that do have regulatory laws.

Credit Counseling Sylvia and Daryl prefer the final method they have found, credit counseling, as the first step toward regaining financial stability. They have recommended it to their fathers. To begin with, they have been told, a person seeking counseling should contact his or her local bank. Many banks are now offering credit counseling to customers who find themselves in difficulty.

Another resource that is available is the National Foundation for Consumer Credit, which is located at 1819 H Street, N.W., Washington, D.C. 20006. The foundation will provide you with the address of one of the consumer credit counseling services near you. These services are nonprofit organizations, backed by local banks, merchants, and educators. They are set up to provide financial counseling to anyone.

Another agency that offers help is the Family Service Agency. Hundreds of Family Service Agencies offer either financial counseling or can refer you to an agency that does. If you do not know which agency offers such help in your area, write to the Family Service Association of America at 44 East Twenty-Third Street, New York, New York 10010.

ASK YOURSELF

1. What action might a creditor use if debtors do not pay their debts?
   Answers may vary. (Ex.: garnishment)
2. What are three signs of carrying too much debt?
   See Additional Answers.
3. What is a safe way to get out of debt?
   Answers may vary. (Ex.: Chapter XIII, bankruptcy...)

ADDITIONAL ANSWERS (ASK YOURSELF)

2. 1) continually lengthening repayment periods; 2) balances on revolving accounts go up; 3) bills begin to pile up

(ALGEBRA REVIEW)

10.

Jan.	21.73%	Feb.	1.85%
Mar.	5.01%	Apr.	10.65%
May	5.75%	Jun.	17.56%
Jul.	0.53%	Aug.	7.96%
Sep.	0.05%	Oct.	26.9%
Nov.	2.00%	Dec.	0%

ALGEBRA *REVIEW*

Solve for x. Round your answers to the nearest hundredth.

1. $x = 12.5\%$ of \$435.60
   \$54.45
2. $x = 0.2\%$ of \$333.99
   \$0.67
3. $x = 202.2\%$ of \$677.98
   \$1370.88
4. $32 = x\%$ of \$250.88
   12.76%
5. $12 = x\%$ of \$1543.00
   0.78%
6. $0.3 = x\%$ of 123.45
   0.24%
7. $75 = 0.2\%$ of x
   37,500
8. $0.05 = 35\%$ of x
   0.14
9. $444 = 102\%$ of x
   435.29
10. The Wolk family saved the following amounts of money:

Jan.	\$412	Feb.	\$35
Mar.	\$95	Apr.	\$202
May	\$109	Jun.	\$333
Jul.	\$10	Aug.	\$151
Sep.	\$1	Oct.	\$510
Nov.	\$38	Dec.	\$0

Determine the percent of the savings that occurred in each month to the nearest hundredth of a percent.

SHARPEN YOUR SKILLS

SKILL 1

INSTRUCTION

TEACHING THE LESSON

Spend one day on this lesson. Have students read the opening text and the sections on debt collection practices, debt problems and solutions to debt problems. Encourage a discussion about debt problems and their solutions. Use *Algebra Review* as a springboard for Example 1. Make sure that students understand which entries in a budget count as credit payments. Go over Example 2, pointing out that larger pie pieces correspond to larger percentages.

FOCUS ON ALGEBRA

In this lesson students practice computing percents and work with pie graphs. (*NCTM Standard 4, p. 146.*)

CRITICAL THINKING

Suppose the family in Example 1 had an income of $5000, then assume an income of $4000. How do these different incomes affect the family's credit picture?

EXAMPLE 1 Daryl and Sylvia have learned that a family that spends more than 20% of their take-home pay on installment payments and credit accounts is in danger of credit overload. They have decided to look at a specific family budget to determine how well the family is handling credit. The family's budget for this month is shown.

Income	
Take-home pay	$4550
Expenses	
Mortgage Payment	462
Utilities	185
Telephone	45
Cable Television	53
Gasoline and Car Repairs	220
Food	650
Savings	450
*Car Payment	225
*Credit Union Loan	75
*MasterCard	120
*VISA	95
Charge Accounts:	
*Sears	135
*Foley's Garage	25
*American Express	140
Everything Else	1670

*The entries that count as credit payments are marked with an asterisk. (A mortgage is not counted as part of the credit expenses.)

QUESTIONS What percent of the family's take-home pay goes toward paying credit bills? Is this family spending too much of its salary on credit payments?

SOLUTIONS

The family members' total credit payments are

$$225.00 + 75.00 + 120.00 + 95.00 + 135.00 + 25.00 + 140.00 = 815.00$$

They are spending

$$815.00 \div 4550 = 0.179, \text{ or } 17.9\%$$

of their take-home pay on credit payments. They are not spending too much, but they are approaching the 20% limit.

SKILL 2

It is often helpful to have a pictorial representation of data.

EXAMPLE 2 The pie graph below represents the Taylor family's monthly take-home pay budget.

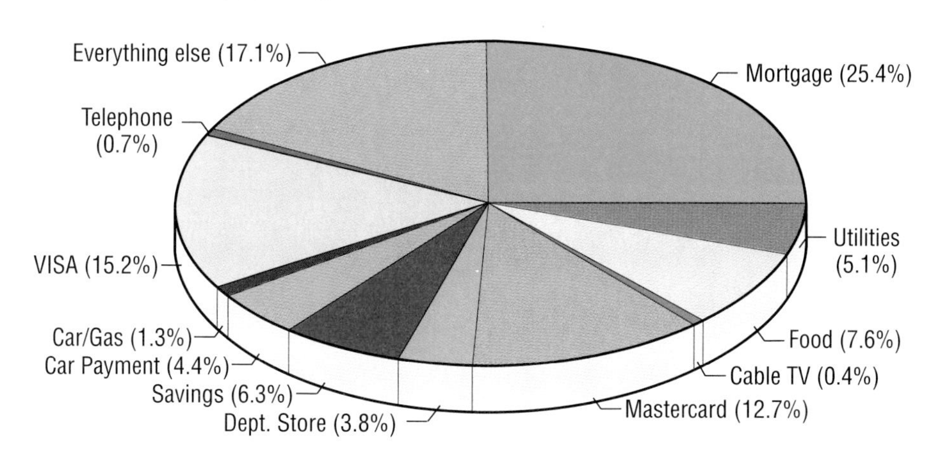

QUESTIONS Approximately what percent of their take-home pay are the family members spending on credit? Are they in danger of credit overload?

SOLUTIONS

They are making the following credit payments.

VISA	15.2%
Car Payment	4.4%
Dept. Store	3.8%
MasterCard	12.7%
Total	36.1%

They are spending approximately 36.1% of their take-home pay on credit payments. They are significantly over the 20% limit and are in great danger of credit overload.

TECHNOLOGY HINT

Spreadsheet programs generate excellent pie charts like the one in Example 2. As an additional exercise, have students create a spreadsheet from the data in Example 1 in order to generate the pie chart.

COMMON ERRORS

Students may include entries which should not be regarded as credit payments or may exclude entries which are credit payments. Discuss the criteria for categorizing budget entries.

RETEACHING

In Example 1 have students compute the percentages for each entry, then add the percentages for credit payments. Assume that the family in Example 1 has a monthly income of $5100. Have students use the pie graph to compute actual expenditures based on the given percentages. Tell students to compare the two cases first based on payments, then on percentages.

ENRICHMENT

Challenge students to determine the minimum income required in Example 1 to keep the credit payments shown to no more than 20% of the family's income. Repeat for Question 1 in the *Try Your Skills* section.

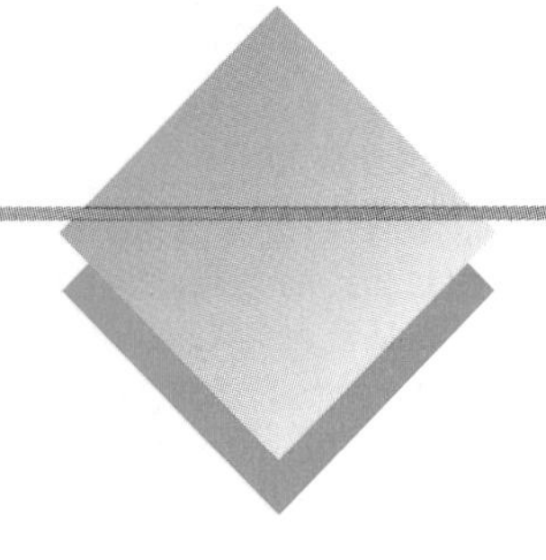

TRY YOUR SKILLS

GUIDED PRACTICE

Have students work in small groups on the *Try Your Skills* exercises. Tell them to make a pie graph for Exercise 1.

ADDITIONAL ANSWERS

1. 21.6%; Yes, they are over the 20% limit.
2. 19%; No, but they are close.

1. The Irrizarry family's budget for this month follows. To the nearest tenth of a percent, how much of their take-home pay are they spending on credit payments? Are they in danger of credit overload?

Income	
Take-home pay	$5300
Expenses	
Mortgage Payment	775
Utilities	225
Telephone	85
Cable Television	45
Gasoline and Car Repairs	125
Food	800
Savings	250
*Car Payment	200
*MasterCard	115
*VISA	500
*Neiman Marcus	155
*Macy's	100
*American Express	75
Everything Else	1850

2. The pie graph below represents the Schmidt family's monthly take-home pay budget. To the nearest tenth of a percent, how much of their take-home pay are they spending on credit? Are they in danger of credit overload?

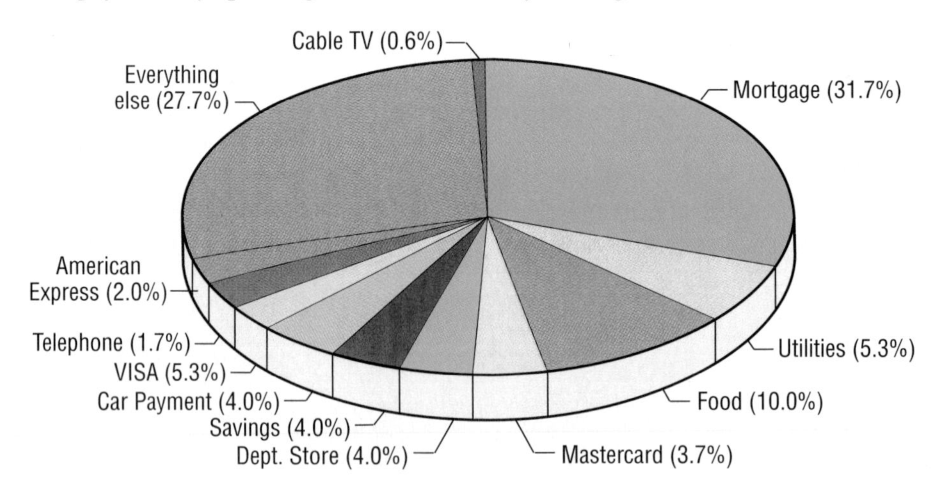

EXERCISE YOUR SKILLS

1. Why do you think the law prohibits an employer from firing an employee because of garnishment of wages for indebtedness?
2. Why is a compulsive shopper at greater risk than a normal shopper of encountering financial problems? Answers may vary.

3. Why would you recommend credit counseling as the first step toward regaining financial stability? Answers may vary.

The following pie graphs represent the monthly take-home pay budgets of four families. To the nearest tenth of a percent, how much of its take-home pay is each family spending on credit? Are any in danger of credit overload?

4.

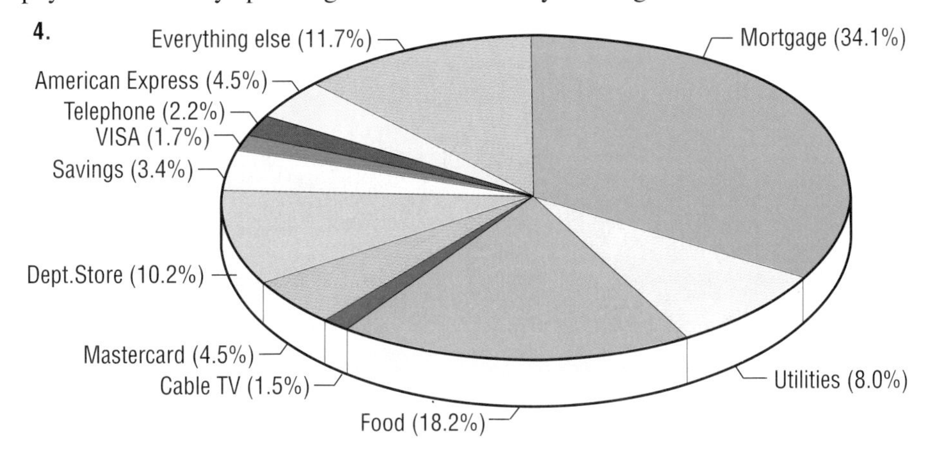

KEY TERMS

bankruptcy
collection agencies
credit counseling
Fair Debt Collection Practices Act
garnishment
loan sharks
pawnbrokers
Wage Earner Plan-Chapter XIII

INDEPENDENT PRACTICE

ASSIGNMENT GUIDE

Exercises 1-3 check students' understanding of the opening text.

Exercises 4-7 ask students to interpret data from a pie graph.

Exercises 8-11 direct students to compute percentage of income spent on credit payments.

ADDITIONAL ANSWERS

1. Answers may vary.
4. 20.9%; Yes, they are in credit overload.

CLOSURE

Have students write a paragraph discussing debt collection practices, danger signals for too much debt and methods for ending perpetual debt. Tell them to give examples of maximum credit payments for families of varying incomes within the 20% limit.

LESSON QUIZ

1. If a family has a monthly income of $3700 and spends $720 per month on credit payments, determine to the nearest tenth of a percent how much it is spending on credit payments. (**19.5%**)
2. A family's monthly income is $2540. Determine the maximum amount it can spend on credit payments to stay within the recommended 20% limit. (**$508.00**)

5.

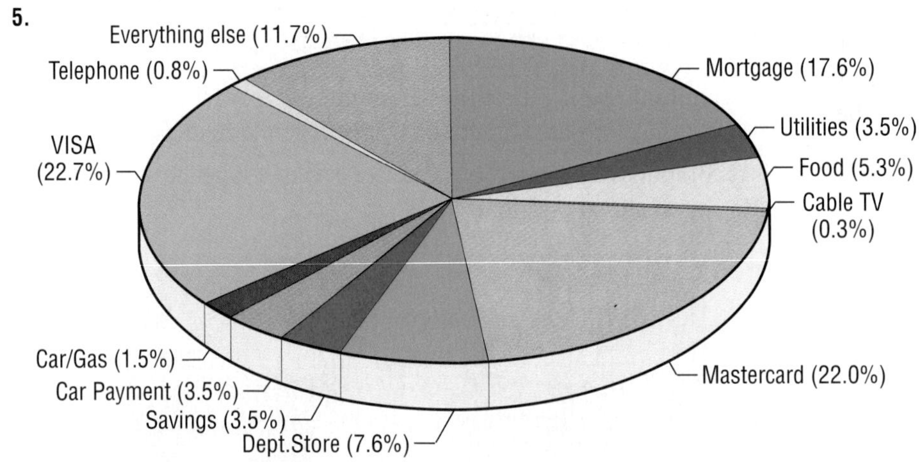

55.8%; Yes, they are in credit overload.

6.

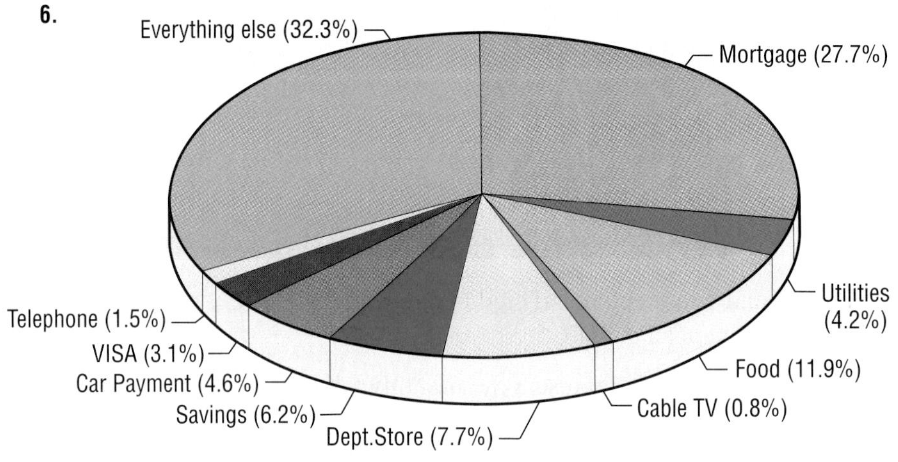

15.4%; They are not in credit overload.

7.

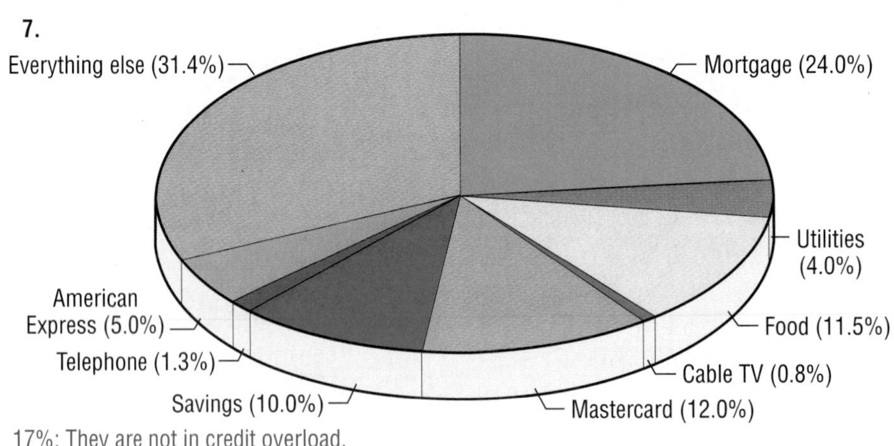

17%; They are not in credit overload.

The monthly budgets for four families follow. Their monthly incomes range from $1850.00 (Family D) to $5400.00 (Family B). Determine, to the nearest tenth of a percent, how much each family is spending on credit payments. Then state whether or not each family is in danger of credit overload.

8. Family A 14.7%; no

Income	
Take-Home Pay	$3290.00
Expenses	
Mortgage Payment	$650.00
Utilities	148.00
Telephone	43.00
Cable Television	53.00
Gasoline and Car Repairs	183.50
Food	580.00
Savings	330.00
*Car Payment	165.00
*MasterCard	56.00
*American Express	140.00
*VISA	45.00
Charge Accounts:	
*Sears	77.00
Everything Else	819.50

9. Family B 11.4%; no

Income	
Take-Home Pay	$5400.00
Expenses	
Rent	$852.00
Utilities	118.70
Telephone	43.60
Sports Club Fee	53.00
Gasoline and Car Repairs	126.00
Food	580.00
Savings	540.00
*Car Payment	237.00
*MasterCard	89.00
*American Express	140.00
*VISA	102.00
Charge Accounts:	
*Hardware Store	34.00
*Penneys	15.00
Everything Else	2469.70

10. Family C 11.2%; no

Income	
Take-Home Pay	$3210.00

Expenses	
Mortgage Payment	$640.00
Utilities	235.00
Telephone	74.00
Sports Club Fee	53.00
Gasoline and Car Repairs	54.00
Food	620.00
Savings	321.00
*Car Payment	0.00
*MasterCard	79.00
*American Express	140.00
*VISA	42.00
Charge Accounts:	
*Pharmacy	34.00
*Montgomery Ward	65.00
Everything Else	853.00

11. Family D 18.3%; no

Income	
Take-Home Pay	$1850.00

Expenses	
Mortgage Payment	$150.00
Utilities	83.00
Telephone	26.00
Cable Television	33.00
Gasoline and Car Repairs	68.50
Food	225.00
Savings	185.00
*Car Payment	233.00
*MasterCard	26.00
*VISA	35.00
Charge Accounts:	
*Sears	45.00
Everything Else	740.50

MIXED REVIEW

6-2 **1.** Debra has a balance of $1325 on her VISA card, which carries an APR of 16%. She makes monthly payments of $150. If she makes no further purchases with this card, how many months will it take her to pay off the balance? 10 months

3-2 **2.** Use the Rule of 72 to determine how long it will take Lou's $4400 savings account to double if it is growing at a rate of 3.5%. 21 y

6-1 **3.** Determine the effective rate of interest for an APR of 17.2%. 18.62%

5-1 **4.** Determine Ryan's monthly payment on a loan of $9875.00 if he borrows the money at an annual interest rate of 10% for 15 years. $106.12

3-2 **5.** How much will $7200 be worth at the end of 3 years if it earns 5% interest, compounded semiannually? $8349.79

1-1 **6.** If Sandra works more than 25 hours in one week, she receives $1\frac{1}{2}$ times her regular wage rate of $6.00 per hour for each of the extra hours. One week she worked 33 hours and received $26 in tips. Find her earnings for the week. $248

5-5 **7.** How much money would you save if you financed an $1800 item at 8% for 3 years rather than accepting a rent-to-own plan for which you pay $60 per month for 3 years? $129.40

6-4 **8.** A credit card account has a balance of $1000 and no further purchases are made. The monthly interest charge is 1.5% and $100 of the amount owed is paid off each month. Find the interest paid after two months.

ADDITIONAL ANSWERS

8. $28.73

5-4 Yumi's parents are purchasing a refrigerator for $420. They finance the purchase with no money down at 15% per year. They pay in 12 equal installments at a store that applies the Rule of 78. Determine the prepayment savings under each condition.

9. It is prepaid after 9 months. $4.85

10. It is prepaid after 5 months. $22.62

6-3 Complete the chart below. Assume that the credit card carries an APR of 11% and that 10% of the amount owed to the nearest dollar is paid off each month.

	Month	Balance	Interest	Amount Owed	Payment
11.	1	$1438.00	$13.18	$1451.18	$145
12.	2	1306.18	11.97	1318.15	132

2-1 A bank offers monthly interest of 0.6% on its checking accounts but charges $0.10 per written check for balances under $500. There is no monthly service charge. Find the amount added or subtracted to the following checking accounts.

13. an account with a balance of $650 and 28 checks written $3.90 added

14. an account with a balance of $480 and 19 checks written $0.98 added

CHAPTER 6 REVIEW

See Additional Answers.

1. Use a spreadsheet program to determine the previous balance, new charges, payment received, finance charges, and new balance that would be shown on Bruce's first 12 MasterCard statements after a purchase of $2315 if he makes monthly payments of $250. The card has a 1-month grace period and an APR of 12.9%. Assume that he has never used the card and that he makes no further charges on it during this time period.
2. Determine the effective rate of interest on a credit card with an APR of 18.5%. Give answer to the nearest hundredth of a percent. 20.15%
3. How long will it take to pay off a credit balance of $2275 with monthly payments of $225 if the card carries a monthly interest rate of 1.25%? 11 mo
4. Determine the monthly payment that must be made to pay off a VISA balance of $12,500 in 30 months if there is an APR of 18%. $520.49
5. Make a chart with the headings below to determine the interest charges over the next 4 months on a balance of $875.00 if your credit card has an APR of 12% and you make monthly payments of 5% of the amount due, rounded to the nearest whole dollar. Assume that you make your payment on the last day of the month. See Additional Answers.

Month	Balance	Interest	Amount Owed	Payment

6. What is the total interest in Exercise 5? $32.94
7. Make a chart with the headings below to determine the interest charges over the next 4 months on a balance of $875.00 if your credit card has an APR of 12% and you make monthly payments of 10% of the amount due, rounded to the nearest whole dollar. Assume that you make your payment on the last day of the month.

Month	Balance	Interest	Amount Owed	Payment

8. What is the total interest in Exercise 7? $30.53
9. Compare your results from Exercises 6 and 8 to determine how much more interest you pay if you pay 5% of the amount owed each month instead of 10% of the amount owed. $2.41
10. How much more interest would you have paid in Exercise 8 if the APR on the card had been 21%? $23.38
11. Use a spreadsheet program with the following headings to work through the first 8 months of Rebecca's VISA account. The monthly interest rate is 1.5%, and she makes monthly payments of 10% of the amount owed. Her beginning balance was $980.12 and she made purchases of $87.35 in month 2, $25.50 in month 5, $100.00 in month 6 and $53.33 in month 7.

Month	New Charges	Balance	Interest	Amount Owed	Payment

12. In Exercise 11, does Rebecca owe more or less in month 8? less

13. Your balance from 9/1 to 9/19 was $400. You made a payment of $50.75 on 9/20. You made no further charges or payments for the remainder of the month. Determine your average daily balance for the month of September. How much did you pay in interest if your card carries a monthly finance charge of 1.75%? $381.39; $6.67

14. How much money would you have saved if you had made your payment on 9/1 instead of on 9/20 in Exercise 13? $0.56

15. Your balance from 8/1 to 8/15 was $780. You made a payment of $100 on 8/16. You made a purchase of $220 on 8/25. You made no further charges or payments for the remainder of the month. Determine your average daily balance for August. How much did you pay in interest if your card carries a monthly finance charge of 1.25%? $778.06; $9.73

16. Use the credit scoring system in Lesson 6-6 to find Julie's credit score and the probability that she will repay a loan. Score: 114; 94%

- She is 32 years old. 0
- She has lived at her current address for 11 years. 21
- Her current car is 2 years old. 16
- Her monthly auto payment is $200. 0
- Her monthly housing costs are $525. 12
- She has a checking account and a savings account. 15
- She has not been referred to a finance company. 15
- She has ten major credit cards. 15
- Her ratio of debt to income is 15%. 20

17. The Garcia family's budget for this month follows. To the nearest tenth of a percent, how much of their take-home pay are they spending on credit payments if their monthly income is $6000? Are they in danger of credit overload? 19.1%; No, they are not in credit ove.load but are approaching the 20% limit.

Expenses	
Mortgage Payment	2000.00
Utilities	300.00
Telephone	95.00
Cable Television	50.00
Food	800.00
Savings	350.00
*Car Payment	250.00
*MasterCard	215.00
*VISA	400.00
*Macy's	180.00
*American Express	100.00
Everything Else	1260.00

CHAPTER 6 TEST

249

255

256

1. Determine the effective rate of interest on a credit card with an APR of 16.5%. Give the answer to the nearest hundredth of a percent. 17.81%
2. How long will it take to pay off a credit balance of $1675 with monthly payments of $180 if the card carries a monthly interest rate of 1.75%? 11 mo
3. Determine the monthly payment that must be made to pay off a VISA balance of $13,100 in 28 months if there is a finance charge of 12%. $538.73
4. Make a chart with the headings below to determine the interest charges over the next 4 months on a balance of $795.00 if your card has an APR of 15% and you make monthly payments of 5% of the amount due, rounded to the nearest whole dollar. Assume you make your payment on the last day of the month. See Additional Answers.

Month	Balance	Interest	Amount Owed	Payment

5. What is the total interest in Exercise 4? $37.54
6. Make a chart with the headings below to determine the interest charges over the next 4 months on a balance of $795.00 if your card has an APR of 15% and you make monthly payments of 10% of the amount due, rounded to the nearest whole dollar. Assume you make your payment on the last day of the month. See Additional Answers.

Month	Balance	Interest	Amount Owed	Payment

7. What is the total interest in Exercise 6? $34.79
8. Compare your results from Exercises 5 and 7 to determine how much more interest you pay if you pay 5% of the amount owed each month instead of 10% of the amount owed. $2.75
9. Your balance from 4/1 to 4/18 was $600. You made a payment of $75 on 4/19. You made no further charges or payments for the remainder of the month. Determine your average daily balance for the month of April. How much did you pay in interest if your card carries a monthly finance charge of 1%? $570; $5.70
10. How much money would you have saved if you had made your payment on 4/1 instead of on 4/19 in Exercise 9? $0.45
11. Your balance from 3/1 to 3/16 was $560. You made a payment of $80 on 3/17. You made a purchase of $160 on 3/22. You made no further charges or payments for the remainder of the month. Determine your average daily balance for March. How much did you pay in interest if your card carries a monthly finance charge of 1.5%? $572.90; $8.59
12. What is a credit rating? See Additional Answers.
13. Would a family with a monthly income of $3000 and monthly credit payments of $350 be in danger of credit overload? Why or why not? No, since it is only 11.7% of their income.

ALTERNATIVE ASSESSMENT

Ask students to write a report on credit cards, discussing their advantages and disadvantages to both user and lender. Have them explain how one might qualify for a credit card for the first time and how to evaluate an APR. Tell students to explain how average daily balances are computed and how much one should pay monthly to avoid debt problems. Remind them to display some information using a spreadsheet program.

P 288

ADDITIONAL ANSWERS

12. An indication of a person's ability to secure goods, services, and money in return for the promise to pay.

CUMULATIVE REVIEW

1-1 **1.** You earn \$7.25 per hour. You worked 46 hours last week. You receive $1\frac{1}{2}$ times your hourly rate for any hours over 40. Find your total wages for last week. \$355.25

1-3 **2.** You are a real-estate agent and you sold a house for \$210,000. You earn 5% commission on the first \$100,000 and 6% commission on any amount over \$100,000. How much commission did you earn? \$11,600

1-2 **3.** You work 35 hours per week for 50 weeks. You get 2 weeks of paid vacation plus 25% of your base salary in fringe benefits. You earn \$4500 per month. Determine your yearly earnings, including benefits. \$67,500

2-1 **4.** The bank charges \$0.02 for the first 20 checks you write each month and \$0.09 for each check over 20. Determine the charges if you write 25 checks each month. \$0.85

2-2 **5.** If you write a check for \$110.23, how would you write that amount in words? One hundred ten and $\frac{23}{100}$

3-2 **6.** Determine the interest for one year on \$600 at 10.5%, compounded semiannually. \$64.65

4-3 **7.** The fixed costs for your business are \$200 per week. The variable costs are \$5.10 per item. What is the cost if you produce 80 items in one week?

4-3 **8.** Using the information in Exercise 7, determine your profit or loss for one week if each item sells for \$9.50. Profit: \$152

5-1 **9.** You borrowed \$3400 for 3 years. You pay \$111 per month. What is the deferred payment price? \$3996

5-1 **10.** In Exercise 9, how much interest did you pay? \$596.00 or 17.5%

5-2 **11.** You have a choice of paying \$450 per month for 3 years or \$300 per month for 5 years on a \$13,500 loan. How much will you save if you choose the 3-year option? \$1800

6-4 **12.** Your credit card carries a monthly finance charge of 1.5%. You pay 10% of the amount owed each month. Your beginning balance is \$540. Complete a table with the following column heads for 3 months.

Month	Balance	Interest	Amount Owed	Payment

5-5 **13.** Determine the plan that would provide the lowest total financed price for a \$16,750 car. Assume the loan in Plan 1 is from a bank or other lending institution and the loans in Plans 2 and 3 are from the car manufacturer's financing subsidiary.

	Plan 1	Plan 2	Plan 3
Rebate	\$1200.00	\$600.00	0
Rate/Time	10.75%/36 months	7.5%/36 months	3.9%/36 months

5-5 **14.** Determine which plan yields the highest profit to the company in Exercise 13. Plan 3

ADDITIONAL ANSWERS

7. \$608.00

13. Plan 3; Plan 1: \$18,260.93; Plan 2: \$18,085.16; Plan 3: \$17,776.15

PROJECT 6–1: **Shopping on Credit**

1. Call or visit 25 different merchants in your area, including at least 3 department stores.
2. Find out which credit cards are accepted by these stores. Try to find at least 5 places that accept credit cards other than MasterCard or VISA.
3. Make a poster presenting the information you gathered, including the name and address of each store and the hours they are open.
4. Write a 1-page report on what you learned through this project. Include comments about the following:
   - Why do so many merchants accept MasterCard, VISA, and Discover cards?
   - What, if any, are the differences among MasterCard, VISA and Discover?
   - What advantage do merchants gain from accepting credit cards?
   - What are the disadvantages of credit cards for merchants and consumers?
   - What department stores issue their own credit cards? What are the advantages and disadvantages of these cards to the merchants and the consumers?

PROJECT 6–2: **Audio-Visual Presentations**

Use the kinds of techniques you see in advertising to point out some of the disadvantages or advantages to the use of credit cards. First, discuss with a small group of your classmates how you feel about using credit cards. Then choose one of the following activities to complete independently or with a group of classmates.

1. Create a poster presenting one or more of the following:
   - Advantages of using credit cards.
   - Disadvantages of using credit cards.
   - A collage of available credit cards—include a wide variety, such as diner's cards, oil company cards, department store cards, and bank cards.

   Illustrate the poster and make it eye-catching or humorous.
2. Draw a cartoon illustrating some negative aspect or disadvantage of credit buying. Exaggeration is permissible in cartoons to emphasize your point.
3. Write a humorous skit to illustrate the negative aspects of using credit cards and perform your skit for your class.

A *random experiment* is the occurrence of something that has an uncertain result. For example, this is the set of all possible results, or *outcomes* of the random experiment of tossing a die:

$\{1, 2, 3, 4, 5, 6\}$

This list of all possible outcomes is the *sample space* of the experiment. An *event* is a subset of a sample space. For example, "rolling an even number" is $\{2, 4, 6\}$ since 2, 4, and 6 are the only outcomes that are even.

List each event. The sample space is the set of six possible outcomes of rolling a die.

Example Rolling a prime number

Solution $\{2, 3, 5\}$

1. Rolling an odd number
   $\{1, 3, 5\}$
2. Rolling a perfect square
   $\{1, 4\}$
3. Rolling a number less than 5
   $\{1, 2, 3, 4\}$
4. Rolling a number greater than π
   $\{4, 5, 6\}$

The *probability* of an event is the ratio

$$P = \frac{\text{number of outcomes in the event}}{\text{number of outcomes in the sample space}}$$

Find the probability of obtaining the given outcome on the roll of a die. The sample space is the set of all possible rolls.

Example Rolling a perfect square

Solution

$$\begin{aligned} P &= \frac{m}{n} && \frac{\text{number of outcomes in the event}}{\text{number of outcomes in the sample space}} \\ &= \frac{2}{6} && \frac{\text{2 perfect squares, 1 and 4}}{\text{6 die faces}} \\ &= 0.33333\ldots, \end{aligned}$$

5. Rolling an odd number
   0.5
6. Rolling an even number
   0.5
7. Rolling a number less than 5
   $0.6\overline{6}$
8. Rolling a multiple of 3
   $0.3\overline{3}$

The following are useful properties of probability.

The probability of an impossible event is 0.
The probability of a certain event is 1.
The probability of any event is a number between 0 and 1, inclusive.

Tell whether the probability of the event is 0, 1, or a number between 0 and 1.

9. Rolling a 7 on a die toss
   0
10. Winning money in a lottery
    $0 < P < 1$
11. Living to the age of 98
    $0 < P < 1$
12. Tossing a coin heads or tails
    1

CHAPTER 7

Planning

FOR TEENAGERS THE FUTURE IS NOW. They often consider it a waste of time to think about matters such as life insurance or retirement planning that have nothing to do with today. Indeed, it is true that a young person without family responsibilities rarely needs life insurance. It is also understandable that for someone with more immediate financial concerns, such as how to pay the rent, retirement is something to worry about later–much later. However, this attitude is probably a mistake. Young people should prepare themselves at an early age for the time down the road when such matters will be very important.

A young person who does devote some time to learning about life insurance and retirement faces some difficult decisions. For example, a young person's life insurance premium is much lower than the premium for someone who is much older, but is that enough of a reason to buy the insurance? Some families have purchased cash-value life insurance so that they can borrow on the policy when it comes time to pay for college. Is that the best way to accomplish this goal?

The earlier retirement is planned for, the more secure and comfortable the retirement years are likely to be. Unfortunately, people from earlier generations have not always learned this lesson. Many older citizens have found themselves trying to get by on fixed-income pensions that do not adequately cover expenses. As a result, they sometimes have had to rely on their grown children for support at a time when those children need to conserve their financial resources to raise and educate their own children.

If the teenagers in this chapter take a closer look at the financial decisions that their parents are making, they may become a bit more understanding when those parents have to turn down a request that would strain the family budget. Who knows? The teenagers might even learn some pointers from the chapter that they can pass on to their parents to assist them in their retirement planning.

Ahead

7-1 LIFE INSURANCE: WHO NEEDS IT?

QUICK REFERENCE

TEACHER'S RESOURCES AND ANSWER KEY

Lesson Quiz Answers 7-1, p. 64

Reteaching Activity Answers 7-1, p. 104

Enrichment Activity Answers 7-1, p. 105

EXTENSION ACTIVITIES

Reteaching Activity 7-1, p. 55

Enrichment Activity 7-1, p. 58

Lily never used to care about life insurance. As a topic of conversation she ranked it just below her grandmother's cataract surgery and slightly above her cousin's discussion of his recent vacation illustrated with 243 slides. Her mother tried once to discuss the subject of insurance with her, but Lily didn't really pay close attention.

Lily's attitude about life insurance quickly changed last year after the father of her very close friend, Annika, was killed in an automobile accident. Of course, Annika and her mother found it very difficult to deal with the shock of the unexpected death. Then, as if the death itself were not bad enough, their adjustment to their new circumstance became complicated by an unpleasant fact: Annika's father had not purchased enough life insurance to help support the family after his death. Annika's mother had to struggle to pay the bills with the income from her job.

She had hoped to find a better-paying job without having to move, since she knew how much it would mean to Annika to stay with her friends and finish her senior year at the same high school. After eight months, however, Annika and her mother finally had to move to another city, where a higher-paying job finally did become available.

As a result of all this, Lily began to wonder what it was about life insurance that could so dramatically affect the quality of a person's life. Her friend Manuel has offered to help her learn more about life insurance—the various types of policies and how much they cost.

OBJECTIVES: *In this lesson, we will help Lily and Manuel to:*

- *Understand what life insurance is.*
- *Decide who needs life insurance.*
- *Determine how much life insurance a family should buy.*
- *Examine the kinds of life insurance that are available.*
- *Calculate the cost of various amounts of life insurance.*

ALGEBRA CONNECTIONS

In the equation $A = rs$, A is an amount depending on a rate r, and a salary, s. Suppose that $s = \$35{,}000$ and r is some number such that $0.5 \leq r \leq 1$. What can you say about A?

LIFE INSURANCE

Life insurance is a contract to pay a specified amount of money to a designated person upon the death of the policyowner. If Annika's father had purchased adequate life insurance, his wife would have received enough money to replace the income lost when he died.

How Does Insurance Work? All types of insurance are based on two ideas: risk sharing and statistical probability. Every person faces the possibility of financial disaster caused by an unpredictable event such as an accident, fire, flood, illness, or the death of the principal wage earner. These risks can be lessened if they are shared by a large group of people paying money into a central pool. Then those who have contributed to the pool have the right to call on that reserve when they suffer a loss. Insurance companies manage such pooled money, called **premiums,** that they have collected from their policyowners. In the event of the of the policyowner's death, the company disburses money from the pool to the **beneficiaries,** those people named in the policy as recipients of the benefits.

Who Needs Life Insurance? When a family has only one provider, it is placing its financial security in the hands of that person. If that person should die, then the family may face serious economic difficulty. Although Annika's father was not the sole provider (her mother also worked), his family did depend upon his income. He should have bought a life insurance policy that was large enough to replace all or most of the income that his **dependents** (his wife and children) needed to survive after he died.

In today's families, women are increasingly responsible for providing all or part of the family income. For that reason, women as well as men must decide how much life insurance they need. If a wife and mother were to die, her husband and children might have to not only replace her income but also pay someone to provide child care and housework.

Remember that life insurance is purchased to protect the dependent members of a family. Therefore it is usually a mistake to buy life insurance to cover the life of anyone other than the principal breadwinners. In particular, the money that might be spent on a child's life insurance policy can probably be put to better use, even though the child's premiums are very low. If you are a high school or college student and have no dependents, you probably do not need life insurance.

How Much Insurance Is Needed? An insurance company will generally sell any amount of insurance that a person wants to buy. However, there are some general guidelines that may be helpful in determining how much life insurance a family should purchase.

An ideal goal is to have the insurance provide enough income for the family to continue its current standard of living. According to one major financial services company, this can usually be done without maintaining 100% of current gross pay. Instead, you should attempt to replace your family's usual *net income,* that is, the amount that remains after income and FICA taxes have been deducted. You can probably accomplish this goal by purchasing insurance that provides 75% of the previous gross income. If the premiums for doing this are too high for the family budget, you should at least aim at a replacement income of 60% of the current gross income to avoid a serious lowering of your family's standard of living.

LIFE INSURANCE: TWO TYPES

Manuel's older brother, Martin, is married to Rachel, his childhood sweetheart. The couple has a young child. Manuel's research will be useful to Martin and Rachel as they decide what kind of insurance they need. Manuel has learned that they can choose from two basic types of insurance. The first type, **term insurance,** offers pure protection. The second, called **cash-value insurance,** offers less protection for the same money. However, it also builds up a cash value that the policyowner can draw on in later years.

Term Insurance If Martin and Rachel choose term insurance, they will be protected for a fixed period of time, or *term.* The term is usually five or ten years. If the insured person dies during the term, the beneficiary will receive the **face value of the policy.** If Annika's father had had $100,000 worth of term insurance, the beneficiary, her mother, would have received $100,000 on his death. If the insured person does not die during the term, the policy either expires or is renewed for another term. Term insurance generally has the lowest premiums of any insurance available until approximately the age of 50. Because the amount of coverage can be increased or decreased at the end of each term, term insurance is a very flexible option. Manuel discovered three common types of term insurance.

Group life insurance, which people can buy through their place of employment, provides term insurance for a large number of people under a single policy without the need for a medical examination.

A policy of the second type, **renewable convertible term insurance** covers a person for a period of time such as one, five, or ten years and can be renewed without a medical examination. In addition to being renewable, renewable convertible term insurance is convertible into a different form of insurance.

The third alternative is **decreasing term insurance.** Under this option the amount of benefits that an insurance company will award the beneficiaries decreases over time as the children grow up and cease being dependents. After this happens, the parents no longer need as much insurance as they did in earlier years. The appeal of decreasing term insurance is that the premiums are lower than those for renewable term insurance. Often decreasing term insurance is used to provide cash to pay off a mortgage. Policies have been designed so that the face value declines exactly with a person's outstanding mortgage balance.

Cash-Value Insurance If Martin and Rachel choose cash-value insurance, they will be buying protection plus savings. As they pay the premiums, the policy will build up cash value much like a savings account.

Manuel learned about three traditional types of cash-value insurance. The first is **whole life insurance,** also called *straight life* or *ordinary life.* With this kind of insurance policy, you pay premiums for your entire life that are more than enough for the insurance coverage you need. The remaining part of your premium is invested for you by the insurance company, building up the cash value of the policy. You always have the option of cashing in the policy, that is, taking out the money and the interest it has earned. In addition, since the policy has a cash value, you can borrow part of the cash value. If you should die with an outstanding loan, the amount of benefit due to your beneficiaries is reduced. However, the rate of return tends to be low in comparison to the return that you would get investing on your own.

Limited payment life insurance has a fixed number of premium payments that stop after a certain number of years; for example, 20 years. The purchaser pays higher premiums, but the policy is paid up to its cash value in fewer years than is the case for whole life insurance.

If the couple wants a policy in which the cash value buildup is accomplished especially quickly, they might choose an **endowment policy.** An endowment policy, like other cash value insurance, is protection plus savings, but the emphasis is on the savings. Because of this rapid accumulation over a short period of time, the premiums on endowment policies are very high.

Universal life insurance is a relatively new type of cash value insurance. As with whole life insurance, it gives both protection and savings. Two basic types of universal life differ in the death benefit. In Type A the death benefit remains constant; in Type B the death benefit varies with the cash value so that as the cash value increases, the death benefit also increases. Usually, the insurer guarantees a minimum rate of interest throughout the duration of the policy. As with whole life, you can borrow or withdraw a part of the cash value.

CRITICAL THINKING

Ask students how life insurance is like health insurance and how it is different. Which one would a 16-year-old be more likely to have?

ADDITIONAL ANSWERS

1. It is a contract to pay a specified amount of money to a designated person upon the death of the policy owner.

The types of life insurance and their characteristics are summarized in the following chart.

LIFE INSURANCE	
Term Insurance	
Group life	Purchased through an employer; low rates
Renewable convertible term	One-, five-, or ten-year terms; renewable and convertible
Decreasing term	Like renewable convertible term but less expensive and with a gradually decreasing death benefit
Cash-Value Insurance	
Whole life	Combines savings with insurance; can be cashed in for its accumulated cash value
Limited payment	Higher premiums for a fixed number of years, such as 20; builds up cash value more quickly than whole life
Endowment	Emphasis on high cash buildup; very high premiums
Universal life	Combines savings with insurance; can be cashed in for its accumulated cash value

ALGEBRA *REVIEW*

A thrown dart may land anywhere on its target. So the probability of landing in the top half of the square is, $P = 0.5$.

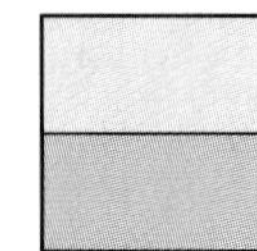

Determine the probability of a dart landing in each region.

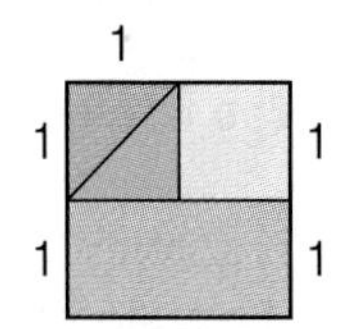

1. Lower rectangle 0.5
2. Upper-right square 0.25
3. Upper-left triangle 0.125

Determine the probability of a dart landing in each region.

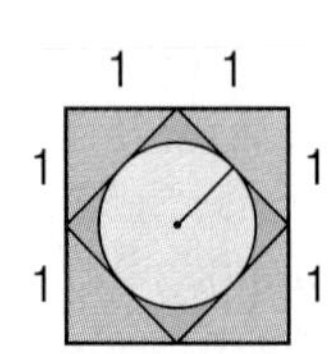

4. Inner circle 0.39
5. Inner diamond 0.5
6. Lower-left triangle 0.125

What Does it Cost? The cost of life insurance varies widely. The range between the most expensive and the least expensive can be as much as 100%. To find the best price for both term insurance and cash-value life insurance, you should shop around and compare the prices that different insurance companies offer.

How About Taxes? The proceeds paid to a beneficiary from some kinds of life insurance policies are generally free from federal income taxes but may be subject to other taxes, such as an estate tax. Tax laws change from time to time, so you should obtain up-to-date information from a tax adviser.

Ask Yourself

1. What is life insurance?
   See Additional Answers.
2. Which member(s) of a family should have life insurance?
   the ones that have dependents
3. How much life insurance should a family assume?
   75% of gross income
4. What are the two basic kinds of life insurance?
   term and cash value
5. What is a beneficiary?
   The recipient of the benefit of a life insurance policy
6. What advantage does decreasing term insurance have over regular term insurance?
   Answers may vary. (Ex.: premiums are lower)

SHARPEN YOUR SKILLS

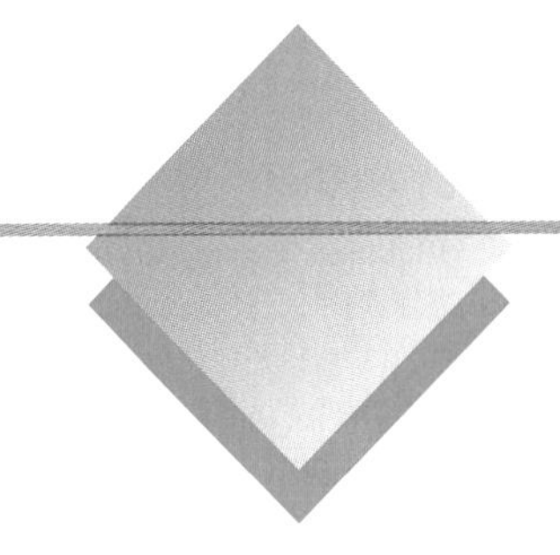

SKILL 1

To determine how much insurance to buy to replace either 75% or 60% of gross income, you can use the following chart and formula. This chart is also in the Reference Section.

MULTIPLES-OF-SALARY CHART

	Current Age							
	25 Years		**35 Years**		**45 Years**		**55 Years**	
Current Gross Earnings	**75%**	**60%**	**75%**	**60%**	**75%**	**60%**	**75%**	**60%**
$ 7,500	4.0	3.0	5.5	4.0	7.5	5.5	6.5	4.5
9,000	4.0	3.0	5.5	4.0	7.5	5.5	6.5	4.5
15,000	4.5	3.0	6.5	4.5	8.0	6.0	7.0	5.5
23,500	6.5	4.5	8.0	5.5	8.5	6.5	7.5	5.5
30,000	7.5	5.0	8.0	6.0	8.5	6.5	7.0	5.5
40,000	7.5	5.0	8.0	6.0	8.0	6.0	7.0	5.5
65,000	7.5	5.5	7.5	6.0	7.5	6.0	6.5	5.0

Replacement Life Insurance Formula

$R = mS$ where R = required replacement insurance
m = multiple from the table
S = original gross salary or wage

EXAMPLE 1 Annika's father was earning $30,000 per year when he was 35 years old.

QUESTION How much insurance should he have bought to replace 75% of his regular gross income?

SOLUTION

Look at the Multiples-of-Salary Chart. Find the income level of $30,000 in the left-hand column. Then read across to the column for 35 years and 75%. The number 8.0 appears in that space. Use the replacement life insurance formula.

$R = mS$ — The multiple of salary m is 8.0.
$= 8 \cdot 30{,}000$
$= 240{,}000$

Annika's father should have purchased $240,000 worth of life insurance at age 35 to replace 75% of his gross income.

INSTRUCTION

TEACHING THE LESSON

Explain the difference between term and whole life insurance and why one or the other might be preferred in different circumstances. When doing the skill examples, point out the importance of carefully reading the headings on tables so that you can find needed information. These tables differ from spreadsheets in that no computations are done within the table. Rather, like a mileage chart, the tables give information defined by the headings.

FOCUS ON ALGEBRA

When discussing the economy and power of mathematical notation, we usually think of equations and formulas. But tables also are highly economical in the amount of precise information they carry. (*NCTM Standard 2, p. 140*)

ESL STUDENTS

The introductory scenario can be used with non-native speakers to increase comprehension both in vocabulary and in consumer awareness. You might pair an ESL student with a first-language student. Have them read alternating paragraphs and discuss the meaning.

COMMON ERRORS

Some students are likely to mix up the new terms. Review the meaning and use of "premium." If possible use an overhead projector to show how to read the tables.

INTERDISCIPLINARY INVESTIGATION

Language Arts: Have students interview an adult—an insurance agent or someone else with a better than average knowledge of insurance—and then write a report about the value of life insurance.

AT-RISK STUDENTS

The mathematics in this lesson is quite elementary. Some students who have difficulty working with equations will be able to use the tables easily. The lesson may be confidence booster for some at-risk students.

RETEACHING

Since the tables are based on differences between term and whole life rates, it might be necessary to review these concepts if they are not understood.

ENRICHMENT

Have students find an actuarial table in a magazine, newspaper, or text. Ask them to study it and then explain its meaning to the class.

SKILL 2

The following table shows the comparative premium rates for two of the major types of insurance. This table is also in the Reference Section. *Note:* To use the table to find the cost of $50,000 worth of life insurance, divide the $100,000 premium by 2. To find the cost of $200,000 worth, multiply by 2.

COMPARISON TABLE FOR TERM AND WHOLE LIFE PREMIUMS

Policy face value is $100,000

Age	Five-Year Renewable Term	Whole Life	First-Year Difference
20	$205	$ 775	$ 570
25	207	918	711
30	218	1112	894
35	254	1374	1120
40	363	1729	1366
45	562	2127	1565
50	878	2689	1811

EXAMPLE 2 Ramón is a 30-year-old father who is comparing the premiums for different types of life insurance.

QUESTIONS

1. How much would five-year term insurance for $200,000 cost Ramón per year?
2. How much would the same amount of whole life insurance cost him?
3. In one year, how much would he save by buying term insurance instead of whole life?

SOLUTIONS

Use the Comparison Table for Term and Whole Life Premiums.

1. For age 30 the number in the "Five-Year Renewable Term" column is $218. Ramón's annual premium would be 2 • 218, or $436.
2. For age 30, the number in the "Whole Life" column is $1112. Ramón's annual premium would be 2 • 1112, or $2224.
3. The difference is 2224 − 436, or $1788. Ramón would save $1788 by buying term insurance.

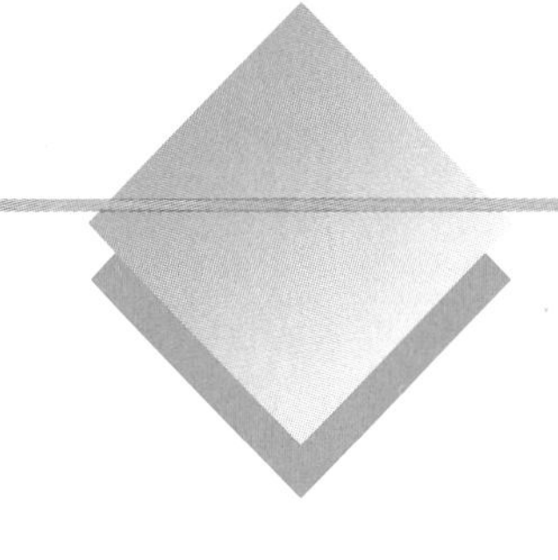

TRY YOUR SKILLS

Use the Multiples-of-Salary Chart and the Replacement Life Insurance Formula in Skill 1 to find the amount of life insurance each of these income earners should buy under the conditions described.

	Current Earnings	Age	Income Replacement	Amount of Insurance
1.	$65,000	35	75%	$487,500
2.	65,000	45	60%	390,000
3.	40,000	55	75%	280,000

Use the Comparison Table for Term and Whole Life Premiums in Skill 2 to find the yearly premium for the amounts of insurance listed below at the ages shown.

	Amount of Insurance	Age	Type	Premium
4.	$100,000	25	Term	$ 207
5.	100,000	25	Whole life	918
6.	200,000	35	Term	508
7.	200,000	35	Whole life	2748
8.	300,000	45	Term	1686
9.	300,000	45	Whole life	6381

GUIDED PRACTICE

The *Try Your Skills* exercises can best be done by students working on their own. Then review each answer, going back to the tables as needed when students have made mistakes. This will further reinforce the use of the tables.

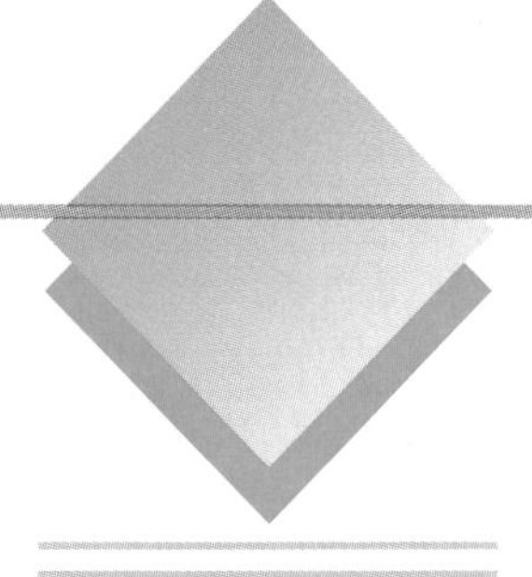

EXERCISE YOUR SKILLS

1. Why do people need life insurance? to replace lost income
2. When would it be a good idea for both a husband and a wife to have a life insurance policy? if they are both employed
3. If an aged grandmother were living with a family, would it be a good idea to insure that grandmother's life? Why or why not? No, she probably does not support the family with income.

4. Why would "death insurance" be a more accurate term than "life insurance" to describe the kind of policy that has been discussed in this lesson? It is only payable upon death.
5. Why is group life insurance cheaper than life insurance bought as an individual policy? Answers may vary.
6. What is the advantage of limited payment life insurance? See Additional Answers.

Use the Multiples-of-Salary Chart to find the amount of life insurance each of these income earners should buy under the conditions described.

	Current Earnings	Age	Income Replacement	Amount of Insurance
7.	$40,000	25	60%	$200,000
8.	30,000	35	75%	240,000
9.	30,000	45	60%	195,000
10.	23,500	55	75%	176,250
11.	23,500	25	60%	105,750
12.	15,000	35	75%	97,500
13.	15,000	45	60%	90,000

KEY TERMS

beneficiaries
cash-value insurance
decreasing term insurance
dependents
endowment policy
face value of the policy
group life insurance
life insurance
limited payment life insurance
premiums
renewable convertible term insurance
term insurance
universal life insurance
whole life insurance

INDEPENDENT PRACTICE

ASSIGNMENT GUIDE

Exercises 1-6 review insurance concepts. They can be answered in writing, and then discussed. Exercises 7-22 employ skills related to the use of tables.

ADDITIONAL ANSWERS

6. The policy is paid up to its cash value in fewer years than whole life insurance.

Use the Comparison Table for Term and Whole Life Insurance to find the yearly premium for the amounts of insurance listed below at the ages shown.

	Amount of Insurance	Age	Type	Premium
14.	$ 50,000	30	Term	$ 109
15.	50,000	30	Whole life	556
16.	150,000	40	Term	544.50
17.	150,000	40	Whole life	2593.50
18.	250,000	50	Term	2195
19.	250,000	50	Whole life	6722.50

20.–22. For each of the three pairs of policies (term and whole life) in Exercises 14–19, which is the less expensive? How much cheaper is the less expensive choice? Term; $447; $2049; $4527.50

MIXED REVIEW

1-1 Assume that a person receives $1\frac{1}{2}$ times the usual hourly rate for each hour or part of an hour beyond 40 hours per week. Find the total wages earned in one week in each case.

1. Hourly wage rate: $8.25; total hours worked: 50 $453.75

2. Hourly wage rate: $11.00; total hours worked: $47\frac{1}{2}$ $563.75

5-2 **3.** Suppose that you can pay off a $12,500 loan either by paying $395 per month for 3 years or by paying $260 per month for 5 years. How much will you save if you pay the loan off in 3 years? $1380

6-3 Complete the chart below to determine how much the interest cost will be if you pay 10% of the amount due to the nearest dollar each month and your credit card carries an APR of 9%.

	Month	Balance	Interest	Amount Owed	Payment
4.	1	$880.00	$6.60	$886.60	$89.00
5.	2	797.60	5.98	803.58	80.00

6. What is the total interest for the two months? $12.58

1-4 **7.** Tom's weekly gross pay is $115. The amount withheld for taxes is $2 and 7.65% of the gross pay is withheld for FICA taxes. Find Tom's take-home pay for the week. $104.20

4-5 **8.** Write an inequality for each of the following 3 constraints: you want to work no more than 15 hours per week, you want to earn at least $100 a week, and you won't get a job paying more than $9 an hour. Let x represent the number of hours you work in 1 week and y represent your weekly salary. $x \leq 15$; $y \geq 100$; $9x \leq y$

CLOSURE

Have students study the table in Skill 2 and write as many things as they can that are not in the table but can be inferred from it.

LESSON QUIZ

1. Find the replacement income for someone earning $40,000 who wants 75% replacement. (**$30,000**)

2. The premium for term insurance is $250 for $100,000. Under the same conditions, what will be the premium for $1,000,000? (**$2500**)

7-2 SPREADING THE RISK: HOW INSURANCE WORKS

QUICK REFERENCE

TEACHER'S RESOURCES AND ANSWER KEY
Lesson Quiz Answers 7-2, p. 66
Reteaching Activity
Answers 7-2, p. 104
Enrichment Activity
Answers 7-2, p. 105

EXTENSION ACTIVITIES
Reteaching Activity 7-2, p. 56
Enrichment Activity 7-2, p. 59

Martin and Rachel finally decided to purchase term insurance on Martin's life rather than the other types of life insurance that were available. Since Martin is still in his twenties, he was able to get a policy with a relatively low premium. The oldest brother, Benjamin, has also started a family recently. Because he is in his thirties, the term insurance that Benjamin purchased is almost $50 more per year than Martin has to pay for the same $100,000 policy. Manuel knew that insurance premiums increase as the age of the insured person increases, but he wondered why the amount of the increase was $50. Why not a $5 increase—or $500?

Lily has recently had a similar experience. Her father told her that he had rejected an insurance salesman's argument that Lily's life should be insured "because her premiums would be so low." She and her father both know that she doesn't need life insurance because no one depends on her for support. Nevertheless, like Martin, she wants to know why her premiums would be so low.

Manuel decided to talk with his cousin Roberto, who works for a major insurance company. One of Roberto's duties is to help prepare new tables that show the premiums that people have to pay for insurance. Manuel hoped that Roberto could explain to him and Lily how the numbers in the tables are computed.

OBJECTIVES: *In this lesson, we will help Lily and Manuel to:*

- *Understand how life-expectation tables are used to estimate the probability that an individual will die within one year.*
- *Learn how an insurance company determines its premium schedule to make a reasonable profit.*

STATISTICAL TABLES

Manuel and Lily visited Roberto at his office. As they were talking, Manuel noticed that on Roberto's desk was a statistical table entitled Expected Deaths per 100,000 Alive at Specified Age. Part of the table is shown below. This table is also in the Reference Section.

EXPECTED DEATHS PER 100,000 ALIVE AT SPECIFIED AGE		
Age	**Expected Deaths Within 1 Year**	**Expected to be Alive in 1 Year**
15	63	99,937
16	79	99,921
17	91	99,909
18	99	99,901
19	103	99,897
20	106	99,894
21	110	99,890
22	113	99,887
23	115	99,885
24	117	99,883
25	118	99,882
26	120	99,880
27	123	99,877
28	127	99,873
29	132	99,868

FOCUS

ALGEBRA CONNECTIONS

Have students write down a number from 1 to 9 without showing it. Ask several student volunteers to guess other students' numbers. Clearly they will not have much luck. Then say that you will guess the average of all the numbers. Guess 5. The average will not be far from this.

Roberto explained that information in the table enabled him to calculate the probability that a person of a given age will die or not die sometime during the next 12-month period. This information helps him to decide how large a life insurance premium should be for each age.

For example, the probability that a 16-year-old person will still be alive 1 year from today is found by using the formula for the **probability of an event.** This probability is written as $P(E)$, which is read "P of E."

Probability of an Event

$P(E) = \frac{m}{n}$ where $P(E)$ = the probability of an event E
m = the number of times the event occurs
n = the number of all possible outcomes

CRITICAL THINKING

Ask students to sketch a graph of the Expected Deaths table for a population of 100,000 from ages 0 to 100. This type graph is commonly drawn with numbers of people on the vertical axis and ages on the horizontal axis. The graph will start at 100,000 on the vertical axis, decline slowly at first, then its negative slope will increase, especially in the years from 65 to 80.

Recall that the probability of an event is always a number between 0 and 1, inclusive.

From the chart, Roberto is able to determine the probability that a 16-year-old person will be alive 1 year from today.

$$P(E) = \frac{99{,}921}{100{,}000} \quad \frac{\text{Number of 16-year-old people alive 1 year later}}{\text{Total number of 16-year-old people}}$$

$$= 0.99921$$

The probability that a 16-year-old will be alive 1 year from today is 0.99921, that is, almost 1. The event E is practically certain.

The probability that a 16-year-old will die within one year is

$$P(E') = \frac{79}{100{,}000}$$

$$= 0.00079$$

This probability is 0.00079, or almost zero. The event E' ("E prime") is very unlikely.

In the above formula, E' is the event "E does not occur," or "not E." For this reason it is called the **complement** of E. Notice that

$$P(E) + P(E') = 0.99921 + 0.00079$$

$$= 1.00000$$

In words, *the sum of the probabilities of an event and its complement is 1.* For example, it is certain (a probability of 1) that a 16-year-old will either be dead or alive 1 year from today.

Manuel found the discussion about probabilities very interesting, but he still did not have an answer to his question: How is the size of an insurance premium determined?

ALGEBRA *REVIEW*

At Central High there are 240 juniors. Of these, 110 are girls, 180 study math, 60 take Spanish, 40 take French, and 100 go to first lunch. No students take both Spanish and French. A student's name is to be drawn for a prize. Each name has an equal chance of being drawn.

Find the following probabilities to the nearest thousandth for the student whose name is selected.

1. Be a girl 0.458
2. Study Spanish 0.25
3. Study French 0.167
4. Study either Spanish or French 0.417
5. Study neither Spanish nor French 0.583
6. Go to first lunch 0.417
7. Not go to first lunch 0.583
8. Be a boy 0.542
9. Study both Spanish and French 0
10. Win the prize 0.004

Ask Yourself

1. How do you think insurance companies can obtain information about life expectancy?
   through statistics
2. If a person's life expectancy is very high, does that mean that he or she can safely drive without a seat belt at high speeds? Explain your answer.
   No; each individual affects his/her lift expectancy.
3. In earlier centuries, life expectancy for infants and very young children was much lower than it is today. Why do you think that the life expectancy for children has improved in recent years?
   Better medical care is more accessible now.

SHARPEN YOUR SKILLS

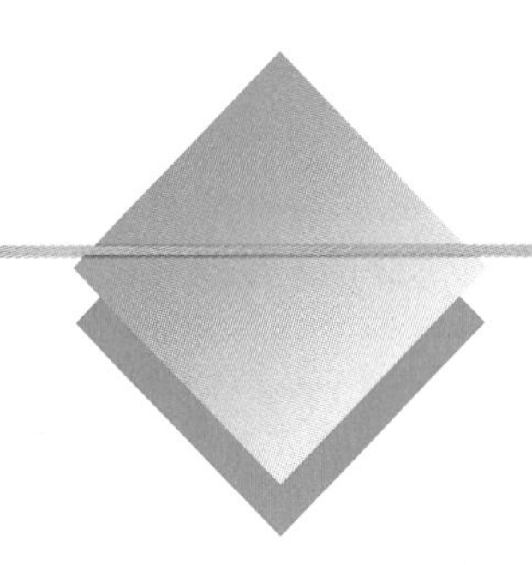

SKILL 1

The **expected value** is the amount of money to be won or lost in the long run. If an event can assume two values, then the *expected value* of the event is the sum of the product of each value and its probability.

> **Expected Value Formula**
>
> $E = P_1v_1 + P_2v_2$ where v_1 and v_2 are values and P_1 and P_2 are the corresponding probabilities

EXAMPLE 1 In a certain game of chance, you win \$4 when a coin shows heads and \$1 when it shows tails.

QUESTION What is the expected value of money that you could win?

SOLUTION

The probability of obtaining heads and the probability of obtaining tails are each 0.5.

$E = P_1v_1 + P_2v_2$ Expected value formula

$E = 0.5(4) + 0.5(1)$ $v_1 = 4$, $v_2 = 1$, $P_1 = 0.5$, $P_2 = 0.5$

$= 2.50$

INSTRUCTION

TEACHING THE LESSON

Be sure that students understand the reason why insurance works. It is related to the number guessing exercise in the *Algebra Connections*. While the time of an individual's death cannot be known, the numbers for a large population can be predicted quite accurately. This knowledge makes it possible for insurance companies to establish rates.

Expected value is a subtle and abstract notion. It exists only "in the long run" or "in a large population."

FOCUS ON ALGEBRA

This lesson models real-world phenomena and also bridges statistics and probability. (*NCTM Standard 10, p. 167.*)

For the "privilege" of tossing the coin you should pay no more than \$2.50. Notice that on a single toss you will win either \$1 or \$4, never \$2.50. For a large number of tosses you can expect to gain about \$2.50.

SKILL 2

Roberto's insurance company must make a reasonable profit on the insurance policies that it writes. Otherwise, it will not be able to pay benefits on the policies or even to survive as a company. Roberto uses the idea of expected value to find the premium that gives the **break-even value** for the insurance company; that is, the value of the premium that gives zero profit after paying for all expenses.

COMMON ERRORS

Some students will be able to follow the explanation of expected value but not be able to develop expected value equations on their own. Reviewing the E and "not E" probabilities should help.

EXAMPLE 2 The direct and indirect expenses for each policy that the insurance company writes are about \$20 per policy.

QUESTION How can Roberto use probability to determine the proper premium for a \$50,000 insurance policy on a 28-year-old person?

SOLUTION

The possible outcomes in the next year for a 28-year-old person are as shown in the table on page 331. The variable x represents the break-even value for the premium. Notice that the probability that a person dies is 1 minus the probability that a person lives.

COMPANY'S GAIN OR LOSS ON ONE POLICY (before deducting expenses)

Possible Outcome	Probability of Outcome	One-year Gain/Loss
The person lives.	$\frac{99,873}{100,000}$, or 0.99873	x (gain)
The person dies.	$\frac{127}{100,000}$, or 0.00127	$x - 50,000$ (loss)

$$E = P_1v_1 + P_2v_2 \quad \text{Use the expected value formula.}$$
$$E = 0.99873x + 0.00127(x - 50,000)$$
$$0 + 20 = 0.99873x + 0.00127x - 63.5 \quad \text{Profit: 0; expenses: 20}$$
$$83.5 = 1.00000x$$
$$x = 83.5$$

COOPERATIVE LEARNING

If needed, give an alternate Example 2: What should be the break-even premium for a \$75,000 policy for a 21- year-old, including \$25 expenses? Assign groups of threes with the following roles: (1) sets up a gain or loss table; (2) sets up and solves equations; (3) observes, helps as needed, checks all work. (**\$107.50**)

The break-even premium for one year of life insurance on a 28-year-old person is \$83.50. The company will have to charge more than \$83.50 to make a reasonable profit.

Roberto explained to Manuel and Lisa that this example applies to a *one-year term* life insurance policy, not to a whole life policy. A whole life policy has additional savings features that would raise the premiums significantly. He also mentioned that most term policies are for five or ten years, not one year, and that five-year and ten-year life expectation tables can be used to prepare premium schedules for such policies.

SKILL 3

The profit that an insurance company makes on policies is determined by the amount of revenue (money) that is received as premiums minus any benefits paid out and any other expenses such as overhead.

Profit on Insurance

$P = R - B - C$ where P = profit
R = revenue received as premiums
B = benefits paid out
C = costs or expenses

TECHNOLOGY HINT

A calculator with a screen will be helpful in doing the calculations for Skills 2 and 3. In this way the decimals and large numbers can be checked for accuracy.

RETEACHING

Review the basic rules for working on both sides to solve a linear equation.

ENRICHMENT

Challenge students to write a calculator program for using the general expected value formula, and then use it to create a table of premiums for ages 20-29.

EXAMPLE 3 Roberto told Lisa and Manuel that his company charges a 20-year-old person \$110 for a one-year term \$50,000 life insurance policy.

QUESTION What profit does the company expect to make on each of 100,000 such policies if the expenses for each policy are \$20?

SOLUTION

Let x represent the profit for each policy. Then $100{,}000x$ represents the profit for 100,000 policies. Use the table of expected deaths on page 331. There are 106 expected deaths for 20-year-old people. The death benefit is \$50,000. So, the benefits paid out are $B = 106(50{,}000)$. The cost is \$20 per policy, so $C = 20(100{,}000)$. The revenue is \$110 per policy, so $R = 110(100{,}000)$.

$P = R - B - C$ Use the profit on insurance formula.
$100{,}000x = (110)100{,}000 - 106(50{,}000) - 20(100{,}000)$
$100{,}000x = 11{,}000{,}000 - 5{,}300{,}000 - 2{,}000{,}000$
$100{,}000x = 3{,}700{,}000$
$x = 37$ Divide each side by 100,000.

The company will make \$37 profit on each policy and 37(100,000) or \$3,700,000 on 100,000 policies.

EXAMPLE 4 To use the table of expected deaths on page 331 for another amount of people, write a proportion.

QUESTION What is the number of expected deaths for 5000 20-year-olds?

SOLUTION

From the table on page 331 the expected deaths for 100,000 20-year-olds is 106. Let d represent the expected deaths for 5000 20-year-olds. Then

$$\frac{106}{100{,}000} = \frac{d}{5000}$$ Multiply both sides by 5000.

$$\frac{106(5000)}{100{,}000} = d$$

$$5.3 = d$$

The expected deaths for 5000 20-year-olds is 5.3.

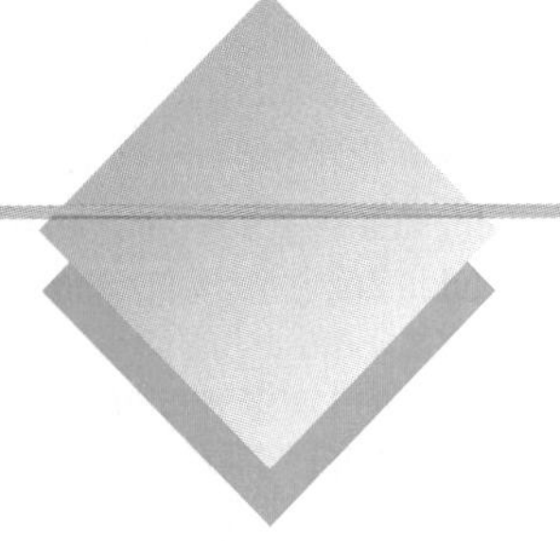

TRY YOUR SKILLS

Find the expected value.

1. The expected gain from a coin toss that pays $3 for heads and $2 for tails $2.50
2. The expected gain from a roll of a die that pays $10 for a 6 or a 2 and $1 for any other result $4.00

A person's life expectancy for one year is shown in the following table. She is contemplating the purchase of a one-year term policy with a face value of $60,000. Refer to the table to answer Exercises 3–5.

COMPANY'S GAIN OR LOSS ON ONE POLICY (before deducting expenses)		
Possible Outcome	**Probability of Outcome**	**One-Year Gain/Loss**
The person lives.	0.99906	x
The person dies.	?	? – ?

3. Find the probability that the person will die within one year. 0.00094
4. What is the algebraic expression that represents the company's one-year gain or loss (before expenses) if the person dies? $x - 60{,}000$
5. What is the algebraic expression that represents the company's expected gain or loss (before expenses) on the policy? See Additional Answers.
6. The company has direct and indirect expenses of $25 for each policy that it issues. Find the premium that the company must charge to break even; that is, neither to make nor to lose money on this policy. $81.40

Use the formula for profit on insurance $P = R - B - C$ and the table of expected deaths to calculate the profit that a company makes on one-year term life insurance policies for the policies and populations described in Exercises 7–10. Begin by letting x represent the money made on each policy to cover profit. The cost for each policy is $25.

7. 1000 19-year-olds; face value: $100,000; annual premium: $200
8. 1000 23-year-olds; face value: $40,000; annual premium: $100
9. 5000 28-year-olds; face value: $135,000; annual premium: $350
10. 50,000 25-year-olds; face value: $45,000; annual premium: $125

7. $72,000; $1000x = 200(1000) - 1.03(100{,}000) - 25(1000)$
8. $29,000; $1000x = 100(1000) - 1.15(40{,}000) - 25(1000)$
9. $767,750; $5000x = 350(5000) - 6.35(135{,}000) - 25(5000)$
10. $2,345,000; $50{,}000x = 125(50{,}000) - 59(45{,}000) - 25(50{,}000)$

GUIDED PRACTICE

The guided practice takes students step-by-step through the skills of the lesson. Have students do these alone or in pairs. Stop and review the correct solutions after Exercises 2, 6, and 10.

ADDITIONAL ANSWERS

5. $E = 0.99906x + 0.00094(x - 60{,}000)$

EXERCISE YOUR SKILLS

1. Why do you think that the probability of a 16-year-old being alive one year from now is higher than that of a 25-year-old? See Additional Answers.
2. Why don't insurance companies charge the same life insurance premium for everyone regardless of his or her age? See Additional Answers.
3. Is there any age group for which an insurance company could not issue a life insurance policy? Explain your answer. See Additional Answers.
4. Suppose that two insurance companies both sell five-year term life insurance policies but that one of the companies charges an annual premium that is $30 higher than the other company's. What are some possible explanations for this? Answers may vary.
5. From time to time, insurance companies update the tables that they use to help them set premium schedules. Why do you think that they have to update the tables? life expectancy changes over the years
6. Because they deal with large populations of people, life insurance companies can use principles of probability to help them set profitable premium schedules. Does this mean that an insurance company can never suffer a loss from selling insurance policies? Explain your answer. No; unlikely events can occur.

A winning lottery ticket pays $200. Find the expected value of the lottery for each number of sold tickets.

7. 50 tickets $4.00
8. 100 tickets $2.00
9. 500 tickets $0.40
10. In a certain store the probability that a person makes one purchase is 0.7; the probability that a person makes two purchases is 0.3. Find the expected number of purchases. 1.3

Find the break-even premium for a one-year term insurance policy for each indicated individual. In each case, assume that the direct and indirect expenses for issuing one policy are $30. Use the table of expected deaths on page 331.

11. a $50,000 policy for a man or woman of 24 $88.50
12. a $100,000 policy for a man or woman of 29 $162.00
13. Find the break-even premium for a one-year term insurance policy for $200,000 for a man or woman of 35. The insurance company expects 99.825% of all 35-year-old people to live at least one more year. $380.00
14. Find the break-even premium for a one-year term insurance policy for $150,000 for a man or woman of 38. The insurance company expects 99.765% of all 38-year-old people to live at least one more year. $382.50
15. Find the break-even premium for a one-year-term insurance policy for $250,000 for a man or woman of 39. The insurance company expects 99.725% of all 38-year-old people to live at least one more year. $717.50

KEY TERMS
break-even value complement expected value probability of an event

ASSIGNMENT GUIDE

Exercises 1-6 ask about the practices of insurance companies and the related mathematics. To answer these, students must apply the information in the lesson to situations beyond the lesson. Exercises 7-10 review expected value. Exercises 11-21 evaluate understanding of break-even and profit as explained in Skills 2 and 3. Exercises 22-25 are more challenging but can be solved using the information and skills from the lesson.

ADDITIONAL ANSWERS

1. The older a person becomes the more likely death may be.
2. A different probability is related to every age.
3. anyone who is alive past his/her life expectancy

CLOSURE

Ask for volunteers to explain probability, expected value, and profit as related to life insurance. Sometimes students will explain a concept in a way that you had not considered, and that is helpful to others in the class.

LESSON QUIZ

1. Find the premium for a one-year term policy of $10,000 for a person 50 years of age if there are 10 expected deaths per 1000 for one year, for a person that age. Allow $40 for expenses and profit. (**$140**)

What profit does an insurance company expect to make for each one-year term insurance policy? Assume that the direct and indirect expenses for each policy are $25.

16. Face value: $50,000; age of insured: 25; annual premium: $135 $51

17. Face value: $150,000; age of insured: 22; annual premium: $390 $195.50

18. Face value: $290,000; age of insured: 26; annual premium: $800 $427.00

19. Face value: $250,000; age of insured: 29; annual premium: $650 $295

20 Face value: $75,000; age of insured: 24; annual premium: $200 $87.25

21. Face value: $125,000; age of insured: 27; annual premium: $325 $146.25

An insurance company employee has the responsibility of determining the company's schedule of life insurance premiums. She is reviewing the following portion of a life-expectancy table.

EXPECTED DEATHS PER 100,000 ALIVE AT SPECIFIED AGE		
Age	**Expected Deaths Within 1 Year**	**Expected to be Alive in 1 Year**
45	315	99,685
46	341	99,659
47	371	99,629
48	405	99,595
49	443	99,557

The company's marketing department has told her that the public loses interest in buying life insurance whenever the annual premium is greater than $800. The company's current expenses for each policy issued are $35 and it aims at making a profit of $50 on each policy.

22. At what age will a person first be inclined to resist purchasing one-year term life insurance with a face value of $200,000? 47 years

23. To appeal to the person of Exercise 22, the company manages to reduce its expenses to $20 per policy. Will this be enough to enable the employee to lower the premium to $799 without cutting into the company's profits? If not, how much of a reduction in profits would the company have to accept for the employee to be able to lower the premium to $799? no; $13 reduction

24. The company wants to sell a one-year term policy to a 49-year-old person and keep the premium below $800. Assume that the expenses have been reduced to $20 per policy and that the profit on the policy has been raised to $55. What would be the largest possible face value of such a policy? $163,656.88

25. The company wants to sell a one-year term policy to a 45-year-old person and keep the premium below $500. Asume that the expenses are $20 per policy and that the profit on the policy is $60. What would be the largest possible face value of such a policy? $133,333.33

MIXED REVIEW

2-2 Suppose that you write a check for the amount shown. Write the amount in words.

1. \$500 Five hundred and $\frac{00}{100}$ **2.** \$10,100.10 See Additional Answers.

5-3 **3.** Which approach will save you more money; to make a 20% down payment on a purchase, or to save up until you can afford a 30% down payment? Explain your answer. See Additional Answers.

6-1 **4.** Determine the effective rate of interest if the APR is 14.75%. 15.79%

6-3 Complete the following chart to determine how much the interest cost will be if you pay 5% of the amount due to the nearest dollar each month and your credit card carries an APR of 9%.

	Month	Balance	Interest	Amount Owed	Payment
5.	1	\$1000.00	\$7.50	\$1007.50	\$50.00
6.	2	957.50	7.18	964.68	48.00

7. What is the total interest for the two months? \$14.68

3-3 **8.** What happens to the nation's money supply when the Federal Reserve System increases the percent of required reserves that banks must hold? money supply decreases

1-1 **9.** You earn \$7.25 per hour. You worked 46 hours last week. You receive $1\frac{1}{2}$ times your hourly rate for any hours over 40. Find your total wages for last week. \$355.25

6-7 **10.** Last month the Wilson family spent \$205 on the car payment, \$506 on credit card payments, \$850 on utilities and rent, \$560 on food, and \$2,000 on everything else. To the nearest tenth of a percent, what percent of their take-home pay did the family spend on credit payments? 17.3%

6-2 **11.** Use the Time-to-Pay-Off formula to find the number of months required to pay off a credit card balance of \$1,850. The APR is 15% and the monthly payment is \$120. 18 mo

5-4 **12.** You have a loan of \$25,000 at 10.5% for 5 years for which you have been making monthly payments of \$537.35. How much money will you save if you prepay the loan at the end of 40 months? \$927.08

3-3 **13.** Suppose that a bank's reserve requirement is 25%. How much new money can be created from a deposit of \$2000? \$6000

4-3 Some students purchased a number of damaged tote bags for \$1.00 each and repaired them for \$0.50 each. They plan to sell the repaired tote bags.

14. What is the unit cost for purchasing and repairing the tote bags? \$1.50

15. Find the total cost for 56 tote bags. \$84

16. The fixed costs are \$62. What are the total costs? \$146

ADDITIONAL ANSWERS

2. Ten thousand one hundred and $\frac{10}{100}$

3. 30% down payment; the less owed, the less interest owed.

7-3 VALUE FOR THE FUTURE

QUICK REFERENCE

TEACHER'S RESOURCES AND ANSWER KEY

Lesson Quiz Answers 7-3, p. 66

Reteaching Activity Answers 7-3, p. 105

Enrichment Activity Answers 7-3, p. 105

EXTENSION ACTIVITIES

Reteaching Activity 7-3, p. 57

Enrichment Activity 7-3, p. 60

While Lily and Manuel are taking a look at life insurance, their friend Eleanor has other concerns—the financial well-being of her grandparents. Her grandmother, who retired earlier this year, has had to cut back on some expenses to help her family make ends meet.

Eleanor's grandfather retired three years ago. Since then, he has had some costly health problems that are not completely covered by Medicare. In addition to dealing with these past cost burdens, both grandparents have had to purchase supplementary health insurance to cover some of their future medical expenses that will not be covered by Medicare.

Eleanor is not certain of the details, but she does know that the retirement pensions that her grandparents receive are just enough to keep them comfortable. She also knows that her grandparents have had a little difficulty adjusting to the fact that they do not have as much income as they did when they were both working. Eleanor's mother is ready to help her parents in case of a financial emergency but hopes that she will not have to step in. Their circumstances have caused her to begin reevaluating her own retirement plans, and even Eleanor has suddenly become interested in topics such as annuities and Individual Retirement Arrangements.

OBJECTIVES: *In this lesson, we will help Eleanor to:*

- *Examine reasons for investing in retirement plans that are tax deferred, such as annuities and Individual Retirement Arrangements (IRAs).*
- *Calculate the future value of regular payments invested at compound interest.*
- *Compare the future value of cash-value life insurance with the future value of the same amount invested at compound interest.*
- *Compare the difference in accumulated cash value between investing directly and saving indirectly through whole life insurance.*

FOCUS

ALGEBRA CONNECTIONS

Substitute values and construct a table of values for the function $y = 1.2^x$. Plot points and connect them to form a graph. (Do this without using the graphing calculator.)

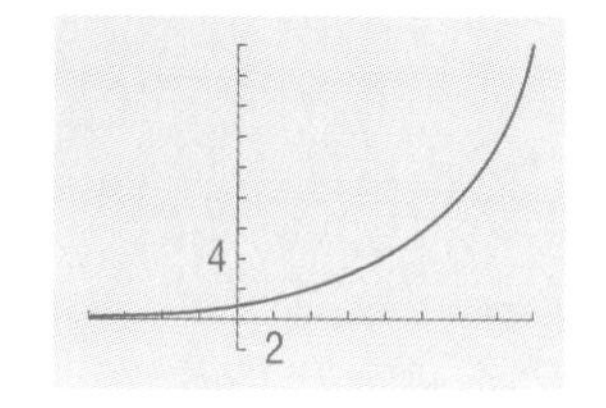

SOCIAL SECURITY IS NOT ENOUGH

Everyone faces the fear of outliving one's pension. Eleanor's grandparents have some degree of financial security because they planned ahead for their retirement. They knew that they could not rely on their monthly Social Security checks, since that system was never intended to provide one's sole source of retirement income, even though the system is adjusted upward periodically for inflation. Instead, Eleanor's grandparents learned to take advantage of supplementary sources of retirement income such as annuities and Individual Retirement Arrangements.

ANNUITIES

Years ago, Eleanor's grandmother placed some of her income from college teaching in a tax-deferred annuity. An **annuity** is an investment plan that provides income upon retirement. Annuities, which are usually purchased through insurance companies, offer two advantages—forced savings and tax deferral. Grandmother chose to make payments into her annuity through a payroll deduction. The amount that she requested was automatically deducted from her paycheck each month.

Now that she has retired, Grandmother withdraws the money that she originally contributed as well as interest on that money. She doesn't pay taxes on the returned contributions, but she does pay taxes on the interest income from the annuity.

Grandmother's annuity has provided her with two major tax advantages. First, she has never had to pay any taxes on any of the interest that has been accumulating for all those years—until now. During this long period of time, that interest has been earning more interest. She would never have received all that additional money if her tax obligation had not been delayed until her retirement years. Second, even though she now has to pay the taxes, they are less than they would have been if she had paid them in earlier years. The reason? Grandmother's tax *rate* on her pension income is lower than it was on her much higher working income. The lower your income, the lower your tax rate.

INDIVIDUAL RETIREMENT ARRANGEMENTS (IRAS)

An **Individual Retirement Arrangement (IRA),** sometimes called an individual retirement *account,* has the same **tax shelter** advantages just described for annuities. It allows employees (and some self-employed people) to put up to $2000 a year into such a plan to shelter the accumulated interest from taxes. If the employee's income is less than a certain amount, then the contribution can also be fully deducted from the adjusted-gross-income line on the federal tax return. The maximum income for a full deduction is, at this writing, $40,000 for a married couple and $25,000 for an individual. (A partial deduction is allowed on up to $10,000 above these levels.) Employees who have no employer pension plan are exempt from these income maximums.

You may withdraw money from an IRA, but you will pay a tax penalty and possibly an interest penalty unless you are either disabled or over the age of $59\frac{1}{2}$. You may choose the kind of institution that you want to handle your IRA, for example, a bank, a brokerage firm, or a mutual fund group. You may also set up more than one IRA among a variety of institutions. It is important never to combine your IRA account with non-IRA funds.

EMPLOYER PENSION PLANS

Some employers offer pension plans for their employees. If such a plan is approved by the Internal Revenue Service (IRS) as a *qualified plan* or as a **401(k) plan,** then an employer will make tax-sheltered contributions to the plan that are larger than an employee could get through a regular IRA. In some such plans, the employee may also contribute to the plan, but whether or not those contributions are tax deductible depends on the exact nature of the plan. Unlike its rules for IRAs, there are no IRS-imposed income maximums on participants in these employer plans. A self-employed person may set up a similar tax-sheltered plan, called a **Keogh plan.**

These employer and Keogh plans have the disadvantage that, unlike IRAs, they are somewhat complicated to set up and to administer. To avoid these complications, some employers or self-employed people prefer to establish a **Simplified Employee Pension Plan (SEP)** that allows a person to contribute to an IRA without all of the usual IRA rules and limitations.

BALANCING INSURANCE AND INVESTMENTS

Eleanor is especially curious about cash-value life insurance as an investment vehicle compared to a direct investment such as a mutual fund.

The CVLI/BTID Controversy A controversy has arisen in the last few years over the merits of cash-value life insurance compared to term insurance. Consumers have assumed, perhaps because of persuasive insurance salespeople, that cash-value insurance is a good way to combine adequate life insurance coverage with a means for saving money. However, some people in the insurance industry have begun to advise their customers that they can save much more money by buying term insurance and investing the money that they save in instruments such as certificates of deposit, mutual funds or even stocks or bonds. These opposing viewpoints are expressed in the code phrase CVLI/BTID, which stands for *Cash-Value Life Insurance/Buy Term, Invest the Difference.*

If your employer offers you the opportunity to buy life insurance at work, you may be able to save significantly on the annual premium. Another way of saving on the cost of the premium is to find a company that sells **low-load insurance policies,** policies with a very low commission. Over a long period the annual savings can be very important. One insurance expert has concluded that over 20 years, a full-commission cash-value insurance policy would have to give a yield that was $1\frac{1}{2}$ percentage points higher than a low-commission insurance policy to match the cash value of the low-load policy. This could amount to a difference of several thousand dollars over 20 years.

The most difficult choice in your planning may be the selection of interest rates that you can reasonably expect to earn over the next 30 years. Since even professional economists cannot accurately predict future interest rates, your best course may be to start with the current rates for government bonds, both short-term (six-month Treasury bills) and long-term (30-year bonds) and make your own projection about rates in the future. Since government bonds are the safest investments of all, you have a right to expect higher returns from any investment that is not as safe as they are.

Do not have unrealistic expectations about the 30-year performance of your own investments. If you can obtain a long-term average of five percentage points more than the yearly return on a 30-year government bond, then you will be doing reasonably well. You will be doing very well indeed if you consistently average ten points a year above such bonds.

On the basis of your research, you can decide whether it is better for you to save through an insurance company or to use the insurance company solely for protecting your family's earning power and do your investing directly. Careful decision making on your part may lead to high long-term yields. Even our teenagers, who see retirement as something many years in the future, know the difference between having $20,000 at retirement and having $200,000!

Ask Yourself

1. What are tax-deferred annuities?
   Investment plans that provide income upon retirement
2. What are Individual Retirement Arrangements?
   See Additional Answers
3. What is the earliest age at which you can normally withdraw money from an IRA without penalty?
   $59\frac{1}{2}$

ADDITIONAL ANSWERS: Accounts in which money is invested for retirement and taxes are deferred until retirement

ALGEBRA *REVIEW*

Find the value of *A* or *B* in the following formulas for the values given.

$$B = \frac{(1.01)^n - 1}{0.01}$$

$$A = \frac{p[(1 + r)^n - 1]}{r}$$

1. $B;\ n = 3$
   3.03
2. $B;\ n = 30$
   34.78
3. $A;\ n = 4,\ r = 0.03,\ P = 1000$
   4183.63
4. $A;\ n = 10,\ r = 0.065,\ P = 1500$
   20,241.63
5. $A;\ n = 20,\ r = 0.09,\ P = 2000$
   102,320.24
6. $A;\ n = 40,\ r = 0.10,\ P = 750$
   331,944.42
7. $A;\ n = 20,\ r = 0.15,\ P = 2000$
   204,887.17

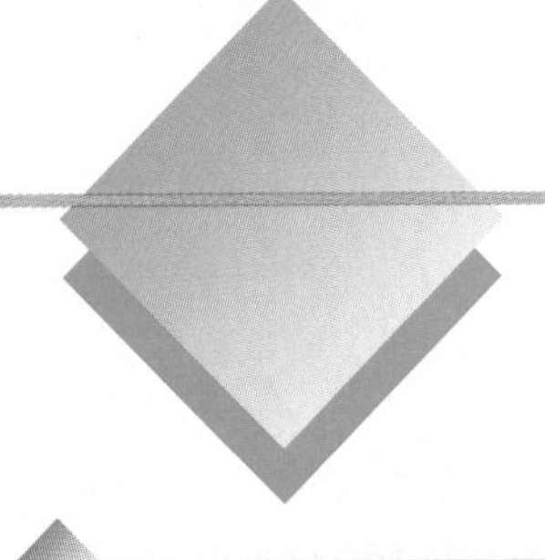

SHARPEN YOUR SKILLS

SKILL 1

To determine the future value of money that is deposited into an interest-bearing account on a regular basis, use the formula for the **future value of a periodic investment.**

Future Value of a Periodic Investment Formula

$$A = \frac{p[(1 + r)^n - 1]}{r}$$

where A = the future value of the investment
p = the investment made at the end of each period
r = the interest rate for the period
n = the number of periods

EXAMPLE 1 Eleanor's father and mother are both 43. They have just opened IRA accounts that they hope will supply them with extra money when they retire 22 years from now at the age of 65. Each year, they will deposit $2000 dollars into each account, which they are assuming will pay 8% interest, compounded annually.

QUESTION By the time they are 65, how much money will be in each parent's account?

SOLUTION

$$A = \frac{p[(1 + r)^n - 1]}{r} \quad \text{Future value of a periodic investment}$$

$$= \frac{2000[(1 + 0.08)^{22} - 1]}{0.08} \quad p = 2000,\ r = 0.08,\ n = 22$$

$$= 110{,}913.51$$

When they retire, Eleanor's parents will each have $110,913.51 in their accounts. This is $66,913.51 more than the $44,000 that they each contributed to their accounts over the 22 years.

SKILL 2

EXAMPLE 2 Suppose Eleanor's grandfather purchased $50,000 worth of whole life insurance when he was 25.

QUESTIONS

1. How much would his annual premium be?
2. How much would he pay for the insurance over 40 years?
3. If he had invested the premium money every year in an annuity that paid $8\frac{1}{2}\%$ interest, compounded annually, how much would his savings be worth at the end of 40 years?

INSTRUCTION

TEACHING THE LESSON

The main point of the lesson is the application of exponential growth to regular savings. This an extension of compound

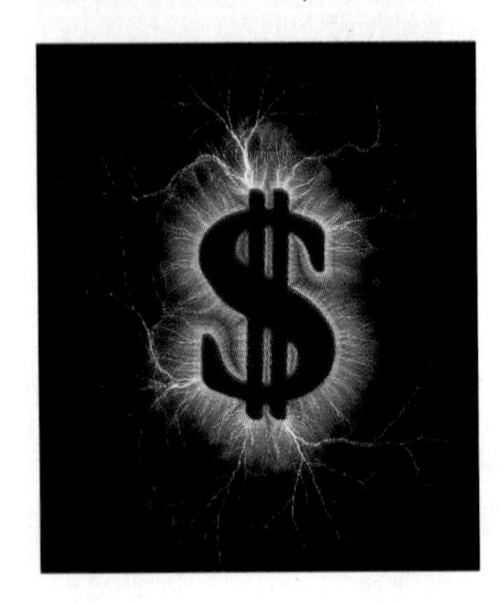

interest and students should see the connection.

The derivation of the future value formula involves the use of a geometric series and so is not given. But students should understand the meaning of each variable in the formula. Review on the chalkboard the addition of $1 + r$ to give, for example, 1.08. Also review the definition and laws of exponents.

TECHNOLOGY HINT

It would be useful to review keying the formula in Skill 1 for different calculators. In particular, the correct use of parentheses is essential.

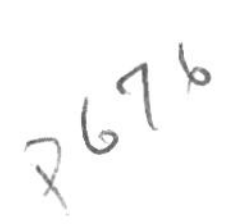

SOLUTIONS

1. See the Comparison Table for Term and Whole Life Premiums in the Reference Section. The premium for \$50,000 of whole life insurance is half of the premium for \$100,000; that is, \$918 ÷ 2, or \$459.
2. The total paid in 40 years would be

   $459 \cdot 40$, or \$18,360

   He would have paid \$18,360 in premiums over 40 years.
3. Use the formula for the future value of a periodic investment.

$$A = \frac{p[(1 + r)^n - 1]}{r}$$

$$= \frac{459\,[(1 + 0.085)^{40} - 1]}{0.085} \qquad p = 459,\ r = 0.085,\ n = 40$$

$$= 135{,}718.28 \qquad \text{To the nearest cent}$$

His annuity would be worth about \$135,718, or about \$85,000 more than the cash value of the insurance policy after 40 years. This money would remain untaxed until he began to receive his periodic payments during retirement. If Eleanor's grandfather had chosen to use the \$18,360 to buy the annuity, then he would have been without the \$50,000 insurance for 40 years.

CRITICAL THINKING

The lesson shows the possible long-term benefit in purchasing term life insurance and investing the savings more productively. Ask students why, in spite of this, many people purchase whole life with its guaranteed payoff. (**Two possible reasons: 1. Fluctuating interest rates make a guaranteed rate impossible. 2. The terms of an insurance contract "force" the savings that might be difficult on a "volunteer" basis.**)

SKILL 3

The following table shows the cash value (accumulation of savings) of a whole life insurance policy of \$100,000 over the first 20 years. This table is also in the Reference Section.

ACCUMULATED CASH VALUE OF \$100,000 WHOLE LIFE POLICY
AGE OF ISSUE: 25

Year	Person's Age	Cash Value	Year	Person's Age	Cash Value
1	25	\$ 0	11	35	\$10,187
2	26	700	12	36	11,501
3	27	1500	13	37	12,860
4	28	2300	14	38	14,246
5	29	3100	15	39	15,667
6	30	4020	16	40	17,094
7	31	5158	17	41	18,555
8	32	6349	18	42	20,014
9	33	7538	19	43	21,563
10	34	8898	20	44	23,197

FOCUS ON ALGEBRA

The power and elegance of mathematical notation is evident in the future value formula. It compresses a process that takes many words to explain. *(NCTM, Standard 2, p. 140)*

COMMON ERROR

Many errors will be made using the calculator. Explain to students that they must read the documentation for each calculator and computer program. Future business and science courses will be easier if students are ready to learn how to learn with calculators and computer programs.

EXAMPLE 3 Eleanor was interested in knowing whether her grandfather could have been better off financially by buying term insurance instead of whole life at age 25 and investing the money saved in an IRA at $8\frac{1}{2}$%.

INTERDISCIPLINARY INVESTIGATION

Have interested students read about the growth of bacteria and prepare a report showing the similarities in the formulas used to describe such growth and the future value of money invested.

RETEACHING

Have students evaluate each of these expressions, then find the sum.

$\$100(1.08)^4 = (\mathbf{\$136.04})$
$100(1.08)^3 = (\mathbf{125.97})$
$100(1.08)^2 = (\mathbf{116.64})$
$100(1.08)^1 = (\mathbf{108})$
$100(1.08)^0 = (\mathbf{100})$
Sum = (**$586.65**)

Then have them compare the sum with the result of evaluating the future value formula when $p = 100$, $r = 0.08$, and $n = 5$. (**$586.66**)

ENRICHMENT

Challenge students to graph the future value formula using the graphing calculator. Let them select which values to fix and which to be the independent variable.

QUESTIONS

1. If Eleanor's grandfather had bought $50,000 worth of whole life cash-value insurance at age 25, what would the cash value have been when he was 44?
2. How much money could a 25-year-old man accumulate between the ages of 25 and 30 by buying a $50,000 term policy instead of a whole life policy and investing the difference in the two premiums in an IRA at $8\frac{1}{2}\%$?
3. How does the money accumulated in Question 1 compare with the cash value that would have accumulated in the whole life policy in the same five-year period?

SOLUTIONS

1. Looking at the table, you can see that the value at age 44 would have been $23,197 ÷ 2, or $11,598.50.
2. Eleanor used the Comparison Table for Term and Whole Life Premiums in the Reference Section to find the difference between the premiums for the two kinds of policies at age 25. To obtain the premiums for a $50,000 policy, she divided each table entry by 2.

   Whole life premium − term premium = annual saving
   459 − 103.50 = 355.50

   Next, Eleanor found how much an annual investment of $355.50 will grow to in five years compounded annually at $8\frac{1}{2}\%$. She used the formula for the future value of a periodic investment.

   $$A = \frac{p[(1 + r)^n - 1]}{r}$$

   $$A = \frac{355.50[(1.085)^5 - 1]}{0.085} \qquad p = 355.50,\ r = 0.085,\ n = 5$$

   $$A = 2106.47$$

   The annual payments will grow to about $2106.
3. From the table entitled Accumulated Cash Value of $100,000 Whole Life Policy, Eleanor found that the cash value of the policy would be worth about $3100 ÷ 2, or $1550 at the end of 5 years. By buying a term policy instead of a whole life policy and investing the difference at $8\frac{1}{2}\%$, Eleanor's grandfather would have accumulated

   2106 − 1550

   or an extra $556 at the age of 30 to reinvest tax-sheltered for another 30 or 35 years; that is, until he retired. When he renewed his term policy at age 30, 35, 40, and so on, his premium would gradually rise. This would reduce the size of the premium savings that he realized in the early years.

   If Eleanor's grandfather had bought term insurance instead of whole life, he would have been better off *provided* that he could be reasonably sure of averaging an $8\frac{1}{2}\%$ return or better. Otherwise, the whole life option might have been a better choice after all.

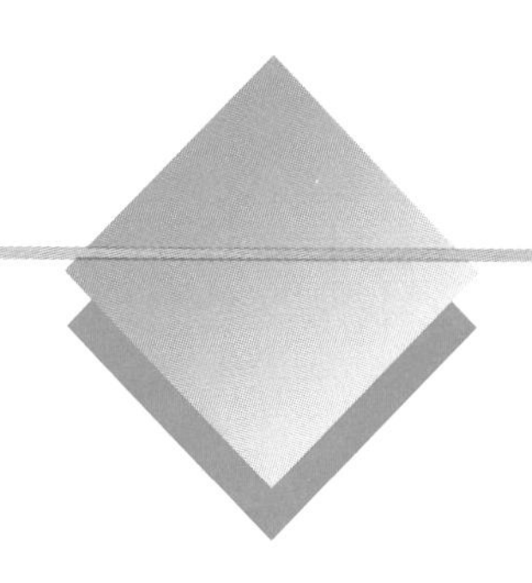

TRY YOUR SKILLS

Use the formula for the future value of a periodic investment to answer Exercises 1–4.

1. If you begin to contribute \$2000 a year at age 45 to an IRA account that you expect to pay 8% per year, how much will you have in your account at age 65? \$91,523.93
2. If you begin contributing the same amount at age 35, how much will you have at age 60? \$146,211.88
3. Suppose you purchase \$100,000 worth of insurance when you become 25. Use the Comparison Table for Term and Whole Life Insurance in the Reference Section to find the premium per year. Then calculate how much you will pay in premiums in 40 years. term: \$8280; whole: \$36,720
4. If you invested the amount of the insurance premium in Exercise 3 in an annuity at a return of $8\frac{1}{2}\%$, how much would your annuity be worth at the end of 40 years? Term: \$61,206.29; whole: \$271,436.57

GUIDED PRACTICE

In this lesson, even the guided practice exercises are rather complex. So you might want to read these exercises and discuss the method for doing each one before students do them. After the first four, have students place their results on the chalkboard and explain the procedures they followed.

Use the table in this lesson entitled Accumulated Cash Value of \$100,000 Whole Life Policy Issued at Age 25 to find the cash value of a \$100,000 policy at the following ages. The policy is bought at age 25.

5. 30 \$4020
6. 35 \$10,187
7. 40 \$17,094
8. 42 \$20,014

Use the Comparison Table for Term and Whole Life Premiums in the Reference Section, the table entitled Accumulated Cash Value of \$100,000 Whole Life Policy in the Reference Section, and the formula for the future value of a periodic investment to answer Exercises 9–14. \$218

9. What is the annual premium for a \$100,000 five-year renewable term life insurance policy for a 30-year-old person? \$218
10. What is the annual premium for a \$100,000 whole life insurance policy for a 30-year-old person? \$1112
11. How much money is saved each year if a 30-year-old person buys the term policy instead of the whole life policy? \$894
12. Use the formula for the future value of a periodic investment to find how much tax-deferred money a 30-year-old person could accumulate between the ages of 30 and 35 by buying a \$100,000 term policy instead of a whole life policy and then investing the money saved because of the lower premium in an IRA that gives an annual return of $7\frac{1}{2}\%$. \$5192.70
13. What would be the cash value of the whole life insurance policy for a 30-year-old person after 5 years? \$3999
14. Compare the cash value of the insurance policy with the value of the IRA at the end of 5 years. Which is the more profitable investment?
IRA is more profitable

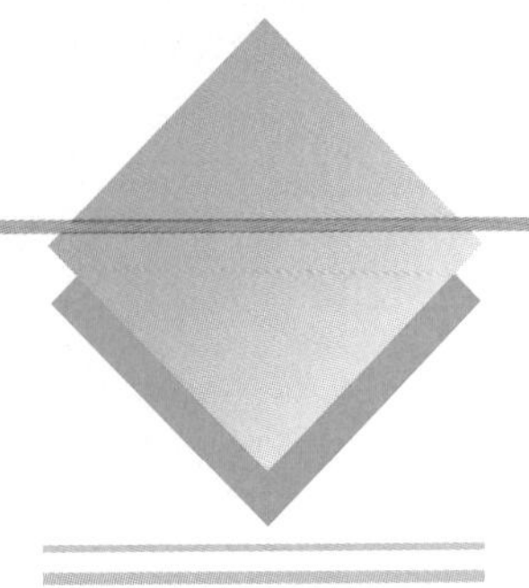

EXERCISE YOUR SKILLS

1. Why should people invest in retirement plans such as tax-deferred annuities and Individual Retirement Arrangements? See Additional Answers.
2. What are two advantages of purchasing tax-deferred annuities through payroll deductions? See Additional Answers.
3. Under what circumstances may a person not defer the taxes on his or her IRA contribution? early withdrawal

Use the formula for the future value of a periodic investment to answer Exercises 4–6.

4. Suppose that at the age of 35 you begin contributing $2000 a year to an IRA account that you think will pay an average of 10% a year. How much can you expect to have in your account at age 65? $328,988.05
5. If you begin contributing the same amount at age 35, how much more will you have at age 65 than if you had begun at age 45? $214,438.05
6. If you begin contributing the same amount at age 35, how much more will you have at age 65 than if you had begun at age 40? $132,293.93
7. Suppose you purchase $100,000 worth of whole life insurance when you become 30. Use the Comparison Table for Term and Whole Life Insurance in the Reference Section to find the premium per year. Then calculate how much you will pay in premiums over 35 years. term: $7630; whole: $38,920

Suppose that you wanted to know how well you would do at the end of 35 years if you had invested the insurance premium in Exercise 7 in an annuity at a return of 9% per year.

8. Use the formula for the future value of a periodic investment to find the value of the annuity in 35 years. term: $47,024.94; whole: $239,870.36
9. Explain why the results of Exercise 8 do not necessarily imply that the annuity investment is the better choice. A person cannot borrow from an annuity.

Use the Comparison Table for Term and Whole Life Premiums in the Reference Section, the table entitled Accumulated Cash Value of $100,000 Whole Life Policy in the Reference Section, and the formula for the future value of a periodic investment to answer Exercises 10–13.

10. How much money is saved each year if a 35-year-old person buys a term policy instead of a whole life policy? $1120
11. Determine how much tax-deferred money a 35-year-old person could expect to accumulate between the ages of 35 and 40 by buying a $100,000 term policy instead of a whole life policy and investing the money that is saved because of the lower premium in an IRA that will earn $7\frac{3}{4}$% per year. $6537.92
12. What would be the cash value of the whole life insurance policy for a 35-year-old person after 5 years? $5151

KEY TERMS

- annuity
- 401(k) plan
- future value of a periodic investment
- Individual Retirement Arrangement (IRA)
- Keogh plan
- low-load insurance policies
- Simplified Employee Pension Plan (SEP)
- tax shelter

INDEPENDENT PRACTICE

ASSIGNMENT GUIDE

Exercises 1-3 ask about the benefits associated with annuities. Exercises 4-7 require the application of the future value formula in different situations. Exercises 8-12 require comparisons of different kinds of regular investments, particularly term insurance, whole life, and regular tax deferred savings. Exercises 13-15 ask students to make judgments based on the information learned and the exercises completed.

ADDITIONAL ANSWERS

1. no taxes paid on income invested

13. Compare the cash value of the insurance policy with the value that the person expects the IRA to have at the end of 5 years. Which seems to be the more profitable investment? IRA

14. Under what circumstances would it be better for a young person to choose a term policy instead of a whole life policy and then use the money saved because of the lower premium to invest into an annuity or IRA?

15. Under what circumstances would it be better for a young man or woman to purchase a whole life policy rather than a term policy, even though the premium money that would be saved might be more profitably invested in an annuity or IRA? See Additional Answers.

CLOSURE

Review the formula for the future value of a periodic investment explaining again the meaning of each variable.

LESSON QUIZ

1. Find the future value of $1000 per year invested at 7% for 20 years. (**$40,995.49**)
2. Find the future value of the investment described in Exercise 1 at 9%. (**$51,160.12**)

MIXED REVIEW

1-3 Each of two real-estate salespeople sold a house for $180,000. Find the commission earned by each person.

1. Rafael gets a straight 5% commission on the entire sales price. $9000
2. Sally gets a $4\frac{1}{2}$% commission on the first $100,000 and a 6% commission on any amount over $100,000. $9300

3-2 3. Benjamin Franklin once remarked, "Money makes money, and the money that money makes makes more money." In two words or fewer, what money concept was Franklin referring to? compound interest

6-4 4. For an all-purpose credit account, set up a chart to show the first 3 monthly payments, and find the total interest paid over 3 months at 1.5% interest per month. The beginning balance is $720, and you make monthly payments of 10% of the amount owed to the nearest dollar. During the second month you make an additional purchase of $125. Total interest: $33.27. See Additional Answers.

6-7 5. In the Larsen family, $190 is spent on the car payment, $400 on credit card payments, $980 on utilities and rent, $760 on food, and $1,170 on everything else. To the nearest tenth of a percent, what percent of their take-home pay is the family spending on credit payments? 16.9%

7-2 6. The probability that a person will be alive in 1 year is 0.99894. What is the probability that the person will die within one year? 0.00106

5-5 7. Ralph is considering renting a computer on a rent-to-own credit plan that allows him to rent the computer for $90 per month for 24 months. If he decides to buy the computer, the rental fees will be applied to cover the entire purchase price. How much will the computer cost him under this plan? $2160

5-5 8. Suppose that Ralph purchases the computer in Exercise 7 for $1,800 with an installment loan with monthly payments at 12% annual interest over 2 years. Is this a better deal than the rent-to-own plan? Explain your answer. $2160 − $2033. 57 = $126.43; It's a better deal.

ADDITIONAL ANSWERS

14. when it is unlikely he/she will need to use the cash before retirement
15. when it is unlikely he/she will need to borrow the money in the future

CHAPTER 7 REVIEW

1. Why is it important for people to have adequate life insurance?
2. How does a person decide whether to purchase term insurance or cash-value insurance? by projecting into the future
3. Why might young families be wise to purchase decreasing term insurance?
4. Why do people need to supplement their Social Security pensions with their own retirement plans? Social Security is not enough to be one's only income.

ADDITIONAL ANSWERS

1. to protect dependent family members
3. Premiums are lower than renewable term.

Use the Multiples-of-Salary Chart in the Reference Section to find the amount of life insurance that each of these income earners should buy.

5. Juanita is 35 years of age and has gross earnings of $40,000. She needs 60% income replacement. $240,000
6. Joshua is 45 years of age and has gross earnings of $30,000. He needs 75% income replacement. $255,000

Use the Comparison Table for Term and Whole Life Premiums in the Reference Section to find the yearly premiums for the amounts of insurance given below.

7. $50,000 whole life insurance at 25 years of age $459
8. $150,000 term insurance at 45 years of age $843
9. $200,000 term insurance at 40 years of age $726
10. $250,000 whole life insurance at 35 years of age $3435

John Andrews is considering the purchase of a one-year term policy with a face value of $90,000. The partially completed table below shows how an insurance company evaluates John's ability to remain alive for one year.

COMPANY'S GAIN OR LOSS ON ONE POLICY (before deducting expenses)		
Possible Outcome	**Probability of Outcome**	**One-Year Gain/Loss**
John Andrews lives.	0.99906	x
John Andrews dies.	?	$x - 90{,}000$

11. Find the probability that John Andrews will die in the next year. 0.00094
12. The company has direct and indirect expenses of $20 for each policy it issues. Find the premium that the company must charge in order to break even; that is, neither to make nor to lose money on this policy. $104.60

Use the table entitled Expected Deaths per 100,000 Alive at Specified Age on page 331 to find the break-even premium for a one-year term insurance policy for each indicated individual. In each case, assume that the direct and indirect expenses for issuing one policy are $20.

13. An $80,000 policy for a man or woman of 28 $121.60
14. A $250,000 policy for a man or woman of 20 $285.00

Use the table entitled Expected Deaths per 100,000 Alive at Specified Age on page 331. What profit does an insurance company expect to make for each one-year term insurance policy? Assume that the direct and indirect expenses for each policy are $20.

15. Face value: $60,000; age of insured: 21; annual premium: $120 $34

16. Face value: $175,000; age of insured: 19; annual premium: $310 $109.75

Use the formula for the future value of a periodic investment to answer Exercises 17–18.

17. At the age of 30, André begins to contribute $2000 a year to an IRA account that he expects to pay 11% a year. How much does he expect to have in the account at the age of 42? $45,426.37

18. If André had begun his IRA contributions at the age of 20, how much more would he expect to have in the account at age 42 than he actually expects because of his beginning at the age of 30? $117,002.25

In Exercises 19–20, use the Comparison Table for Term and Whole Life Premiums on page 326, and the formula for the future value of a periodic investment.

19. Miguel purchases $150,000 worth of whole life insurance when he is 35 years old. How much will he pay in premiums over 30 years? $61,830

20. If Miguel had invested the amount of the insurance premium in Exercise 19 in an annuity that gave him an annual compounded return of 7% interest, how much would the account be worth at the end of 30 years? $194,683.68

Use the Comparison Table for Term and Whole Life Premiums in the Reference Section, the table entitled Accumulated Cash Value of $100,000 Whole Life Policy in the Reference Section, and the formula for the future value of a periodic investment in Lesson 7–3 to answer Exercises 21–24.

21. How much money does a 40-year-old person save each year if he buys the term policy instead of the whole-life policy? $1366

22. Determine how much tax-deferred money a 40-year-old person could accumulate between the ages of 40 and 45 by buying a $100,000 term policy instead of a whole life policy and investing the difference between the two premiums in an IRA account that is expected to return $7\frac{1}{2}\%$ per year. $7934.26

23. What would be the cash value of the whole life insurance policy for a 40-year-old person after 5 years? $6367

24. Compare the cash value of the insurance policy with the value of the IRA at the end of 5 years. Which is the more profitable investment? IRA is more profitable

CHAPTER 7 TEST

1. Melanie is 45 years old and has gross earnings of $65,000. Use the Multiples-of-Salary Chart for Net Income Replacement on page 325 to find the amount of life insurance that she should buy. She needs 60% income replacement. $390,000

2. Use the Comparison Table for Term and Whole Life Premiums on page 326 to find the yearly premium for a $100,000 whole life insurance policy purchased at the age of 20. $775

ALTERNATIVE ASSESSMENT

Ask students to find a table with information about insurance rates. It will not have the same headings as those in the text. Students should write several paragraphs explaining the information in the table and how it is used. They should include an example of the table's use in a specific situation.

3. Use the table entitled Expected Deaths per 100,000 Alive at Specified Age on page 331 of Lesson 7–2 to find the break-even premium for a one-year $150,000 term insurance policy for a man or woman of 18. Assume that the direct and indirect expenses for issuing one policy are $20. $168.50

Use the table entitled Expected Deaths per 100,000 Alive at Specified Age on page 331. What profit does an insurance company expect to make for each one-year term insurance policy? Assume that the direct and indirect expenses for each policy are $20.

4. Face value: $90,000
Age of insured: 20
Annual premium: $170 $54.60

5. Face value: $350,000
Age of insured: 29
Annual premium: $980 $498.00

Use the formula for the future value of a periodic investment to answer Exercises 6 and 7.

Ask students to explain why insurance companies include certain risk factors when computing premiums. Have interested students report on two or three factors, and include life expectancy tables and graphs as they explain how and why premiums differ.

6. At the age of 25, Marie began contributing $1500 a year to an annuity paying 10.5% annually. How much will she have in the account at the age of 40? $49,590.05

7. If Marie had begun her annuity contributions at the age of 20, how much more would have been in the account at age 40 than was there because she began at the age of 25? $41,356.16

Use the Comparison Table for Term and Whole Life Premiums in the Reference Section, the table entitled Accumulated Cash Value of $100,000 Whole Life Policy in the Reference Section, and the formula for the future value of a periodic investment to answer Exercises 8–11.

Have students use the computer to generate spreadsheets showing the growth in investment of $1000 invested annually at interest percents varying from 5-10%, and from 10 to 40 years.

8. How much money is saved each year if a 45-year-old person buys the term policy instead of the whole life policy? $1565

9. Find how much tax-deferred money a 45-year-old person could accumulate between the ages of 45 and 50 by buying a $100,000 term policy instead of a whole life policy and investing the difference between the two premiums in an IRA at 8% per year. $9181.23

10. What would be the cash value of the whole life insurance policy for a 45-year-old person after 5 years? $7737

11. Compare the cash value of the insurance policy with the value of the IRA at the end of 5 years. Which is the more profitable investment? IRA is more profitable

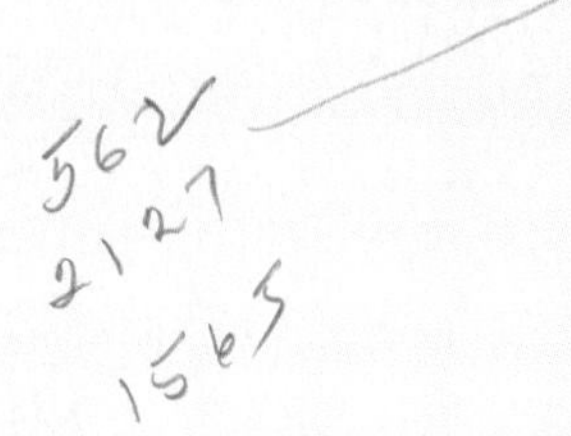

CUMULATIVE REVIEW

1-2 A company has the following fringe benefits policy:

Employee	Salary	Fringe Benefits Policy
Bank vice-president	$51,000	7.65% of gross for FICA taxes
		3% of gross for retirement
		$160 per month for life and health insurance
		25 nonworking days @ 212.50 per day

1. Find the amount that the company must pay annually for extra taxes, retirement, life and health insurance, and other indicated fringe benefits.
2. Find how much salary is paid for the indicated number of nonworking days in a total of 240 operating days. $5312.50

1-3 3. Frank's earnings as a real-estate salesman are based on a graduated commission. He receives 3.5% on each sale up to $150,000 and 5% of the amount above $150,000. How much would he earn for selling a $215,000 house? $8500

2-1 4. Suppose you wrote 28 checks in June. Your bank charges $0.03 per check for the first 20 checks and $0.15 for each check over 20. How much were your June bank charges? $1.80

3-2 5. What would be the interest on $950 at 9.5% for one year on an investment that is compounded semiannually? $92.39

5-5 6. Find the total financed price, the total interest, and the dealer's profit for each car-sale plan for a car that is priced at $12,900.

	Plan A	Plan B
Rebate	0	$1500
Rate	4.5%	10.5%
Time	48 months	48 months
Monthly payment	$294.16	$291.88
Total financed price	$14,119.92	$14,010.17
Total interest	$1219.92	$2610.17
Dealer's profit	$2719.92	$2610.17

6-3 7. Suppose that your credit card charges 1.75% interest per month. You pay 10% of the balance to the nearest dollar each month. Complete a table using the following columns for 3 months. Your beginning balance is $690.
See Additional Answers.

Month	Balance	Interest	Amount Owed	Payment

7-1 8. Use the Comparison Table for Term and Whole Life Premiums in the Reference Section to find the yearly premium for $200,000 of whole life insurance purchased at the age of 35. $2748

ADDITIONAL ANSWERS

1. $7351.50

PROJECT 7–1: **Comparing Types of Available Insurance**

Work with a group of your classmates to complete this activity.

1. Each member of the group should choose two insurance companies that sell policies in your area.
2. Find out what types of life insurance policies each company sells.
3. Obtain a list of the premiums for each type of policy.
4. Create a bulletin board to display your findings.
5. Discuss your findings with the other members of your group.
6. Evaluate the various policies and premiums. Decide which company and which policy provide the best investment at the lowest cost.
7. Prepare a brochure stating which policy is the best and why. Make the brochure colorful and eye-catching so it might be used as an advertisement for the policy.

PROJECT 7–2: **How are Insurance Premiums Determined?**

1. Contact various insurance companies to find out what statistics they use to determine the premiums for different types of life insurance.
2. Contact people (actuaries) who compile the statistics used by insurance companies to determine premiums. Find out how they compile the statistics.
3. Write a report summarizing your findings.
4. Use reference books such as a world almanac to study the changes in the life expectancy of men and women in the last century.
5. Compare the changes in life expectancies with the changes in life insurance premiums over the years for which you can find figures. How do they compare?

A *scatter plot* is a graph of several points. Four scatter plots are shown below. A *line of best fit* is superimposed on each graph.

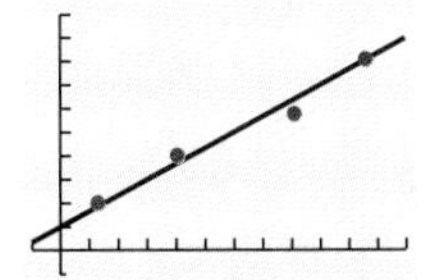 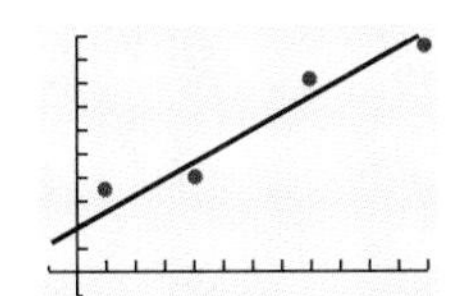 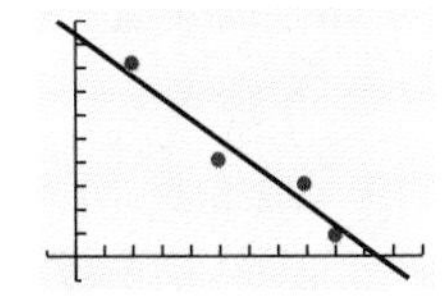 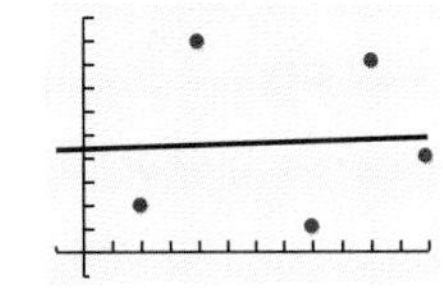

A *line of best fit* is the straight line that is closer to all the points than is any other line you might draw. The *correlation coefficient* r is a number between -1 and 1 that indicates how good the fit is. If r is 1 or -1, then the fit is perfect. If r is close to 1 or -1, the fit is good. If r is 0 or near 0, then the fit is poor.

Graph the table of values at the right as a scatter plot. Then graph the equations of Exercises 1–3 over the scatter plot and tell whether you think the fit of each line is *poor, good,* or *excellent.* See Additional Answers.

x	0	1	3	5	7
y	0	2	4	6	7

1. $y = 0.5x + 2$
good

2. $y = -0.2x - 2.5$
poor

3. $y = x + 0.5$
excellent

A graphing calculator can be used to find an equation of a line that best fits a scatter plot. The calculator will give you values for a and b in the *linear-regression equation* $y = a + bx$. It will also give you the correlation coefficient for the line.

Write an equation for a linear-regression equation that corresponds to the given calculator display. Round the constants to the nearest hundredth.

Example $a = -0.136986$
$b = 3.260274$
$r = 0.973955$

Solution $y = 3.26x - 0.14$

4. $a = 0.486765$
$b = 2.242647$
$r = 0.837326$
$y = 2.24x + 0.50$

5. $a = 1.186441$
$b = -3.067797$
$r = -0.993112$
$y = -3.07x + 1.19$

6. $a = 0.612805$
$b = 0.484756$
$r = 0.474731$
$y = 0.48x + 0.61$

7. $a = 4.254237$
$b = 1.271186$
$r = 0.159854$
$y = 1.27x + 4.25$

8. In Exercises 4–7, which line is the best fit for its scatter plot? Which line is the poorest fit?
$y = -3.07x + 1.19$ is the best. $y = 1.27x + 4.25$ is the poorest.

9. Use a graphing calculator to find the linear-regression equation for the scatter plot of Exercises 1–3.
$y = 0.98x + 0.66$

CHAPTER 8

investments IN STOCKS AND BONDS

WALL STREET IS MORE THAN A PLACE, FOR MANY OF us, Wall Street represents wealth. It is an arena in which millions of dollars circulate every day. Portions of companies, even whole companies, are bought and sold to other companies and to wealthy investors, and the money flashes by as numbers on computer screens.

Many companies, however, are also owned by thousands of individual stockholders, including wage-earning investors. Companies with over 100,000 stockholders include Westinghouse, Atlantic Richfield, Eastman Kodak, Texaco, Chrysler, Occidental Petroleum, Mobil, Ford, Sears, and Southwestern Bell, to name a few.

Some of the students whom we have met in earlier chapters will investigate various aspects of stock investments. They will study the differences among stocks, bonds, and mutual funds and how these investments compare with money that is kept in a savings account. They will also examine some of the ways to avoid risky investment schemes.

York Stock
WALL
ST
Volume
NOV. DEC. JAN. FEB. MAR. APR. MAY JUNE JULY AUG. SEPT. OCT. NOV. DEC. JAN. FEB. MAR. APR. MAY JUNE
12-Month
High Low Stock
Yld P.E Sales
Div % Ratio hds.
Week's
High Low Last Chg
Net
Yld P.E
Div %

8-1 STOCKING UP

QUICK REFERENCE

TEACHER'S RESOURCES AND ANSWER KEY

Lesson Quiz Answers 8-1, p. 68

Reteaching Activity Answers 8-1, p. 106

Enrichment Activity Answers 8-1, p. 107

EXTENSION ACTIVITIES

Reteaching Activity 8-1, p. 61

Enrichment Activity 8-1, p. 65

Jeff owns some stock that his grandfather bought for him when Jeff was three years old. Granddad retired seven years ago from a large engineering firm that offered stock ownership in the company as part of the employee benefit plan. Granddad had also been able to buy some of the shares for Jeff as a way of saving toward his college costs.

Jeff didn't know much about stocks when Granddad gave him the stock certificates on his 16th birthday, but now that the time is approaching for him to go to college, he has become more interested. Four times a year, the engineering company sends him payments that his father told him are called *dividend checks.* Although the amounts have varied over the years, Jeff notices that they have added significantly to his new checking account.

Stocks are not Jeff's only investment. Ever since he was five years old, Jeff's parents have bought him a $50 U.S. savings bond every month. Jeff is not sure that these investments will cover the high costs of his college education. He is asking himself questions such as the following.

Will he need to continue working this summer tutoring algebra?

Should he find a job that pays more than tutoring?

Should he cash in the savings bonds and invest the money in mutual funds?

Jeff has become interested in mutual funds because his father owns some shares in a mutual fund that he buys through an employee benefit plan. Dad says he is more than willing to let the mutual fund manage his investments, since the daily and weekly changes in the world of business and finance are so varied and complex that he prefers not to be bothered with them.

OBJECTIVES: *In this lesson, we will help Jeff to:*

- *Examine several types of investments, including stocks, bonds, and mutual funds.*
- *Calculate the number of bonds or shares of stock an investor can buy at a specified price.*
- *Read and understand daily stock and bond transactions in the newspaper.*

STOCKS

Jeff's stock certificates, called simply **shares** or **stock**, represent ownership in a **corporation**, a form of business organization that people beginning a business very often choose because it limits the liability of its owners, the **shareholders**. If the corporation is successful, Jeff will take part in its success through **dividends** (the periodic payments that a corporation distributes to its stockholders). The dividends from Granddad's company are declared four times a year.

Granddad and Jeff may also make money on their stock through an increase in the market value of the shares of stock. Such an increase in value, called a **capital gain**, will normally occur if a company increases its profits. On the other hand, if the corporation is not successful, it may have to reduce or eliminate its dividends, and the value of the stock may decrease. Then all

FOCUS

ALGEBRA CONNECTIONS

Ask students which of the graphs below could represent the cost of postage (y-axis) for a given number of ounces that a letter weighs (x-axis). **(Graph 1)**

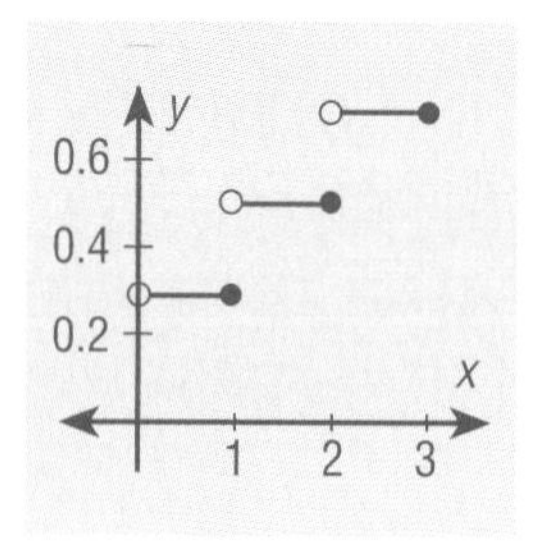

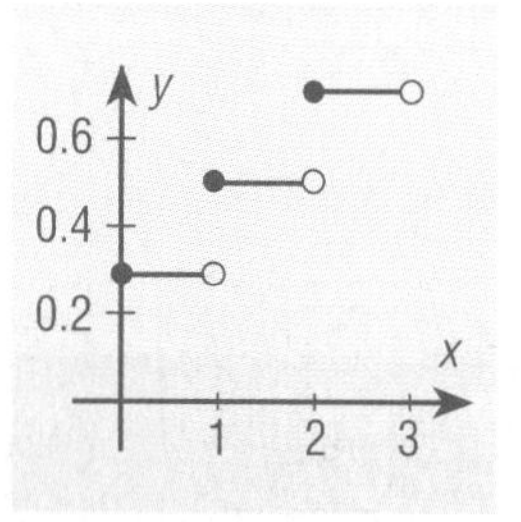

Point out that both graphs represent step functions. The second graph represents the greatest integer function, that is, the function that associates any number with the greatest integer that is less than or equal to that number. Explain that the notation $[x]$ is used to refer to " the greatest integer less than or equal to x." (This concept can be used with Skills 1 and 2, as explained later in the *Focus On Algebra*.)

of the shareholders such as Granddad and Jeff could even lose part or all of their investment.

A corporation always issues *common stock* (like Jeff's) and may also issue *preferred stock.*

Common Stock The owners of **common stock** of a corporation are entitled to elect a board of directors and to receive any dividends that the board declares. However, a corporation's board can choose not to declare a dividend. This may occur if a corporation has poor earnings. Sometimes, even a profitable corporation will pay little or no dividends in order to have funds for research or expansion. In that case, shareholders will expect to see the value of their shares increase to compensate for the lack of dividends.

Preferred Stock Like owners of common stock, owners of **preferred stock** have a share in the ownership of a corporation. However, they usually do not have a right to vote in the election of corporate directors. Unlike the holders of common stock, they are entitled to a specified, fixed dividend. If the company encounters financial difficulty, all preferred shareholders must be paid before any dividends are declared on the common stock.

Newspaper Stock Listings Newspapers print closing stock and bond prices for the previous trading day along with other information about the stock or bond, such as the volume of trading. A typical line from a stock table is shown below in simplified form. It shows how the stock price for Growth Company of America performed the previous day.

52-Week							
High	**Low**	**Stock**	**Div**	**High**	**Low**	**Last**	**Change**
$35\frac{3}{4}$	$19\frac{3}{4}$	GrowCo	1.10	$22\frac{1}{8}$	$21\frac{5}{8}$	22	$+\frac{1}{4}$

The abbreviated name of this company, GrowCo, appears in column three. The first two columns show the highest and lowest prices at which GrowCo traded during the previous 52 weeks ($35.75 and $19.75, respectively). After the name of the stock comes the annual per-share dividend of $1.10. Yesterday's high and low price are shown next. Finally, the day's closing price appears along with the price change for the day. In this case the stock price for GrowCo varied between $21\frac{5}{8}$ and $22\frac{1}{8}$ ($21.625 and $22.125), ending the trading session at $22; this was up $\frac{1}{4}$, or 25 cents, from the previous day's closing price of $21.75.

BONDS

When you buy a **bond,** you are lending money to either a corporation or a government body, such as a city, state, county, or independent agency. The **face value of a bond** is the original loan amount that appears on the bond, usually a multiple of $1000. In return for your money you get three things:

1. The bond issuer's promise to pay you back the entire face value
2. A maturity date, which is the earliest date when the bond will be redeemed
3. A promise to pay interest at a specified interest rate on the face value

For example, if you buy a "\$1000 Joy $8\frac{1}{8}$ 14" 20-year bond from Joy Motor Company in 1994, you will be paid an annual interest rate of $8\frac{1}{8}\%$, or \$81.25 per \$1000, until the year 2014. On the date of maturity (sometime in 2014), Joy will pay back the \$1000. If you decide to sell the bond before the maturity date, it will probably be worth either more or less than the face value, depending on the prevailing interest rate for all similar bonds at the time you choose to sell.

Corporate Bonds Private companies issue **corporate bonds** to raise money for plant expansion or other company purposes. If a corporation's earnings are not great enough to cover its obligations to both stockholders and bondholders, then the bondholders have priority. All bond interest must be paid before any dividends can be paid to the stockholders.

Municipal Bonds States, cities, counties, school districts, and other governmental bodies issue **municipal bonds** to raise money for schools, hospitals, streets, and so forth. They are usually partially or totally exempt from local and federal income taxes.

U.S. Savings Bonds Jeff's parents have purchased over \$6000 in **U.S. savings bonds** for him. Like many other Americans, they take advantage of the payroll savings plan that allows individuals to buy a bond each month through a payroll deduction.

U.S. savings bonds are backed by the power of the federal government to tax and the country's faith in the government to pay its debts. However, such bonds are sometimes criticized because bondholders have to pay for their exceptional safety by accepting a relatively low rate of interest. If a savings bond is lost, stolen, or destroyed, the bondholder may be able to have it replaced by the government if he or she can supply evidence of actually owning the bond.

U.S. savings bonds have one additional important advantage; there are no transaction costs such as commissions to a broker. Unless you are purchasing a large number of bonds that you plan to hold for many years, the large commission that you pay may severely reduce or even wipe out the interest that you get from a bond that is not a savings bond.

MUTUAL FUNDS

The company at which Jeff's father works assists its employees in buying mutual funds. A **mutual fund** is a means of pooling funds with thousands of other people to acquire a wide variety of stocks, bonds, and other types of investments. This way of investing appeals particularly to small investors, such

CRITICAL THINKING

Ask students why it might be more prudent for a person with $12,000 to invest in stocks to do so systematically, investing $500 or $1,000 a month for 24 months or 12 months rather than to invest the entire $12,000 at one time. This procedure, known as "dollar cost averaging," will reduce the danger of losing a significant portion of one's capital caused by "buying at the top of the market."

as Jeff's father, who do not have enough money to **diversify** (buy a well-balanced assortment of investments) or enough knowledge to make sound stock and bond investment decisions.

When investors buy shares from a mutual fund company, the company uses that money to purchase stocks, bonds, and similar **securities.** Thus, instead of investing $1000 in a particular corporation, Jeff's father can buy $1000 worth of shares in a mutual fund and have a stake in perhaps 100 or more different corporations.

BEFORE YOU INVEST . . .

Jeff wants to invest in stocks either directly or indirectly through a mutual fund. However, he also remembers something his grandfather told him: "Buying stocks or mutual fund shares should be part of your long-term investment program, but first be sure that you have a savings account or a reputable money-market account that equals at least six months of your living expenses." Jeff has a savings account at his local bank but was unfamiliar with money-market funds. Jeff's father told him that a **money-market account** was similar to a NOW account that combines the features of a savings and checking account.

A money-market account can be opened with a mutual fund group or with a bank. The interest received on a money-market account is higher than on a savings account, but the account usually has certain restrictions such as requiring any withdrawal to be at least $500 or limiting the number of withdrawals that may be made in a month. If you open such an account at a bank insured by the Federal Deposit Insurance Corporation (FDIC), your account will be insured up to $100,000. However, the interest will be less than you get from a mutual fund group because the money-market fund at the mutual-fund group is not insured by the federal government. These uninsured money-market funds are generally regarded as reasonably safe, but you should investigate the background of the mutual fund group before sending it any money.

Jeff decided to follow his Granddad's advice. In addition to the savings account he already has, he opened a money-market account at the bank and will bring his combined savings in the two accounts to a half year of living expenses before he invests in stocks or a mutual fund.

ALGEBRA *REVIEW*

Solve for x.

1. $2650 = 42.50x$ $x = 62.35$
2. $19.875x = 1700$ $x = 85.53$
3. $25.375x = 8764$ $x = 345.38$
4. $112\frac{1}{2} = x + 108\frac{3}{4}$

   $x = 3.75$ or $3\frac{3}{4}$
5. $x - \frac{7}{8} = 5\frac{1}{4}$

   $x = 6.125$ or $6\frac{1}{8}$
6. $x + 34\frac{5}{8} = 35\frac{1}{2}$

   $x = 0.875$ or $\frac{7}{8}$

Ask Yourself

1. What is your relationship to a company when you buy its stock?
   You share ownership in that corporation.
2. What is your relationship to a company when you buy its bond?
   You are lending money to the company.
3. What is the main difference between buying a mutual fund and buying a stock? A stock is a share in an individual company and a mutual fund is an investment in a wide variety of companies.

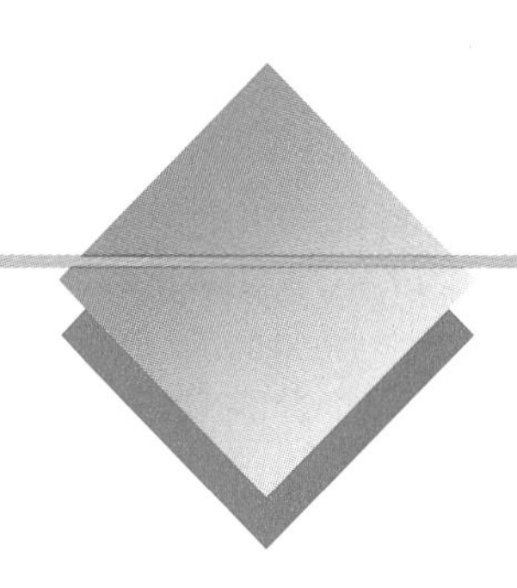

SHARPEN YOUR SKILLS

SKILL 1

Stock prices are quoted in fractional parts of a dollar. This means that

$$\$0.25 = \frac{1}{4} \qquad \$0.50 = \frac{1}{2} \qquad \$0.75 = \frac{3}{4}$$

To determine what $\frac{1}{8}$, $\frac{3}{8}$, $\frac{5}{8}$, and $\frac{7}{8}$ are equivalent to in dollars, divide the numerator by the denominator.

$$\frac{1}{8} = 0.125 \qquad \frac{3}{8} = 0.375 \qquad \frac{5}{8} = 0.625 \qquad \frac{7}{8} = 0.875$$

EXAMPLE 1 Jeff would like to cash in his U.S. savings bonds. He has $2200 in savings bonds that have matured.

QUESTION How many shares could he buy of a stock that is selling for $33\frac{3}{8}$ dollars per share? Disregard transaction costs such as commissions.

SOLUTION

The price of one share of the stock is $33\frac{3}{8}$, or $33.375. To find how many shares of stock Jeff can buy, divide the amount that he has ($2220) by the price per share ($33.375).

$$2200 \div 33.375 = 65.917603$$

Jeff knows that he cannot buy a part of a share of stock, so he drops the decimal portion of the number. Jeff can buy 65 shares of the stock (before transaction costs). The shares will cost 65 • 33.375 = $2169.38. Notice that you must use the fractional portion of a cent to find the cost that is rounded to the nearest cent.

SKILL 2

Bond prices that appear in the newspaper bond tables are somewhat misleading, since they show the true price *divided by 10.* Thus a price of $104\frac{1}{2}$, or 104.5, actually refers to a bond that has a market price of $1045.

EXAMPLE 2 The bond table in the newspaper shows that a high quality corporate bond is selling at $104\frac{1}{2}$.

QUESTION How many corporate bonds could Jeff buy with the $2200 from his U.S. savings bonds? Disregard transaction costs such as commissions.

INSTRUCTION

TEACHING THE LESSON

After reading the opening scenario, students will need help in distinguishing between relative attractiveness of stocks, bonds, and mutual funds. For most people, mutual funds are the best choice; these are covered more thoroughly in Lesson 8-3.

Mention that preferred stocks are rarely suitable for private investors. However, this kind of stock is often bought by corporations because of special tax advantages that only corporations are able to obtain from preferred stocks.

FOCUS ON ALGEBRA

Use the greatest integer function illustrated earlier in *Algebra Connections* to explain Example 1. The notation $[x]$ can be used to emphasize the process. Point out that the answer, $[65.917603]$ is the desired number, 65. Sixty-five is the greatest integer less than or equal to 65.917603. (*NCTM Standard 1, p. 137*)

COMMON ERRORS

In Example 1, some students may unthinkingly round up 65.917603 to 66 "because 65.917603 is closer to 66 than 65." Point out that there are times, as here, when rounding down to the lesser integer is the appropriate step, regardless of whether or not it is the closer integer.

INTERDISCIPLINARY INVESTIGATION

Students should explore the social effects of investing in stocks. They may be surprised to learn that institutions such as pension funds, not individuals, constitute the major influence in today's securities markets.

RETEACHING

Have students solve a few equations of the form $an = b$. (You can do this by having them solve *Algebra Review* exercises 1-3.) Then lead students to see that equations of this form can be used to solve the problems of Skills 1 and 2. Here, a is the price per share, n is the number of shares, and b is the total value of the shares. Similarly, *Algebra Review* exercises 4-6 represent equations of a type that could be used to solve some of the questions of Skill 3.

SOLUTION

To find how many bonds Jeff can buy, divide the amount that he has by the price of one bond.

$2200 \div 1045 = 2.105$ $\quad 104\frac{1}{2} \bullet 10 = 1045$

Jeff cannot buy a part of a bond, but he can buy two bonds. The two bonds will cost $2 \bullet 1045 = \$2090$. Jeff probably would not purchase the bond because of the high commission cost.

SKILL 3

EXAMPLE 3 Jeff has been reading the financial section of the newspaper to learn how the prices of stocks are reported. Following is a partial listing of stock transactions for one day.

52-Week							
High	Low	Stock	Div	High	Low	Last	Change
$53\frac{1}{4}$	$39\frac{1}{2}$	BlcInc	1.25	$41\frac{1}{2}$	$40\frac{3}{4}$	41	$-\frac{1}{4}$
163	$135\frac{1}{4}$	BxCo	2.00	$137\frac{1}{4}$	$136\frac{3}{4}$	$137\frac{1}{4}$	$+\frac{3}{4}$
$4\frac{3}{4}$	$2\frac{1}{4}$	Maldmer		$2\frac{5}{8}$	$2\frac{1}{2}$	$2\frac{5}{8}$	$+\frac{1}{8}$
$12\frac{3}{4}$	$7\frac{5}{8}$	TemCont	0.75	$11\frac{1}{8}$	11	$11\frac{1}{8}$	$+\frac{1}{4}$

QUESTIONS

1. What was the closing price for TemCont?
2. What was the highest price at which BlcInc traded?
3. How much did BxCo stock vary during trading?
4. How does BxCo's closing price compare with its closing price for the previous trading session?
5. What was the closing price of BxCo at the end of the previous trading session?
6. What is the annual dividend per share paid by BlcInc?
7. What would your annual dividend income be from 100 shares of BlcInc?
8. Which companies closed at their high prices for the day? Which did not?
9. What is the highest price at which Maldmer has traded in the previous 52 weeks? What is the lowest price?

SOLUTIONS

1. $11\frac{1}{8}$ (or \$11.125)
2. $41\frac{1}{2}$ (or \$41.50)
3. $137\frac{1}{4} - 136\frac{3}{4} = \frac{1}{2}$
4. $\frac{3}{4}$ of a point (75 cents) higher
5. $137\frac{1}{4} - \frac{3}{4} = 136\frac{1}{2}$
6. \$1.25
7. $100 \bullet 1.25 = \$125$
8. TemCont, BxCo, and Maldmer closed at their high for the day; BlcInc did not.
9. $4\frac{3}{4}$ is the highest price at which Maldmer has traded in the previous 52 weeks; $2\frac{1}{4}$ is the lowest.

In this lesson, the effect of transaction costs has been ignored. These include transfer fees, which are usually very small, and the commission that you pay the broker when you buy or sell. The effect of commissions on your stock price is considered in the next lesson.

ENRICHMENT

Ask interested students to find out the difference between a "closed-end fund" and an "open-end," or mutual fund. (**They will discover that an important difference is that closed-end funds, unlike mutual funds, have a fixed number of shares that can be traded in over-the-counter market or at a stock exchange, just like ordinary stock. Mutual funds continually distribute new shares to its customers who send it money to invest.**)

TRY YOUR SKILLS

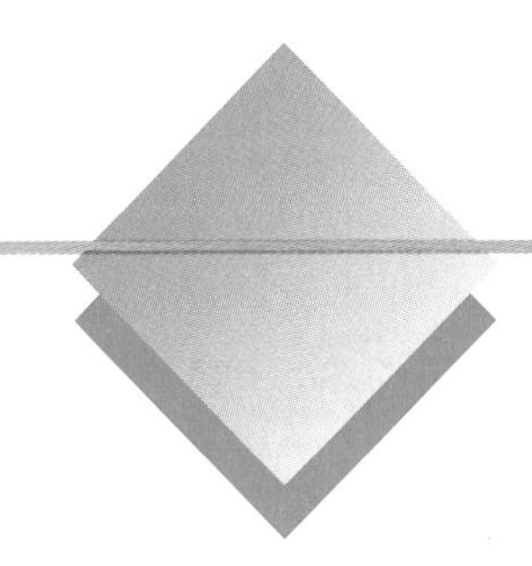

Use the following prices per share to find the number of shares of stock you can afford to purchase if you have \$50,000 to invest in each of the following companies. Also, find the total cost of the shares. Remember that you cannot purchase part of a share of stock. Ignore transaction costs such as brokerage commissions.

	Company	Price per Share	Number of Shares	Total Cost
1.	AdoreMin	$6\frac{1}{2}$	7692	\$49,998.00
2.	ATR	$24\frac{7}{8}$	2010	\$49,998.75
3.	AcnInc	$9\frac{1}{8}$	5479	49,995.88
4.	AcmeAl	$10\frac{3}{4}$	4651	49,998.25

5. How many bonds can you buy with \$50,000 if the price of a bond is $97\frac{1}{2}$?
6. AdoreMin's closing price for the day is $6\frac{5}{8}$, $\frac{1}{2}$ point lower than yesterday. What was yesterday's closing price? $7\frac{1}{8}$
7. The annual dividend for ATR is \$1.33 per share. What is the dividend income on 490 shares? \$651.70

GUIDED PRACTICE

Students are asked to calculate the number of shares of stock that they can afford to buy with a fixed amount of money. Transaction costs, which are disregarded in the exercises of this lesson, will be taken into account in Lesson 8-2.

ADDITIONAL ANSWERS

5. 51

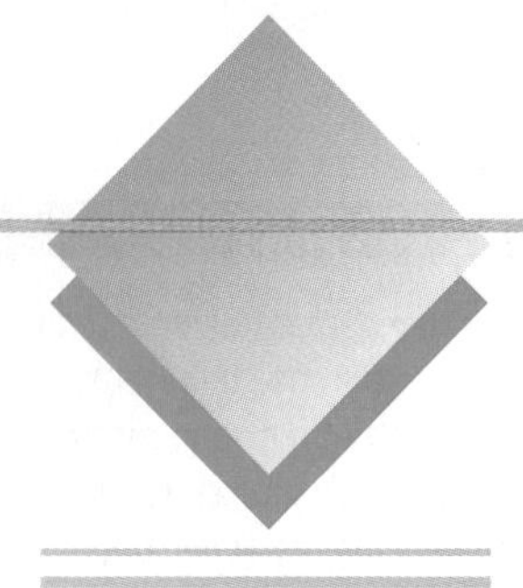

EXERCISE YOUR SKILLS

1. What is the risk involved in buying stocks? See Additional Answers.
2. Who is more likely to realize a capital gain, the owners of common stock or the owners of preferred stock? Why? preferred stock, they get their dividends first
3. For what purposes might your city or town raise money by issuing municipal bonds for sale? schools, hospitals, streets, and so on

Use the following prices per share to find the number of shares of stock you can afford to purchase if you have $50,000 to invest in each of the following companies. Also find the total cost of the shares. Ignore transaction costs such as brokerage commissions.

	Company	Price per Share	Number of Shares	Total Cost
4.	Aber&Son	56	892	$49,952.00
5.	AceText	$8\frac{7}{8}$	5,633	49,992.88
6.	AmerScCo	$10\frac{5}{8}$	4,705	49,990.63
7.	AmerTool	90	555	49,950.00
8.	AyerHrd	$3\frac{1}{4}$	15,384	49,998.00
9.	Babbage	$74\frac{1}{2}$	671	49,989.50
10.	BLTInc	$53\frac{5}{8}$	932	49,978.50

11. Jeff has $3400 that he can invest in either a growth stock that is trading at $70 per share or a high-quality bond that pays annual interest of 5.8% and matures in the year 2010. Which would be a more suitable investment for him? Explain your answer. See Additional Answers.

Use the following portion of a table of stock transactions to answer Exercises 12–20.

52-Week							
High	**Low**	**Stock**	**Div**	**High**	**Low**	**Last**	**Change**
$57\frac{3}{4}$	$40\frac{3}{4}$	LollPp	1.42	$41\frac{3}{8}$	41	$41\frac{1}{4}$	$-\frac{1}{8}$
$49\frac{7}{8}$	$29\frac{3}{8}$	LukInd	0.97	30	$29\frac{5}{8}$	$29\frac{5}{8}$	—
$91\frac{1}{8}$	67	MamaMi	2.24	$81\frac{3}{4}$	$80\frac{1}{4}$	$81\frac{1}{4}$	$-\frac{1}{4}$
24	$15\frac{5}{8}$	NajiCo	0.46	$17\frac{3}{4}$	$17\frac{3}{8}$	$17\frac{5}{8}$	$-\frac{1}{8}$
$36\frac{3}{8}$	22	NolInc	0.64	$24\frac{1}{4}$	$23\frac{7}{8}$	$24\frac{1}{4}$	$+\frac{1}{4}$

12. Which stock finished up for the day? NolInc.
13. Which stock closed at the same price at which it closed at the previous trading session? LukInd
14. How much did NajiCo vary during the day? $\frac{3}{8}$
15. Which stock varied by exactly $1\frac{1}{2}$ during the day? MamaMi
16. What is the difference between the closing price of NajiCo for the day and the closing price for the previous trading session? down 0.125 or $-\frac{1}{8}$

KEY TERMS

bond
capital gain
common stock
corporate bonds
corporation
diversify
dividends
face value of a bond
money-market account
municipal bonds
mutual fund
preferred stock
securities
shares
shareholders
stock
U.S. savings bonds

INDEPENDENT PRACTICE

ASSIGNMENT GUIDE

Exercises 1-3 are intended to focus students' attention on the role of stocks and bonds on individuals and society in general. Exercises 4-10 and 12-20 provide practice in calculating the monetary results of several kinds of stock transactions. Exercise 11 asks students to evaluate the appropriateness of a particular investment decision.

ADDITIONAL ANSWERS

1. If the company is not successful, part or all of the investment could be lost.
11. The bond would be a safer, long-term investment if he is a beginning investor.

17. By how much does LollPp's closing price differ from its high for the last 52 weeks? $16\frac{1}{2}$

18. Which stock closed exactly 2 points above its low for the last 52 weeks?

19. If you had 100 shares of NolInc, how much would you receive in dividends in 1 year? $64

20. Which stock pays the highest dividend? MamaMi

MIXED REVIEW

1-1 A restaurant is advertising for a cashier in the local newspaper. The job is for an 8-hour day, 5 days per week. There is a 2-week paid vacation each year. Find the weekly and yearly earnings for the given hourly rate.

1. $7.50 $300; $15,600
2. $8.25 $330; $17,160
3. $9.00 $360; $18,720

3-1 4. Find the simple interest earned in 3 years on $890.00 at the annual rate of 7%. $186.90

6-4 Use a spreadsheet program to fill in the missing entries from the table below that shows the activity in Victor's VISA account over a period of 3 months. The account carries a monthly interest charge of 1.5% and Victor makes monthly payments of 10% of the amount owed to the nearest dollar.

	Month	Purchases	Balance	Interest	Amount Owed	Payment
5.	1	0.00	$624.50	$9.37	$633.87	$63
6.	2	$92.00	662.87	9.94	672.81	67
7.	3	0.00	605.81	9.09	614.90	61

8. Find the total interest paid in the 3-month period. $28.40

5-5 9. Alice is considering renting a piano on a rent-to-own credit plan that allows her to rent the piano for $200 per month for 24 months. If she decides to buy the piano, the rental fees will be applied to cover the entire purchase price. How much will the piano cost under this plan?

5-5 10. Suppose that Alice purchases the piano of Exercise 9 for $4,000 on an installment loan with monthly payments at 10% annual interest over 2 years. Is this a better deal than the rent-to-own plan? Explain.

1-4 Use the Income Tax Withholding Table in the Reference Section to find the take-home pay for the given wages and withholding allowances. Assume that the person is single. The FICA withholding is 7.65% of gross pay.

	Monthly Salary	Withholding Allowances	Take-home Pay
11.	$1540	0	$1223.19
12.	2750	1	2112.63
13.	4130	3	3105.06

CLOSURE

Ask students what they think might be the special difficulties involved in investing in stocks and bonds. They should understand that at some brokerage houses, the commission cost on the purchase of a small number of bonds may severely reduce the expected return. They should also know that when directly buying common stocks, an investor should usually reduce his or her risk by diversifying purchases among a number of different companies that he or she knows well.

LESSON QUIZ

The opening price of a certain stock was $35\frac{1}{2}$. It reached a daily high of $36\frac{3}{8}$ and closed the day at its low price of $35\frac{1}{8}$.

1. What is the most number of shares that could have been bought with $20,000? Ignore transaction costs. (**569**)
2. Why is it incorrect to conclude that the daily change in the stock's price is $35\frac{1}{2} - 35\frac{1}{8}$, or a drop of $\frac{3}{8}$? (**Yesterday's closing price is not necessarily the same as today's opening price.**)

ADDITIONAL ANSWERS (EXERCISE YOUR SKILLS)

18. NajiCo

(MIXED REVIEW)

9. $4800
10. yes, total cost is $4429.91 which is less than $4800

8-2 COMMISSIONS AND STOCKS, BONDS, AND MUTUAL FUNDS

QUICK REFERENCE

TEACHER'S RESOURCES AND ANSWER KEY

Lesson Quiz Answers 8-2, p. 68

Reteaching Activity Answers 8-2, p. 106

Enrichment Activity Answers 8-2, p. 107

EXTENSION ACTIVITIES

Reteaching Activity 8-2, p. 62

Enrichment Activity 8-2, p. 66

Maria has continued to save money. Recently, she has been wondering whether she should invest in the stock market or a mutual fund to make more money than she can get in a savings account. Maria has a good friend, Cal, whose father, Clarence Sr., has owned thousands of shares of stock in various companies.

Sometimes Clarence Sr. has done really well in the market, and sometimes he hasn't. Cal still talks about October 19, 1987, known as Black Monday in business circles. The Dow Jones average dropped 508 points that day, losing 22.6% of its value. Cal's father lost money in that crash. The market did rebound eventually and had record highs within a couple of years. But that was too late for Clarence Sr.

As a result of financial difficulties caused by the crash, Cal's family had to sell its large home in the expensive part of town and buy a much smaller house in a suburb farther away from the city. Clarence Sr. also sold his luxury car and bought a used one inexpensively from the car dealership that he owned. Cal himself had to give up his vision of having his own car when he reached 17.

Maria wonders how investing in stocks can cause such an upheaval in a family. She is aware of how stock prices and mutual fund prices can vary but believes that she could choose some fine companies or a good fund to invest in if she had the money to invest. Maria will find out how stocks and mutual funds are bought and sold as well as other things that she needs to know if she is considering investing real money.

OBJECTIVES: *In this lesson, we will help Maria to:*

- *Discover what factors must be considered before deciding to buy stocks or a mutual fund.*
- *Understand the process that takes place when stocks are traded.*
- *Calculate profit or loss from buying and selling shares of stock.*
- *Understand how transaction costs affect the value of stock or mutual fund investments.*

FOCUS

ALGEBRA CONNECTIONS

Ask students how well off an investor will be who loses 50% of the initial value of an investment, reinvests what he has left in a new investment, and then manages to increase that amount by 50%. (**Observe students who think that two 50% changes offset one another and that the investor has thereby regained all of his previous loss. Point out that to regain his loss, the investor will need to increase the shrunken capital by 100%.**)

SHOULD I INVEST IN THE STOCK MARKET?

Clarence Sr. watched his acquaintances acquire large sums of money by "playing the market." Unfortunately for him, somewhere along the line he crossed over the line between *investing* in stocks and *gambling* with money his family needed to secure its financial independence.

Cal's family always knew that stock ownership involved some risk. Perhaps Clarence Sr. was careful at first with his choice of stocks. Then, as those choices paid well and he was able to make bigger profits, he was swept along by the tide of his increasing fortune, and he began to take bigger risks. When the market dropped so suddenly in October 1987, some of the riskier companies lost so much of their value that he was forced to sell many of his holdings at a loss.

It is hard to define the line between investing safely and gambling with money needed for essentials; this line is defined differently in every family. You should be aware that carefully conducted studies have shown that very few people successfully "play the market," that is, very few people can *consistently* buy stocks for a few weeks or months and then sell the stocks at a profit.

On the other hand, if you carefully choose a group of stocks for the fundamental value of the underlying companies or the ability of those companies to sustain long-term growth, you may, *if you hold such stocks over several years,* be able to build up your family's wealth, especially if the overall economy is healthy. In the past, stocks in well-managed companies have also been a good buffer against inflation when held for many decades. Moreover, stocks, unlike some other forms of investment, are easy to liquidate if you need cash.

Clarence Sr. was lucky that he could still afford to buy a house and car. He would say that you should consider all of the following points before you invest in stocks.

1. Be sure you can afford to lose what you risk.
2. Determine specific long-term and short-term investment goals.
3. Ask yourself if you have the emotional temperament suited to handling the ups and downs of investing.
4. Plan to spend time as well as money on the companies you invest in.
5. Choose an experienced, reputable broker to give you guidance and advice.
6. Do not expect too much too soon.
7. After you have set your goals and objectives, stick to them.
8. Be aware that you are buying part of a particular company, not a lottery ticket on the next stock market move.
9. Have an overall family investment plan to protect you from falling into hit-or-miss investing.

CRITICAL THINKING

Ask students what they think might be the difference between the following two basic approaches to investing in stocks;

a. investing for growth

b. investing for value

These two approaches were briefly alluded to in the scenario. Companies with good growth prospects are expected to have steadily rising earnings over several years. Value investing concentrates on companies that are selling below their true worth, often because of difficulties that are temporary in nature.

THE PROCESS OF BUYING AND SELLING

If Maria decides that she can follow the above guidelines, then she may open an account with a **broker,** that is, a salesperson who specializes in buying and selling stocks and bonds. When Maria buys or sells a stock, she must pay the broker a **commission.** Commission rates vary according to the brokerage house and the size of the trade.

If her brokerage firm is a **full-service firm,** then it will issue reports on individual companies, and its brokers will offer opinions on individual stocks. If she has an account with such a firm, she can get its reports and study them. However, the quality of advice in these reports and from individual brokers will vary from very good to poor. To find a competent broker, she should do what she would do to find a competent physician or lawyer: obtain some names from a person whose advice she trusts. In this case, Maria decides to contact an older friend who has years of investing experience.

If Maria had preferred to do her own research, she could have chosen to open an account with a **discount broker,** who charges much lower commissions than a full-service broker but provides no investment advice at all.

Many investors purchase a hundred shares of a particular stock, which is called a **round lot.** A trade of fewer than 100 shares is called an **odd lot.** Because odd lots can be inconvenient to trade, Maria may have to pay a higher broker commission for an odd lot than for a round lot.

Stocks are traded on national and regional stock exchanges. The largest of these is the *New York Stock Exchange,* where over 84 percent of all listed securities are bought and sold. The *American Stock Exchange* is the other national exchange in the United States. Many stocks trade not on an exchange but rather in the **over-the-counter market,** a national network of dealers and brokers who trade among themselves by telephone, telegraph, or teletype.

Stocks are bought and sold on a security exchange by the auction method. A stock may be offered for sale at a certain price, or someone may bid for the same stock at a lower price. A sale is made when either the buyer or the seller meets the other's price. Maria can instruct her broker to purchase or sell shares of stock for her with a limit on the price that she will accept.

Ask Yourself

1. How does investing in stocks differ from "playing the market"?
   Investing does not involve gambling.
2. What are some guidelines that you should follow before you decide to invest in stocks? Which guideline do you think is the most important for you? Be sure you can afford to lose what you invest, have investment goals, and don't expect too much too soon.
3. Why should investors be careful about following the advice of a stockbroker? The quality of advice can vary, and a broker works on commission.

ALGEBRA *REVIEW*

One way to think about **absolute value** is to look at the number line. The distance from zero to any point on the number line is positive.

Find each distance.

Examples: From 0 to A is 4
From 0 to B is 4

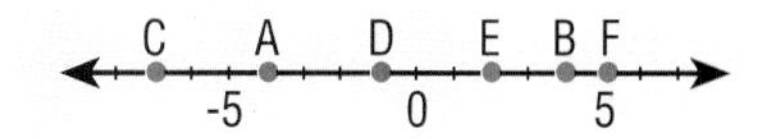

1. From 0 to C
   7
2. From 0 to D
   1
3. From 0 to E
   2
4. From 0 to F
   5

Find the absolute value of each change in price.

5. Old price: \$100
   New price: \$85
   \$15
6. Old price: \$85
   New price: \$100
   \$15
7. Old price: $65x$
   New price: $95x$
   $30x$
8. Old price: $5a$
   New price: $1.5a$
   $3.5a$

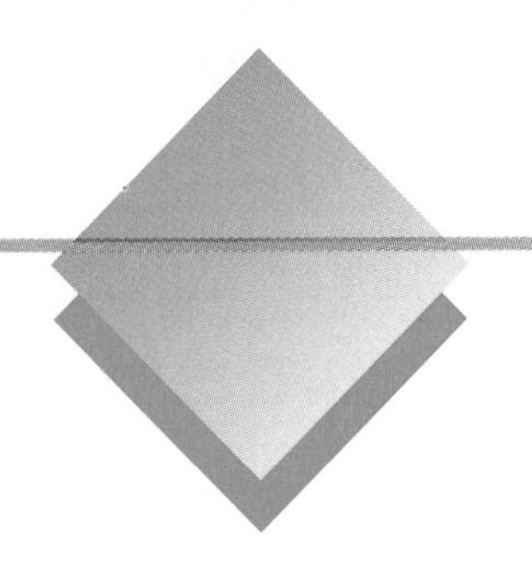

SHARPEN YOUR SKILLS

SKILL 1

Commission rates are charged on a percent-of-sale basis. In other words, in addition to the cost of the stocks or bonds, you must pay a percentage of the cost for commission. To determine the cost of a commission, you multiply the number of shares, the price of one share, and the commission rate.

INSTRUCTION

TEACHING THE LESSON

Encourage students to study major points of the opening scenario carefully. Of particular importance is point 8, that investing money is buying part of an actual company with some combination of positive and negative attributes; it is not gambling on an abstraction that rises or falls in value for no particular reason. Some very successful investors have suggested that beginners pay particular attention to products and services that they are quite familiar with in their own lives and to learn as much as possible about the companies behind those products and services. (Most companies will send financial and other information to anyone who phones in a request.)

> **Commission Cost Formula**
>
> $c = npr$ where c = cost of commission
> n = number of shares
> p = price of one share
> r = commission rate

EXAMPLE 1 Maria would like to buy 250 shares of International Enterprises, Inc. at $17\frac{3}{4}$ per share. She wants to compare the commissions charged by Osgood and Company, a full-service broker with a commission rate of 2%, and Willcox and Company, a discount broker with a commission rate of 1%.

QUESTION How much commission will each broker charge, and what will the total cost be?

SOLUTION

Osgood and Company

$c = npr$
$c = 250 \bullet 17.75 \bullet 0.02$
$c = 88.75$

The commission is $88.75.

Willcox and Company

$c = npr$
$c = 250 \bullet 17.75 \bullet 0.01$
$c = 44.375$

The commission is $44.38.

The purchase price of the stock is 250 • 17.75 = 4437.50. Add the commission to $4437.50 to find the total cost.

$88.75 + 4437.50 = 4526.25$ $\qquad$ $44.38 + 4437.50 = 4481.88$

The total cost at Osgood is $4526.25, and the total cost at Willcox is $4481.88. Maria will pay 88.75 − 44.38 = $44.37 more at Osgood.

If you decide to sell a stock, you will pay a brokerage commission, just as you did when you bought the stock. However, the selling commission and buying commission will normally not be equal since the selling price usually differs from the purchase price.

SKILL 2

When you sell stock that you have owned for a period of time, your stock may have increased in value. This increase is called a *capital gain.* Usually, you will have to pay income taxes on the capital gain on a stock that you have sold. If the value of your stock decreases and you sell, you will have a loss.

EXAMPLE 2 Maria would like to know what the capital gain would be on the 250 shares of International Enterprises. Assume that she purchased the stock from the full-service broker.

QUESTION What would be Maria's capital gain if the market price of the stock increases to $72\frac{1}{4}$ over the next 10 years?

SOLUTION

Multiply the new price of one share by the number of shares.

$72.25 \bullet 250 = 18{,}062.50$

The capital gain is found by subtracting the total purchase price from the total value of the stock after 10 years.

$18{,}062.50 - 4526.25 = 13{,}536.25$

Maria's capital gain after 10 years is $13,536.25. If she does not sell, her capital gain may increase or decrease in value. If she does decide to sell, then the amount that will be subject to tax is the capital gain reduced by both her buying commision and her selling commission.

COMMON ERRORS

In Example 2 and the related exercises, there is no selling commission because the stock is not sold. If the stock were sold, the commission would, of course, be subtracted from the selling price to obtain the net price. A common careless error is to add the commission to the selling price, unthinkingly following the operation used in finding the net cost.

SKILL 3

The rate of increase or decrease is the *change in price* divided by the original price. A formula for the rate of increase or decrease involves *absolute value.* To find the percent of increase or decrease, divide the rate by 100.

Rate-of-Change Formula

$$r = \frac{|P_n - P_o|}{P_o}$$

where r = rate of increase or decrease
P_o = original price
P_n = new price

FOCUS ON ALGEBRA

The Rate-of-Change Formula can aid students in calculating the perçent increase or decrease in the value of an investment. Mention that the denominator is always the original price, regardless of whether it is greater than or less than the new value. (*NCTM Standard 5, p. 150*)

EXAMPLE 3 Maria would like to know the percent of decrease or increase when the price of a share of stock varies.

QUESTIONS

1. If the market price of a share of stock drops from $45 to $40, what is the percent of decrease?
2. If the market price of a share of stock increases from $45 to $60, what is the percent of increase?

AT-RISK STUDENTS

Have students obtain commission schedules from several full-service brokers and discount brokers. Then have them compare the actual commissions that would be paid to each broker for a fictitious purchase of an actual company, such as "100 shares of XYZ company at $50 a share."

SOLUTIONS

1. $r = \frac{|P_n - P_o|}{P_o}$ Use the rate-of-change formula.

$= \frac{|40 - 45|}{45}$ $P_n = 40$, $P_o = 45$

$= \frac{5}{45}$ $|40 - 45| = |-5| = 5$

$= 0.11$

The percent of decrease is 11%.

2. $r = \frac{|P_n - P_o|}{P_o}$ Use the rate-of-change formula.

$= \frac{|60 - 45|}{45}$ $P_n = 60$, $P_o = 45$

$= \frac{15}{45}$ $|60 - 45| = |15| = 15$

$= 0.33$

The percent of increase is 33%.

SKILL 4

Maria has learned that commission rates can vary widely on the purchase of stocks or mutual funds. Some brokers charge as little as 1%, while mutual funds have commissions that vary from 0 to over 8%.

EXAMPLE 4 Maria decided to compare how various commission rates affect an investment of $1000.

QUESTION How does the commission rate affect the purchase price of a stock or mutual fund?

SOLUTION

She wrote an equation for the relationship between the total cost T of a stock or mutual fund purchase and the commission rate r.

$$\text{Total price} = \text{price} + \text{commission}$$
$$T = 1000 + 1000r$$

Next, she used a spreadsheet program to calculate the total price T for commission rates r ranging between 0 (no commission) and 8%.

r	0	0.005	0.01	0.02	0.03	0.04	0.05	0.06	0.07	0.08
T	1000	1005	1010	1020	1030	1040	1050	1060	1070	1080

Using the spreadsheet's graphing program, Maria graphed the values with r on the horizontal axis and T on the vertical axis. The graph showed clearly that the commission rate and total purchase price had a *linear* relationship. Whenever the commission rate increases by 1 percentage point, her cost also increases, by $10.

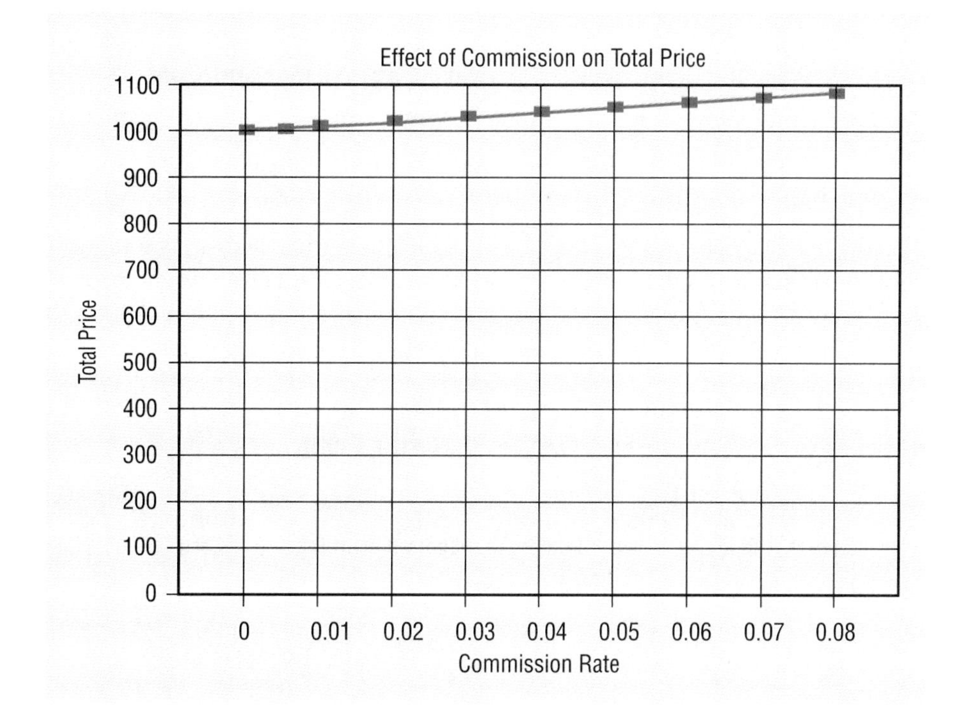

TECHNOLOGY HINT

The linear relationship illustrated in Example 4 of Skill 4 can also be shown using a graphing calculator. To bring out the linearity more clearly, have students set the vertical range between 1000 and 1100.

RETEACHING

In reteaching Skill 2, consider the case of both a capital loss and also the case of no change at all in the market price. Some students may not fully appreciate that, because of the buying commission, there will be a loss even if the market price remains unchanged after the purchase.

ENRICHMENT

Some brokers have a commission schedule that involves a formula such as "$22 + 1.4% of the purchase price." In addition, there is usually an overriding minimum commission. Have students find the commission and percent commission they would pay on the following stock purchases at a firm that uses the above formula and has a minimum commission charge of $35.

a. $250 (**$35; 14%**)
b. $1000 (**$36; 3.6%**)
c. $2500 (**$57; 2.3%**)

TRY YOUR SKILLS

GUIDED PRACTICE

Exercises 1-11 are similar to some of the exercises of Lesson 8-1 except that the holding period is much longer and the cost of commission is now taken into account.

ADDITIONAL ANSWERS

5. $95.06; $6432.56
6. $213.47; $14,444.72
7. $21.94; $1484.44

Find the cost of these shares of stock before commission.

1. 300 shares at $21\frac{1}{8}$ $6337.50
2. 225 shares at $63\frac{1}{4}$ $14,231.25
3. 45 shares at $32\frac{1}{2}$ $1462.50
4. 40 shares at $16\frac{5}{8}$ $665.00

5.–7. For Exercises 1–3, find the commission and the total cost if the commission rate is 1.5%. See Additional Answers.

Suppose that the 40 shares of Exercise 4 are held for 5 years. Find the capital gain or capital loss in each case. Ignore the effect of commissions.

8. Market price is $\$42\frac{7}{8}$ $1050 gain
9. Market price is $14 $105 loss

10.–11. Find the percent gain or loss in market price for the stocks of Exercises 8 and 9. **10.** 157.9% gain **11.** 15.79% loss

Use a spreadsheet graphing program to show how the total cost of a mutual fund purchase is related to the commission rate for commission rates of 0 to 8% when the amount invested is as given.

12. $500 See Additional Answers.
13. $2000

EXERCISE YOUR SKILLS

KEY TERMS

broker
commission
discount broker
full-service broker
odd lot
over-the-counter market
round lot

ADDITIONAL ANSWERS

1. without patience, money may foolishly be lost
2. money not needed for living expenses
3. Answers may vary.

1. How does temperament affect the way a person handles investments?
2. How should an investor determine how much money to invest in stocks?
3. Why might it be wiser for some people to invest in a completely safe security, such as U.S. savings bonds, rather than stocks?

Find the purchase price of these shares of stock at the prices shown. Ignore the effect of commissions. Find the capital gain or capital loss at the current market value one year later by subtracting the purchase price from the current market value. Note that if the difference is a negative number, then there is a loss. Calculate the percent of increase or decrease. Put parentheses around a decrease.

			Original Purchase		Current Market			
	Company	Number of Shares	Price/Share	Total Price	Price/Share	Total Price	Gain or Loss	Percent of Change
4.	AP&P	306	$41\frac{7}{8}$	$12,813.75	$43\frac{3}{8}$	$ 13,272.75	$459 gain	3.6%
5.	Zola	215	$10\frac{1}{4}$	2,203.75	$16\frac{1}{2}$	3,547.50	1343.75 gain	61%
6.	ABM	400	$42\frac{1}{2}$	17,000.00	$65\frac{7}{8}$	26,350.00	9350 gain	55%
7.	Lands	6000	20	120,000.00	$18\frac{1}{4}$	109,500.00	10,500 loss	8.8%
8.	Dr. Pop	150	$29\frac{3}{8}$	4,406.25	$26\frac{3}{8}$	3,956.25	450 loss	10.2%
9.	Doledo	310	$9\frac{7}{8}$	3,061.25	$17\frac{1}{4}$	5,347.50	2286.25 gain	74.7%
10.	Binbury	55	$65\frac{5}{8}$	3,609.38	$74\frac{1}{4}$	4,083.75	474.37 gain	13.1%

For the following companies from Exercises 4–10, find the commission cost and the total original purchase cost given the commission rates.

11. 306 AP&P, 2% $256.28; $13,070.03 **12.** 215 Zola, 1% $22.04; $2225.79

13. 400 ABM, 1.5% $255; $17,255 **14.** 6000 Lands, 0.5% $600; $120,600

Maria is reviewing the prospectuses (informational booklets) of two mutual funds: a "load" fund and a "no-load" fund. The load fund charges a commission of 8%; the no-load fund charges no commission. The following table assumes that a $1000 investment in each fund will vary in value as shown over the next 10 years and end up with a value of $4050.

				Percent Return	
Elapsed Time	**Cost of No-Load Fund**	**Cost of Load Fund**	**Current Value of Fund**	**No-Load Fund**	**Load Fund**
0 years	$1000	$1080	$1000	0%	−7%
$\frac{1}{2}$ year	1000	1080	1100	10%	2%
1 year	1000	1080	950	−5%	−12%
2 years	1000	1080	1300	30%	20%
4 years	1000	1080	1750	75%	62%
6 years	1000	1080	2300	130%	113%
8 years	1000	1080	3200	220%	196%
10 years	1000	1080	4050	305%	275%

15. Find the missing table values. Round percents to the nearest whole percent.

16. Draw a line graph that shows the 10-year performance of the no-load fund. Use the x axis for the number of years from 0 to 10, and use the y axis for the percent of gain or loss from 10% loss to 310% gain.

17. Draw a graph that shows the ten-year performance of the load fund on the same x and y axes as the graph of Exercise 16. See Additional Answers.

18. Which fund had the better 10-year performance? How much better did that fund perform in 10 years? The No-Load Fund performed better at 305% increase over a 275% increase for the Load Fund

ASSIGNMENT GUIDE

Exercises 1-3 can be used to initiate a discussion about who should consider investing in stocks and who should not. Exercises 4-18 consider the kinds of results that can actually occur when a person buys stocks or a mutual fund. In particular, Exercises 15-18 illustrate the drag that an initial load, or commission can have on the long-range performance of an investment (see Exercise 18).

Ask students whether they would prefer to buy stocks directly or indirectly through mutual funds. What element of one's personality would one have to consider when choosing between the two approaches to stock investing?

LESSON QUIZ

One hundred shares of a stock were bought 5 years ago at $39.50 a share. The purchase commission was 1.6% of the price. Find the current capital gain or loss in each case.

1. The price is now $62\frac{7}{8}$. (**$2274.30 gain**)
2. The price is now $38. (**$213.20 loss**)

MIXED REVIEW

1-4 Find the amount that each company must pay each year for the indicated fringe benefits.

1. Salary: $38,000
7.65% for FICA taxes
4% of gross pay for retirement
$1800 per year for training $6227

2. Salary: $25,000
7.65% for FICA taxes
$75 per month for health insurance
$80 per month for life insurance $3772.50

3-2 Find the new balance B for the indicated investments. Use the compound interest formula $B = P(1 + r)^n$, where P is the initial principal, r is the interest rate for the period, and n is the number of periods.

ADDITIONAL ANSWERS

3. $1469.33

3. Starting principal: $1000; annual compounding at 8% for 5 years

4. Starting principal: $12,500; semiannual compounding at 7% per year for 12 years $28,541.61

5-1 **5.** Use the monthly payment formula $M = \dfrac{Pr(1 + r)^n}{(1 + r)^n - 1}$ to find the monthly payment for a car loan of $8000 at 9% a year for 4 years. In the formula, r represents the monthly interest rate and n represents the total number of payments. $199.08

6-2 **6.** How long will it take to pay off a credit card balance of $2500 with monthly payments of $150 if the APR is 9%? Express your answer as a whole number of months. 18 mo

6-5 The following chart shows the payments in a credit card account during the month of April. The monthly interest rate is 1.5%.

Dates	Payment	Balance at End of Day	Number of Days	Sum of Daily Balances
4/1–4/29	0.00	$406.00	29	$11,774.00
4/30	$95.00	311.00	1	311.00
		Total	30	$12,085.00

7. Determine the average daily balance. $402.83

8. Determine the finance charge on April 30. $6.04

9. Determine the ending balance for April. $317.04

6-5 **10.** Your balance from 1/1 to 1/15 was $450. Then you made a payment of $100 on 1/16. You made no further charges or payments for the remainder of the month. Determine your average daily balance for January. How much did you pay in interest if your card carries a monthly finance charge of 1.75%? $398.39; $6.97

8-3 INVESTMENT PROS AND CONS

QUICK REFERENCE

TEACHER'S RESOURCES AND ANSWER KEY

Lesson Quiz Answers 8-3, p. 70

Reteaching Activity Answers 8-3, p. 106

Enrichment Activity Answers 8-3, p. 107

EXTENSION ACTIVITIES

Reteaching Activity 8-3, p. 63

Enrichment Activity 8-3, p. 67

During the many months that Maria was saving up for her car, Nelson kept track of her savings account and noticed how steadily it was growing. After hearing his explanation of the effects of compound interest, she became as intrigued as he was by the idea that compound interest can make invested money appear to take on a life of its own and seemingly grow without any outside help.

Nelson knows that Maria is learning about the stock market and mutual funds and that she is thinking about the risks involved in investing real money in one of these areas. They have spent many hours discussing the problem Clarence Sr. has been having since the value of his investments took such a beating on Black Monday.

Having watched what happened to Cal and his family, Nelson and Maria understand that a wise investor must not view investments as a gamble and must not jump into any investment scheme that promises large, fast profits. If such an "opportunity" seems too good to be true, it probably is!

OBJECTIVES: *In this lesson, we will help Nelson and Maria to:*

- *Examine ways to avoid risky investment schemes.*
- *Graph the change in the market value of an investment over a long period of time.*
- *Use a line of best fit to predict the approximate future value of an investment.*

INVESTMENT BOOBY TRAPS

Nelson has seen ads such as these in the newspaper:

"Worried about inflation? Diamonds are an investor's best friend."

"Make quick money! For only \$1000, you can earn \$16,000."

"Let me show you how I can quadruple the value of your stock portfolio in 18 months!"

Investment come-ons such as these operate on the principle that greed will overcome common sense. They often separate the investor from his or her money without anything of value being provided in return. However, in the past the situation was far worse. For example, unlike today, stock prices were frequently manipulated by those in a position to distribute misleading information about stocks.

After the stock market crash of 1929 the U.S. Congress established the **Securities and Exchange Commission (SEC),** an independent federal agency whose purpose is to prevent the return of the unsound stock selling practices and schemes that were common before the crash. The commission has two basic responsibilities.

- To require companies that offer securities for sale in more than one state to file with the commission and to make complete and accurate information about the company available to investors
- To protect investors against misrepresentation and fraud in the issuance and sale of securities

The SEC requires companies, including mutual fund companies, to disclose facts that are essential for an investor to make an informed investment decision. However, this agency cannot protect the public from making poor investment decisions. To become an investor who is not taken advantage of, you should take to heart the following tips from the SEC.

- Do not deal with security firms or salespeople with whom you are not personally familiar. Consult people whose knowledge about such matters you trust; perhaps your local banker can play this role.
- Be sure you understand the risk that you are assuming when you buy stock or mutual fund shares. Risk cannot be eliminated, but it can be managed; be sure that the level of risk is not too great for the gain you expect to achieve.
- Tell any salesperson who is promising a specific investment result to put the recommendations and expectations in writing.
- Give at least as much consideration to your investments as you would to any other valuable asset or property.
- Do not play the market, that is, do not buy a stock with the intention of selling with a profit in only a few weeks or months.

FOCUS

ALGEBRA CONNECTIONS

Ask students for what value of x the graph of $y = 2x + 1$ intersects the y-axis ($x = 0$). Mention that the y-value at that point of intersection is called the y-intercept. If time allows, this example can lead to a general discussion of the slope-intercept form of a linear equation (as background for Skill 2).

CRITICAL THINKING

After students have read the opening scenario, ask them what additional steps they can think of to avoid being caught by an "investment booby trap." Remind them that even after choosing a broker, they should exercise caution in managing their financial affairs. Examine monthly or quarterly statements for accuracy.

- Do not listen to high-pressure sales talk.
- Beware of tips, rumors, and promises of spectacular profits. A deal that sounds too good to be true probably *is* too good to be true.

DOES THE BROKER HAVE A DEAL FOR YOU!

While the great majority of brokers are very honest, you may occasionally run into a con artist who is interested only in taking your money. Be very wary of a salesperson who does not have a reputation that you can check on or who does any one of the following.

- Plugs one certain stock or mutual fund and refuses to sell you anything else
- Promises a quick, sure profit
- Claims to have inside information
- Urges you to hurry "before the price goes up"

If you wind up falling for a slick operator's con game at some time during your life, take comfort from the fact that almost everyone gets stung in that way at least once. Finally, when you do decide to invest, never risk more than you could feel comfortable losing. Otherwise, you may lose money even from a well-chosen investment. This might very well happen if a temporary (but severe) drop in its price were to frighten you into selling a basically sound investment at a loss.

Ask Yourself

1. What is the Securities and Exchange Commission?
   See Additional Answers.
2. What are the basic responsibilities of the U.S. Securities and Exchange Commission?
   To require companies that sell securities in more than one state to file with the commission; to protect investors against fraud

ADDITIONAL ANSWERS

1. An independent federal agency to protect against unsound stock selling

ALGEBRA *REVIEW*

Select the equation that *best fits* the points in each of the following graphs. Explain why the equation fits best and the others do not fit. Remember that none of the equations will fit perfectly.

1. a. $y = x + 1$
   b. $y = 2x$
   c. $y = x + 2$
   c

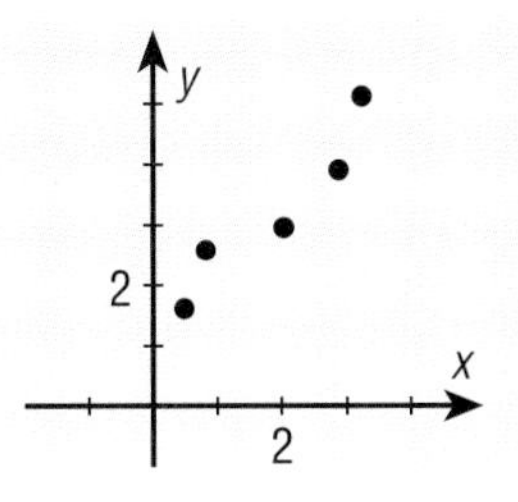

2. a. $y = x$
   b. $y = 0.5x - 1$
   c. $y = x - 1$
   b

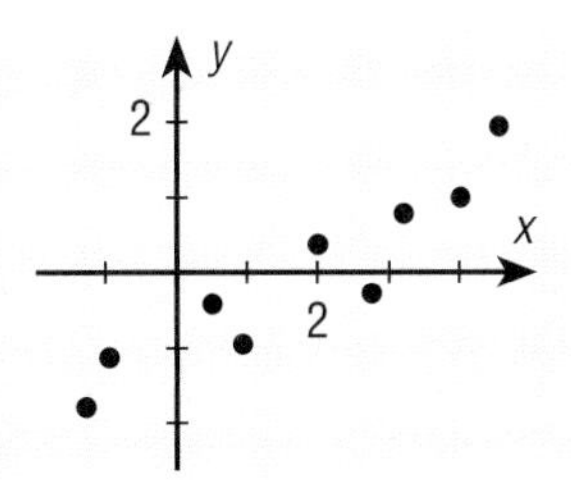

3. a. $y = -x$
   b. $y = x$
   c. $y = -x + 1$
   a

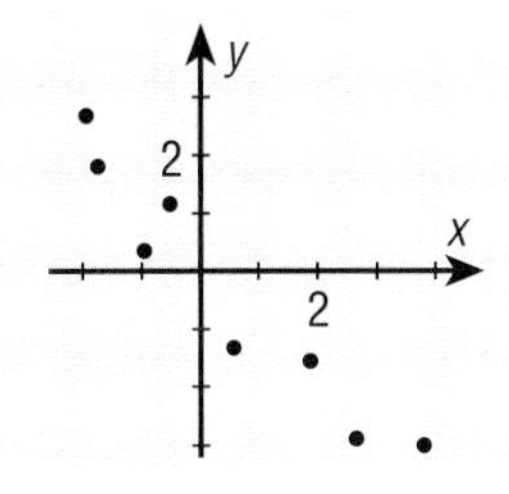

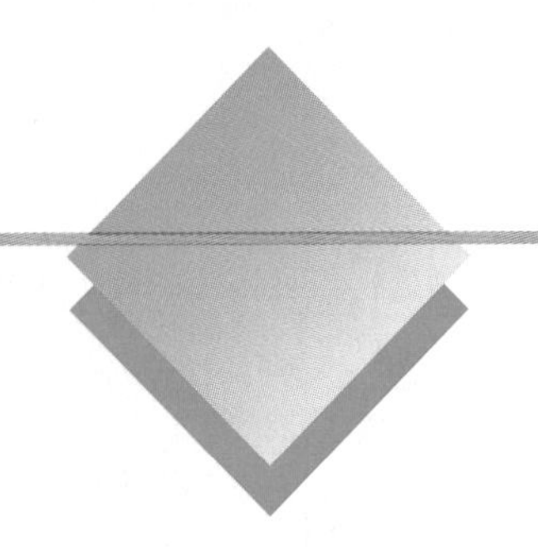

SHARPEN YOUR SKILLS

SKILL 1

EXAMPLE 1 Nelson has consulted a mutual funds reference manual at his neighborhood library. From the manual he recorded the 5-year price performance of the Vantage Balanced Mutual Fund, a fund recommended by Maria's investor friend. The prices were taken at 6-month intervals from July 1, 1988, to July 1, 1993, as shown.

Date	Years After Start	Price of Vantage Balanced Mutual Fund
July 1, 1988	0.0	$10.00
Jan. 1, 1989	0.5	10.60
July 1, 1989	1.0	11.80
Jan. 1, 1990	1.5	11.60
July 1, 1990	2.0	12.20
Jan. 1, 1991	2.5	12.40
July 1, 1991	3.0	13.60
Jan. 1, 1992	3.5	14.80
July 1, 1992	4.0	15.50
Jan. 1, 1993	4.5	16.20
July 1, 1993	5.0	16.80

INSTRUCTION

TEACHING THE LESSON

This lesson concentrates on mutual funds, probably the most suitable investment for the average young investor with a long-term time horizon. You can mention that the 5-year records for the funds in the text, while fictitious, are adapted from actual average price changes as provided by the CDA/Wiesenberger Mutual Funds Update of May, 1993.

FOCUS ON ALGEBRA

This lesson uses statistical data to make predictions about future performance of stocks. *(NCTM Standard 10, p. 167)*

ESL STUDENTS

Some students from other lands may be interested in learning that international investment opportunities have recently become more accessible to the average investor through *mutual funds.* Have such students report to the class on two kinds of mutual funds: international funds, which invest only in markets outside the United States, and global funds, which invest in both United States companies and companies from other countries.

QUESTION How can Nelson graph this price information?

SOLUTION

Nelson enters the data into his graphing calculator with the years after the start as x values and the prices as y values (11 ordered pairs of numbers starting with $x_1 = 0$, $y_1 = 10.0$). He chooses a range.

Xmin:	−0.5	Ymin:	−2
Xmax:	6	Ymax:	20
Xscl:	0.5	Yscl:	2

He plots all of the price/date information first as a **scatter plot** (left below), then as a **broken line graph** (right below). The broken line graph shows how the prices of the Vantage Balanced Mutual Fund changed over 5 years.

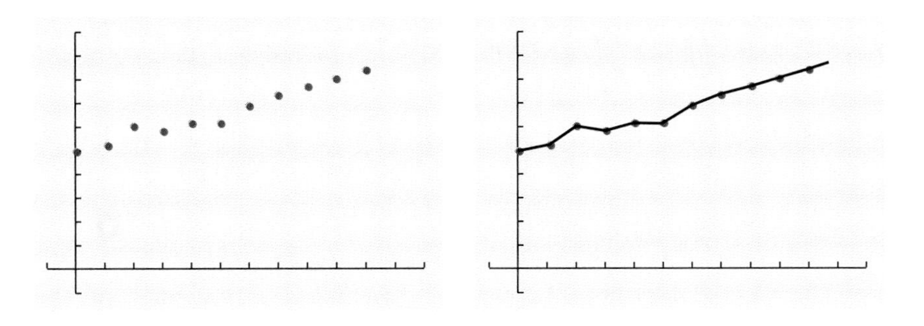

SKILL 2

EXAMPLE 2 Nelson wants to find an equation that will fit the points he has graphed.

QUESTION How can Nelson use a linear regression equation to find a line of best fit?

SOLUTION

Nelson decides to use his graphing calculator to find the *line of best fit* for the scatter plot. A linear equation of the form

$$y = a + bx$$

is called the **linear regression equation.** To obtain a and b, Nelson chooses "linear regression equation" from a menu on the graphing calculator. (Some graphing calculators use a code word for "linear regression equation," for example "LinReg.") As soon as Nelson chooses "linear regression equation," the calculator displays the following:

$a = 9.804545455$
$b = 1.369090909$
$r = 0.9850588294$

COMMON ERRORS

As mentioned in Skill 2, some calculators use the abbreviation "LinReg" for "linear regression." Caution students not to confuse this abbreviation with "LnReg," the abbreviation often used for "logarithmic regression," an option that is NOT used for any work in this text.

TECHNOLOGY NOTE

Many popular graphing calculators use the parameters a and b to represent the y-intercept and slope, respectively. However, the linear equation represented in mathematics texts is $y = mx + b$, where m and b are the slope and y-intercept, respectively. It is understandable that this inconsistent use of symbols will confuse students who are familiar with the slope-intercept equation from an earlier course.

COOPERATIVE LEARNING

Have a committee of students find out what an *index fund* is from their local library and report to the class on the role that such a mutual fund might play in a person's investment portfolio.

Using the values for a and b rounded to the nearest hundredth, Nelson writes an equation for the line of best fit

$$y = 9.80 + 1.37x$$

Nelson can graph the linear regression equation on the same set of axes as the original scatter plot as shown below.

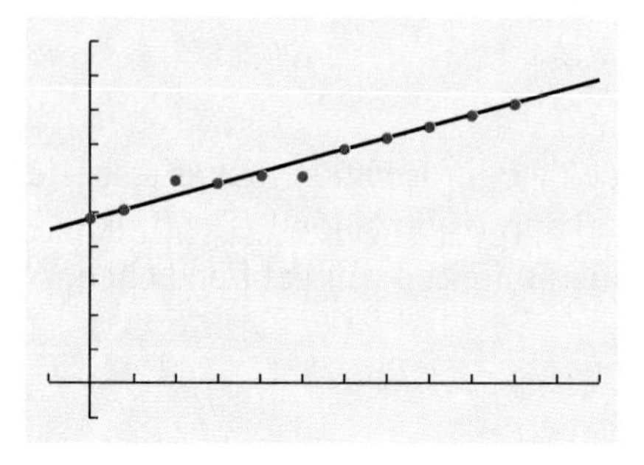

RETEACHING

For simplicity, the x-value in the tables of Skill 1 and the exercises are the years after the start, rather than the years themselves. In the reteaching, the years themselves could also be used, with similar results.

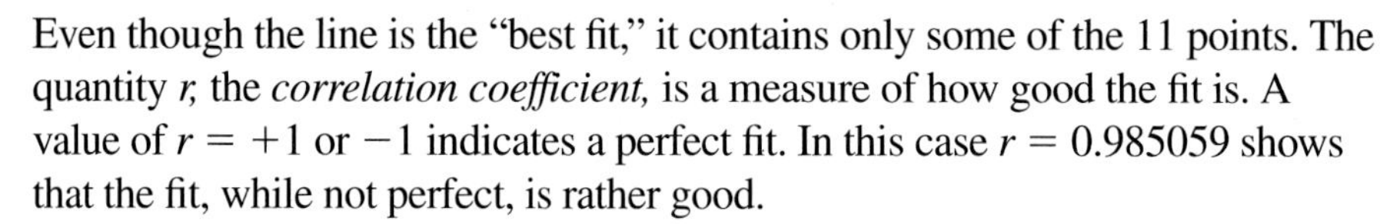

Even though the line is the "best fit," it contains only some of the 11 points. The quantity r, the *correlation coefficient,* is a measure of how good the fit is. A value of $r = +1$ or -1 indicates a perfect fit. In this case $r = 0.985059$ shows that the fit, while not perfect, is rather good.

Nelson has another reason for knowing that the linear regression equation is not perfect, since if $x = 0$ then

$$y = 9.80 + 1.37(0)$$
$$y = 9.80$$

Nelson already knows that when $x =$ "0 years after the start," the actual price of the mutual fund is \$10.00, not \$9.80. He concludes that if he uses the equation to predict future prices of the mutual fund, then he cannot expect the predictions to be completely accurate.

ENRICHMENT

Have students who are curious about the complete range of graphing-calculator capabilities choose a regression model other than the linear one to represent the data of Skill 1. Do any of the other three models have a better correlation coefficient than the linear one that was chosen? If so, what conclusions can be drawn from the fact? (**The exponential model, which reflects the effect of compounding, may better fit the data. See Skill 3 of Lesson 8-4.**)

EXAMPLE 3 Nelson wants to predict future prices of the Vantage Balanced Mutual Fund.

QUESTION What price does Nelson's linear regression equation predict for the mutual fund in July 1994?

SOLUTION

Use Nelson's linear regression equation. July 1, 1994, is 6.0 years after the start.

$$y = 1.37x + 9.80$$
$$y = 1.37(6.0) + 9.80 \quad \text{Substitute 6.0 for } x.$$
$$y = 18.02$$

The predicted price for July 1994 is a bit more than \$18. As a check, Nelson uses the trace function on his graphing calculator and notices that the point that corresponds to the ordered pair (6, 18) is almost on his line of best fit, so the algebraic solution and the graphical solution agree closely.

Nelson is happy with the results that he has obtained using the linear regression model. He believes that the model, while not perfect, is good enough to help him understand how the mutual fund may perform in the future. The model will provide a basis for advising Maria on whether she should invest in the fund.

TRY YOUR SKILLS

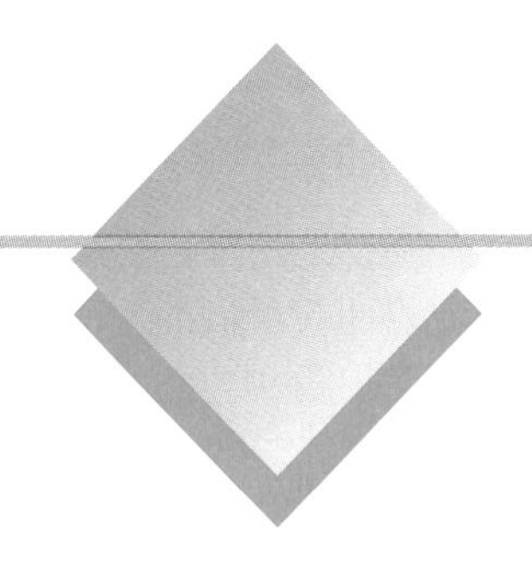

Nelson's friend Maria is interested in taking a bit more risk than Nelson to have a greater chance of creating a higher value for her investment. So instead of looking at the Vantage Balanced Fund, she decides to study the 5-year performance of the Vantage Maximum Capital Gain Fund, shown below.

Date	Years After Start	Price of Vantage Maximum Capital Gain Fund
July 1, 1988	0	$10.00
Jan. 1, 1989	0.5	10.40
July 1, 1989	1.0	12.40
Jan. 1, 1990	1.5	11.50
July 1, 1990	2.0	12.60
Jan. 1, 1991	2.5	11.50
July 1, 1991	3.0	13.70
Jan. 1, 1992	3.5	16.20
July 1, 1992	4.0	15.30
Jan. 1, 1993	4.5	17.50
July 1, 1993	5.0	18.00

GUIDED PRACTICE

The *Try Your Skills* exercises allow students to examine the 5-year behavior of a mutual fund that increases in value a bit more than the fund of Skill 1. Students should realize that with the increase in risk, the fund could very well have performed worse than the more conservative fund, rather than better.

1. Make a scatter plot for the above data beginning with July 1, 1988.
2. Make a broken line graph for the data. See Additional Answers.
3. To find a line of best fit using a graphing calculator, what should be your values for x_1 and y_1? (0, 10)
4. For the graph of the line of best fit, what would be a good range?
5. Which of the following equations could represent a linear equation for the line of best fit? c.

   a. $y = bx^2 + a$ **b.** $y = ax^b$ **c.** $y = a + bx$ **d.** $y = ab^x$
6. Use a graphing calculator to find the linear regression equation that is the line of best fit for the data. How good is the fit? See Additional Answers.
7. Use the equation that you found in Exercise 6 to predict a possible value for the Vantage Maximum Capital Gain Fund in July 1995. $20.70

ADDITIONAL ANSWERS

4. $0 \leq x \leq 6$; $0 \leq y \leq 20$
6. $y = 9.57 + 1.59x$; good, $r = 0.9417$

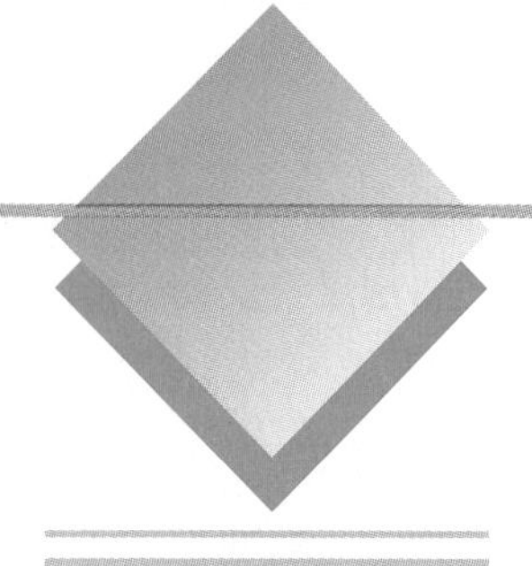

EXERCISE YOUR SKILLS

1. Why was the Securities and Exchange Commission established?
2. Why do some people lose money on investments despite the protection provided by the Securities and Exchange Commission?
3. What signals should alert you to suspect that a broker might not be acting in your best interest? See Additional Answers.

Maria's sister, Sarita, is interested in taking a bit more risk than she would have to assume in a balanced mutual fund but not as much as in a fund that tries to attain a maximum of capital gains. She decides to study the 5-year performance of the Vantage Growth and Current Income Fund, shown below.

Date	Years After Start	Price of Vantage Growth and Current Income Fund
July 1, 1988	0	$20.00
Jan. 1, 1989	0.5	21.60
July 1, 1989	1.0	24.80
Jan. 1, 1990	1.5	23.80
July 1, 1990	2.0	25.60
Jan. 1, 1991	2.5	25.20
July 1, 1991	3.0	28.40
Jan. 1, 1992	3.5	31.20
July 1, 1992	4.0	32.10
Jan. 1, 1993	4.5	34.00
July 1, 1993	5.0	35.00

4. Make a scatter plot for the above data beginning with July 1, 1988.
5. Make a broken line graph for the data. See Additional Answers.
6. Which of the following equations could represent a linear regression equation for the line of best fit? **b.**

   a. $y = x^b + a$ **b.** $y - bx = a$ **c.** $y = b^x + a$ **d.** $axy = b$
7. Use a graphing calculator to find the linear regression equation that is the line of best fit for the data. For your first data points, use $x_1 = 0$ and $y_1 = 20.00$. How good is the fit? $y = 2.98x + 19.97$; good, $r = 0.9797$

Use the equation that you found in Exercise 7 to predict a possible value for the Vantage Growth and Current Income Fund on the following dates.

8. January 1, 1995 $39.34
9. January 1, 1998 $48.28
10. January 1, 1999 $51.26
11. July 1, 1999 $52.75
12. How well does the equation of Exercise 7 predict the actual value of the Vantage Growth and Income Fund on July 1, 1988? $19.97; quite well, only off by $0.03

KEY TERMS

broken line graph
linear regression equation
scatter plot
Securities and Exchange Commission (SEC)

INDEPENDENT PRACTICE

ASSIGNMENT GUIDE

Exercises 1-3 test what the students learned about the SEC. The fund of Exercises 4-6 is neither as conservative nor as aggressive as the funds analyzed earlier. Exercises 7-15 reinforce the fact that the line of "best" fit is not necessarily a perfect representation of the data, merely the best that can be obtained under the given restraints.

ADDITIONAL ANSWERS

1. to ensure sound trading practices
2. The SEC cannot protect people from making investments with little or no value.
3. if the broker will sell only one stock or fund, makes promises for a quick, sure profit, has inside information, and urges you to buy before "prices go up"

For the 5-year performance of Vantage Growth and Current Income Fund, find an equation for the line of best fit for the first ten data points, omitting the data for July 1, 1993. $y = 2.97x + 19.99$

13. What price does the equation predict for the value of the fund on July 1, 1993? $34.84

14. How does the predicted price compare with the actual price?

15. What do you conclude from this result? See Additional Answers.

MIXED REVIEW

8-2 **1.** Find the total cost of 100 shares of Daring Ventures, Inc. at $63\frac{1}{2}$. The commission cost is 1.5%. $6445.25

5-3 Determine the total savings on a loan of $49,800 at 8.5% for 25 years if you make a down payment of 20% or 30% instead of 10%.

	Percent Down	Down Payment, D	Loan Amount, A	Monthly Payment, M	Total Amount, T	Savings Over 10% Down
2.	10%	$4,980	$44,820	$360.90	$113,250.83	—
3.	20%	9,960	39,840	320.80	106,200.74	$ 7,050.09
4.	30%	14,940	34,860	280.70	99,150.65	14,100.18

6-6 **5.** Why do some large lending institutions use a credit scoring system such as the one that appears on page 293 of Lesson 6–6?

7-3 **6.** Use the formula for the Future Value of a Periodic Investment on page 344 of Lesson 7–3 to find the amount that will be in an IRA account after 25 years if the account consistently pays a compounded rate of 7.5% per year and $1,500 is contributed to the IRA every year.

2-3 **7.** Suppose that your name is Charles Lewis and your uncle sends you a check made out to "Chuck Lewis." Show how you would endorse that check. Chuck Lewis (Charles Lewis)

7-2 **8.** A certain insurance company expects 99.830% of all 34-year-old people to live at least one more year. What is the break-even premium for a $50,000 policy issued to such a person by the company if the direct and indirect expenses for issuing the policy amount to $15? $100

4-1 **9.** If you earn $65,000 per year, how much do you pay in Social Security and Medicare taxes? $4513.70

4-1 **10.** If you earn $159,000 per year, how much do you pay in Social Security and Medicare taxes? $5528.70

5-2 **11.** Suppose that your family's take-home pay is $3500 per month. How much can you afford to spend for credit card payments each month? $700.00

CLOSURE

Ask students which kind of mutual fund they would consider for an investment: (1) a fund with moderate risk and moderate expectations of reward, or (2) a fund with conservative expectations and conservative risks.

LESSON QUIZ

On January 1 for 5 years in a row, a fund had the following values: $15.60, $15.95, $16.82, $17.80; $19.30.

Find each of the following.

1. a linear regression equation for the data (**$y = 0.925x + 15.244$**)

2. the correlation coefficient for the equation of Exercise 1. (**0.9760816643**)

ADDITIONAL ANSWERS (EXERCISE YOUR SKILLS)

14. $0.16 difference

15. The line of best fit is accurate in predicting stock prices.

(MIXED REVIEW)

5. Helps to determine the credit risk of a borrower in an objective way that does not discriminate.

6. $101,966.79

8-4 STOCK PRICES AND INFLATION

QUICK REFERENCE

TEACHER'S RESOURCES AND ANSWER KEY

Lesson Quiz Answers 8-4, p. 70

Reteaching Activity Answers 8-4, p. 106

Enrichment Activity Answers 8-4, p. 107

EXTENSION ACTIVITIES

Reteaching Activity 8-4, p. 64

Enrichment Activity 8-4, p.68

Maria read an article in the newspaper that referred to some investors who feared that a "bear market" might be developing. She was not sure what bears had to do with stock prices. Nelson told her that in a **bear market,** most stocks go down and that in a **bull market,** most stocks go up. He also told her that not all the stocks go up or down at the same time.

Maria also remembers from her study of the Federal Reserve System that the U.S. economy is affected by inflation. She also knows that though the rates of inflation have varied from year to year, inflation has affected salaries and prices for a number of years. One means of measuring the effects of this inflation is the *Consumer Price Index.*

Maria has often heard her parents speak of nickel postage stamps and 10-cent newspapers as though their childhood was not that long ago. To Maria, 35 years might as well be 100, but she does know that most adults talk that way. She decides that when she is an adult, she will remember what she thought when she was a teenager and not tell her children about how low prices were when she was a kid.

OBJECTIVES: *In this lesson, we will help Maria to:*

- *Understand the effects of inflation as charted by the Consumer Price Index.*
- *Use the Consumer Price Index to calculate changes in prices of products.*
- *Use a regression model to illustrate changes in the Consumer Price Index.*
- *Use the concept of probability to estimate whether to invest money in a mutual fund or stocks.*

THE CONSUMER PRICE INDEX

Inflation affects the cost of every product and service that we purchase because it cheapens the value of the dollar. Even though salaries may also have increased, the purchasing power of those salaries may be less than the salaries of 10 years ago. The **Consumer Price Index (CPI)** is an economic yardstick that can help Maria to judge the changes in the buying power of the dollar. Economists use the CPI to measure **inflation** (a general increase in prices) or **deflation** (a general decrease in prices). Periods of deflation are usually accompanied by high unemployment.

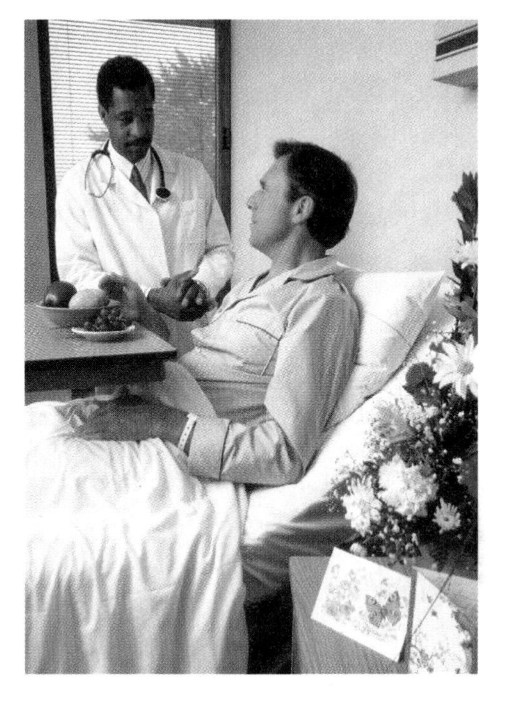

The Bureau of Labor Statistics compiles and publishes two CPIs. The CPI-W represents the consumption experience of urban wage earners and clerical workers; this represents approximately 40% of the population. The CPI-U is a broader index that covers all urban consumers, about 80% of the population. The CPI-U includes the consumption habits of diverse groups such as salaried workers, the unemployed, the retired, and the self-employed.

The following chart shows the breakdown of the CPI for all urban consumers in various categories. The categories listed are very general; more detailed categories are available from the Bureau of Labor Statistics and other sources. The chart illustrates the changes in the cost of all items and also the costs for four major categories for the years 1960–1992. These special categories are food, shelter, apparel, and medical care. The year 1984 is the base year, the year in which a **market basket** of items is assigned an arbitrary value of $100. Each separate category, such as Shelter or Apparel, is also assigned a value of $100 in 1984. As a result, all years that precede or follow 1984 have CPI

CONSUMER PRICE INDEX (1960–1992)

Year	All Items	Food	Shelter	Apparel	Medical Care
1960	29.6	30	25.2	45.7	22.3
1970	36.8	39.2	35.5	59.2	34
1975	53.8	59.8	48.8	72.5	47.5
1980	82.4	86.8	81	90.9	74.9
1984	100	100	100	100	100
1985	107.6	105.6	107.7	105	113.5
1986	109.6	109	115.8	105.9	122
1987	113.6	113.5	121.3	110.6	130.1
1988	118.3	118.2	127	115.4	138.6
1989	124	125.1	132.8	118.6	149.3
1990	130.7	132.4	140	124.1	162.8
1991	136.2	135.8	146.3	128.8	177
1992	140.7	138.3	152.2	131	189.4

FOCUS

ALGEBRA CONNECTIONS

Quickly review the compound-interest formula $B = P(1 + r)^n$ from Chapter 3. Then ask students whether they think that the variable P in the formula could stand for any quantity other than invested money. If a hint is needed, mention the phrase "population explosion." In Skill 3, values of the Consumer Price Index will be revealed as another possible interpretation of P. The general principle is that of exponential growth represented by the equation $y = ab^x$.

CRITICAL THINKING

Ask students whether there are any other indexes for measuring the cost of goods and services. Mention the Wholesale Price Index that reflects costs to producers of goods and suggest that students inquire about other indexes at their local library.

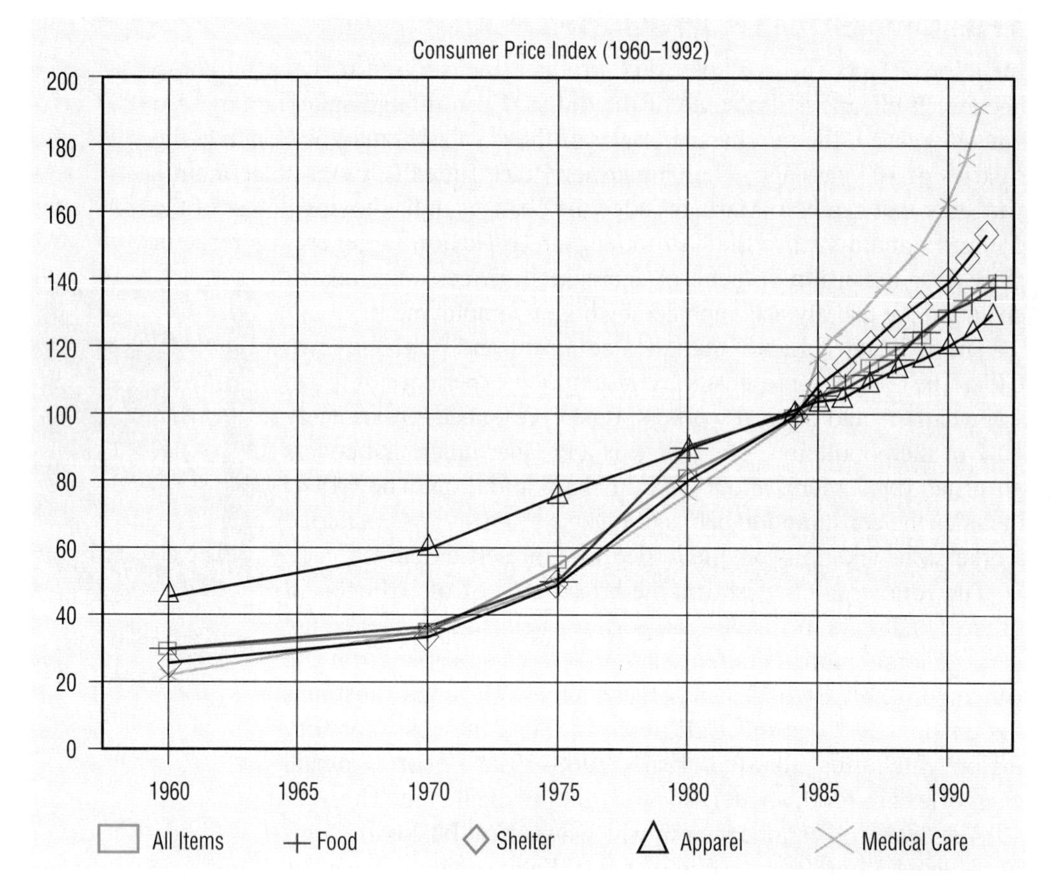

values that are proportionally related to the base year. Notice that in the combined category, "All Items," a basket of goods that cost $100 in 1984 would have cost $29.60 in 1960 and $140.70 in 1992.

Maria wanted to see the numbers displayed in a graph for easy comparison among the categories. She entered the numbers for each category in a spreadsheet and created the graph shown above. From the graph, Maria could see that the rate of price increase was not the same for every category. For example, the rate of increase for the price of shelter was greater than the rate for all items.

ALGEBRA *REVIEW*

Solve each proportion.

1. $\frac{x}{3} = \frac{8}{2}$
   12
2. $\frac{x+1}{5} = \frac{x-1}{4}$
   9
3. $\frac{2.5}{a} = \frac{1}{2}$
   5
4. $\frac{3}{15} = \frac{b-1}{10}$
   3
5. $\frac{7}{4x+2} = \frac{1}{2x-4}$
   3

Ask Yourself

1. What is a bear market?
   the market when most stock prices go down
2. What is a bull market?
   the market when most stock prices go up
3. What is the Consumer Price Index?
   measurement of the effects of inflation
4. Why do you think a period of deflation is usually accompanied by unemployment?
   a decrease in prices could force companies to close

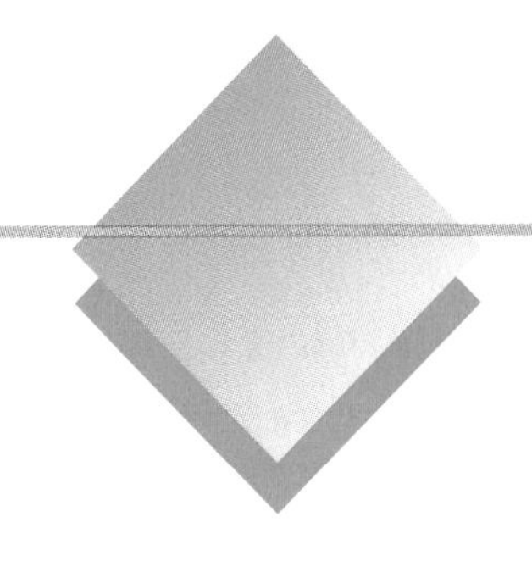

SHARPEN YOUR SKILLS

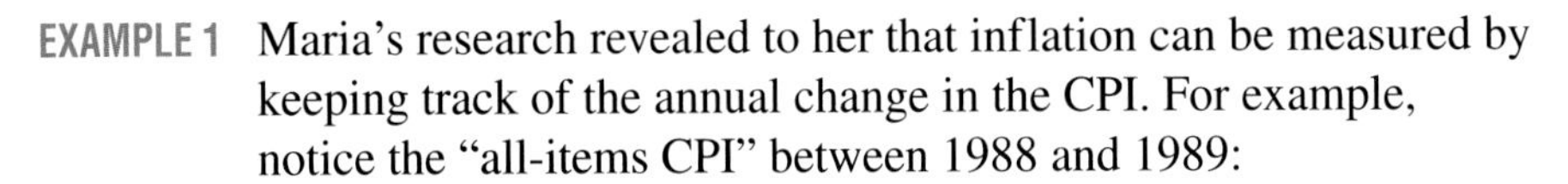

SKILL 1

EXAMPLE 1 Maria's research revealed to her that inflation can be measured by keeping track of the annual change in the CPI. For example, notice the "all-items CPI" between 1988 and 1989:

CPI (all items) for 1988: 118.3
CPI (all items) for 1989: 124

QUESTION What is the percent of increase in the CPI between 1988 and 1989?

SOLUTION
Use the rate-of-change formula from Lesson 8–2. The rate of change in the CPI is the change divided by the CPI for the earlier year.

$$r = \frac{|P_n - P_o|}{P_o}$$ Rate-of-change formula

$$r = \frac{|124 - 118.3|}{118.3}$$

$$= 0.04818$$

Since the CPI increased, the rate of inflation is 4.8%.

SKILL 2

EXAMPLE 2 Since 1984 is the base year for the CPI, Maria would like to compare her parents' spending power in 1984 with what they make now.

QUESTION Maria's parents spent $20,000 in 1984. How much would they have needed in 1992 just to buy the same things?

SOLUTION
Write a proportion. Let x represent the unknown 1992 expenditure.

$$\frac{\text{1992 expenditure}}{\text{1992 CPI}} = \frac{\text{1984 expenditure}}{\text{1984 CPI}}$$

$$\frac{x}{140.70} = \frac{20{,}000}{100}$$ Next, use the *Rule of Proportions.*

$$x \cdot 100 = 140.70 \cdot 20{,}000$$ If $\frac{a}{b} = \frac{c}{d}$, then $ad = bc$.

$$100x = 2{,}814{,}000$$

$$x = 28{,}140$$

In 1992 it would have taken about $28,000 to buy the same market basket of goods that $20,000 would have bought in 1984.

INSTRUCTION

TEACHING THE LESSON

The Consumer Price Index is the most widely followed measure of the general price level for goods and services. Mention that in addition to the national Consumer Price Index, Consumer Price Indexes are prepared for various regions of the country to reflect local variation in the cost of living.

FOCUS ON ALGEBRA

Example 2 of Skill 2 provides another opportunity to apply the Rule of Proportions. Curve fitting is extended to include the exponential-regression model in Skill 3. Probability concepts, including that of expected value, are applied in Skill 4. (*NCTM Standard 1, p. 137; Standard 5, p. 150; Standard 10, p. 167; Standard 11, p. 171*)

COMMON ERRORS

Students are sometimes inclined to apply a mathematical method without keeping its limitations in mind. Remind them that the prices of goods and services do not change uniformly over time.

EXAMPLE 3 Maria would like to estimate the cost of living for the year 2000.

QUESTION How can Maria use the CPI table for 1960–1992 to estimate the cost of living for the year 2000?

SOLUTION

Maria knew from talking with Nelson that she can use a graphing calculator to find a line of best fit. She decides to use the data points from the "All Items" column of the 1960–1992 CPI table. To make the data easier to work with, she simplifies the values for x that she will have to enter. She lets x represent the number of years after 1960 and y represent the CPI for each year in the table beginning with 1960 as shown below.

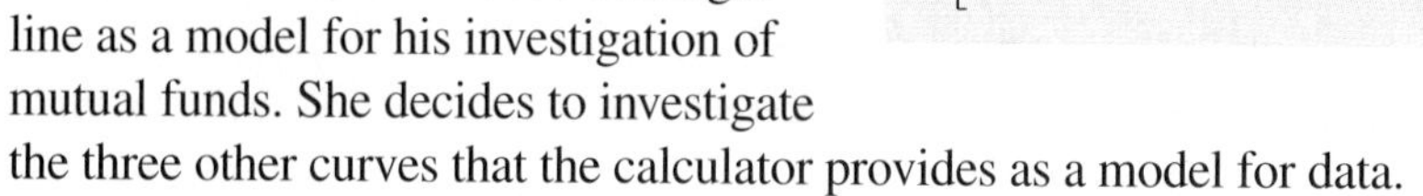

x	0	10	15	20	24	25	26	27	28	29	30	31	32
y	29.6	36.8	53.8	82.4	100	107.6	109.6	113.6	118.3	124	130.7	136.2	140.7

Maria enters the data in a graphing calculator and creates a scatter plot using the following range values.

Xmin: -2	Ymin: -10
Xmax: 33	Ymax: 150
Xscl: 2	Yscl: 20

She knows that Nelson used a straight line as a model for his investigation of mutual funds. She decides to investigate the three other curves that the calculator provides as a model for data.

Two of the other three curves cannot use her data (they will not take 0 as a value for x). The two models that do accept all of her data are the *linear regression* equation that Nelson used and another curve called the **exponential regression equation.** The graphing calculator provides a correlation coefficient r for each model and values for two constants, a and b, that appear in the formulas for the models.

Regression Model	Formula	Equation	Correlation Coefficient
Linear	$y = a + bx$	$y = 10.73788254 + 3.850867094x$	$r = 0.973223596$
Exponential	$y = ab^x$	$y = 26.19388895(1.05527341)^x$	$r = 0.9876350425$

Maria uses her graphing calculator to graph each model on top of the scatter plot.

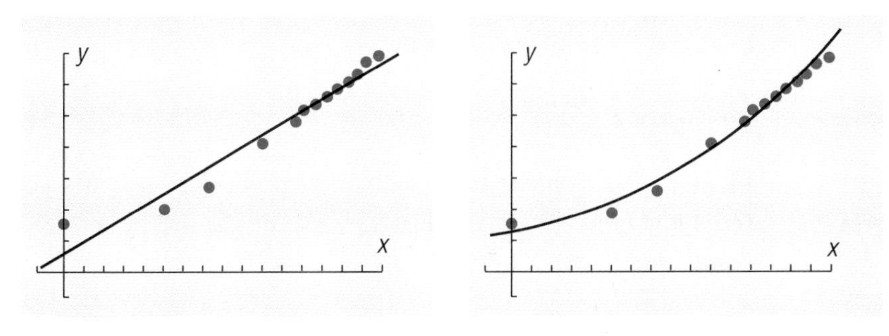

She decides to use the exponential model instead of the linear model for two reasons.

1. The exponential graph seems to fit the scatter points better.
2. She knows that the closer the correlation coefficient is to 1, the better the fit. The correlation coefficient for the exponential curve is closer to 1 than is the correlation coefficient for the straight line.

To estimate the cost of living for the year 2000, Maria notes the year 2000 is 40 years after 1960. Maria uses the ZOOM and TRACE features of the graphing calculator to find that y is about 225 when x is close to 40. To check, Maria substitutes into the exponential equation.

$y = 26.19388895(1.05527341)^{40}$ Substitute 40 for x.
$y = 225$ To the nearest dollar

According to the exponential model, the CPI will be about 225 in the year 2000.

EXAMPLE 4 Maria would like to predict the rate of inflation for the year 2000.

QUESTION If the exponential model is correct, what will be the rate of inflation?

SOLUTION

Maria notices that the exponential equation is similar to the compound interest formula (see page 107 in Lesson 3–2).

$B = p(1 + r)^n$ where p = initial investment
r = interest rate for the period
n = number of periods

RETEACHING

In Example 3, have students use the linear-regression model instead of the exponential-regression model and compare the results of the two models.

ENRICHMENT

Ask interested students who have studied logarithms to find out why two of the four regression models available in a graphing calculator's memory cannot take 0 as a value for x (see Example 3 of Skill 3). For the power model $y = ax^b$, they will probably need to consult the calculator's manual. (**The logarithmic model $y = a + b \ln x$ fails because the domain of the logarithmic function is the set of all positive numbers. The power model with equation $y = ab^x$ would accept $x = 0$ (unless $b = 0$) but, as a good manual will point out, the statistical results are calculated within the calculator using not x and y but $\ln x$ and $\ln y$. (This is done in order to transform the power curve to more convenient linear form, i.e., to $\ln y = \ln a + b \ln x$, which does have the linear form $Y = A + bX$.) Thus, $x = 0$ still cannot be used.**)

She compares the two formulas

$$B = p(1 + r)^n \qquad\qquad y = 26.19(1 + 0.055)^x$$

and notices that the compound interest formula is also an exponential model. She writes 1.055 as 1 + 0.055 to sharpen the comparison and correctly concludes that the predicted annual increase in the CPI (the rate of inflation) is 0.055, or about 5.5%. She also notices that the model is not perfect, since in 1960 when x was 0,

$$\begin{aligned} y &= 26.19(1.055)^0 \\ &= 26.19 \bullet 1 \qquad\qquad \text{Recall } b^0 = 1 \text{ when } b \neq 0. \\ &= 26.19 \end{aligned}$$

which is not the true value for the CPI in 1960. (The true value was 29.6.) Maria suspects that the predicted inflation rate, 5.5%, might also be inaccurate. In a few more years she will know.

SKILL 4

Maria asks her friend who knows about investments what she should expect her fund to be worth in 5 years if inflation continues at a low rate and what her fund might be worth if deflation occurs at a low rate. Her friend's opinion is that her \$1000 might drop to about \$700 if deflation were to occur. Her friend agrees with Maria that if a mild inflation occurs, the fund will probably be worth at least \$1800. Her friend also believes that the Federal Reserve will do whatever it can to avoid a deflation, since a deflation is usually accompanied by a recession and above-average unemployment. However, a deflation might occur anyway. The investor friend thinks that the probability of a deflation is only about 1 in 5; that is, about 0.20.

EXAMPLE 5 How can Maria use her new knowledge about inflation to help her with her decision about investing in a mutual fund?

QUESTION What is the expected value of her portfolio in 5 years?

SOLUTION

Maria decides to calculate the expected value of her portfolio in 5 years. If the probability of a deflation is 0.20 and inflation is the only other choice, then the probability of inflation can be found by solving the following probability formula for P(inflation).

$$\begin{aligned} P(\text{inflation}) + P(\text{deflation}) &= 1 \\ P(\text{inflation}) &= 1 - P(\text{deflation}) \\ &= 1 - 0.20 \\ &= 0.80 \end{aligned}$$

The expected value of her fund is found by using the expectation formula.

$$\begin{aligned} E &= a_1P_1 + a_2P_2 \\ &= 1800 \cdot 0.80 + 700 \cdot 0.20 \\ &= 1440 + 140 \\ &= 1580 \end{aligned}$$

The expected value of Maria's fund in 5 years is $1580. She knows that this value is useful as a guide in deciding whether to invest in a fund or to stick to a savings account. She also knows that the actual value of the fund is likely to be around either $1800 (if there is some inflation) or $700 (if there is deflation), not around $1580. In deciding whether to invest in the fund, Maria has to face the possibility that the fund will drop in value to $700 in 5 years.

TRY YOUR SKILLS

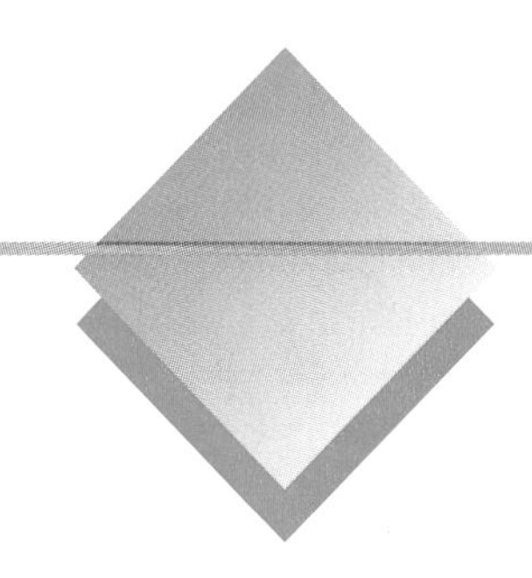

Use the table of the Consumer Price Index (1960–1992) on page 389 to find the percent of increase in the CPI (all items) for the following time periods. Round your answers to the nearest tenth of a percent.

1. From 1989 to 1990 5.4%

2. From 1984 to 1985 7.6%

Use the table of the Consumer Price Index (1960–1992) on page 389 to write and solve a proportion that can be used to find how much money a family would need to buy a market basket of goods in the given year that was worth $10,000 in 1987.

3. 1984 See Additional Answers.

4. 1990

Suppose that an exponential regression model gives the following equations as a description of the cost of living over a five-year period. What would be the expected rate of inflation?

5. $y = 100(1 + 0.04)^x$ 4%

6. $y = 53.6(1.062)^x$ 6.2%

7. Suppose that the probability of low inflation is 0.15 and the only other possibility is moderate deflation. Find the probability of moderate deflation.

8. Suppose that there are three possibilities: low inflation with a probability of 0.3, severe deflation with a probability of 0.1, and perfect price stability (neither inflation or deflation). Find the probability of perfect price stability. 0.6

GUIDED PRACTICE

Exercises 1-4 provide practice using algebraic techniques with problems involving the Consumer Price Index. Students who have difficulty with Exercise 6 should rewrite the equation of that exercise in the expanded form of Exercise 5.

ADDITIONAL ANSWERS

3. $\frac{10{,}000}{113.6} = \frac{x}{100}$

$x = \$8{,}802.82$

4. $\frac{10{,}000}{113.6} = \frac{x}{130.7}$

$x = \$11{,}505.28$

7. 0.85

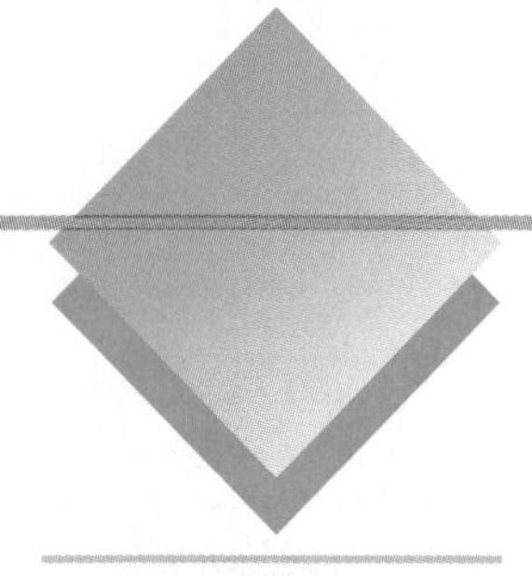

EXERCISE YOUR SKILLS

1. How do economists use the Consumer Price Index? See Additional Answers.
2. In the table for the CPI, which category had the largest increase between 1984 and 1992? Which had the least increase?
3. Which category in the CPI table most nearly matched the "all items" column? Why do you think this happened?
4. The "all items" column in the table for the CPI consists of a market basket of 400 items. Why is this a better indication of how prices are behaving than the cost of a few selected items would be?

The prices given are for 1984. Find the cost of the indicated items in 1992. Use the table of the Consumer Price Index (1960–1992) on page 389.

5. $500 worth of food $691.50
6. $1000 in medical care $1894.00
7. A $25 market basket $35.18

Find the percent of change in the CPI for the indicated item.

8. Food between 1984 and 1985
9. Shelter between 1989 and 1990
10. Apparel between 1987 and 1988 4.3%
11. Medical care between 1985 and 1986 7.5%

Use a graphing calculator to make an exponential regression model for the cost of medical care from 1960 to 1992 as shown in the table on page 389. Then answer Exercises 12–16.

12. What is the equation for the model? $y = 18.87(1.074)^x$
13. How accurately does the model agree with the actual cost of medical care in 1960? 18.87 compared to 22.3
14. What value does the model suggest for the average annual rate of increase in the cost of medical care between 1960 and 1992? 7.4% increase
15. Use the graphing calculator to make a scatter plot of the data, and on the same calculator screen, graph the exponential regression equation that you found. See Additional Answers.
16. Do you think that a linear regression model would have given a more accurate picture than the exponential regression model for the increase in the cost of medical care from 1960 to 1992? Explain your answer.

Maria's sister, Sarita, has become interested in a mutual fund that stresses both growth and current income. She believes that if there is inflation in the next 5 years, then her $1000 investment will grow to about $1750 but that if there is a deflation, then the fund may drop to $800. She thinks that the next 5 years will be either inflationary or deflationary with a 20% probability that the economy will be deflationary.

17. What is the probability that the economy will be inflationary? 80%
18. What is the expected value of Sarita's fund in 5 years? $1560

KEY TERMS

bear market
bull market
Consumer Price Index (CPI)
deflation
exponential regression equation
inflation
market basket

INDEPENDENT PRACTICE

ASSIGNMENT GUIDE

Exercises 1-4 probe students' general understanding of the Consumer Price Index. The remaining exercises test their ability to use concepts of algebra and probability to solve everyday problems dealing with inflation and with investment decisions affected by deflation and inflation.

ADDITIONAL ANSWERS

1. to measure inflation or deflation
2. Largest: Medical Care Least: Apparel
3. Food; explanations may vary.
4. Answers may vary.
8. 5.6%
9. 5.4%
16. exponential: $r = 0.99$; linear: $r = 0.94$; exponential is more accurate

MIXED REVIEW

4-3 The following table shows cost and sales figures for a company that sells board games. Use a computer spreadsheet to fill in the missing information.

	Unit Cost	Number Produced	Fixed Cost	Total Cost	Unit Price	Revenue	Profit (Loss)
1.	$8.93	15	$310.00	$ 443.95	$25.00	$ 375	($68.95)
2.	7.50	125	310.00	1247.50	25.00	3125	$1877.50

7-1 Use the following table for Exercises 3–5.

Comparison Table for Term and Whole Life Annual Premiums for a Policy with a Face Value of $100,000			
Age	**5-year Renewable Term**	**Whole Life**	**First-year Difference**
20	$205	$ 775	$570
25	207	918	711
30	218	1112	894

3. What is the annual premium for a $100,000 whole life policy for a 20-year-old person? $775

4. What is the annual premium for a $200,000 term policy for a 30-year-old person? $436

5. Determine the difference that a 25-year-old person will pay in annual premiums the first year if he or she takes a term policy with a face value of $250,000 instead of a whole life policy with the same face value.

2-2 Enter the following transactions in a check register form. Make up your own form with the following headings: Check Number, Date, Checks/Deposits, Amount, and Balance. Under "Checks/Deposits" have two lines for each entry, the top one labeled "To:" and the bottom one labeled "For:." Find the new balance after each transaction. The starting balance is $450.

6. Check 201, September 27, to Aaron Jones, $158.50 for car repair

7. Deposit on September 29 of $130 See Additional Answers.

8. Check 202, September 29, to Superior Supermarket, $87.63 for groceries

7-2 **9.** The probability that a person will be alive in 1 year is 0.98329. What is the probability that the person will die within 1 year? 0.01671

6-7 **10.** Last month, the Antonelli family spent $241 on the car payment, $452 on credit card payments, $841 on the mortgage, $86 on utilities, $475 on food, and $1500 on everything else. To the nearest tenth of a percent, what percent of their take-home pay did the family spend on credit payments? 19%

CLOSURE

Ask students how they should take inflation into account when trying to estimate the value of their investment over a 30-year period. Point out that it would be prudent to subtract their long-term inflation estimate (for example, 3%) from their expected average annual return before figuring the calculation of the future value of their investments.

LESSON QUIZ

Use a graphing calculator to make an exponential regression model for the cost of food from 1960 to 1992 as shown in the table on page 389.

1. What is an equation for the model? (**$y = 27.92(1.053)^x$**)

2. What is the percent increase of food costs between 1960 and 1992? (**361%**)

ADDITIONAL ANSWERS

5. $1777.50

CHAPTER 8 REVIEW

1. How can you make money from stocks? through careful investments
2. How is buying bonds the same as lending money to a municipality or a corporation? See Additional Answers.
3. Why are mutual funds suitable for small investors? See Additional Answers.
4. What do you think is the most important factor in deciding whether or not to buy stocks? being able to afford the loss
5. What method is used by brokers to buy and sell stocks on a security exchange? The auction method
6. How does the Securities and Exchange Commission help to protect investors?
7. Why is it important to adopt a long-term point of view when buying stocks whether directly in the stock market or indirectly through a mutual fund?

ADDITIONAL ANSWERS

2. Bonds are loans to companies or municipalities, in return the company or municipality must pay interest to the investor.
3. It is a way to have a stake in many different corporations, rather than risking all your money on the future of just one or two corporations.
6. It requires companies to provide complete and accurate information.
7. Stocks are not instant profit and take time to develop.

Use the prices per share shown below to find the number of whole shares of stock that you can afford to purchase if you have $20,000 to invest in each of the following companies. Then find the total cost of the shares. Ignore brokerage commissions.

	Company	Price per Share	Number of Whole Shares	Total Cost
8.	NunnInc	42	476	$19,992
9.	OctOfAm	$7\frac{5}{8}$	2622	19,992.75
10.	OscarsPies	$18\frac{3}{4}$	1066	19,987.50
11.	OwlInd	150	133	19,950

Find the capital gain or loss from buying 2500 shares at the given purchase price and selling them at the current market value. Ignore the effect of commissions.

	Company	Buy	Total Value	Sell	Total Value	Gain/Loss
12.	Disnel	$63\frac{1}{2}$	$158,750	$68\frac{1}{2}$	$171,250	$ 12,500
13.	AP & P	$18\frac{1}{2}$	46,250	$30\frac{1}{4}$	75,625	29,375
14.	Banter	$24\frac{3}{8}$	60,937.50	$19\frac{5}{8}$	49,062.50	(11,875)
15.	Zola	$43\frac{1}{2}$	108,750	89	222,500	113,750

The 5-year performance of the Vantage Long-Term Growth Fund is shown below.

Date	Number of Years After Start	Price of Vantage Long-Term Growth Fund
July 1, 1988	0	$10.00
Jan. 1, 1989	0.5	10.70
July 1, 1989	1.0	12.90
Jan. 1, 1990	1.5	11.80
July 1, 1990	2.0	13.10
Jan. 1, 1991	2.5	12.50
July 1, 1991	3.0	14.30
Jan. 1, 1992	3.5	16.10
July 1, 1992	4.0	17.60
Jan. 1, 1993	4.5	18.90
July 1, 1993	5.0	20.10

16. Make a scatter plot for the above data beginning with July 1, 1988.

17. Make a broken line graph for the data. See Additional Answers.

18. Use a graphing calculator to find the linear regression equation that is the line of best fit for the data. $y = 9.49 + 1.95x$

19. Use the equation that you found in Exercise 18 to predict a possible value for the Vantage Long-Term Growth Fund in July 1995. $23.14

Use a graphing calculator to make an exponential regression model for the cost of food from 1960 to 1992 as shown in the table on page 389. Then answer Exercises 20–22.

20. What is the equation for the model? $y = 27.916\ (1.053)^x$

21. How accurately does the model agree with actual cost of food in 1960?

22. What value does the model suggest for the average annual rate of increase in the cost of food between 1960 and 1992? 5.3%

ADDITIONAL ANSWERS

21. close; model is $27.916 and actual is $30.00

CHAPTER 8 TEST

For the companies listed below, find the number of whole shares that you can buy with $100,000. Ignore commissions.

1. Z Mart at $65\frac{1}{4}$ per share 1532

2. Disnel at $38\frac{3}{4}$ per share 2580

3. Zaxta at $87\frac{1}{2}$ per share 1142

4. Algin at $25\frac{7}{8}$ per share 3864

5.–8. Find the total value of the shares of stock in Exercises 1–4.

For the following companies, find the commission cost and the total purchase cost (price plus commission) for the given information. See Additional Answers.

9. 306 Cts&Dgs; $20/share; 2%

10. 215 Elmo; $33.50/share; 1%

11. 400 SuprCo; $100.375/share; 1.5% $C = 602.25$; $40,752.25

12. 6000 Langly; $7/share; 0.5% $C = 210$; $42,210

The value of the Vantage Conservative Growth Fund grew annually for 5 years as follows, beginning on January 1, 1988: $15.00 (starting value), $15.70, $16.80, $17.80, $19.00, $20.60. See Additional Answers.

13. Make a scatter plot for the above data beginning with January 1, 1988.

14. Make a broken line graph for the data.

15. Use a graphing calculator to find the linear-regression equation that is the line of best fit for the data. $y = 14.70 + 1.11x$

16. Use the equation that you found in Exercise 15 to predict a possible value for the Vantage Conservative Growth Fund in January 1996. $23.58

Use the Consumer Price Index (1960–1992) on page 389 in Lesson 8–4 to find the approximate cost of each of the following items in 1989. Choose from the categories Food, Shelter, and Apparel. For each item the 1984 price is given.

17. Woman's jacket; $50.75 $60.19

18. House; $77,000 $102,256

19. Rent; $435 $577.68

20. Pizza slice; $0.70 $0.88

An investor believes that if there is inflation in the next 5 years, then her $10,000 investment will grow to about $18,000 but that if there is a deflation, then the investment may drop to $7500. She thinks that the next 5 years will be either inflationary or deflationary with a 30% probability that the economy will be deflationary.

21. What is the probability that the economy will be inflationary? 70%

22. What is the expected value of the investment in 5 years? $14,850

ALTERNATIVE ASSESSMENT

Ask students to write a paragraph describing how they would advise a friend to choose between the following two investment choices: the stock of a large, established company that is not expanding but that pays a good dividend and a small, well-regarded company that has growing earnings and that pays no dividend. Have students comment on the following remark by a young person contemplating what to do with his savings. "Stocks are too risky and mutual funds are not much better. I'm never going to put my money in anything else but a savings account or a money-market mutual fund." Is this the most conservative approach to managing one's own money?

ADDITIONAL ANSWERS

5. $99,963

6. $99,975

7. $99,925

8. $99,981

9. $C = 122.40$; $6242.40

10. $C = 72.03$; $7274.53

CUMULATIVE REVIEW

1-1 **1.** If Phyllis works more than 25 hours in one week, she receives $1\frac{1}{2}$ times her regular wage rate of $4.80 for each of the extra hours. One week she worked 32 hours and received $68.75 in tips. How much money did Phyllis earn that week? $239.15

1-3 **2.** Angela earns a weekly salary of $150 at the real estate agency where she works and gets a 4% commission for each house that she sells. In a 4-week period her sales were $100,000. What were her total earnings? $4600

2-1 **3.** Suppose you wrote 30 checks in June. Your bank charges $0.040 per check for the first 25 checks and $0.10 for each check over 25. How much were your June bank charges? $1.50

3-1 Andy can save $35 a week. He wants to buy a 25-inch remote stereo color TV that sells regularly for $419.99 but will be on sale for 10 weeks at $369.99.

4. Find how much he can save between now and the end of the sale. Will he be able to buy the set on sale? Explain your answer. See Additional Answers.

5. How long will he have to save in order to be able to afford the TV at the regular price? 12 weeks

3-2 Use a calculator to find the amount of interest and the new balance that will accumulate over 2 years on the following principal amounts at the given interest rate compounded as shown.

	Principal	Interest Rate	How Often Compounded	Interest Earned		New Balance
6.	$5000	8%	Annually	1st period:	$400	$5400
				2nd period:	432	5832
7.	$9000	9.5%	Semiannually	1st period:	427.50	9427.50
				2nd period:	447.81	9875.31

4-3 The following table shows cost and sales amounts for a company that sells sports equipment. Use a computer spreadsheet to fill in the missing information.

	Unit Cost	Number Produced	Fixed Cost	Total Cost	Unit Price	Revenue	Profit (Loss)
8.	$6.50	20	$550.00	$ 680	$86.00	$ 1,720	$ 1,040
9.	6.50	295	550.00	2467.50	86.00	25,370	22,902.50

5-1 **10.** Use the monthly payment formula $M = \frac{Pr(1 + r)^n}{(1 + r)^n - 1}$ to determine how much money will be saved by purchasing a $950 computer with monthly payments at an interest rate of 12% rather than 18%. The loan is to be repaid in 2 years. $64.99

6-1 **11.** Determine the effective interest rate on a credit card with an APR of 14%. 14.93%

ADDITIONAL ANSWERS

4. $350; no, he would need $369.99 to buy the TV on sale.

PROJECT 8–1: Companies Listed on the Major Stock Exchanges or Traded over the Counter

1. Find several companies of medium or large size that make high-quality products with which you are familiar.
2. In the financial section of your local public library, find your chosen companies' names in a standard reference manual such as *Standard & Poor's.* Make a note of the the telephone numbers of the four companies that appeal to you the most.
3. Phone each company. Ask the person responsible for shareholder relations to send you a copy of the most recent annual report and any other available information about the company. The vast majority of companies will respond favorably to your request.
4. The material from the companies may tell you whether any of the companies has a local branch. Find out what you can about that branch, such as the role it plays in its community and its employees' attitudes.
5. Track your companies' stock prices for several weeks in the newspaper.
6. Share with the rest of the class all of the information that you have about your companies.

Extension

On the basis of what you have learned, select the company that you think would be the best investment for at least the next 5 years. Share your reasons for this choice with the rest of the class.

A *number line* assigns each real number a point on the line. A number line is helpful in comparing numbers and examining some of the relationships between numbers, especially inequalities.

Graph each number on a number line and describe the position of each number.

Example 2.6

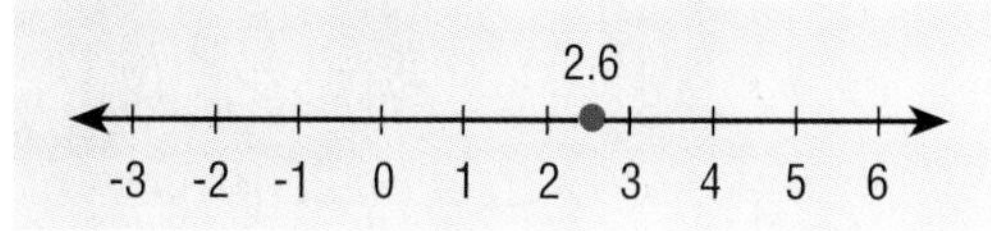

The number 2.6 is a little more than half way between 2 and 3.

1. -1.5 **2.** $6\frac{1}{3}$ **3.** $-\frac{8}{3}$ **4.** π **5.** $\sqrt{2}$ **6.** $(-2.5)^2$

See Additional Answers.

Inequalities can be represented on the number line. A *solid circle* at an end point means that the point is included in the set. An *open circle* means the number is excluded.

Graph each inequality on a number line.

Example $x \leq 7$

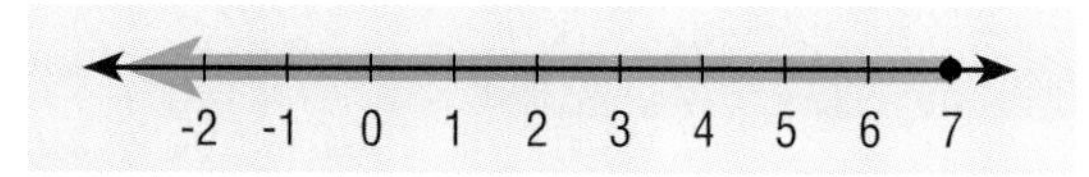

7. $x > 1$ **8.** $x < -2$ **9.** $-2 \leq x$ **10.** $5 \geq x$

See Additional Answers.

A *compound inequality* satisfies more than one condition. When two conditions are linked by "and," the solution set is the *intersection* of the sets satisfying the conditions. A number in both sets is in the solution set. Sometimes "and" is not written but it is clear that two conditions are given. When two conditions are linked by "or," the solution is the *union* of two sets. A number is in the solution set if it satisfies either one or the other condition or both conditions.

Graph the solution set on the number line.

Example: **a.** $-3 \leq x \leq 5$ **b.** $x < -1$ or $x \geq 2$

a. $-3 \leq x \leq 5$ means $-3 \leq x$ and $x \leq 5$

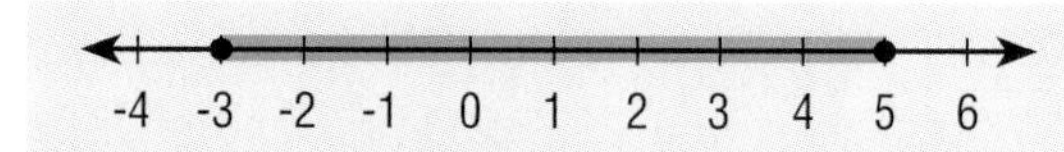

b. $x < -1$ or $x \geq 2$

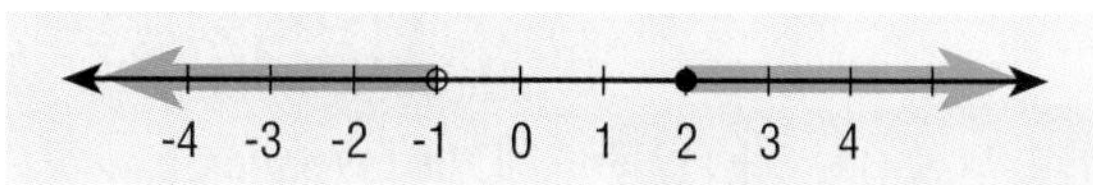

11. $1 \leq x < 10$ **12.** $x \leq -2$ or $x > 3$ **13.** $x \leq 5$ and $x > 0$

14. $1 \geq x$ or $1 \leq x$ **15.** $-5 < x < 0$ **16.** $0 < x$ and $x < 1$

See Additional Answers.

CHAPTER 9

Federal Paying the Price

WHETHER OR NOT WE AGREE WITH HOW our tax dollars are spent, most of us would not want to do without at least some of the services that our federal, state, and local governments provide.

Taxes are one of the largest items in the family budget. The average taxpayer works every year from January 1 to around early May just to cover tax obligations. However, most of the tax that we pay never goes through our hands; our employer withholds some of our pay for taxes and sends that withheld money directly to the Internal Revenue Service (IRS). In most communities, part of the withheld money goes also to the state and local governments for its taxes. If it were not for the discipline of having the government regularly withholding part of our pay, many of us might find it difficult to meet our tax obligation at all.

Even so, before every April 15, each of us must calculate our income for the previous year, and if it is above a certain minimum amount, we must file a tax form with the IRS. Since there are so many kinds of income, the IRS has devised forms to accommodate the various possibilities. These forms do take some getting used to.

From time to time, attempts have been made to simplify the forms for the benefit of the taxpayer, especially the taxpayer who does not have a complicated tax situation.

Income Tax:

Three of our high school students will examine the three tax forms that individuals most frequently encounter. The students will demonstrate how, for many individuals, the forms are fairly easy to complete. They will also come in contact with sources of assistance in filling out the forms. Unfortunately, these sources do not extend their assistance to making the tax payments themselves; for that, we are on our own!

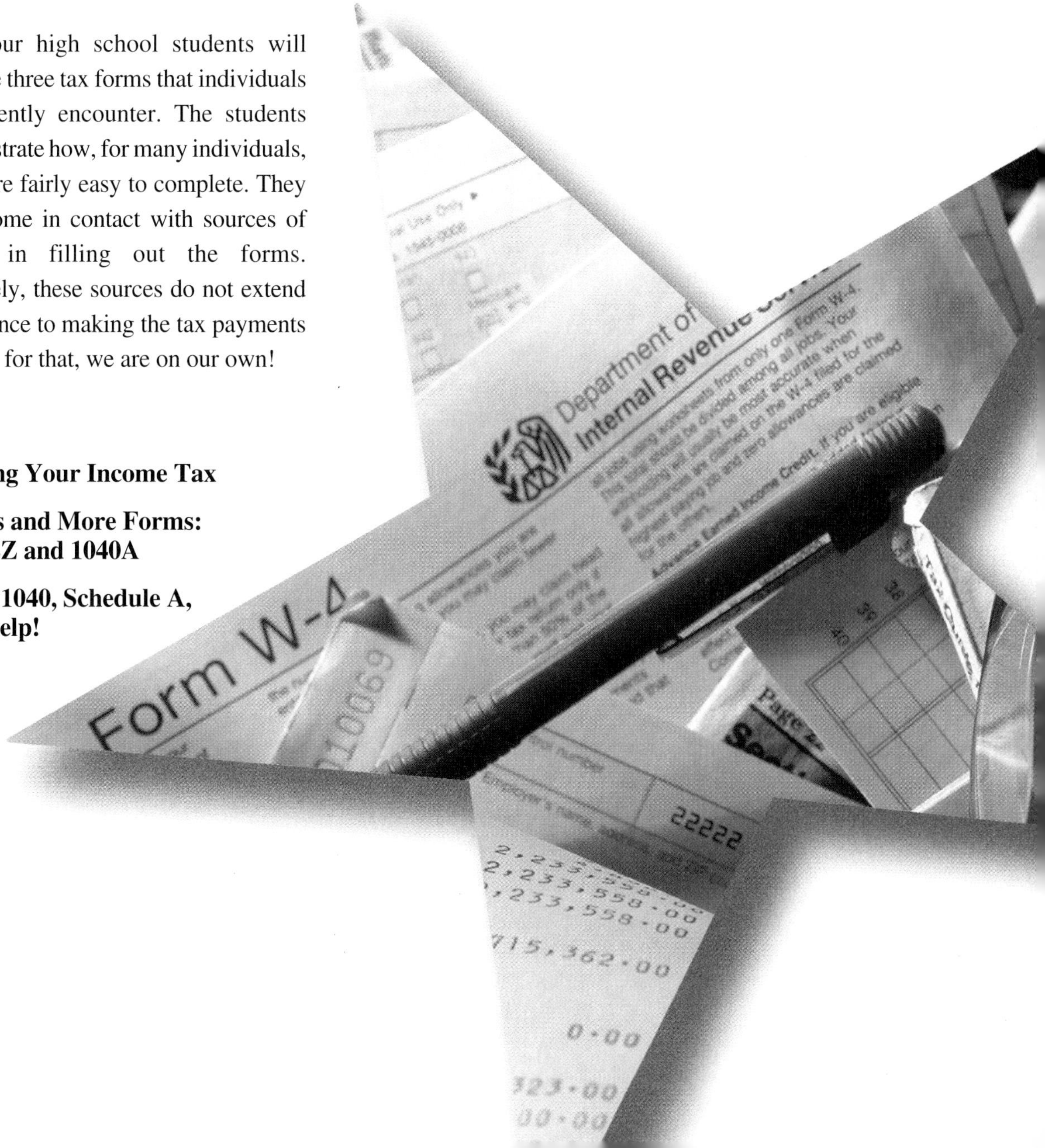

9–1 Finding Your Income Tax

9–2 Forms and More Forms: 1040EZ and 1040A

9–3 Form 1040, Schedule A, and Help!

9-1 FINDING YOUR INCOME TAX

QUICK REFERENCE

TEACHER'S RESOURCES AND ANSWER KEY

Lesson Quiz Answers 9-1, p. 72

Reteaching Activity Answers 9-1, p. 108

Enrichment Activity Answers 9-1, p. 108

EXTENSION ACTIVITIES

Reteaching Activity 9-1, p. 69

Enrichment Activity 9-1, p. 72

Luis had big plans for his first paycheck. His car needed some tires, and the amount that he expected to receive for his first week's work should have been just enough. For that reason, he was quite surprised when he received the check—$11.89 was missing! He examined the stub more closely and noticed that deductions had been made from his salary.

The largest deduction caught his attention: "Federal Income Tax Withheld." He remembered the mass of papers that his father and mother shuffle and sort once a year during the first week of April. The activity is usually accompanied by intense discussions about depreciation, Schedule A deductions, and other equally strange notions. In addition, there are invariably many unkind comments about wasteful government spending. Now it is Luis's turn to deal with income taxes.

He is determined to figure out what it all means. He received his W-2 form from his employer and knows that he must file a federal income tax return. He hopes to receive a refund but wonders whether the amount withheld from his paycheck will *really* be enough to cover his taxes. He plans to compare the withheld amount with the amount of the tax given in the federal tax table. If he gets back as much as he thinks, he will be able to buy a whole new set of tires for his car.

OBJECTIVES: *In this lesson, we will help Luis to:*

- *Realize how much of the year a person works to pay income taxes.*
- *Discover some of the characteristics of the federal income tax system.*
- *Find how much income tax will be owed in specific cases.*
- *Compare the tax owed with the amount withheld by the employer.*

THE PRICE TAG OF GOVERNMENT

Taxes are what we pay for civilized society.
Oliver Wendell Holmes, Jr.

In this world nothing is certain but death and taxes.
Benjamin Franklin

Since Luis first realized that all his earnings were being taxed by the federal government, he has noticed newspaper and magazine articles about taxes. One that recently caught his attention was entitled "Tax-Freedom Day." As he read, he discovered that the Tax Foundation Incorporated, a private organization, calculates an annual Tax Freedom Day, the day each year when the average American will stop working to pay for government services. To Luis this meant that if he were working full time and receiving a full salary, the amount of money that he would pay the government in taxes for the year would be equivalent to what he would earn from the beginning of January to May 1.

The table that follows shows how the cost of government as measured by the number of working days has increased considerably between 1930 and the present. The Tax Foundation estimates that typical workers pay approximately one third of their earnings into local, state, and federal taxes. Luis certainly hopes that these governments are using his money wisely.

TAX FREEDOM DAYS

Year	Date
1930	February 13
1940	March 8
1950	April 3
1955	April 9
1960	April 17
1965	April 15
1970	April 28
1975	April 28
1980	April 28
1985	May 1
1990	May 5
1991	May 8

ALGEBRA CONNECTIONS

Jason earns $400 monthly working at a restaurant. He puts 50% in the bank for college and spends $120 on car-related expenses. He uses 70% of the remainder for going out with his friends. How much money is left each month for Jason to spend on other things? (**$24**)

HOW DID THIS HAPPEN?

Luis wondered whether governments in our country have always needed so much money. As he looked into the matter, he discovered that for its first 100 years the United States covered its expenses by imposing taxes on certain goods manufactured here and on products brought in from other countries. Then our country changed from a society that was mainly rural and agricultural to one that was mainly industrial. This change was accompanied by a growth in the role played by the federal government. As a consequence, the federal government's need for income increased. The 16th Amendment, adopted in 1913, gave Congress the power to pass income tax laws. We now have taxes on income as well as on sales and property.

UNITED STATES INDIVIDUAL INCOME TAX

Luis also discovered some features of the elaborate process that we go through to supply funds to our government. The **individual income tax** is imposed on an individual's earnings from wages, salary, tips, interest, rents, dividends, and capital gains. It is the largest revenue-producing tax for the federal government. Each year, on or before April 15, U.S. citizens and other legal residents determine their tax liability and usually either pay the amount that is due or request a refund. The only other possibility is to allow the government to apply the refund toward next year's tax liability.

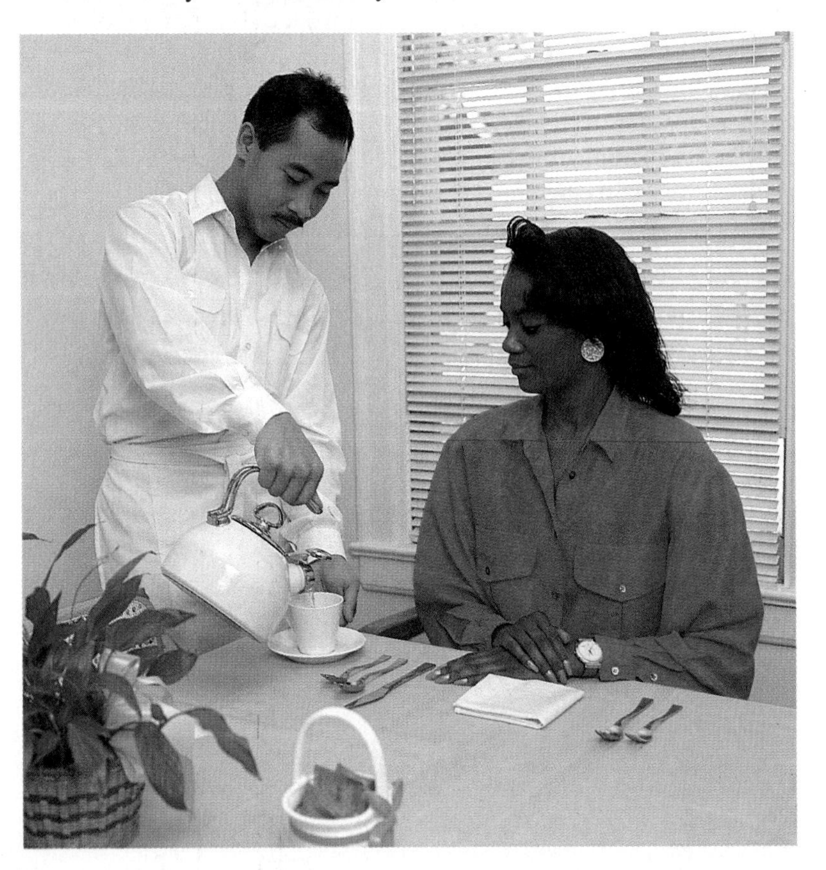

Here are some of the characteristics of our income tax system.

Ability to Pay The federal income tax rules are designed to match one's ability to pay. To accomplish this goal, the IRS has established a **progressive tax schedule,** one that takes an increasingly higher share of any additional income that you earn. That explains why Luis's parents will pay a higher percent of their income for taxes than he does. A person with a very low income may even owe no federal income tax at all.

Voluntary Compliance Another characteristic of the federal personal income tax is voluntary compliance. The Internal Revenue Service relies on Luis and the rest of us to meet our tax responsibilities faithfully. However, any U.S. citizen or other legal resident who does not pay taxes voluntarily can be fined or even imprisoned; tax evasion is illegal.

Pay-as-You-Earn Luis especially appreciates a feature of the federal tax system known as **pay-as-you-earn.** As we receive types of income that are subject to a tax, money is held back to pay the tax. These types of income include tips, pensions, gambling winnings, and, in Luis's case, wages. This system helps prevent Luis and other taxpayers from owing large amounts of taxes on April 15 of the following year. Luis's employer is responsible for withholding tax from Luis's paycheck and depositing it with the IRS. Luis learned how much money had been withheld from his paycheck during the year when he received his Form W-2 from his employer early in the following year.

1 Control number	22222	For Official Use Only ▶ OMB No. 1545-0008	
2 Employer's name, address, and ZIP code		6 Statutory employee ☐ Deceased ☐ Pension plan ☐ Legal rep. ☐ 942 emp. ☐ Subtotal ☐ Deferred compensation ☐ Void ☐	
		7 Allocated tips	8 Advance EIC payment
		9 Federal income tax withheld	10 Wages, tips, other compensation
3 Employer's identification number	4 Employer's state I.D. number	11 Social security tax withheld	12 Social security wages
5 Employee's social security number		13 Social security tips	14 Medicare wages and tips
19a Employee's name (first, middle initial, last)		15 Medicare tax withheld	16 Nonqualified plans
19b Employee's address and ZIP code		17 See Instrs. for Form W-2	18 Other
20	21	22 Dependent care benefits	23 Benefits included in Box 10

24 State income tax	25 State wages, tips, etc.	26 Name of state	27 Local income tax	28 Local wages, tips, etc.	29 Name of locality

Copy A For Social Security Administration Cat. No. 10134D Department of the Treasury—Internal Revenue Service

Form **W-2 Wage and Tax Statement 19--**

For Paperwork Reduction Act Notice and instructions for completing this form, see separate instructions.

FORM W-2

Every employer for whom Luis works during the year must provide him with a W-2 form for the period of time during which he worked. This form provides a record of wages earned as well as federal, state, and other taxes withheld during the year. When Luis files his tax return, he must submit a copy of each of his W-2 forms at the same time.

TAX TABLES

You can find your **tax liability,** the amount you owe, in the **tax tables,** which you can obtain from the IRS. (A typical set of these tables are in the Reference Section at the back of the book.) Before you can use the tables, however, you need to find your *taxable income.* **Taxable income** is the amount of money that you actually make during the year *minus* certain adjustments, deductions, and exemptions that the government allows.

After you know your taxable income, you use the column in the table that is appropriate for your filing status. A portion of a tax table is shown in Skill 1. There are columns for single people, married people filing jointly, married people filing separately, and heads of households. Read down the income column until you find your taxable income. Then read across to the column that applies to you. The amount shown is your tax. (For married couples, filing a **joint return** often results in lower taxes, especially if one income is lower than the other.)

ALGEBRA *REVIEW*

Evaluate each expression for the indicated value of *I*.

1. $0.15I$ for $I = 5{,}600$
   840
2. $1200 + 0.25\,(I - 7800)$ for $I = 8000$
   1250
3. $0.15I$ for $I = 21{,}430$
   3214.50
4. $3217.50 + 0.28(I - 21{,}450)$ for $I = 21{,}470$
   3223.10
5. $3217.50 + 0.28(I - 21{,}450)$ for $I = 51{,}890$
   11,740.70
6. $11{,}743.50 + 0.31(I - 51{,}900)$ for $I = 51{,}910$
   11,746.60
7. $11{,}743.50 + 0.31(I - 51{,}900)$ for $I = 75{,}750$
   19,137.00
8. $11{,}743.50 + 0.31(I - 51{,}900)$ for $I = 105{,}380$
   28,322.30

Ask Yourself

1. What is an income tax?
   A tax on an individual's income.
2. How long does the average person have to work in a year to earn the amount of money that he or she pays to the government in taxes?
   January 1 to early May
3. What are two characteristics of our income tax system?
   ability to pay; voluntary compliance; pay as you earn

SHARPEN YOUR SKILLS

SKILL 1

EXAMPLE 1 Elizabeth's taxable income for the past year was $2,275.

QUESTION How much tax does Elizabeth, who is single, owe?

SOLUTION

Use the portion of the Tax Table shown below. Read down the income column headed "If line 5 (Form 1040EZ), line 22 (Form 1040A), or line 37 (Form 1040) is—" to find 2,275.

Notice that 2,275 is listed twice, once under the column marked "At least" and once under "But less than." Use the line where 2,275 is shown under the "At least" column. Read across to the column headed "Single." The amount in that column is 343.

Elizabeth owes $343.00 in income tax for the year.

Tax Table

If line 5 (Form 1040EZ), line 22 (Form 1040A), or line 37 (Form 1040) is–		And you are—			
At least	But less than	Single	Married filing jointly •	Married filing sepa-rately	Head of a house-hold
		Your tax is—			
1,900	1,925	287	287	287	287
1,925	1,950	291	291	291	291
1,950	1,975	294	294	294	294
1,975	2,000	298	298	298	298
2,000					
2,000	2,025	302	302	302	302
2,025	2,050	306	306	306	306
2,050	2,075	309	309	309	309
2,075	2,100	313	313	313	313
2,100	2,125	317	317	317	317
2,125	2,150	321	321	321	321
2,150	2,175	324	324	324	324
2,175	2,200	328	328	328	328
2,200	2,225	332	332	332	332
2,225	2,250	336	336	336	336
2,250	2,275	339	339	339	339
2,275	2,300	343	343	343	343
2,300	2,325	347	347	347	347
2,325	2,350	351	351	351	351
2,350	2,375	354	354	354	354
2,375	2,400	358	358	358	358

If line 5 (Form 1040EZ), line 22 (Form 1040A), or line 37 (Form 1040) is–		And you are—			
At least	But less than	Single	Married filing jointly •	Married filing sepa-rately	Head of a house-hold
		Your tax is—			
3,400	3,450	514	514	514	514
3,450	3,500	521	521	521	521
3,500	3,550	529	529	529	529
3,550	3,600	536	536	536	536
3,600	3,650	544	544	544	544
3,650	3,700	551	551	551	551
3,700	3,750	559	559	559	559
3,750	3,800	566	566	566	566
3,800	3,850	574	574	574	574
3,850	3,900	581	581	581	581
3,900	3,950	589	589	589	589
3,950	4,000	596	596	596	596
4,000					
4,000	4,050	604	604	604	604
4,050	4,100	611	611	611	611
4,100	4,150	619	619	619	619
4,150	4,200	626	626	626	626
4,200	4,250	634	634	634	634
4,250	4,300	641	641	641	641
4,300	4,350	649	649	649	649
4,350	4,400	656	656	656	656

SKILL 2

Taxpayers whose taxable income is over $100,000 do not use tax tables. Instead, they use the **Tax Rate Schedules** X, Y-1, Y-2, or Z. Schedule X, for filers who are single, is shown below.

Tax Rate Schedules

Caution: *Use **only** if your taxable income (Form 1040, line 37) is $100,000 or more. If less, use the **Tax Table.** Even though you cannot use the tax rate schedules below if your taxable income is less than $100,000, all levels of taxable income are shown so taxpayers can see the tax rate that applies to each level.*

Schedule X—Use if your filing status is **Single**

If the amount on Form 1040, line 37, is: *Over—*	*But not over—*	Enter on Form 1040, line 38	*of the amount over—*
$0	$21,450	.... **15%**	**$0**
21,450	51,900	**$3,217.50 + 28%**	**21,450**
51,900	....	**11,743.50 + 31%**	**51,900**

INSTRUCTION

TEACHING THE LESSON

Spend two days on this lesson.

On the first day, have students read the opening text. Discuss characteristics of the federal income tax system. After reviewing Skills 1 and 2, tell students to practice using tables and tax formulas by making up other examples (incomes of $4100; $23,050; $96,000) and determining the tax liability in each case.

On the second day, explain Skill 3 and vary Luis' income to give added practice in computing. Make sure students understand the concepts listed in *Exercise Your Skills*.

You can construct a formula for any tax that is calculated using the Tax Rate Schedules. (Recall that the symbol $<$ means "is less than" and the symbol $\leq$ means "is less than or equal to.")

Formulas for Tax Rate Schedule X (for *single* taxpayers)

i. $t = 0.15I$ if $I \leq 21{,}450$

ii. $t = 3{,}217.50 + 0.28(I - 21{,}450)$ if $21{,}450 < I \leq 51{,}900$

iii. $t = 11{,}743.50 + 0.31(I - 51{,}900)$ if $51{,}900 < I$

where I = taxable income
t = tax on the income

Although only taxpayers with income over $100,000 use the Tax Rate Schedules, the above formulas can be used by *any* single taxpayer as a rough check of the tax obtained from the Tax Table. The difference between the two methods of finding the tax may be about 5 or 10 dollars, although for someone with a high income the difference can be almost 20 dollars. The Tax Table, the Tax Rate Schedules, and other tax-related forms and schedules are those that were in effect for 1992. From time to time, tax laws are changed so that the particular details shown here may differ from those currently in effect. However, the overall principles and procedures affecting your tax obligations have not significantly changed.

EXAMPLE 2 Marilyn's taxable income is $102,000.

QUESTION How much does Marilyn, who is single, pay in taxes?

SOLUTION

Since $51{,}900 < I$, use the third Schedule X formula.

$$t = 11{,}743.50 + 0.31(102{,}000 - 51{,}900)$$
$$t = 27{,}274.50$$

Marilyn's tax is $27,274.50.

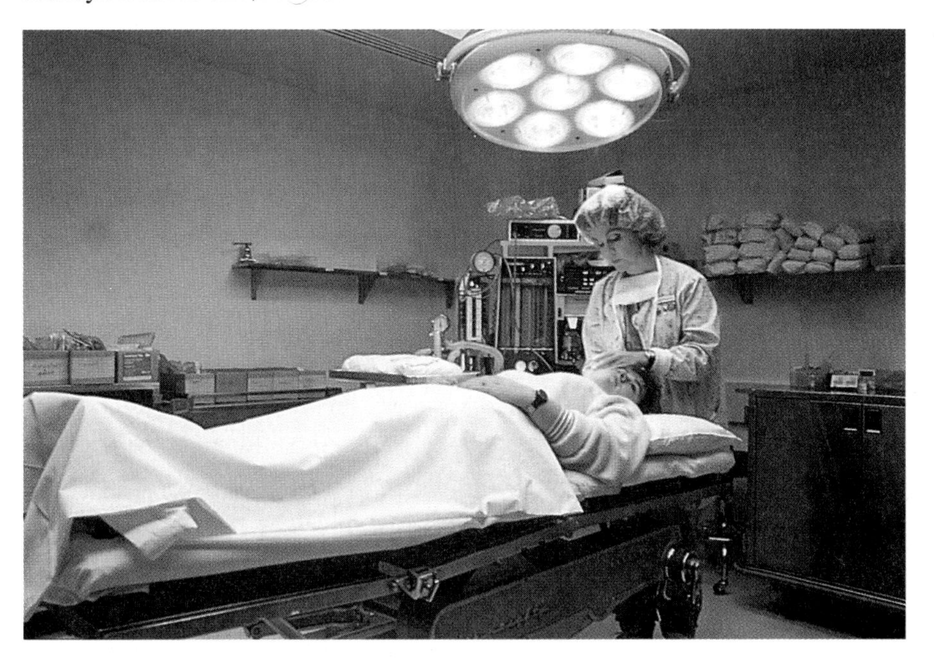

SKILL 3

To find taxable income I, you must reduce the total income T by any *deductions D* and any *exemptions E.*

Taxable income = total income − (deductions + exemptions)

$$I = T - (D + E)$$

Deductions and exemptions will be explained more fully in later lessons.

EXAMPLE 3 Luis's monthly income for this year was $675. He had a deduction of $3600 and an exemption of $2300. He has one withholding allowance, and he is single.

QUESTIONS

1. What was Luis's total income for the year?
2. What was his taxable income?
3. How much income tax does he owe?
4. How much was withheld from his paychecks?
5. Does he have a refund? If so, how much is it?

FOCUS ON ALGEBRA

In this lesson, students work with linear functions and with inequalities. *(NCTM Standard 6, p. 154)*

CRITICAL THINKING

Ask students to consider the purpose of "number of exemptions" and "withholding allowances." Have them compare the tax owed by a single person and a married person with four exemptions and a withholding allowance of 5 if the taxable income in each case is $42,000.

COOPERATIVE LEARNING

Have students work in pairs as partners. Tell each student to make up a profile of a taxpayer with a filing status and income. Have each student work out the tax owed for his partner's hypothetical taxpayer.

INTERDISCIPLINARY INVESTIGATION

History: Have students research the events leading to the passage of the Sixteenth Amendment.

COMMON ERRORS

Students may use monthly incomes to look up tax rates on the Tax Table. Point out that monthly income is used on the withholding table but that annual income is used on the Tax Table.

RETEACHING

Students may need additional practice in determining which table or schedule to use in different situations. Provide more examples for them to look up. Do problems involving withholding allowances and numbers of exemptions as well.

ENRICHMENT

Challenge students to rewrite the formulas for Tax Rate Schedule X to reflect the rates of 15%, 28%, and 31% on different portions of a taxpayer's income (hint: $.15 \cdot 21{,}450 = 3{,}217.50$).

SOLUTIONS

1. His total yearly income was

   $675 \cdot 12$, or $8100

2. To find Luis's taxable income, you must reduce his total income by a *deduction* of $3600 and an *exemption* of $2300. Luis's taxable income is

   $I = T - (D + E)$
   $I = 8100 - (3600 + 2300)$
   $I = 2200$

   Luis's taxable income is $2200.

3. Use the Tax Table in the Reference Section or the abbreviated table in this section. The tax on $2200 is $332.

4. To find the amount that would have been withheld from Luis's check, look at the withholding tables for Single Persons—Monthly Payroll Period. A portion of that table is shown below.

SINGLE Persons—MONTHLY Payroll Period

If the wages are—		And the number of withholding allowances claimed is—					
		0	1	2	3	4	5
At least	But less than	The amount of income tax to be withheld is—					
500	520	45	16	0	0	0	0
520	540	48	19	0	0	0	0
540	560	51	22	0	0	0	0
560	580	54	25	0	0	0	0
580	600	57	28	0	0	0	0
600	640	61	32	3	0	0	0
640	680	67	38	9	0	0	0
680	720	73	44	15	0	0	0
720	760	79	50	21	0	0	0
760	800	85	56	27	0	0	0
800	840	91	62	33	3	0	0
840	880	97	68	39	9	0	0
880	920	103	74	45	15	0	0
920	960	109	80	51	21	0	0
960	1,000	115	86	57	27	0	0

   In the table, locate Luis's monthly income of $675. It falls on the line that reads "At least 640, But less than 680." Use the column for one withholding allowance, the number that Luis chose when he completed his Form W-4 (see Lesson 1–4). Under that column, find the amount withheld, $38 per month. To find the yearly withholding, multiply the monthly amount by 12: $38 \cdot 12 = 456$. The withheld amount was $456.

5. The amount owed is less than the amount already withheld. The difference is $456 - 332$, or 124. Luis can expect a refund of $124.

TRY YOUR SKILLS

Use the Tax Table in this lesson or in the Reference Section to find the taxes owed on each of the following incomes. (In Exercise 4 a "head of household" could be a single parent or a person caring for an elderly parent.)

	Filing Status	Taxable Income	Tax Owed
1.	Single	$ 2,050	$ 309
2.	Married, filing jointly	2,200	332
3.	Married, filing separately	2,268	339
4.	Head of household	2,132	321
5.	Single	11,000	1,654
6.	Married, filing jointly	27,500	4,129
7.	Head of household	42,000	8,030
8.	Head of household	52,600	10,998

GUIDED PRACTICE

Have students work in small groups on the *Try Your Skills* exercises. Tell them to compare results and to make sure that each group member understands how to use the tables and schedules.

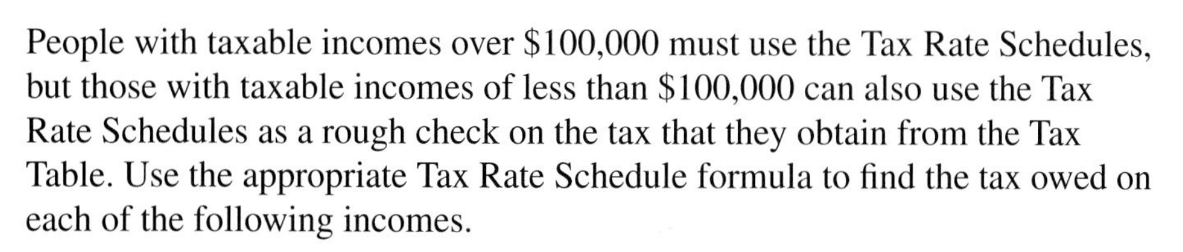

People with taxable incomes over $100,000 must use the Tax Rate Schedules, but those with taxable incomes of less than $100,000 can also use the Tax Rate Schedules as a rough check on the tax that they obtain from the Tax Table. Use the appropriate Tax Rate Schedule formula to find the tax owed on each of the following incomes.

	Filing Status	Taxable Income	Formula	Tax Owed
9.	Single	$ 2,245	$t = 0.15I$	$ 336.75
10.	Single	23,000	$t = 3217.50 + 0.28(I - 21,450)$	3,651.50
11.	Single	104,000	$t = 11,743.50 + 0.31(I - 51,900)$	27,894.50
12.	Single	151,900	$t = 11,743.50 + 0.31(I - 51,900)$	42,743.50
13.	Single	132,000	$t = 11,743.50 + 0.31(I - 51,900)$	36,574.50
14.	Single	185,000	$t = 11,743.50 + 0.31(I - 51,900)$	53,004.50
15.	Single	200,000	$t = 11,743.50 + 0.31(I - 51,900)$	57,654.50

16. The taxpayers of Exercises 9 and 10 are not permitted to use the Tax Rate Schedule. Why not? If they were, which taxpayer would have a lower tax by using the Tax Rate Schedule than by using the Tax Table?
The tax rate schedule is for use only for incomes over $100,000; the tax on an income of $23,000 is $7.50 less using the Tax Rate Schedule.

EXERCISE YOUR SKILLS

1. Why is it necessary for people to pay federal income taxes? See Additional Answers.
2. What is the date by which personal income tax returns must be filed for the previous year? April 15
3. Does the voluntary compliance nature of our income tax system mean that you legally have the right not to pay if you wish? If not, what does it mean?

Use the Tax Table in the Reference Section of the book to find the taxes owed on each of the following incomes.

	Filing Status	Taxable Income	Tax Owed
4.	Married, filing jointly	$ 8,870	$ 1,331
5.	Married, filing separately	47,850	11,217
6.	Head of household	7,280	1,091
7.	Single	10,170	1,526

Use Tax Rate Schedule X or Y-1 in the Reference Section to find the tax owed on each of the incomes given in Exercises 8–12. (For taxpayers with taxable incomes less than $100,000, this method serves only as a rough check on the actual tax found in the tax table.) Use the two sets of formulas shown below, where I = taxable income and t = tax on the income.

Formulas for Tax Rate Schedule X (for single taxpayers)

i.	$t = 0.15I$	if $I \leq 21{,}450$
ii.	$t = 3{,}217.50 + 0.28(I - 21{,}450)$	if $21{,}450 < I \leq 51{,}900$
iii.	$t = 11{,}743.50 + 0.31(I - 51{,}900)$	if $51{,}900 < I$

Formulas for Tax Rate Schedule Y-1 (married filing jointly or qualifying widow or widower)

i.	$t = 0.15I$	if $I \leq 35{,}800$
ii.	$t = 5{,}370.00 + 0.28(I - 35{,}800)$	if $35{,}800 < I \leq 86{,}500$
iii.	$t = 19{,}566.00 + 0.31(I - 86{,}500)$	if $86{,}500 < I$

KEY TERMS

- individual income tax
- joint return
- pay-as-you-earn
- progressive tax schedule
- tax liability
- Tax Rate Schedules
- tax tables
- taxable income

ASSIGNMENT GUIDE

Exercises 1-3 ask students to display their understanding of the opening text.

Exercises 4-7 concentrate on using a Tax Table to determine tax owed based on filing status and income.

Exercises 8-12 use Tax Rate Schedules X and Y-1 to determine tax owed.

Exercises 13-15 ask students to compare results using the Tax Table and the Tax Rate Schedule.

Exercises 16-27 require students to calculate yearly income, taxable income, taxes owed and amount withheld.

Exercise 28 challenges students to write formulas for Tax Rate Schedule Y-2.

Exercise 29 asks students to apply the formulas from Exercise 28.

ADDITIONAL ANSWERS

1. Income tax is the largest revenue-producing tax the government has.
3. No; instead of the government telling us what we owe, we inform them

	Filing Status	Taxable Income	Tax Owed
8.	Single	$107,000	$28,824.50
9.	Married, filing jointly	138,500	35,686.00
10.	Qualifying widow	30,750	4,612.50
11.	Single	42,654	9,154.62
12.	Qualifying widow	68,756	14,597.68

For the taxpayers of Exercises 10–12, find their actual tax using the Tax Table in the Reference Section. Then answer Exercises 13–15.

13. Which of the three taxpayers has the greatest dollar difference between the two methods of calculating the tax? Taxpayer #11

14. Which of the three differences is the greatest when expressed as a percent of the tax found by using the Tax Table? Taxpayer #10

15. Which, if any, of the taxpayers of Exercises 10–12 would pay a lower tax if he or she were allowed actually to use the Tax Rate Schedule? all three

Find the yearly income, the taxable income, the taxes owed, and the amounts to be withheld annually for the incomes described below. Use a deduction of $3600 for a single person, $6000 for a married couple, and $5250 for a head of household. Each exemption is $2300. Assume that the married couples are filing a joint return and that the head of household is single. Use the Tax Table where possible. Otherwise, use the Tax Rate Schedules.

	Filing Status	Monthly/ Yearly Income	Number of Exemptions	Taxable Income	Tax Owed	Withholding Allowances		Refund (+) or Money Owed (−)
						Number	Annual Amount	
16.	Single	$ 825/9900	1	$ 4,000	$ 604	1	$744	+$140
17.	Single	1555/18,660	1	12,760	1,916	1	2,040	+ 124
18.	Single	785/9420	1	3,520	529	0	1,020	+ 491
19.	Single	2378/28,536	0	24,936	4,191	0	4,572	+ 381
20.	Single	4634/55,608	0	52,008	11,782	1	11,556	− 226
21.	Married	3500/42,000	4	26,800	4,024	4	3,960	− 64
22.	Married	3777/45,324	5	27,824	4,174	6	3,756	− 418
23.	Married	5300/63,600	2	53,000	10,193	2	10,260	+ 67
24.	Married	4200/50,400	5	32,900	4,939	4	5,316	+ 377
25.	Head of household	1780/21,360	3	9,210	1,384	3	1,764	+ 380
26.	Head of household	2116/25,392	2	15,542	2,329	2	2,700	+ 371
27.	Head of household	3209/38,508	4	24,058	3,611	6	3,444	− 167

CLOSURE

Have students write a paragraph discussing characteristics of the federal income tax system and explaining how to compute tax owed. Tell them to take into account variations in filing status and income.

LESSON QUIZ

1. Use the Tax Table in the Reference Section of the book to determine taxes owed by a single person with a taxable income of $26,140. (**$4527**)
2. Use the Tax Rate Schedule X or Y-1 in the Reference Section of the book to find the tax owed by a married couple filing jointly with a taxable income of $106,250. (**$25,688.50**)

ADDITIONAL ANSWERS

28. i. $t = 0.15I$ if $I \leq 17{,}900$

ii. $t = 2685 + 0.28(I - 17{,}900)$ if $17{,}900 < I \leq 43{,}250$

iii. $t = 9783 + 0.31(I - 43{,}250)$ if $43{,}250 \leq I$

p 671

28. Write a set of algebraic formulas for Tax Rate Schedule Y-2 (for married couples filing separately), which is in the Reference Section.

29. Because of unusually large medical expenses, a married woman finds that it is to the family's advantage for her and her husband to file separately one year rather than jointly. Her taxable income that year is $111,500. Use one of the formulas that you wrote in Exercise 28 to calculate her income tax. $30,940.50

MIXED REVIEW

1-3 **1.** Daniel has a small business grooming dogs. He charges $20 for each dog that he grooms but gives a 10% discount to each customer who brings him a new customer. One week he had 27 customers, six of whom brought in a new customer. What were Daniel's gross earnings that week? $528

4-1 Use the table for weekly withholding amounts for single persons in the Reference Section and the fact that 7.65% of gross pay is withheld for FICA taxes to find the missing entries in the payroll summary below. Assume that the number of claimed withholding allowances is 1.

	Week Ending	Gross Pay	Withholding	FICA	Take-home Pay
2.	5/17	$115.00	$ 4	$ 8.80	$102.20
3.	5/17	$287.00	29	21.96	236.04

6-5 **4.** Your credit card balance from 4/1–4/12 was $750. You made a payment of $125 on 4/13. You made no further charges or payments for the remainder of the month. The monthly finance charge is 1.25% and is calculated on the last day of the month. Determine your average daily balance for April. $675

3-3 Make a table to show the multiplier effect on the amounts below if the Federal Reserve requirement is 20%. Show the first five levels and find the total extra money that is generated at the fifth level. See Additional Answers.

5. Initial deposit: $1,200 **6.** Initial deposit: $10,000

7.–8. For the initial deposits of Exercises 5–6, find the total amount of new money created by the multiplier effect. **7.** $4,800 **8.** $40,000

5-4 **9.** You have a loan of $17,500 at 9.5% for 5 years. Your monthly payments are $367.53. Determine how much money you will save if you prepay the loan at the end of 30 months. $1245.80

4-3 Some students purchased some canvas bags for $2.00 each and painted them for $1.50 each. They plan to sell the painted canvas bags.

10. What is the unit cost for purchasing and painting the canvas bags? $3.50

11. Find the cost for purchasing and painting 56 canvas bags. $196.00

12. The fixed costs are $75. What are the total costs? $271.00

9-2 FORMS AND MORE FORMS: 1040EZ AND 1040A

QUICK REFERENCE

TEACHER'S RESOURCES AND ANSWER KEY

Lesson Quiz Answers 9-2, p. 71

Reteaching Activity Answers 9-2, p. 108

Enrichment Activity Answers 9-2, p. 109

EXTENSION ACTIVITIES

Reteaching Activity 9-2, p. 70

Enrichment Activity 9-2, p. 73

Larry learned his lesson about automated teller machines the hard way. He still remembers the Saturday afternoon when he took so much cash from half a dozen different ATMs around town that he had to rush to the bank on Monday to put some of the money back in. He even had to borrow $100 from his mother to cover some of the checks that he had already written on his checking account. Since that weekend, he has been working to pay her back. She has hired him to do everything from mowing the lawn to doing the laundry.

Larry also has an outside job as a clerk at a sporting goods store. When he received his W-2 form in January, Larry was amazed that his earnings had been so high. But where had all that money gone? He knew that he had spent quite a lot on his car. Then, of course, he and Lorrie Anne had gone out most weekends. Larry had bought a few gifts for her, too. Even as he was spending most of his paycheck, Larry knew that he was supposed to be saving for college.

With all of his failure to control his spending, Larry has one consolation; the income taxes on his earnings have been withheld throughout the year. He would certainly be in debt to his mother if he had to borrow $309 to pay the taxes. As it is, Larry will actually get a refund. All he has to do now is fill out the proper form. Larry isn't sure which form to use, the 1040EZ or the 1040A. He wonders whether his choice of a form will affect the time he will have to wait for his refund.

OBJECTIVES: *In this lesson, we will help Larry and his parents to:*

- *Choose one of the two forms 1040EZ or 1040A.*
- *Fill out Form 1040EZ for Larry.*
- *Fill out Form 1040A for Larry's parents.*

FOCUS

ALGEBRA CONNECTIONS

Rachel has four hours 20 minutes to do her homework. She allots $\frac{1}{4}$ of her time to history and 40% of the remaining time to math. After completing these two subjects she splits the rest of her homework time evenly between science, English and Spanish. What percentage of Rachel's homework time does she spend on science? (**15%**)

WHICH FORM IS THE RIGHT ONE?

Larry would like to have his refund as soon as possible. For one thing, he has not saved very much of the income from his job to use at college next year. He imagines that if he puts his tax refund into a savings account, his mother will notice that he is at least making an effort to save! On the other hand, he really could use a jacket to go with the new skis that he just bought.

Larry knows that the IRS will not send his refund at all if he does not send in a properly completed form. The question is, which form? He can choose from the regular 1040, the 1040A, or the 1040EZ. Larry hopes that he can use the 1040EZ, since it sounds "E-Z" to fill out.

Form 1040EZ As Larry will see when he reads the instructions, if he uses **Form 1040EZ,** he must meet all the following requirements:

1. His filing status is "single."
2. He does not claim any dependents.

3. He was under 65 for the entire year and not blind at the end of the year.
4. His taxable income was less than $50,000.
5. His income included only wages, salaries, tips, taxable scholarship or fellowship grants, and taxable interest of $400 or less.
6. He did not receive any advance earned-income credit payments.
7. He was not a nonresident alien at any time during the year.
8. His wages were not over $55,000 if he had more than one employer.

If he does not meet these requirements, he must use Form 1040 if he has deductions to itemize (see Lesson 9-3) or Form 1040A if he has none.

Form 1040A If Larry chooses Form 1040A, he can still get his refund. The **Form 1040A** is longer and more complex than Form 1040EZ because it can be used by people who have a more complicated tax situation. It is the form that his parents will use. To use it, you must know a little more about *exemptions, dependents,* and *deductions.*

The filer is the person whose name first appears at the beginning of a joint return. For a nonjoint return, it is the only person whose name appears at the beginning of the return. The IRS allows you to subtract $2300 from your gross income for each exemption that you can claim. An **exemption** may normally be claimed for each family member, including the filer. However, except for the filer's spouse, a family member may be counted as an exemption only if he/she meets the IRS's tests for being a **dependent.** If you are a dependent of your parents (or of someone else), then you cannot claim an exemption for yourself.

The qualifications for being a dependent are fairly complicated. For a complete account of the rules, you should obtain an IRS publication such as *Your Federal Income Tax, for Individuals* (Publication 17). (All IRS publications are free.) However, the basic requirements, all of which must be satisfied, are that a dependent must:

- Be a relative of the filer or a member of the filer's household
- Be a citizen or resident of the United States
- Have a gross income (before adjustments) of less than $2300
  unless he or she is under 19 and is a child of the filer, or
  unless he or she is under 24 and qualifies as a student and is a child of the filer
- Have received more than half of his or her support from the filer

Every person filing a return may subtract a number called a **deduction.** People with expenses such as large medical bills, mortgage interest, and charitable gifts may be able to list, or itemize, these expenses and subtract them from their income as **itemized deductions** on Schedule 1040, as described in Lesson 9–3. The IRS allows every filer to choose the larger of his or her itemized deductions or a fixed number called a **standard deduction.** Most people choose the standard deduction, since it is usually larger than the itemized deduction.

People who use Form 1040A can find the standard deduction that they are allowed by referring to the standard deduction charts in the booklet that accompanies the form. They contain the same information as shown in the following charts, which have been adapted from IRS Publication 17.

Standard Deduction Tables

Standard Deduction Chart for Most People*

If Your Filing Status is:	Your Standard Deduction Is:
Single	$3,600
Married filing joint return or Qualifying widow(er) with dependent child	6,000
Married filing separate return	3,000
Head of household	5,250

* DO NOT use this chart if you were 65 or older or blind, OR if someone can claim you as a dependent.

Standard Deduction Chart for People Age 65 or Older or Blind*

Check the correct number of boxes below. Then go to the chart.

You — 65 or older ☐ Blind ☐

Your spouse, if claiming spouse's exemption — 65 or older ☐ Blind ☐

Total number of boxes you checked ☐

If Your Filing Status is:	And the Number in the Box Above is:	Your Standard Deduction is:
Single	1	$4,500
	2	5,400
Married filing joint return or Qualifying widow(er) with dependent child	1	6,700
	2	7,400
	3	8,100
	4	8,800
Married filing separate return	1	3,700
	2	4,400
	3	5,100
	4	5,800
Head of household	1	6,150
	2	7,050

* If someone can claim you as a dependent, use the worksheet in Table 20-3, instead.

Caution: If you are married filing a separate return and your spouse itemizes deductions, or if you are a dual-status alien, you cannot take the standard deduction even if you were 65 or older or blind.

Standard Deduction Worksheet for Dependents*

If you were 65 or older or blind, check the correct number of boxes below. Then go to the worksheet.

You	65 or older ☐	Blind ☐
Your spouse, if claiming spouse's exemption	65 or older ☐	Blind ☐

Total number of boxes you checked ☐

1. Enter your **earned income** (defined below). If none, go on to line 3	**1.**________
2. Minimum amount	**2.** $600
3. Compare the amounts on lines 1 and 2. Enter the **larger** of the two amounts here	**3.**________
4. Enter on line 4 the amount shown below for your filing status. • Single, enter $3,600 • Married filing separate return, enter $3,000 • Married filing jointly or Qualifying widow(er) with dependent child, enter $6,000 • Head of household, enter $5,250	**4.**________
5. Standard deduction. **a.** Compare the amounts on lines 3 and 4. Enter the **smaller** of the two amounts here. If under 65 and not blind, stop here. This is your standard deduction. Otherwise, go on to line 5b	**5a.**________
b. If 65 or older or blind, multiply $900 ($700 if married or qualifying widow(er) with dependent child) by the number in the box above. Enter the result	**5b.**________
c. Add lines 5a and 5b. This is your standard deduction for 1992.	**5c.**________

Earned income *includes wages, salaries, tips, professional fees, and other compensation received for personal services you performed. It also includes any amount received as a scholarship that you must include in your income.*

* Use this worksheet ONLY if someone can claim you as a dependent.

Ask Yourself

1. What is the maximum taxable income that a person may have and still be allowed to use Form 1040EZ?
   $49,999.99
2. What basic requirements must a person meet to be claimed as a dependent?
   See Additional Answers.
3. What kinds of income besides salaries must be reported on federal tax forms?

ALGEBRA *REVIEW*

Graph the function described by the given function rule.

Example $y = 0.50x$ if $x < 4$
$y = 1 + 0.25x$ if $x \geq 4$

Solution

Make a table of ordered pairs. Then graph.

x	−2	0	1	2	3	4	5	6	8
y	−1	0	0.50	1	1.5	2	2.25	2.5	3

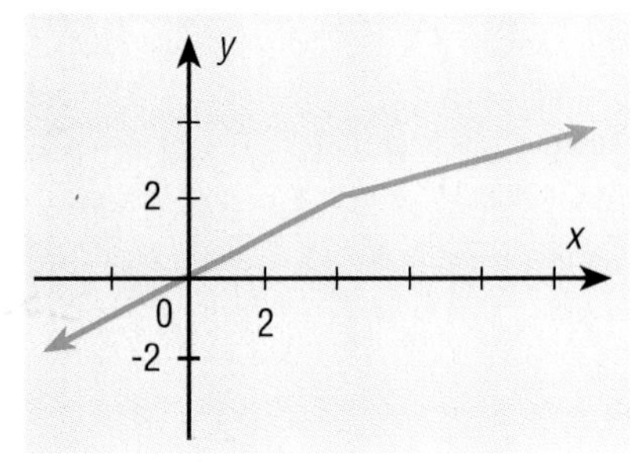

1. $y = 0.3x$ for all x

2. $y = 0.3x$ if $x < 0$
   $y = x$ if $x \geq 0$

3. $y = x$ if $x < 2$
   $y = 2x - 2$ if $x \geq 2$

4. $y = -1$ if $x < 0$
   $y = x^2 - 1$ if $x \geq 0$

5. $y = -x$ if $x \leq -1$
   $y = x + 2$ if $-1 < x \leq 3$
   $y = 2x - 1$ if $x > 3$

6. $y = 0.2x$ if $0 \leq x \leq 10$
   $y = 2 + 0.5(x - 10)$ if $10 < x \leq 16$
   $y = 5 + 0.8(x - 16)$ if $x > 16$
   See Additional Answers.

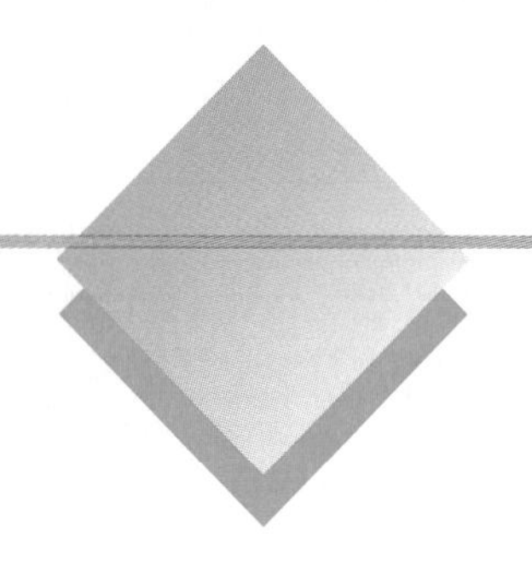

SHARPEN YOUR SKILLS

SKILL 1

EXAMPLE 1 Larry meets the requirements for using Form 1040EZ.

QUESTION How will he fill out the form to determine the tax he owes?

SOLUTION

Larry completed the form as shown on page 426. He used the Tax Table and the instructions on the back of the form. Then he followed these steps.

Step 1 Larry filled in his name, address, and Social Security number.

Step 2 Larry reported all of his wages on line 1 and the interest from his savings account on line 2. He then added lines 1 and 2 to get his **adjusted gross income** on line 3.

Step 3 Larry used a worksheet to enter "Yes" and "3600" on line 4 of the form. The worksheet is shown below.

Standard Deduction Worksheet for Dependents*

If you were 65 or older or blind, check the correct number of boxes below. Then go to the worksheet. You 65 or older ☐ Blind ☐ Your spouse, if claiming spouse's exemption 65 or older ☐ Blind ☐ **Total number of boxes you checked** ☐	
1. Enter your **earned income** (defined below). If none, go on to line 3	1. 5,460
2. Minimum amount	2. $600
3. Compare the amounts on lines 1 and 2. Enter the **larger** of the two amounts here	3. 5,460
4. Enter on line 4 the amount shown below for your filing status. • Single, enter $3,600 • Married filing separate return, enter $3,000 • Married filing jointly or Qualifying widow(er) with dependent child, enter $6,000 • Head of household, enter $5,250	4. 3,600
5. Standard deduction. **a.** Compare the amounts on lines 3 and 4. Enter the **smaller** of the two amounts here. If under 65 and not blind, stop here. This is your standard deduction. Otherwise, go on to line 5b	5a. 3,600
b. If 65 or older or blind, multiply $900 ($700 if married or qualifying widow(er) with dependent child) by the number in the box above. Enter the result	5b.
c. Add lines 5a and 5b. This is your standard deduction for 1992.	5c.
Earned income *includes wages, salaries, tips, professional fees, and other compensation received for personal services you performed. It also includes any amount received as a scholarship that you must include in your income.*	

* Use this worksheet ONLY if someone can claim you as a dependent.

INSTRUCTION

TEACHING THE LESSON

Spend two days on this lesson.

On the first day have students read the opening text. Review with them the requirements for qualifying for a 1040 EZ form. Go over Larry's tax form step-by-step and discuss the checklist for Form 1040 EZ.

On the second day tell students to do exercises in the *Algebra Review* section. Introduce the 1040A form and go over each line of the Lender's form carefully.

FOCUS ON ALGEBRA

In this lesson students must pick up information from tables and use a calculator to perform arithmetic operations. They also use inequalities to determine appropriate categories. (*NCTM Standard 10, p. 167*)

COMMON ERRORS

Students may find the tax tables difficult to decipher initially, inadvertently picking up incorrect entries. Practice looking up specific information with them.

He then subtracted line 4 from line 3 to get line 5, his **taxable income.**

Step 4 On line 6, Larry entered the federal income tax withheld shown on his W-2 form. Then, to find his tax on line 7, Larry first found the portion of the Tax Table shown on page 411 of Lesson 9–1. Since his taxable income was between $2,050 and $2,075, he selected the tax that he found to the right of these numbers under the column headed "Single." He entered this tax, $309, on line 7 of the form.

Step 5 Since the tax is less than the $478 that was withheld from his paycheck, Larry was entitled to a refund. He subtracted line 7 from line 6: 478 − 309, or 169. His refund was $169.

To verify his work, Larry used the checklist on the back of Form 1040EZ. The checklist is similar to the following.

Checklist for Form 1040EZ

1. Did you check your computations (additions, subtractions, and so on), especially when calculating your taxable income, federal income tax withheld, and your refund or the amount you owe?
2. Did you check the "Yes" box on line 4 if your parents (or someone else) can claim you as a dependent on their return, even if they chose not to claim you? If no one can claim you as a dependent, did you check the "No" box?
3. Did you enter an amount on line 4? If you checked the "Yes" box on line 4, did you fill in the worksheet to figure the amount to enter? If you checked the "No" box, did you enter 5,900?
4. Did you use the amount from line 5 to find your tax in the Tax Table? Did you enter the correct tax on line 7?
5. If you didn't get a label, did you enter your name, address (including ZIP Code), and Social Security number in the spaces provided on Form 1040EZ?
6. If you got a label, does it show your correct name, address, and Social Security number? If not, did you enter the correct information?
7. Did you attach your W-2 form(s) to the left margin of your return? And did you sign and date Form 1040EZ and enter your occupation?

AT-RISK STUDENTS

Generate interest by asking students who have or are planning to get part-time jobs to fill out a 1040EZ form using their (assumed) earnings and estimating their interest income. Remind them that they probably can be claimed as dependents on a parent's or guardian's tax return.

Notice that Larry used whole dollar amounts when filling out his income tax return. In Example 2, Larry's parents use dollars and cents when filling out their income tax return. The IRS allows you to round to the nearest dollar, if you prefer.

Form **1040EZ**

Department of the Treasury—Internal Revenue Service

Income Tax Return for Single Filers With No Dependents **19--**

OMB No. 1545-0675

Name & address

Use the IRS label (see page 10). If you don't have one, please print.

LABEL HERE

Print your name (first, initial, last): LARRY L. LENDER

Home address (number and street). If you have a P.O. box, see page 10. Apt. no.: 1040 LESSON ST.

City, town or post office, state, and ZIP code. If you have a foreign address, see page 10.: OMAHA, INDIANA 45533

Please print your numbers like this: 9876543210

Your social security number: 345 45 6789

Please see instructions on the back. Also, see the Form 1040EZ booklet.

Presidential Election Campaign (See page 10.) Do you want $1 to go to this fund? **Note:** *Checking "Yes" will not change your tax or reduce your refund.* ✓

		Dollars	Cents
Report your income			
	1 Total wages, salaries, and tips. This should be shown in box 10 of your W-2 form(s). Attach your W-2 form(s). **1**	5,460	00
Attach Copy B of Form(s) W-2 here. Attach tax payment on top of Form(s) W-2.	**2** Taxable interest income of $400 or less. If the total is more than $400, you cannot use Form 1040EZ. **2**	200	00
	3 Add lines 1 and 2. This is your **adjusted gross income. 3**	5,660	00
Note: *You **must** check Yes or No.*	**4** Can your parents (or someone else) claim you on their return? ☒ **Yes.** Do worksheet on back; enter amount from line E here. ☐ **No.** Enter 5,900.00. This is the total of your standard deduction and personal exemption. **4**	3,600	00
	5 Subtract line 4 from line 3. If line 4 is larger than line 3, enter 0. This is your **taxable income. 5**	2,060	00
Figure your tax	**6** Enter your Federal income tax withheld from box 9 of your W-2 form(s). **6**	478	00
	7 **Tax.** Look at line 5 above. Use the amount on **line 5** to find your tax in the tax table on pages 22-24 of the booklet. Then, enter the tax from the table on this line. **7**	309	00
Refund or amount you owe	**8** If line 6 is larger than line 7, subtract line 7 from line 6. This is your **refund. 8**	169	00
	9 If line 7 is larger than line 6, subtract line 6 from line 7. This is the **amount you owe.** Attach your payment for full amount payable to the "Internal Revenue Service." Write your name, address, social security number, daytime phone number, and "1992 Form 1040EZ" on it. **9**		

Sign your return

Keep a copy of this form for your records.

I have read this return. Under penalties of perjury, I declare that to the best of my knowledge and belief, the return is true, correct, and complete.

Your signature: X Larry Lender

Date: 4-12--

Your occupation:

For IRS Use Only — Please do not write in boxes below.

For Privacy Act and Paperwork Reduction Act Notice, see page 4 in the booklet. Cat. No. 11329W Form 1040EZ (19--)

SKILL 2

EXAMPLE 2 Larry's parents must complete the more complicated Form 1040A.

QUESTION What steps must they take to complete the form?

SOLUTION

The Lenders' completed Form 1040A appears on pages 428–429. An explanation of some of the lines follows.

Step 1 The Lenders have filled in their names, address, and Social Security numbers.

Step 2 Mr. and Mrs. Lender are married filing a joint return.

Step 3 On lines 6a and 6b the Lenders claim three exemptions, one for each of them, and one for Larry. On line 6c Larry is claimed as a dependent.

Step 4 Line 7 is the total of Mr. and Mrs. Lender's income from their salaries. These two amounts were shown on their W-2 forms:

Mr. Lender earned $17,600.
Mrs. Lender earned $31,000.

Line 8a shows the interest income received on a savings account that they have at their credit union.
Lines 8b, 9, 10, 11, 12, and 13 are left blank. If the Lenders had received any dividends from stock or money-market investments, they would have included these payments on line 9.
Line 14, the **total income,** is the sum of lines 7, 8a, and 9–13.

Step 5 Since the Lenders have no IRA deduction, line 15 is left blank. For the Lenders, line 16, the **adjusted gross income,** is the same as line 14.

Step 6 Line 17 at the top of page 2 of the form is the same as line 16. None of lines 18a, 18b, or 18c applies to the Lenders, so they are left blank.
Line 19, the **standard deduction,** is $6,000, since the Lenders are married and filing jointly.
Line 20 is the adjusted gross income reduced by the standard deduction.
In line 21 the Lenders multiply $2,300 by 3, the number of exemptions, to get their total exemption allowance of $6,900.
After subtracting their exemptions from line 20, the Lenders arrive at their **taxable income** in line 22.

Step 7 The Lenders find their tax on lines 23 and 27 by using the Tax Table in the Reference Section.
Lines 24, 25, and 26 are left blank.
On lines 28a and 28d the Lenders write the sum of the amounts shown as withheld on their two W-2 forms. This is their **total payment.**
Lines 28b and 28c are left blank.

Step 8 The Lenders complete the form by subtracting their total tax (line 27) from their total payments (line 28d) to get line 30, the amount of their **refund,** $1,783.80. They both sign and date the form and mail it to the IRS.

CRITICAL THINKING

Have students fill out a 1040A form for Larry. Ask them to compare the results with those on his 1040EZ form.

RETEACHING

Students may find the 1040A form confusing. Do extra examples with them to help solidify their ability to fill out the form.

ENRICHMENT

Suggest that students write for a state tax form with the accompanying booklet. Tell them to see if they understand its filing directions and terminology.

Form **1040A** Department of the Treasury—Internal Revenue Service
U.S. Individual Income Tax Return (0) **19--** IRS Use Only—Do not write or staple in this space.

OMB No. 1545-0085

Label (See page 14.)

Use the IRS label. Otherwise, please print or type.

LABEL HERE

Your first name and initial	Last name	Your social security number
NATHAN S.	LENDER	432 : 12 : 7777
If a joint return, spouse's first name and initial	Last name	Spouse's social security number
MAUREEN T.	LENDER	459 : 38 : 4141
Home address (number and street). If you have a P.O. box, see page 15.	Apt. no.	
1040 LESSON ST.		
City, town or post office, state, and ZIP code. If you have a foreign address, see page 15.		
OMAHA, INDIANA 45533		

For Privacy Act and Paperwork Reduction Act Notice, see page 4.

Presidential Election Campaign Fund (See page 15.)

	Yes	No
Do you want $1 to go to this fund?	✓	
If a joint return, does your spouse want $1 to go to this fund?	✓	

Note: *Checking "Yes" will not change your tax or reduce your refund.*

Check the box for your filing status (See page 15.)

Check only one box.

1 ☐ Single
2 ☑ Married filing joint return (even if only one had income)
3 ☐ Married filing separate return. Enter spouse's social security number above and full name here. ▶ ______
4 ☐ Head of household (with qualifying person). (See page 16.) If the qualifying person is a child but not your dependent, enter this child's name here. ▶ ______
5 ☐ Qualifying widow(er) with dependent child (year spouse died ▶ 19 ____). (See page 17.)

Figure your exemptions (See page 18.)

If more than seven dependents, see page 21.

6a ☑ **Yourself.** If your parent (or someone else) can claim you as a dependent on his or her tax return, **do not** check box 6a. But be sure to check the box on line 18b on page 2.

b ☑ **Spouse**

No. of boxes checked on 6a and 6b: 2

c **Dependents:**

(1) Name (first, initial, and last name)	(2) Check if under age 1	(3) If age 1 or older, dependent's social security number	(4) Dependent's relationship to you	(5) No. of months lived in your home in 1992
LARRY L. LENDER		345 : 45 : 6789	SON	12

No. of your children on 6c who:
- lived with you: 1
- didn't live with you due to divorce or separation (see page 21): ____

No. of other dependents on 6c: ____

d If your child didn't live with you but is claimed as your dependent under a pre-1985 agreement, check here . . . . ▶ ☐

e Total number of exemptions claimed.

Add numbers entered on lines above: 3

Figure your total income

Attach Copy B of your Forms W-2 and 1099-R here.

If you didn't get a W-2, see page 22.

Attach check or money order on top of any Forms W-2 or 1099-R.

7	Wages, salaries, tips, etc. This should be shown in box 10 of your W-2 form(s). Attach Form(s) W-2.	7	48,600 00
8a	**Taxable** interest income (see page 24). If over $400, also complete and attach Schedule 1, Part I.	8a	300 00
b	**Tax-exempt** interest. DO NOT include on line 8a. 8b		
9	Dividends. If over $400, also complete and attach Schedule 1, Part II.	9	
10a	Total IRA distributions. 10a — **10b** Taxable amount (see page 25).	10b	
11a	Total pensions and annuities. 11a — **11b** Taxable amount (see page 25).	11b	
12	Unemployment compensation (see page 29).	12	
13a	Social security benefits. 13a — **13b** Taxable amount (see page 29).	13b	
14	Add lines 7 through 13b (far right column). This is your **total income.** ▶	14	48,900 00

Figure your adjusted gross income

15a	Your IRA deduction from applicable worksheet. 15a		
b	Spouse's IRA deduction from applicable worksheet. **Note:** *Rules for IRAs begin on page 31.* 15b		
c	Add lines 15a and 15b. These are your **total adjustments.**	15c	
16	Subtract line 15c from line 14. This is your **adjusted gross income.** If less than $22,370, see "Earned income credit" on page 39. ▶	16	48,900 00

Cat. No. 11327A — 19-- **Form 1040A page 1**

Name(s) shown on page 1 | Your social security number 432 12 7777

Figure your standard deduction, exemption amount, and taxable income

Line	Description	Ref	Amount	Cents
17	Enter the amount from line 16.	17	48,900	00
18a	Check if: ☐ **You** were 65 or older ☐ Blind; ☐ **Spouse** was 65 or older ☐ Blind. **Enter number of boxes checked** ▶	18a	☐	
b	If your parent (or someone else) can claim you as a dependent, check here . . . ▶	18b	☐	
c	If you are married filing separately and your spouse files Form 1040 and itemizes deductions, see page 35 and check here ▶	18c	☐	
19	Enter the **standard deduction** shown below for your filing status. **But if you checked any box on line 18a or b,** go to page 35 to find your standard deduction. **If you checked box 18c,** enter -0-. • Single—$3,600 • Head of household—$5,250 • Married filing jointly or Qualifying widow(er)—$6,000 • Married filing separately—$3,000	19	6,000	00
20	Subtract line 19 from line 17. (If line 19 is more than line 17, enter -0-.)	20	42,900	00
21	Multiply $2,300 by the total number of exemptions claimed on line 6e.	21	6,900	00
22	Subtract line 21 from line 20. (If line 21 is more than line 20, enter -0-.) This is your **taxable income.** ▶	22	36,000	00

Figure your tax, credits, and payments

If you want the IRS to figure your tax, see the instructions for line 22 on page 36.

Line	Description	Ref	Amount	Cents
23	Find the tax on the amount on line 22. Check if from: ☒ Tax Table (pages 48–53) or ☐ Form 8615 (see page 37).	23	5,433	00
24a	Credit for child and dependent care expenses. Complete and attach Schedule 2.	24a		
b	Credit for the elderly or the disabled. Complete and attach Schedule 3.	24b		
c	Add lines 24a and 24b. These are your **total credits.**	24c		
25	Subtract line 24c from line 23. (If line 24c is more than line 23, enter -0-.)	25		
26	Advance earned income credit payments from Form W-2.	26		
27	Add lines 25 and 26. This is your **total tax.** ▶	27	5,433	00
28a	Total Federal income tax withheld. If any tax is from Form(s) 1099, check here. ▶ ☐	28a	7,216	80
b	1992 estimated tax payments and amount applied from 1991 return.	28b		
c	**Earned income credit.** Complete and attach Schedule EIC.	28c		
d	Add lines 28a, 28b, and 28c. These are your **total payments.** ▶	28d	7,216	80

Figure your refund or amount you owe

Attach check or money order on top of Form(s) W-2, etc., on page 1.

Line	Description	Ref	Amount	Cents
29	If line 28d is more than line 27, subtract line 27 from line 28d. This is the amount you **overpaid.**	29	1,783	80
30	Amount of line 29 you want **refunded to you.**	30	1,783	80
31	Amount of line 29 you want **applied to your 1993 estimated tax.**	31		
32	If line 27 is more than line 28d, subtract line 28d from line 27. This is the **amount you owe.** Attach check or money order for full amount payable to the "Internal Revenue Service". Write your name, address, social security number, daytime phone number, and "1992 Form 1040A" on it.	32		
33	Estimated tax penalty (see page 41).	33		

Sign your return

Keep a copy of this return for your records.

Under penalties of perjury, I declare that I have examined this return and accompanying schedules and statements, and to the best of my knowledge and belief, they are true, correct, and complete. Declaration of preparer (other than the taxpayer) is based on all information of which the preparer has any knowledge.

Your signature	Date	Your occupation
Nathan Lender	4-13—	ACCOUNTANT
Spouse's signature. If joint return, BOTH must sign.	**Date**	**Spouse's occupation**
Maureen Lender	4-13—	SYSTEMS ANALYST

Paid preparer's use only

Preparer's signature	Date	Check if self-employed ☐	Preparer's social security no.
Firm's name (or yours if self-employed) and address		E.I. No.	
		ZIP code	

*U.S. Government Printing Office: 1992—315-155

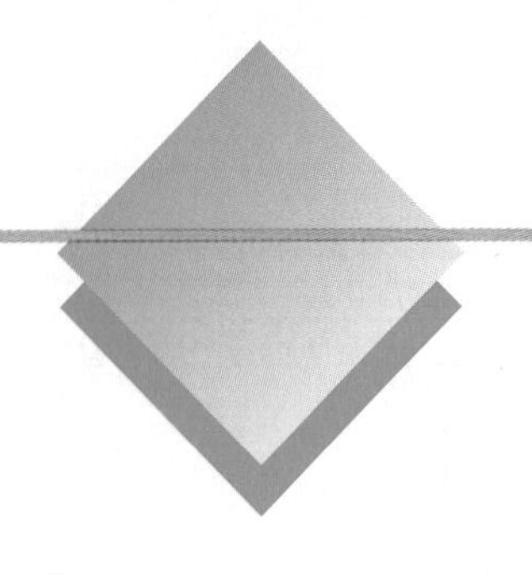

TRY YOUR SKILLS

GUIDED PRACTICE

Have students work in groups of three or four on the *Try Your Skills* exercises. After they have compared and corroborated results, tell them to repeat 1-4 assuming $210 taxable interest in each case.

Fill out a 1040EZ form for each of the following incomes to determine the amount owed or refund due to the nearest dollar. Assume there is no taxable interest. All of the people are single; make up names and addresses. You may contribute to the election campaign fund; this choice will not affect anything else. The standard deduction amount for line 4 will be $3,600 unless the adjusted gross income from line 3 is less than that.

	Income	Amount Withheld	Can Be Claimed as a Dependent by Someone?	
1.	$ 2,400	$ 225	Yes	Refund: $225
2.	9,020	1,200	No	Refund: $731
3.	3,800	420	Yes	Refund: $388
4.	46,000	8,748	No	Refund: $301

Fill out a 1040A form for each of the following couples to determine the amount owed or refund due to the nearest dollar. Their Social Security numbers and those of their dependents are shown in parentheses.

5. (689-78-6523) (658-31-4353)
Fred and Fawn Turner
10 Springfield Ave.
Austin, Texas 78749
Filing a joint return
Wages: Fawn, $9,700
Fred, $18,000
Tax owed: $1114

Dependent:
Angela, age 7 (632-29-4168)

Withheld: from Fawn, $0
from Fred, $1,200
Interest income: $350
Dividend income: $298

6. (652-81-3099) (618-39-4788)
Sten and Greta Larson
9500 Crosswicks Drive
Orlando, Florida 32819
Filing a joint return
Wages: Sten, $37,500
Greta, $29,320
Tax owed: $3911

Dependents:
Karl, age 10 (399-68-9915)
Kristina, age 8 (015-33-2751)
Withheld: from Sten, $4,512
from Greta, $1,504
Interest income: $287
Dividend income: $155

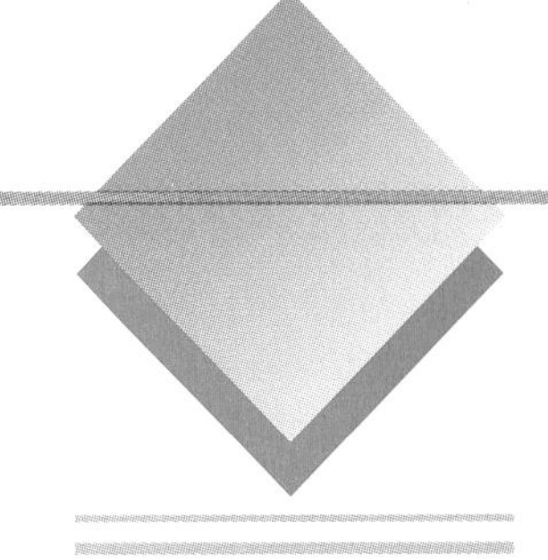

EXERCISE YOUR SKILLS

1. Why do you think the government does not require everyone to file the same income tax form? See Additional Answers.
2. Why do you think most married people file joint returns?
3. Why do you think people are allowed to subtract standard deductions to find their taxable income?

Fill out a 1040EZ form for each of the following incomes to determine the amount owed or refund due to the nearest dollar. For Exercises 4–8, make up a name and address. For Exercises 9–12, use the given information. All of the people are single. You may choose whether or not to contribute to the election campaign fund; this choice will not affect anything else you fill out. The standard deduction amount for line 4 will be $3,600 unless the adjusted gross income from line 3 is less than that.

	Wages, Salaries, and Tips	Interest	Amount Withheld	Can Be Claimed as a Dependent by Someone?
4.	$ 1,900	$100	$ 340	No
5.	4,200	50	600	Yes
6.	8,500	0	1,400	Yes
7.	18,600	350	2,040	No
8.	Two jobs: $15,400	200	3,700	No
	$ 2,200		0	

9. Daryl Hawkins
   1234 Sesame Street
   Phoenix, Arizona 85442
   S.S. No. 551-76-3333
   Wages: $20,650
   Withheld: $2,400
   Interest income: $50.00
   Daryl will be claimed as a dependent on another's return.

10. Brent Poulson
    1415 16th Ave.
    Dallas, Texas 75277
    S.S. No. 432-55-7777
    Wages: $6,200
    Withheld: $192
    Interest income: $246.80
    Brent will be claimed as a dependent on another's return.

11. Yazmin Rogers
    2020 Twentieth Ave.
    Seattle, Washington 98043
    S.S. No. 456-78-4321
    Wages: $39,400
    Withheld: $7,008
    Interest income: $150
    Yazmin will not be claimed as a dependent elsewhere.

12. Renata Taylor
    2323 Ranger Street
    Cincinnati, Ohio 45208
    S.S. No. 543-77-2222
    Wages: $35,170
    Withheld: $5,796
    Interest income: $310
    Renata will not be claimed as a dependent elsewhere.

KEY TERMS

adjusted gross income
deduction
dependent
exemption
Form 1040A
Form 1040EZ
itemized deductions
refund
standard deduction
taxable income
total income
total payment

INDEPENDENT PRACTICE

ASSIGNMENT GUIDE

Exercises 1-3 check students' understanding of the opening text.

Exercises 4-12 provide practice in filling out a 1040 EZ form.

Exercises 13-16 ask students to fill out a 1040A form.

ADDITIONAL ANSWERS

9. Tax owed: $169
10. Tax owed: $234
11. Refund: $367
12. Refund: $303

CLOSURE

Have students write a paragraph to explain who qualifies for the 1040 EZ form and for the 1040A form. Tell them to discuss itemized and standard deductions, exemptions, total income, and taxable income.

LESSON QUIZ

Use the Tax Tables in the Reference Section of the book and a 1040 EZ form in conjunction with the given information to answer Questions 1 and 2.

Wages, Salaries and Tips	$5600
Interest	$109
Amount Withheld	$420
Filing Status	Single
Can Be Claimed as a Dependent by Someone	Yes

1. Determine the taxable income. (**$2109**)
2. Determine the amount of tax owed or the refund due. (**$103 refund**)

Fill out a 1040A form for each of the following couples or individuals to determine the amount owed or refund due to the nearest dollar. Their Social Security numbers are shown in parentheses. See Additional Answers.

13. (421-32-4587) (476-58-9023)
Flora and Ernest Jones
10101 Barley Street
Anaheim, California 98043
Filing a joint return
Wages: Flora, $19,400
Ernest, $21,620
Withheld: from Flora, $2,350
from Ernest, $1,836

Dependents:
Bea, age 6 (452-79-5644)
Phillip, age 14 (751-44-9086)
Earl, age 17 (555-44-9922)

Interest income: $200
Dividend income: $180

14. (458-29-7011) (492-71-6322)
Emilio and Rosa Ortiz
13422 Euclid Avenue
Des Moines, Iowa 55105
Filing a joint return
Wages: Emilio, $48,750
Rosa, $53,790
Withheld: from Emilio, $9,212
from Rosa, $8,268

Dependents:
Rosita, age 4 (711-42-1195)
Julio, age 8 (329-44-7890)
Carlos, age 12 (666-21-4434)

Interest income: $280
Dividend income: $195

15. (463-44-9292)
Janice Parker
9286 Spring Blvd.
Two Egg, Florida 32304
Filing as head of household
Wages: $38,450
Withheld: $3,800

Dependents:
Dimitri, age 12 (428-77-4083)
Newton, age 14 (523-97-6702)
Sophia, age 16 (452-95-2965)

Interest income: $40
Dividend income: $300

16. (716-43-9026)
John Madison, Jr.
1409 S. Alta Vista
San Remo, Texas 77566
Filing as head of household
Wages: $47,060
Withheld: $8,952

Dependent:
John Madison, Sr., age 87
(490-62-6710)

Interest income: $390
Dividend income: $309

ADDITIONAL ANSWERS

13. Refund: $597
14. Tax owed: $1813
15. Refund: $151
16. Refund: $2070

MIXED REVIEW

Determine the monthly payment and total payments (deferred payment price) for each car loan. See Additional Answers.

5-1 **1.** \$15,500 at $6\frac{1}{2}\%$ for 5 years **2.** \$9,000 at 10% for 4 years

7-1 **3.** Use the Comparison Table for Term and Whole Life Premiums on page 326 to find the yearly premiums for \$100,000 of whole life insurance and of 5-year term insurance, each purchased at the age of 45. Which is the less expensive insurance and what is the first-year saving if the less expensive insurance is purchased? 5-year term; \$1565

8-1 **4.** How many whole shares of Truffles Company stock can you buy if you have \$37,665 to invest and the stock is selling for $102\frac{1}{4}$ per share? Ignore commission costs. 368 shares

6-6 Use the credit scoring table on page 293 of Lesson 6–6 to determine the credit score for each family described below. Then determine the probability that the head of the household will repay the loan.

5. Family 1

- Age of the head of household: 40
- Lived at current address: 6 years
- Age of their current car: 2 years
- Monthly car payment: \$190
- Housing cost: \$800 per month
- They have a checking account and a savings account
- They have not been referred to a finance company
- They have 4 major credit cards
- Their ratio of debt to income is 15% 120; 95.5%

6. Family 2

- Age of the head of household: 24
- Lived at current address: 3 years
- Age of their current car: 4 years
- Monthly car payment: \$90
- Housing cost: \$350 per month
- They have a checking account but no savings account
- They have not been referred to a finance company
- They have 8 major credit cards
- Their ratio of debt to income is 15% 98; 91%

8-4 Suppose that an exponential regression model gives the following equations as a description of the cost of living over a 5-year period. What would be the expected rate of inflation?

7. $y = 150(1 + 0.06)^x$ 6% **8.** $y = 83.5(1.027)^x$ 2.7%

ADDITIONAL ANSWERS

1. \$303.28; \$18,196.52
2. \$228.26; \$10,956.64

9-3 FORM 1040, SCHEDULE A, AND HELP!

QUICK REFERENCE

TEACHER'S RESOURCES AND ANSWER KEY

Lesson Quiz Answers 9-3, p. 74
Reteaching Activity
Answers 9-3, p. 108
Enrichment Activity
Answers 9-3, p. 109

EXTENSION ACTIVITIES

Reteaching Activity 9-3, p. 71
Enrichment Activity 9-3, p. 74

Evelyn went into business with her friends selling T-shirts and sweatshirts decorated with hand-painted designs. After a few months the partners had their production process fairly well under control, and the enterprise began to show a profit. Now it is February; the business has been in operation since August. Evelyn knows that she must file an income tax return for the income she has received from the business.

She wants her tax return to be completed very carefully. She heard some of her friends talking about the problems of the father of one of their classmates. They were discussing several of his visits to the IRS office. Apparently, the IRS examined some of his tax returns from earlier years. The way Evelyn understood the matter, he had to pay over $12,000 in back taxes that the IRS said he owed.

Evelyn knew she couldn't begin to scrape together $12,000 to pay the government, especially not at any one time. Her parents assure her that unless her profits are over $60,000 in one year, she will not have to pay that much. They have had years when their tax bill was that high, however, so Evelyn knows that the federal income tax is not an insignificant amount.

If Evelyn decides that she cannot fill out the proper forms and needs some help, several sources of assistance are available. Evelyn will explore some of these sources.

OBJECTIVES: *In this lesson, we will help Evelyn to:*

- *Find where a taxpayer can go for help in filing a tax return.*
- *Complete Form 1040 and Schedule A.*

TAX HELP: WHERE TO GET IT

When Evelyn first looked at the array of different tax forms that are available at the local IRS office, she was not sure which ones to pick up. She finally settled on Form 1040, a booklet of instructions for Form 1040, and a few other IRS publications that looked as if they might have something to do with business.

Evelyn's parents have been filing the 1040 form every year since they bought their first house. The interest that they pay on the mortgage is deductible. However, to take advantage of this deduction, they must use the 1040 form.

Evelyn, who cannot tell a deduction from a dependent, looks at the 1040 form and wonders whether it is truly written in English. She recognizes the words, but making sense out of what they say is a formidable task. Fortunately for Evelyn, there are people who make their livings interpreting all the rules set up by the IRS.

> **FOCUS**
>
> **ALGEBRA CONNECTIONS**
>
> Nicole wants to invest $20,000, some at 4% annual interest and some at 8% annual interest. She wants to invest no more than $16,000 at 8% and no more that $8,000 at 4%. Find out how much money she should invest at each rate to receive maximum interest. (**$16,000 at 8%, $4000 at 4%**)

In fact, by the middle of the 1980s, over one third of all taxpayers sought help from the IRS, and millions of others paid from ten dollars to thousands of dollars for tax help from other sources. For Evelyn and millions like her, four major sources of tax help are available: the IRS itself, professional tax preparers including mass-market tax preparation services, CPA tax specialists, and tax lawyers. The first three of these sources are described below.

Internal Revenue Service The IRS has a toll-free telephone answering service and nearly 1000 tax assistance centers located throughout the United States. Under certain circumstances the IRS will figure your tax for you on Form 1040, Form 1040A, or Form 1040EZ. Of course, you must provide complete information.

Relying on an IRS employee's opinion about a tax issue is not always a good idea. The IRS has been known to give taxpayers wrong forms, answer questions incorrectly, make mathematical errors, and, very rarely, dispense misleading or incorrect advice in its publications. Moreover, if you make a mistake because of incorrect advice from the IRS, you will probably have little success in using this fact as an excuse for the mistake. In other words, claiming that you were misinformed will probably not help if the IRS selects your return for examination.

CRITICAL THINKING

Ask students to consider the advantage of using Schedule A. Tell them to hypothesize what type of deduction is most commonly used and to explain their reasoning.

Professional Tax Preparation Services There are tax preparation services that will prepare your 1040A forms for a fee, sometimes as low as $10 to $25. However, since you have to find and provide all the essential records, which is most of the work, you may be able to do the whole job yourself. The IRS has exerted enormous effort to simplify the forms so that you can fill them out without help and thus save the money you would otherwise pay a tax preparer. If you feel that a preparation service is necessary, however, you should talk personally with the person who works on the return to find out something about the preparer's background.

Certified Public Accountants For complicated tax work, a certified public accountant (CPA) who specializes in taxes is the best source of tax help.

Do-It-Yourself Being in business for herself with her friends has given Evelyn more confidence, especially since the business has started to grow and make a profit. Evelyn found in the early stages of the business that visits to her local bookstore and library were very beneficial. She did notice in the business section of the bookstore a whole shelf of books offering advice on how to fill out tax returns. Evelyn's parents have read a few of these tax guides, and every year they study the new changes in the tax law.

Another source of help that has gained acceptance in recent years is **tax preparation software** that can be used by any person who has a personal computer. Programs for both federal taxes and state taxes are available. These programs may cost over $50 (less for state tax programs) and cannot normally be reused for the following year's taxes. However, after your initial purchase, you can usually buy the tax program for later years at a discount from the original price.

ALGEBRA *REVIEW*

Evaluate each expression for the indicated values of *x* and *y*.

1. $x + 0.025y$
   $x = 80, y = 10{,}000$
   330
2. $x(1 + 0.025)y$
   $x = 80, y = 10{,}000$
   820,000
3. $x - 0.075y$
   $x = 10{,}000, y = 30{,}000$
   7750
4. $x - 0.075y$
   $x = 2250, y = 30{,}000$
   0
5. $x - 0.075y$
   $x = 4250, y = 44{,}000$
   950

Graph each inequality on a number line. See Additional Answers.

6. $x \geq -2$
7. $0 \leq x$
8. $x < 5$
9. $x > -4$
10. $2 \leq x < 10$
11. $x < -1$ or $3 < x$
12. $-3 < x \leq 5$
13. $x > -2$ and $x < 1$
14. $x < 0$ or $x \geq 2$

Ask Yourself

1. What are four major sources of help in completing income tax forms? the IRS, professional tax preparers, CPA tax specialists, tax lawyers
2. What is the best source of tax help for someone who has a complicated return? CPA tax specialist
3. What sources of help are available for someone who wants to prepare his or her own tax return? books, tax guide publications, tax preparation software for the personal computer

SHARPEN YOUR SKILLS

SKILL 1

EXAMPLE 1 Mr. and Mrs. Enterprise file the 1040 form each year mainly because of the mortgage interest deduction, which requires filling out Schedule A. Many parts of the 1040 form are similar to those on the 1040A form. Quite a few of the lines on the 1040 form do not apply to the Enterprises.

QUESTION How should the Enterprises complete page 1 of Form 1040?

SOLUTION

To fill out Schedule A, the Enterprises first have to find their adjusted gross income on line 31 of Form 1040. This line is at the bottom of the first page of the form, which is shown on page 438. Here is an explanation of some of the lines of page 1.

Lines 1–5	Filing status of taxpayer; similar to 1040A
Line 6	Exemptions; similar to 1040A
Lines 7–9	Income from wages, salaries, interest, and dividends; similar to 1040A
Lines 10–22	Other kinds of income that an individual or family might have
Line 23	The family's total income
Lines 24a, 24b	The IRA contribution of $2000 from each of the Enterprises' two salaries. Neither of Evelyn's parents is covered at work by a retirement plan, so each is allowed a full deduction of the $2000 contribution.
Lines 25–29	Other adjustments to income that do not apply to the Enterprises
Line 30	The total adjustments. For the Enterprises this is just the IRA contributions.
Line 31	The adjusted gross income

INSTRUCTION

TEACHING THE LESSON

Spend two days on this lesson.

On the first day, discuss the opening reading and have students write out the answers to the *Ask Yourself* questions in their notes. Direct students to complete the *Algebra Review* section. Go over Example 1, giving examples of other kinds of incomes and explaining what types of employment may or may not offer retirement plans.

On the second day discuss Examples 2 and 3 in detail, making sure that students understand how to find the information requested on specific lines of both forms.

Form **1040** Department of the Treasury—Internal Revenue Service
U.S. Individual Income Tax Return 19-- (0) IRS Use Only—Do not write or staple in this space.

For the year Jan. 1–Dec. 31, 1992, or other tax year beginning , 1992, ending , 19 OMB No. 1545-0074

Label (See instructions on page 10.) **Use the IRS label.** Otherwise, please print or type. LABEL HERE

Your first name and initial: FREEMONT O. Last name: ENTERPRISE — Your social security number: 526 31 7055

If a joint return, spouse's first name and initial: ELAINE P. Last name: ENTERPRISE — Spouse's social security number: 475 42 9266

Home address (number and street). If you have a P.O. box, see page 10.: 300 MAIN STREET — Apt. no.

City, town or post office, state, and ZIP code. If you have a foreign address, see page 10.: MUNCIE, INDIANA 47304

For Privacy Act and Paperwork Reduction Act Notice, see page 4.

Presidential Election Campaign (See page 10.)
Do you want $1 to go to this fund? ✓ Yes No
If a joint return, does your spouse want $1 to go to this fund? ✓ Yes No
Note: *Checking "Yes" will not change your tax or reduce your refund.*

Filing Status (See page 10.) Check only one box.
1 Single
2 ✓ Married filing joint return (even if only one had income)
3 Married filing separate return. Enter spouse's social security no. above and full name here. ▶
4 Head of household (with qualifying person). (See page 11.) If the qualifying person is a child but not your dependent, enter this child's name here. ▶
5 Qualifying widow(er) with dependent child (year spouse died ▶ 19). (See page 11.)

Exemptions (See page 11.)
6a ☒ **Yourself.** If your parent (or someone else) can claim you as a dependent on his or her tax return, **do not** check box 6a. But be sure to check the box on line 33b on page 2
b ☒ **Spouse**
No. of boxes checked on 6a and 6b: 2

c **Dependents:**

(1) Name (first, initial, and last name)	(2) Check if under age 1	(3) If age 1 or older, dependent's social security number	(4) Dependent's relationship to you	(5) No. of months lived in your home in 1992
EVELYN ENTERPRISE		237 11 8642	DAUGHTER	12

If more than six dependents, see page 12.

No. of your children on 6c who: • lived with you: 1 • didn't live with you due to divorce or separation (see page 13): — No. of other dependents on 6c: —

d If your child didn't live with you but is claimed as your dependent under a pre-1985 agreement, check here ▶ ☐
e Total number of exemptions claimed — Add numbers entered on lines above ▶ 3

Income — Attach Copy B of your Forms W-2, W-2G, and 1099-R here. If you did not get a W-2, see page 9. Attach check or money order on top of any Forms W-2, W-2G, or 1099-R.

7	Wages, salaries, tips, etc. Attach Form(s) W-2	7	47,650 00
8a	**Taxable** interest income. Attach Schedule B if over $400	8a	350 00
b	**Tax-exempt** interest income (see page 15). DON'T include on line 8a — 8b		
9	Dividend income. Attach Schedule B if over $400	9	
10	Taxable refunds, credits, or offsets of state and local income taxes from worksheet on page 16	10	
11	Alimony received	11	
12	Business income or (loss). Attach Schedule C or C-EZ	12	
13	Capital gain or (loss). Attach Schedule D	13	
14	Capital gain distributions not reported on line 13 (see page 15)	14	
15	Other gains or (losses). Attach Form 4797	15	
16a	Total IRA distributions 16a — b Taxable amount (see page 16)	16b	
17a	Total pensions and annuities 17a — b Taxable amount (see page 16)	17b	
18	Rents, royalties, partnerships, estates, trusts, etc. Attach Schedule E	18	
19	Farm income or (loss). Attach Schedule F	19	
20	Unemployment compensation (see page 17)	20	
21a	Social security benefits 21a — b Taxable amount (see page 17)	21b	
22	Other income. List type and amount—see page 18	22	
23	Add the amounts in the far right column for lines 7 through 22. This is your **total income** ▶	23	48,000 00

Adjustments to Income (See page 18.)

24a	Your IRA deduction from applicable worksheet on page 19 or 20	24a	4,000 00
b	Spouse's IRA deduction from applicable worksheet on page 19 or 20	24b	
25	One-half of self-employment tax (see page 20)	25	
26	Self-employed health insurance deduction (see page 20)	26	
27	Keogh retirement plan and self-employed SEP deduction	27	
28	Penalty on early withdrawal of savings	28	
29	Alimony paid. Recipient's SSN ▶	29	
30	Add lines 24a through 29. These are your **total adjustments** ▶	30	4,000 00

Adjusted Gross Income
31 Subtract line 30 from line 23. This is your **adjusted gross income.** *If this amount is less than $22,370 and a child lived with you, see page EIC-1 to find out if you can claim the "Earned Income Credit" on line 56* ▶ 31 — 44,000 00

Cat. No. 11320B Form **1040** (19--)

Form 1040 (19--) Page 2

Tax Computation

(See page 22.)

Line	Description		Line	Amount
32	Amount from line 31 (adjusted gross income)		32	44,000 00
33a	Check if: ☐ **You** were 65 or older, ☐ Blind; ☐ **Spouse** was 65 or older, ☐ Blind. Add the number of boxes checked above and enter the total here ▶ 33a ☐			
b	If your parent (or someone else) can claim you as a dependent, check here ▶ 33b ☐			
c	If you are married filing separately and your spouse itemizes deductions or you are a dual-status alien, see page 22 and check here ▶ 33c ☐			
34	Enter the **larger** of your: **Itemized deductions** from Schedule A, line 26, **OR** **Standard deduction** shown below for your filing status. **But if you checked any box on line 33a or b,** go to page 22 to find your standard deduction. **If you checked box 33c,** your standard deduction is zero. • Single—$3,600 • Head of household—$5,250 • Married filing jointly or Qualifying widow(er)—$6,000 • Married filing separately—$3,000		34	9,210 00
35	Subtract line 34 from line 32		35	34,790 00
36	If line 32 is $78,950 or less, multiply $2,300 by the total number of exemptions claimed on line 6e. If line 32 is over $78,950, see the worksheet on page 23 for the amount to enter		36	6,900 00
37	**Taxable income.** Subtract line 36 from line 35. If line 36 is more than line 35, enter -0-		37	27,890 00
38	Enter tax. Check if from **a** ☒ Tax Table, **b** ☐ Tax Rate Schedules, **c** ☐ Schedule D, or **d** ☐ Form 8615 (see page 23). Amount, if any, from Form(s) 8814 ▶ **e** ______		38	4,181 00
39	Additional taxes (see page 23). Check if from **a** ☐ Form 4970 **b** ☐ Form 4972		39	
40	Add lines 38 and 39 ▶		40	4,181 00

If you want the IRS to figure your tax, see page 23.

Credits

(See page 23.)

Line	Description		Line	Amount
41	Credit for child and dependent care expenses. Attach Form 2441	41		
42	Credit for the elderly or the disabled. Attach Schedule R	42		
43	Foreign tax credit. Attach Form 1116	43		
44	Other credits (see page 24). Check if from **a** ☐ Form 3800 **b** ☐ Form 8396 **c** ☐ Form 8801 **d** ☐ Form (specify)______	44		
45	Add lines 41 through 44		45	0 00
46	Subtract line 45 from line 40. If line 45 is more than line 40, enter -0- ▶		46	4,181 00

Other Taxes

Line	Description	Line	Amount
47	Self-employment tax. Attach Schedule SE. Also, see line 25	47	
48	Alternative minimum tax. Attach Form 6251	48	
49	Recapture taxes (see page 25). Check if from **a** ☐ Form 4255 **b** ☐ Form 8611 **c** ☐ Form 8828	49	
50	Social security and Medicare tax on tip income not reported to employer. Attach Form 4137	50	
51	Tax on qualified retirement plans, including IRAs. Attach Form 5329	51	
52	Advance earned income credit payments from Form W-2	52	
53	Add lines 46 through 52. This is your **total tax** ▶	53	4,181 00

Payments

Attach Forms W-2, W-2G, and 1099-R on the front.

Line	Description			Line	Amount
54	Federal income tax withheld. If any is from Form(s) 1099, check ▶ ☐	54	5,100 00		
55	1992 estimated tax payments and amount applied from 1991 return	55			
56	**Earned income credit.** Attach Schedule EIC	56			
57	Amount paid with Form 4868 (extension request)	57			
58	Excess social security, Medicare, and RRTA tax withheld (see page 26)	58			
59	Other payments (see page 26). Check if from **a** ☐ Form 2439 **b** ☐ Form 4136	59			
60	Add lines 54 through 59. These are your **total payments** ▶			60	5,100 00

Refund or Amount You Owe

Attach check or money order on top of Form(s) W-2, etc., on the front.

Line	Description			Line	Amount
61	If line 60 is more than line 53, subtract line 53 from line 60. This is the amount you **OVERPAID** ▶			61	919 00
62	Amount of line 61 you want **REFUNDED TO YOU** ▶			62	919 00
63	Amount of line 61 you want **APPLIED TO YOUR 1993 ESTIMATED TAX** ▶	63			
64	If line 53 is more than line 60, subtract line 60 from line 53. This is the **AMOUNT YOU OWE.** Attach check or money order for full amount payable to "Internal Revenue Service." Write your name, address, social security number, daytime phone number, and "1992 Form 1040" on it			64	
65	Estimated tax penalty (see page 27). Also include on line 64	65			

Sign Here

Keep a copy of this return for your records.

Under penalties of perjury, I declare that I have examined this return and accompanying schedules and statements, and to the best of my knowledge and belief, they are true, correct, and complete. Declaration of preparer (other than taxpayer) is based on all information of which preparer has any knowledge.

Your signature	Date	Your occupation
Freemont Enterprise	4/12/--	Production Manager
Spouse's signature. If a joint return, BOTH must sign.	Date	Spouse's occupation
Elaine Enterprise	4-12--	Travel Agent

Paid Preparer's Use Only

Preparer's signature ▶	Date	Check if self-employed ☐	Preparer's social security no.
Firm's name (or yours if self-employed) and address ▶		E.I. No.	
		ZIP code	

*U.S. Government Printing Office: 19-- — 315-148

FOCUS ON ALGEBRA

In this lesson students compute with linear functions and translate from verbal situations to algebraic formulas. (*NCTM Standard 6, p. 154*)

COMMON ERRORS

Students may be unsure about the placement of many pieces of information on a 1040 form (such as a self-employed person's income). Added practice should alleviate this confusion.

ESL STUDENTS

Explain to non-native speakers the meaning of *gross adjusted income, charitable contributions, dividends* and *retirement plan.*

RETEACHING

Vary details of the Enterprises' financial picture (medical and dental expenses = $3500, gifts to charity = $300, taxable interest income = $500) and have students rework the family's 1040 and Schedule A forms. Ask them to comment on their findings.

ENRICHMENT

Tell students to prepare a list of questions for the IRS toll-free answering service. Have them record the responses and compare the answers to check for consistency.

SKILL 2

EXAMPLE 2 With page 1 of Form 1040 now completed, the Enterprises can now fill out Schedule A.

QUESTION How should the Enterprises use Schedule A to itemize their deductions?

SOLUTION

Mr. and Mrs. Enterprise have paid $4500 in mortgage interest on their home and also have several other deductible expenses, such as state income taxes and charitable contributions. Here is an explanation of the Enterprises' Schedule A, which appears on page 442.

Line 1	Medical expenses: $4250.00
Lines 2–4	This amount from line 31 (or line 32) of Form 1040 is used to reduce the amount of the medical deduction from $4250 to $950.
Line 5	During the tax year the Enterprises paid $1320 to the state of Indiana in income taxes. Of this, $1200 was money withheld by the two employers during the tax year. The balance was paid by Mr. Enterprise himself in April of the year when he sent the State of Indiana a check for $120 to complete his income tax payment for the year before that.
Line 6	Real-estate taxes on their home: $1240
Line 8	Total deductible taxes: $2560
Lines 9a–12	Interest payments paid on their home mortgage: $4500
Lines 13–16	Contributions to a local charity: $1200
Line 26	Totals from lines 4, 8, 12, and 16: $9210. The Enterprises enter this amount on line 34 of Form 1040 shown on page 439.

SKILL 3

EXAMPLE 3 The Enterprises can now complete Form 1040 shown on page 439.

QUESTION How will the Enterprises use Schedule A to complete Form 1040?

SOLUTION

Line 32	This is the same as line 31, the adjusted gross income.
Lines 33–34	These lines determine your deduction. They are similar to the lines of Form 1040A except that here there is a choice of taking an itemized deduction based on Schedule A.
Line 35	The gross adjusted income reduced by the itemized deductions: 44,000 − 9210 = $34,790.

Line 36	The exemptions: 3 exemptions • \$2300 = \$6900.
Line 37	The taxable income: line 35 − line 36, that is, 34,790 − 6900, or \$27,890.
Lines 38–40	Since the taxable income is less than \$100,000, the Enterprises use the Tax Table to find their tax. To the right of the line that reads "27,850 27,900" they find their tax, \$4,181, in the second column. If the income had been greater than \$100,000, they would have used the Tax Rate Schedule instead.
Lines 41–52	The Enterprises have no credits to subtract from their tax on line 40 or other taxes to pay.
Line 53	The total tax. Since there are no credits or other taxes, the total tax is the same as line 40 above.
Lines 54–60	Payments already made on the tax. For the Enterprises this is just line 54, the payments withheld on the W-2 forms.
Lines 61–62	Since line 60 is greater than line 53, the Enterprises have a refund: \$5100 − \$4181 = \$919.

The Enterprises both sign their Form 1040. Then, before the filing deadline, they attach the W-2 forms to the form and mail it together with Schedule A to the IRS in an envelope that the IRS has provided.

If line 5 (Form 1040EZ), line 22 (Form 1040A), or line 37 (Form 1040) is—		And you are—			
At least	But less than	Single	Married filing jointly •	Married filing sepa-rately	Head of a house-hold
		Your tax is—			
26,000					
26,000	**26,050**	4,499	3,904	4,960	3,904
26,050	**26,100**	4,513	3,911	4,974	3,911
26,100	**26,150**	4,527	3,919	4,988	3,919
26,150	**26,200**	4,541	3,926	5,002	3,926
26,200	**26,250**	4,555	3,934	5,016	3,934
26,250	**26,300**	4,569	3,941	5,030	3,941
26,300	**26,350**	4,583	3,949	5,044	3,949
26,350	**26,400**	4,597	3,956	5,058	3,956
26,400	**26,450**	4,611	3,964	5,072	3,964
26,450	**26,500**	4,625	3,971	5,086	3,971
26,500	**26,550**	4,639	3,979	5,100	3,979
26,550	**26,600**	4,653	3,986	5,114	3,986
26,600	**26,650**	4,667	3,994	5,128	3,994
26,650	**26,700**	4,681	4,001	5,142	4,001
26,700	**26,750**	4,695	4,009	5,156	4,009
26,750	**26,800**	4,709	4,016	5,170	4,016
26,800	**26,850**	4,723	4,024	5,184	4,024
26,850	**26,900**	4,737	4,031	5,198	4,031
26,900	**26,950**	4,751	4,039	5,212	4,039
26,950	**27,000**	4,765	4,046	5,226	4,046
27,000					
27,000	**27,050**	4,779	4,054	5,240	4,054
27,050	**27,100**	4,793	4,061	5,254	4,061
27,100	**27,150**	4,807	4,069	5,268	4,069
27,150	**27,200**	4,821	4,076	5,282	4,076
27,200	**27,250**	4,835	4,084	5,296	4,084
27,250	**27,300**	4,849	4,091	5,310	4,091
27,300	**27,350**	4,863	4,099	5,324	4,099
27,350	**27,400**	4,877	4,106	5,338	4,106
27,400	**27,450**	4,891	4,114	5,352	4,114
27,450	**27,500**	4,905	4,121	5,366	4,121
27,500	**27,550**	4,919	4,129	5,380	4,129
27,550	**27,600**	4,933	4,136	5,394	4,136
27,600	**27,650**	4,947	4,144	5,408	4,144
27,650	**27,700**	4,961	4,151	5,422	4,151
27,700	**27,750**	4,975	4,159	5,436	4,159
27,750	**27,800**	4,989	4,166	5,450	4,166
27,800	**27,850**	5,003	4,174	5,464	4,174
27,850	**27,900**	5,017	4,181	5,478	4,181
27,900	**27,950**	5,031	4,189	5,492	4,189
27,950	**28,000**	5,045	4,196	5,506	4,196

SCHEDULES A&B (Form 1040)

Department of the Treasury Internal Revenue Service (0)

Schedule A—Itemized Deductions

(Schedule B is on back)

▶ Attach to Form 1040. ▶ See Instructions for Schedules A and B (Form 1040).

OMB No. 1545-0074

19--

Attachment Sequence No. **07**

Name(s) shown on Form 1040: FREEMONT AND ELAINE ENTERPRISE

Your social security number: 526 : 31 : 7055

Section	Line	Description		Amount	Line	Amount
Medical and Dental Expenses		**Caution:** *Do not include expenses reimbursed or paid by others.*				
	1	Medical and dental expenses (see page A-1)	1	4,250 00		
	2	Enter amount from Form 1040, line 32. [2] 44,000				
	3	Multiply line 2 above by 7.5% (.075)	3	3,300 00		
	4	Subtract line 3 from line 1. If zero or less, enter -0- ▶			4	950 00
Taxes You Paid (See page A-1.)	5	State and local income taxes	5	1,320 00		
	6	Real estate taxes (see page A-2)	6	1,240 00		
	7	Other taxes. List—include personal property taxes ▶	7			
	8	Add lines 5 through 7 ▶			8	2,560 00
Interest You Paid (See page A-2.)	9a	Home mortgage interest and points reported to you on Form 1098	9a	4,500 00		
	b	Home mortgage interest not reported to you on Form 1098. If paid to an individual, show that person's name and address. ▶	9b			
Note: Personal interest is not deductible.	10	Points not reported to you on Form 1098. See page A-3 for special rules	10			
	11	Investment interest. If required, attach Form 4952. (See page A-3.)	11			
	12	Add lines 9a through 11 ▶			12	4,500 00
Gifts to Charity (See page A-3.)		**Caution:** *If you made a charitable contribution and received a benefit in return, see page A-3.*				
	13	Contributions by cash or check	13	1,200 00		
	14	Other than by cash or check. If over $500, you **MUST** attach Form 8283	14			
	15	Carryover from prior year	15			
	16	Add lines 13 through 15 ▶			16	1,200 00
Casualty and Theft Losses	17	Casualty or theft loss(es). Attach Form 4684. (See page A-4.) ▶			17	
Moving Expenses	18	Moving expenses. Attach Form 3903 or 3903F. (See page A-4.) ▶			18	
Job Expenses and Most Other Miscellaneous Deductions (See page A-5 for expenses to deduct here.)	19	Unreimbursed employee expenses—job travel, union dues, job education, etc. If required, you **MUST** attach Form 2106. (See page A-4.) ▶	19			
	20	Other expenses—investment, tax preparation, safe deposit box, etc. List type and amount ▶	20			
	21	Add lines 19 and 20	21			
	22	Enter amount from Form 1040, line 32. [22]				
	23	Multiply line 22 above by 2% (.02)	23			
	24	Subtract line 23 from line 21. If zero or less, enter -0- ▶			24	
Other Miscellaneous Deductions	25	Other—from list on page A-5. List type and amount ▶			25	
Total Itemized Deductions	26	Is the amount on Form 1040, line 32, more than $105,250 (more than $52,625 if married filing separately)? • **NO.** Your deduction is not limited. Add lines 4, 8, 12, 16, 17, 18, 24, and 25. • **YES.** Your deduction may be limited. See page A-5 for the amount to enter. ▶			26	9,210 00
		Caution: *Be sure to enter on Form 1040, line 34, the* ***LARGER*** *of the amount on line 26 above or your standard deduction.*				

For Paperwork Reduction Act Notice, see Form 1040 instructions. Cat. No. 11330X **Schedule A (Form 1040) 19--**

TRY YOUR SKILLS

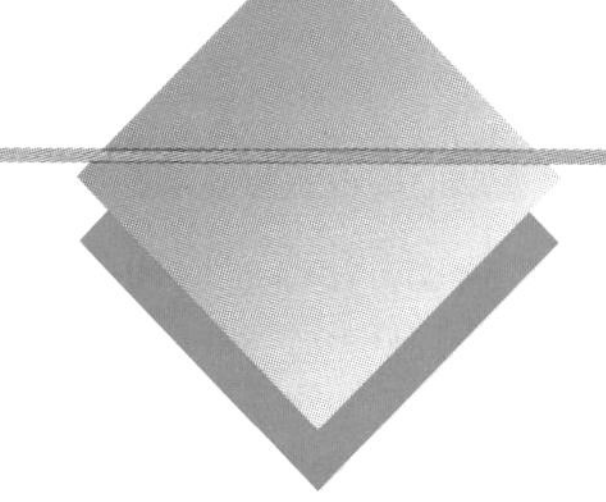

Suppose that the Enterprises had made no contribution to their IRAs. Find what effect, if any, that would have had on each of the following.

1. Their adjusted gross income
2. Their medical deduction −$300
3. Their mortgage deduction None
4. Their total deductions −$300
5. Their taxable income +$4300
6. Their tax +$645
7. Their total payments (line 60)
8. Their refund −$645

ADDITIONAL ANSWERS

1. +$4000
7. no effect

EXERCISE YOUR SKILLS

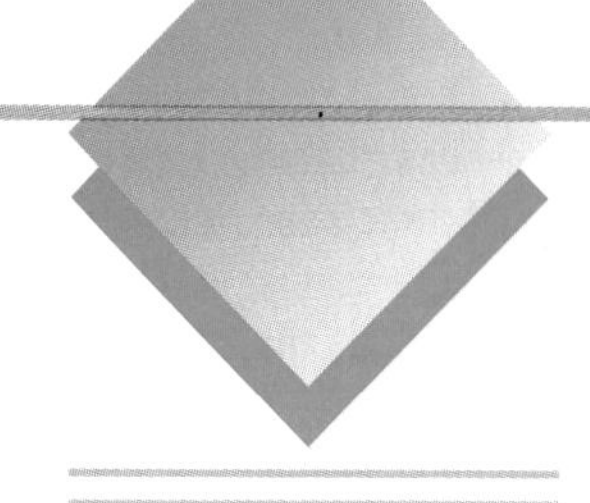

1. Why should taxpayers be informed about income tax laws even when they choose to have someone fill out their forms for them?
2. Why should taxpayers with complicated tax work use a certified public accountant who specializes in taxes? See Additional Answers.
3. When should taxpayers file a Schedule A instead of claiming the standard deduction? When the taxpayers' deductions are larger than the standard deduction.

KEY TERM
tax preparation software

Use the information provided for the two families below to answer Exercises 4–16. The numbers in parentheses are Social Security numbers.

(465–88–1033) (543–66–7121)
Howard and Florence Charisma
4565 Gulfstream
Panama Beach, Florida 32233
Wages: Howard, $37,800
Florence, $21,866
Withheld: from Howard, $4020
from Florence, $1992
Interest income: $250
Dividend income: $300
IRA for Florence: $2000

Filing Status: Married filing jointly
Dependents:
Loren, age 5 (305–67–7868)
Charlie, age 18 (690–13–9870)

Itemized deductions:
Medical expenses: $6500
Real-estate taxes: $1200
Mortgage interest on home: $6420
Contributions: $1300

(143–68–2569) (390–22–1378)
Macon and Milly Logan
12322 Commonwealth Avenue
Boston, Massachusetts 02108
Wages: Macon, $44,000
Milly, $31,700
Withheld: from Macon, $5440
from Milly, $4200
Interest income: $375
Dividend income: $200
IRA for Macon and Milly: $4000

Filing Status: Married filing jointly
Dependents:
Molly, age 17 (531–22–3657)
Mark, age 17 (531–22–3656)

Itemized deductions:
Medical expenses: $700
Real-estate taxes: $2253
Local income taxes: $4000
Mortgage interest on home: $7564
Contributions: $1600

(Neither Macon nor Milly has a pension plan at work.)

GUIDED PRACTICE

Have students work in small groups on the *Try Your Skills* exercises. Tell them to repeat 1-8 assuming that the Enterprises had $200 taxable interest income.

INDEPENDENT PRACTICE

ASSIGNMENT GUIDE

Exercises 1-3 check students' understanding of the opening text.

Exercises 4-16 ask students to fill out 1040 and Schedule A forms for the given families.

CLOSURE

Have students write a paragraph discussing available forms of tax help and their advantages and disadvantages. Suggest that students list reasons for using a Schedule A form.

LESSON QUIZ

1. Name four categories of deductions itemized on a Schedule A form. **Answers may vary.** (**Ex.: medical expenses, state income tax, real-estate taxes, interest on mortgage payments, charitable contributions, moving expenses**)
2. What is the maximum number of exemptions that can be taken by a married couple filing jointly with four children? What is the minimum? (**6, 2 if the children are over the age of 24**)

ADDITIONAL ANSWERS

12. Adjusted gross income: $47,764; Itemized Deductions: $11,838; Tax owed: $3780

4. Complete the first page of Form 1040 for Howard and Florence Charisma to find their adjusted gross income to the nearest dollar. $58,216
5. Complete Schedule A for the Charismas to find their total itemized deductions to the nearest dollar. $11,054
6. Complete page 2 of Form 1040 for the Charismas to find their tax owed or refund due to the nearest dollar. Refund: $33.00
7. Complete the first page of Form 1040 for Macon and Milly Logan to find their adjusted gross income to the nearest dollar. $72,275
8. Complete Schedule A for the Logans to find their total itemized deductions to the nearest dollar. $15,417
9. Complete page 2 of Form 1040 for the Logans to find their tax owed or refund due to the nearest dollar. Refund: $945.00

Mr. Charisma expects to be self-employed next year, earning the same amount of money as this year. He must fill out a Schedule C form for self-employed persons. However, you are not expected to fill out Schedule C to answer questions 10–13.

10. If he earns $37,800 next year, on which line of Form 1040 will he enter his business income as a self-employed person? Line 12
11. If Mr. Charisma becomes self-employed, he will be able to start a Keogh plan (self-employment retirement plan) by contributing 20% of his earnings to the plan each year. On which line of Form 1040 will he enter his contribution? Line 27
12. Prepare a Form 1040 and a Schedule A for the Charisma family to show their adjusted gross income, their total itemized deductions, and their tax owed or refund due to the nearest dollar next year when Mr. Charisma is self-employed with a Keogh plan. Assume that all of the earnings and expenses are the same as for this year. Note that Mr. Charisma is both an employer and an employee, so his FICA, or self-employment tax (line 47) is calculated at the rate of 2 • 7.65%, or 15.3%. One-half of his self-employment tax (line 25) is half of the line 47 entry. You may assume that Mr. Charisma's estimated tax payments (line 55) are the same as what his withheld taxes had been in Exercises 4–6. See Additional Answers.
13. How much money will the Charisma family save on taxes next year because of the Keogh plan? $222
14. Suppose that Mr. and Mrs. Logan had not contributed to their IRAs this year. How much more would they have paid in taxes? $1120
15. Suppose that instead of owning their own home, the Logans were renting a house and paying $9817 in rent. (This is the total of their current real-estate taxes and mortgage interest.) Would they find it worthwhile to use Schedule A? No. The standard deduction would be higher.
16. Use the assumptions of Exercise 15, with the IRA deduction, to find out how much more or how much less the Logans would pay in taxes if they were renting instead of owning. $2632 more

MIXED REVIEW

6-3 **1.** At the beginning of a monthly billing period, Albert's Mastercard balance was $385. The monthly finance charge was 1.7% of the amount owed. His bank required him to pay at least 15% of his unpaid balance or $25, whichever was larger. If he always paid the minimum required amount and made no additional purchases, in how many months did Albert reduce his balance to below $200? 5 mo

TAX Table p659

7-2 Find the break-even premium for a 1-year term insurance policy for each indicated individual. Assume that the direct and indirect expenses for issuing one policy are $20.

2. A $100,000 policy for a person with a life expectancy of 99.750% $270

3. A $200,000 policy for a person with a life expectancy of 98.950% $2120

8-1 Use the prices per share shown below to find the number of shares of stock you can afford to purchase if you have $50,000 to invest in each of the following companies. Also, find the total cost of the shares in each company. Round all money amounts to the nearest penny. Remember that you cannot purchase part of a share of stock. Ignore transaction costs such as brokerage commissions.

	Company	Price per Share	Number of Shares	Total Cost
4.	BullyPulp	$ 97	515	$49,955.00
5.	ElbGrs, Inc.	$18\frac{3}{4}$	2666	49,987.50
6.	Strm&Drng	$210\frac{3}{8}$	237	49,858.88

9-1 Use the Tax Table in the Reference Section to find the taxes owed on each of the following incomes.

	Filing Status	Taxable Income	Tax Owed
7.	Married, Filing Jointly	$68,500	$14,533
8.	Single	17,800	2,674

5-2 **9.** Suppose that your family's take-home pay is $4100 per month. How much can you afford to spend for credit-card payments each month?

7-2 **10.** The company wants to sell a 1-year term policy to a 45-year-old person and to keep the premium below $600. Assume that the expenses are $30 per policy and that the profit on the policy is $75. What would be the largest possible face value of such a policy? $157,143

4-1 **11.** If you earn $83,500 per year, how much do you pay in Social Security and Medicare taxes? $4781.95

4-1 **12.** If you earn $142,500 per year, how much do you pay in Social Security and Medicare taxes? $5528.70

ADDITIONAL ANSWERS

9. $820

CHAPTER 9 REVIEW

1. What does the term *filing status* mean? See Additional Answers.
2. May a person with taxable income over $100,000 use the tax table? If not, how does such a person find out how much tax he or she owes?
3. How does having dependents affect the amount of taxes that people pay?

Use the Tax Table in the Reference Section to find the tax owed each of the following incomes.

	Filing Status	Taxable Income	Tax Owed
4.	Single	$26,450	$ 4,625
5.	Married, filing jointly	87,000	19,729
6.	Married, filing separately	8,490	1,271
7.	Head of household	38,400	7,022
8.	Single	1,256	189
9.	Married, filing jointly	19,500	2,929

Use Tax Rate Schedules X and Y-1 in the Reference Section and the formulas on page 416 of Lesson 9–1 to find the tax owed on each of the following incomes.

	Filing Status	Taxable Income	Tax Owed
10.	Single	$150,000	$42,154.50
11.	Married, filing jointly	108,000	26,231.00
12.	Married, filing jointly	121,347	30,368.57
13.	Single	106,734	28,742.04

Find the yearly income, the taxable income, the taxes owed, and the amounts to be withheld annually for the incomes described below. Use a deduction of $3600 for a single person, $6000 for a married couple, and $5250 for a head of household. Each exemption is $2300. Assume that the married couple is filing a joint return and that the head of household is single. Use the Tax Table in the Reference Section.

	Filing Status	Monthly/ Yearly Income	Number of Exemptions	Taxable Income	Tax Owed	Withholding Allowances		Refund (+) or Money Owed (−)
						Number	Annual Amount	
14.	Single	$ 655/ 7,860	1	$1960	$ 294	1	$ 456	+$ 162
15.	Married	$ 940/ 11,280	3	0	0	2	60	+ 60
16.	Single	$1000/ 12,000	1	6,100	919	2	756	− 163
17.	Head of Household	$4450/ 53,400	4	38,950	7176	4	8928	+ 1752

ADDITIONAL ANSWERS

1. how you submit your claim form—single, married filing jointly, married filing separately, head of household
2. No. He/She must use the Tax Rate Schedules.
3. It lowers the amount of taxes.

Fill out a 1040EZ form for each of the following incomes to determine the amount owed or refund due to the nearest dollar. Assume that there is no taxable interest. For each exercise, make up a name and address. All of the people are single. You may choose whether or not to contribute to the election campaign fund; this choice will not affect anything else you fill out. The standard deduction amount for line 4 will be $3600 unless the adjusted gross income from line 3 is less than that.

	Wages, Salaries, and Tips	Interest	Amount Withheld	Can Be Claimed as a Dependent by Someone?
18.	$22,400	$350	$2460	No
19.	5,671	97	823	No
20.	12,500	290	2528	Yes
21.	18,600	350	3440	Yes

ADDITIONAL ANSWERS

18. Tax owed: $71

19. Refund: $823

20. Refund: $1152

21. Refund: $1134

22. Fill out a 1040A Form for the following married couple filing jointly to determine the amount owed or refund due to the nearest dollar. Their Social Security numbers are shown in parentheses. Refund: $4362

(321–45–3660) (544–36–1565)
Frederick and Patricia Moore
222 Beachfront Lane
Memphis, Tennessee 38132
Wages: Frederick, $23,026.75
Patricia, $22,617.16
Withheld: from Frederick, $4217
from Patricia, $4431

Dependents:
William, age 3 (534–56–8567)
Arthur, age 12 (574–24–6149)
David, age 15 (422–22–7469)
Interest income: $254
Dividend income: $195

23. Fill out Form 1040 for the following married couple filing jointly to determine their adjusted gross income, their total itemized deductions, and their tax owed or refund due to the nearest dollar. The numbers in parentheses are Social Security numbers. Use Schedule A for itemized deductions. Assume no tax penalty is owed.

(284–11–6997) (492–22–6451)
Alex and Jennifer Robertson
496 Jefferson Ave.
Longmont, Colorado 80502
Wages: Alex, $43,400
Jennifer, $22,000
Withheld: from Alex, $5020
from Jennifer, $2392
Interest income: $250
Dividend income: $385
IRA for Alex: $2000

Dependents:
Dorry, age 7 (551–33–1092)
Lorry, age 9 (493–44–0122)
Adjusted gross income: $64,035
Itemized deductions: $11,475
Tax owed: $79
Itemized Deductions:
Medical expenses: $3500
Real-estate taxes: $2900
Mortgage interest on home: $7595
Contributions: $980

CHAPTER 9 TEST

Use the Tax Table in the Reference Section to find the tax owed by each of the following taxpayers.

1. A single person with taxable income of $25,600 $4387
2. A married couple, filing jointly, with taxable income of $36,450 $5559
3. A married person, filing separately, with taxable income of $38,725 $8516
4. A head of a household with taxable income of $2600 $392

Use the Tax Rate Schedules in the Reference Section to find the tax owed by each of the following taxpayers.

5. A single person with taxable income of $122,400 $33,598.50
6. A married couple, filing jointly, with taxable income of $178,650 $48,132.50
7. Write a set of algebraic formulas for Tax Rate Schedule Z (for heads of households), which is found in the Reference Section. See Additional Answers.
8. Use one of the formulas from Exercise 7 to find the tax on a single parent with two young children. The family's total taxable income is $107,550. $27,378.50
9. Fill out a Form 1040EZ for the taxpayer described below to determine the amount owed or refund due to the nearest dollar. He cannot be claimed as a dependent on another taxpayer's return.

(354–20–8238)	Refund: $153
Nick Piocci	Salary: $16,780 per year
3821 Oak Avenue	Interest income: $235
Berwyn, Illinois 60402	Withheld taxes: $1822

10. Fill out 1040A form for the following married couple who file jointly and have four dependents to determine the amount owed or refund due to the nearest dollar. Fiala is blind; neither Miguel nor Fiala is over 65. To make a decision concerning the deduction on line 34, use the chart on page 422.

(102–00–9812) (324–57–3214)	
Miguel and Fiala Ferguson	Refund: $1693
2325 Ski Slope Drive	Interest income: $250
Red River, New Mexico 84222	Dividend income: $320
Miguel earns $34,222	Withheld taxes: $3564
Fiala does not work.	IRA deduction (Miguel): $1800

11. Suppose that in Exercise 10, the Fergusons have the following expenses.

Medical expenses: $8240	State income tax payments: $2400
Real estate taxes: $1500	Charitable contributions: $750

    Find the Fergusons' adjusted gross income, total itemized deductions, and tax owed or refund due to the nearest dollar using Form 1040 and Schedule A. To compute the standard deduction on line 19, use the chart on page 422. (You may ignore the directions concerning page 35 of the tax booklet.) Does the family benefit by using Form 1040 instead of Form 1040A? If so, by how much? See Additional Answers.

ALTERNATIVE ASSESSMENT

Have students create a hypothetical taxpayer with at least one dependent, assigning a wage, taxable interest income and Schedule A deductions. Direct them to fill out 1040 and Schedule A forms using the monthly withholding table and the Tax Table or Tax Rate Schedules. Have students explain the advantage of Schedule A in their taxpayer's case.

ADDITIONAL ANSWERS

7. i. $t = 0.15I \quad I \leq 28{,}750$
   ii. $t = 4312.50 + 0.28(I - 28{,}750)$ if $28{,}750 < I \leq 74{,}150$
   iii. $t = 17{,}024.50 + 0.31(I - 74{,}150)$ if $I > 74{,}150$

11. Adjusted gross income: $32,992
    Itemized deductions: $10,416
    Refund: $2248
    Yes, by $555.

CUMULATIVE REVIEW

1-2 Molly is considering each of the following employment opportunities.

Company	A	B	C
Salary	$42,500	$48,000	$38,800
Retirement benefits	4.0% of gross pay	3% of gross pay	4.5% of gross pay
Medical insurance	$60/month	$50/month	$56/month
Educational expense	$450/year	$900/year	$1400/year
Travel allowance	$1900/year	$200/month	$450/month

1. Calculate the total amount of money being offered, including fringe benefits for each position. A: $47,270; B: $53,340; C: $48,018
2. Determine which of the three positions offers the most money. Company B

4-3 Three students plan to sell souvenir umbrellas at the school football game. Their cost for 30 umbrellas is as follows.

Labor	25 hours at $4.50 per hour
Materials	umbrellas at $5.50 each
Advertising, energy, and transportation	$36.50

3. Find the cost of producing the 30 umbrellas. $314
4. What price should they put on the umbrellas in order to make $100 in profit? $13.80

6-5 Complete the chart below that shows one month's activity in a MasterCard account. Then find the quantities asked for in Exercises 5–7. Assume that the monthly finance charge is 1.5%.

Dates	Payment	Balance at End of Day	Number of Days	Sum of Daily Balances
1/1–1/8	0.00	$1549.50	8	$12,396.00
1/9	$183.00	1366.50	1	1,366.50
1/10–1/31	0.00	1366.50	22	30,063.00

5. Average daily balance 6. Finance charge 7. Ending balance

7-3 Use the formula for the Future Value of a Periodic Investment on page 344 of Lesson 7–3 to answer Exercises 8–9.

8. If you contribute $2000 to an IRA account that pays 11% a year beginning at the age of 20, how much will you have in your account at age 60?
9. If you begin contributing to your IRA at the age of 30, how much less will you have in your account at the age of 60 than you would have if you had begun your contributions at the age of 20?

9-1 Use the Tax Table or the Tax Rate Schedule in the Reference Section to find the taxes for the following taxpayers.

10. A single taxpayer with a taxable income of $151,000 $42,464.50
11. A head of household with a taxable income of $92,500 $22,721

ADDITIONAL ANSWERS

5. $1413.73
6. $21.21
7. $1387.71
8. $1,163,652.10
9. $1,163,652.10 − $398,041.76 = $765,610.34

PROJECT 9–1: **State Income Taxes**

1. Use reference books (such as a world almanac) to find the methods different states use to determine how much state income tax their residents pay.
2. Evaluate the methods of determining state income taxes to find out in which states people pay the highest taxes.
3. Find out in which states people pay the lowest taxes.
4. Devise a way to report your findings in an interesting way. Show the five highest, the five lowest, the most unusual, and so on. Or make a bar graph comparing the income tax rates in different states.

Extensions

1. Make up some problems based on your findings to calculate the state income taxes for people making the same salaries but living in different states. Have your classmates solve some of the problems.
2. Find which cities have income taxes and what their rates of taxation are.
3. Discuss with your classmates some reasons why some states have higher income taxes than others.

PROJECT 9–2: **The History of Income Taxes**

Work with a group of classmates to complete this activity. Use reference books (such as encyclopedias, history books, and political science books) to find out when federal income taxes were first introduced in the United States.

1. What is the name of the law that made federal income taxes legal?
2. Why and how was it passed?
3. When was the Internal Revenue Service established?
4. Who administers the Internal Revenue Service?
5. How are changes in tax laws made?
6. Do income taxes in fact strictly conform to the "earn more, pay more" principle?

Extensions

1. Organize a debate with your classmates to argue the pros and cons of federal income taxes. Also include how the taxes could be made more equitable.
2. Investigate income tax laws of other countries in the world.
3. Calculate what percent of income the average American pays in taxes—include federal, state, city, county, and social security taxes.

Very often we want to summarize a set of numbers or scores with just one number. The mean, median, and mode are *statistics* used to do this.

The *mean* is the sum of the scores divided by the total number of scores.

The *median* is the score that lies in the middle when the scores are ordered from least to greatest. If there are an even number of scores then the median is the average of the two middle scores.

The *mode* is the score that occurs the most times. If no score occurs most frequently then the set of numbers has no mode.

Find the mean, median, and mode. Round to the nearest tenth.

Example {70, 70, 70, 74, 75, 76, 82, 91, 97}

Mean: $\frac{70 + 70 + 70 + 74 + 75 + 76 + 82 + 91 + 97}{9} = 78.3$ The mean is 78.3.

Median: 70, 70, 70, 74, **75,** 76, 82, 91, 97 The median is 75.
Mode: **70, 70, 70,** 74, 75, 76, 82, 91, 97 The mode is 70.

1. {60, 68, 68, 78, 78, 78, 89, 90, 90}
   mean: 77.7; median: 78; mode: 78
2. {56, 89, 23, 48, 56, 91, 184, 56}
   mean: 75.4; median: 56; mode: 56
3. {200, 220, 220, 220, 240, 400, 800, 1000, 1000, 2000}
   mean: 630; median: 320; mode: 220
4. {4.1, 4.5, 4.5, 5.3, 5.3, 5.3, 6.5, 6.5, 6.9, 8.2, 8.3}
   mean: 5.9; median: 5.3; mode: 5.3

Sometimes when many scores are repeated, frequencies are used to express the results. Frequency means how many times a score is repeated.

Find the mean, median, and mode. Round to the nearest tenth.

Example Thirty-two students took a test. The scores are shown in the table.

Score	65	76	82	83	84	87
Frequency	1	3	4	8	9	7

Mean: $\frac{65 + 3(76) + 4(82) + 8(83) + 9(84) + 7(87)}{32} = 82.8$ The mean is 82.8.

Median: There are 32 scores. The first score is 65. The 16th score is 83 and the 17th score is 84. The median is $(83 + 84) \div 2 = 83.5$.

Mode: The score that occurs most often is 84. The mode is 84.

5.

Score	60	65	74	75	82	85	86	90	92
Frequency	1	3	7	10	9	7	8	4	2

mean: 80.2; median: 82; mode: 75

6.

Score	205	214	254	301	420	475	500
Frequency	6	9	14	60	34	18	54

mean: 382.6; median: 420; mode: 301

Owning a Car

IF YOU DON'T ALREADY DRIVE YOU ARE probably looking forward to the day when you will. Your parents and teachers have probably already told you that driving is not just for fun, and a car is most certainly not a toy. Driving is a big responsibility. Unless you are a very careful driver, you might injure yourself and others.

A car is one of the most expensive items you will ever purchase. After the initial purchase of a car, there are also fixed ownership costs and variable operating costs. Even though you take good care of it, a car gradually loses its resale value.

The young people in this chapter explore some of the differences between buying a new car and buying a used car and between owning and leasing a car. They learn about estimating and budgeting the different operating costs and what protection different kinds of insurance offer. Maria suffers "sticker shock" when she first sees the prices at a new car dealership. She learns, however, that dealers are willing to negotiate.

With his father, Alex learns that it can be wise to buy a used car. Freda looks closely at the mileage her car gets as she does the driving for the T-shirt business. She and Hari discover that leasing a car can make sense, particularly for a business. Finally, Trevor learns about automobile insurance. In particular, he researches the different kinds of auto insurance available and how the rates are determined for a driver and a car.

10-1 NEW CARS: SELECTING AND FINANCING

QUICK REFERENCE

TEACHER'S RESOURCES AND ANSWER KEY

Lesson Quiz Answers 10-1, p. 74

Reteaching Activity Answers 10-1, p. 109

Enrichment Activity Answers 10-1, p. 110

EXTENSION ACTIVITIES

Reteaching Activity 10-1, p. 75

Enrichment Activity 10-1, p. 79

When Maria's father told her that it is expensive to own and maintain a car, she thought that he just didn't want her driving too much. Then he told her that if she could save half of her weekly paycheck for six months, he would help her buy a car. That was a challenge. Much to her surprise, she was able to set a goal and achieve it.

Maria wanted a new car, so she and her father visited several new car dealers. At first, Maria was shocked by what she found. Each new car had a sheet of paper stuck to the inside of one of the windows. On the paper was a list of various features of that car: automatic transmission, power steering, air conditioning, AM/FM stereo with cassette deck, cruise control—and on and on. Many of the features had prices listed beside them.

The total price at the bottom of the list surprised Maria the most. She wondered how ordinary people could buy new cars. Maria's father explained that the sticker price was not what the dealer actually expected to receive for the car. Then Maria was really confused. How was she to know what price to pay for a car?

Maria wondered whether there were other things that she should know before spending her hard-earned money for a car.

OBJECTIVES: *In this lesson, we will help Maria to:*

- *Find out how to shop for a new car.*
- *Calculate the percent of markup.*
- *Calculate monthly car payments.*
- *Compare terms for financing a car.*

SHOP BEFORE GOING TO A CAR DEALER

Maria has been reading consumer magazines such as *Car and Driver, Motor Trend, Consumer Reports,* and *Changing Times.* She was able to find many of these in bookstores, in newsstands, and in her local and school library. From these sources she has learned a lot about the cost and quality of different cars.

Maria still does not understand why there is such a difference between the **sticker price** that appears on a car window and what the consumer finally pays. But she found that books such as *Edmund's New Car Prices* and the *Kelley Blue Book New Car Price Manual* give useful information about what dealers pay for cars and options. The sticker price is the suggested retail price of the car. The range between the dealer's cost and the sticker price is where Maria can negotiate.

Markup is the amount that a dealer adds to his cost to arrive at the sticker price. Maria noticed that some cars and some options have a higher markup than others. She also knows that the purchase price is not everything. She will try to choose a reputable dealer with a good service department, one that will be convenient for her when service is required.

FOCUS

ALGEBRA CONNECTIONS

A car selling for $9000 was marked up 25% by the dealer. Write and solve a one-variable equation for the dealer's cost. (**$1.25x = 9000$; $x = 7200$**)

DEALING WITH DEALERS

Maria has shopped around and is prepared to bargain. She knows that she must be able to walk away from a car and go to another dealer if the price does not suit her. "Falling in love" with a car will often lead to paying too much. Maria has learned three essentials for bargaining:

1. Use competitive conditions to your advantage.
2. Know the dealer's costs.
3. Allow the dealer a reasonable profit.

What is a reasonable profit? You and the dealer might have different ideas about that. There are, of course, many other factors involved. But most of them come down to the amount of the original markup and the laws of supply and demand. In other words, if many people want a particular model car, then the dealer will not be motivated to reduce the price very much. Following are some of the tactics that car dealers might use in their efforts to attract buyers and sell cars.

CRITICAL THINKING

Organize a brief debate with two or three students on each team to present reasons for and against: (1) placing a large down payment; (2) obtaining the longest-term loan possible. Be sure that students state the mathematical and economic consequences of their recommendations.

Bait and Switch Maria saw an advertisement in the newspaper that seemed too good to be true. It was. When Maria and her father visited the dealer, they found that the advertised car was not available. But the salesperson tried to interest them in another, more expensive car. This was Maria's first encounter with the bait-and-switch technique.

Financing Maria's father explained that when the time comes for **financing**, that is, borrowing the money needed for a car, they will obtain a loan from their own bank or credit union rather than from a company recommended by the auto dealer. This is because sometimes the dealer, in cooperation with the finance company, tries to add special fees or insurance charges to the purchase price. The terms of financing include the amount to be borrowed, the purchaser's down payment, the interest rate, and the number of months of the loan. Maria will consider all of these and weigh them in light of her own financial resources. The **finance charge** is the amount that the credit will cost.

Trade-In If Maria has a car to **trade in,** that is, to sell to the dealer when she buys a new car, then her negotiations will be even more complicated, and she will have to be even more careful. Sometimes, to get a buyer's interest, a dealer will talk about paying a high price for a trade-in and later lower the amount. In this case the dealer is using a highball strategy. In a similar way a dealer might quote you an unrealistically low price for a new car and then, when you return after shopping around, explain that, on review, the original quote must be raised. This approach is called the lowball. In most cases it is better not to mention that there is a trade-in until the price of the new car has been settled. If the price offer for your old car seems too low, you can always sell it yourself.

ALGEBRA *REVIEW*

Solve each equation or, if needed, select a variable, write an equation, and then solve.

1. $0.2 \cdot \$5647 = c$
   $1129.40
2. $7.5\% \cdot \$7800 = c$
   $585
3. Find 2.3% of $4568.
   $105.06
4. Five is what percent of 15?
   33.3%
5. Ten is what percent of 25?
   40%
6. What percent of $7560 is $529.20?
   7%
7. Add $4500 to 9% of $4500.
   $4905
8. Find 1.12% of $9800.
   $109.76
9. $585.12 is 12% of what number?
   $4876
10. $500 is 10% of what number?
    $5000

Ask Yourself

1. What should you do before shopping for a car?
   See Additional Answers.
2. What are some of the tactics a car dealer might use to interest a potential buyer?
   bait-and switch, highball trade-in, lowball price quote
3. What does financing mean?
   borrowing money to purchase an expensive item

ADDITIONAL ANSWERS

1. read consumer magazines and gather information about the car(s) you may be interested in

INSTRUCTION

TEACHING THE LESSON

Review the meaning of percent, the method for decimal-percent conversion, and three basic percent cases.

Following up on the *Algebra Connections* and the use of percent in other chapters, you may want to have students add markup percent to 100%, giving, for example, 118% or 1.18. In this way some problems can be done in one rather than two steps.

Most students will be familiar with monthly payments. Be sure to emphasize that the monthly interest rate is used in the formula. Demonstrate how it is obtained from the annual rate.

FOCUS ON ALGEBRA

This lesson applies an integrated approach to the solution of real world problems, moving from familiar skills such as percent to more complex formulas. (*NCTM, Standard 1, p. 137*)

AT-RISK STUDENTS

Some students, generally not mathematically inclined, will have an interest in cars. Try to use this interest to motivate students and lead them gradually to the related mathematical ideas.

SKILL 1

EXAMPLE 1 The sticker price for a Starfire two-door convertible is $14,255. Using her car books, Maria finds the dealer's cost is $12,450.

QUESTION What is the percent of the dealer's markup?

SOLUTION

The markup m is the difference between the sticker price s and the dealer's cost d.

$$m = s - d$$
$$m = 14{,}255 - 12{,}450$$
$$m = 1805$$

The markup is $1805.

To find p the percent of markup, Maria writes an equation using the following question to help her.

What percent of 12,450 is 1805?

$$p(12{,}450) = 1805$$

$$p = \frac{1805}{12{,}450} \quad \text{Divide both sides by 12,450.}$$

$$p = 0.145 \quad \text{To the nearest thousandth}$$

The percent of markup is 14.5%.

COMMON ERRORS

Students may be uncertain whether to use the dealer's cost or sticker price of a car as the base when working with percent. Reinforce the idea that the *base* in percent is the number on which a percent is based.

TECHNOLOGY HINT

Help students become familiar with how to use a formula repeatedly on the graphing calculator. It saves time and possible mistakes to copy a formula already entered and checked, and then to change numbers as needed. This method can be shown in Skill 3, and used in some of the exercises.

RETEACHING

If your class requires reteaching, go back to the basic meaning of percent as parts out of one hundred. Then explain the meaning of interest. Finally, using round numbers, explain down payment and monthly payment.

ENRICHMENT

Ask students to use the monthly payment formula to devise a formula for calculating loan payments on an annual basis. Then apply this to the figures in Skill 3 to find an annual amount. How does this compare with the amount paid on a monthly basis? (**For 10% down for 3 years, the annual payment is $6562.26, slightly higher than twelve monthly payments because the principal is paid back more slowly.**)

SKILL 2

EXAMPLE 2 Maria compares the prices of the following options listed on the sticker with dealer's costs she finds in her car books.

Option	Dealer's Cost	Sticker Price
Air conditioning	$356	$420
Roof rack	116	138
Automatic transmission	645	774
AM/FM radio, tape deck	440	506

QUESTION How can Maria use a spreadsheet to show the markup and percent of markup for the options?

SOLUTION

Maria uses the formula +C3−B3 for cell D3 and +D3/B3 for cell E3. To find the totals she uses the sum function of the spreadsheet. Her spreadsheet looks like the following. Notice that she formatted column E for percent.

	A	B	C	D	E
1		Dealer	Sticker		Percent of
2	Option	Cost	Price	Markup	Markup
3	Air conditioning	356	420	64	18%
4	Roof rack	116	138	22	19%
5	Automatic transmission	645	774	129	20%
6	AM/FM radio, tape deck	440	506	66	15%
7	Total	1557	1838	281	18%

SKILL 3

Recall the monthly payment formula for repaying a loan from Lesson 5–1.

$$M = \frac{\left(Pr(1+r)^n\right)}{\left((1+r)^n - 1\right)}$$

where M = monthly payment
P = amount of loan
r = monthly interest rate
n = number of payment periods

EXAMPLE 3 Maria's friend Bryan plans to buy a new Moonbeam car with a list price of $19,490 and finance it through his credit union. He can make a down payment of 10% or 20% and then pay off the loan at a 6% annual interest rate over a period of 3 years or 5 years.

QUESTION What will be the monthly payments and the total cost of the car under the different payment plans?

SOLUTION

Bryan has four ways to pay:

10% down for 3 years
10% down for 5 years
20% down for 3 years
20% down for 5 years

He decides to use a spreadsheet to compare costs. The loan amount is the original cost minus the down payment. Use the monthly payment formula to calculate the payments for each plan. Since the APR is 6%, the monthly interest rate is $0.06 \div 12 = 0.005$. To find the total payments, remember to multiply the unrounded amount by the number of payments. The total cost includes the down payment. The finance charge is the difference between the total cost and the original cost.

	A	B	C	D	E	F	G	H	I
1	Original	Percent	Down	Loan		Monthly	Total	Total	Finance
2	Cost	Down	Payment	Amount	Years	Payment	Payment	Cost	Charge
3	19490	10%	1949	17541	3	533.63	19210.72	21159.72	1669.72
4	19490	10%	1949	17541	5	339.12	20347.00	22296.00	2806.00
5	19490	20%	3898	15592	3	474.34	17076.20	20974.20	1484.20
6	19490	20%	3898	15592	5	301.44	18086.22	21984.22	2494.22

From the spreadsheet Bryan concludes that if he can afford to make higher monthly payments he will save money.

TRY YOUR SKILLS

A car costing the dealer $12,800 has a sticker price of $15,232.

1. Find the markup. $2432
2. Find the percent of markup. 19%

A car costing the dealer $11,250 is marked up 18.5%.

3. Find the markup. $2081.25
4. Find the sticker price. $13,331.25

A car is to be sold for $13,200. The annual interest rate is 8.4% with a 20% down payment for 3 years.

5. Find the monthly interest rate. 0.007
6. Find the monthly payment. $332.86
7. Find the total cost of the car. $14,623.11
8. Find the finance charge. $1423.11

GUIDED PRACTICE

The *Try Your Skills* exercises take students through each of the subskills learned in the lesson. After most students have completed the first four exercises, review these on the chalkboard. Similarly, review numbers 5-8 allowing for questions.

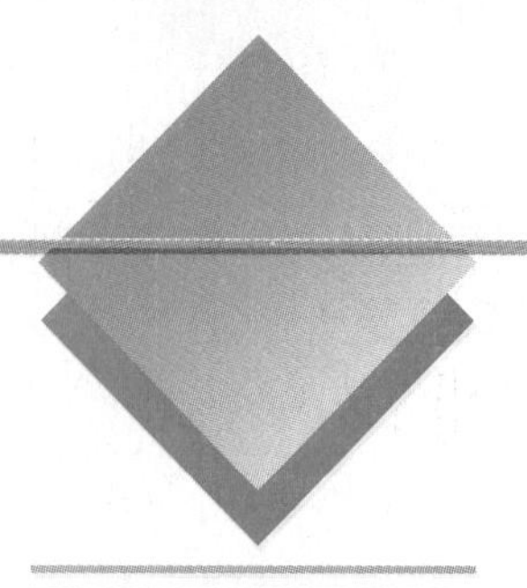

EXERCISE YOUR SKILLS

KEY TERMS

finance charge
financing
markup
sticker price
trade-in

INDEPENDENT PRACTICE

ASSIGNMENT GUIDE

Exercises 1-8 ask students to find percent. In each case they must: find the amount, identify the base, and correctly use the method to find the percent.

Exercises 9-12 present slightly more complex situations, but they are still directed toward finding percent.

Exercises 13-15 ask for the calculation of monthly payment.

Exercises 16-26 review monthly payment and ask about the costs of different financing arrangements.

ADDITIONAL ANSWERS

7. 17.4%

The following is a portion of the Moonbeam price list, which also appears in the Reference Section of the book.

Moonbeam	Dealer Cost	Sticker Price
Model A, 4-door wagon, 8 cylinder	$12,725	$14,722
Model B, 4-door sedan, 6 cylinder	14,062	16,295
Model C, 4-door sedan, 4 cylinder	11,053	12,755
Options		
Conventional spare tire	$ 62	$ 73
Air conditioning	695	817
Electronic climate control	850	1000
Antilock brake system	838	985
Rear window defroster	136	160
AM/FM radio and cassette player	132	155
Rear-facing third seat	132	155
Cruise control	178	210
Stripe, painted	51	61

Find the percent of markup for each of the following. Round to the nearest tenth of a percent.

1. Model A 15.7%
2. Model B 15.9%
3. Model C 15.4%
4. Air conditioning 17.6%
5. Electronic climate control 17.6%
6. Antilock brake system 17.5%
7. AM/FM radio and cassette player
8. Rear facing third seat 17.4%

Find the total dealer cost and total sticker price for each of the following.

9. Model B with air conditioning, rear window defroster, AM/FM radio and cassette player, cruise control $15,203; $17,637
10. Model A with electronic climate control, anti-lock brakes, rear window defroster, rear-facing third seat $14,681; $17,022
11. Model C with air conditioning, rear window defroster, AM/FM radio and cassette player, stripe $12,067; $13,948
12. Find the percent of markup for each of the total suggested retail prices in Exercises 9, 10, and 11. 16.0%; 15.9%; 15.6%
13. Find the monthly payment for Model B at the sticker price with air conditioning, rear window defroster, AM/FM radio and cassette player, and cruise control. Financing terms are 6% annual interest rate, 20% down, 3 years to pay. $429.24

14. Using the sticker price, find the monthly payment for the car and options described in Exercise 10. Terms are 8.4% annual interest rate, 15% down, 4 years to pay. $355.95

15. Using the sticker price, find the total amount paid for the car and options described in Exercise 11. Terms are 6% annual interest rate, 20% down, 5 years to pay. $15,732.99

Exercises 16–22 are based on the purchase of Model A at the sticker price with air conditioning, anti-lock brake system, rear window defroster, AM/FM radio and cassette player, rear-facing third seat, and cruise control.

16. Find the total dealer cost. $14,836

17. Find the total sticker price. $17,204

18. Find the percent of markup. Round to the nearest tenth of a percent. 16.0%

19. Find the monthly payment at 6% APR for 6 years with 20% down. $228.10

20. Find the monthly payment at 6% APR for 4 years with 10% down. $363.63

21. In Exercises 19 and 20, which of the terms of financing has the lower monthly payment? 6% APR for 6 years with 20% down

22. In Exercises 19 and 20, which of the terms of financing has the lower total cost? 6% APR for 4 years with 10% down

23. Compare the terms of financing, and discuss reasons why a person might wish to select one or the other. See Additional Answers.

24. The cash price for certain used cars is shown in the following table. The annual interest rate is 9%. The payment period is 4 years. The down payment can be 10% or 20%. Use a spreadsheet program or calculator to complete the table.

Original Cost	Percent Down	Down Payment	Loan Amount	Years	Monthly Payment	Total Payment	Total Cost	Finance Charge
$7500	10%	$ 750	$6750	4	$167.97	$8062.75	$8812.75	$1312.75
7500	20%	1500	6000	4	149.31	7166.89	8666.89	1166.89
8000	10%	800	7200	4	179.17	8600.27	9400.27	1400.27
8000	20%	1600	6400	4	159.26	7644.69	9244.69	1244.69

25. For the used car in Exercise 24 with an original cost of $7,500, how much will be saved by making a 20% down payment? $145.86

26. Suppose the dealer agrees to finance the $8000 in Exercise 24 at a 5% annual interest rate if you make a down payment of 25%. How much will you save over putting 20% down at the 9% rate? $612.25

ADDITIONAL ANSWERS

23. In Exercise 19, although the high down payment decreased the amount financed, the increased term of the loan increased the total finance charge. The less expensive financing is in Exercise 20 because the term of the loan is decreased to 4 years. You would choose the financing in Exercise 19 if you could not afford a high monthly payment. If you could afford the higher monthly payments, it would be better to choose the financing in Exercise 20, and lower the total finance charge.

CLOSURE

Review the ideas of interest earned and interest paid. Ask a student to explain to the class the meaning of a finance charge and the factors that determine finance charges: interest rate, down payment, and length of loan.

LESSON QUIZ

1. Find the monthly interest rate for a 9% annual rate. (**0.75%**)
2. How many monthly payments will be made on a 4-year loan? (**48**)
3. Find the monthly payment on a 4-year loan of $15,000 with a 9% annual interest rate. (**$373.28**)

MIXED REVIEW

3-2 **1.** Find the value of $2500 after it has been invested at a 7% annual return and compounded annually for 5 years. $3506.38

3-2 **2.** How long will it take for an investment to double if it increases in value by 10% a year? 8 years

6-1 **3.** If 1400 out of 5000 people interviewed pay their credit card bills in full each month, what percent of the sample does not pay their bills in full each month? 72%

7-3 **4.** Use the formula for the Future Value of a Periodic Investment to find the value of an IRA into which annual contributions of $2000 have been made for 20 years. Assume that the investment has had an annual compounded return of 8% a year over the entire period. $91,523.93

5-1 **5.** If Marci can afford a monthly payment of $360, how much can she afford to borrow at a yearly interest rate of 9.5% for 5 years? $17,141.34

4-4 **6.** The cost function for a small business is given by the equation $c = 2.5n + 125$, where n is the number of items produced. The revenue function for the business is given by the equation $r = 6n$. Use a graphing calculator to find the intersection of the graphs of the functions, that is, the break-even point of the business. 36 items

8-3 **7.** Use a graphing calculator to find the linear regression equation that is the line of best fit for the following table values. $y = 0.998x + 8.942$

Number of Years After Start	0	0.5	1.0	1.5	2.0	2.5	3.0	3.5
Value of Mutual Fund (in $)	9	9.6	9.8	10.3	10.9	11.4	12.0	12.5

9-2 **8.** Use IRS Form 1040A to find the refund or payment due for the following married couple. Social Security numbers are shown in parentheses. Payment due: $12,721

(012–46–3774) (012–49–1132)
Alex and Nina Black
9500 Banyon Drive
Orlando, Florida 32819
Filing a joint return
Wages: Alex, $57,500
Nina, $28,000

Dependents:
Hal, age 16 (422–55–9031)
Gladys, age 17 (422–55–9032)
Withheld: from Alex, $3200
from Nina, $950
Interest income: $4,100
Dividend income: $2,455

1-3 **9.** You are a real-estate broker who has just sold a house for $325,000. You earn 5% on the first $125,000 and 6% on any amount over $125,000. Determine your total commission. $18,250

2-3 **10.** Your checkbook shows a balance of $285.35. Your bank deducts a service charge of $2.50 each month but does not pay interest and does not charge for each check. The only check that you wrote during the month was for $185.44 and was mailed the day before the bank's monthly statement arrived. What will the bank show as the closing balance? $468.29

10-2 EQUITY AND DEPRECIATION FOR NEW AND USED CARS

QUICK REFERENCE

TEACHER'S RESOURCES AND ANSWER KEY

Lesson Quiz Answers 10-2, p. 76

Reteaching Activity Answers 10-2, p. 109

Enrichment Activity Answers 10-2, p. 110

EXTENSION ACTIVITIES

Reteaching Activity 10-2, p. 76

Enrichment Activity 10-2, p. 80

Alex thought it would be fun to help his father pick out a used van for the family. His father took him to quite a few used-car lots, where they looked at a variety of vans before they finally made the deal for the van they bought. At first, Alex did not understand why his father inspected the vans so closely and insisted on driving them, sometimes even onto the freeway and back.

Alex's father told him about the first used car that he had bought. It was only two years old and showed 14,000 miles on the odometer. Alex's father took a short test drive and didn't even notice that the odometer was not working. In fact, he didn't notice a number of things about the car. It turned out that the car was a repainted taxi with 90,000 miles on it. During the first year the car needed a new radiator, an alternator, a fuel pump, and a rebuilt transmission. These repairs were very costly.

Alex now understands why the van that his father finally purchased was thoroughly checked by their mechanic before his father bought it. He also knows that there is a lot more to buying a used car than first meets the eye.

OBJECTIVES: *In this lesson, we will help Alex to:*

- *Learn how to select a used car.*
- *Estimate the value of a used car.*
- *Understand equity and depreciation.*

ALGEBRA CONNECTIONS

A car has a value of $10,000. With each passing year it loses 10% of this original value. Write an equation for the value of the car as a function of years.
(**$v = 10{,}000 - 1000t$**)

EQUITY AND DEPRECIATION

When you take out a loan to buy a car, the car dealer is being paid mostly by the bank. The bank really owns the car until you finish paying for it. If for some reason you cannot continue your car payments, you will have to sell the car to pay the remainder of loan, or the bank will repossess the car—that is, take it from you. As you pay off the loan, you increase your equity in the car. **Equity** is your part of the financial worth of the car. If your car could be resold for $8000 but you still owe $3000 on a loan, then your equity in the car is $5000.

But what is a car worth at any time? You know that a car loses value as it gets older. It can be sold for less and less, until it is worth nothing. This loss in the value of the car is called **depreciation.** For a new car this loss begins immediately after you purchase the car. For a used car, depreciation is more gradual.

Depreciation varies from one car to another. Some cars have a reputation for lasting and therefore hold their value better than others. Also, inflation means that the price of everything goes up a little bit each year. Inflation slightly reduces the effects of depreciation. Straight-line depreciation means that a car, or whatever is being discussed, depreciates an equal amount each year. For example, if a car costs $20,000 and depreciates over a straight line for 10 years, it loses 10% of its original cost or $2000 each year.

But cars, especially new cars, do not depreciate in a straight line. They tend to lose more of their resale value in the first years and less in the later years. A car might depreciate by 25–30% in the first year and 20–25% in the second year. To Alex, this does not seem "right." An "almost new car" should be worth almost as much as a new car. In fact, people now watch their finances more closely and are often interested in buying almost new cars. Consequently, newly purchased cars do not depreciate as rapidly as brand-new cars.

WHERE TO BUY

Alex has learned that while it is fairly easy to compare the prices of new cars at different dealers, it is difficult to compare the prices of used cars. This is because you can rarely find two used cars of the same year and model with the same number of miles and the same amount of wear and tear. Fortunately, there are lists that give the book value for used cars. **Book value** is a standard estimate of the resale value for an average used car of a particular year and model. Current used car prices are found in *Edmund's Used Car Prices* and in the *Official Used Car Guide* published monthly by the National Automobile Dealer's Association. The latter publication is also know as the *Blue Book.* The book value is sometimes used for tax purposes. In practice, you can expect to pay more or less than the book value depending on the condition of the car.

Dealers New car dealers generally keep the best used cars traded in, so you can expect to find good, but fairly expensive, used cars at new car dealers. There are many different kinds of salespeople at a used-car dealership. Some will help you find the best car to suit your needs; others just want to make a sale as quickly as possible. Until you get some experience, it is important to go slowly when looking at cars. Talk to your friends and relatives about different dealers. Ask a knowledgeable friend to go with you when you first visit car dealers. The Federal Trade Commission (FTC) has written consumer-protection regulations concerning the selling of used cars. The FTC encourages the buyer to have all verbal claims put in writing and to have a mechanic inspect the car before it is purchased.

Private Sales Cars for sale are listed in the newspaper. If a buyer knows what he or she wants, what questions to ask, and how to look at a car, then a private sale can be a good deal. It is always important to read the car's **title** carefully to be sure that there is no mistake about the car or owner. FTC regulations do not apply to used cars if they are purchased from an individual.

CRITICAL THINKING

Ask what factors seem to determine whether something will appreciate or depreciate. What are three kinds of things that appreciate and three that depreciate? Do some cars appreciate?

Car Rental Company Sales and Auctions Sometimes good used cars are available when rental agencies or other companies sell their cars. Rental cars are generally kept in excellent condition, but a buyer should always think twice about buying a car that has been driven by many people. Cars bought at auction or from a car-leasing agency require extra investigation because maintenance information may not be available.

Inspecting a Used Car It should be clear that staring at an engine, kicking the tires, or driving a car around the block will not tell you what you need to know about the condition of a used car. A smart buyer conducts three types of inspections: on the lot, on the road, and in the garage.

On the lot, you should ask questions about the history of the car including replacement of parts, accidents, and the number of previous owners. On the road, you should pay attention to how the car feels and sounds. Trust your experience, and ask more questions if anything does not seem right. Finally, pay a mechanic to examine the car. You will do this only with a car that you are almost ready to buy.

In addition, the name of the previous owner must be given to you with a proof of odometer reading. It is also a good idea to call the National Highway Traffic Safety Administration (1-800-424-9393) to see if there are any safety recalls on the car.

ALGEBRA *REVIEW*

Set up an appropriate coordinate system and plot the points.

1. (1, 1000), (2, 600), (3, 400)
   See Additional Answers.
2. (1, 10,000), (2, 8000), (3, 6000)
3. (1, 10.5), (2, 21), (3, 42)

Find the value of *t*. Round to the nearest hundredth.

4. $t = 1.05^{10}$
   1.63
5. $t = (1000)(1.006)^{24}$
   1154.39
6. $t = (5000)(1.005)^{36}$
   5983.40

Evaluate each for $x = 3$.

7. $3.5x - 1$
   9.5
8. $1.05x + 2.007$
   5.157
9. $1.05x + 2.25$
   5.4
10. $1.005x$
    3.015

Ask Yourself

1. What is equity? Equity is the buyer's part of the financial worth of the car.
2. What is depreciation?
   the loss in value of the car
3. Which depreciates more rapidly in the early years, a new car or a used car?
   a new car
4. What is book value? the standard estimate of the resale value for an average used car of a particular year and model
5. Why do new car dealers tend to charge more for their used cars than other dealers? New car dealers generally keep the best used cars traded in.

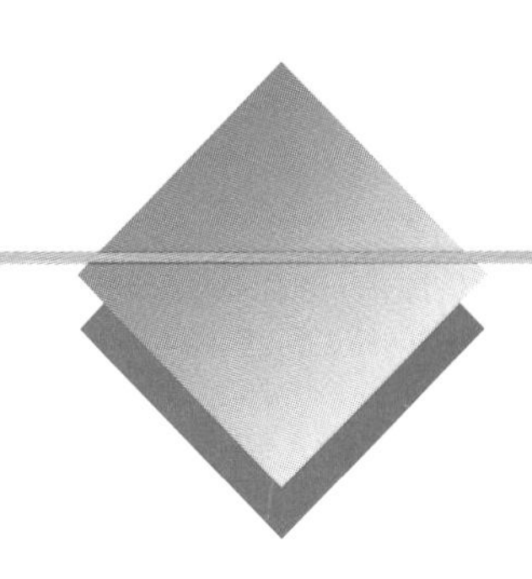

SHARPEN YOUR SKILLS

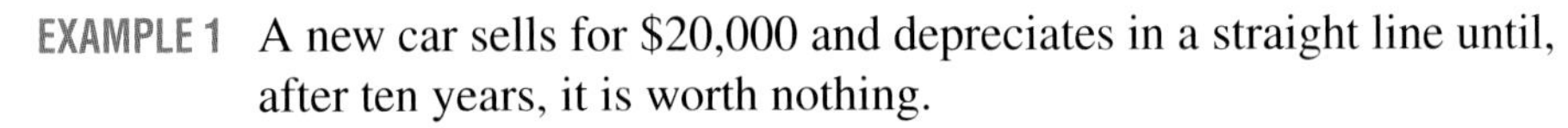

SKILL 1

There are several different ways to calculate depreciation. The method depends on the rate at which a car, or something else of value, tends to lose its value.

EXAMPLE 1 A new car sells for \$20,000 and depreciates in a straight line until, after ten years, it is worth nothing.

QUESTION How would you construct a table and draw a graph to show the straight-line depreciation?

SOLUTION

The amount of depreciation is the same for each year. Depreciation begins as soon as the car is purchased and continues at the rate of 10% per year. At the end of the first year the car has depreciated \$2000.

$20{,}000 \bullet 0.10 = 2000$

You can use a spreadsheet to calculate the depreciation and graph the results. Choose the number of years for x values and the resale values for y values. The graph shows the resale value as a function of time.

	A	B	C	D
1		Percent of	Amount of	Resale
2	Year	Depreciation	Depreciation	Value
3	0			20000
4	1	10%	2000	18000
5	2	10%	2000	16000
6	3	10%	2000	14000
7	4	10%	2000	12000
8	5	10%	2000	10000
9	6	10%	2000	8000
10	7	10%	2000	6000
11	8	10%	2000	4000
12	9	10%	2000	2000
13	10	10%	2000	0

+D3–C4

The graph of the resale value of the car is a straight line that slopes downward and to the right. This slope accounts for the name "straight-line" depreciation. The graph appears at the top of the next page.

INSTRUCTION

TEACHING THE LESSON

Be sure to spend time explaining depreciation and appreciation. Students will be familiar with the fact that some things, like cars, wear out while others, like houses and jewelry, increase in value.

Skill 2 uses the formula for monthly payment, from Lesson 10-1, to calculate the amount of the first and second payments that are paid toward principal and interest. Use the example to point out that interest is paid only on the unpaid balance, not on the original amount borrowed.

The graph in Skill 2 contains a good deal of information and will require a thorough explanation. The practical significance is that the early sale of a car may not bring in enough money to repay what is left of the loan.

FOCUS ON ALGEBRA

In discussing mathematics as communication, the Standards note that students should be encouraged to ask clarifying and extending questions about what they have learned. This lesson affords several opportunities in this area. (*NCTM Standard 2, p. 140*)

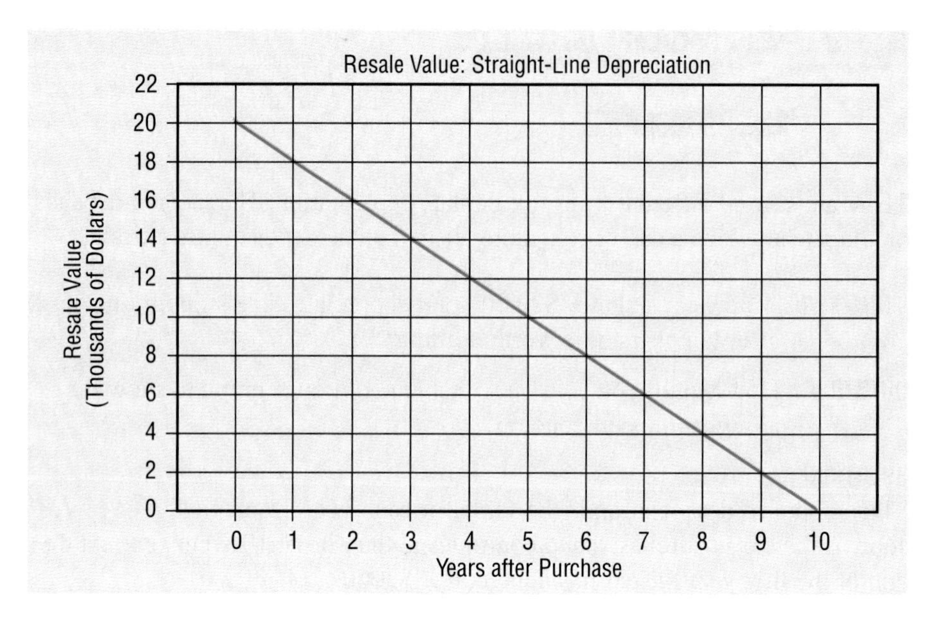

EXAMPLE 2 Records for one standard model of car indicate that during 10 years it depreciates by the following percent of the original cost:

25, 20, 12, 10, 8, 7, 6, 5, 4, 3.

QUESTION How can these percents be used to form a table and a graph to represent the depreciation?

SOLUTION

We start with the original price of $20,000 and subtract the depreciation amount for each year. Use a spreadsheet as in Example 1. Again, the graph shows the resale value of the car as a function of time. In the formula for cell C4, be sure to use an *absolute* address for cell D3 so the address stays the same when the formula is copied to another cell.

COMMON ERRORS

Students may confuse "resale value" and "outstanding balance" because they are mentioned in the same skill and because they are new terms. Explain that the first means, "the amount you could get for the car if you sold it," and the second means, "the amount you have left to pay."

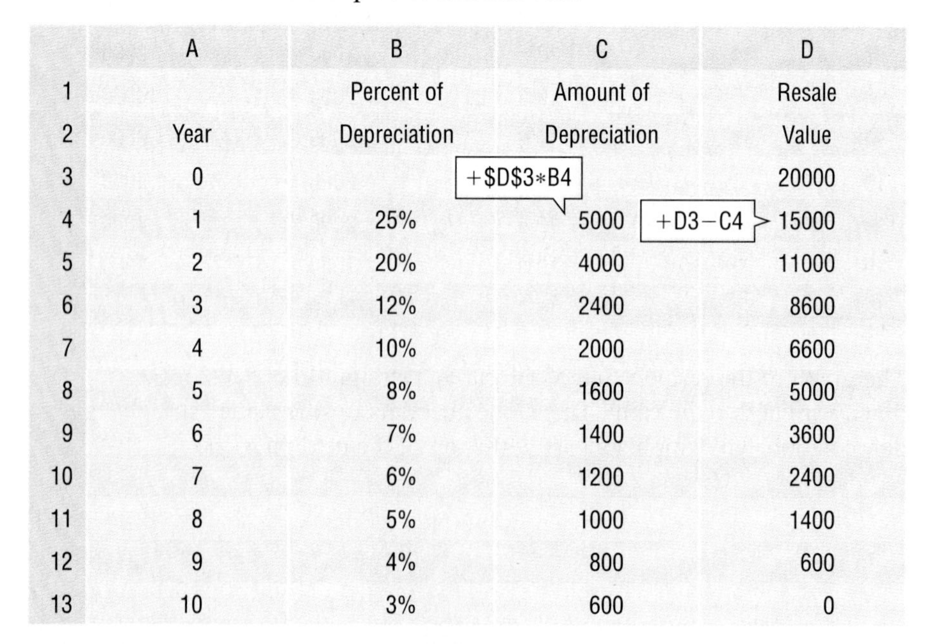

	A	B	C	D
1		Percent of	Amount of	Resale
2	Year	Depreciation	Depreciation	Value
3	0		+$D$3*B4	20000
4	1	25%	5000	+D3−C4 15000
5	2	20%	4000	11000
6	3	12%	2400	8600
7	4	10%	2000	6600
8	5	8%	1600	5000
9	6	7%	1400	3600
10	7	6%	1200	2400
11	8	5%	1000	1400
12	9	4%	800	600
13	10	3%	600	0

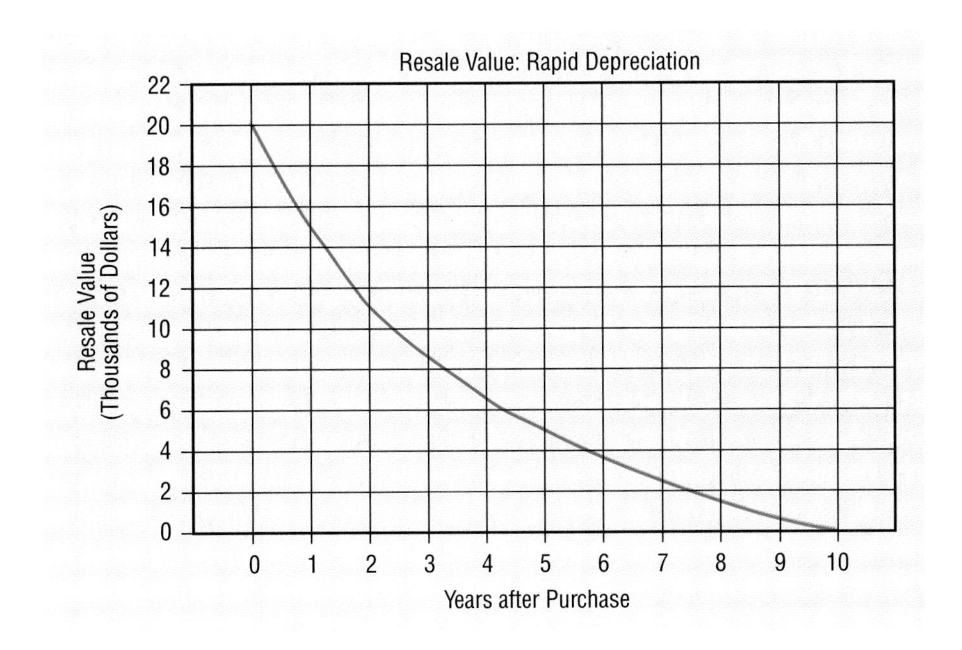

SKILL 2

Comparing depreciation with loan repayment can help you understand what your car is really worth and how to use your money.

EXAMPLE 3 Bryan decides to buy the Moonbeam car discussed in Lesson 10–1, Example 3, for $19,490. He makes a down payment of 20% and obtains a 3-year loan at an APR of 6%.

QUESTION How does Bryan's equity in the car vary during the life of the car?

SOLUTION

In Lesson 10–1, Example 3, Bryan found that the monthly payment is $474.34. He uses a spreadsheet to make an amortization schedule as in Lesson 5–4, Example 2. Bryan includes an additional column to calculate the depreciated value of the car. He assumes straight-line depreciation at 10% per year, that is 19,490(0.10 ÷ 12), or $162.42 per month.

He also includes a column to calculate the equity, which is the part of the car he owns. The equity is the difference between the depreciated value and the amount you owe on the car; that is, the unpaid balance of the loan. Notice that the final payment is adjusted for a zero unpaid balance. Although the loan is repaid in 36 months, Bryan extends the spreadsheet to 60 months so he can find out his equity after 5 years.

Bryan's spreadsheet is shown on pages 470 and 471. He used formulas as shown.

ESL STUDENTS

Review key words with non-native speakers: *loan, payment, depreciation, equity, book value, principal, balance.*

COOPERATIVE LEARNING

Exercises 18-21 are based on a scenario that describes two different financial situations. These could be done in cooperative groups with four students in a group. The exercises require mathematical calculations and non-mathematical reasoning. Students should distribute the mathematical tasks but discuss and try to agree on the decisions.

+F3−E3

@ROUND(E3*0.005, 2)

+B4−C4

+E3−D4

+F3−162.42

	A	B	C	D	E	F	G
1	Payment	Payment	Interest	Note	Unpaid	Depreciated	
2	Number	Amount	Due	Reduction	Balance	Value	Equity
3	0				15592.00	19490.00	3898.00
4	1	474.34	77.96	396.38	15195.62	19327.58	4131.96
5	2	474.34	75.98	398.36	14797.26	19165.16	4367.90
6	3	474.34	73.99	400.35	14396.91	19002.74	4605.83
7	4	474.34	71.98	402.36	13994.55	18840.32	4845.77
8	5	474.34	69.97	404.37	13590.18	18677.90	5087.72
9	6	474.34	67.95	406.39	13183.79	18515.48	5331.69
10	7	474.34	65.92	408.42	12775.37	18353.06	5577.69
11	8	474.34	63.88	410.46	12364.91	18190.64	5825.73
12	9	474.34	61.82	412.52	11952.39	18028.22	6075.83
13	10	474.34	59.76	414.58	11537.81	17865.80	6327.99
14	11	474.34	57.69	416.65	11121.16	17703.38	6582.22
15	12	474.34	55.61	418.73	10702.43	17540.96	6838.53
16	13	474.34	53.51	420.83	10281.60	17378.54	7096.94
17	14	474.34	51.41	422.93	9858.67	17216.12	7357.45
18	15	474.34	49.29	425.05	9433.62	17053.70	7620.08
19	16	474.34	47.17	427.17	9006.45	16891.28	7884.83
20	17	474.34	45.03	429.31	8577.14	16728.86	8151.72
21	18	474.34	42.89	431.45	8145.69	16566.44	8420.75
22	19	474.34	40.73	433.61	7712.08	16404.02	8691.94
23	20	474.34	38.56	435.78	7276.30	16241.60	8965.30
24	21	474.34	36.38	437.96	6838.34	16079.18	9240.84
25	22	474.34	34.19	440.15	6398.19	15916.76	9518.57
26	23	474.34	31.99	442.35	5955.84	15754.34	9798.50
27	24	474.34	29.78	444.56	5511.28	15591.92	10080.64
28	25	474.34	27.56	446.78	5064.50	15429.50	10365.00
29	26	474.34	25.32	449.02	4615.48	15267.08	10651.60
30	27	474.34	23.08	451.26	4164.22	15104.66	10940.44
31	28	474.34	20.82	453.52	3710.70	14942.24	11231.54
32	29	474.34	18.55	455.79	3254.91	14779.82	11524.91
33	30	474.34	16.27	458.07	2796.84	14617.40	11820.56
34	31	474.34	13.98	460.36	2336.48	14454.98	12118.50
35	32	474.34	11.68	462.66	1873.82	14292.56	12418.74
36	33	474.34	9.37	464.97	1408.85	14130.14	12721.29

	A	B	C	D	E	F	G
1	Payment	Payment	Interest	Note	Unpaid	Depreciated	
2	Number	Amount	Due	Reduction	Balance	Value	Equity
37	34	474.34	7.04	467.30	941.55	13967.72	13026.17
38	35	474.34	4.71	469.63	471.92	13805.30	13333.38
39	36	474.28	2.36	471.92	0.00	13642.88	13642.88
40	37					13480.46	13480.46
41	38					13318.04	13318.04
42	39					13155.62	13155.62
43	40					12993.20	12993.20
44	41					12830.78	12830.78
45	42					12668.36	12668.36
46	43					12505.94	12505.94
47	44					12343.52	12343.52
48	45					12181.10	12181.10
49	46					12018.68	12018.68
50	47					11856.26	11856.26
51	48					11693.84	11693.84
52	49					11531.42	11531.42
53	50					11369.00	11369.00
54	51					11206.58	11206.58
55	52					11044.16	11044.16
56	53					10881.74	10881.74
57	54					10719.32	10719.32
58	55					10556.90	10556.90
59	56					10394.48	10394.48
60	57					10232.06	10232.06
61	58					10069.64	10069.64
62	59					9907.22	9907.22
63	60					9744.80	9744.80

Notice what happens in row 39 of the spreadsheet. Bryan's last payment amount was $474.28, which he inserts in cell B39. Now his unpaid balance is zero. Also, the depreciated value and the equity are equal.

Since his loan is paid off, he only needs to calculate the depreciated value and the equity in row 40. He adjusts row 40 to calculate only the depreciated value and the equity, and copies row 40 onto rows 41–63.

RETEACHING

Review as often as necessary: the meaning of exponents, exponential growth, graphs, and how to enter formulas in the calculator and computer.

ENRICHMENT

Challenge students to continue the principal and interest table for a loan on \$6,000 over 2 years at 6% interest. This table will have 24 entries with interest gradually decreasing and principal increasing.

Bryan graphs the unpaid balance, the depreciated value, and the equity on the same graph. He notices that the graphs intersect and observes that after 36 months, the depreciated value is equal to his equity in the car. His equity in the car increases and then decreases when the depreciated value is less than the amount of money he has spent for the car. Although the graphs of the unpaid balance and the equity appear to be a straight lines, they are not. These graphs are slightly curved.

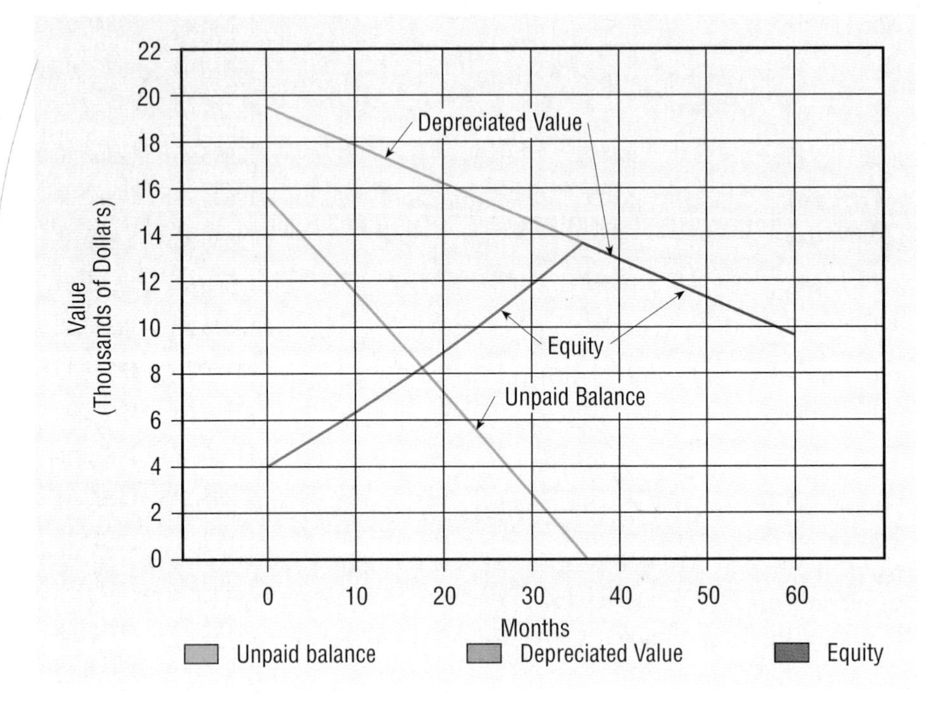

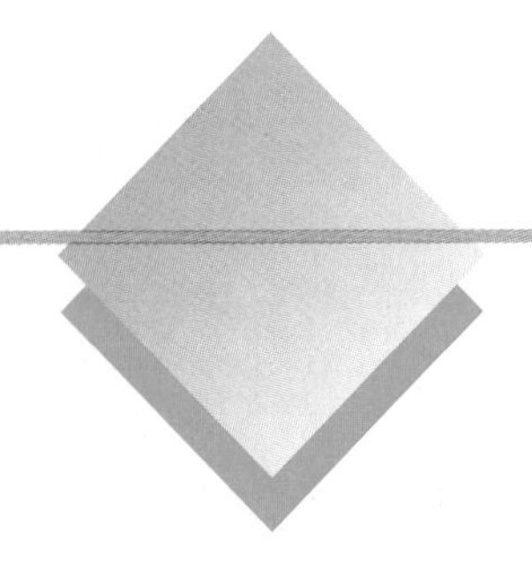

TRY YOUR SKILLS

A new car is purchased for \$16,000, with a 20% down payment, and is financed for 3 years at a rate of 7.2% per year.

1. Find the amount of the monthly payment. \$396.40
2. Find the amount of interest paid in the first payment. \$76.80
3. Find the amount of principal paid in the first payment. \$319.60
4. Find the amount of interest and principal paid in the second payment.
5. If the car depreciates 30% the first year and 20% the second year, what is its resale value at the beginning of the third year? \$8000
6. Since the car depreciates 30% during the first year, what is the percent of depreciation during the first month? $30\% \div 12 = 2.5\%$
7. Find the owner's equity in the car at the end of 1 year. \$2364.34

GUIDED PRACTICE

After students work through the *Try Your Skills* exercises, review each one on the chalkboard and answer questions. This will review the lesson's procedures, some of which are challenging.

ADDITIONAL ANSWERS

4. \$74.88; \$321.52

EXERCISE YOUR SKILLS

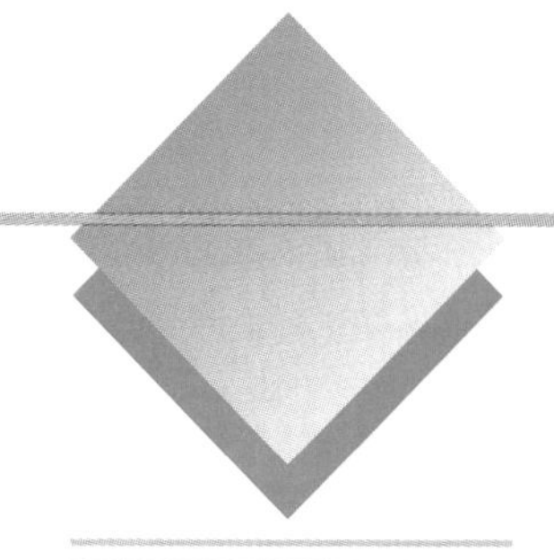

1. Why do some buyers prefer "young" used cars to new cars?
2. How can a used-car buyer reduce the chances of making a mistake when purchasing a used car? See Additional Answers.

Alex's uncle purchased a new car for $24,000. Terms of financing were 15% down, 6% annual interest, and 36 months to pay. Annual depreciation on the car as a percent of the purchase price is likely to follow the pattern 30%, 20%, 15%, 10%, 8%, 7%, 5%, 3%, 2%.

3. Find the amount of the monthly payment. $620.61
4. Find the amount of interest paid in the first payment. $102
5. Find the amount of principal paid in the first payment. $518.61
6. What is the estimated resale value (depreciated value) of the car at the start of the fourth year after purchase? $8400
7. Draw a graph showing the unpaid balance as a function of time. Example 3 will remind you of how this is done. (A straight line will approximate this graph.) See Additional Answers.
8. On the same axes as the graph for Exercise 7, graph the resale value of the car based on depreciation.
9. Draw a graph showing equity as a function of time.
10. At approximately what point in time does Alex's uncle have the least amount of equity in the car? 1 year
11. At approximately what point in time does Alex's uncle have the greatest amount of equity in the car? 3 years

Alex's father purchases a used van for $12,000. He pays $4000 down and finances the car for 3 years at an annual rate of 9%. He estimates that, because it is used, the van will depreciate in a straight line over 6 years.

12. Find the monthly payment for the van. $254.40
13. Find the amount of interest paid in the first payment. $60
14. Find the amount of principal paid in the first payment. $194.40
15. Graph the outstanding balance of the loan as a function of time.
16. On the same axes, graph the resale value as determined by depreciation.
17. What is the approximate equity in the car after 3 years? $6000

KEY TERMS

book value
depreciation
equity
title

ASSIGNMENT GUIDE

Exercises 1 and 2 provide a general review of the guidelines for selecting a car.

Exercises 3-11 take students, step-by-step, through all of the skills learned in the lesson.

Exercises 12-17 review the same skills for a used vehicle.

For Exercises 18-21, see the *Cooperative Learning* note on page 469.

ADDITIONAL ANSWERS

1. For a new car, depreciation is greater and begins immediately after you purchase the car. For a used car, depreciation is more gradual.
2. by conducting three types of inspection 1) on-the-lot; 2) on-the-road; and 3) in-the-garage

CLOSURE

Ask for student explanations of: the reason why principal and interest change as loan payments are made; the meaning of depreciation, including a graph; the meaning of equity.

LESSON QUIZ

1. Calculate the interest and principal for the first month on a $4000 loan for 3 years at 6% annual interest. (**$101.69, $20**)
2. Calculate the resale value of a car as it depreciates according to the following percents of the original $8000 price: 40%, 30%, 20%, 10%. (**$4800, $2400, $800, $0**)

ADDITIONAL ANSWERS (EXERCISE YOUR SKILLS)

18. Bob should choose 20% down at 5.5% annual interest over 3 years if he can afford the high monthly payment.
19. Dena will need to choose 10% down at 6.5% annual interest over 5 years to make her monthly payments as low as possible.

ADDITIONAL ANSWERS (MIXED REVIEW)

4. $105,750

Bob and Dena are both planning to purchase used cars for $8000. Financing can be selected from the following terms: Pay 10% down and finance the remainder at a 6.5% annual interest rate or pay 20% down and finance the remainder at a 5.5% annual interest rate; the time of the loan can be 3, 4, or 5 years. Bob has $3000 in the bank and a good job and would like to purchase a new car in 3 years. Dena has $500 saved and can borrow a few hundred dollars from her parents. She has a part-time job and hopes to keep the car for as long as possible. See Additional Answers.

18. Which financing terms would be best for Bob and why?
19. Which financing terms would be best for Dena and why?
20. According to the terms you selected, who pays the greater total amount for the car? Dena
21. If the car depreciates in a straight line over 5 years, what amount is Bob likely to get when he sells it? $3200

MIXED REVIEW

1-1 **1.** Jason works from 10:00 A.M. until 12:30 P.M., and then from 1:00 P.M. to 5:00 P.M. He earns $9.60 an hour. What is his daily pay? $62.40

2-2 **2.** Emily has $568.21 in her checking account. What is the balance in the account after she pays bills totaling $178.96? $389.25

3-1 **3.** Find the interest earned on $790.30 for one month at a monthly interest rate of 0.45%. $3.56

7-1 **4.** Use the Multiples-of-Salary Chart in the Reference Section to find the amount of life insurance a 25-year-old who has gross earnings of $23,500 would need in order to provide 60% income replacement.

8-2 **5.** If you buy 680 shares of stock at a price of $21\frac{1}{4}$ per share, what is the total cost of the stock? The commission rate is 1.5%. $14,666.75

7-2 What profit does an insurance company expect to make for each 1-year term insurance policy described in Exercises 6 and 7? Assume that the direct and indirect expenses for each policy are $20. Refer to the following table. Write and solve an equation.

EXPECTED DEATHS PER 100,000 ALIVE AT SPECIFIED AGE		
Age	**Expected Deaths Within 1 Year**	**Expected to be Alive in 1 Year**
45	315	99,685
46	341	99,659
47	371	99,629
48	405	99,595
49	443	99,557

6. Face value: $100,000; Age of insured: 47; annual premium: $500 $109
7. Face value: $180,000; Age of insured: 45; annual premium: $860 $273

10-3 OWNING AND OPERATING COSTS

QUICK REFERENCE

TEACHER'S RESOURCES AND ANSWER KEY

Lesson Quiz Answers 10-3, p. 76
Reteaching Activity
Answers 10-3, p. 110
Enrichment Activity
Answers 10-3, p. 111

EXTENSION ACTIVITIES

Reteaching Activity 10-3, p. 77
Enrichment Activity 10-3, p. 81

When Freda became a salesperson in the business started by her friends Evelyn and Greg, she began to use her car for business. She distributed posters and flyers and also delivered the finished T-shirts and sweatshirts.

Freda soon found that Hari, who kept the company's financial records, wanted an accurate report on the car expenses. So Freda began keeping all of her receipts for the purchase of gas and for car maintenance. Hari said that the business will at least pay for her gas. Freda also has monthly payments on her car loan and insurance premiums to pay. She asked Hari whether the company could pay part of these bills. Hari said he would speak to an accountant to find out more about personal and business expenses.

After learning more about business expenses for a car, Hari and Freda looked into car leasing. They found that in some circumstances it makes sense financially for a business to lease, rather than own, a company car.

OBJECTIVES: *In this lesson, we will help Freda to:*

- *Calculate car mileage per gallon of gasoline.*
- *Compute operating costs of a car and total costs per mile.*
- *Compute the costs of leasing a car.*

MILEAGE

Hari tells Freda that she should keep all receipts for the purchase of gas and also keep track of miles driven for business. Each time she uses the car for business, she is to write down the beginning and ending odometer readings. The **odometer** indicates the number of miles a car has been driven. Freda also is to calculate **miles per gallon,** the average number of miles her car travels on each gallon of gas.

MAINTENANCE

Since Freda's car has only 12,000 miles on it, expenses for tune-ups, parts, and repairs should be low. Her car is still under **warranty,** which means that the manufacturer pays for many repairs. The warranty expires after a certain number of miles or years.

To maintain her car in good condition, Freda has the oil changed every 3000 miles. She has a tune-up every 10,000 miles and regularly checks the air pressure in the tires and the levels of engine coolant.

REPAIRS

Freda has talked with mechanics and has learned that she will eventually need a new battery, new tires, and brake linings. She will also need to replace filters and hoses. She may need replacements for or repairs on a variety of other parts including the fuel pump, alternator, and muffler. Freda plans to set aside an amount of money each month so that she will be able to pay her bills for repairs as they come up.

LEASING A CAR

Because car expenses are often high and because cars depreciate, car leasing has become popular. **Leasing** is an arrangement by which a car is rented for a monthly fee under a contract that extends for several years. The person leasing the car pays for gas; the company that leases the car pays some maintenance and repair expenses.

ASK YOURSELF

1. What is an odometer? an instrument that indicates the number of miles a car has been driven
2. What is the difference between maintenance and repairs?
   See Additional Answers.
3. What is meant by "miles per gallon?" the average number of miles that a car travels using a gallon of gas
4. What does leasing mean?

ALGEBRA *REVIEW*

1. Find the average of 4, 7, and 9.
   $6\frac{2}{3}$
2. Jack walked 15 miles in 5 hours. What was his average speed in miles per hour?
   3 miles per hour
3. Pat ran 8 miles in 70 minutes. Find her average rate in minutes per mile.
   8.75 minutes per mile
4. Ms. Gonzales drove 140 miles in 3.5 hours. Find her average speed in miles per hour.
   40 miles per hour
5. Mr. Oliver used 12 gallons of gas to drive 156 miles. What was his average miles per gallon?
   13 miles per gallon

Find the solution of each system of equations.

6. $x + y = 10$
   $y = 5$
   (5, 5)
7. $y = 2x + 1$
   $y = 15$
   (7, 15)
8. $y = 2x - 3$
   $x = 2.5$
   (2.5, 2)

SHARPEN YOUR SKILLS

SKILL 1

The number of miles that you can drive for each gallon of gas depends on the kind of driving you do. Steady driving at a moderate speed gives the best miles per gallon. Idling or moving at very low speeds in city traffic and driving at high speeds use more gas and lower the miles per gallon rate.

Miles per gallon of gas is like miles per hour except that you divide total miles by the number of gallons of gas used rather than by the hours driven.

Average Miles per Gallon

$a = m \div g$ where a = average miles per gallon
m = total miles driven
g = total gallons purchased

Freda learned from a friend the way to determine the average miles per gallon of gas.

1. Record the mileage when the gas tank is full.
2. Fill the tank the next time you purchase gas; then record the mileage, gallons purchased, and amount spent.
3. Follow this procedure several times so that you have enough miles and gallons for a valid average.

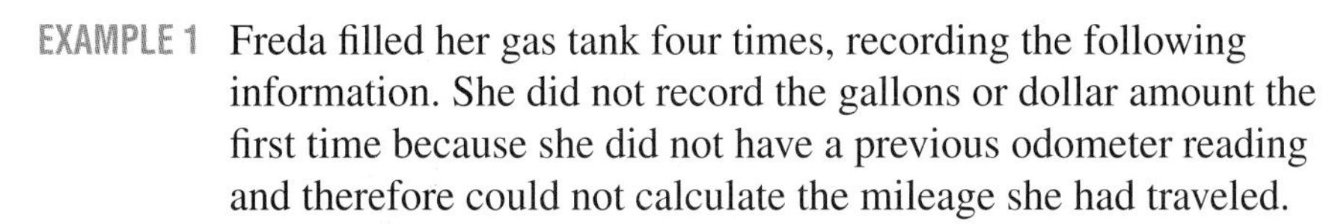

EXAMPLE 1 Freda filled her gas tank four times, recording the following information. She did not record the gallons or dollar amount the first time because she did not have a previous odometer reading and therefore could not calculate the mileage she had traveled.

Odometer Reading	Gallons of Gas	Amount
11,400	——	——
11,604	9.7	$12.03
11,842	11.2	13.89
12,027	8.6	10.67

QUESTION For the given information, what is the average miles per gallon?

SOLUTION

Freda can find the answer by using the formula $a = m \div g$.

Total miles driven: $m = 12{,}027 - 11{,}400 = 627$
Total gallons purchased: $g = 9.7 + 11.2 + 8.6 = 29.5$
Average miles per gallon: $a = 627 \div 29.5 = 21.3$

Freda averages 21.3 miles per gallon of gas.

FOCUS

ALGEBRA CONNECTIONS

Beverly does fitness exercises in 40-minute intervals. First she rows 1 mile in 40 minutes. Then she bikes 9 miles in 40 minutes. Then she runs 4 miles in 40 minutes. Taking all these together, what is her average speed in miles per hour? Write a formula for this. (***$a = \frac{m}{t}$; $a = \frac{14}{2} = 7$ mph***)

INSTRUCTION

TEACHING THE LESSON

Review the introductory information by having students answer the *Ask Yourself* questions. They provide a transition to the Skill activities.

Review the fact that an average is not a fixed number. Miles per hour and cost per mile are not given, but must be computed based on other data.

If students have trouble with Example 1, calculate the miles driven for each odometer reading, add them, and then divide to compute the average miles per gallon.

In Skills 1 and 2 point out that car costs per mile are calculated based on three different kinds of data: gasoline usage only ($0.058); total costs including loan payment ($0.544); costs without including loan payment ($0.239).

EXAMPLE 2 Freda has purchased gas three times to drive 627 miles in Example 1.

QUESTION What is Freda's cost per mile for gas?

SOLUTION

The cost per mile, like gallons per mile, is an average. It is found by dividing total cost by the number of miles. Since we have used a for average miles per gallon, we will use b for the cost per mile.

> **Average Cost per Mile**
>
> $b = c \div m$ where b = average cost per mile
> c = total cost
> m = total miles driven

According to Freda's records,

Total cost:	$12.03 + $13.89 + $10.67 = $36.59
Total miles driven:	627
Average cost per mile:	$b = 36.59 \div 627 = 0.058$

The average cost per mile for gas is $0.058 or 5.8 cents. This does not seem like a lot of money, but, as we shall see, the total cost per mile of owning and operating a car is much higher.

SKILL 2

Freda is driving about 1000 miles per month. Using this fact and the information from Example 2, she calculates that her yearly expense for gas will be 12 • 1000 • 0.058 = $696.

In addition to the purchase of gas, Freda has to pay the following.

Regular maintenance: tune-up, oil change, filters, wheel alignment. This will cost about $125 per year.

Repairs: Some can be anticipated, such as new tires, but most cannot be known in advance. Freda can be reasonably sure of some repair expenses, and she knows that these will increase with time. Freda feels that $240 per year will be a safe estimate for now.

Monthly loan payment: $304.72

Insurance premium: $428.60, four times per year

Taxes and fees: $94 per year

EXAMPLE 3 Using the information given, Freda decides to set up a budget.

QUESTION How should Freda set up a budget for her car expenses?

SOLUTION

All expenses must be annualized, that is, the total amount per year must be calculated for each expense. Gasoline, maintenance, repairs, and taxes/fees are already stated as annual. The loan payment is monthly. Insurance is paid four times a year. We multiply to annualize these expenses.

Loan payment: $12(304.72) = 3656.64$
Insurance: $4(428.60) = 1714.40$

Annual expenses are shown in the following table.

Annual Expenses (Dollars)					
Gasoline	Maintenance	Repairs	Loan Payment	Insurance	Taxes/Fees
696	125	240	3656.64	1714.40	94

After looking at the yearly budget, Freda decides that she needs a monthly budget to know how much money to put aside for monthly bills each month and to get a better idea of what the car is costing her. She divides each annual amount by 12 to find the monthly amounts.

Monthly Expenses (Dollars)					
Gasoline	Maintenance	Repairs	Loan Payment	Insurance	Taxes/Fees
58	10.42	20	304.72	142.87	7.83

Freda was surprised when she saw what her car was costing her each month. She hadn't realized how much it was because she pays the bills at so many different times. She used her spreadsheet to make a pie graph of the costs.

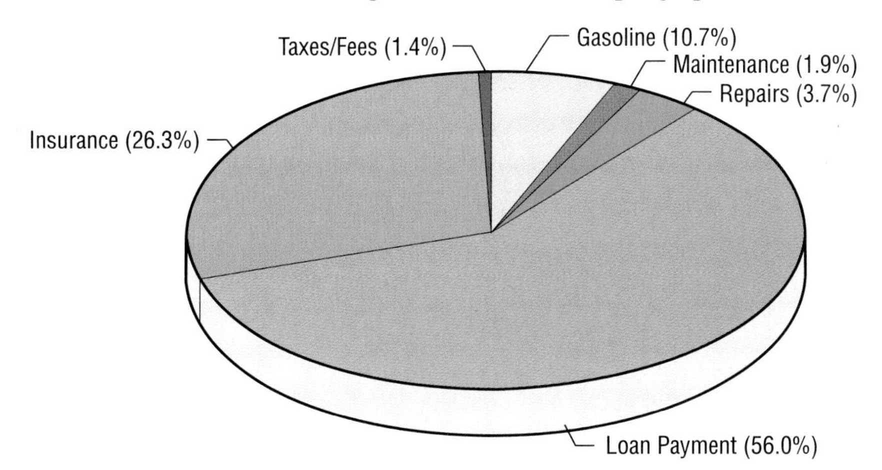

FOCUS ON ALGEBRA

This lesson follows up on mathematical processes previously learned, taking them an additional step so that students can, "reflect upon and clarify their thinking about mathematical ideas and relationships." *(NCTM, Standard 2, p. 140)*

COMMON ERRORS

Students may demonstrate some carelessness with the information to be entered in spreadsheets. Information is deliberately given in different bases: monthly, quarterly, semiannually, or annually, and must be converted to a common basis.

CRITICAL THINKING

Ask students to use data from the lesson to create a formula for the hourly cost of running a car. Have them clearly identify the data and the procedures that they use.

EXAMPLE 4 Freda totaled her monthly expenses and found that they came to \$543.84.

QUESTION What is Freda's average cost per mile to own and operate her car?

SOLUTION

Freda drives an average of 1000 miles per month, so the average cost per mile can be found by dividing her total costs by 1000.

$543.84 \div 1000 = 0.544$

Freda can't believe that it costs her more than 50 cents for every mile that she drives. She shows her figures to Hari. He has been investigating car costs and tells her that the business should pay some of her operating expenses but not the loan payment. He says that she is buying the car, so it will belong to her, not to the business.

Freda then calculates her car expenses, leaving out the loan payment.

Expenses: $58 + 10.42 + 20 + 142.87 + 7.83 = 239.12$

Average per mile: $239.12 \div 1000 = 0.239$

Freda's operating expenses are \$0.239 or 23.9 cents per mile.

Hari and the others agree that it is fair to pay Freda 24 cents a mile for the business use of her car. They later discover that this is about the amount that most employees are allowed to charge when using their cars for business.

SKILL 3

Freda's cousin, Vanessa, drives for a delivery company. At first the company pays her 25 cents a mile. Then, as business increases, the company considers leasing a car. It can lease a car for \$300 a month. The leasing agency pays for all expenses (except gas) and repairs. Leasing would save wear and tear on Vanessa's car, but would it save money for the company?

EXAMPLE 5 Vanessa's company asks her to prepare figures and graphs showing the costs of leasing a car at \$300 a month and \$0.058 per mile for gas compared with paying \$0.25 per mile for the use of her car.

QUESTION For the stated rates, how does the cost of paying Vanessa to use her car compare with the cost of leasing?

SOLUTION

First we compare the total monthly costs. We use x for the number of miles driven and y for the total cost. The equations are

$y = 0.25x$ — Cost of using Vanessa's car

$y = 300 + 0.058x$ — Cost of leasing plus 0.058 per mile

$0.25x = 300 + 0.058x$ — Substitute $0.25x$ for y in the second equation.

$0.192x = 300$ — Combine x terms.

$x = 1562.5$

INTERDISCIPLINARY INVESTIGATION

Have students interview a salesperson and prepare a report about the arrangements followed in using a car for business. This could be a personal car that is also used for business, or a company car that also receives personal use.

The costs are equal at 1562.5 miles per month. Since some rounding is involved in these calculations, you can say that if Vanessa drives more than 1560 miles per month, the company will save by leasing a car. Graphing will give a better idea of the comparison.

Using a graphing calculator, you can graph these two equations as shown.

Range:

Xmin:	0	Ymin:	0
Xmax:	2000	Ymax:	500
Xscl:	100	Yscl:	50

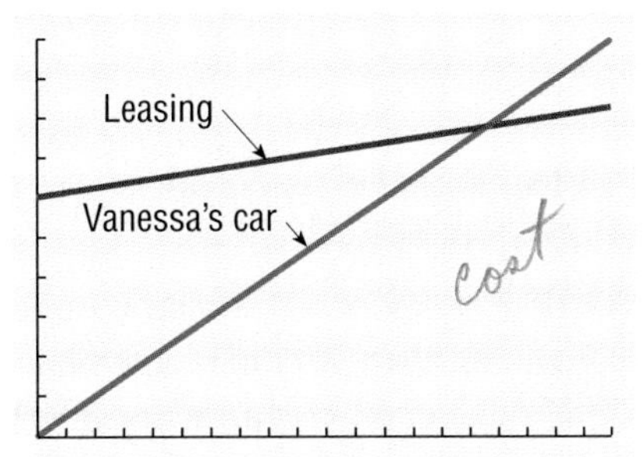

EXAMPLE 6 Vanessa's company would like to compare the cost per mile of leasing with paying \$0.25 for using Vanessa's car.

QUESTION How do the costs per mile compare?

SOLUTION

In this case the cost of using Vanessa's car is fixed at 25 cents a mile. To find the cost per mile of leasing c, you must divide the monthly leasing cost by the number of miles x, and then add the cost per mile for gas.

The cost per mile equations are

$c = 0.25$ Cost per mile of using Vanessa's car

$c = \frac{300}{x} + 0.058$ Cost per mile of leasing a car

Solve these equations for x.

$0.25 = \frac{300}{x} + 0.058$ Substitute 0.25 for c in the second equation.

$0.192 = \frac{300}{x}$ Subtract 0.058 from both sides.

$0.192x = 300$

$x = 1562.5$

Graph the cost per mile equations on a graphing calculator using the following range values. Use X and Y for the variables x and c in the equations.

Xmin:	0	Ymin:	0
Xmax:	2000	Ymax:	1
Xscl:	100	Yscl:	0.1

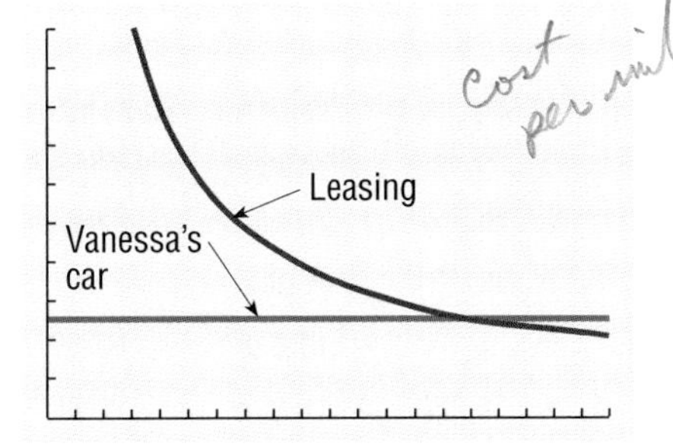

TECHNOLOGY HINT

Students who have been using the graphing calculator and computer spreadsheet regularly through the course should now be proficient. You might wish to have a special mastery test on one or both of these.

RETEACHING

With students who have difficulty in Skill 1, review the meaning of average using round numbers.

For students who do not fully grasp the comparative costs graphed, start over with the equations and do paper and pencil graphs rather than with the graphing calculator.

ENRICHMENT

Ask students to gather information from whatever sources they can about the fuel efficiency of a particular car at different speeds, and then draw a graph showing miles per gallon as a function of speed in miles per hour.

The tracing feature on the calculator can be used to find the cost per mile for different total monthly amounts. For example, the cost of leasing at 1000 miles per month is 35.8 cents per mile, and the cost at 1600 miles per month is 24.6 cents per mile.

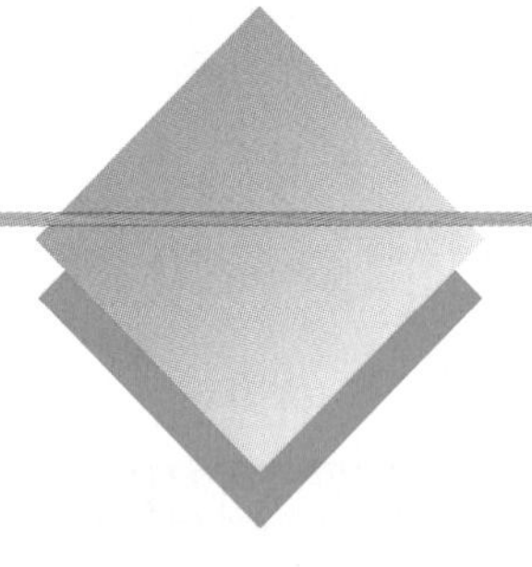
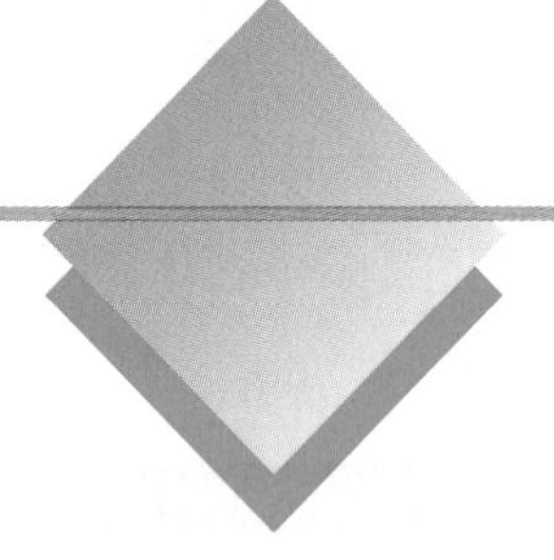

TRY YOUR SKILLS

GUIDED PRACTICE

The *Try Your Skills* can be done individually and then reviewed in pairs.

Mark drives 537 miles on 23.7 gallons of gas. He pays $30.57 for the gas. He calculates his yearly car expenses, including loan payments of $2560, to be $5470. He can lease a car for $300 per month.

1. Find Mark's average miles per gallon. 22.7 mpg
2. Find his average cost per gallon. $1.29
3. Find his average monthly expense. $455.83
4. Find Mark's average cost per mile if he drives 12,000 miles per year. $0.456
5. Find the cost of leasing for a month and driving 1350 miles while paying an average of 5.7 cents per mile for gas. $376.95
6. Write equations for the cost of paying Mark a per mile amount for operating expenses (not counting the loan) and for the leasing cost.
   $y = 0.24x$; $y = 300 + .057x$

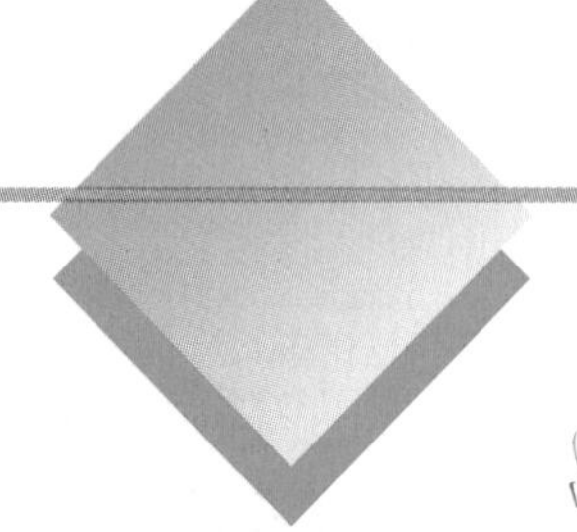

EXERCISE YOUR SKILLS

1. Car companies advertise the miles per gallon that can be expected from a car. Why should this be considered when buying a car? See Additional Answers.
2. Why is it important to take care of minor repairs promptly?
3. Why should a car owner estimate repair expenses when preparing an annual budget? in order to put money aside to pay for the repairs needed
4. What are some of the differences between owning and leasing a car?

Use the information in the table to find the distance traveled, the miles per gallon, and the cost per mile for the trip.

	Odometer First	Odometer Last	Gallons of Gas	Cost of Gas	Distance	Miles per Gallon	Cost per Mile
5.	28,431	28,848	18.4	$25.15	417	22.7	$0.060
6.	38,715	39,326	15.7	18.95	611	38.9	0.031
7.	11,477	11,628	10.1	14.00	151	15.0	0.093
8.	18,388	19,374	40.6	63.74	986	24.3	0.065
9.	15,428	15,639	10.9	14.38	211	19.4	0.068

KEY TERMS

leasing
miles per gallon
odometer
warranty

ADDITIONAL ANSWERS

1. the miles per gallon affect the annual operating cost
2. to decrease the need for major repairs and to increase efficiency
4. When a car is leased all expenses except gas are covered by the leasing company.

Isabel drives 15,000 miles a year and has the following real and estimated car costs:

Maintenance: \$150 per year
Repairs: \$300 per year
Loan payments: \$401.30 per month
Insurance: \$603.27 twice per year
Taxes: \$114 per year
Gasoline: \$0.06 per mile

Find each of the following for Isabel.

10. Her annual expense for gasoline \$900

11. Her monthly expense for insurance \$100.55

12. The total of her annual car expenses \$7486.14

13. Her average cost per mile to own and operate her car \$0.499

14. Her annual cost to operate the car not including the monthly loan payment

15. Create a pie graph to show her annual expenses for her car. See Additional Answers.

Mark works for a company that will make a decision about whether to pay him 20 cents per mile to drive his car for the business or to lease a car for \$200 per month and 6 cents per mile.

16. Write equations for the total monthly cost under each arrangement.

17. Solve the equations simultaneously to find the number of miles per month for which the costs will be equal. 1428.6

18. Graph the two equations to show the relationship. See Additional Answers.

19. Which is the better arrangement for 1000 miles per month? Mark's car

20. What is the difference in costs for a 1000-mile month? \$60

21. Which is the better arrangement for 2000 miles per month? lease

22. What is the difference in costs for a 2000-mile month? \$80

MIXED REVIEW

5-1 **1.** Jacob borrowed \$10,000 and repaid it at the rate of \$200 per month for 5 years. Find the total interest paid over the 5 years. \$2000

8-1 Phyllis bought 1200 shares of stock at $37\frac{3}{8}$ per share. In Exercises 2–4, ignore the commission costs.

2. Find the cost of the shares. \$44,850

3. After 2 years, Phyllis sold the shares at $45\frac{5}{8}$ per share. What was the total sale price? \$54,750

4. What was the capital gain or loss on the shares? Gain: \$9900

9-2 **5.** Use IRS Form 1040EZ to find the amount of the refund or payment due to the nearest dollar for a single taxpayer who had \$1,600 withheld on his \$13,900 income. The taxpayer has no taxable interest and cannot be claimed as a dependent by someone else. Refund: \$396

INDEPENDENT PRACTICE

ASSIGNMENT GUIDE

Exercises 1-4 should help students understand that, although automobile repairs and the related expenses often come unexpectedly, they can, in general, be anticipated.

Exercises 5-9 concern miles per gallon and costs per mile.

Exercises 10-15 ask for overall and average expenses based on those listed.

Exercises 16-22 compare two different car-for-business arrangements.

CLOSURE

Ask questions to be sure that students understand each of the key ideas of the lesson: miles per gallon, cost per mile, simultaneous solution to equations, and graphing equations.

LESSON QUIZ

A person drove 364 miles using 16 gallons of gas at an average price of \$1.21 a gallon.

1. What was the average number of miles per gallon? (**22.8**)

2. What was the average cost per mile for gas? (**\$0.053**)

3. If the person was reimbursed at the rate of 30 cents a mile, how much was left to cover operating expenses other than gasoline? (**\$89.84**)

ADDITIONAL ANSWERS

14. \$2670.54

16. $y = 0.20x$; $y = 200 + 0.06x$

10-4 INSURANCE COSTS

QUICK REFERENCE

TEACHER'S RESOURCES AND ANSWER KEY

Lesson Quiz Answers 10-4, p. 78

Reteaching Activity Answers 10-4, p. 110

Enrichment Activity Answers 10-4, p. 111

EXTENSION ACTIVITIES

Reteaching Activity 10-4, p. 78

Enrichment Activity 10-4, p. 82

Trevor, who is 17, is allowed to drive the family car. Sometimes, he drives his 11-year-old sister Tracey and her friends to a weekend event. Last Saturday, Trevor drove Tracey and her friends, Roy and Leah, to a movie. As he drove, another car came speeding past, rapidly changing lanes and weaving back and forth among the other cars. Just as the car passed, Trevor was distracted for an instant by the younger children in the back seat, who were wrestling over a piece of candy.

As Trevor looked up, the car that was speeding collided with another car. Both cars seemed to be heavily damaged as they skidded off the road. Trevor had to change lanes quickly and slow down to avoid another car that had moved into his lane because of the accident. His sister and her friends had their seat belts on, but they were shaken up.

As a result of seeing the accident, Trevor asked his parents to tell him about insurance. He wanted to know what it is, why it is needed, and why costs are higher for young people. His parents explained that by collecting money from thousands of car owners, the insurance company can pay for car repairs and medical bills after accidents.

Compared with the many thousands of people who purchase insurance, only a few actually have accidents. But no one knows which ones will have the accidents, so everyone should purchase insurance. Furthermore, in most states, the law requires that you have auto insurance.

OBJECTIVES: *In this lesson, we will help Trevor to:*

- *Examine several types of insurance that are available for car owners.*
- *Compare rates of auto insurance for different drivers on the basis of the age, sex, training, and marital status of the driver and several other factors.*
- *Understand how insurance companies use accident statistics to estimate their costs.*

AUTOMOBILE INSURANCE

Most states now require that car owners purchase insurance. The reason for this is that any driver might have an accident. Depending on the circumstances, the driver might then be responsible for the financial consequences of the accident. The legal term for this responsibility is **liability.** Even if not legally liable, a driver would still want to have his or her car repaired or replaced. There are a number of different kinds of automobile insurance dealing with liability.

Bodily Injury Liability This coverage is protection against claims or lawsuits that are brought against you by pedestrians, riders in your car, and people in other cars whose injuries were caused by an accident for which you are responsible. The amount of protection can vary from $10,000 to $1,000,000 or more. Some people choose high amounts because in serious accidents the medical bills can be enormous.

Property Damage Liability Coverage for property damage liability is paid for the damage done to other people's property, such as a car, a building, or a fence. Property damage insurance might range from $5,000 to $50,000. It also protects you when you are driving someone else's car.

Medical Payments Medical payment insurance covers the person driving the car and others riding in it, whether or not that driver is at fault. In Trevor's case this coverage protects him and the other members of his family when they ride in their own car or in another person's car.

Collision Coverage for **collision** pays for damage to Trevor's family's car if it is in an accident. Some people decline this coverage if their cars are not worth much. For example, if a car is worth less than $1000, many people would not want to pay several hundred dollars a year for collision insurance.

Comprehensive Physical Damage Coverage for comprehensive physical damage pays for damages that result from a fire, falling object, theft, flood, earthquake, and the like. The accident that Trevor witnessed was not caused by any of these events.

FOCUS

ALGEBRA CONNECTIONS

If risk factor means likelihood of being the driver in an automobile accident, assign a risk factor between 0 and 1 to each: (a) a 17-year-old city driver; (b) a 50-year-old country driver; (c) an 80-year-old suburban driver. Explain the reasons for the risk factor. Write an equation using the risk factor.

CRITICAL THINKING

Ask students to think of two objections that a person might have when assigned a certain risk factor. (**My driving record shows that the risk does not apply to me. The insurance company should create a different population to even out the risk, for example, by combining urban and suburban groups.**)

Uninsured and Underinsured Motorist Protection Coverage for uninsured and underinsured motorist protection pays for personal expenses due to an accident if it can be shown that the driver of the other car was at fault and if the other driver has little or no bodily injury liability coverage.

THE COST OF INSURANCE

Naturally, it costs more to insure a car that is worth $40,000 than one that is worth $8000. In addition to the value of cars, insurance companies keep records of all kinds of driving-related accidents. They have found that accidents are more likely to occur in some places than in others. These companies also keep track of driver characteristics and their relationship to accidents. Insurance companies charge higher rates for people and places that they find are more often involved in accidents.

In the Reference Section of the text you will find a table entitled Driver Rating Factors. The **driver rating factor** is based on characteristics such as age, sex, whether the person is married, the kind of driving done, and whether the driver has been through a driver training program. The base premium is multiplied by the appropriate factor to find the premium for a specific individual.

Ask Yourself

1. What is liability?
   See Additional Answers.
2. What is bodily injury liability insurance?
3. What is property damage liability insurance?
4. What does comprehensive physical damage insurance cover?

ADDITIONAL ANSWERS

1. Liability is the legal term for the responsibility of the financial consequences of an accident.
2. coverage that protects the driver against claims or lawsuits that are brought against him by pedestrians, riders in his car or persons in other cars
3. Coverage for property damage liability is paid for damage done to other people's property, such as a car, building or fence.
4. Coverage for comprehensive physical damage pays for damage that results from fire, falling objects, theft, flood or other natural causes.

ALGEBRA *REVIEW*

Given the equation $p = rb$

1. Find p when $r = 1.8$ and $b = 525$.
   945
2. Find p when $r = 2.25$ and $b = 980$.
   2205
3. Find p when $r = 4.4$ and $b = 1200$.
   5280
4. Find r when $p = 2015$ and $b = 3.1$.
   650
5. In general, what happens to the value of p if r is unchanged and b decreases?
   p decreases

Given the equation $p = \frac{e}{n} + h$

6. Find p if $e = 568{,}500$, $n = 10{,}000$, and $h = 90$.
   146.85
7. Find e if $p = 52.80$, $n = 10{,}000$, and $h = 30$.
   228,000
8. Find n if $p = 825.86$, $e = 72{,}586{,}000$, and $h = 100$.
   100,000
9. Find h if $p = 591$, $e = 526{,}000$, and $n = 65$.
   −7501.31
10. In general what happens to p if e and h are unchanged and n increases?
    p decreases

SHARPEN YOUR SKILLS

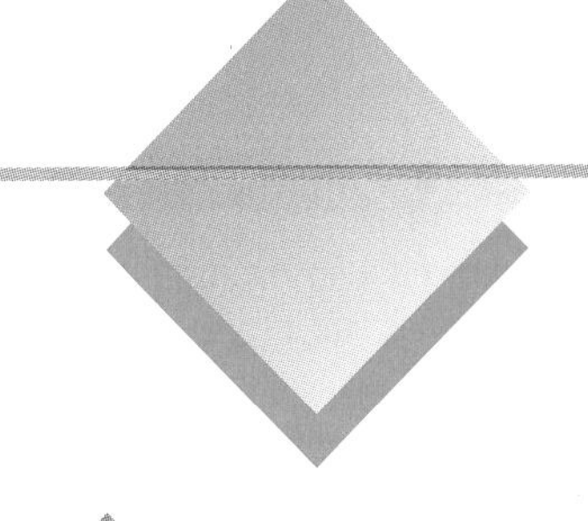

Trevor is now looking at actual insurance costs. Through a friend who works for an insurance company, he has obtained the following table of basic rates.

AUTOMOBILE INSURANCE, SIX-MONTH BASIC RATE SCHEDULE

Car Class Rating	Collision Deductible			Comprehensive Deductible		
	$100	**$250**	**$500**	**$50**	**$250**	**$500**
1–10	$ 500	$ 430	$ 370	$140	$130	$110
11–20	820	750	690	350	310	280
21–30	1140	1040	940	450	410	370
31–40	1280	1190	1100	670	610	560

Trevor noticed that several factors affect basic rates. One is the car class rating. This rating reflects the resale value of the car. Another factor is the amount of the **deductible,** the first expenses that are not covered by insurance. The deductible is a way of reducing insurance rates. In the case of bills resulting from an accident or injury, the insured person pays the deductible amount before the insurance company begins to pay. The higher the deductible, the more you pay yourself and the lower your **premium,** that is, the amount you pay for your insurance. For example, if repairs after an accident cost $1000 and your deductible is $250, then you pay $250, and the insurance company pays $750.

The insurance premium is dependent on the driver rating factor r and the basic rate b for the car. The six-month basic rate is found in the table at the top of this page or in the Reference Section. The driver rating factor is found in a Driver Rating Factor table in the Reference Section. Note there are three tables, one for each of these categories: married youths, unmarried youths, and all adults.

Annual Insurance Premium Formula

$p = 2rb$ where p = the annual premium paid to the company
r = the driver rating factor
b = the six-month basic rate for the car

INSTRUCTION

TEACHING THE LESSON

Use the *Ask Yourself* questions to review key concepts. They must be thoroughly understood before a student can do the lesson's application exercises.

The table at the opening of the *Sharpen Your Skills* section contains a good deal of information. After explaining it, ask for student questions.

RETEACHING

For students who have trouble with the idea that insurance depends on ratings of cars and drivers, use analogies from sports or school. Ratings are something like grades or scores given for gymnastic performances. In some sports, such as wrestling, students compete according to weight classes.

FOCUS ON ALGEBRA

The use of mathematics to model real-world problems is evident through much of this text. Students should begin to appreciate the practical value of mathematics. *(NCTM Standard 1, p. 137)*

COMMON ERROR

It may be necessary to review the terms of the lesson several times. Also, until they have had practice, students may make mistakes using the tables.

COOPERATIVE LEARNING

The *Try Your Skills* exercises are not difficult but require some checking and consequently could be done in pairs, with students checking each others' work.

INTERDISCIPLINARY INVESTIGATION

Business: Have students write to different insurance companies requesting information about how different populations of drivers and risk factors are determined. Ask also about how tables are constructed to reflect the factors.

ESL STUDENTS

Ask non-native speakers to write a paragraph using the terms *deductible, liability,* and *collision.*

SKILL 1

EXAMPLE 1 Trevor is 17 years old. He has had driver training. He is single. He is not the owner or principal operator of the car that he drives. He does errands in the car. He does not use the car to drive to work.

QUESTION According to the Driver Rating Factor tables, what is Trevor's rating factor?

SOLUTION

First find the table, Part II, Unmarried Youths. Trevor is 17, so you use the top half of the table and the section marked M. Trevor is not the owner or usual driver, but he has had driver education and does not drive to work. Thus, working your way across the table, you find that his rating is 2.45.

SKILL 2

EXAMPLE 2 Trevor is driving his family's car, which has a car class rating of 8. He plans to purchase collision insurance with a deductible of $500.

QUESTION How can you use the premium equation to find Trevor's annual premium?

SOLUTION

The basic rate for Trevor is found in the Automobile Insurance table at the top of the previous page. Look in the collision half of the table (left side) in the column under $500 and the row 1–10. Thus the basic rate is $430.

Using the equation

$$p = 2rb$$

gives

$$p = 2(2.45)(430) = 2107$$

The annual premium is $2107.

SKILL 3

EXAMPLE 3 One insurance company is considering a new policy for young people aged 14 and 15. The policy will insure them for injuries sustained as passengers riding in a car. Some parents are interested in this because their children in this age range are often passengers in cars driven by new, 16- or 17-year old drivers who lack adequate insurance.

The insurance company must decide on the price of a premium. Their records show that for every 10,000 14- and 15-year-old passengers, there are 425 accidents resulting in personal injuries. The medical costs of these injuries average $3800 per person. The company must also receive $35 per policy to cover overhead and profit.

QUESTION What should be the premium for the new policy?

SOLUTION
The premium can be found by considering the total estimated expense for the population, then dividing the amount by the total number of policies paying premiums. Then the overhead must be added.

Algebraically, the cost of the premium can be expressed as

$$p = \frac{e}{n} + o$$

where p = premium
e = expense paid out
n = number paying premiums
o = overhead per premium

$$p = \frac{425(3800)}{10{,}000} + 35$$
$$= 196.50$$

The company's figures suggest a premium of $196.50.

ENRICHMENT

Ask interested students to obtain records of numbers of accidents for people of different age groups, different residential areas, and different models of cars. A librarian could help point them toward the appropriate resources. Then have the students create their own premium tables based on the accident rates.

TRY YOUR SKILLS

1. Use the Driver Rating Factor tables to find the rating for a married female, 20 years old, who owns her own car, has not had driver training, and drives the car to work. 2.10
2. Use the Automobile Insurance table to find the annual premium for comprehensive coverage, with a $250 deductible, on a car with a class rating of 12. $700
3. Write an algebraic equation for finding the annual insurance premium based on the six-month basic rate for the car and the driver rating factor. Explain each part of the equation. See Additional Answers.

Insurance records show that of 10,000 drivers in a certain class, 1378 were in accidents with an average cost to the insurance company of $6089.

4. What is the average cost per person insured? $839.06
5. If the company overhead is $95 per policy, what should the company charge as a premium? $934.06

GUIDED PRACTICE

After students do the practice exercises alone or in pairs, check their work and review the use of the tables as needed.

$\frac{1378(6089)}{10{,}000}$

ADDITIONAL ANSWERS

3. $p = 2rb$; p represents the annual insurance premium; r represents the driver rating factor; b represents the six-month basic rate for the car

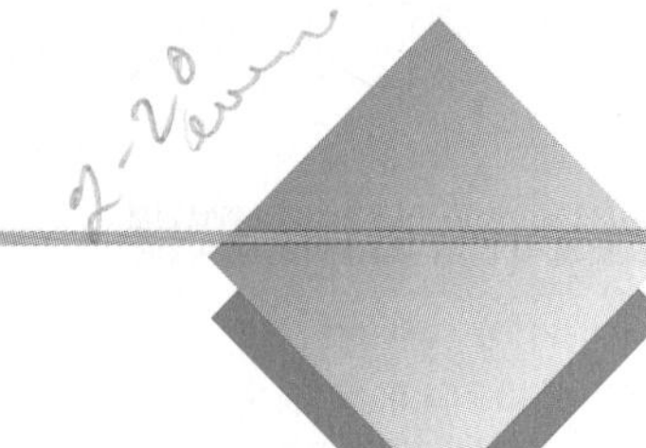

EXERCISE YOUR SKILLS

1. Why do laws require drivers to be financially responsible for damage to other people or their property? Any driver might have an accident.
2. Why is it important for drivers to have adequate medical payment coverage? In serious accidents the medical bills can be quite high.
3. What is the difference between a premium and a deductible?
4. What is the advantage in having a deductible in insurance coverage?

Use the Driver Rating Factor tables in the Reference Section to find the driver rating factor for each of the following individuals.

5. A 19-year-old female without driver training who drives to work, and is not the owner or usual driver of the car. married or unmarried: 1.65
6. A 19-year-old unmarried male without driver training who drives to work and is not the owner or usual driver. 2.30
7. A 17-year-old female with driver training who drives to work and is the owner. married or unmarried: 2.40
8. A 22-year-old female with or without driver training who drives to work and is the owner. 1.50
9. The usual driver of the car, 33 years of age, who drives to work. 1.25
10. The usual driver of the car, 65 years of age, who does not drive to work.

Use the Automobile Insurance, Six-Month Basic Rate Schedule to find the total six-month premiums in Exercises 11–15.

	Driver Rating Factor	Car Class Rating	Collision Deductible	Comprehensive Deductible	Six-Month Premium
11.	2.25	15	$500	$ 50	$2340
12.	1.00	32	250	250	1800
13.	2.40	9	500	500	1152
14.	0.80	23	100	500	1208
15.	1.10	10	500	250	550

16. An insurance company's records show that for every 10,000 drivers in a certain category there are 1067 accidents with an average cost to the company of $12,400 each. The company charges an overhead amount of $125 per premium. On the basis of these figures, what premium should be charged for a policy in this category? $1448.08
17. The AVR insurance company's records show that for every 100,000 drivers in a certain category there are 2311 accidents with an average cost to the company of $18,200. For this category the company charges an overhead of $190 per premium. What should be the premium to cover costs? $610.60

KEY TERMS

- collision
- deductible
- driver rating factor
- liability
- premium

INDEPENDENT PRACTICE

ASSIGNMENT GUIDE

Exercises 1-4 review the reasoning behind the classifications and terms that insurance companies use.

Exercises 5-15 require the use of tables and risk factors in determining premiums.

Exercises 16-17 ask for the determination of premiums based on given accident and payout numbers.

Exercises 18-19 ask for some reflection on the meaning of insurance to different parties.

ADDITIONAL ANSWERS

3. The insured person pays the deductible amount of the bill resulting from the accident before the insurance company begins to pay. The premium is the amount one pays for the insurance coverage.
4. the higher the deductible, the lower the insurance rate
10. 0.80

18. What are some of the differences between the ways in which an insurance company and an individual look at an insurance policy?

19. What are some of the factors that lead an insurance company to raise or lower the premiums every year? See Additional Answers.

MIXED REVIEW

1-3 **1.** You are a real-estate agent and you sold a house for $120,000 earning a commission of 3% on the first $100,000 and 6% on the amount above $100,000. What was your commission? $4200

5-1 A loan of $6000 is repaid one year later in a single payment of $6600.

2. What was the interest paid on the loan? $600

3. What was the rate of interest on the loan? 10%

7-3 **4.** A person buys a $100,000 whole life insurance policy at age 30. Use the table entitled Accumulated Cash Value of $100,000 Whole Life Policy in the Reference Section to find the cash value of the policy when the person retires 20 years later. $27,339

8-2 Find the cost of these shares of stock. The cost includes a commission of 1.5%.

5. 50 shares at $35\frac{1}{2}$ a share $1801.63 **6.** 1,000 shares at $9\frac{7}{8}$ a share

7-3 Use the Comparison Table for Term and Whole Life Premiums on page 326, the table showing Accumulated Cash Value of $100,000 Whole Life Policy on page 345, and the formula for the Future Value of a Periodic Investment to answer Exercises 7 and 8.

7. Find how much tax-deferred money a 25-year-old person could expect to accumulate between the ages of 25 and 30 by buying a $100,000 term policy instead of a more expensive whole life policy. The money saved will be invested into an IRA that should earn 8% a year. $4171.15

8. For the person in Exercise 5, compare the cash value of a whole life insurance policy after 5 years with the value of the IRA at the end of 5 years. Which seems to be the more profitable investment?

9-3 The Tierney family consists of Mr. and Mrs. Tierney and their three dependent children. Last year, the family had salary income of $47,500, interest income of $1457, and dividend income of $2869. Their IRA contribution was $1750. The allowable deductions were real-estate taxes of $1690, local income taxes of $2100, and mortgage interest on their home of $5980.

9. Determine whether the family is better off itemizing their deductions or using the standard deduction. What is the deduction that the family should take? itemizing; $9770

10. Use IRS Form 1040 and Schedule A to find the Tierneys' income tax for the previous year. $4324

CLOSURE

The main idea of the lesson is that the factors related to automobile accidents, such as age and location of residence, can be quantified so that the premium amount is greater or less depending on the category. Ask questions to be sure students grasp this concept.

LESSON QUIZ

1. An insurance company found that in a certain population of 1000 there were 24 automobile accidents with an average payout of $3000. What should be the premium for people in a population like this one, including $65 per premium for expenses and profit? (**$137**)
2. Why might the premium determined for this population not work so well for a large population? (**The sample of 1000 is small and therefore may not be a good predictor.**)

ADDITIONAL ANSWERS (EXERCISE YOUR SKILLS)

18. An insurance company is trying to maximize profits and pay out as little as possible. It would be conservative on who and how much it insures. An individual wants as much coverage as possible for the money. The individual wants coverage sufficient to pay all bills related to an accident.

19. expenses paid out (claims); number of people paying premiums; and overhead costs

(MIXED REVIEW)

6. $10,023.13

8. the IRA; cash value of policy: $3100

CHAPTER 10 REVIEW

1. When financing a car, what factors cause monthly payments to be higher and what factors cause them to be lower? See Additional Answers.
2. What is meant by straight-line depreciation? equal depreciation each year
3. If you purchase a new car, what will probably be the most costly factor in operating the car? loan payments followed by insurance costs
4. How would you describe the depreciation of most cars?
5. What are some of the causes contributing to the variation in insurance rates? the driver rating factor, the car being insured, coverage and deductible

ADDITIONAL ANSWERS

1. factors that raise monthly payments: higher interest, longer term, higher amount financed
   factors that lower monthly payments: lower interest, shorter term, smaller amount financed
4. greater depreciation in early years
8. Air conditioning $63; 15%
   AM/FM radio, tape deck $127.50; 17%
   Full-size spare tire $40; 33.3%
9. $2922.50; 19.8%

The following is a portion of the price list for a new car.

Item	Dealer's Cost	Sticker Price
4-door sedan	$13,460	$16,152
Air conditioning	420	483
AM/FM radio, tape deck	750	877.50
Full-size spare tire	120	160

6. Find the markup for the car without options. $2692
7. Find the percent of markup of the car without options. 20%
8. Find the markup and percent of markup for each of the options.
9. Find the markup and percent of markup for the car with all options.

A car with a selling price of $16,800 can be financed for

(a) 20% down and 6% annual interest rate with three years to pay

(b) 10% down and 7.2% annual interest rate with four years to pay

10. Find the monthly interest rate for an annual rate of 6%. 0.5%
11. Find the monthly payment under terms (a). $408.87
12. Find the monthly payment under terms (b). $363.47
13. Find the amount of interest and principal in the first month's payment under terms (a). Interest $67.20; Principal $341.67
14. Find the amount of interest and principal in the first month's payment under terms (b). Interest $90.72; Principal $272.75
15. Find the annual depreciation of the car for 10 years with straight-line depreciation. $1680
16. Draw a graph showing the resale value of the car with the following annual depreciation percents, starting with the purchase of the car: 25, 20, 15, 12, 10, 8, 5, 3, 2. See Additional Answers.
17. With the depreciation given in Exercise 16, what will be the resale value of the car at the beginning of the fifth year? $4704

The odometer of a car read 10,356 at the start of a trip and 12,621 at the end of the trip. During the trip, 92.1 gallons of gas were purchased at a total cost of $114.20.

18. Find the average miles per gallon for the trip. 24.6 miles per gallon

19. Find the average cost per mile for the trip. $0.05 per mile

Betty drives an average of 900 miles a month. Her average cost per mile for gas is 5.5 cents. In addition to the gas expense, she estimates $140 per year for maintenance and $200 per year for repairs. Her monthly loan payment is $186.74, her insurance premium is $715.50 every six months, and she pays $60 a year in taxes.

20. Find the estimated amounts that Betty spends on gas per month and per year. $49.50; $594

21. How much does Betty pay per year in insurance premiums? per month?

22. Set up a yearly budget showing Betty's car expenses. See Additional Answers.

23. Make a pie graph of Betty's yearly car expenses.

24. Find Betty's per mile expense for owning and operating her car. $0.43

25. On the same coordinate system, graph Betty's per mile cost of operating her car (not including the loan expense) and the cost of leasing a car for $200 a month and paying 5.5 cents a mile for gas.

26. When would it be more cost efficient for Betty to lease?

27. Use the Driver Rating Factor table to find the rating of a 21-year-old unmarried male who owns his own car, has not had driver training, and uses the car for work. 3.35

28. Use the Driver Rating Factor table to find the rating of a 17-year-old unmarried female who does not own her own car, has had driver training, and does not use the car for work. 1.75

29. Find the annual insurance premium for a person with a driver rating factor of 2.10 and a six-month base rate of $675. $2835

30. Suppose the driver in Exercise 29 has a six-month base rate of $710. Find her annual insurance premium. $2982

31. Use the Driver Rating Factor table to find the rating of a 25-year-old female who drives to work. 1.65

32. Find the annual insurance premium for the woman in Exercise 31 if the six-month base rate is $525. $1732.50

ADDITIONAL ANSWERS

21. $1431; $119.25

22.

Gas	$ 594
Maintenance	140
Repairs	200
Loan	2241
Insurance	1431
Taxes	60
TOTAL	$4666

26. when mileage is over 533.3 miles per month

CHAPTER 10 TEST

A car with a base price of \$12,200 has a sticker price of \$14,030.

1. Find the markup. \$1830
2. Find the percent of markup. 15%

Kayla borrows \$12,000 at a monthly rate of 0.5% for 36 months.

3. Find the amount of the monthly payment. \$365.06
4. Find the amount of interest paid in the first month. \$60

Sam records the following odometer readings: start, 34,560; end, 36,723. While traveling this distance, he uses 82.6 gallons of gas and spends \$106.65.

5. How many miles did he travel? 2163 miles
6. Find his average number of miles per gallon of gas. 26.2 miles per gallon
7. Find his average cost per mile of driving. \$.049 per mile

The following table shows Sandra's monthly car budget in dollars. She drives an average of 1000 miles per month. She has also found that she can lease a car for \$240 a month and pay 6 cents per mile for gas.

Gasoline	Maintenance	Repairs	Loan Payment	Insurance	Taxes
\$52	\$15	\$20	\$245.30	\$150.85	\$11

8. Make a pie graph of Sandra's annual costs. See Additional Answers.
9. How much does Sandra pay in a year on her car loan? \$2943.60
10. Find her average cost per mile of owning and operating the car. \$0.494
11. Find her average cost per mile of operating the car without the purchasing cost. \$0.249 per mile
12. Write an equation for the monthly cost of operating the car. $y = 0.249x$
13. Write an equation for the monthly cost of leasing a car. $y = 240 + 0.06x$
14. Find the solutions common to the equations in Exercises 12 and 13.
15. What does the common solution mean to Sandra? See Additional Answers.
16. Find the annual insurance premium for a situation in which a driver with a rating of 2.55 purchases insurance with a six-month base rate of \$627.80. \$3201.78

ALTERNATIVE ASSESSMENT

Ask for a table of monthly payments for loans of \$5000 to \$12,000 at \$500 increments and at interest rates from 4% to 10%. Ask students preparing the table to explain how it might be used to make quick estimates of financing costs.

Have students create their own premium tables based on factors of their own choosing. Score them on how well their figures reflect the information contained in the lesson.

ADDITIONAL ANSWERS

14. 1269.8 miles
15. At mileage over 1269.8 it will be cheaper to lease.

CUMULATIVE REVIEW

5-2 **1.** Your family's take-home pay is $3950 a month. How much can you afford to spend for credit card payments each month? $790

3-2 **2.** Use the Rule of 72 to determine how long it will take your $2300 savings account to double in value if it is growing at a rate of 4.5%. 16 years

5-3 **3.** Boris intends to buy a computer system for $1300. At one store he can finance the purchase in monthly installments at 12% over 2 years. At another store he can finance the purchase at 9% over 2 years. How much money will he save by financing the purchase at 9%? $43.33

6-3 **4.** Complete the following chart that shows the interest cost for two months on a credit card when you pay 10% of the amount due to the nearest dollar each month and your card carries an APR of 9%.

Month	Balance	Interest	Amount Owed	Payment
1	$1000.00	$7.50	$1007.50	$100.00
2	907.50	6.81	914.31	91.00

6-3 **5.** Find the total interest paid in Exercise 4. $14.31

7-1 **6.** Use the Comparison Table for Term and Whole Life Premiums in the Reference Section to find the amount paid in premiums in 30 years for a whole life insurance policy of $200,000 purchased at the age of 25. $55,080

9-1 Use the Tax Rate Schedule in the Reference Section to find the tax owed to the nearest dollar by each of the following taxpayers.

7. a single person with a taxable income of $105,950 $28,499

8. a married couple, filing jointly, with a taxable income of $137,846 $35,483

9-2 **9.** Use IRS Form 1040EZ and the Tax Table in the Reference Section to find the tax paid by a single person with a salary of $15,950 and interest income of $380. The person cannot be claimed as a dependent by anyone else. $1564

10-1 **10.** A car with a base price of $6200 has a sticker price of $7500. Find the amount and percent of markup. $1300; 21%

10-2 **11.** A used car purchased for $6000 is expected to depreciate in a straight line over five years. Find its resale value after two years. $3600

10-3 **12.** A car uses 43.5 gallons of gasoline to cover 1155 miles. Find the average number of miles traveled on one gallon of gas. 26.6 miles per gallon

8-3 **13.** Which equation or equations could represent a linear regression equation for the line of best fit? **b** and **d**

a. $y = x^a + b$

b. $-a = bx - y$

c. $y = b^x + a$

d. $y = a + bx$

9-1 **14.** Use the Tax Tables in the Reference Section to find the tax owed by a head of household with a taxable income of $20,300. $3049

PROJECT 10–1: **Investigating Percent of Markup**

Work with a group of four or five of your classmates to complete this activity.

1. Use *Edmund's New Car Prices* or the *Kelley Blue Book New Car Price Manual* as a reference.
2. Each student in the group selects ten automobiles, representing two or three manufacturers. Choose a wide range of models and prices.
3. Each student chooses a list of options for his or her automobiles.
4. Calculate the percent of markup for the automobiles and for each of the options.
5. Prepare a table showing the markup for different models and calculate averages for the different manufacturers.
6. Prepare a similar table for the options by manufacturer.

PROJECT 10–2: **Percent of Markup for Used Cars**

Work with a group of four or five classmates.

1. Have each member of a group select 20 cars advertised in the newspaper or on a used-car lot.
2. Consult a reference book such as *Edmund's* or the *Blue Book* for used-car prices to use in calculating the trade-in value of the cars.

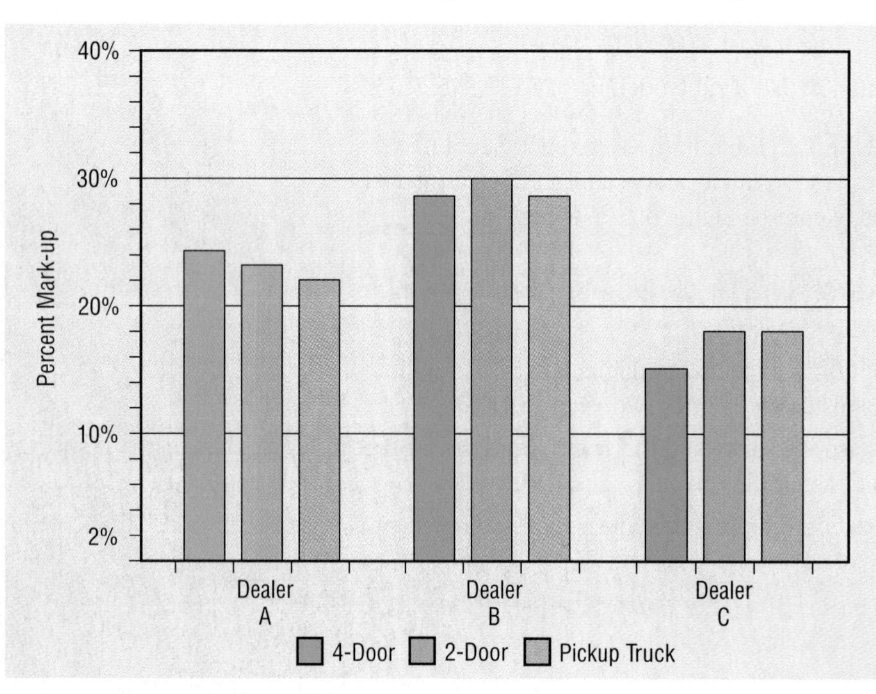

3. Use the prices listed in the newspaper or on the cars in the lot to calculate the percent of markup over the book value the dealer is using.
4. Prepare a table of the trade-in values and the percent of markup for the cars you researched.
5. Draw bar graphs, similar to the one shown, indicating the percent of markup by dealer and car model.

Some equations and their graphs belong to families. The equations in the same family all have some common characteristics. Recall that the slope-intercept form of the equation of a line is $y = mx + b$, where m is the slope and b is the y-intercept. To determine slope from a graph of a line, remember that slope $= \frac{\text{change in } y}{\text{change in } x}$. The y-intercept is the point where the graph intersects the y-axis.

Find the equation of each graph then note the family characteristic.

Example

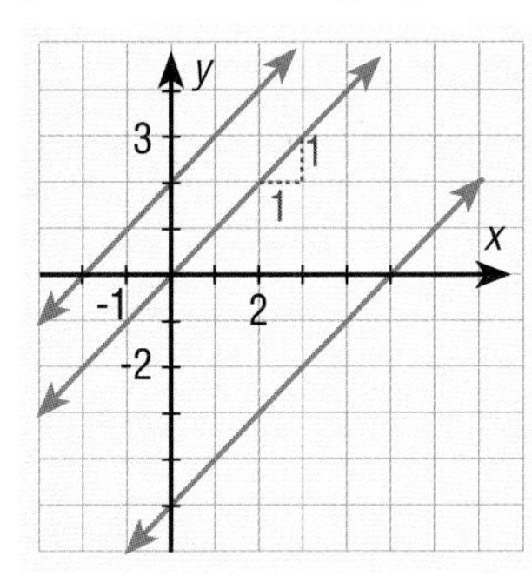

From the graph, the slope of each line is 1, so $m = 1$ in each case. The y-intercepts are 2, 0, and -5. Therefore the equations are

$$y = x + 2$$
$$y = x$$
$$y = x - 5$$

The family characteristic is that they all have the same slope.

1.

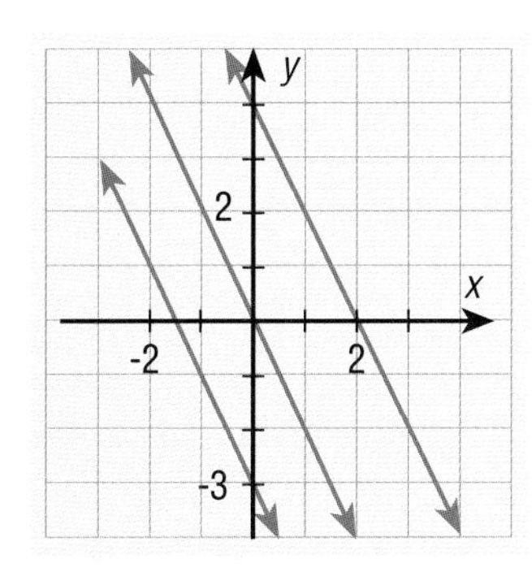

$y = -2x - 3$; $y = -2x$; $y = -2x + 4$; $m = -2$

2.

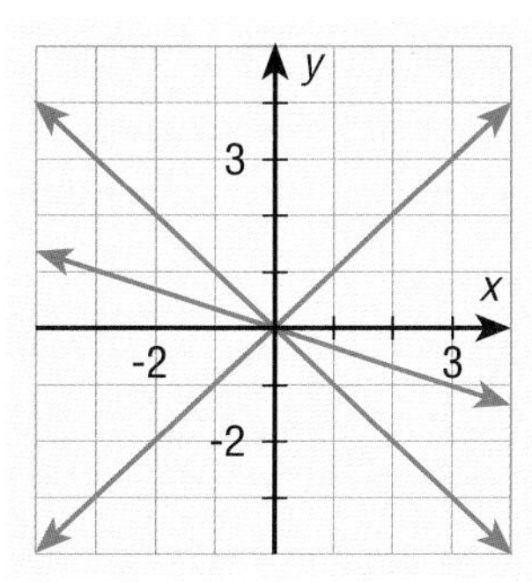

$y = x$, $y = -x$, $y = -\frac{1}{3}x$; $b = 0$

Graph the equations and note the family characteristics.

Example $y = x^2 - 3x$
$y = -x^2 + 3x$

The second is the reflection of the first in the x-axis. Note that

$$-(x^2 - 3x) = -x^2 + 3x$$

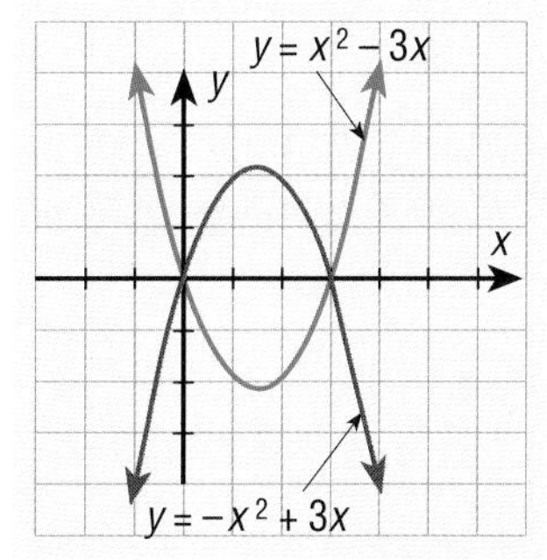

3. $y = x^2$
$y = 3x^2$
$y = 0.5x^2$

See Additional Answers.

4. $y = x^2$
$y = x^2 + 2$
$y = x^2 - 3$

Travel Plans

AMERICANS LIKE TO TRAVEL, ESPECIALLY AT VACATION time. We visit relatives, beaches, mountains, and cities. We play and watch all sorts of games and sports. We visit places that were important in our nation's history and marvel at the natural beauty of different parts of the country. There really is a lot to see and do. For some, getting away from home and the everyday routine is in itself reason to travel.

In any case, travel is a big business. The travel and tourism industry is happy to cater to our needs and desires. Places to stay, amusements, souvenirs, good things to eat – these and much more call out to us from brochures, from roadside signs, and in all kinds of advertising.

But travel is not free. Whether reasonable or expensive, it always costs money. In this chapter we consider some travel costs. We work with Betty and her family, reading maps, figuring distances, and calculating the costs of automobile travel.

We consider Sylvia and her mother, who travel not only for fun but also because Sylvia's mother is a travel agent and wants first-hand experience about different places. Sylvia helps her mother by using spreadsheets to budget travel costs and a graphing calculator to compare the expenses and earnings from tours.

Reading airline schedules, with the various abbreviations, takes some skill. With Ramón, a college student, we compare the costs of travel by air and by a car.

11–1 Making Travel Plans

11–2 Travel Costs: Different Perspectives

11–3 Flying Saves Time and Sometimes Money

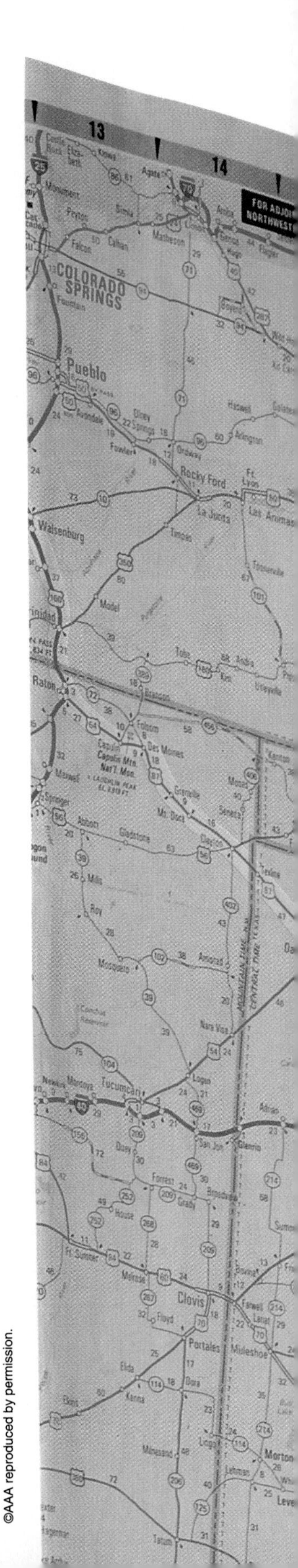

16
17
18
19
20
21
22
23
24
25
26
KANSAS CITY
Topeka
Lawrence
Manhattan
Salina
Hays
Russell
Great Bend
Hutchinson
McPherson
WICHITA
Garden City
Dodge City
Liberal
Pratt
Kingman
Emporia
El Dorado
Winfield
Arkansas City
Joplin
SPRINGFIELD
Sedalia
Enid
Ponca City
Stillwater
Bartlesville
TULSA
OKLAHOMA CITY
Norman
Shawnee
Lawton
Duncan
Ardmore
Wichita Falls
Amarillo
Pampa
Borger
Plainview
LUBBOCK
Fort Smith
Fayetteville
Harrison
Russellville
Hot Springs
OUACHITA MOUNTAINS
Texarkana
Sherman
Denison
Paris
McAlester
Muskogee
Gainesville
Vernon
Childress
Guymon
Woodward
Elk City
Clinton
Chickasha
Ada

11-1 MAKING TRAVEL PLANS

QUICK REFERENCE

TEACHER'S RESOURCES AND ANSWER KEY

Lesson Quiz Answers 11-1, p. 78

Reteaching Activity Answers 11-1, p. 111

Enrichment Activity Answers 11-1, p. 112

EXTENSION ACTIVITIES

Reteaching Activity 11-1, p. 83

Enrichment Activity 11-1, p. 86

Betty likes to travel. She and her family take vacation trips almost every summer. One of her favorite trips was to the West Coast. Betty remembers seeing giant sequoia trees that are the oldest living things on earth, more than 5000 years old. Betty's father bought her a small redwood box and a model truck with tiny redwood logs on the back.

On other trips, Betty saw the Liberty Bell in Philadelphia and stood at the top of the Empire State Building in New York. In Boston she really liked the science museum. She also remembers how her little brother left his teddy bear in a taxicab.

Betty helps her family plan their trips. They usually drive between 300 and 500 miles a day for two days. Then they stay for a few days to relax, see the sights, and visit. They usually take a different way home to see other sights.

Betty's mother teaches history at the high school and frequently knows interesting stories about the people who have lived in the area they are visiting. Betty's father makes up riddles and games for them to play as they drive along. She expects to enjoy the trip this year more than ever.

OBJECTIVES: *In this lesson, we will help Betty to:*

- *Use map-reading skills to plan a trip within the United States.*
- *Use a mileage chart to determine distances.*
- *Determine the cost of driving different distances.*

WHICH WAY TO GO

After deciding where you are going, the next big travel decision is selecting the route, that is, which roads you will take. There are almost always several ways to drive to any given place. Each route has advantages and disadvantages.

If you want to get to your destination in the least amount of time, then you will probably take the big **interstate highways.** These do not have stop lights and with two or three lanes in each direction, traffic generally keeps moving. But sometimes it is more interesting, at least for a time, to take what are called **secondary roads.** These are well-marked federal and state highways that pass through towns and let you see more of what an area is really like. But traveling a distance on secondary roads can make a trip take a very long time.

Betty's family has found that it helps to have an interstate atlas along on any trip. This is a booklet with detailed road maps of every state. The **mileage chart** printed in the atlas gives the distances between many cities. Other mileage distances either are marked on the map or can be estimated on the basis of the map distance and the scale showing miles as distances.

When Betty and her family plan a trip, they have a lot to consider. Betty now helps with some of the driving. They usually switch drivers after one or two hours. When they stop to change drivers, they stretch and sometimes have a snack.

The family has learned to have a destination but also to be flexible when traveling. They sometimes find the most interesting places unexpectedly and stay longer than they had planned. On other occasions they find that a city or attraction is not what they had hoped, and they move along sooner than planned. There are also difficulties beyond their control. Once they had to wait for four hours while the car was repaired. Another time they traveled back roads for two hours to get around rush hour traffic. They have learned to turn problems into adventures and disappointments into opportunities.

Ask Yourself

1. How do you decide which routes to take when traveling?
   See Additional Answers.
2. Why would people take secondary roads instead of interstate highways? if they had the time and wanted to see what the area is like
3. What are two unexpected events that might change the timing of a trip?
   Answers may vary. (Ex.: car trouble, heavy traffic)

ALGEBRA CONNECTIONS

A treasure is buried one mile from the old oak tree and one-half mile from a north-south road located one-half mile east of the old oak tree. Draw a diagram showing the three points where you should dig for the treasure.

ALGEBRA *REVIEW*

If $d \div a = g$ and $g \cdot c = t$, find each of the following.

1. g when $d = 500$ and $a = 25$
   $g = 20$
2. d when $g = 22.5$ and $a = 19.6$
   $d = 441$
3. a when $d = 300$ and $g = 28$
   $a = 10.71$
4. t when $g = 52$ and $c = 1.35$
   $t = 70.2$
5. c when $t = 127$ and $g = 105$
   $c = 1.21$
6. g when $t = 10.56$ and $c = 1.28$
   $g = 8.25$

For four numbers a, b, c, and d,

7. Write an expression for the average of a and b.
   $(a + b) \div 2$
8. Write an expression for the average of a, b, c, and d.
   $(a + b + c + d) \div 4$
9. If the average of a, b, and c is d, what is the average of a, b, c, and d?
   d
10. If the average of a and b is 10, the average of c and d is 12, find the average of a, b, c, and d.
    11

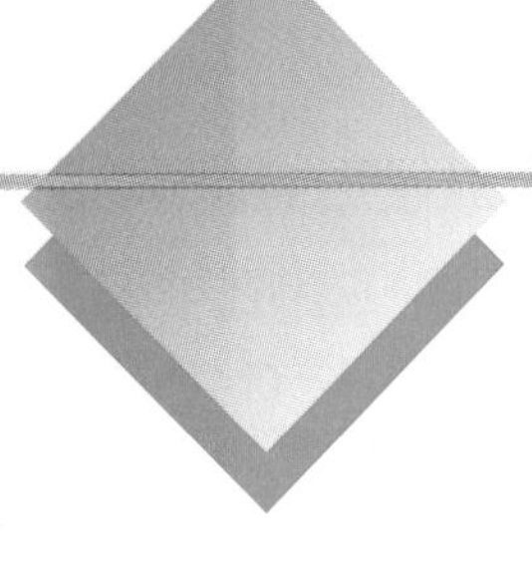

SHARPEN YOUR SKILLS

EXAMPLE 1 Betty and her family have decided to drive from their home in Indianapolis, Indiana, to visit their cousins in Milledgeville, Georgia.

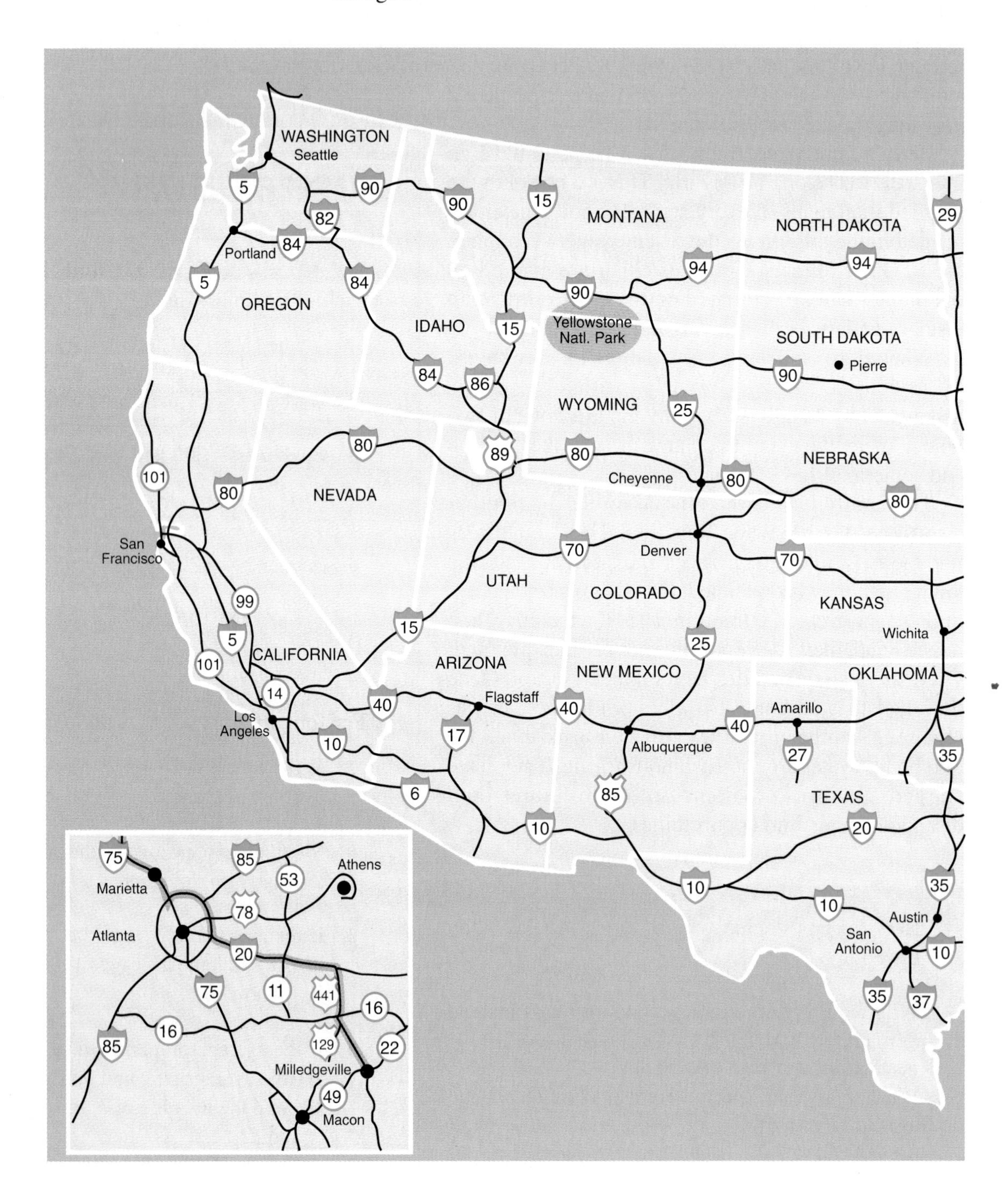

QUESTION What highways can they take to drive from Indianapolis to Milledgeville?

SOLUTION

The map below shows the interstate highways in the United States. The route that they will take is highlighted: first, south on 65 through Louisville to Nashville; then southeast on 24 to Chattanooga and then 75 to Atlanta. The trip from Atlanta to Milledgeville is highlighted on the inset map of the area around Atlanta. They first take Interstate 20 east and then 441, a secondary road, south.

CRITICAL THINKING

Have students list some of the differences between planning for a business trip and planning for a pleasure trip. Ask, in particular, how time, expenses, and mode of travel might be handled in each case.

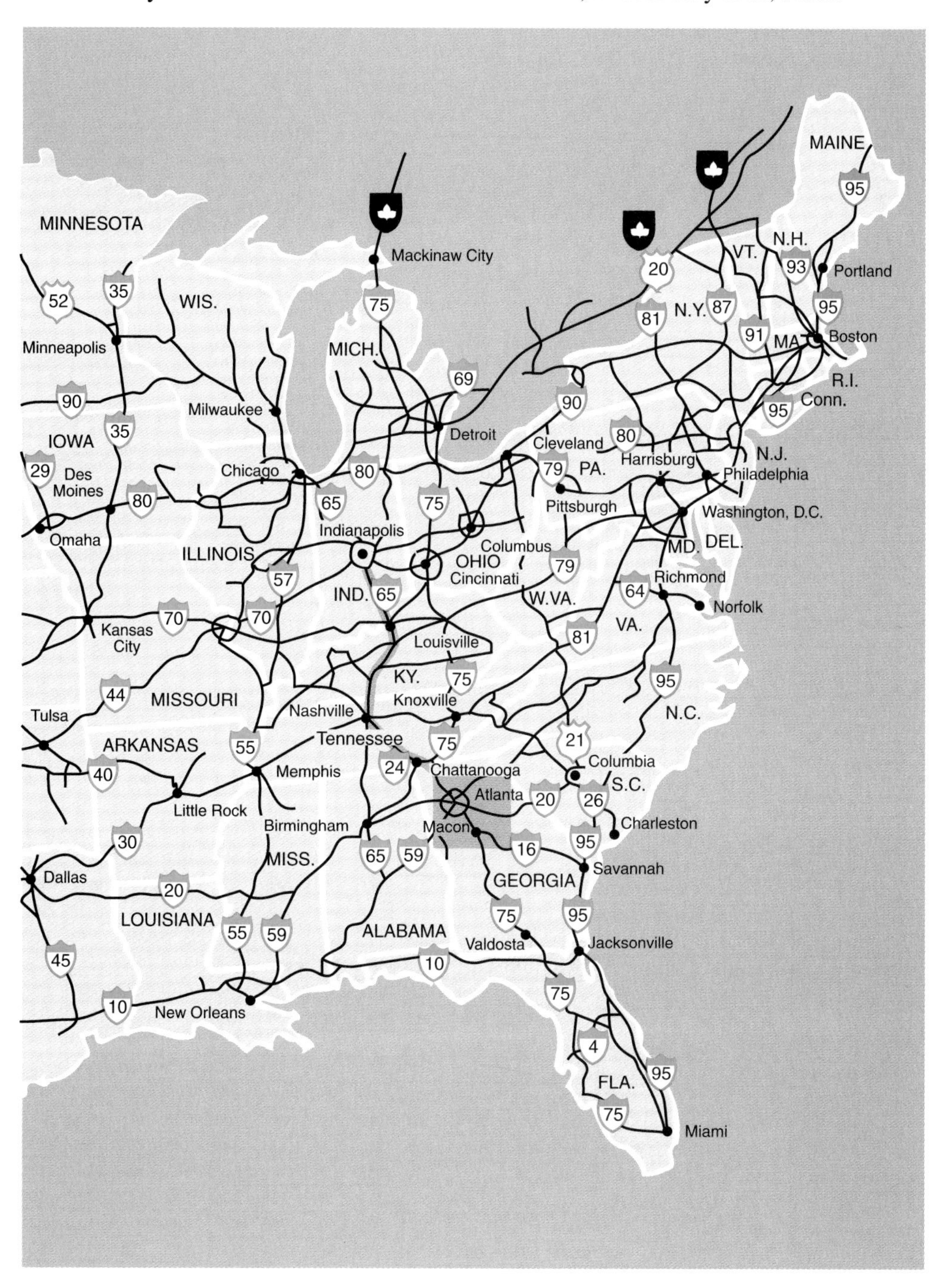

ESL STUDENTS

The names of cities and states should be reviewed with non-native speakers. Many take their origin from Native American words or from places or people not originally from the United States.

SKILL 2

EXAMPLE 2 The mileage chart shows the distances between many major cities.

QUESTION How far is it from Indianapolis, Indiana, to Atlanta, Georgia?

SOLUTION

To answer the question, we use the mileage chart. Atlanta is the third city down the left-hand column. We read across the Atlanta row to the Indianapolis column. The cell where this row and column intersect shows the mileage between the two cities: 493 miles.

United States Mileage Chart

	Atlanta, GA	Boston, MA	Cheyenne, WY	Chicago, IL	Cincinnati, OH	Cleveland, OH	Dallas, TX	Denver, CO	Des Moines, IA	Detroit, MI	Indianapolis, IN	Kansas City, MO	Louisville, KY	Memphis, TN	Milwaukee, WI	Minneapolis, MN	New Orleans, LA	Omaha, NE	Philadelphia PA	Pittsburgh, PA
Albuquerque, NM	1381	2172	517	1281	1372	1560	638	417	977	1525	1266	782	1301	1010	1319	1190	1134	858	1899	1619
Amarillo, TX	1097	1897	511	1043	1096	1285	358	423	742	1269	991	547	1019	726	1084	975	850	643	1624	1344
Atlanta, GA		1037	1442	674	440	672	795	1398	870	699	493	798	382	371	761	1068	479	986	741	687
Austin, TX	919	1911	994	1110	1083	1327	193	906	877	1315	1037	682	982	615	1184	1129	517	837	1615	1367
Birmingham, AL	150	1165	1347	642	465	709	645	1286	787	724	475	697	364	246	728	1006	342	898	869	741
Boston, MA	1037		1907	963	840	628	1748	1949	1280	695	906	1391	941	1296	1050	1368	1507	1412	296	561
Charleston, SC	289	929	1722	877	603	730	1072	1678	1150	842	696	1078	591	660	964	1282	720	1266	633	666
Cheyenne, WY	1442	1907		954	1174	1279	869	100	627	1211	1068	650	1161	1101	987	788	1361	495	1678	1390
Chicago, IL	674	963	954		287	335	917	996	327	266	181	499	292	530	87	405	912	459	738	452
Cleveland, OH	672	628	1279	335	244		1159	1321	652	170	294	779	101	468	374	692	786	693	567	287
Columbus, OH	533	735	1235	308	108	139	1028	1229	618	192	171	656	209	576	395	713	894	750	462	182
Dallas, TX	795	1748	869	917	920	1159		781	684	1143	865	489	819	452	991	936	496	644	1452	1204
Denver, CO	1398	1949	100	996	1164	1321	781		669	1253	1058	600	1120	1040	1029	841	1273	537	1691	1411
Des Moines, IA	870	1280	627	327	571	652	684	669		584	465	195	566	599	361	252	978	132	1051	763
Detroit, MI	699	695	1211	266	259	170	1143	1253	584		278	743	360	713	353	671	1045	716	573	287
Flagstaff, AZ	1704	2495	757	1604	1695	1883	961	657	1300	1848	1589	1105	1624	1333	1642	1481	1457	1171	2222	1942
Harrisburg, PA	700	373	1579	639	468	314	1383	1592	952	474	534	1019	569	931	726	1044	1142	1084	102	189
Indianapolis, IN	493	906	1068	181	106	294	865	1058	465	278		485	111	435	268	586	796	587	633	353
Jackson, MS	391	1406	1257	742	655	899	404	1169	809	914	646	644	554	212	824	1036	178	845	1110	939
Kansas City, MO	798	1391	650	499	591	779	489	600	195	743	485		520	451	537	447	806	201	1118	838
Knoxville, TN	193	911	1372	527	253	485	837	1328	800	512	346	728	241	385	614	932	596	916	615	511
Louisville, KY	382	941	1161	292	101	345	819	1120	566	360	111	520		367	379	697	685	687	668	388
Mackinaw City, MI	935	916	1291	387	495	439	1261	1341	673	284	460	864	562	880	368	508	1247	805	842	556
Miami, FL	655	1504	2097	1329	1095	1264	1300	2037	1525	1352	1148	1448	1037	997	1416	1723	856	1641	1208	1200
Minneapolis, MN	1068	1368	788	405	692	740	936	841	252	671	586	447	697	826	332		1214	357	1143	857
New Orleans, LA	479	1507	1361	912	786	1030	496	1273	978	1045	796	806	685	390	994	1214		1007	1211	1070
Norfolk, VA	540	558	1764	831	604	508	1329	1758	1141	666	700	1162	642	877	918	1236	1019	1273	263	384
Pierre, SD	1361	1726	434	763	1050	1098	943	518	492	1029	944	592	1055	1043	690	394	1394	391	1501	1215
Pittsburgh, PA	687	561	1390	452	287	129	1204	1411	763	287	353	838	388	752	539	857	1070	895	288	
Portland, ME	1139	106	1986	1042	942	707	1850	2028	1359	775	1001	1486	1043	1398	1129	1447	1609	1491	398	663
Portland, OR	2601	3046	1159	2083	2333	2418	2009	1238	1786	2349	2227	1809	2320	2259	2010	1678	2505	1654	2821	2535
San Antonio, TX	983	1988	1027	1187	1160	1404	270	939	954	1392	1114	759	1059	692	1261	1206	550	914	1692	1444
San Francisco, CA	2496	3095	1188	2142	2362	2467	1753	1235	1815	2399	2256	1835	2349	2125	2175	1940	2249	1683	2866	2578
Seattle, WA	2618	2976	1228	2013	2300	2348	2078	1307	1749	2279	2194	1839	2305	2290	1940	1608	2574	1638	2751	2465
Tulsa, OK	772	1537	765	683	736	925	257	681	443	909	631	248	659	401	757	695	647	387	1264	984
Washington, DC	608	429	1611	671	481	346	1319	1616	984	506	558	1043	582	867	758	1075	1078	1116	133	221
Wichita, KS	903	1587	583	696	787	975	365	509	392	940	681	197	710	532	734	644	815	298	1314	1034

SKILL 3

EXAMPLE 3 Betty estimates that the driving distance from Atlanta to Milledgeville is 115 miles. The family car gets an average of 20 miles per gallon of gas. They expect to pay about $1.35 a gallon for gas while on the trip.

QUESTION How much should they budget for gas between Indianapolis and Milledgeville?

SOLUTION

The total distance d is the distance from Indianapolis to Atlanta plus the distance from Atlanta to Milledgeville

$$493 + 115 = 608 \text{ miles}$$

To find the number of gallons of gas to be used, divide the total distance by the average miles per gallon.

$g = \frac{d}{a}$ where d = total distance in miles
a = average miles per gallon
g = number of gallons needed

$$g = \frac{608}{20} = 30.4 \text{ gallons}$$

Next, we find the total cost.

$t = gc$ where g = number of gallons
c = average cost per gallon
t = total cost for gas

$$t = 30.4(1.35) = 41.04$$

They should budget about $41 for gas.

INSTRUCTION

TEACHING THE LESSON

This lesson can be a practical, hands-on change of pace for many students. If possible encourage them to bring in maps and mileage charts from home.

The lesson provides the opportunity for a brief review of United States geography. Students will all have done some work with maps but may not be familiar with the use of route numbers. Explain as much as you wish about how federal, state, and local roads are identified.

To test understanding of the mileage chart, ask why some of the cells are blank.

FOCUS ON ALGEBRA

The lesson gives a chance to use mathematics in every-day problems arising outside the normal realm of mathematics. *(NCTM Standard 1, p. 137)*

COMMON ERRORS

When going from map to map—for example, using the two maps for Skill 1—it is easy to make mistakes about distances because the scales are so different. Even in a United States atlas, the scales for different states can be very different.

COOPERATIVE LEARNING

The exercises on planning trips could profitably be done by groups of two or three students working together.

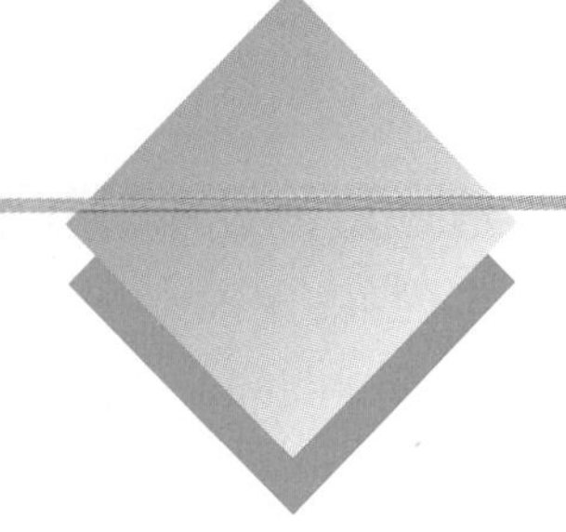

TRY YOUR SKILLS

See Additional Answers.

1. Use the map on pages 502–503 to find the quickest and most direct route to drive from New Orleans, Louisiana, to Jacksonville, Florida.

GUIDED PRACTICE

Let students do all of the *Try Your Skills* exercises individually. Then ask individuals to explain their results to the class. See if there are alternatives that might be defended.

RETEACHING

To reteach the map skills and the mileage chart, begin with a given distance and ask students to find two places that are about that far apart.

ENRICHMENT

Have students give directions for someone from a town or city in another state to get exactly to the student's house. Why is it that the last few miles are usually the most complicated with regard to directions?

2. Use the map on pages 502–503 to find the fastest and most direct route to drive from Columbia, South Carolina, to Birmingham, Alabama.
3. Use the mileage chart on page 504 to find the distance from Milwaukee, Wisconsin, to Portland, Oregon. 2010 miles
4. Use the mileage chart on page 504 to find the distance from Chicago, Illinois, to Cleveland, Ohio. 335 miles
5. If you drive 387 miles and get an average of 21.5 miles per gallon of gas, how many gallons of gas do you use? 18 gallons
6. If you used 24 gallons of gas and got an average of 18.5 miles per gallon of gas, how many miles did you drive? 444 miles
7. If you use 29.6 gallons of gas and pay an average of $1.39 a gallon, what is the total cost of the gas? $41.14
8. If you purchase 13.8 gallons of gas at a total cost of $15.85, what is the average price per gallon? $1.15
9. Write an equation for finding gallons of gas when you know total miles and miles per gallon. Explain what each variable in the equation stands for.
   Answers may vary. (Ex.: $g = d \div a$; g = gallons of gas, d = total distance in miles, a = average miles per gallon)

ADDITIONAL ANSWERS

1. North on 59, East on 10
2. Interstate 20 West

EXERCISE YOUR SKILLS

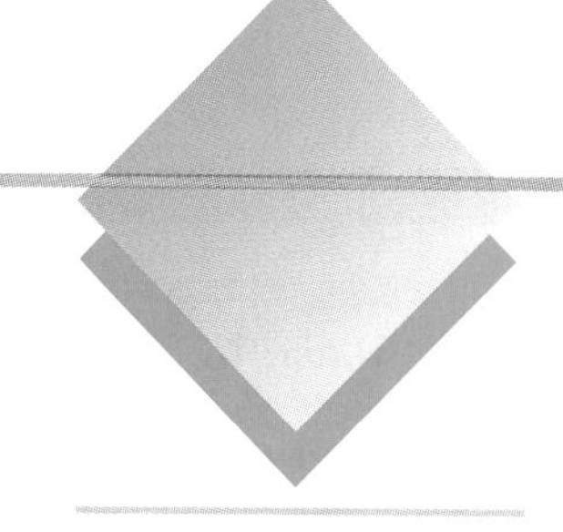

1. What are three factors that might affect your travel plans? Answers may vary.
2. What are some advantages of traveling by car? Answers may vary.
3. Why should you use a map to plan before you begin a car trip? Answers may vary.

Use the maps provided in the lesson to find a direct route for each trip.

4. Norfolk, Virginia, to Valdosta, Georgia See Additional Answers.
5. Macon, Georgia, to Boston, Massachusetts east on 16; north on 95
6. Miami, Florida, to Savannah, Georgia north on 95
7. Dallas, Texas, to Nashville, Tennessee northeast on 30; east on 40
8. Use maps to find a direct route from where you live to Macon, Georgia.

Use the mileage chart in the lesson or the one in the Reference Section of the book to find the following distances.

9. Tulsa, Oklahoma, to Denver, Colorado 681 mi
10. Albuquerque, New Mexico, to Cheyenne, Wyoming 517 mi
11. Mackinaw City, Michigan, to Boston, Massachusetts 916 mi
12. Flagstaff, Arizona, to Cleveland, Ohio 1883 mi
13. Minneapolis, Minnesota, to Tulsa, Oklahoma 695 mi
14. Pierre, South Dakota, to Dallas, Texas 943 mi

Find the cost of driving the following distances with the miles per gallon and the cost of gasoline as shown. Round each number of gallons to the nearest tenth and each cost to the nearest cent.

	Miles	Miles per Gallon	Number of Gallons	Cost per Gallon	Total Cost
15.	540	20.5	26.3	$1.29	$ 33.93
16.	2110	17.3	122.0	1.19	145.18
17.	750	26.4	28.4	1.24	35.22
18.	1235	22.6	54.6	1.15	62.79
19.	1827	19.7	92.7	1.35	125.15
20.	926	27.5	33.7	1.27	42.80

A family from Philadelphia, Pennsylvania, plans a 9-day vacation. Their destination is Nashville, Tennessee, but they plan to stay overnight three times on the way and twice on their return trip. Work in groups of two or three to do the following exercises. Use information from travel magazines and talk with experienced travelers, as needed, to help you answer the following.

21. Plan the trip, indicating the miles to be covered each day, where they would stay each night, and what things they might do and see along the way. Answers may vary.

KEY TERMS

interstate highways
secondary roads
mileage chart

INDEPENDENT PRACTICE

ASSIGNMENT GUIDE

Exercises 1-3 review the process of planning for a trip.

Exercises 4-8 are answered using the maps.

Exercises 9-14 are answered using the mileage chart.

Exercises 15-20 are on gasoline usage.

Exercises 21-24 bring together the skills and concepts of the lesson. These may be completed by students working in small groups.

ADDITIONAL ANSWERS

4. northwest on 64; south on 95; west on 16; south on 75
8. Answers may vary.

CLOSURE

To complete the instructional part of the lesson, have an oral quiz. Using the maps and then the mileage chart, quickly ask students for the distance between different cities, and then for cities that are a given distance apart.

LESSON QUIZ

1. Use the first map to find two cities that are about 600 miles apart. (**Answers will vary.**)
2. Use the mileage chart to find three cities that are abut 700 miles from Detroit. (**Answers will vary. Possibilities: Boston, Kansas City, Memphis, Omaha**)

ADDITIONAL ANSWERS (EXERCISE YOUR SKILLS)

22. Answers may vary.
23. Answers may vary.
24. Answers may vary.

22. Plan the trip back using a different route but also indicating distances, places of interest, and where they would stay overnight. See Additional Answers.

23. Assuming that they drive an average of about 80 miles a day while in the Nashville area, make an estimate of their total miles for the trip.

24. Estimate miles per gallon and cost per gallon, and on the basis of these estimates, determine an approximate cost for gasoline for the trip.

MIXED REVIEW

2-2 **1.** With $778.22 in your checking account you write a check for $105.89 to Grocery Mart for your weekly groceries. Show how you would enter this check in your check register. See Additional Answers.

6-5 Your credit account balance from the first of the month through the 12th was $1298. You made a payment of $325 on the 13th. You made no further charges or payments for the remaining 17 days of the month. The monthly finance charge is 1.5%.

2. What was your average daily balance? $1103.00

3. What was the finance charge for the month? $16.55

8-4 The following table shows the Consumer Price Index for selected years between 1984 and 1992. The prices given in Exercises 4–6 are for 1984. Find the approximate cost that you might expect to pay for each item in 1992.

CONSUMER PRICE INDEX (1984–1992)

Year	All Items	Food	Shelter	Apparel	Medical Care
1984	100	100	100	100	100
1985	107.6	105.6	107.7	105	113.5
1986	109.6	109	115.8	105.9	122
1987	113.6	113.5	121.3	110.6	130.1
1988	118.3	118.2	127	115.4	138.6
1989	124	125.1	132.8	118.6	149.3
1990	130.7	132.4	140	124.1	162.8
1991	136.2	135.8	146.3	128.8	177
1992	140.7	138.3	152.2	131	189.4

4. a $100,000 house $152,200

5. $250 in clothes $327.50

6. $2500 in medical care $4735

9-1 Use the Tax Table in the Reference Section to find the tax owed by each of the following taxpayers.

7. a single person with taxable income of $35,900 $7271

8. a married couple, filing jointly, with taxable income of $48,548 $8933

11-2 TRAVEL COSTS: DIFFERENT PERSPECTIVES

QUICK REFERENCE

TEACHER'S RESOURCES AND ANSWER KEY

Lesson Quiz Answers 11-2, p. 80

Reteaching Activity Answers 11-2, p. 111

Enrichment Activity Answers 11-2, p. 112

EXTENSION ACTIVITIES

Reteaching Activity 11-2, p. 84

Enrichment Activity 11-2, p. 87

A perspective is a point of view. A city looks quite different from the window of a bus than from an airplane. Similarly, the cost of a restaurant meal appears different, depending on whether you are the person having dinner or the one who owns the restaurant. In this lesson we will consider, from different perspectives, some of the costs of vacationing.

Ever since her parents were divorced, Sylvia has helped her mother with the cooking and cleaning at home. Consequently she and her mother really enjoy a vacation when they can eat out and let someone else wash the dishes. They also like camping out. There is a great feeling of freedom and relaxation that comes from being out of doors all day long.

Sylvia's mother has another reason for traveling. She is a travel agent and needs first-hand information about some of the accommodations and attractions that she recommends to her clients. She and Sylvia watch their costs very closely, both for themselves and also so that they can tell clients of the travel agency how to budget for a trip.

OBJECTIVES: *In this lesson, we will help Sylvia and her mother to:*

- *Compare and calculate the costs of food and lodging.*
- *Prepare a vacation budget.*
- *Compute costs for tour vacations.*

FOCUS

ALGEBRA CONNECTIONS

Draw a graph comparing time and distance for a person who travels as follows: 1 hour at 50 miles an hour; stops 1.5 hours for lunch; travels 2 hours at 20 miles an hour; stops 0.5 hours; travels 1 hour at 40 miles an hour.

MAKING CHOICES ON A BUDGET

Sylvia and her mother have discovered that no one who travels likes expensive surprises. For example, many people do not mind paying a little extra for a good meal or accommodations with special features. But they want to know in advance what to expect. For that reason, Sylvia helps her mother prepare a detailed budget before a trip, and they also make reservations to be sure that they will not be disappointed. A **reservation** is an establishment's assurance that a motel room or a table in a restaurant will be held for you at a particular date or time. Sometimes you must use your credit card to guarantee a reservation. Even campgrounds accept reservations, and with so many people now using these facilities it is important to reserve a spot.

When Sylvia and her mother go camping or stay in a cabin with kitchen facilities, they purchase groceries and cook some meals for themselves. However, they prefer restaurants and have become skilled in predicting the prices of meals. They set up budgets for people with different tastes and means so that Sylvia's mother can make recommendations. They are always collecting brochures about restaurants and points of interest.

Everyone who travels says, "What shall we do today?" Sylvia has learned that the costs of entertainment vary greatly. Walking in the woods is free; the fee to enter a zoo or museum is usually reasonable; a special evening tour can be costly. Going to the beach is much cheaper than skiing. Gifts and souvenirs are always fun to buy, but once again, the costs can add up. Sylvia's mother says that they must stick to their budget and be careful to avoid impulse buying.

Ask Yourself

1. How can you plan your lodging expenses when you travel? make reservations in advance and ask what to expect and how much it will cost
2. How can you save money on food costs when you travel? stay where there are kitchen facilities, purchase groceries, and cook for yourselves
3. What are several less expensive forms of entertainment that you enjoy? Answers may vary. (Ex.: hiking, walking tours, beachcombing)

ALGEBRA *REVIEW*

For numbers x, y, and z,

1. If $x = 10.50$ and $y = 11.90$, find the average of x and y.
   11.2
2. If $x = 13.20$ and the average of x and y is 15.00, find y.
   16.8
3. If $x = 8.76$, $y = 9.49$, and $z = 11.75$, find the average of x, y, and z.
   10.0
4. If $x = 45.60$, $y = 58.20$, and the average of x, y, and z is 51.42, find z.
   50.46
5. If $a = b + 1$ and $b = c + 1$, find the average of a, b, and c.
   $c + 1$

Graph each system of equations and find the coordinates of the intersection of the graphs.

6. $y = 10$ and $y = x + 2$
   (8, 10)
7. $y = 30 + 3x$ and $y = 5x$
   (15, 75)
8. $y = 1000 + 40x$ and $y = 75x$
   (28.6, 2142.9)

SHARPEN YOUR SKILLS

SKILL 1

Sylvia has found the following lodging information in a travel guide.

LOTUS INN
LOTUS, IOWA

I-123 at 33rd Place, S.W.
Exit 3 555-1234
Location: 3322 South Street,
Southwest Lotus 50010
Downtown: 3.5 mi
Ball park: 1 mi
Zoo: 2 mi.
Airport: 7 mi
Restaurants & shopping nearby.

Rates: Tax 8%

1 person single $37.00	2 people double bed $43.00
2 people 2 dble bds $48	4 people 2 dble bds $58

PLEASURE LODGE
MONMOUTH, IOWA

Near International Airport
Indoor Pool HBO
Health Club Sauna
Location: 1001 Sandwich Road
Exit Stream Road,
Follow Airport signs.
Complimentary: airport limo,
light breakfast, cable TV.

Rates: Tax 8%. XP $8
Teens free. RB $8

1 person	Standard: $65 Luxury: $90
2 people	Standard: $75 Luxury: $105

All rooms have two double beds.

(XP means extra person)
(RB means roll-away bed)

INSTRUCTION

TEACHING THE LESSON

You will be helping students if you instill the realization that spending money is not something that just happens to us. By thinking and planning, we can control what we do with our money.

As in some earlier lessons, the easy to understand table of expenses leads to a spreadsheet which is slightly more generalized.

Skill 3 offers another opportunity to review the concept of break-even.

EXAMPLE 1 Sylvia and her mother want to plan their lodging costs.

QUESTION What would be the cost for Sylvia and her mother to stay at each of these motels for 2 nights? They would take the least expensive room with two double beds.

SOLUTION

At Lotus Inn, two beds for two people cost $48 per night.

$2(48) = 96$ Rate for two nights
$0.08(96) + 96 = 103.68$ The tax is 8%.

The cost at Lotus Inn is $103.68.

At Pleasure Lodge teens can stay free of charge and the least expensive room costs $65.

$2(65) = 130$ Rate for two nights
$1.08(130) = 140.40$ Multiply by 1.08 for room and tax (100% + 8% = 108%).

The cost at Pleasure Lodge is $140.40.
Sylvia and her mother would stay at Lotus Inn.

FOCUS ON ALGEBRA

Students comfortable with the spreadsheet and the use of the graphing calculator will be much better equipped to apply their mathematical learning in more advanced courses and in real-life situations. *(NCTM Standard 1, p. 137)*

EXAMPLE 2 Sylvia and her mother want to use the actual cost of their meals while traveling as a reasonable basis for a budget for clients.

QUESTION How can they arrive at a reasonable estimate of the cost for meals?

SOLUTION

The best method is to record a number of prices for meals and then find the average. The following are the amounts spent by Sylvia and her mother. Sylvia used her spreadsheet skills to write formulas that would give totals and averages.

	A	B	C	D	E	F	G
1		Monday	Tuesday	Wednesday	Thursday	Total	Average
2	Breakfast	7.50	12.75	6.28	11.32	37.85	9.46
3	Lunch	12.85	14.50	6.70	9.43	43.48	10.87
4	Dinner	52.45	32.60	44.76	18.40	148.21	37.05
5	Total	72.80	59.85	57.74	39.15	229.54	57.39

+F2/4

@SUM(B2..B4)

On the basis of this record of expenses, Sylvia's mother will recommend to her clients that, for a similar trip, they plan to budget \$10 per day for breakfast, \$11 per day for lunch, and \$37 per day for dinner, or \$58 total per day.

TECHNOLOGY HINT

The formula for cell F2 can be written in two ways. Students can use +B2+C2+D2+E2 or @SUM(B2..E2) to get \$37.85.

Notice that \$229.54 is the total of all the daily totals and also the total of all the meal totals. These two totals are equal because they are simply different ways of adding the same numbers. If you were preparing this with a calculator rather than a spreadsheet, this total would provide a check on your calculations.

SKILL 2

COMMON ERROR

Be sure that students use the correct divisor when calculating averages, and that they are careful to round correctly.

EXAMPLE 3 Her mother was so impressed with Sylvia's computer skills that she asked Sylvia to prepare a table in which to record the expenses for their trip.

QUESTION How can a spreadsheet be used to record travel expenses?

SOLUTION

Sylvia set up a budget as follows. Her mother suggests that her clients prepare a budget like this as a target, using their own numbers, and then take a blank form on the trip and fill in the actual amounts.

	A	B	C	D	E	F	G
1		Day 1	Day 2	Day 3	Day 4	Total	Average
2	Food	72.80	59.85	57.74	39.15	229.54	57.39
3	Lodging	56.70	56.70	28.65	31.50	173.55	43.39
4	Car	19.20	0.00	15.60	10.00	44.80	11.20
5	Entertainment	32.00	29.00	8.50	14.75	84.25	21.06
6	Gifts	0.00	13.00	21.68	8.90	43.58	10.90
7	Total	180.70	158.55	132.17	104.30	575.72	143.93

Sylvia's mother sponsors tours for people who like someone else to do the planning. Some of these trips include costs for travel, accommodations, meals, entrance to some attractions, and guided tours. Some costs are fixed, such as chartering a plane. Other costs, such as meal expenses, vary with the number of people on the tour. You can use break-even analysis to determine how many people are needed to make a profit.

EXAMPLE 4 Sylvia's mother has found that for a certain tour, the fixed cost is \$10,000, and the cost per person is \$150. She charges \$720 per person for the tour.

QUESTION How can Sylvia help her mother determine the number of people needed to break even?

SOLUTION

Graphing equations for income (revenue) and expenses (cost) will show profit and loss.

The equation for income is

$y = 720x$ where x = number of people
y = total revenue

The equation for expenses is

$y = 10{,}000 + 150x$ where $10{,}000$ = fixed expense
x = number of people
y = total expenses

Graph the equations with the following range values.

Xmin: −2	Ymin: −2500
Xmax: 30	Ymax: 20000
Xscl: 2	Yscl: 2500

X = 17.536842 Y = 12626.526

Using the trace function shows a break-even point at about 17.5.

Solve the system of two equations using substitution.

$$y = 720x$$
$$y = 10{,}000 + 150x$$
$$720x = 10{,}000 + 150x$$
$$570x = 10{,}000$$
$$x = 17.5$$ To the nearest tenth

To break even, Sylvia's mother will need 18 people on the tour.

CRITICAL THINKING

Ask students to write a paragraph about two meanings of the expression, "saving money." It can mean putting money away for future use, or it can mean buying an item less expensive than another. Ask how the second meaning can lead to problems.

INTERDISCIPLINARY INVESTIGATION

Economics: Have students research the costs associated with starting a small business, particularly a restaurant or inn. Ask them to find statistics on the number of new businesses that are successful and the reasons for success or failure.

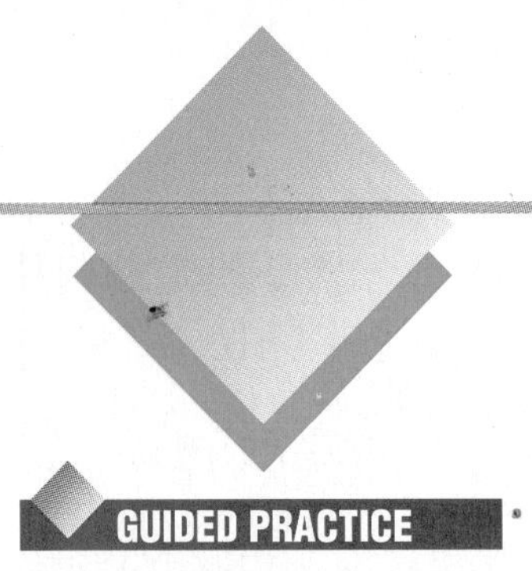

TRY YOUR SKILLS

GUIDED PRACTICE

The *Try Your Skills* exercises are structured to review the examples. Have students read them, review the examples as needed and ask questions before doing the work.

The Downtown Motor Inn charges the following: Single, $47; double, $57; tax, 6.5%.

1. Find the cost of a single for 1 night, with tax. $50.06
2. Find the cost of a double for 3 nights, with tax. $182.12
3. The table in Example 2 shows amounts spent for restaurant meals by Sylvia and her mother. Find the average amount per person for each meal. $4.73, $5.44, $18.53
4. On the basis of the information in Example 4, what will be Sylvia's mother's profit if 20 people purchase the tour? $1400

EXERCISE YOUR SKILLS

KEY TERM
reservation

The following summarizes the costs at two different hotels.

Milledgeville Lodge	Georgia Inn
Rates: single, $48	Rates: single, $39
2 adults, $56	2 adults, $48
Teens, free	Teens, $10
Tax: 8% on room only	Tax: 8% on room only
Free in-room movies	Video rental: $7.00
Game machines	Miniature golf, $3.00

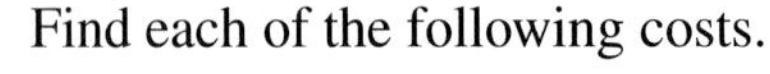

Find each of the following costs.

1. Two adults at the Milledgville Lodge for 3 nights $181.44
2. Single at the Georgia Inn for 2 nights, with 1 video rental $91.24
3. Two adults and 2 teens at the Georgia Inn for 4 nights, with 2 rounds of miniature golf per person and 2 video rentals $331.76

RETEACHING

It is always beneficial to review basic operations with decimals. As a check on comprehension, ask students to estimate the answers before they add the weekly expenses, or compute averages.

Use a calculator to find the totals for each day, the totals for each meal, and the average cost for each meal; each cost shown is for a family of four.

		Monday	Tuesday	Wednesday	Thursday	Total	Average
4.	Breakfast	$17.50	$22.75	$19.54	$23.32	$ 83.11	$ 20.78
5.	Lunch	28.75	24.50	15.27	17.43	85.95	21.49
6.	Dinner	78.45	32.60	58.36	98.63	268.04	67.01
7.	Total	124.70	79.85	93.17	139.38	437.10	109.28

ENRICHMENT

Have students find out the expenses of staying in several hotels or motels in your area, and also the costs of meals at several different kinds of restaurants.

8. How can you double check your totals? add row total and compare to column total
9. Using these averages and the rates listed above, prepare a food and lodging budget for a family of four (2 adults, 2 teens) to stay at the Milledgeville Lodge for 3 days. Answers may vary.

A travel agent prepares a tour package, charging \$560 per person. The tour has fixed costs of \$7500 and a per person cost of \$180.

10. Write an equation for the total revenue of the travel agency. $y = 560x$

11. Write an equation for the total expenses of the travel agency. $y = 7500 + 180x$

12. Set up a range and use a graphing calculator to graph the equations.

13. Use the graphs to find the break-even point for the tour. 20 people

14. Find the break-even point by solving the system of equations. 20 people

15. What will be the profit or loss if 15 people take the tour? loss: \$1800

16. What will be the profit or loss if 28 people take the tour? profit: \$3140

John and Harriet Gomez operate a restaurant. Their average costs per day are \$500 fixed, \$10 per person having dinner. The average amount spent by a person having dinner is \$20.

17. How many dinner patrons are needed for the Gomezes to break even? 50

18. What is the financial result of having only 35 dinner patrons? loss: \$150

19. How does the Gomezes' thinking about the price of dinner differ from the thinking of one of their patrons coming for dinner? See Additional Answers.

20. How might lowering the average price of dinner lead the Gomezes to earn less money from their restaurant?

21. How might lowering the average price of dinner lead the Gomezes to earn more money from their restaurant?

22. Without actually lowering the prices, how might the Gomezes determine whether this would be a good strategy?

A tour company arranges a bus tour. The advertised price is \$445 per person. The tour has fixed costs of \$9100 and a per person cost of \$225.

23. Write an equation for the amount paid in to the tour company.

24. Write an equation for the amount paid out by the tour company.

25. The number of people able to take the tour is limited by the fact that the bus has a maximum capacity of 45 passengers. Will the tour company be able to make a profit? Yes, if more than 41 people take the tour.

26. Suppose 50 people sign up for this tour. Accommodating the extra people will raise the fixed costs to \$12,350 and lower the cost per person to \$210. The price of the tour must remain at \$445 per person. Should the tour company accept reservations from 50 people, or maintain a limit of 45 people? Explain your answer. maintain the limit of 45 people because the break-even point for more than 45 people is 53 people

ASSIGNMENT GUIDE

Exercises 1-3 ask for answers based on lodging information similar to that given in Skill 1.

Exercises 4-7 require the completion of a table for meal expenses. Exercises 8 and 9 ask for the interpretation of a table.

Exercises 10-16, 17-22, and 23-26 provide scenarios and step-by-step questions about break-even.

CLOSURE

Ask students to explain why average meal costs are important to a consumer and why they are important to a restaurant owner.

Ask how a family might control expenses while on a vacation trip without making up a budget for every meal.

Ask how the slopes of the graphs for costs and sales relate to break-even.

LESSON QUIZ

1. The Murphy family budgets \$600 for lodging and meals for a 5-day vacation. They will have to pay \$350 for lodging. How much will they have per day for meals? (**\$50**)
2. Sam's Hotdog Stand sells only hotdogs. He pays \$30 a day to rent his stand. He buys each hotdog and roll for thirty-five cents and sells them for sixty cents each. How many must he sell to break even? (**120**)

MIXED REVIEW

1-1 **1.** Marsha's job at the library pays her $12.50 plus $1\frac{1}{2}$ times her regular wage rate for each hour over 35 hours. How much did she earn in a week during which she worked 42 hours? $568.75

1-2 **2.** Keisha earns $9.50 per hour. She works $37\frac{1}{2}$ hours per week for 49 weeks and received 3 weeks of paid vacation. Her other benefits cost her employer $4000 per year. What is the annual cost to the company of her salary and benefits? $22,525

2-1 Liam's bank pays interest of 0.25% per month on his checking account. It imposes a monthly service charge of $6 for balances under $200, $5 for balances between $200 and $499, and no service charge for balances over $499. For balances under $500 there is a charge of $0.02 for each check. Find the new balance in a checking account with the given balance and number of checks written.

3. $501; 22 checks $502.25 **4.** $400; 15 checks $395.70

ADDITIONAL ANSWERS

5. Loan **b**

a. $16,753.14

b. $18,608.78

5-3 **5.** Determine which of the following loans has a greater total cost.

a. $12,000 at 14% for 5 years **b.** $15,000 at 11% for 4 years

10-2 The annual depreciation on a new $20,000 car is 25% immediately upon purchase. At the end of each year that follows, the percent of depreciation from the original price changes as follows: 20%, 12%, 10%, 8%, 7%, 6%, 5%, 4%, 3%. Find the value of the car at the times indicated in Exercises 6–8.

6. Immediately after purchase $15,000

7. In 1 year $11,000 **8.** In 5 years $3600

3-2 Use the Rule of 72 to find how many years it will take for $3000 to grow to $24,000 at the given rates of interest.

9. 5.25% 42 years **10.** 7.5% 30 years **11.** 9% 24 years

8-1 José bought 1500 shares of stock at $43\frac{1}{8}$ per share. In Exercises 12–14 ignore the commission costs.

12. Find the cost of the shares. $64,687.50

13. After three years, José sold the shares at $57\frac{3}{8}$ per share. What was the total sale price? $86,062.50

14. What was the capital gain or loss on the shares? Gain: $21,375

5-1 A loan of $10,000 is repaid one year later in a single payment of $10,900.

15. What was the interest paid on the loan? $900

16. What was the rate of interest on the loan? 9%

11-3 FLYING SAVES TIME AND SOMETIMES MONEY

QUICK REFERENCE

TEACHER'S RESOURCES AND ANSWER KEY

Lesson Quiz Answers 11-3, p. 80

Reteaching Activity Answers 11-3, p. 112

Enrichment Activity Answers 11-3, p. 112

EXTENSION ACTIVITIES

Reteaching Activity 11-3, p. 85

Enrichment Activity 11-3, p. 88

Ramón, like Sylvia and Betty, has taken family vacations by car. He has always enjoyed visiting new places, especially those that played a part in history, such as the Alamo, and those that have great natural beauty, such as Niagara Falls. He loves art and frequently sketches what he sees.

Now Ramón goes to college and sometimes flies to school. For this reason he has become familiar with airline timetables and air fares. He lives in Chicago and goes to school in New Orleans. He occasionally visits friends in other parts of the country.

It takes some practice to read airline schedules. Comparing times and connections can be tricky. In addition, there is quite a range in air fares. Ramón, like many people, depends on a travel agent to make arrangements. But there are always several options, and planning ahead can often save time and money.

OBJECTIVES: *In this lesson, we will help Ramón to:*

- *Understand what is involved in air travel.*
- *Make travel plans based on airline schedules.*
- *Compare the costs of traveling by air and by car.*

FOCUS

ALGEBRA CONNECTIONS

Boston time is three hours later than Seattle. A plane leaves Boston for Seattle at 10 A.M. Boston time. At the same time a plane leaves Seattle on a 5-hour trip to Boston. In local time, when will the plane reach Boston? (**3 P.M.**)

MAKING AIR TRAVEL PLANS

Ramón likes to fly because it saves time. When school is out, he wants to get home. When he is home on vacation, he likes to stay until the last possible minute. Nevertheless, he has learned to plan ahead. He realizes that a lot of students go back to school at the same time, so it is important to buy a ticket early. This saves money and helps him get the most convenient flight.

As Ramón became more involved in travel matters, checking flight times and rates became a hobby. He was interested not just in his own trips, but also in how the airlines operate. He has learned to consider time zones. Airlines always list departure and arrival times according to local time. This is helpful because most people want to know the local time when they will arrive.

People also want to know the length of a trip. The airline schedule doesn't tell you this. You have to figure it out yourself, making sure to include the change in time from the place of departure to that of arrival. A **time zone** is a geographic area that has the same time. Dallas and Chicago are in the same time zone. Dallas and Los Angeles are in different time zones.

Ramón knows that several major airlines fly many times a day to major cities. Smaller cities are often connected by smaller regional airlines. As he started checking schedules, Ramón was surprised to learn how many ways there are to get from one place to another.

The **advance purchase** of a ticket, that is, buying a ticket 14–60 days ahead can sometimes save money. On a round trip, staying over Saturday night can also save money. One-way fares are usually priced higher than round-trip fares. Some tickets that are purchased under discount arrangements are **nonrefundable,** that is, you cannot get your money back or use the tickets on another flight. Tickets for which you pay full fare can be used on flights other than the one specified on the ticket. Sometimes an airline will have special rates to attract customers.

Ask Yourself

1. What is the obvious advantage of air travel?
   Answers may vary. (Ex.: faster than driving)
2. What is a disadvantage of air travel?
   Answers may vary.
3. Why is it important to check schedules and costs before taking a trip?
   Answers may vary.

ALGEBRA *REVIEW*

The following are distances along a line:

***A* to *B*, 300 miles**
***B* to *C*, 400 miles**
***C* to *D*, 1000 miles**

1. What is the least possible distance from *A* to *C*?
   100 mi
2. What is the greatest possible distance from *A* to *C*?
   700 mi
3. Find the distance from *B* to *D*, passing through *C*.
   1400 mi
4. Find the least possible distance from *D* to *B*.
   600 mi
5. Draw a diagram showing the least possible distance from *A* to *D*.
   See Additional Answers.
6. What is the least distance from *A* to *D*?
   300 mi
7. If the distance from *A* to *C* is 700 miles and the distance from *A* to *D* is 1700 miles, draw a diagram showing the arrangement of *A*, *B*, *C*, and *D*.
   See Additional Answers.

SHARPEN YOUR SKILLS

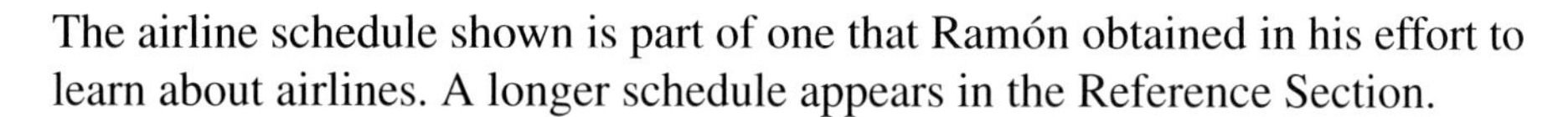

SKILL 1

The airline schedule shown is part of one that Ramón obtained in his effort to learn about airlines. A longer schedule appears in the Reference Section.

WINDWARD AIRLINES SCHEDULE

Leave	Arrive	Stops/Via	Rmks
From Chicago, IL			
To Muscle Shoals, AL			491 mi
1:00p0	5:10p	ATL	L
4:55p0	9:00p	ATL	D X6
To Myrtle Beach, SC			743 mi
8:10a0	1:18p	ATL	B
9:54a0	3:15p	ATL	S
Eff. May 15			
1:00p0	6:59p	ATL	L
6:44p0	11:45p	ATL	D X6
To Naples, FL			1136 mi
8:15a0	1:48p	MCO	B
2:10p0	7:25p	MCO	S
To Nashville, TN			401 mi
5:30a0	8:45a	CVG	S X67
6:15a0	9:24a	CVG	S X67
7:40aM	10:55a	CVG	X67
9:57a0	12:59p	CVG	X67
11:35aM	2:50p	CVG	X67
3:20pM	6:55p	CVG	X6
3:50p0	6:55p	CVG	
5:35p0	8:43p	CVG	X6
5:40pM	8:43p	CVG	X67
Eff. May 1			
To Nassau, Bah			1301 mi
8:10a0	1:55p	ATL	
To New Orleans, LA			831 mi
6:15a0	10:14a	CVG	B/S X67
8:10a0	12:11p	ATL	B
9:25aM	1:35p	CVG	L6
Eff. May 6			
9:57a0	1:35p	CVG	L
10:18a0	2:45p	DFW	X67
11:35a0	3:55p	ATL	L
1:00p0	5:20p	ATL	L
3:14p0	7:20p	ATL	S6
3:20pM	7:45p	CVG	DX6
3:50p0	7:45p	CVG	D
6:44p0	10:35p	ATL	D
8:45p0	12:59a	DFW	S
To New York, NY/ Newark, NJ			734 mi
L-LaGuardia; J-Kennedy; E-Newark			
7:40aM	12:25pL	CVG	X67
11:35aM	4:40pL	CVG	X67
11:35aM	4:55pE	CVG	X67
3:20pM	7:45pE	CVG	D X6
5:35p0	10:20pL	CVG	
5:40pM	10:20pL	CVG	X67
Eff. May 1			
To Norfolk/Virginia Beach/ Williamsburg, VA			707 mi
6:30a0	12:29p	ATL	B X7
9:54a0	3:10p	ATL	S
11:35aM	3:55p	CVG	X67
1:00p0	6:25p	ATL	L
5:35p0	9:40p	CVG	
5:40pM	9:40p	CVG	X67
To Oakland, CA			1843 mi
Also see San Francisco, San Jose			
8:00a0	12:05p	SLC	B
11:45a0	3:45p	SLC	
3:10p0	7:10p	SLC	D
6:15p0	9:55p	SLC	
To Oklahoma City, OK			695 mi
8:30a0	12:35p	DFW	B
10:18a0	2:15p	DFW	X67
12:10p0	4:15p	DFW	X7
3:50p0	9:25p	DFW	S
8:45p0	12:35a	DFW	S
To Ontario, CA			1707 mi
Also see Los Angeles, Burbank, Long Beach and Orange County			
8:00a0	11:55a	SLC	B
8:30a0	1:10p	DFW	
11:45a0	3:25p	SLC	
12:10p0	4:20p	DFW	X7
3:50p0	8:10p	DFW	
6:15p0	9:45p	1	
To Orange County, CA			1732 mi
Also see Los Angeles, Burbank, Long Beach and Ontario			
6:15a0	12:05p	CVG	X67
8:00a0	12:05p	SLC	B
8:30a0	1:15p	DFW	
9:25aM	3:40p	CVG	6
Eff. May 6			
11:45a0	3:40p	SLC	
3:10p0	7:35p	SLC	S
3:50p0	8:10p	DFW	
To Orlando, FL			995 mi
5:30a0	10:00a	CVG	B X67
8:15a0	11:42a		B
9:25aM	2:35p	CVG	L6
Eff. May 6			
9:45a0	2:54p	ATL	S
9:57a0	2:35p	CVG	L

Ramón has learned that each commercial airport in the world is assigned a three-letter code. Usually, the code makes it easy to recognize the name of the city or airport that it represents. However, sometimes the reference is not so obvious. Following are codes from the schedule and the cities that they represent. In the schedule, the codes refer to places at which the flight stops on its way to the final destination. A flight that does not make any stops is called a **non-stop flight.** A **connection** is a stop at which you must change planes in the course of your trip. If your flight is a **direct flight,** the plane will have one or more intermediate stops but you will not change planes.

INSTRUCTION

TEACHING THE LESSON

Remind students about time zones. Picturing the United States on a map or globe can help. The earth turns from the west toward the east. Morning comes first to the east coast; when dawn comes to the west it is later in the east. When it is 5 A.M. in California, it is 8 A.M. in Virginia.

Airline schedules can be tricky to read. Explain to students that flight times and other information are usually given and explained by a travel agent.

Skill 3 covers the practical decision of whether flying or driving is most economical under particular conditions.

FOCUS ON ALGEBRA

Students are now in the final chapters of the text. They should have increased their proficiency in reading mathematics and interpreting mathematical situations. *(NCTM Standard 2, p. 140)*

ATL	Atlanta, Georgia	CVG	Cincinnati, Ohio
DFW	Dallas/Ft. Worth, Texas	MCO	Orlando, Florida
SLC	Salt Lake City, Utah		

Other symbols in the schedule and their meanings are as follows.

B, L, S, D	Breakfast, Lunch, Snack, Dinner
X6 and X7	The flight travels every day except 6 (Saturday) and 7 (Sunday).

EXAMPLE 1 Ramón is making travel arrangements to go back to school. He wants information about flights from Chicago, Illinois, to New Orleans, Louisiana. Chicago has two airports: O'Hare (O) and Midway (M).

QUESTIONS

1. Which flight will get him to New Orleans the earliest in the day? At what time and from which Chicago airport does the flight leave? Where does it stop along the way?
2. Find the same information for the latest flight he can take.
3. About how long is the time for the trip?

SOLUTIONS

By examining the schedule we obtain the information.

1. The flight with the earliest arrival time leaves O'Hare at 6:15 A.M. and arrives in New Orleans at 10:14 A.M. It passes through Cincinnati. This flight does not travel on Saturday or Sunday.
2. The latest flight leaves O'Hare at 8:45 P.M. and arrives in New Orleans at 12:59 A.M. the following morning. It stops at the Dallas/Ft. Worth airport.
3. Calculating several of the times shows that the average length of the trip is about 4 hours. Since Chicago and New Orleans are both on Central Time, it is not necessary to adjust the time as calculated from the schedule.

SKILL 2

You cannot always calculate the length of a flight directly from the departure and arrival times. Use the time zone map as needed to determine the duration or length of flights.

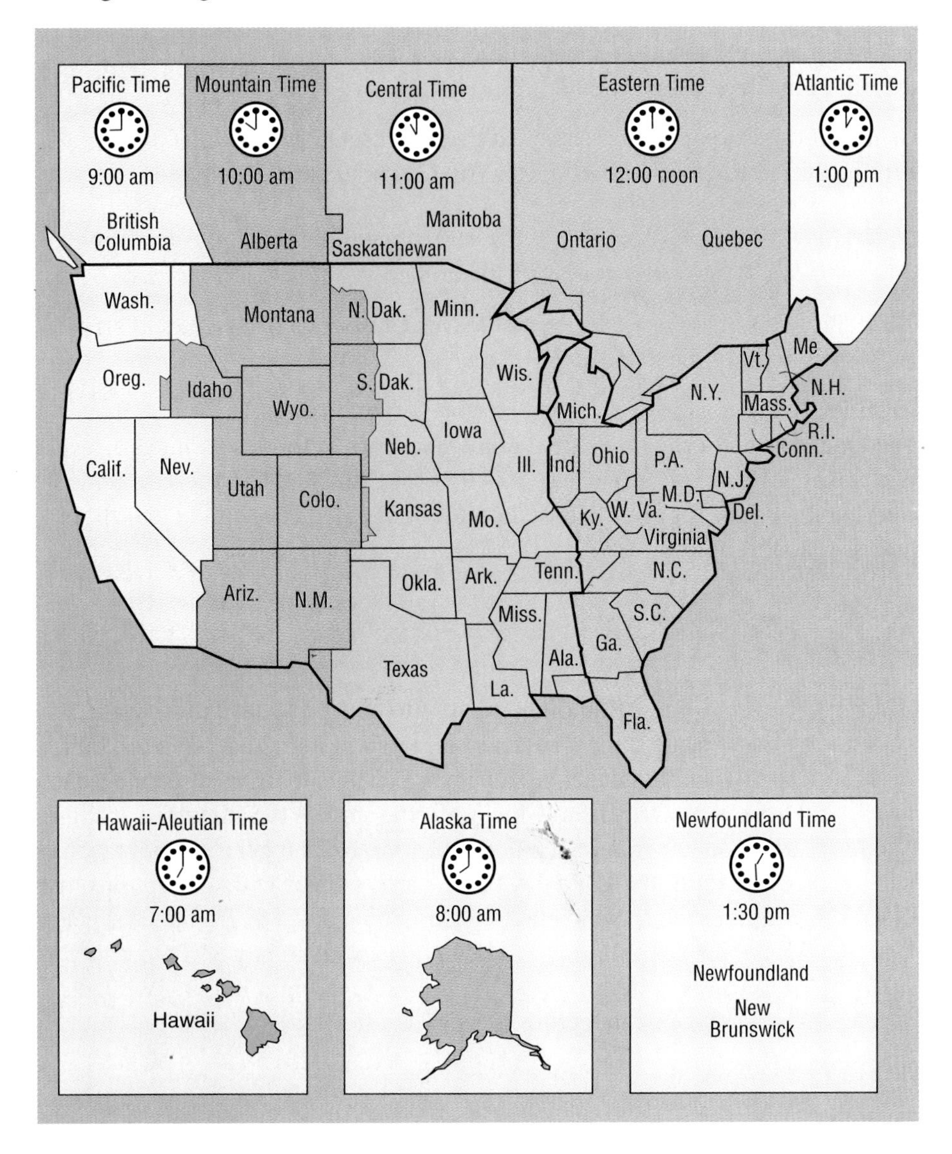

EXAMPLE 2 Ramón wants to find out how to adjust the travel time for flights traveling from west to east.

QUESTION How long is the first flight listed from Chicago to Norfolk/Virginia Beach?

CRITICAL THINKING

Ask students to consider whether, within a time zone, there might be a significant difference in the amount of daylight in the morning or evening. (**Although the time throughout a zone is the same, the sun reaches the west side later. Therefore, people on the west side of a time zone have nearly an hour less light in the morning and an hour more light in the evening.**)

AT-RISK STUDENTS

Some students without a strong inclination toward mathematics can be motivated in the lesson by appealing to their interest in airplanes and air travel.

INTERDISCIPLINARY INVESTIGATION

Geography: Obtain maps of airline routes. Have students select a country and report back on the flying times, time changes, as well as tourist and business travel to that country.

RETEACHING

If reteaching is needed, begin with a time-zone chart. Quiz students on the times in different parts of the world. Then use flights with whole-number times to discuss arrival times. Finally, move to more complex situations.

ENRICHMENT

Pose the following problem to interested students: The circumference of Earth at the equator is approximately 25,000 miles. Suppose that you could fly from east to west above the equator at a speed of a little more than 1000 miles per hour. Write a report about your experiences as you cross time zones. (**The time of day will remain the same. When you cross the international date line it will be one day later.**)

SOLUTION

To determine the effect of time change on the length of flight, it is necessary first to find out the time zones of the various cities. Use the time zone map to do this.

Chicago is in the Central Time Zone. Norfolk is in the Eastern Time Zone. The clocks in Norfolk are an hour later than those in Chicago. To make the arrival time correspond to Chicago time, subtract 1 hour from the length of the trip according to the schedule.

Using Chicago time, you can see that the trip lasts from 6:30 A.M. until 11:29 A.M. Subtracting 1 hour gives a trip of 4 hours and 59 minutes.

EXAMPLE 3 Ramón now wants to work with travel time for flights traveling west from one time zone to another.

QUESTION How long is the flight that leaves Chicago at 3:10 P.M. and arrives in San Francisco at 7:10 P.M.?

SOLUTION

The clocks in San Francisco are 2 hours earlier than those in Chicago. When it is 7:10 in San Francisco, it is 9:10 in Chicago. Therefore you must add 2 hours to the flight time based on the schedule. The total actual flight time is 6 hours.

SKILL 3

EXAMPLE 4 Several times, Ramón has driven to college to save money and also to bring things, such as his stereo system, that were too bulky to take on the plane. Now he has a chance to drive someone's car from New Orleans to Chicago at the start of the summer vacation. He wonders how much it would save. He can get a special rate flight for $300. Each day that he arrives later will keep him from a summer job at which he will earn $65 a day. His friend's car gets 20 miles per gallon of gas. Ramón expects to pay $1.25 per gallon for gas. He plans to drive for 2 days. Overnight lodging will cost $40 and he is budgeting $25 per day for food.

QUESTION How much will he save by driving?

SOLUTION

From the mileage chart in Lesson 11–1, Ramón finds the distance from New Orleans to Chicago is 912 miles. Driving 912 miles will take $912 \div 20 = 45.6$ gallons of gas.

Cost of gas: $1.25(45.6) = \$57$
Lodging for one night: $40
Meals for two days: $2(25) = \$50$
Total cost of driving: $57 + 40 + 50 = \$147$

Since the cost of flying is \$300, it seems that Ramón will save 300 − 147 = \$153 by driving. However, he will lose two days pay which is 2(65) = \$130. His actual cost of driving is 147 + 130 = \$277. So he will save only 300 − 277 = \$23. Ramón decides saving \$23 is not worth a tiring trip by car.

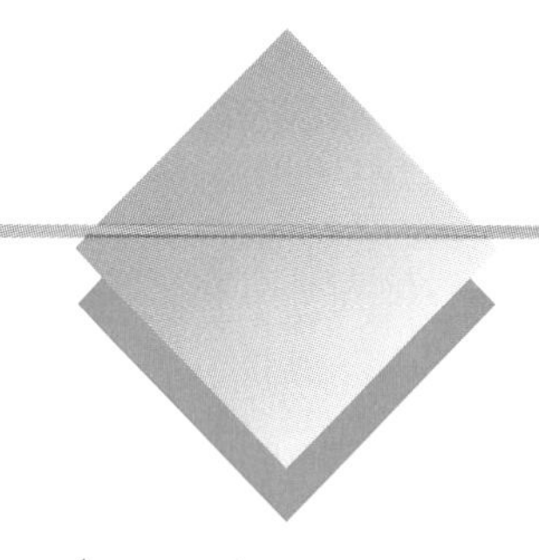

TRY YOUR SKILLS

Ramón plans to take a flight to visit a friend in Orlando, Florida. His friend can pick him up at about noon. Ramón's ticket costs \$625. Use the Windward Airlines Schedule in the Reference Section.

1. At what time does the most convenient flight leave Chicago? 8:15 A.M.
2. From what airport does it leave? O'Hare
3. Is it a direct flight? Yes
4. How long is the flight? 2 h, 27 min
5. Does it seem likely that Ramón could save money by driving? Why or why not? \$625 is an expensive flight, so Ramón would likely save money by driving, given he has a car to use.

GUIDED PRACTICE

The *Try Your Skills* exercises might be done cooperatively by students with two or three to a group.

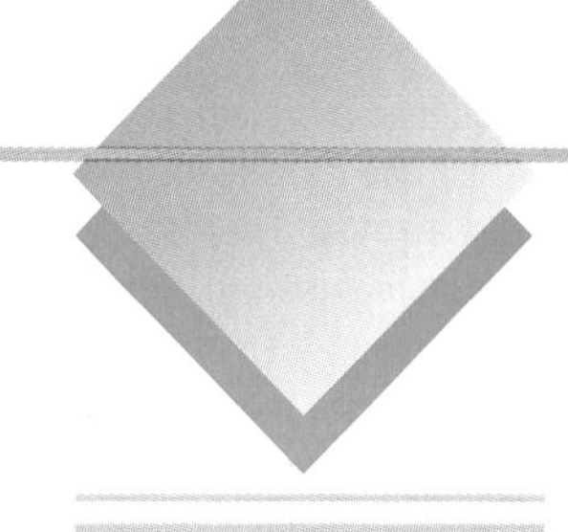

EXERCISE YOUR SKILLS

Ramón's sister is planning a trip from Chicago to Williamsburg, Virginia. She wants to arrive between 2:00 and 4:00 in the afternoon. Use the Windward Airlines Schedule in the Reference Section.

1. Which departure times could she take? 9:54 A.M. or 11:35 A.M.
2. From which airport does the earlier flight leave? O'Hare
3. From which airport does the later flight leave? Midway

KEY TERMS

advance purchase
connection
direct flight
nonrefundable
non-stop flight
time zone

INDEPENDENT PRACTICE

ASSIGNMENT GUIDE

Exercises 1-6 and 10-16 require reading a flight schedule and some interpretation.

Exercises 7-9 ask about the airline procedures and comparative costs.

Exercises 17-22 offer practice in reading time zones.

Exercises 23-26 compare air and car travel.

CLOSURE

Ask students to explain why the departure and arrival times for flights are not sufficient to calculate the time of the trip. Have them document their answers with reference to the airline schedules.

LESSON QUIZ

1. A flight leaves Chicago at 3:00 P.M. and arrives at San Diego at 5:00 P.M., local time. How long was the flight? (**4 hours**)
2. Why are "per mile costs" not used for air fares? (**There are many different fares for the same distance, depending on time of trip and other factors. An average is not dependable.**)

4. How long is each flight? 4 h 11 min; 3 h 20 min
5. Why might she choose the first flight? See Additional Answers.
6. Why might she choose the second flight? It takes less time.
7. If you want to compare the cost of air travel with car travel, what are some of the costs that must be included in each? See Additional Answers.
8. Why do you think airlines often charge people less if they pay for a particular flight weeks or months in advance? It ensures that the flight will be full.
9. What is one disadvantage of buying an airline ticket months in advance?

Use the Windward Airlines Schedule in the Reference Section to do the following.

10. Find the shortest flight from Chicago to San Francisco. 6:15 P.M.
11. Find the earliest flight from Chicago to San Diego that goes through Dallas/Ft. Worth. 8:30 A.M.
12. Find the latest flight from Chicago to Portland, Oregon. 6:15 P.M.
13. Find the shortest flight from Chicago to Ontario, California. 6:15 P.M.
14. Estimate the average flight time from Chicago to Phoenix. 282 min
15. Estimate the average flight time from Chicago to San Diego. 381 min
16. What is the air mileage for a trip from Chicago to Palm Springs, California? 1658 mi

Name the time zone in which each city is located.

17. Nashville, Tennessee Central
18. Newark, New Jersey Easterm
19. El Paso, Texas Mountain
20. Oakland, California Pacific
21. Denver, Colorado Mountain
22. Boston, Massachusetts Eastern

Ramón's friend Michael is planning a trip from Chicago to Portland, Oregon. Use the information in the Reference Section as needed. If he drives, he expects that it will take 4 days. He will have to spend $70 a day for meals and lodging. His car gets 20 miles to the gallon of gas, and he expects to pay an average of $1.30 per gallon. He can get an airline ticket for $780.

23. What is the air mileage for the trip? 1749 mi
24. What is the driving mileage for the trip? 2083 mi
25. What will be the approximate cost of the trip by car? $415.40
26. Does he save by traveling by car or flying? How much? by car; $364.60

ADDITIONAL ANSWERS

5. If she's traveling on the weekend, the second flight is not available.
7. Car travel costs include gas, lodging and meals; air travel costs include ticket and transportation to and from airport.
9. The ticket is probably not refundable so you can't change your plans.

MIXED REVIEW

6-4 Your credit card carries a monthly finance charge of 1.25%. You pay 10% of the amount owed to the nearest dollar each month. Your beginning balance is $1676. Complete the following table for 3 months. Assume that you make no further purchases during that time.

	Month	Balance	Interest	Amount Owed	Payment
1.	1	$1676	$20.95	$1696.95	$170
2.	2	1526.95	19.09	1546.04	155
3.	3	1391.04	17.39	1408.43	141

9-1 Use the Tax Table in the Reference Section to determine the tax owed by each of the following taxpayers. See Additional Answers.

4. A married person, filing separately, with $45,750 in taxable income

5. A head of household with $72,000 in taxable income

ADDITIONAL ANSWERS

4. $10,566

5. $16,430

10-3 Use the information in the table below to find the distance traveled, the miles per gallon, and the cost per mile of the trip.

	Odometer		Gallons of Gas	Cost of Gas	Distance (miles)	Miles per Gallon	Cost per Mile
	First	Last					
6.	26,580	27,000	21	$24.50	420	20	$0.058
7.	250	866	19.5	27.56	616	31.6	0.045
8.	86,290	87,326	51.6	49.75	1036	20.1	0.048

1-3 **9.** Cheryl prepares flower arrangements for Niven's Flower Store. She receives $7.50 for each budget arrangement, $12.25 for each regular arrangement, and $18.50 for each deluxe arrangement. How much did she earn in a week in which she assembled 30 budget arrangements, 15 regular arrangements, and 12 deluxe arrangements? $630.75

1-3 **10.** Myrna sells annual memberships in an exercise club. Her boss has offered her a choice between Plan 1, no salary but a straight $40 for each membership sold, and Plan 2, a weekly base salary of $200 plus $30 for each membership sold. How many memberships would Myrna have to sell to make the same amount of money under either plan? 20 memberships

4-1 **11.** Joseph's annual salary is $68,250. How much does Joseph pay for FICA taxes each year? $4560.83

4-1 **12.** Heidi's annual salary is $137,500. How much does Heidi pay for FICA taxes each year? $5528.70

4-1 **13.** Oshi's annual salary is $52,500. How much does Oshi pay for FICA taxes each year? $4016.25

CHAPTER 11 REVIEW

1. What is an advantage in traveling by interstate highways? faster
2. What is an advantage in traveling by secondary roads? more scenic, interesting
3. Why should plans for travel by car be somewhat flexible? unexpected events
4. What is an advantage of air travel over travel by car? faster
5. What is an advantage in traveling by car? often less expensive, more flexible
6. In addition to times of departure and arrival, what must you check when calculating the length of time for a flight? time zones

Use the U.S. interstate map in the Reference Section to identify which routes you would take to travel from Dallas, Texas, to the following cities. Answers may vary. See Additional Answers.

7. New Orleans, Louisiana
8. Richmond, Virginia
9. Boston, Massachusetts
10. Little Rock, Arkansas
11. Chicago, Illinois
12. Kansas City, Missouri

ADDITIONAL ANSWERS

7. 20 east, 55 south, 10 east
8. 20 east, 95 northeast
9. 20 east, 95 northeast
10. 30 northeast
11. 30 northeast, 40 east, 55 north, 57 north
12. 30 northeast, 40 east, 55 north, 70 west
17. \$81.14, \$104.59, \$89.35, \$41.22

Use the mileage chart in the Reference Section to find the distances by road between the following.

13. Austin, Texas, and Detroit, Michigan 1315 mi
14. Cincinnati, Ohio, and Flagstaff, Arizona 1695 mi
15. Kansas City, Missouri, and Miami, Florida 1448 mi
16. Louisville, Kentucky, and Philadelphia, Pennsylvania 668 mi
17. Find the cost of traveling the distances for Exercises 13–16 if you get 23.5 miles per gallon and gas costs an average of \$1.45 per gallon.
18. A family of five wants to share a room at the Honolulu Royale. The rates are \$86.00 for two people, \$10 for each additional person, and \$4.00 for a rollaway bed; the tax is 9.6%. Assuming that they need one rollaway bed, how much will it cost the family to stay for 5 nights? \$657.60

Use the following for Exercises 19–21.

ATLANTA PLAZA Rates: 1 pers, \$65–83; 2 people, dbl, \$73–91
Tax 8%, Addit. Pers \$8. Children free. Rollaway \$6

19. What is the least that it would cost two adults and one child to stay for 4 nights, assuming that each has his or her own bed? \$341.28
20. What is the least that it would cost one person to stay for 2 nights? \$140.40
21. What is the most that it would cost two people to stay 2 nights? \$196.56

Fnd the total amount spent each day, the total for each meal, and the average cost of a meal.

		Monday	Tuesday	Wednesday	Thursday	Total	Average
22.	Breakfast	$15.20	$22.75	$12.28	$23.78	$ 74.01	$ 18.50
23.	Lunch	25.47	30.80	11.45	17.56	85.28	21.32
24.	Dinner	89.63	71.29	67.40	38.59	266.91	66.73
25.	Total	130.30	124.84	91.13	79.93	426.20	106.55

Harvey's Tours arranges guided tours to different parts of the country. The 4-day bus tour of New England has a fixed cost of $4200 and a per person cost of $160. The charge for the tour is $450 per person.

26. Write an equation for the expenses of the tour. See Additional Answers.

27. Write an equation for the revenue received by Harvey's Tours.

28. Graph the equations. See Additional Answers.

29. Find the approximate break-even point. 15 people

30. Find the break-even point algebraically. 15 people

31. How much profit will Harvey's make if 20 people go on the tour? $1600

ADDITIONAL ANSWERS

26. $y = 4200 + 160x$, where $x =$ the number of people and $y =$ the total expenses

27. $y = 450x$, where $x =$ the number of people and $y =$ the total revenue

Find the time zone for each city.

32. Washington, D.C. eastern

33. Eugene, Oregon pacific

34. Butte, Montana mountain

35. Tulsa, Oklahoma central

Use the Windward Airlines Schedule in the Reference Section to do the following.

36. Estimate the average length of a trip from Chicago to San Antonio. 241 min

37. Find the departure time of the earliest flight from Chicago to Oklahoma City, Oklahoma. 8:30 A.M.

38. Find the arrival time of the latest flight from Chicago to Roanoke, Virginia. 9:55 P.M.

39. Find the length of the first flight from Chicago to Raleigh/Durham, North Carolina. 230 min

Consuela is planning a trip from Chicago to San Antonio, Texas. Use the information in the Reference Section as needed. If she drives, she expects that it will take 3 days. She will have to spend $80 a day for meals and hotels. Her car gets 22 miles to the gallon of gas, and she expects to pay an average of $1.35 per gallon. She can get an airline ticket for $660.

40. What is the air mileage for the trip? 1040 mi

41. What is the driving mileage for the trip? 1187 mi

42. What will be the approximate cost of the trip by car? $312.84

43. About how much money does she save traveling by car? $347.16

CHAPTER 11 TEST

1. What are two possible causes of delays in traveling by car?
2. Why should travelers estimate costs before starting on a car trip?
3. Use the interstate map to plan the most direct route from Indianapolis, Indiana, to Yellowstone National Park in Wyoming.

Complete the table to find the number of gallons and total cost.

	Miles	Miles per Gallon	Number of Gallons	Cost per Gallon	Total Cost
4.	1768	22.6	78.23	$1.47	$115.00
5.	2591	25.6	101.21	1.29	130.56

Use the following descriptions to answer Exercises 6 and 7.

HUMPHREY LODGE
Rates: Double $56
Addit. Pers: $8
Tax: 9%

McCARTHY INN
Rates: Double $86
Children free
Tax: 8%

6. Find the cost of three adults sharing a room at the Humphrey Lodge for 2 nights. $139.52
7. Find the cost of two adults and two children for 3 nights at the McCarthy Inn. $278.64

Alex and Alicia run nature tours to different parts of the country. Three days in the Everglades, leaving from Miami, cost them a flat rate of $1500 and $125 per person. The charge per person is $225.

8. Write an equation for the cost of the tour to Alex and Alicia. See Additional Answers.
9. Write an equation for the income for the tour.
10. Graph the equations. See Additional Answers.
11. Use the graph to find the break-even point for the tour. 15 people
12. Find the break-even point algebraically. 15 people
13. What is the amount of profit or loss if ten people go on the tour? loss: $500

Use the Windward Airlines schedule in the Reference Section to do the following.

14. Find the length of time of the first flight from Chicago to Myrtle Beach.
15. Find the length of time of the last flight from Chicago to Norfolk, Virginia.

ALTERNATIVE ASSESSMENT

Have students plan a class trip to another city. They should include the length of the trip, and expenses for transportation, meals, and lodging.

Have students obtain flight information and tourist information about a destination of their choice. Have them make a list of the questions they would ask a travel agent, and the follow-up questions they would have ready depending on the answers. But do not have them actually contact a travel agent.

ADDITIONAL ANSWERS

1. Answers may vary. (Ex.: heavy traffic, car repairs)
2. So they can budget their money to last throughout the trip.
3. 65 northwest, 90 west
8. $y = 1500 + 125x$, where $x =$ the number of people and $y =$ the total cost
9. $y = 225x$, where $x =$ the number of people and $y =$ the total income
14. 3 h, 8 min
15. 3 h

CUMULATIVE REVIEW

1-1 Assume that 40 hours = 1 workweek, 52 weeks = 1 year, and 12 months = 1 year. Find the hourly, weekly, monthly, and yearly salaries for the following pay rates.

1. $17,680 per year **2.** $2000 per month **3.** $5.50 an hour

2-3 **4.** Your checkbook shows a balance of $178.64. You have outstanding checks that total $53.21. The bank statement shows a balance of $229.85. Find the amount of a bank fee that has not yet been recorded in your checkbook. $2.00

3-2 You have a savings account that contains $6000. Your bank pays interest at an annual rate of 5.5% compounded as given. Find the value of the account at the end of 3 years.

5. annually $7045.45 **6.** quarterly $7068.41

4-3 Jackie and Joshua have gone into business selling doughnuts. They charge 75¢ for each doughnut. Their fixed costs are $400. Each doughnut costs $0.36.

7. How many doughnuts must they sell before they begin to make a profit?

8. How much profit will they have after selling 1500 doughnuts? $185

6-1 **9.** The previous balance of a credit card statement is $930. The monthly interest rate is 1.5% and you pay $125 plus all of the interest that has accumulated for that month. What is the new balance? $805

6-6 **10.** Use the Credit Scoring Table on page 293 of Lesson 6-6 to find the probability that a person with a score of 115 will repay a loan. 0.95

7-2 An insurance company expects 99.804% of all 36-year-old people to live at least one more year. The direct and indirect expenses for issuing one policy are $25.

11. Find the break-even premium for a 1-year $150,000 term insurance policy on a 36-year-old person. $319

12. What profit does the company expect to make on a $200,000 1-year term policy for a 36-year-old person who is charged with a premium of $550?

8-1 **13.** How many bonds can you buy with $20,000 given that the price of the bond is quoted at $101\frac{3}{4}$, that is, at $1017.50 for each bond? Ignore the effect of commissions. 19

8-2 **14.** Find the loss incurred in buying 100 shares of stock at $63\frac{3}{8}$ and then selling them 1 year later at $57\frac{5}{8}$. The commission rate on each transaction is 2%.

9-1 **15.** Use the Tax Tables in the Reference Section to find the tax to be paid by a single person who has $29,760 of taxable income. $5549

10-1 A car is purchased by a dealer for $12,600 and sold for $16,000.

16. Find the markup. $3400 **17.** Find the percent of markup. 27%

10-2 **18.** If a used car is purchased for $8000 and depreciates in a straight line for 8 years, what is its value after 3 years? $5000

ADDITIONAL ANSWERS

1. $8.50/h; $340/wk; $1473.33/mo
2. $11.54/h; $461.54/wk; $24,000/y
3. $220/wk; $953.33/mo; $11,440/y
7. 1026
12. $133
14. $586.50

PROJECT 11–1: **Planning a Camping Trip**

Work with a group of classmates to plan a 2-week camping trip for a family of four.

1. Decide where you want to go and what sights you want to see.
2. Plan a driving route to include campsites along the way.
3. Estimate the cost of gasoline, lodging, and food for the trip, including the cost of camping at the sites you choose.
4. Find the approximate cost of renting a bus with a driver for a camping trip similar to the one you planned above.
5. Estimate the fixed and variable costs.
6. Set a fair per-person price for the trip.
7. Write cost and revenue equations.
8. Graph the cost and revenue equations.
9. Find the break-even point for the trip.

PROJECT 11–2: **Planning an Airline Trip**

Work with a group of classmates to find bargain air fares.

1. Have each person choose five cities to visit.
2. Get prices from various airlines and travel agents.
3. Find the savings achieved by purchasing in advance and by staying particular lengths of time.
4. Have individuals present their findings and discuss different arrangements with the group.
5. Prepare a list of tips to travelers about how to save money on air fare.

ALGEBRA *REFRESHER*

The algebraic expressions x, $x^2 + x$, and $2x^2 - 3x + 1$ are called *polynomials.* Polynomials can be multiplied by using rules such as the distributive property.

Multiply.

Example $(x - 1)(x - 2)$

$$(x - 1)(x - 2) = (x - 1)x - (x - 1)2 \quad \text{Distributive property}$$
$$= x^2 - x - 2x + 2 \quad \text{Distributive property}$$
$$= x^2 - 3x + 2 \quad \text{Collect like terms.}$$

1. $(x - 1)(x + 2)$
$x^2 + x - 2$

2. $(x + 3)(x - 2)$
$x^2 + x - 6$

3. $(x + 3)(x + 3)$
$x^2 + 6x + 9$

4. $(x - 2)(x + 2)$
$x^2 - 4$

5. $(x + 1)(x - 5)$
$x^2 - 4x - 5$

6. $(x - 3)(x - 2)$
$x^2 - 5x + 6$

Notice in your answers to Exercises 1–6 that the third term is equal to the product of the two constant terms, and the coefficient of the x term is equal to the sum of the two constant terms. You can use this pattern to factor some polynomials, that is, to express them as the product of two binomials.

Factor.

Example $x^2 - x - 2$

$$x^2 - x - 2 = (x + 1)(x - 2)$$

The constant terms required must have a product of -2 and a sum of -1. Two numbers with these properties are 1 and -2.

7. $x^2 + x - 2$
$(x - 1)(x + 2)$

8. $x^2 + 5x + 6$
$(x + 3)(x + 2)$

9. $x^2 - 4x + 3$
$(x - 1)(x - 3)$

10. $x^2 + 2x + 1$
$(x + 1)(x + 1)$

11. $x^2 + x - 20$
$(x + 5)(x - 4)$

12. $x^2 - 1$
$(x - 1)(x + 1)$

The method of factoring can be used to find the points at which the graph of a quadratic function crosses the x-axis. By setting the value of the function equal to 0 and factoring you can find the x values for which the value of the function (y value), is 0.

Find the x values that make the function equal 0.

Example: $f(x) = x^2 - 6x$

$$x^2 - 6x = 0 \quad \text{Set the function equal to 0.}$$
$$x(x - 6) = 0 \quad \text{Factor.}$$
$$x = 0 \text{ or } x - 6 = 0$$
$$x = 0 \text{ or } x = 6$$

13. $y = x^2 + 5x + 4$
$x = -4$ or $x = -1$

14. $f(x) = x^2 + 2x - 3$
$x = 1$ or $x = -3$

15. $f(x) = x^2 - x - 6$
$x = 3$ or $x = -2$

16. $y = x^2 - 9$
$x = 3$ or $x = -3$

CHAPTER 12

HOME OWNERSHIP

The American Dream

THE SELF-CONTAINED SINGLE-FAMILY home with its nuclear family unit—Mom, Dad, two children, and a dog—is portrayed as every middle-class American's fondest wish. The fact that not every family even wants children, a dog, or a house of its own is overlooked when the message is broadcast. Of the families that do wish to pursue this "American dream," many will find home ownership to be expensive and laden with responsibilities. Even if they can afford to buy a house, there is always something in it that needs maintenance, repair, or updating.

Because many families still want to own a house no matter what the cost, they must turn to banks and other lending institutions for financing. At a time when a single-family, three-bedroom house in a moderate neighborhood carries a price tag of over $100,000 in many cities, most buyers must obtain a mortgage to be able to afford to buy a house.

The teenagers in this chapter will examine the process of buying a house and some of the many responsibilities associated with home ownership.

Vernon's family decided to buy a house when they moved from Chicago to Dallas. They considered the advantages of buying as compared with renting and the differences between new and older houses.

When Olivia's family is thinking about moving, she becomes aware of all the costs associated with buying a house, such as the fees and points. She also learns about making a down payment and the total financed price.

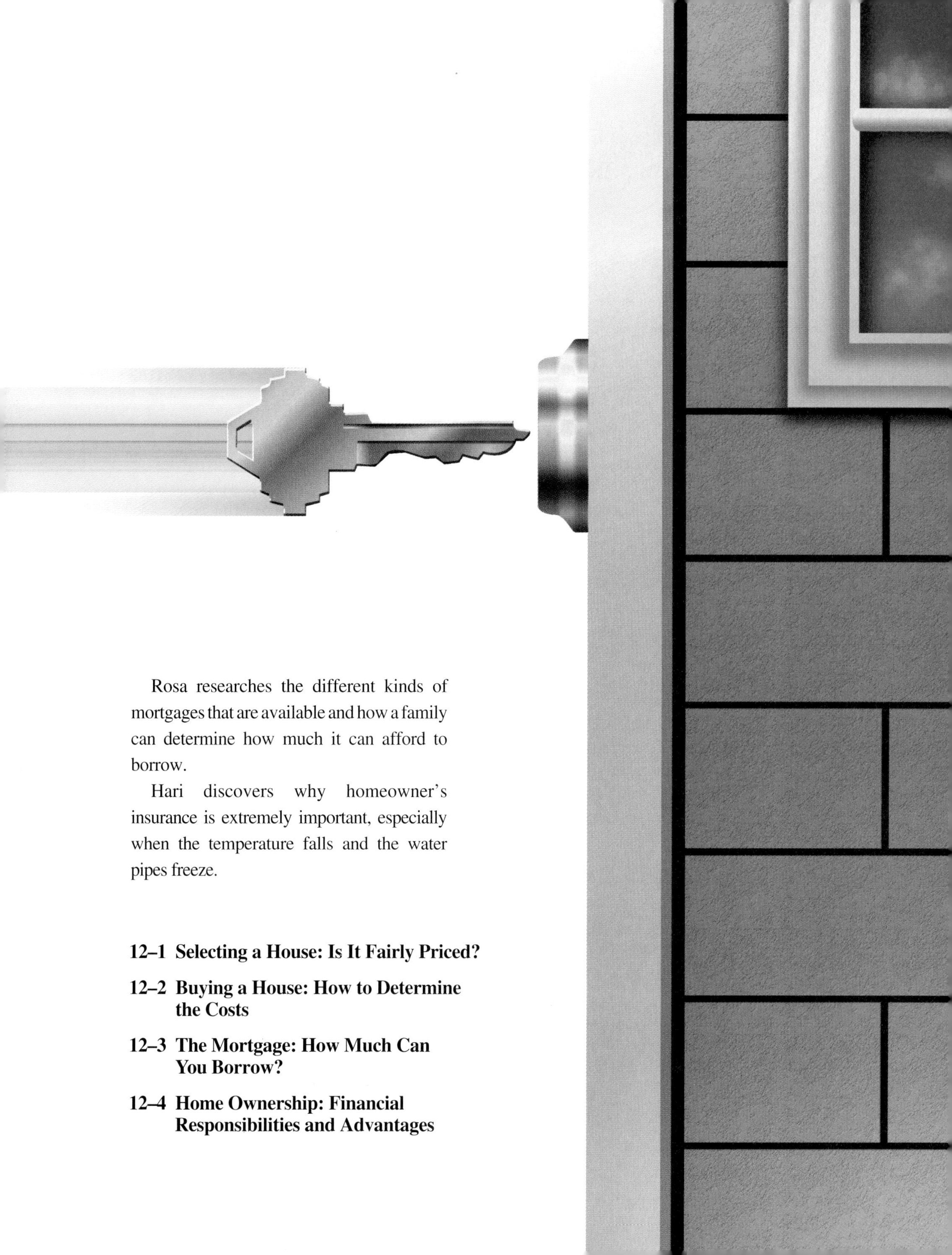

Rosa researches the different kinds of mortgages that are available and how a family can determine how much it can afford to borrow.

Hari discovers why homeowner's insurance is extremely important, especially when the temperature falls and the water pipes freeze.

12-1 SELECTING A HOUSE: IS IT FAIRLY PRICED?

QUICK REFERENCE

TEACHER'S RESOURCES AND ANSWER KEY

Lesson Quiz Answers 12-1, p. 82

Reteaching Activity Answers 12-1, p. 113

Enrichment Activity Answers 12-1, p. 114

EXTENSION ACTIVITIES

Reteaching Activity 12-1, p. 89

Enrichment Activity 12-1, p. 93

When Vernon's family moved to Dallas from Chicago, where they had lived in a three-bedroom apartment, they decided to consider buying a house. They were moving because Vernon's mother was transferred to Dallas by her company. Vernon's father, who is an accountant, was confident that he could find another accounting job in their new city. It seemed unlikely that the company would move Vernon's mother again soon, so they thought that it was an excellent time to buy a house.

Vernon was very excited about having his own room in a new house. He had shared a bedroom with his younger brother Kev since their sister needed a room of her own. The brothers never liked sharing the bedroom. Kev always complained about Vernon's choice of music, so Vernon could not play his CD player while Kev was studying. Vernon finally bought a pair of headphones to solve that problem. The neighbors were grateful, too!

Vernon also liked the idea that they might have space for a real recreation room. It would allow him to have more than one friend at a time over to his house. His mother and father were pleased that they could decide what kinds of carpeting, drapes, and appliances they would have. They also liked controlling the temperature themselves. In rented apartments it had always seemed too hot or too cold.

The family knew that buying a house would certainly cost more money than renting an apartment. A lot of responsibility comes with home ownership, too. Vernon and Kev would have to help care for the yard. They might all have to help make repairs and improvements.

To get a whole closet to himself and to have a place where he could close the door and get away from his brother, Vernon was willing to do a lot of painting and lawn mowing. He decided to find out all that he could about home ownership in order to help his parents make the buying decision.

OBJECTIVES: *In this lesson, we will help Vernon to:*

- *Consider some of the advantages and disadvantages of owning a house.*
- *Understand the advantages and disadvantages of buying a new house.*
- *Recognize the relationship between the selling price of a house and its size, location, age, and amenities.*

FOCUS

ALGEBRA CONNECTIONS

How long would it take to pay back a mortgage on a house for $288,000, if the monthly payment is $800? (**30 years**)

CHOOSING A PLACE TO LIVE

Vernon's family was ready to buy a house. Like many, they chose to buy a house partly for noneconomic reasons—such as the freedom to have pets, the space to plant a garden, and the ability to make home improvements. Vernon and Kev were eager to have bedrooms of their own, and this was now possible.

Home ownership usually costs more than renting, but it does allow you to build up equity. **Equity** is the difference between what the house (or other property) is worth and what is owed on the mortgage. The monthly payments decrease the debt and increase the amount of equity the buyer has in the property. When their house is paid for in full, Vernon's family will have 100% equity in it.

Home ownership has been a good investment in recent years. Many people have been able to experience a large **capital gain,** or increase in the value of their property, when selling their houses. Vernon's parents do not intend to sell this house until their three children are through college and perhaps have homes of their own. By then, their equity in the house will be substantial.

ADVANTAGES OF OWNING A HOME

Home ownership often provides a feeling of security and independence.
Home ownership provides a means of forced savings by building up equity.
Home ownership has proven to be a good investment over the years because of increasing property values.
Interest paid on a mortgage is tax deductible.

DISADVANTAGES OF OWNING A HOME

The money invested in the house is not readily available for other purchases.
An owner assumes responsibility for financing and maintaining a house; renters leave those responsibilities to the owner.
Home ownership makes moving more difficult because the old house must be sold.
Home ownership may require larger monthly payments (including utilities, property taxes, and insurance) than renting requires.

Vernon's parents felt that they were ready for home ownership because they met the following conditions.

- They could make the monthly payments out of their current income.
- They planned to live in the same place for three years or longer.
- They were prepared to take on the responsibility for the financing, repair, and maintenance of a house.

Owning a house often provides individuals and family members with a sense of pride, enjoyment, and satisfaction. Vernon was not sure he would actually enjoy mowing the lawn, but he would derive a lot of satisfaction from being able to have his own room!

When they examined the housing market in their new city, Vernon and his family realized that the next major decision they would have to make would be whether to buy a new house or an older one. A new house often costs more than an older one, but the repairs are usually less expensive. However, the buyer may have to make a substantial investment in landscaping, window treatments, carpeting, and major appliances.

Although older houses are usually less expensive than new ones, they frequently need repairs such as a new roof, a new furnace, new plumbing, or electrical work. Sometimes the buyer will want to add a room or modify existing ones. Many of the improvements or changes that the buyer makes may not appeal to the next owner. Therefore these improvements may not increase the cash value of the house.

CRITICAL THINKING

After the students have answered the questions about buying houses, have them consider owning a co-op, or a condominium.

ALGEBRA *REVIEW*

Solve for y if $x = 10.18$. Round your answers to the nearest hundredth.

1. $y = 177.23 + 0.76x$
   184.97
2. $y = 5.787 - 0.0065x$
   5.72
3. $y = \frac{109.75}{x}$
   10.78
4. $y = \frac{0.098}{x}$
   0.01
5. $x = 15.666 + 0.123y$
   −44.60
6. $x = 14.75y - 12.7$
   1.55
7. $x = \frac{1.65y}{43.47}$
   268.20
8. $x = \frac{177.711}{y}$
   17.46
9. $x = \frac{3.81y}{542.6}$
   1449.78
10. $x = \frac{2301.46}{y}$
    226.08

Ask Yourself

1. What are some noneconomic reasons people may choose to buy a house?
   See Additional Answers.
2. What is equity?
3. What are some of the advantages of buying a new house instead of an older one? What are the disadvantages?
4. What conditions should you meet before buying a house?

ADDITIONAL ANSWERS

1. freedom to have pets, make home improvements, and plant a garden
2. difference between what the house is worth and what is owed on the mortgage
3. Advantages: fewer repairs; repairs needed are less expensive. Disadvantages: more expensive to buy; need to spend more on landscaping, carpeting, appliances, etc.
4. Conditions: I can make the monthly payments on current income; I plan to live in the same place for three or more years; I can take on the responsibility for financing, repairing, and maintaining the house.

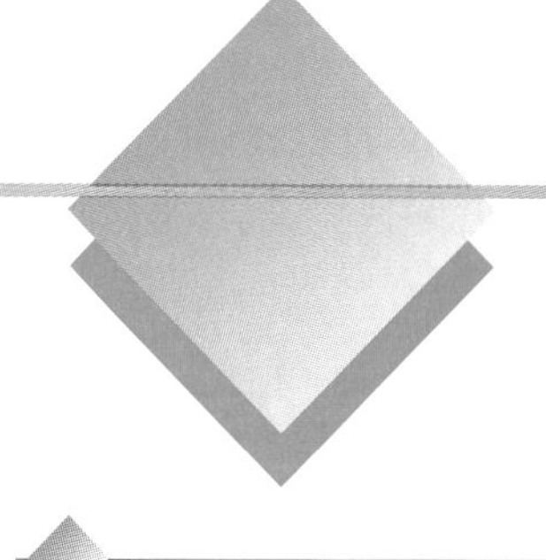

SHARPEN YOUR SKILLS

SKILL 1

The price of a house is related to many factors, including size, age, location, and amenities. You can use *regression analysis* to determine a correlation between the price and one of these factors. Remember that a graphing calculator can automatically calculate a, the y intercept, and b, the slope of a line that best fits a set of data. The calculator also determines a value for r, the *correlation coefficient,* which is a measure of how good the fit is. If r is close to $+1$ or -1, the fit is very good.

EXAMPLE 1 The most frequently used measure of the size of a house is its floor area. Vernon collected the information shown about nine houses that are for sale in his new city.

Price (in thousands of dollars)	Area (in square feet)
$348	4350
91	2800
260	3600
150	3100
164	3100
185.9	3400
102.4	2850
375	4375
136.3	3000

QUESTION What is a linear model of the form $y = a + bx$ that will allow Vernon to predict the number of square feet in a house in his city if he knows the selling price? Is there a good correlation between price and area?

SOLUTION

Use the linear regression function on your graphing calculator to find the line that best fits the data. Begin by entering the ordered pairs (x, y) of data, where x represents the price and y represents the area in each house. Your calculator will display the following information:

$a = 2235.432486$
$b = 5.768568698$
$r = 0.9916637943$

The linear equation that best relates the price and the number of square feet is

$$y = 2235.432486 + 5.768568698x$$

The graph of the equation is shown in a scatter plot of the data.

Xmin: -25	Ymin: -500
Xmax: 400	Ymax: 5000
Xscl: 50	Yscl: 1000

Since the value of r is close to 1, there is an excellent linear correlation between the price and the number of square feet in the house. This means that as the price increases, the area increases.

INSTRUCTION

TEACHING THE LESSON

Have students read the opening scenario. Use the *Algebra Connections* to illustrate how long it would take to pay back a mortgage on a house. Make sure students understand the factors involved in choosing a place to live by reading the advantages and disadvantages of owning a house.

After completing Skills 1 and 2, have students try prices of homes over $500,000. Have them determine if there is a significant jump in the area of a house and whether those houses are fairly priced.

TECHNOLOGY HINT

Skill 1 emphasizes the use of the regression function on a graphing calculator to determine a correlation between various factors. In entering different ordered pairs, students should be careful in identifying the values that represent the independent and the dependent variables.

FOCUS ON ALGEBRA

In this lesson, using regression analysis gives students the opportunity to explore ways that help them make decisions that apply to their daily lives. *(NCTM Standard 1, p. 137)*

SKILL 2

You can use a regression equation to determine missing information. You can use regression analysis to make a prediction about a house, such as its square footage, if you know its price.

EXAMPLE 2 Vernon found an advertisement for a house in his new city that appealed to him. Although the price of the house is listed at \$290,000, the number of square feet is misprinted in the ad.

QUESTION What should the area of this house be?

SOLUTION

Use the regression equation that you found in Example 1. Since the original data had only four digits, round a and b to four significant digits. Remember to use 290 for \$290,000 since the original data gave the prices of houses in thousands.

$y = 2235 + 5.769x$ — Substitute 290 for x.

$y = 2235 + 5.769(290)$ — Solve for y.

$y = 3908.01$

Vernon estimates that there are approximately 3900 ft^2 in this house.

SKILL 3

You can use a regression equation to evaluate information such as whether a house is appropriately priced.

EXAMPLE 3 Vernon's family looked at a house with 2700 ft^2 that was priced at \$150,000.

QUESTION Is this house fairly priced?

SOLUTION

Use the regression equation that you used in Example 2.

$y = 2235 + 5.769x$ — Substitute 2700 for y.

$2700 = 2235 + 5.769x$ — Solve for x.

$2700 - 2235 = 5.769x$ — Subtract 2235 from both sides.

$\frac{2700 - 2235}{5.769} = x$ — Divide both sides by 5.769.

$80.603 = x$

Remember the price of houses in the original data was given in thousands so the result 80.603 gives a price of \$80,603. At \$150,000, this house appears to be significantly overpriced. A price of \$80,603 seems more appropriate.

ESL STUDENTS

Non-native speakers may have difficulty with the meaning of the terms *equity, capital gain, linear regression,* and *correlation coefficient.*

SKILL 4

Since the price of a house is related to many factors, you can determine many different correlations between various factors.

EXAMPLE 4 Priscilla has collected the information shown about the average prices of houses in towns outside a major business financial center to which people commute each day.

Average Price (in thousands of dollars)	Distance from Financial Center (in miles)
$400	24
375	27
360	29
325	33
295	36
280	42
200	50
125	60

QUESTIONS What is a linear model of the form $y = a + bx$ that will allow Priscilla to predict the distance from the financial center to a house in her county if she knows the selling price of the house? Is there a good correlation between the price and the distance from the financial center?

SOLUTION

Use the linear regression function on your graphing calculator to find the linear equation that best relates the price (in thousands) and the location.

Your calculator will display the following values for a, b, and r.

$a = 76.62623558$
$b = -0.1322075783$
$r = -0.995827639$

The linear equation that best relates the price and the distance from the financial center to four significant digits is

$y = 76.63 - 0.1322x$

A graph of the linear regression equation is shown on a scatter plot of the data.

Xmin: -25	Ymin: -5
Xmax: 500	Ymax: 75
Xscl: 50	Yscl: 10

Notice that the slope of the line is a negative number. Therefore the line falls downward to the right. From the graph, you can see that the distance y decreases as the price x increases.

Since the value of r is -0.995827639, there is a very good *negative* linear correlation between price and the distance from the financial center. This means that as the price increases, the distance decreases.

COMMON ERRORS

Students may confuse the linear model $y = a + bx$ with the slope-intercept form $y = mx + b$. Point out the difference.

INTERDISCIPLINARY INVESTIGATION

Economics: Have students research the establishment of the Federal Housing Administration. Have them report on how the Federal Housing Administration's policies affect the housing market.

RETEACHING

Some students might benefit from solving linear equations in two variables using decimals. To reteach the lesson, concentrate on solving for one variable in terms of the other.

ENRICHMENT

Challenge students to write a computer program using the linear regression equation.

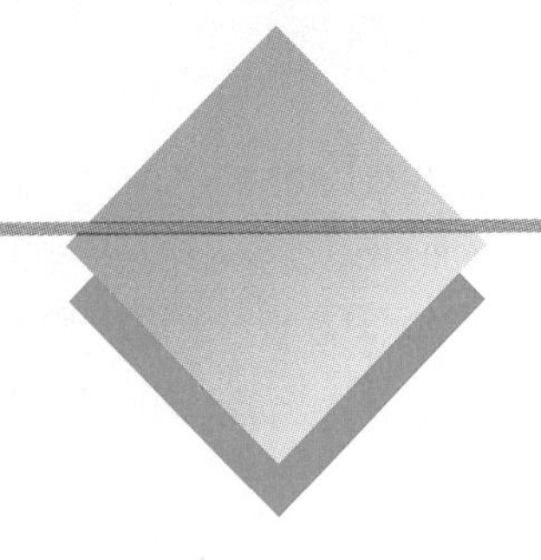

TRY YOUR SKILLS

GUIDED PRACTICE

You may wish to have students work in groups. Observe each group as they gather information, providing guidance whenever necessary.

ADDITIONAL ANSWERS

2. yes, $r = 0.9838$
5. yes

Carla noticed that the following amenities are used as selling points in advertisements for homes: hardwood floors, walk-in closets, professional landscaping, fireplace, island kitchen, central air conditioning, skylight, underground sprinkler system, security system, patio/deck, pool, and spa/jacuzzi. She collected the information shown about houses selling in her town. Give answers to four significant digits.

Price (in thousands of dollars)	Number of Amenities
$200	10
102	3
111	4
88	1
120	4
145	7
125	5
162	8
175	9
130	6
173	8
210	11

1. What is a linear model of the form $y = a + bx$ that will allow Carla to predict the number of amenities in a house in her town if she knows the selling price of the house? $y = -4.74 + 0.0763x$
2. Is there a good correlation between the price and the number of amenities?
3. Graph the linear regression equation on a scatter plot of the datas. See Additional Answers.
4. Estimate the number of amenities in a house in Carla's town that is selling for $190,000. 10
5. Do you think that a $120,000 house with five amenities is fairly priced?

EXERCISE YOUR SKILLS

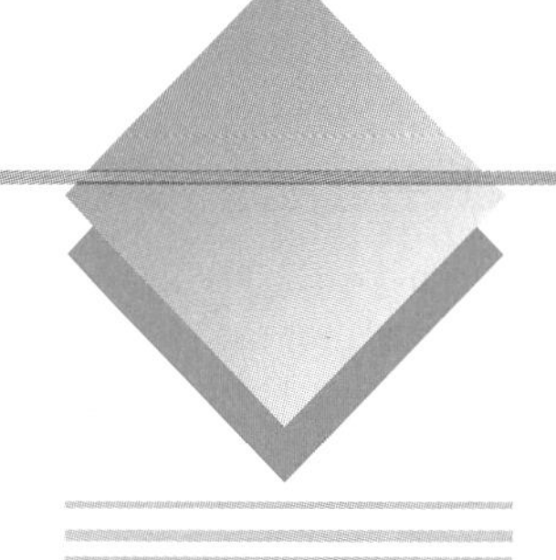

1. What are some of the disadvantages of owning a house? Answers may vary.
2. In your opinion, do the advantages of owning a house outweigh the disadvantages or vice versa? Explain. Answers may vary.
3. Why does owning a house usually cost more than renting an apartment?

Juanita collected the information shown about houses selling in her town. Give answers to four significant digits.

Price (in thousands of dollars)	Age (in years)
$108	47
200	10
95	50
150	28
125	44
215	4
240	0*
255	0*
225	2
140	35

*Brand new house

4. What is a linear model of the form $y = a + bx$ that will allow Juanita to predict the age of a house in her town if she knows the selling price of the house?
5. Is there a good correlation between the price and the age of a house? yes; $r = -0.9873$
6. Graph the linear regression equation on a scatter plot of the data. See Additional Answers.
7. Estimate the age of a house in Juanita's town that is selling for $185,000. 19 years
8. Do you think that a $160,000 house that is 40 years old is fairly priced?

Daniel collected the information shown about houses selling in his town. Give answers to four significant digits.

Price (in thousands of dollars)	Number of Bathrooms
$188	1
200	1
290	2.5
210	1.5
310	2.5
325	3
380	4
250	2
350	3.5

9. What is a linear model of the form $y = a + bx$ that will allow Daniel to predict the number of bathrooms in a house in his town if he knows the selling price of the house? $y = -1.8969 + 0.0151x$
10. Is there a good correlation between the price and the number of bathrooms in a house?
11. Graph the linear regression equation on a scatter plot of the data. See Additional Answers.
12. Estimate the number of bathrooms in a house in Daniel's town that is selling for $225,000. Round your answer to the nearest half bathroom. about 1.5
13. Do you think that a $300,000 house with 1.5 bathrooms is fairly priced?
no; $220,000 would be fair; $x = 223.11$

KEY TERMS

capital gain
equity

INDEPENDENT PRACTICE

ASSIGNMENT GUIDE

Exercises 1-3 ask students to weigh the advantages and disadvantages of owning a house.

Exercises 4-18 assess students' ability to apply the linear regression equation taught in the lesson.

ADDITIONAL ANSWERS

3. Answers may vary.
4. $y = 84.07 - 0.3541x$
8. No, about $124,000; $x = 124.45$
10. yes; $r = 0.9891$

CLOSURE

Ask students to define the variables in the linear regression equation, $y = a + bx$.

LESSON QUIZ

1. Ellen collected the information shown about houses selling in her town. What is a linear model of the form $y = a + bx$ that will allow Ellen to predict whether a house has a 1- or 2-car garage if she knows the selling price of the house?

Price (in thousands of dollars)	1- or 2-Car Garages
$200	1
225	1
300	2
285	2
195	1
325	2
350	2
380	2

(**$y = -0.2680412371 + 0.0067010309x$**)

2. Ellen found an advertisement for a house in her town that was appealing to her. The price of the house was $250,000, and the advertisement did not mention a 1- or 2-car garage. Use the regression equation in Exercise 1 to estimate whether the house has a 1- or 2-car garage. (**one**)

ADDITIONAL ANSWERS (EXERCISE YOUR SKILLS)

14. $y = 6.649 - 0.0172x$

15. yes; $r = -0.9627$

(MIXED REVIEW)

2. 24 years

A realtor conducted a study to determine whether the price of houses in his city is related to traffic patterns on the street on which the house is located. He used the traffic scoring system shown. He then made the chart shown to organize the data. Give answers to four significant digits.

1. Quiet cul-de-sac
2. Very little traffic
3. Moderate traffic
4. Heavy traffic

Price (in thousands of dollars)	Traffic Patterns
$200	3
150	4
300	1
350	1
275	2
250	2
225	3
175	4

14. What is a linear model of the form $y = a + bx$ that will allow the realtor to predict the traffic patterns on the street on which a house is located if he knows the selling price of the house?

15. Is there a good correlation between the price and the traffic patterns on the street on which a house is located?

16. Graph the linear regression equation on a scatter plot of the data. See Additional Answers.

17. Estimate the traffic pattern score on the street on which a house selling for $325,000 is located. about 1; $y = 1.059$

18. Do you think that a $280,000 house with a traffic pattern score of 4 is fairly priced? no; $x = 154.01$

MIXED REVIEW

2-2 **1.** With $987.44 in your checking account, you write a check for $122.87. What is the new balance? $864.57

3-2 **2.** Use the Rule of 72 to determine how long it will take Elliott's $2400 savings account to double in value if it is growing at a rate of 3.0%?

6-1 **3.** Determine the effective rate of interest if the APR is 16.5%. 17.8%

6-2 **4.** Marcia has a balance of $1232 on her VISA card, which carries an APR of 15%. If she makes no further purchases with this card, how many months will it take her to pay off the balance? She can afford monthly payments of $160. 9 months

6-7 **5.** Last month, the Reynolds family spent $250 on the car payment, $315 on credit card payments, $400 on utilities, $620 on rent, $810 on food, and $2256 on everything else. To the nearest tenth of a percent, what percent of their take-home pay did the family spend on credit payments? 12.1%

11-1 **6.** Use the United States Mileage Chart in the Reference Section to find the distance from Columbus, Ohio, to Atlanta, Georgia. 533 mi

11-2 **7.** The Uptown Motor Inn charges $42 for a single room and $52 for a double room. The tax is 5%. Find the cost paid by Mr. and Mrs. Engel for a stay of four nights. $218.40

12-2 BUYING A HOUSE: HOW TO DETERMINE THE COSTS

QUICK REFERENCE

TEACHER'S RESOURCES AND ANSWER KEY

Lesson Quiz Answers 12-2, p. 82

Reteaching Activity Answers 12-2, p. 113

Enrichment Activity Answers 12-2, p. 114

EXTENSION ACTIVITIES

Reteaching Activity 12-2, p. 90

Enrichment Activity 12-2, p. 94

Olivia's parents would like to buy another house. If the family lived a little closer to the school, they would allow Olivia to drive her brother Orson and herself to school. Olivia's mother has been working for several years and recently received a substantial increase in her salary, so this may be a good time to make a move. They have often talked about wanting more space, especially since Olivia and Orson became teenagers. The children are both pleased about the prospect of having groups of friends over on the weekends.

Olivia has noticed that adults become very secretive when they talk about how much things cost—except, of course, when they talk about how little things used to cost! Olivia will ask Vernon whether he knows about the kinds of costs they will encounter if they do decide to buy another house. From what Vernon has already said, Olivia knows that her parents might be in for quite a surprise. If they think the cost of a bicycle has gone up in 35 years, just wait until they see what has happened to housing prices!

OBJECTIVES: *In this lesson, we will help Olivia to:*

- *Recognize the major costs to consider when purchasing a house, including the price, the mortgage loan, points, down payment, closing costs, and the cost of new furnishings.*
- *Compute the monthly payments and total financed price of a house.*

THE COST OF OWNING YOUR HOME

ALGEBRA CONNECTIONS

What is the total amount to be paid on a 10-year loan if the monthly payment is $700? (**$84,000**)

Since 1980 the prices of houses have consistently outpaced consumer incomes. Since Olivia has studied the effects of inflation and the changes in prices as measured by the Consumer Price Index, she understands why close to two thirds of the families buying houses have two incomes. Six out of ten first-time buyers are couples in which both partners hold jobs.

Because of the high cost of new houses, most first-time buyers must make compromises such as the following:

- Buy a condominium as a first home.
- Commute longer distances to buy less expensive housing.
- Live in smaller and older houses that may need repairs.
- Buy a manufactured house, generally called a mobile home or prefabricated home.
- Accept a more moderate life-style to pay for shelter.

How Much to Spend Olivia's family banker told her parents that there is a rule of thumb that a family should look for housing priced within $2\frac{1}{2}$ times their annual income. Since Olivia's mother and father are both working, they can spend a large sum of money on a house. However, they certainly cannot pay that much in cash.

The Mortgage The family will take a **mortgage,** which is a loan on which both principal (the amount borrowed) and interest are paid back over a fixed number of years, usually 15 to 30 years. The amount of the mortgage is based on the **appraised value** of the house. A mortgage contract is signed by the lender and the borrower. The amount of the monthly mortgage payment is based on the amount of the loan, the interest rate, and the amortization period of the loan. The lender may include property taxes and the cost of insurance in the mortgage payment. Over the course of the loan, the borrower may end up making total payments that equal between 2 and 4 times the purchase price of the house. However, the alternative of saving until they can pay cash for a house would prevent many families from ever being able to purchase a house.

Another important factor that they must consider when borrowing for a real-estate purchase are the points that the lender may charge. **Points** are a one-time charge that is paid at the time of the mortgage closing. One *point* is equal to 1% of the value of the mortgage. Points provide additional revenue to the bank, and some banks require a larger number of points to be paid in exchange for a lower interest rate over the term of the loan.

The Down Payment When Olivia's parents buy the house, they will make a **down payment** (pay part of the purchase price). Their down payment could range from 0 to 30% or more of the purchase price. The higher their original down payment, the lower their monthly mortgage expense will be. When they sell their other house, they can use the equity that they had in it as the down payment on the new house.

The Closing Costs The family will also have to have enough cash available for various *closing costs* that occur at the time the sale is completed. **Closing costs** are the charges and fees associated with the transfer of ownership of a home to a new buyer. These costs are incurred by the buyer for the work involved in preparing and processing all of the documents needed to complete the purchase of the home. A *buyer's costs* may include the following:

- Title search
- Mortgage preparation
- Appraisal fee
- Recording fees
- Credit report
- Legal fees

The *seller's costs* include the realtor's commission, the preparation of the deed, and any real estate taxes that are due. These costs can range from several hundred to several thousand dollars. The realtor who is handling a sale can estimate the closing costs when final arrangements for the sale are being made.

Additional Costs Olivia's family knows that their expenses do not end at the closing. When they move into the house, certain other expenses will be incurred. The costs of curtains, draperies, floor coverings, and landscaping are in this category. Also, the family may be required to submit deposits to the utility companies that supply electricity, water, and telephone service. If they are not changing utility companies, they may be able to apply the deposits made when they bought their first house. They do know that they will want new curtains, bedspreads, draperies, and several new carpets for the new house. They may even want a few new pieces of furniture. They would like to get a new bedroom suite for Olivia and replace the couch that the dog chewed. Therefore they must also plan to have the resources available to meet these costs.

Ask Yourself

1. What are points? a one-time charge paid at time of mortgage closing. 1 point = 1% of the value of the mortgage
2. What is the name given to costs incurred by the buyer for work involved in preparing and processing all the documents for a house purchase? closing costs
3. Name three compromises that first-time buyers might make because the prices of new houses are so high. Answers may vary.

ALGEBRA *REVIEW*

The following table shows average home prices in thousands of dollars in two counties at different times.

	1978	1983	1988	1993
Bell	43	49	61	52
Mann	25	34	47	40

1. When was the highest average price reached in Bell County? 1988
2. What year shows the greatest difference in prices for the two counties? 1978
3. Draw a double bar graph to represent the information given in the table. See Additional Answers.

SHARPEN YOUR SKILLS

SKILL 1

To find the monthly mortgage payment, use the monthly payment formula that you used in Chapter 5.

Monthly Payment Formula

$$M = \frac{Pr(1 + r)^n}{(1 + r)^n - 1}$$

where P = amount of loan
r = monthly interest rate
n = number of payment periods
M = monthly payment

The **total financed price** is the total purchase price plus the financing costs and the down payment. The total financed price T is found by multiplying the monthly payment M by the number of months n over the course of the loan and then adding the down payment d.

$$T = nM + d$$

INSTRUCTION

TEACHING THE LESSON

Encourage the students to read the opening scenario. Use the *Algebra Connections* to point out the different choices that are available for obtaining a mortgage on a house. Have students read the section on the cost of owning a home. Make sure students understand how to apply the monthly payment formula where the rate has to be divided by 12.

In Skill 2, when discussing the fact that Olivia's parents pay back significantly more than they borrow, point out that a dollar paid back 30 years from now may be worth far less than the dollar borrowed today, depending on the inflation rate.

Before proceeding to Skill 5, have students try different amounts of loan, rate, and length of repayment to determine the monthly payment.

FOCUS ON ALGEBRA

In this lesson, formulas for monthly payments and total financed price give students the opportunity to explore mathematical functions that apply to their daily lives. *(NCTM Standard 6, p. 154)*

Recall that the calculation of the monthly payment is given to many decimal places in your calculator. Use that answer to calculate the total financed price by using the answer key [ANS] on your calculator. Remember that the last monthly payment is adjusted to the actual total.

Olivia has noticed that house listings give the purchase price of the house. However, the listings do not tell her how much the monthly mortgage payments will be. The house listing shown is very appealing to her.

GIANT TEEN ROOM ON 3RD FLOOR!
Teens can entertain their guests in their own suite. Parents will enjoy the built-in sauna, spa, pool, and large paneled billiard room. A steal at $765,000.

EXAMPLE 1 Olivia knows that her parents plan to make a down payment of 20% on the house that they purchase. They want to amortize their loan over 30 years, so they will apply for a 30-year mortgage. The prevailing interest rate on a home loan is 12%.

TECHNOLOGY HINT

In using a calculator to determine the monthly payment on a mortgage, be sure students know the proper use of the exponential function key [y^x].

QUESTIONS What are the monthly payment and the total financed price of the house for $765,000 under these conditions? What is the ratio of the total financed price to the purchase price?

SOLUTIONS

Determine the amount that they will borrow.

$0.20(765{,}000) = 153{,}000$ They make a down payment of 20%.

Therefore they must borrow $765{,}000 - 153{,}000 = 612{,}000$.

Use the monthly payment formula.

$$M = \frac{612{,}000(0.01)(1 + 0.01)^{360}}{(1 + 0.01)^{360} - 1}$$ $P = 612{,}000$, $r = (0.12 \div 12) = 0.01$, $n = 30 \bullet 12 = 360$

$M = 6295.11$ To the nearest cent

The monthly payment is $6295.11. Olivia is certain that her parents will not share her enthusiasm for a house with such a large monthly payment!

In the course of 30 years the total financed price will be

$T = nM + d$
$T = 360$[ANS]$ + 153{,}000$
$T = 2{,}419{,}239.27$

COMMON ERRORS

Remind students that the down payment should be added to the payments over the course of the loan to determine the total financed price.

The total financed price is $2,419,239.27.

The ratio of the total financed price to the purchase price is

$$\frac{2{,}419{,}239.27}{765{,}000} = 3.2$$ To the nearest tenth

The total financed price of $2,419,239.27 is more than three times the original cost of the house.

COOPERATIVE LEARNING

Have students work in small groups using calculators to determine the total financed price of a house with different combinations of down payments, interest rates, and amortization periods.

SKILL 2

As you learned in Chapter 5, the total payment on a loan for a given amount of money over a specified amortization period is always lower when the interest rate is lower. Olivia has heard that even a *small* decrease in the interest rate on a home loan can create a *large* decrease in the total financed price.

EXAMPLE 2 The mortgage plans that are available to Olivia's family require that they make a down payment of 20% of the purchase price. The house that they want to buy costs $250,000. They have found rates for 30-year fixed mortgages that range from 9% at her mom's credit union to 12% at a local bank.

QUESTIONS How does the total financed price on a $250,000 home with a 20% down payment and a 30-year mortgage differ as interest rates vary from 9% to 12% in increments of 0.5%? In each case, what is the savings over a 12% loan?

SOLUTION

Determine the 20% down payment and the loan amount.

Down payment: 0.2(250,000) = 50,000
Loan amount: 250,000 − 50,000 = 200,000

Use a spreadsheet program to find the total financed price, to the nearest dollar, and the savings over a 12% loan as the interest rate varies from 9% to 12% in increments of 0.5%. For cell B5 the formula is:

200000*(A5/12)*(1+(A5/12))^360/((1+(A5/12))^360−1)

Format column A for three fixed places. Format column B for two fixed places and format columns C and D for zero fixed places. In the formula for cell D5, be sure to use an address for cell C11 that stays the same when the formula is copied, called an *absolute* address.

	A	B	C	D
1		$200,000 mortgage with $50,000 down payment		
2		30-year	30-year	
3	Interest	Monthly	Total Financed	Savings
4	Rate	Payment	Price	Over 12%
5	0.090	1609.25	629328	161273
6	0.095	1681.71	655415	135186
7	0.100	1755.14	681852	108749
8	0.105	1829.48	708612	81989
9	0.110	1904.65	735673	54928
10	0.115	1980.58	763010	27591
11	0.120	2057.23	790601	0

@ROUND(B5*360+50000,0) (cell C5)

+A5+0.005 (cell A6)

+$C$11−C5 (cell D5)

Olivia used the graph function of her computer spreadsheet program to graph the relationship between the interest rate and the total financed price.

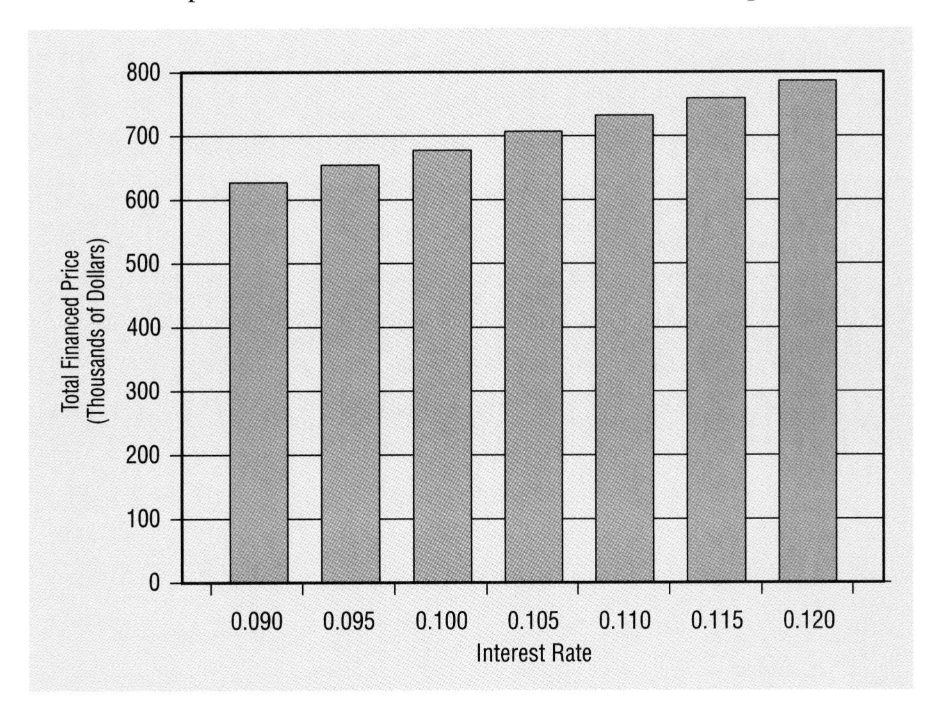

Over the course of 30 years, they will pay $161,273 more with a loan at 12% than at 9%. The savings with a 9% loan over a 12% loan for $200,000 over 30 years is almost as large as the amount that is borrowed!

Note that Olivia's parents will pay back significantly more than they borrowed over the course of 30 years. At 9% they will pay back $629{,}328 \div 250{,}000 = 2.5$ times the cost of the house. At 12% they would pay back $790{,}601 \div 250{,}000 = 3.2$ times the cost of the home.

CRITICAL THINKING

Have students consider the impact of inflation on the value of a house and your equity in it. If prices fall dramatically after you purchase a house, what might be the consequences when you want to sell? (**You may not be able to get enough money from the sale to pay off the outstanding balance of the mortgage loan.**)

SKILL 3

Olivia's parents learned that the total financed price is much smaller on 15-year mortgages, but the monthly payments are larger. For example, the monthly payment on a $200,000 loan at 9% is $1609.25 for 30 years and $2028.53 for 15 years. The 15-year mortgage costs $419.28 more each month.

EXAMPLE 3 Although her family cannot afford the larger monthly payment, Olivia would like to know what the savings over the course of the loan would be if they could afford the monthly payments on a 15-year mortgage.

QUESTION What is the total financed price of a $250,000 home with a loan of $200,000 and a $50,000 down payment for 15 years as interest rates vary from 9% to 12% in increments of 0.5%? In each case, what is the savings over a 30-year loan?

SOLUTION

Olivia used the spreadsheet program from Example 2 to find the total financed price, to the nearest dollar, and the savings over a 30-year mortgage as the interest rate varies from 9% to 12% in increments of 0.5%. She copied and adjusted the formulas as necessary.

	A	B	C	D	E	F
1	$200,000 mortgage with $50,000 down payment					
2		15-year	15-year	30-year	30-year	Savings
3	Interest	Monthly	Total Financed	Monthly	Total Financed	Over
4	Rate	Payment	Price	Payment	Price	30-year
5	0.090	2028.53	415136	1609.25	629328	214192
6	0.095	2088.45	425921	1681.71	655415	229494
7	0.100	2149.21	436858	1755.14	681852	244994
8	0.105	2210.80	447944	1829.48	708612	260668
9	0.110	2273.19	459175	1904.65	735673	276498
10	0.115	2336.38	470548	1980.58	763010	292462
11	0.120	2400.34	482061	2057.23	790601	308540

200000*(A5/12)*(1+(A5/12))^180/((1+(A5/12))^180−1)

@ROUND(B5*180+50000,0)

+E5−C5

She used the graph function of her computer spreadsheet program to graph the relationship between the interest rate and the total financed price for the 15-year and 30-year mortgages.

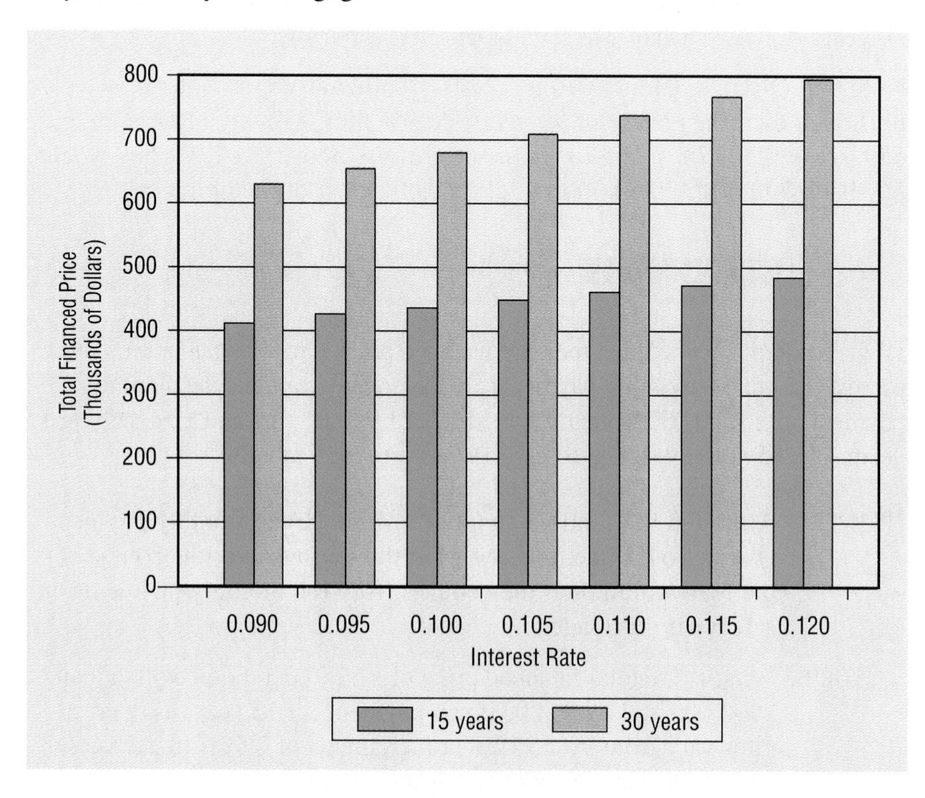

For example, the monthly payment on a $200,000 mortgage at 9% for 15 years would cost Olivia's family over $400 more each month than the same loan for 30 years. However, they would save $214,192 over the course of the loan.

SKILL 4

As you learned in Chapter 5, the higher the down payment on a loan, the lower the total payment.

INTERDISCIPLINARY INVESTIGATION

Finance: Have students research the founding of financial institutions other than banks that provide mortgage money to home buyers. Have them report on differences in lending practices.

EXAMPLE 4 Olivia's family planned $50,000 (20% of the purchase price of $250,000) for the down payment on their home. However, Olivia would like to know what the savings over the course of the loan would be if they made a down payment of $75,000, which is 30% of $250,000, and borrowed only $175,000.

QUESTION What is the total financed price of a mortgage for $175,000 for 30 years with a $75,000 down payment as the interest rate varies between 9% and 12% in increments of 0.5%? In each case, determine the savings over a $200,000 mortgage for 30 years with a $50,000 down payment.

SOLUTION

Olivia used a spreadsheet program to find the total financed priced with a mortgage of $175,000 for 30 years with a $75,000 down payment, to the nearest whole dollar, and the savings over a mortgage of $200,000 for 30 years with a $50,000 down payment as the interest rate varies from 9% to 12% in increments of 0.5%.

Interest Rate	Total Financed Price 30% Down	Total Financed Price 20% Down	Savings Over 20% Down
0.090	$581912	$629328	$47416
0.095	604738	655415	50677
0.100	627870	681852	53982
0.105	651286	708612	57326
0.110	674964	735673	60709
0.115	698884	763010	64126
0.120	723026	790601	67575

Then she used the graph function of her spreadsheet program to graph the relationship between the interest rate and the total financed price for down payments of 20% and 30%.

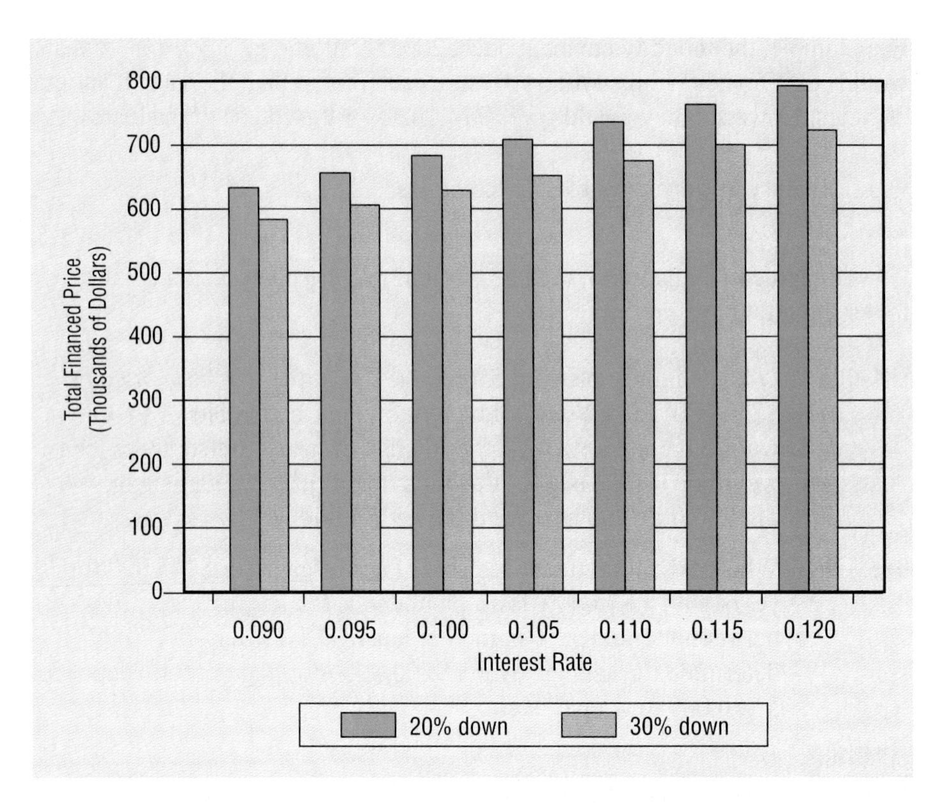

At 30% down the family would make a down payment of $25,000 more than at 20% down. However, at a rate of 9%, for example, the total financed price of the house would be

$$\$629{,}328 - \$581{,}912 = \$47{,}416$$

less than the total financed price with 20% down.

In general, you will obtain the lowest total financed price on a house if you select a mortgage with the *lowest* interest rate, the *shortest* amortization period, and the *largest* down payment.

SKILL 5

Some banks charge *points,* which are a one-time charge that is paid at the time of the mortgage closing. As mentioned earlier in this lesson, one point is equal to 1% of the value of the mortgage. When points are charged, the points p must be added to the other costs to find the *total financed price T.*

$$T = nM + d + p$$

EXAMPLE 5 A local bank is offering a choice of the following 30-year fixed mortgage rates:

A: 10% with 0 points
B: 9.75% with 1 point
C: 9.5% with $1\frac{1}{2}$ points

QUESTION If you would like to buy a house that costs $125,000 with a loan of $100,000 and a down payment of $25,000, what is your lowest-priced option?

SOLUTION

Use the monthly payment formula to find each monthly payment. Then find the total financed price. Remember to use the monthly payment price in your calculator, not the rounded amount.

For plan A:

$$M_A = \frac{100{,}000(0.10 \div 12)(1 + 0.10 \div 12)^{360}}{(1 + 0.10 \div 12)^{360} - 1}$$ $P = 100000$; $n = 30 \cdot 12 = 360$; $r = 0.10 \div 12$

$= 877.57$ To the nearest cent

$T_A = 360(M_A) + 25{,}000 + 0$ $T = nM + d + p$

$= 340{,}926$ To the nearest dollar

For plan B:

$$M_B = \frac{100{,}000(0.0975 \div 12)(1 + 0.0975 \div 12)^{360}}{(1 + 0.0975 \div 12)^{360} - 1}$$

$= 859.15$ To the nearest cent

$T_B = 360(M_B) + 25{,}000 + 0.01(100{,}000)$

$= 335{,}296$ To the nearest dollar

For plan C:

$$M_C = \frac{100{,}000(0.095 \div 12)(1 + 0.095 \div 12)^{360}}{(1 + 0.095 \div 12)^{360} - 1}$$

$= 840.85$ To the nearest cent

$T_C = 360(M_C) + 25{,}000 + 0.015(100{,}000)$

$= 329{,}208$ To the nearest dollar

The lowest-priced option is plan C for $329,208. Therefore if you have the money to pay $1\frac{1}{2}$ points up front, you can save money over the course of the loan.

It is important to explore and evaluate all of your options before selecting a mortgage plan.

RETEACHING

Some students might benefit from a review on decimals and percents. Review the operations with these topics.

ENRICHMENT

Challenge students to write a computer program using the Monthly Payment Formula.

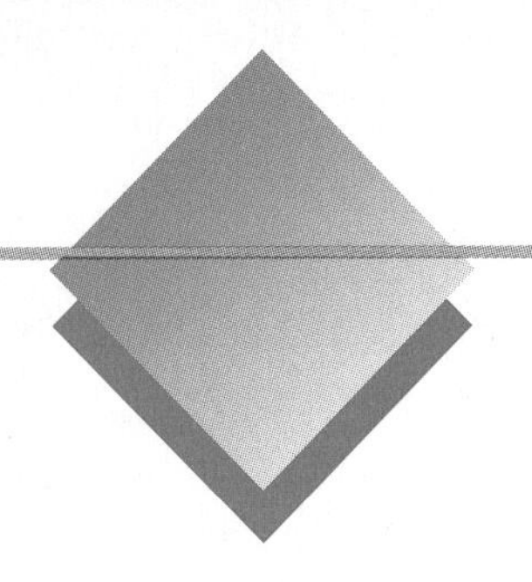

TRY YOUR SKILLS

1. Determine the total financed price on a $180,000 house if you finance it at 15% for 30 years with a 25% down payment. $659,520
2. Calculate the ratio of the total financed price to the purchase price in Exercise 1. 3.7:1

GUIDED PRACTICE

You may wish to have students work in groups to create the spreadsheet in Exercises 3-7. If time permits, have students graph the relationship between the interest rate and total financed price.

Use a spreadsheet program to determine how the total financed price on a $240,000 loan for 15 years with a down payment of $60,000 differs as interest rates vary from 6% to 8% in increments of 0.5%. In each case, what is the savings over an 8% loan?

	Interest Rate	Total Financed Price	Savings Over 8%
3.	6%	$424,546	$48,296
4.	6.5%	436,318	36,524
5.	7%	448,294	24,548
6.	7.5%	460,469	12,373
7.	8%	472,842	0

8. If you would like to buy a house that costs $160,000 with a loan of $125,000 and a down payment of $35,000, which of the following is your lowest-priced option for a 15-year fixed mortgage? Plan C

   A: 12% with 0 points
   B: 11.5% with 1 point
   C: 11% with 2 points

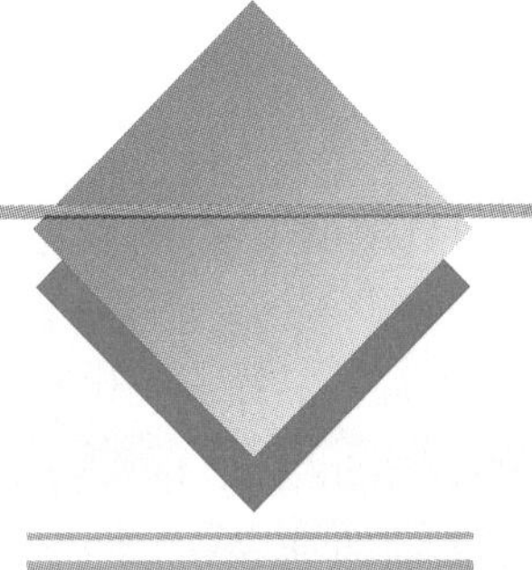

EXERCISE YOUR SKILLS

1. What are the major costs to consider in buying a house?
2. How does the amount of the down payment on a house affect the amount that is borrowed in the form of a mortgage? See Additional Answers.
3. The total financed price for a house is approximately how many times the purchase price? between 2 and 4 times
4. Determine the monthly payment and total financed price on a $188,000 house if you finance it at 9.5% for 30 years with a down payment of 25%.
5. Determine the ratio of the total financed price to the purchase price of the house in Exercise 4. 2.5:1

KEY TERMS

appraised value
closing costs
down payment
mortgage
points
total financed price

ADDITIONAL ANSWERS

1. down payment, closing costs, mortgage, additional costs
2. the higher the down payment, the lower the mortgage
4. $1185.60; $473,817.60

Use a spreadsheet program to determine the total financed price of a $200,000 house with each combination of down payment, interest rate, and amortization period.

	Percent Down	Down Payment	Loan Amount	Interest Rate	Number of Years	Total Financed Price
6.	20%	$40,000	$160,000	10%	15	$349,486
7.	20%	40,000	160,000	10%	30	545,481
8.	20%	40,000	160,000	11%	15	367,340
9.	20%	40,000	160,000	11%	30	588,538
10.	30%	60,000	140,000	10%	15	330,800
11.	30%	60,000	140,000	10%	30	502,296
12.	30%	60,000	140,000	11%	15	346,422
13.	30%	60,000	140,000	11%	30	539,971

14. In Exercises 6–13, which combination of percent down, interest rate, and amortization period yields the lowest total financed price? Which yields the highest total financed price? See Additional Answers.

Use a spreadsheet program to determine the total financed price of a $120,000 home with each combination of down payment, interest rate, and amortization period.

	Percent Down	Down Payment	Loan Amount	Interest Rate	Number of Years	Total Financed Price
15.	25%	$30,000	$90,000	8%	30	$267,740
16.	20%	24,000	96,000	8%	30	277,589
17.	25%	30,000	90,000	9%	30	290,698
18.	20%	24,000	96,000	9%	30	302,078
19.	25%	30,000	90,000	9%	15	194,311
20.	20%	24,000	96,000	9%	15	199,265
21.	25%	30,000	90,000	8%	15	184,816
22.	20%	24,000	96,000	8%	15	189,137

23. In Exercises 15–22, which combination of percent down, interest rate, and amortization period yields the lowest total financed price? Which yields the highest total financed price? See Additional Answers.

24. Use the graph function of your computer spreadsheet program to graph the relationship between interest rate and total financed price on a $150,000 house financed for 15 years with a 25% down payment as interest rates vary between 6% and 10% in increments of 1%.

INDEPENDENT PRACTICE

ASSIGNMENT GUIDE

Exercises 1-3 ask students about factors to consider in buying a house. Exercises 4-26 assess students' ability to apply technology and the formulas taught in the lesson.

CLOSURE

Have students explain the different options available in buying a house. Ask students to define the terminology in the lesson and the variables given in the monthly payment formula.

LESSON QUIZ

1. Sylvia's parents plan to make a down payment of 15% on a house that costs $285,000. They will apply for a 30-year mortgage. The interest rate on a home loan is 9%. Determine the monthly payment and the total financed price. (**$1,949.20, $744,461.38**)
2. Determine the monthly payment and the total financed price on a house that costs $165,000 with a down payment of $30,000 if financed at 8.5% with one point for 15 years. (**$1,329.40, $270,641.71**)

ADDITIONAL ANSWERS

14. lowest: 30% down, 10% interest, 15 years; highest: 20% down, 11 % interest, 30 years

23. lowest: 25% down, 8% interest, 15 years; highest: 20% down, 9 % interest, 30 years

25. If you would like to buy a house that costs $225,000 with a loan of $175,000 and a down payment of $50,000, which of the following is your lowest-priced option for a 15-year fixed mortgage? Plan C

    A: 8% with 0 points
    B: 7.5% with 1 point
    C: 7% with 2 points

26. If you would like to buy a house that costs $185,000 with a loan of $150,000 and a down payment of $35,000, which of the following is your lowest-priced option for a 30-year fixed mortgage? Plan C

    A: 9% with 1 point
    B: 8.5% with 2 points
    C: 8% with 3 points

MIXED REVIEW

2-1 **1.** A bank charges $0.03 for each of the first 25 checks written each month and $0.055 for each check above 25. What is the monthly charge for 33 checks? $1.19

6-2 **2.** Determine the monthly payment that must be made to reduce a $2222.44 MasterCard balance to zero in 3 years. The card carries an APR of 15%. $77.04

6-3 **3.** Your VISA card carries an APR of 21%. Your balance is currently $1200.50 and you pay 10% of the amount owed to the nearest dollar each month. Complete the following table for the next three months. Assume that you make no further purchases during that time.

Month	Balance	Interest	Amount Owed	Payment
1	$1200.50	$21.01	$1221.51	$122
2	1099.51	19.24	1118.75	112
3	1006.75	17.62	1024.37	102

7-1 **4.** Use the Comparison Table for Term and Whole Life Premiums in the Reference Section to find the yearly premium for $350,000 of whole life insurance purchased at the age of 40. $6051.50

8-1 **5.** How many whole shares of stock in DWB company can you purchase for $50,000 at the current price of $65\frac{1}{2}$ per share? Ignore the cost of commissions. 763

10-1 A car that cost the dealer $13,900 has a sticker price of $16,150.

6. Find the mark-up. $2250 **7.** Find the percent of mark-up. 16.2%

11-3 **8.** An airplane departed from Newark Airport in New Jersey at 1:00 p.m. local time and arrived in Austin, Texas, at 3:15 p.m. local time. How long was the flight? 3h 15 min

12-3 THE MORTGAGE: HOW MUCH CAN YOU BORROW?

QUICK REFERENCE

TEACHER'S RESOURCES AND ANSWER KEY

Lesson Quiz Answers 12-3, p. 84

Reteaching Activity Answers 12-3, p. 113

Enrichment Activity Answers 12-3, p. 114

EXTENSION ACTIVITIES

Reteaching Activity 12-3, p. 91

Enrichment Activity 12-3, p. 95

Rosa had never given the price of buying a home much thought until Vernon and Olivia started discussing it. Her family had bought their house when she was much younger, and she and her sister Consuela had always shared a room. It is a large room for a bedroom, with two closets. Rosa and Consuela are good friends as well as sisters, so they have been able to accommodate each other's belongings and attitudes about the room fairly well.

Consuela left for college last fall, so Rosa now has a lot more space and privacy. Consuela left a lot of her posters and other "kid's stuff" at home when she left. She will probably be coming home in the summer to find a job before she goes back to college, but she is talking about spending next year in France if she can arrange the appropriate work-study program.

Rosa herself will be leaving for college next year. Her parents are talking about selling this house and buying one that is more convenient to downtown and their father's job. Although Rosa has not been particularly interested in mortgages, she wants Olivia and Vernon to tell her what they have learned. Rosa's father is a veteran, and the house in which they now live was bought with a VA loan. Rosa has never cared about the family mortgage before, but that attitude may change since she will be relying on her parents to help support her when she goes to college. If the family buys a different house, their whole financial plan may undergo some significant changes.

FOCUS

ALGEBRA CONNECTIONS

What is the monthly payment on $5000 to be paid back in two years, assuming no interest is charged? (**$208.33**)

OBJECTIVES: *In this lesson, we will help Rosa to:*

- *Learn about the types of mortgages that are available to home buyers, including the fixed-rate mortgage, the adjustable-rate mortgage, the balloon mortgage, the Federal Housing Administration loan, and the Veterans Administration loan.*
- *Compute the price of a home loan for which a family can qualify, on the basis of their monthly income.*
- *Understand the importance of knowing whether a mortgage may be prepaid or assumed.*

FINANCING THE PURCHASE OF A HOME

When Rosa's family bought their first house, interest rates and house prices were quite high. As they consider their second house purchase, they will notice that in recent years, interest rates have varied substantially. Over the past several years, interest rates on home loans have ranged from as high as 16% to as low as 6%. Financing for a home purchase is as important as the purchase price or the construction costs.

Rosa's family will consider fixed-rate mortgages and adjustable-rate mortgages. **Fixed-rate mortgages (FRMs)** are amortized with a constant monthly payment over a fixed number of years (usually 15, 20, 25, or 30). If the family selects a fixed-rate mortgage, their monthly mortgage payment will remain the same despite rising (or falling) interest rates. Most conventional fixed-rate mortgages require that the buyer make a down payment of at least 20% of the appraised value of the house and borrow no more than 80%.

At one time, a fixed-rate mortgage was the only choice available. As interest rates began to rise and vary widely, mortgage lenders developed an alternative method of providing mortgages. This alternative is called a variable or adjustable-rate mortgage. An **adjustable-rate mortgage (ARM)** is a loan with an interest rate that can be adjusted up or down by the lending institution over the life of the loan. In most cases the starting interest rate on an ARM is lower than the rates offered on fixed-rate mortgages. Most adjustable mortgages have a **cap** on upward and downward rate adjustments, which specifies the largest and smallest rates that the lending institution can charge the borrower at any time during the course of the loan. Usually, the adjustments are yearly and are based on the **prime interest rate.** This rate varies from bank to bank, but often the rate set by the largest banks is used.

Another type of mortgage is the **balloon mortgage,** in which after a certain time the entire remaining balance becomes due. When a borrower expects to sell the house after a short period of time, a balloon mortgage may be desirable because the interest rate for the years before the balloon payment becomes due may be very low.

GOVERNMENT-INSURED REAL ESTATE LOANS

Some fixed-rate mortgages are backed by insurance provided by the federal government.

Federal Housing Administration Loans In **Federal Housing Administration loans,** the Federal Housing Administration (FHA) of the Department of Housing and Urban Development provides federal insurance on certain loans that are obtained through an approved lending institution such as a bank. A buyer who can make only a small down payment can sometimes obtain an FHA loan. With FHA loans, the borrower can borrow 95–100% of the sales price.

Veterans Administration Loans Rosa's father qualifies for a **Veterans Administration loan** because he is a veteran. People who served in the armed forces during World War II, the Korean War, the Vietnam War, and certain cold war periods receive special loan privileges from the Veterans Administration (VA). If a veteran fails to repay the loan, the Veterans Administration has the right to deduct the amount not repaid from any pension or other compensation received.

An individual with a VA loan must also pay a small monthly charge for insurance in addition to the regular monthly payment. A down payment is recommended, but the VA loan law permits the Veterans Administration to guarantee loans up to a specified maximum with no down payment if the lender is willing to make the loan for the full amount of the purchase price. Rosa's father can qualify for the VA loan only if his monthly payments do not exceed a specified amount of his take-home pay. This standard applies to FHA loans as well.

QUALIFYING FOR A MORTGAGE

Since housing costs have gone up so much since they bought their first house, Rosa's parents will have to determine how much they can borrow. They know that under current mortgage lending standards, buyers will qualify for a mortgage as long as the monthly cost of the loan does not exceed 28% of their gross monthly income. The bank that lends them the money for their new house will want to be sure that they can afford to make the payments each month. Bankers have found that 28% of gross income is a reasonable limit to place on the size of the mortgage payment.

CRITICAL THINKING

Ask students to examine the disadvantages of adjustable-rate mortgages.

PREPAYING A MORTGAGE AND ASSUMING A MORTGAGE

As with installment loans, it is important to know whether your mortgage has a **prepayment clause** that allows you to make additional payments to principal or prepay the balance of the mortgage. Many banks allow you to make additional payments to principal at any time with no penalty. If you can make prepayments without a penalty, you will save a lot of interest if you can pay an extra amount each month. It is possible to pay off a 30-year mortgage in only 20 years by making one extra payment per year.

Sometimes mortgages have an **assumability clause** that enables a buyer to assume your mortgage when you sell your house. This may make it easier for you to sell your house.

ALGEBRA *REVIEW*

Solve for x.

1. $0.3x = 1200$
   4000
2. $0.4x = 2100$
   5250
3. $0.25x = 356$
   1424
4. $0.82x = 3500$
   4268.29
5. $x = 21.6\%$ of \$2500
   \$540
6. $x = 0.29\%$ of \$1030
   \$2.99
7. $x = 0.018(123{,}456.78)$
   2222.22
8. $y = \dfrac{4x(1+q)}{2t^4r}$ $\dfrac{yt^4r}{2(1+q)}$
9. $z = \dfrac{5ab^{36}}{x(c+2)^{36}}$ $\dfrac{5ab^{36}}{z(c+2)^{36}}$
10. $\sqrt{t} = \dfrac{5xy^{42}z}{36}$ $\dfrac{36\sqrt{t}}{5y^{42}z}$

Ask Yourself

1. How do fixed-rate and adjustable-rate mortgages differ?
   A fixed rate does not vary but an adjustable rate does.
2. What is the main advantage of an adjustable-rate mortgage? The starting interest rate on an ARM is usually lower than a FRM.
3. Who is eligible for a VA loan? veterans of WWII, the Korean War, the Vietnam War, and certain cold war conflicts
4. What percent of gross income is the limit that banks place on the size of a borrower's mortgage payment?
   28%

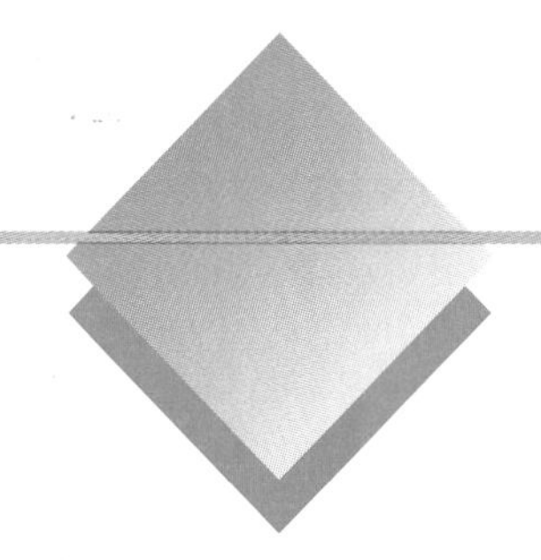

SHARPEN YOUR SKILLS

SKILL 1

As you learned in Chapter 5, the monthly payment formula can be solved for M, giving the amount formula. This formula is used to determine the amount P that can be borrowed if the monthly payment M, the monthly interest rate r, and the number of months n of the loan are all known.

> **Amount Formula**
>
> $P = \dfrac{M[(1 + r)^n - 1]}{r(1 + r)^n}$ where P = amount of loan
> r = monthly interest rate
> n = number of payment periods
> M = monthly payment

EXAMPLE 1 Rosa's parents have a gross monthly income of \$3600. Their bank will lend them the money to buy a house if their monthly mortgage payment is no more than 28% of their monthly income.

QUESTIONS How large can their monthly mortgage payment be? How much money can they borrow at 9% for 30 years?

SOLUTION

The family's monthly mortgage payment can be no more than 0.28(3600) = 1008.

$$P = \frac{M[(1 + r)^n - 1]}{r(1 + r)^n}$$ Use the amount formula.

$$P = \frac{1008[(1 + 0.0075)^{360} - 1]}{0.0075(1 + 0.0075)^{360}}$$ $M = 1008$, $r = 0.09 \div 12 = 0.0075$, $n = 30 \cdot 12 = 360$

$$P = 125{,}276.12$$

They qualify for a loan of \$125,276.12 at 9% for 30 years.

Their loan represents 80% of the purchase price x.

$$0.80x = 125{,}276.12$$
$$x = 125{,}276.12 \div 0.80$$
$$x = 156{,}595.15$$

They can purchase a house priced no more than \$156,595.15.

INSTRUCTION

TEACHING THE LESSON

Have students read the opening scenario. Use the *Algebra Connections* to illustrate the types of mortgages that are available to home buyers. Make sure students understand how to apply the amount formula.

After completing Skills 1 and 2, have the students examine the role of making the monthly payment and the down payment on purchasing a house.

FOCUS ON ALGEBRA

The amount formula in this lesson gives the students the opportunity to explore mathematical functions that apply to their daily lives. *(NCTM Standard 1, p. 137)*

TECHNOLOGY HINT

Remind students to use the exponential function key y^x on the calculator to evaluate the expression $(1 + r)^n$ in the amount formula.

COMMON ERRORS

Be sure that the students follow the order of operations with exponents and know the proper use of the function key y^x.

ESL STUDENTS

Non-native speakers may have difficulty with the terminology used in this lesson. Explain the meaning of key words such as *cap, prime interest rate,* and *amount of loan.*

AT-RISK STUDENTS

Encourage students to gather information on conventional mortgages and VA/FHA mortgages. Have them work in groups to compare rates, points, and down payments.

RETEACHING

A review of math skills in decimals and percents can be helpful in reteaching the lesson. In addition, you may find it helpful to review the monthly payment formula from Chapter 5, comparing it to the amount formula.

ENRICHMENT

Have students research assumable mortgages.

SKILL 2

Even if you qualify, on the basis of your monthly income, to buy a house for a specific amount of money, you must still have the cash for a down payment.

EXAMPLE 2 On the basis of their monthly income, Rosa's family will qualify to purchase a house priced at \$156,595.15. Rosa's parents will have \$28,000 in cash after they sell their current house.

QUESTIONS Will they have enough money to make a 20% down payment on a house selling for \$156,595.15? If not, what is the most expensive house that they can consider?

SOLUTION

$0.20(156{,}595.15) = 31{,}319.03$ Find 20% of the purchase price.

Since they have only \$28,000 for a down payment, they *cannot* afford to buy a house priced at \$156,595.15 because the down payment must be \$31,319.03.

Use the amount of money that they have for a down payment to determine the price x of a house that they can afford.

$$0.20x = 28{,}000 \quad \text{20\% of } x \text{ is \$28,000.}$$
$$x = 28{,}000 \div 0.20$$
$$x = 140{,}000$$

They *can* afford a house priced at \$140,000.

The borrower must meet all of the requirements set out by the bank or financial institution from which mortgage money will be borrowed. Meeting the 28% of gross monthly income requirement is not sufficient if you do not have a down payment of 20% of the purchase price. Similarly, it is not sufficient for you to have 20% of the purchase price. You must also meet the 28% of gross monthly income requirements.

TRY YOUR SKILLS

1. To qualify for a loan at the Scelfo family's local bank, a borrower's monthly mortgage payment must be no more than 28% of the family's gross monthly income. The bank is currently offering mortgages for 30 years at 15%. If the Scelfo family has a monthly income of \$5200, for how large a loan will they qualify? \$115,149.42
2. If their loan in Exercise 1 represents 80% of the purchase price of a house, what is the most expensive house that the family can consider? \$143,936.78
3. If their bank requires a 20% down payment and the family has \$30,000 in cash, will they be able to purchase the house in Exercise 2? yes; 20% is \$28,787.36

EXERCISE YOUR SKILLS

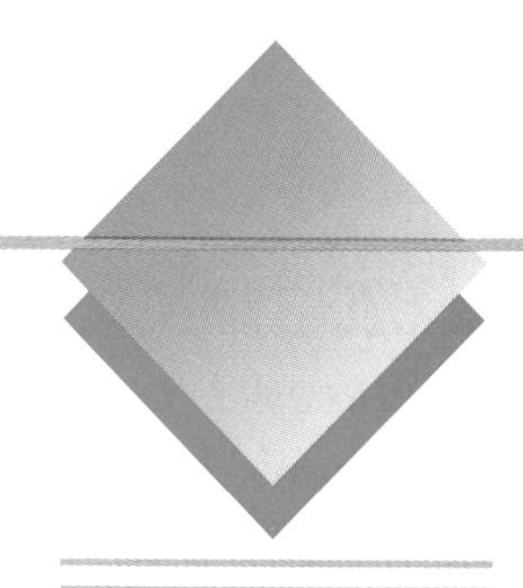

1. Why is the interest rate on a mortgage an important factor in a home purchase? the higher the interest rate, the higher the total cost of the loan
2. In times of inflation, would you rather have an adjustable-rate mortgage or a fixed-rate mortgage? Explain. See Additional Answers.
3. Who provides the insurance for loans made under the FHA and VA programs? federal government
4. Why do lending institutions set a limit on the amount of money that a family can borrow? See Additional Answers.

KEY TERMS

adjustable-rate mortgage (ARM)
assumability clause
balloon mortgage
cap
Federal Housing Administration loans
fixed-rate mortgage (FRM)
prepayment clause
prime interest rate
Veterans Administration loans

GUIDED PRACTICE

You may wish to have students work in groups to calculate the amount of loan and the purchase price that the Scelfo's family can consider. Observe each group as they gather information and provide assistance whenever necessary.

ADDITIONAL ANSWERS (EXERCISE YOUR SKILLS)

2. fixed rate, so your mortgage payments will not go up
4. They do not want to create a situation where the family would be unable to repay the loan.

INDEPENDENT PRACTICE

ASSIGNMENT GUIDE

Exercises 1-4 ask students about factors to consider in purchasing a house. Exercises 5-21 assess students ability to apply the formula taught in the lesson.

CLOSURE

Have students explain options available in financing the purchase of a house. Ask them to define the terminology in the lesson and the variables in the amount formula.

LESSON QUIZ

1. A family has a gross monthly income of $4200. How much money can the family borrow at 8% for 30 years if their monthly mortgage payment is no more than 28% of their monthly income? (**$160,269.39**)
2. If the bank requires that a borrower makes a down payment of 20% on the purchase price of a house and the family in Exercise 1 has $22,000 to use as a down payment, will they qualify for the loan? If not, what is the most expensive house that they can consider? (**No, $110,000**)

To qualify for a loan at the local bank, a borrower's monthly mortgage payment must be no more than 28% of his or her family's gross monthly income. The bank is currently offering mortgages for 30 years at 9%. On the basis of monthly income, for how large a loan will each family qualify? If the loan represents 80% of the purchase price of a house, what is the most expensive house that each family can consider?

	Monthly Income	Mortgage Payment	Loan Amount	House Price
5.	$2500	$ 700	$ 86,997.31	$108,746.64
6.	5900	1652	205,313.64	256,642.05
7.	3500	980	121,796.23	152,245.29
8.	1200	336	41,758.71	52,198.39
9.	2650	742	92,217.14	115,271.43
10.	2899	811.72	100,882.08	126,102.60
11.	5245	1468.60	182,520.35	228,150.44
12.	4350	1218	151,375.31	189,219.14
13.	3495	978.60	121,622.23	152,027.79
14.	1600	448	55,678.28	69,597.85
15.	5777	1617.56	201,033.37	251,291.71

16. If the bank requires that a borrower make a down payment of 20% of the purchase price of his or her house and the family in Exercise 5 has $50,000 to use as a down payment, will they qualify for the loan? If not, what is the most expensive house that they can consider? yes

17. If the bank requires that a borrower make a down payment of 20% of the purchase price of his or her house and the family in Exercise 6 has $32,000 to use as a down payment, will they qualify for the loan? If not, what is the most expensive house that they can consider? no; $160,000

18. If the bank requires that a borrower make a down payment of 20% of the purchase price of his or her house and the family in Exercise 7 has $8000 to use as a down payment, will they qualify for the loan? If not, what is the most expensive house that they can consider? no; $40,000

19. If the bank requires that a borrower make a down payment of 20% of the purchase price of his or her house and the family in Exercise 8 has $20,000 to use as a down payment, will they qualify for the loan? If not, what is the most expensive house that they can consider? yes

20. If the bank requires that a borrower make a down payment of 20% of the purchase price of his or her house and the family in Exercise 9 has $24,000 to use as a down payment, will they qualify for the loan? If not, what is the most expensive house that they can consider? yes

21. If you want to make a down payment of 30% of the purchase price of your house and you have $50,000 in cash, can you purchase a house priced at $165,000? yes

MIXED REVIEW

3-1 **1.** Carin saves $85 per week. She wants to buy a computer that costs $2750. For how many weeks must she save before she can buy the computer? 33 wk

3-2 **2.** How much will $4859 be worth at the end of 3 years, given that the annual interest rate is 6%, compounded semiannually? $5801.90

7-3 **3.** Use the formula for the Future Value of a Periodic Investment to find how much an IRA will be worth in 35 years if the annual contribution is $1,500 and if the compounded annual return is 10.5%. $456,238.19

8-3 The table below shows the recent 5-year record of a mutual fund.

Date	Time After Start (in Years)	Price of Mutual Fund (in dollars)
Jul 1, 1990	0	$30.00
Jan 1, 1991	0.5	31.40
Jul 1, 1991	1.0	35.80
Jan 1, 1992	1.5	36.20
Jul 1, 1992	2.0	34.10
Jan 1, 1993	2.5	38.40
Jul 1, 1993	3.0	41.40
Jan 1, 1994	3.5	40.25
Jul 1, 1994	4.0	45.10
Jan 1, 1995	4.5	46.03
Jul 1, 1995	5.0	47.30

4. Use a graphing calculator to find the linear regression equation that is the line of best fit for the data. $y = 30.165 + 3.424x$

5. Use the equation that you found in Exercise 4 to predict a possible value for the mutual fund in January of 1997. $52.42

9-2 Use IRS Form 1040EZ and the tax table in the Reference Section to find the tax paid by the following single persons with the given salaries and interest income. The person cannot be claimed as a dependent by anyone else.

6. A salary of $12,000 and no interest income $919

7. A salary of $8,800 and interest income of $213 $469

10-3 **8.** Use the odometer readings and gasoline purchases shown below to determine the distance traveled, the number of miles obtained per gallon of gas, and the cost per mile for a weekend trip of the Baxter family.

1st odometer reading: 15,401.8
2nd odometer reading: 15,853.9
16 gallons of gas costing $20.75 452.1 mi; 28.3 mi/gal; $0.046/mi

12-4 HOME OWNERSHIP: FINANCIAL RESPONSIBILITIES AND ADVANTAGES

QUICK REFERENCE

TEACHER'S RESOURCES AND ANSWER KEY

Lesson Quiz Answers 12-4, p. 84

Reteaching Activity Answers 12-4, p. 113

Enrichment Activity Answers 12-4, p. 114

EXTENSION ACTIVITIES

Reteaching Activity 12-4, p. 92

Enrichment Activity 12-4, p. 96

Hari first realized the importance of homeowner's insurance on a cold New Year's Day three years ago. His family got up to find that all the water pipes had frozen. Since it was their custom to spend the day with Hari's grandparents, who live in a nearby town, they decided to leave early and stop for coffee on the way.

Hari's father does not think very clearly without his morning coffee. He opened the cold water tap in the bathtub to help drain the frozen pipes but forgot to open the drain as well. When Hari opened the front door that evening when the family arrived home, he could hear the water splashing over the side of the bathtub. The bathroom floor was a small lake! Water had run down the hall, soaking the carpets in two bedrooms, and was slowly seeping into the dining room.

While Hari was eager to help clean up the mess, his mother suggested that they first call the insurance company to find out what they should do. The insurance agent told her to take photographs of all the damaged areas before moving or cleaning anything. She was also advised to make a list of all damaged items and any parts of the house that needed repair because of the flood. Hari's mother and the agent then agreed on a time at which an insurance adjuster could come to assist the family in filing an insurance claim.

Hari's father pointed out that even though they have a $250 deductible clause in their insurance policy, their settlement from the insurance company would still pay for a new carpet for Hari's room. Hari was thrilled to help pick it out. He chose a color that went well with his computer desk chair and a type that would not build up static electricity when he walked on it.

OBJECTIVES: *In this lesson, we will help Hari to:*

- *Understand why homeowner's insurance is important.*
- *Examine the different categories of coverage in a typical insurance policy.*
- *Learn about two important factors that affect how much an insurance company pays on claims—the deductible and the replacement cost.*
- *Recognize the value of keeping good records to document a claim.*
- *Examine the costs of home maintenance and repairs.*
- *Calculate some of the tax advantages of home ownership.*

FOCUS

ALGEBRA CONNECTIONS

Would you prefer to pay a higher deductible and a lower insurance premium, or a lower deductible and a higher insurance premium?

THE ROLE OF HOMEOWNER'S INSURANCE

A house is the single largest purchase that most families make. Unfortunately, as Hari discovered, unforeseen events may result in damage to this valuable asset and its contents. Incidents such as theft or an injury on their property can also occur. An injured person could hold Hari's family liable and require them to pay medical expenses.

Homeowner's insurance generally covers the following potential disasters: fire and lightning, theft, windstorm, hail, explosion, smoke, cave-ins, frozen pipes, falling objects, and water damage. Sometimes there is an extra charge for certain types of coverage. For example, it may be difficult or very costly to obtain flood insurance in areas that are prone to floods.

Homeowner's insurance protects Hari's family against the financial hardship created by these potentially costly incidents. Requirements for homeowner's insurance vary by state. In the state in which Hari lives, the cost of homeowner's insurance is included in the monthly mortgage payment that his family makes to their bank. The bank then pays the insurance company.

PURCHASING DECISIONS

You may buy insurance from one of three sources: from an independent insurance agent who sells insurance for many companies, from an agent who sells only one company's insurance, or from the insurance company itself. It is often advantageous to find a local representative who can help file claims, clarify questionable points, and accelerate the payment of claims.

Hari's family made choices about the coverage that they wanted when they purchased their homeowner's insurance policy. **Coverage** refers to the items eligible for financial compensation and the dollar limits on that compensation. Decisions that the family had made on coverage affected their premium. The **premium** is the amount that they pay to the insurance company for the coverage that they selected. Some of the options from which they had to choose when they selected their insurance are the following.

Dwelling (Coverage A) Dwelling coverage includes the dwelling described in the policy, any structures connected to it, outdoor equipment, building equipment used to maintain the property, and permanently attached fixtures. Rented or borrowed property is not included.

Other Structures (Coverage B) Structures that are detached from the dwelling are insured for 10% of the amount of the insurance on the dwelling.

Personal Property (Coverage C) Personal property coverage includes all items owned, regardless of where they are located. If they are not located on the premises, they are covered up to a total of 10% of the dwelling coverage. Since many items are excluded, it is important to read this section of a homeowner's policy carefully.

Loss of Use (Coverage D) If the water had damaged Hari's house so much that his family had been unable to live in it, they would have been covered for temporary living expenses up to the limit specified in their policy.

Personal Liability (Coverage E) If a visitor had slipped and fallen on the flooded bathroom floor, Hari's family would have been held liable, or responsible, for any injury that the person suffered as a result. Their insurance company would have been required to defend them in court if the injured person had decided to bring a lawsuit against them.

Medical Payments to Others (Coverage F) If an injured visitor had been admitted to the hospital, Coverage F would have helped pay for medical expenses, even if Hari's family was not held personally liable.

REIMBURSEMENT FACTORS

The amount of the reimbursement that Hari's family will receive for any claim will be determined by the original cost of the damaged items, the deductible amount, and the kind of insurance they have.

The amount that Hari's family pays—in their case, $250—for repairing the eligible damage before the insurance policy begins to reimburse them is the **deductible.** The deductible applies to each accident or loss. If Hari's father makes the same mistake one week later, they will pay the deductible again. In general, the higher the deductible, the lower the insurance premium.

Buildings and other permanent structures (Coverage A and B) automatically have **replacement cost** coverage. If the flooded bathroom floor and part of the walls were completely damaged, the insurance would cover the actual amount that it costs to repair or replace these items. Personal property, however, is not automatically covered at replacement cost. It is covered for the purchase price at the time it was purchased. For example, if the damaged carpeting in Hari's room cost \$675 when it was installed four years ago but costs \$790 to replace at the time of the flood, a standard policy would have reimbursed the family for \$675. However, Hari's family paid an additional premium to ensure that their personal property was covered at replacement cost.

KEEPING RECORDS

When the insurance adjuster arrived to inspect the damage, Hari quickly learned how important it was to keep detailed information about what the family owned. They had kept the receipts that showed where they had purchased the carpeting and exactly what kind it was. This was fortunate because they had purchased high-quality carpeting.

CRITICAL THINKING

After the students have answered the questions, have them consider the impact of inflation on insurance premiums and monthly savings.

OTHER FINANCIAL RESPONSIBILITIES

There are many costs associated with maintaining a home that are not covered by homeowner's insurance. Many of these costs can be anticipated, so you can put money aside to cover them. Hari's family has periodic service check-ups of the following.

- Heating and air-conditioning systems
- All parts of the house for termite inspection and extermination of other pests if necessary
- Gutters, drains, and roof
- Plumbing and electrical systems

Appliances such as the refrigerator, range, freezer, clothes dryer, washing machine, television, hot water heater, dishwasher, and garbage disposal will not last forever and will need to be replaced at some time. The family must also maintain their lawn and the outside of their home.

Property taxes and mortgage interest can be used to reduce your taxable income. These items can be deducted on your federal tax return, saving you a considerable amount of money, particularly in the early years when your mortgage interest is high.

Ask Yourself

1. What is the advantage of buying replacement cost insurance? You will be reimbursed for the current cost of the item, not the price originally paid.
2. What is included in personal property coverage? all items owned
3. Which kinds of coverage protect the family against liability for injury to a nonfamily member? Coverage E (personal liability)

ALGEBRA *REVIEW*

Express each percent as a decimal.

1. 22.7% 0.227
2. 0.96% 0.0096
3. 1.31% 0.0131
4. 107.8% 1.078

Solve for x.

5. $x = (2.8\% \text{ of } \$160{,}980) + \$2512$ \$7019.44
6. $x = (0.85\% \text{ of } \$212{,}212) + \$7579$ \$9382.80
7. $x = \dfrac{1209[(1 + 0.0025)^{72} - 1]}{0.0025}$ 95,244.28
8. $x = \dfrac{1558[(1 + 0.0175)^{108} - 1]}{0.0175}$ 490,729.15
9. $1668 = \dfrac{x[(1 + 0.0125)^{60} - 1]}{0.0125}$ 18.83
10. $12{,}220 = \dfrac{x[(1 + 0.015)^{120} - 1]}{0.015}$ 36.89

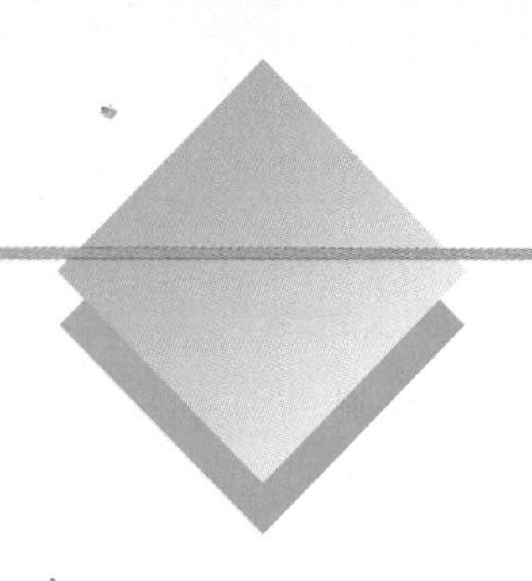

SHARPEN YOUR SKILLS

SKILL 1

EXAMPLE 1 Hari's family pays an additional annual premium of $35 on their homeowner's insurance for replacement cost coverage on their personal property. Their deductible is $150. They filed two insurance claims this year for personal property damage. One was for $200, the other for $800. They estimated that they would have been reimbursed 10% less for their claims if they had not had the replacement cost coverage.

QUESTION Was the extra premium that they paid for replacement cost coverage worth it?

SOLUTION

	With Extra Coverage	Without Extra Coverage	
Claim 1	$200	$180	200 − 0.1(200) = 180
Less deductible	− 150	− 150	
Reimbursement	$ 50	$ 30	
Claim 2	$800	$720	800 − 0.1(800) = 720
Less deductible	− 150	− 150	
Reimbursement	$650	$570	

Total personal property damage claims
With replacement cost coverage: 50 + 650 = $700
Without replacement cost coverage: 30 + 570 = $600
Advantage of extra coverage: 700 − 600 = $100

They paid $35 to receive an extra $100 in reimbursements. The additional coverage was worth it because they had a net savings of 100 − 35 = $65.

SKILL 2

In Chapter 7 you learned to find the future value of money that you are investing periodically. The formula is as follows.

Future Value of a Periodic Investment Formula

$$A = \frac{p[(1 + r)^n - 1]}{r}$$

where A = future value of periodic investment
p = the investment made at the end of each period
r = periodic interest rate
n = number of periods

INSTRUCTION

TEACHING THE LESSON

Ask students to read the opening scenario. Use the *Algebra Connection* to explain the importance of homeowner's insurance. Have students read the different types of coverage that are available to homeowners. Make sure students understand the reimbursement factors.

After completing Skills 1 and 2, have students examine the monthly savings needed to have $3000 at the end of 5 years with an APR of 4%. (**$45.25**)

In Skill 2, the interest on the savings is compounded monthly. Many banks compound interest daily. Students could investigate how their local banks pay interest on savings accounts.

In Skill 3, note that the family saves on its taxes only to the extent that their itemized deductions exceed their standard deduction.

FOCUS ON ALGEBRA

The Periodic Investment Formula in this lesson gives the students the opportunity to explore mathematical functions that apply to their daily lives. *(NCTM Standard 1, p. 137)*

COMMON ERRORS

Remind students to subtract one where appropriate in using the Periodic Investment Formula.

This formula can be solved for p, which will allow you to calculate the periodic investment (or payment) if you know the future value, the periodic interest rate, and number of periods for which the money will be earning interest.

$$A = \frac{p[(1 + r)^n - 1]}{r}$$

$$Ar = p[(1 + r)^n - 1] \quad \text{Multiply both sides by } r.$$

$$\frac{Ar}{(1 + r)^n - 1} = p \quad \text{Divide both sides by } (1 + r)^n - 1.$$

Periodic Investment Formula

$$p = \frac{Ar}{(1 + r)^n - 1}$$

where p = periodic investment
A = future value of periodic investment
r = periodic interest rate
n = number of periods

TECHNOLOGY HINT

For those students who are not using a calculator with a screen, encourage them to write down the proper keystrokes before beginning the calculation.

EXAMPLE 2 Hari's parents would like to set up a savings account from which they can draw money to replace major appliances or make home improvements when it becomes necessary. For purposes of setting up this savings account, they are assuming that they will have to replace their appliances in ten years, although their appliances will probably last longer. They know that appliance costs will probably increase in ten years, so they estimated the cost of each appliance at that time as follows:

Appliance/Home Improvement	Estimated Cost in 10 Years
Refrigerator	$ 1500
Freezer	800
Dryer	450
Washer	600
Television	1250
Disposal	250
Hot water heater	380
Dishwasher	850
Air conditioner	3000
Carpets	6500
Total:	$15,580

COOPERATIVE LEARNING

Have students work in small groups to determine the monthly savings using a combination of cost, APR, and number of periods.

INTERDISCIPLINARY INVESTIGATION

Have students survey local insurance offices and prepare a report on different types of coverage and how they calculate the premiums they charge.

QUESTION How much should Hari's parents save each month to have $15,580 in their account at the end of ten years if their account carries an APR of 6% and interest is paid monthly?

SOLUTION

$p = \dfrac{Ar}{(1 + r)^n - 1}$ Use the periodic investment formula.

$p = \dfrac{15{,}580(0.005)}{(1 + 0.005)^{120} - 1}$ Substitute $A = 15{,}580$, $r = 0.06 \div 12 = 0.005$, and $n = 10 \cdot 12 = 120$.

$p = 95.07$ To the nearest cent

Hari's parents will have to save $95.07 each month to have $15,580 in their savings account at the end of ten years.

RETEACHING

A review of the basic skills on decimals, percents, and evaluating expressions with exponents might benefit some students.

ENRICHMENT

Challenge students to write a computer program using the Periodic Investment Formula.

SKILL 3

EXAMPLE 3 Hari's parents want to know how much they will save in federal taxes this year by deducting their property taxes and the interest charges on their mortgage on their return. Their property taxes are 1.25% of the assessed value of their home, which is $105,000. They will pay $6200 in mortgage interest this year.

QUESTION If they are in a 28% tax bracket, how much can they save in federal income tax this year?

SOLUTION

Property tax cost: $0.0125(105{,}000) = 1312.50$

Save 28% of deduction: $0.28(1312.50 + 6200.00) = 2103.50$

They may save as much as $2103.50 on their federal income tax by deducting their property taxes and mortgage interest on their tax return. If they have no other large deductions, such as medical expenses, these tax savings will be less.

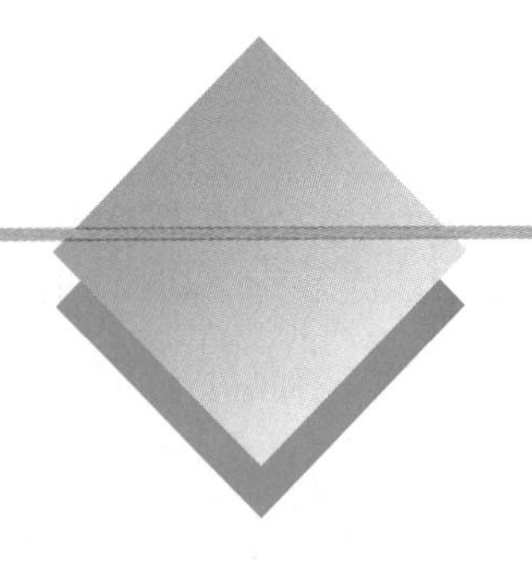

TRY YOUR SKILLS

1. Justin's family has homeowner's insurance for their house. They paid $100 this year for replacement cost coverage on their personal property. Their deductible is $200. They filed an insurance claim for $1500 this year on a stereo system that was stolen and for which they paid $980 some years ago. Was the extra premium that they paid for replacement cost coverage worth it? yes
2. If you estimate that it will cost $8000 to replace all the carpeting in your house in 5 years and you wish to save for that expense now, how much should you place in your savings account each month if it has an APR of 3%? $123.75
3. What would you have to save each month in Exercise 2 if the APR on your account were 4% instead of 3%? $120.67

GUIDED PRACTICE

Have students work in groups to calculate the monthly savings for Exercises 2-3. You might ask students to recall and discuss other investment options for their money.

Calculate the maximum tax savings for the following deductions. Assume that each family is in a 28% tax bracket and that their property tax is the given percent of the assessed value of their home.

	Assessed Value	Property Tax	Amount of Property Tax	Mortgage Interest	Total Deductions	Tax Savings
4.	$150,000	1.8%	$2700	$11,205	$13,905	$3893.40
5.	225,000	0.9%	2025	15,876	17,901	5012.28
6.	85,000	1.2%	1020	6,788	7,808	2186.24

EXERCISE YOUR SKILLS

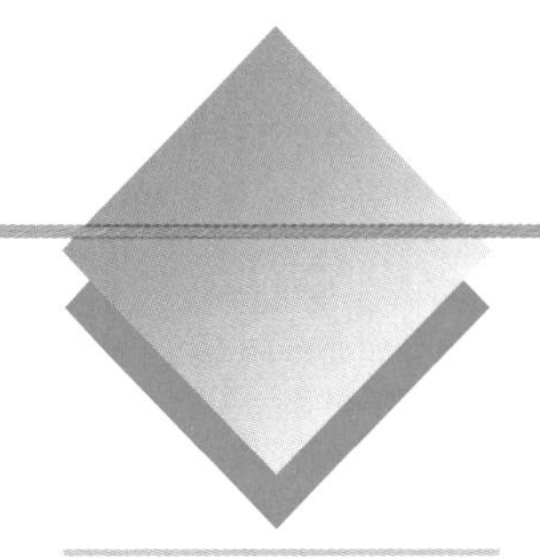

1. Why is it more difficult to obtain flood insurance in areas that are prone to flooding? Answers may vary.
2. Why is it important to keep receipts from household appliances and other furnishings that you purchase?
3. Should you clean up the mess from a disaster before or after the insurance adjuster comes to your home?
4. What costs are not covered by homeowner's insurance?

ADDITIONAL ANSWERS (EXERCISE YOUR SKILLS)

2. for insurance reimbursement
3. document with photos and contact the insurance agent before beginning to clean
4. heating and air-conditioning; extermination control; gutters, drains, roofs; plumbing; maintenance of lawn, and outside of house

KEY TERMS

coverage
deductible
homeowner's insurance
premium
replacement cost

INDEPENDENT PRACTICE

ASSIGNMENT GUIDE

Exercises 1-4 assess students' understanding of homeowner's insurance and financial responsibilities. Exercises 5-10 assess students' ability to calculate amount recovered with and without replacement cost. Exercises 11-22 assess students' ability to apply the formula taught in the lesson. Exercises 23-28 assess students' ability to calculate the tax savings.

CLOSURE

Ask students to explain the importance of homeowner's insurance and the different coverages available. Have students define each of the variables in the Periodic Investment Formula.

LESSON QUIZ

1. Sadie's family has homeowner's insurance for their house. They paid $125 this year for replacement cost coverage on their personal property. Their deductible is $200. They filed an insurance claim for $850 on a stolen television set. What amount did they recover? (**$650**)
2. Find the amount that Jonathan's family must save each month to replace a refrigerator in 5 years that is estimated to cost $1400. Their account carries an APR of 6%. (**$20.07**)

Li-Ming's family has homeowner's insurance for their house. They paid $150 this year for replacement cost coverage on their personal property. Their deductible is $250. They filed an insurance claim for $1275 this year on a piece of furniture that was damaged by a flood and for which they paid $790 some years ago.

5. What amount did they recover? $1025
6. What amount would they have recovered if they had not had replacement cost coverage? $540
7. Was the extra premium that they paid for replacement cost coverage worth it? yes

Phillipa's family has homeowner's insurance for their house. They paid $60 this year for replacement cost coverage on their personal property. Their deductible is $150. They filed an insurance claim for $400 this year on an awning that was blown down by a hurricane and for which they paid $225 some years ago.

8. What amount did they recover? $250
9. What amount would they have recovered if they had not had replacement cost coverage? $75
10. Was the extra premium that they paid for replacement cost coverage worth it? yes

Complete the following chart to find the amount that you must save each month to replace each appliance or make each home improvement in 10 years. Assume that your account carries an APR of 5%.

	Appliance/ Home Improvement	Estimated Cost in 10 Years	Monthly Savings Required
11.	Refrigerator	$1850	$11.91
12.	Water heater	375	2.41
13.	Garage	3500	22.54
14.	Carpeting	8200	52.81
15.	Dishwasher	850	5.47
16.	Kitchen cabinets	8700	56.03
17.	Bedroom furniture	1695	10.92
18.	Garbage disposal	280	1.80
19.	Replacement roof	3600	23.18
20.	Outdoor deck	520	3.35
21.	Large-screen television	3350	21.57
22.	Microwave oven	420	2.70

Calculate the maximum tax savings for the following deductions. Assume that each family is in a 28% tax bracket and that its property tax is the given percent of the assessed value of its house.

	Assessed Value	Property Tax	Amount of Property Tax	Mortgage Interest	Total Deductions	Tax Savings
23.	$140,000	2.40%	$3360	$10,625.00	$13,985	$3915.80
24.	120,000	1.58%	1896	9,850.75	11,746.75	3289.09
25.	205,000	1.25%	2562.50	16,143.75	18,706.25	5237.75
26.	83,500	1.05%	876.75	6,575.63	7,452.38	2086.67
27.	102,000	0.85%	867	8,032.50	8,899.50	2491.86
28.	165,600	0.70%	1159.20	13,041.00	14,200.20	3976.06

MIXED REVIEW

3-2 **1.** Use the Rule of 72 to determine how long it will take Tiffany's $5500 savings account to double. It is growing at a rate of 5.25% per year.

4-3 **2.** The fixed costs of James's business are $275 per week. The variable costs are $5.20 per item. James produces 90 items in 1 week. What is his total cost for the week? $743

5-2 **3.** A family's take-home pay is $4400.00 per month. How much can it afford to spend on credit payments each month, excluding a home mortgage? $880

10-4 **4.** Use the Driver Rating Factors Table in the Reference Section to find the driver rating factor for the usual driver of the car who is 50 years of age and drives to work. 1.15

11-1 Use the United States Mileage Chart in the Reference Section to find the distance between the following pairs of cities.

5. Austin, Texas, and Philadelphia, Pennsylvania 1615 mi

6. Pierre, South Dakota, and New Orleans, Louisiana 1394 mi

11-2 The charges at two motels are as follows.

Cozy Inn	Stay-A-While Lodge
Rates: single, $45	Rates: single, $35
2 adults, $60	2 adults, $45
teens, free	teens, $10 each
Room tax: 6%	Room tax: 6%
Free in-room movies	Game machines
Swimming pool	Free continental breakfast

Find the motel costs for each individual or family.

7. a family of two adults and one teenager at Stay-A-While Lodge for 2 nights $116.60

8. a single person at Cozy Inn from Friday afternoon until Monday morning

ADDITIONAL ANSWERS (MIXED REVIEW)

1. 14 years

8. $143.10

CHAPTER 12 REVIEW

1. What are the advantages and disadvantages of owning a house?
2. How do fixed-rate and adjustable-rate mortgages differ?
3. What percent of gross income is the limit that banks place upon the size of a borrower's mortgage payment? 28%

Karen collected the information shown about houses selling in her town. Give answers to four significant digits. See Additional Answers.

4. What is a linear model of the form $y = a + bx$ that will allow Karen to predict the number of bedrooms in a house in her town if she knows the selling price of the house?
5. Graph the linear regression equation on a scatter plot of the data.
6. Is there a good correlation between the price and the number of bedrooms in a house? yes; $r = 0.9659$
7. Estimate the number of bedrooms in a house in Karen's town that is selling for $260,000. Round your answer to the nearest whole bedroom. 4

Price (in thousands of dollars)	Number of Bedrooms
$150	2
195	3
68	1
210	3
310	5
270	4
135	2
190	2

ADDITIONAL ANSWERS

1. Answers may vary.
2. Adjustable rates vary but fixed rates do not.
4. $y = -0.3500 + 0.0162x$

Use a spreadsheet program to determine how the total financed price on a $210,000 mortgage for 30 years with a down payment of $50,000 differs as interest rates vary from 9% to 11% in increments of 0.5%. In each case, what is the savings over an 11% loan?

	Interest Rate	Total Financed Price	Savings Over 11%
8.	9.0	$513,463	$85,075
9.	9.5	534,332	64,206
10.	10.0	555,481	43,057
11.	10.5	576,890	21,648
12.	11.0	598,538	0

To qualify for a loan at the local bank, a borrower must have a monthly mortgage payment that is no more than 28% of his or her family's gross monthly income. The bank is currently offering mortgages for 30 years at 8%. On the basis of monthly income, for how large a loan will each family qualify? If the loan represents 80% of the purchase price of a house, what is the most expensive house that each family can consider?

	Monthly Income	Mortgage Payment	Loan Amount	House Price
13.	$3200	$ 896	$122,110.01	$152,637.51
14.	2700	756	103,030.32	128,787,90
15.	3300	924	125,925.95	157,407.44
16.	1750	490	66,778.91	83,473.64
17.	6200	1736	236,588.15	295,735.19
18.	4350	1218	165,993.30	207,491.63
19.	2375	665	90,628.52	113,285.65
20.	3000	840	114,478.14	143,097.68
21.	2244	628.32	85,629.65	107,037.06
22.	2890	809.20	110,280.60	137,850.75

23. If the bank requires that a borrower make a down payment of 20% of the purchase price of a house and the family in Exercise 13 has $25,000 to use as a down payment, will they qualify for the loan? If not, what is the most expensive house that they can consider? no; $125,000

24. Avram's family has homeowner's insurance for their house. They paid $125 this year for replacement cost coverage on their personal property. Their deductible is $250. They filed an insurance claim for $1600 this year on some sterling silver flatware that was stolen and for which they paid $1020 some years ago. Was the extra premium that they paid for replacement cost coverage worth it? yes

25. If you estimate that it will cost $7200 to replace all the carpeting in your house in 5 years and you wish to save for that expense now, how much should you place in your savings account each month if it has an APR of 4.5%? $107.23

Calculate the maximum tax savings for the following deductions. Assume that each family is in the 28% tax bracket and that their property tax is the given percent of the assessed value of their home.

	Assessed Value	Property Tax	Amount of Property Tax	Mortgage Interest	Total Deductions	Tax Savings
26.	$165,000	1.78%	$2937	$11,375	$14,312	$4007.36
27.	245,000	0.89%	2180.50	16,988	19,168.50	5367.18
28.	75,000	1.12%	840	5,899	6,739	1886.92

CHAPTER 12 TEST

Ben collected the following information about houses selling in his town. Give answers to four significant digits.

1. What is a linear model of the form $y = a + bx$ that will allow Ben to predict the number of acres of land on which a house stands in his town if he knows the selling price of the house?
2. Is there a good correlation between the price and the number of acres?
3. Graph the linear regression equation on a scatter plot of the data. See Additional Answers.
4. Estimate the number of acres of land to the nearest tenth for a house selling for $220,000. 0.9 acre

Price (in thousands of dollars)	Acres of Land
$145	0.4
190	0.6
65	0.3
205	0.7
315	1.5
275	1.3
140	0.5
185	0.8

ADDITIONAL ANSWERS

1. $y = -0.2215 + 0.005179x$
2. yes; $r = 0.9525$
12. $99.52

ALTERNATIVE ASSESSMENT

Ask students to write a report on insurance premiums that are charged to commercial and industrial institutions. Have them research the differences in APR charged to individuals and institutions.

Find the total financed price of a $270,000 house with each of the following combinations of down payment, interest rate, and amortization period.

	Percent Down	Down Payment	Loan Amount	Interest Rate	Number of Years	Total Financed Price
5.	20%	$54,000	$216,000	10%	15	$471,806
6.	20%	54,000	216,000	10%	30	736,400
7.	30%	81,000	189,000	11%	15	467,670
8.	20%	54,000	216,000	11%	30	794,527

To qualify for a loan at the local bank, a borrower must have a monthly mortgage payment that is no more than 28% of the family's gross monthly income. The bank is currently offering mortgages for 15 years at 7.5%. On the basis of monthly income, for how large a loan will each family qualify? If the loan represents 80% of the purchase price of a house, what is the most expensive house that each family can consider?

	Monthly Income	Mortgage Payment	Loan Amount	House Price
9.	$4300	$1204	$129,879.61	$162,349.51
10.	1950	546	58,898.89	73,623.61
11.	6500	1820	196,329.64	245,412.05

12. If you estimate that it will cost $3800 to stain the exterior of your house in 3 years and you wish to save for that expense now, how much should you place in your savings account each month if it has an APR of 4.0%?
13. Calculate the federal income tax savings for a family in a 28% tax bracket with a house assessed at $162,000 with property tax of 2.1% and mortgage interest of $10,500. $3892.56

CUMULATIVE REVIEW

5-1 **1.** Determine your monthly car payment if you borrow $8,500 at 9.5% annual interest for 5 years. $178.52

5-5 **2.** Determine the plan that would provide the lowest total financed price for a $18,900 car. Assume that the loan in Plan 1 is from a bank or other lending institution and that the loans in Plans 2 and 3 are from the car manufacturer's financing subsidiary. Plan 2 is lowest

	Plan 1	Plan 2	Plan 3
Rebate	$1500	$1000	0
Rate	10.25%	7.9%	4.5%
Time	36 months	36 months	36 months
Loan Amount	$17,400	$17,900	$18,900
Monthly Payment	$563.49	$560.10	$562.22
Total Financed Price	$20,285.77	$20,163.44	$20,239.81

6-3 Your credit card carries a monthly finance charge of 1.75%. You pay 10% of the amount owed to the nearest dollar each month. Your beginning balance is $1,240. Complete a table with the following column heads for 3 months.

	Month	Balance	Interest	Amount Owed	Payment
3.	1	$1240	$21.70	$1261.70	$126
4.	2	1135.70	19.87	1155.57	116
5.	3	1039.57	18.19	1057.76	106

6-5 **6.** Your credit card account balance from 2/1–2/13 was $550. Then you made a payment of $100 on 2/14. You made no further charges or payments for the remainder of the month. Determine your closing balance for February if your card carries a monthly finance charge of 1.25% and it is not a leap year.

7-1 **7.** Use the Multiples of Salary Chart in the Reference Section to find the amount of life insurance a 45-year-old person who has gross earnings of $40,000 would need in order to provide 60% income replacement.

8-2 **8.** If you buy 750 shares of stock at a price of $75\frac{1}{4}$, what will the total cost of the stock be? The commission rate is 1%. $57,001.88

10-3 **9.** Determine the distance traveled, the mileage per gallon of gas, and the cost per mile based on the following information.

1st odometer reading: 35,461.9
2nd odometer reading: 35,953.5
17.4 gallons of gas at a cost of $19.87

10-3 **10.** Use the cost of gasoline per mile found in Exercise 9 to find the total operating expenses and the cost per mile for operating a car for 11,750 miles with $250.00 in maintenance costs and $274.50 in repair costs.

ADDITIONAL ANSWERS

6. $456.21
7. $240,000
9. 491.6 mi; 28.3 mi/gal; $0.04 cost/mi
10. $994.50; $0.08/mi

PROJECT 12–1: **Shopping for a House**

Answers may vary.

1. Find newspaper advertisements for 20 houses that are for sale in your town.
2. For each house, find the number of square feet, the number of bedrooms, and the number of bathrooms.
3. Make a poster presenting the information you gather.
4. Run a regression analysis to find a linear model of the form $y = a + bx$ that will allow you to predict the number of square feet in a house in your town if you know its selling price.
5. Run a regression analysis to find a linear model of the form $y = a + bx$ that will allow you to predict the number of bedrooms in a house in your town if you know its selling price.
6. Run a regression analysis to find a linear model of the form $y = a + bx$ that will allow you to predict the number of bathrooms in a house in your town if you know its selling price.
7. Use your results from Exercise 4 to predict how many square feet are in a house in your town that is selling for $150,000.
8. Use your results from Exercise 5 to predict how many bedrooms are in a house in your town that is selling for $175,000.
9. Use your results from Exercise 6 to predict how many bathrooms are in a house in your town that is selling for $120,000.
10. Select a house that has fewer square feet than your model would predict based on its price. Determine whether the house has other advantages or amenities that warrant the higher price.
11. Select a house that has more bedrooms than your model would predict based on its price. Determine whether the house has any major drawbacks.

Recall that the *principal square root* of a number is either positive or zero.

$\sqrt{9} = 3 \qquad \sqrt{9} \neq -3 \qquad \sqrt{5-5} = 0$

A *radical equation* contains a variable under a radical sign. To solve a radical equation involving one or more square roots, isolate one square root on one side of the equation. Then square each side of the equation. The resulting equation may have solutions that are not solutions of the original equation. Therefore, it is important to check all solutions in the original equation.

Solve.

Examples

a. $2\sqrt{x} + 1 = 3\sqrt{x} - 4$

$1 + 4 = 3\sqrt{x} - 2\sqrt{x}$

$5 = \sqrt{x}$

$25 = x$

Check $2\sqrt{25} + 1 \;?\; 3\sqrt{25} - 4$

$2 \cdot 5 + 1 \;?\; 3 \cdot 5 - 4$

$10 + 1 \;?\; 15 - 4$

$11 = 11$

25 is a solution

b. $\sqrt{x} = x - 2$

$x = x^2 - 4x + 4$ Square both sides.

$0 = x^2 - 5x + 4$

$0 = (x - 4)(x - 1)$ Factor.

$0 = x - 4$ or $0 = x - 1$

$4 = x$ or $1 = x$

Check $\sqrt{4} \;?\; 4 - 2$ $\qquad$ $\sqrt{1} \;?\; 1 - 2$

$2 = 2$ $\qquad$ $1 \neq -1$

4 is a solution. $\qquad$ 1 is not a solution.

1. $\sqrt{x} - 3 = 2$
25

2. $3\sqrt{x} - 2 = \sqrt{x} + 6$
16

3. $\sqrt{x+1} = x - 5$
8

4. $\sqrt{x-1} = 0$
1

5. $\sqrt{4+x} = x - 2$
5

6. $\sqrt{x^2 - 5} = 4$
9, −9

A *rational equation* contains one or more rational expressions. The denominators in a rational equation may be eliminated by multiplying both sides of the equation by each denominator. Check each solution in the original equation.

Solve.

Example $\dfrac{14}{x} = x + 5$

$14 = x^2 + 5x$ Multiply both sides by x.

$0 = x^2 + 5x - 14$ Subtract 14 from both sides.

$0 = (x + 7)(x - 2)$ Factor.

$0 = x + 7$ or $0 = x - 2$

$-7 = x$ or $2 = x$ Check in the original equation.

7. $\dfrac{6}{x} = 3$
2

8. $\dfrac{x}{3} = \dfrac{x-4}{2}$
12

9. $\dfrac{1}{x} = \dfrac{x}{4}$
−2, 2

10. $\dfrac{x}{2} + \dfrac{5}{x} = \dfrac{7}{2}$
2, 5

11. $\dfrac{x}{3} = \dfrac{x+3}{5}$
$\frac{9}{2}$

12. $\dfrac{3}{x+2} = \dfrac{x+2}{3}$
1, −5

CHAPTER 13

Apartments and Other Housing Alternatives

MANY YOUNG PEOPLE FIND THE IDEA of moving away from their parents and establishing a residence of their own to be very appealing. Whether they are going to college, into the work force, or a combination thereof, young people gain a feeling of independence from living on their own. With the expression of that feeling of independence comes responsibility. If they do not go to the grocery store, no food will appear in the refrigerator!

Most young people share an apartment, rented house, or other living quarters with others. Many aspects of sharing living space with another person become immediately apparent. Roommates learn very quickly whether their living habits differ. They must decide who makes sure that the rent and the telephone bill are paid on time, who buys the groceries, who prepares the meals, and how much noise, quiet, clutter, fastidiousness, dirty laundry, television, extra company, immodesty, late-night snacking, and so on each person can tolerate. How does each roommate adjust so that the group can live harmoniously?

After renting for a while, many young people want to purchase a home. A condominium or a mobile home may fit their needs and match their ability to pay.

In this chapter our high school students will learn about the factors involved in renting an apartment or a house. They will study the purchase of smaller-scale housing for people who want the advantages of home ownership without paying the price for a full-sized house.

or Rent

13-1 THE HUNT FOR AN APARTMENT: HOW MUCH CAN YOU AFFORD IN RENT?

QUICK REFERENCE

TEACHER'S RESOURCES AND ANSWER KEY

Lesson Quiz Answers 13-1, p. 86

Reteaching Activity Answers 13-1, p. 115

Enrichment Activity Answers 13-1, p. 115

EXTENSION ACTIVITIES

Reteaching Activity 13-1, p. 97

Enrichment Activity 13-1, p. 100

Evelyn and Freda enjoy being in business together. Their painted T-shirts and sweatshirts are selling very well. They hope that their sales will continue over the summer before they both enroll at the community college in the fall. They are planning to add beach towels with hand-painted beach scenes to their product list for the summer to offset the loss of sweatshirt sales during June, July, and August.

Evelyn is planning to study business management at the community college. Freda will pursue advertising and marketing. They have found that they are quite compatible and have the same attitudes about many things. They have even discussed taking an apartment together when they attend college. The drive from home to the community college takes 45 minutes. Although they can live at home with their parents during college, the commute will be inconvenient. Evelyn is also anxious to be out on her own.

There are several living arrangements from which Evelyn and Freda can choose if they do move away from home. The community college has some very nice dormitories on campus. However, they are not sure that a dormitory is the best choice for them because they will need space in which to continue their business.

There are also a number of reasonably priced apartments near campus. Most of the apartments are furnished, several have access to a swimming pool, and some are conveniently located near shopping areas. As summer approaches, Evelyn and Freda will discover other factors that they must consider before they present their plans to their parents.

OBJECTIVES: *In this lesson, we will help Evelyn and Freda to:*

- *Consider the advantages of rental housing.*
- *Recognize what is involved in sharing living space with a roommate.*
- *Recognize different types of rental housing.*
- *Consider the cost of renting as a percent of income.*

ALGEBRA CONNECTIONS

If a person's monthly rent is multiplied by 52, and 205 is subtracted from the product, the result is 22,675. What is the monthly rent? (**440**)

LOOKING FOR AN APARTMENT

Evelyn and Freda have decided that they want a two-bedroom apartment near campus. It would be ideal if the apartment were well kept, sunny, and had room for their business. They are finding a great variety in number of bedrooms, features, and rents in the newspaper advertisements. They see ads for expensive small apartments in ideal locations and affordable large apartments that are located out of walking distance from campus. They wonder whether apartments are available that are between these two extremes. They notice the abbreviations that appear in these ads. They recognize "bdrms" as bedrooms and "LR" and "K" as living room and kitchen, respectively. They now know that "hdwd flrs," "Ht/wtr," and "Fully crptd" stand for hardwood floors, hot water, and fully carpeted, respectively.

They realize that they will have to base their decision on a number of factors including the **rent,** the location, the number and size of the rooms, the condition of the apartment, the availability of the apartment, whether it is furnished, and whether they will have to pay for **utilities** (heat, water, electricity) separately.

FACTORS TO CONSIDER IN DECIDING TO RENT

Renting an apartment has many advantages over buying a house. They include the following:

- Knowing exactly how much you must spend each month
- Not being tied down to one location
- Not being responsible for maintenance of the grounds
- Getting to know the community well before making a commitment to stay there
- Spending less for housing than buyers do

Evelyn and Freda know that they want to share an apartment. They are already good friends, and sharing will cut their costs in half. However, they are aware that they should discuss the practical arrangements ahead of time. Evelyn's older cousin, Ed, had problems when he and his best friend moved in together. They did not anticipate the kinds of decisions they would have to make to live together amicably. For instance, they had very different opinions about when they could have visitors. Ed's roommate had so many parties that Ed found it very difficult to get all his studying done. They finally became so angry with each other that they are no longer on speaking terms! In order not to endanger their friendship, Evelyn and Freda will discuss some of these potential problems ahead of time.

The issues that they will consider include the following.

- *Living habits* How will they set hours for listening to music, watching TV, and other activities that might interrupt the other person? How will they divide the responsibilities for washing dishes, making beds, cleaning the bathroom, vacuuming, and so on?
- *Independence or companionship* Do they want to spend time together or lead relatively separate lives? Do they want to eat meals together? Do they want to do their laundry at the same time?
- *Sharing expenses* How will they divide the rent, utility bills, food costs, new furniture, household supplies, and the security deposit?
- *Overnight guests* Who may stay, and how often?
- *Ending the arrangement* Can either one move out at any time? How would the remaining roommate replace the one who leaves?

HOUSING ALTERNATIVES

Before making their decision to share an apartment, Freda and Evelyn considered several other possibilities:

Living with Parents Many college students live at home with their parents. This can be a good alternative when the college is nearby and the family enjoys being together. For Evelyn and Freda, however, this alternative was not practical. It would mean a long commute every day, and they would have to own and maintain cars to make the 45-minute drive each way. This would definitely cut into their studying time, the time that they hope to devote to their business, and their sense of belonging to the college community.

Single Rooms Another option that they considered was renting single rooms in a home or hotel. They would want rooms with kitchen and laundry facilities. Evelyn and Freda think that they need more space than single rooms would provide. However, a single room can be a good alternative, especially for someone who does not want to invest in furniture and household items or who is not planning to live in one place for long.

Dormitory The dormitories on campus contain sleeping rooms, usually for two people, with separate desks and closet space for each. Some have eating facilities, and some do not. Evelyn and Freda thought about this alternative because it is generally less expensive than private housing. Their reason for eliminating this option was the same as their reason for eliminating single rooms—they would simply have less space than they would like.

Apartments An apartment building or complex is a multiple-unit dwelling that may contain as few as two living units or as many as hundreds. Apartments can be found in a high-rise building, house, converted warehouse, over the garage of a single-family house, or in the coach house of a large, old estate. Most apartments are rented unfurnished, but some are available with furniture. They seldom have any private yard space, but other conveniences such as a parking space and coin-operated washing machines are often provided.

CRITICAL THINKING

- Students have been introduced to the concept of equity in earlier chapters. Remind them that equity is a major advantage in owning a house. Ask why, in spite of equity, owning a house might not be advantageous for young people, even if it is affordable.

HOW TO CHOOSE AN APARTMENT

Having decided that what they really want is an apartment, Freda and Evelyn now must go about finding one. For them, the most important factor in choosing an apartment is location—they want to live near campus. Next, they know that they want two bedrooms, a common living room and kitchen, and a dining room. They really need only one bathroom. They also need a place to do their laundry and a safe place to keep their bikes.

Other people looking for apartments will have different needs. Elderly people or people with limited mobility may prefer a ground-floor apartment. Commuters may want to be close to train and bus lines. A couple with children may want to be near schools and shopping. A single person may choose a large building or one with a common pool or tennis courts so that he or she can meet new people. Apartments are available for people with special needs. For example, apartments with ramps, low counters, and extra-wide doorways are available for the wheelchair-bound.

ALGEBRA *REVIEW*

Simplify the following expressions.

1. $12\%[52(\$344)]$
   $2146.56
2. $16\%[58(\$508)]$
   $4714.24
3. $22\%[65(\$555)]$
   $7936.50
4. $3(6500 \div 52)$
   375
5. $3 \cdot 6500 \div 52$
   375
6. $3 + 6500 \div 52$
   128

True or false?

7. $700 > 435 + 190 + 230$
   F
8. $580 < 5(105) + 210 + 25$
   T
9. $805 < 4(155) + 150 + 62$
   T
10. $980 > 7(110) + 126 + 22$
   T

Ask Yourself

1. What are four common types of housing arrangements from which college students may choose?
   living with parents; single rooms; dormitory; apartments
2. What are two economic advantages of renting instead of buying?
   less housing cost; not responsible for repairs
3. What are two noneconomic advantages of renting instead of buying?
   See Additional Answers.
4. Why should you discuss potential problems with a prospective roommate before deciding whether or not to share an apartment?
   to avoid endangering a friendship
5. What issues should you discuss with a prospective roommate before you decide whether or not to share an apartment?
   living habits, independence or companionship, sharing expenses, overnight guests, ending the arrangement

ADDITIONAL ANSWERS

3. not tied down to one location; getting to know community well before making a commitment to stay there

SHARPEN YOUR SKILLS

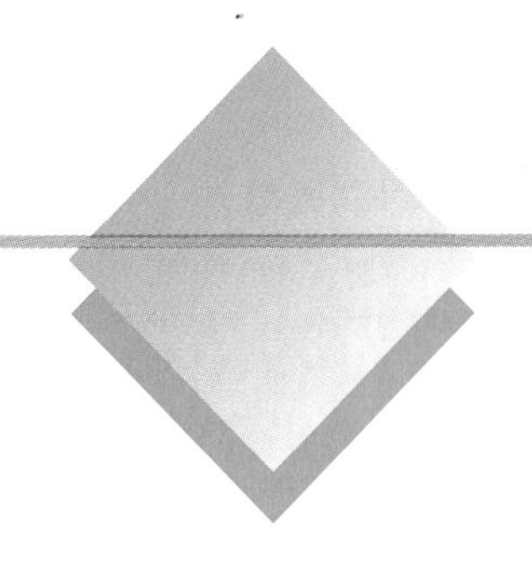

SKILL 1

One guideline that financial advisers provide for renters is that they should spend no more than *one week's gross income* on their monthly rent.

EXAMPLE 1 Evelyn and Freda estimate that the yearly income from their business will be $18,000 next year. Each will also receive $2500 per year from her parents for living expenses at college.

QUESTION How much can they afford to pay in rent each month?

SOLUTION

Annual income: $18{,}000 + 2(2500) = 23{,}000$
Weekly income: $23{,}000 \div 52 = 442.31$

Therefore they should spend no more than $442.31 on their monthly rent.

EXAMPLE 2 Evelyn found an advertisement for an apartment with a monthly rental charge of $550.

QUESTION Can they afford this apartment? If not, how much yearly income would they need to do so?

SOLUTION

They cannot afford this apartment because their weekly income is only $442.31. They would need a gross income for one week of $550 or a yearly income of $52(550) = \$28{,}600$ to afford this apartment.

SKILL 2

There are costs associated with living in an apartment in addition to the rent. If you use a **broker** to help you find an apartment, you may have to pay the broker a fee. When you rent an apartment, you probably will have to pay your landlord a **security deposit,** which will be returned to you when you move out if you have done no damage to the apartment. You must also consider the costs of utilities, which might or might not be included in your rent.

EXAMPLE 3 Miranda used a broker to find an apartment that rents for $525 per month. The broker charges a fee of 10% of one year's rent. When she signs the lease, she must pay a security deposit equal to two month's rent in addition to the first month's rent. If she moves out at the end of one year and does no damage to the apartment, her security deposit will be returned. However, she will be unable to use that money during the year.

QUESTIONS What is Miranda's total cash outlay for the year? How much will she have to pay before she moves into the apartment?

INSTRUCTION

TEACHING THE LESSON

This lesson does not involve any new mathematics and the concepts are not difficult to grasp. It could, therefore, be assigned as a project to be done individually or in small groups. You might have students spend one class reading and working through the *Guided Practice* exercises, and another class working through the *Exercise Your Skills.*

Remind students that real-life money situations often demand reasoning that shifts back and forth between what a person would like and what he or she can afford.

FOCUS ON ALGEBRA

Students will have achieved a great deal if, after completing this text, they are able on their own to read and understand materials that include mathematical presentations. (*NCTM Standard 2, p. 140*)

COMMON ERROR

An error that frequently arises in money math situations is the failure to consider trade-offs and alternatives. This lesson points out that money spent for one thing takes away from another, and money saved can be spent on something else.

AT-RISK STUDENTS

Because the mathematics of the lesson is not new, you may be able to invite less accomplished students to do work at the chalkboard and to explain exercises to the class.

RETEACHING

If the lesson was originally covered using independent or small group work, you may review and reteach common problem areas with the entire class.

ENRICHMENT

Some people own two-family homes, living in one part and renting the other. Ask students to gather information about the advantages and disadvantages of this arrangement.

SOLUTIONS

Total rent for the year:	$12(525) = 6300$
Broker's fee:	$0.1(6300) = 630$
Security deposit:	$2(525) = 1050$
Total cash outlay for the year:	$6300 + 630 + 1050 = 7980$

She must pay the broker's fee of \$630, the security deposit of \$1050, and the first month's rent of \$525 before she moves into the apartment. Therefore she must pay $630 + 1050 + 525 = \$2205$ in advance.

EXAMPLE 4 Jorge pays \$425 in rent and has a high electric bill of approximately \$200 each month because, owing to the hot, sticky climate in which he lives, he uses the air conditioner all the time. His monthly water and gas bills are \$50 and \$30, respectively. He has the opportunity to move to another apartment in which utilities (electricity, water, and gas) are included in the monthly rent of \$600.

QUESTION Will Jorge save money by moving to this new apartment even if he spends \$100 on relocation expenses?

SOLUTION

Current monthly expenses: $425 + 200 + 50 + 30 = 705$

The new apartment will cost him $705 - 600 = \$105$ less each month. His moving expenses will be recovered in the first month. Therefore he will save money by moving.

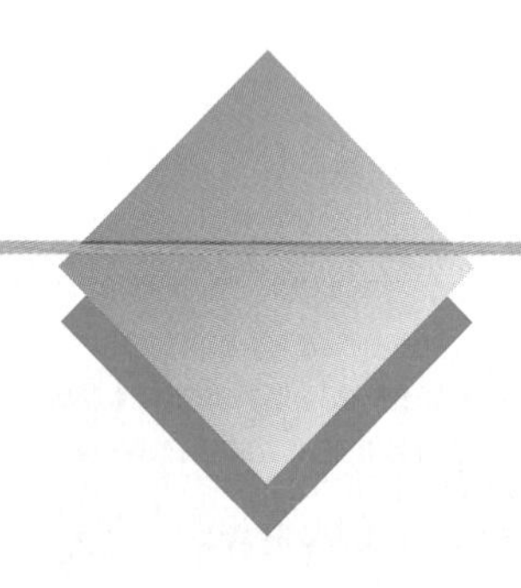

TRY YOUR SKILLS

Determine how much monthly rent you can afford for each gross annual income.

1. \$37,500 \$721.15 **2.** \$52,500 \$1009.62 **3.** \$22,750 \$437.50

Determine the gross annual income that you must earn to be able to afford each monthly rent.

4. \$610 \$31,720 **5.** \$295 \$15,340 **6.** \$805 \$41,860

7. Room and board (food) cost \$6000 in a dormitory on campus for the nine months of the school year. Jacob would like to rent a furnished apartment near campus with his old friend Leo. He found one that rents for \$400 per month, including electricity, water, and gas. However, he must sign a lease for 12 months. He estimates that grocery expenses for two people will be about \$350 per month. Should he rent the apartment? Why or why not?
Yes. The lease is for \$4800; food costs \$4200; $9000 \div 2 = \$4500$. Jacob will save \$1500 over a 12-month period by living in the apartment. Even if they leave the apartment after 9 months, their grocery expenses will only be \$3150, lowering their apartment costs further.

GUIDED PRACTICE

You may want students to do the *Try Your Skills* exercises individually and then check their answers in pairs. Exercise 7 could provide the basis for a small-group or class discussion.

EXERCISE YOUR SKILLS

1. What are some of the advantages of sharing an apartment with another person? sharing expenses, companionship
2. What are some of the disadvantages of sharing an apartment with another person? lack of privacy, many arrangements must be agreed to
3. If you had a choice, would you prefer to buy a house or rent an equally comfortable apartment? Why? See Additional Answers.

Determine how much monthly rent you can afford for each gross annual income.

4. \$66,500
5. \$22,000
6. \$35,000
7. \$29,600
8. \$19,750
9. \$45,095
10. \$33,900
11. \$18,000

Determine the gross annual income that you must earn to be able to afford each monthly rent.

12. \$595 \$30,940
13. \$700 \$36,400
14. \$375 \$19,500
15. \$775 \$40,300
16. \$525 \$27,300
17. \$975 \$50,700
18. \$300 \$15,600
19. \$645 \$33,540

20. Thelma used a broker to find an apartment that rents for \$675 per month. The broker charges a fee of 10% of one year's rent. When she signs the lease, she must pay a security deposit equal to one month's rent as well as the first and last month's rent. What is her total cash outlay for the year? How much will she have to pay before she moves into the apartment? \$9585; \$2835
21. Rosalyn used a broker to find an apartment that rents for \$505 per month. The broker charges a fee of 15% of one year's rent. When she signs the lease, she must pay a security deposit equal to two months' rent as well as the first and last month's rent. What is her total cash outlay for the year? How much will she have to pay before she moves into the apartment? \$7979; \$2929
22. Howard pays \$535 in rent and has monthly electric, water, and gas bills of \$175, \$69, and \$75, respectively. He has the opportunity to move to another apartment in which utilities are included in the monthly rental fee of \$900. His relocation bills will be approximately \$75. Will he save money by moving to this apartment? no
23. Caleb pays \$760 in rent and has monthly electric, water, and gas bills of \$150, \$38, and \$65, respectively. He has the opportunity to move to another apartment in which utilities are included in the monthly rental fee of \$875. His relocation bills will be approximately \$150. Because of the distant location of the new apartment, the monthly transportation cost increases by \$30. Will he save money by moving to this apartment? Yes, in 2 months he will recover his moving cost and start saving \$108/mo.

KEY TERMS

broker
rent
security deposit
utilities

INDEPENDENT PRACTICE

Exercises 1-3 give the opportunity for students to reflect on and discuss some of the non-economic aspects of living on their own.

Exercises 4-19 ask students to work back and forth between affordability and income.

Exercises 20-25 present personal finance cases about living on one's own. Students must work out the mathematical constraints and then make judgements about the advisability of a particular living situation.

ALTERNATIVE ASSESSMENT

Have students collect newspaper ads or read real estate literature to find at least four different ways of calculating the initial payments for an apartment. Have them calculate the different initial costs for one rental rate.

ADDITIONAL ANSWERS

3. Answers may vary. (Ex.: rent—do not have to worry about maintenance)

CLOSURE

Have students explain the method for calculating the amount that an individual or pair can afford to spend on housing. Ask them to make a list of the fees and charges that might be added when a person begins to rent a house or apartment. Ask them to use realistic numbers to calculate the total cost of renting.

LESSON QUIZ

1. Charles and Jon are considering renting a house. Charles earns $15,000 a year. Jon earns $1100 a month. What monthly rent can they afford to pay? (**$542**)
2. An apartment renting for $400 a month requires a broker's fee of 8% of the annual rent, and one month's rent as security. In addition to rent, what is the total cost to begin renting the apartment? (**$784**)

ADDITIONAL ANSWERS (EXERCISE YOUR SKILLS)

25. no; rent is $3480, food for 9 months is $1800; total: $5280

(MIXED REVIEW)

6. the flight that departs at 12:10 P.M.

8. $119,206.58

10. $c = 200 + 3n$; $r = 7n$; 50 T-shirts

24. Room and board (food) cost $6500 in a dormitory on campus for the nine months of the school year. Martha would like to rent a furnished apartment near campus with her friend McKee. She found one that rents for $425 per month, including electricity, water, and gas. Martha expects to pay about $350 per month for their food bills. However, she must sign a lease for 12 months. Should she rent the apartment? Why or why not? yes; rent is $1500, food for 12 months is $4200; total: $9300, or $4650 each

25. Room and board (food) costs $5000 in a dormitory on campus for the nine months of the school year. Damien would like to rent a furnished apartment near campus. He found one that rents for $290 per month, including electricity, water, and gas. Groceries will cost about $200 per month. However, he must sign a lease for 12 months. Should he rent the apartment? Why or why not? See Additional Answers.

MIXED REVIEW

5-1 **1.** If you can afford a monthly payment of $285, how much money can you borrow at 8.75% for 5 years? $13,809.99

10-1 A car that cost a dealer $13,750 is marked up 21%.

2. Find the mark-up. $2887.50

3. Find the sticker price. $16,637.50

10-4 **4.** Use the Driver Rating Factors Table in the Reference Section to find the driver rating factor for a 20-year-old unmarried female without driver training who drives to work and is not the owner or usual driver. 1.65

11-3 Use the Windward Airlines Schedule in the Reference Section to find each of the following:

5. The earliest flight from Chicago to Phoenix, Arizona. 6:15 A.M.

6. The flight from Chicago to Phoenix that takes the least amount of time.

12-3 **7.** A family's monthly income is $4,575. How large can their monthly mortgage payment be? $1281

12-3 **8.** If the prevailing annual interest rate is 10% and the family of Exercise 7 would like a 15-year mortgage, how much can the family borrow?

6-7 **9.** The monthly take-home pay of a family is $3300. Credit-card payments are $540. Determine, to the nearest tenth of a percent, how much the family is spending on its credit payments. Is the family in danger of credit overload? 16.4%; no

4-4 **10.** Students who are selling painted T-shirts have 25 hours of labor costs at $6.00 per hour; advertising, energy, and transportation costs of $50.00; and variable costs of $3.00 for each T-shirt produced. The price of each T-shirt is $7.00. Write the cost function and the revenue function. Determine the break-even point algebraically. See Additional Answers.

13-2 SELECTING AN APARTMENT: WHAT SHOULD YOU DO BEFORE SIGNING A LEASE?

QUICK REFERENCE

TEACHER'S RESOURCES AND ANSWER KEY

Lesson Quiz Answers 13-2, p. 86

Reteaching Activity Answers 13-2, p. 115

Enrichment Activity Answers 13-2, p. 116

EXTENSION ACTIVITIES

Reteaching Activity 13-2, p. 98

Enrichment Activity 13-2, p. 101

Ana is working as a hostess at the seafood restaurant at which Sylvia and her father sometimes dine. The restaurant is near the mall, where Ursula works as a clerk at Joyful Toys. Sometimes the three friends get together after work and do some shopping at the mall or go to the movies.

Ana and Ursula have been thinking about sharing an apartment after high school graduation. Ana will be going on vacation for a month this summer, but the restaurant manager told her that she can have her job back when she returns. Ursula enjoys her job and plans to continue working at the toy store after graduation. She may apply for an assistant manager's position at the end of the year.

Ana and Ursula have been good friends since they were 10 years old and spent two weeks at Girl Scout camp together. They discovered that they could prepare a pretty good stew over an open campfire and they both liked their marshmallows lightly toasted, not burned. Perhaps they would be compatible roommates. Sylvia has promised to show them all the best places to buy towels, dishes, blankets, and sheets for their new apartment.

OBJECTIVES: *In this lesson, we will help Ana and Ursula to:*

- *Investigate potential sources of information on rental housing.*
- *Learn how to inspect an apartment.*
- *Read and comprehend a lease agreement.*
- *Recognize the relationship between the monthly rental charge and the number of amenities, size, location, number of bedrooms, and so on.*

FOCUS

ALGEBRA CONNECTIONS

Match the ordered pairs with the equations they satisfy: (2, 3), (−4, −2), (5, −1)

$y = x - 6$, $2y = x$, $y = 2x - 1$

(**(5, −1)**, **(−4, −2)**, **(2, 3)**)

HELPERS IN THE SEARCH

Ana does not have to move out of her family's house, but she wants to give independence a try. If she shares an apartment with Ursula for a while, she may discover that being an independent adult carries more responsibility than she is prepared to accept. But she will not know that until she tries. Her parents will support her efforts at independence and help her to cope with the adult decisions that she will be making. They have suggested several sources of information that may be useful to her in her search for an apartment.

Friends It is often helpful to ask friends whether they know of any available apartments to rent. If they are currently renting, they can also fill you in on the procedures to follow and any problems that they have had.

Newspapers Reading the advertisements for apartments in the real-estate section of the newspaper will give Ana and Ursula an idea of the rents being charged for assorted types and sizes of apartments in various locations.

Real-Estate Agencies Owners of apartments or apartment buildings often list available rental units with a local real-estate broker. **Real-estate agents** are licensed by the state to help people find an apartment. They may charge the renter a fee based on the amount of rent paid for the apartment they have shown.

INSPECTING THE APARTMENT

When Ana and Ursula find an apartment in which they are interested, they should inspect it carefully and thoroughly. They should look into the closets, peer under the sink, tour the basement, flush the toilet, and test the shower. They should ask themselves whether the cabinet space is sufficient and the refrigerator is large enough. Other questions that they should ask themselves include: Will the size and shape of the rooms fit their needs? Are the locks and security systems adequate? Are storage facilities, laundry facilities, and parking areas available?

If they are seriously considering a particular apartment, they should ask other tenants in the building questions such as the following.

- Is routine maintenance adequate and prompt?
- Are complaints responded to quickly, and is the owner accessible?
- Is the building noisy?
- Have rent increases been excessive?
- Does the building have a high turnover of tenants?

Turnover refers to people moving in and out in rapid succession. In general, it is less desirable to live in a building with a high turnover rate than in one with a more stable population.

READING THE LEASE

When Ana finds an apartment, she probably will be required to sign a **lease,** which is an agreement between the landlord and her in which she agrees to pay for the use of the apartment. The landlord agrees to allow her to occupy the apartment as long as she complies with the provisions of the lease. The lease imposes financial obligations on her as a tenant; therefore she must be sure that she understands all of the clauses in the lease before she signs it. Some of the important clauses of a lease include the following.

Term of Lease This clause sets the length of time your lease will be in force.

Rent You agree to pay a specified amount on a certain day in each month during the term of the lease. You are legally obligated to pay the rent for the entire term of the lease, even if you want to move out before the expiration of the term of the lease.

Security Deposit The security deposit, usually equal to one month's rent, is held by the landlord (the owner of the apartment), to pay for any damages to the apartment caused by the tenant. The security deposit is collected when you sign the lease. It will be returned to you when the lease ends and you leave the apartment in the same condition in which you found it. Therefore, before renting an apartment, you should make a list of existing damages and have the landlord sign it. This will assist you in settling claims for any damage found in the apartment when you leave and ask for your security deposit refund.

Condition of Premises You agree that the premises are acceptable "as is." Sometimes, the landlord may agree to paint the apartment or to fix existing damage before you move in.

Repair You agree not to damage the apartment, which is the landlord's property.

Use; Sublet You agree that only you and your immediate family will live with you. If you are sharing an apartment with a roommate, you both must sign the lease. Some leases may prohibit pets from living with you. In the event that you want to leave the apartment before the expiration of the lease term, you may want to arrange for someone else to move in and assume your obligations to pay rent. This is called a sublet arrangement. Most leases prohibit you from subletting your apartment without the landlord's consent.

Plurals If you have a roommate, both of you share responsibility for obeying the lease. If your roommate fails to pay his or her portion of the monthly rent, you are obligated to pay the entire monthly rent due under the lease. Therefore you should choose your roommates carefully.

Access; Quiet Enjoyment You give the landlord permission to enter your apartment. As long as you obey the terms of the lease, the landlord agrees to allow you to enjoy your apartment and not to disturb your occupancy of the apartment.

Compliance You agree to obey the laws regulating pets, sanitation, noise, and the rights of other tenants in the apartment building.

Default You agree that if you fail to pay your rent or to otherwise comply with the terms of the lease, the landlord has the right to terminate the lease and to commence eviction proceedings. In addition, the landlord may have the right to retain your personal property in your apartment until the rent that is owed is paid in full.

Ana realizes that the terms of a lease are quite serious. She will select her apartment carefully, knowing that if she signs a lease for one year, she must be willing to make that year's commitment as an adult.

ALGEBRA *REVIEW*

Evaluate y for $x = 15.82$. Round answers to the nearest hundredth.

1. $y = 37 - 0.17x$
   34.31
2. $y = \frac{225.72}{x}$
   14.27
3. $x = 49.975 + 0.246y$
   −138.84
4. $x = \frac{4.78y}{98.98}$
   327.59

Determine a linear equation that fits the points.

5.

x	2	4	6
y	7	11	15

$y = 2x + 3$

6.

x	−2	−1	0
y	−5	−4	−3

$y = x - 3$

7. The cost of renting an apartment in a building is \$100 plus \$150 per room. Write an equation for the cost.
   See Additional Answers.

The yearly cost of home heating oil in one neighborhood was \$120 plus \$1.00 per square foot.

8. Find the cost of heating a 1200 ft^2 home.
   \$1320.00
9. Find the cost of heating an 850 ft^2 home.
   \$970.00
10. Write an equation for the yearly cost of heating oil.
    See Additional Answers.

Ask Yourself

1. What are three sources of information that you can consult if you are looking for an apartment?
   friends, newspapers, real estate agencies
2. What should you inspect in any apartment before renting it? look in closets, peer under sink, tour the basement, flush the toilet, test the shower
3. What is a lease? Why does a person who is renting an apartment have to sign one?
   See Additional Answers.

SHARPEN YOUR SKILLS

SKILL 1

Ana noticed that the following amenities are used as selling points in advertisements for apartments: extra storage, attached parking, fireplace, exercise facility, spa, pool, sauna, playground, tennis courts, clubhouse, cable TV or satellite hookup, washer/dryer, central air conditioning, and patio.

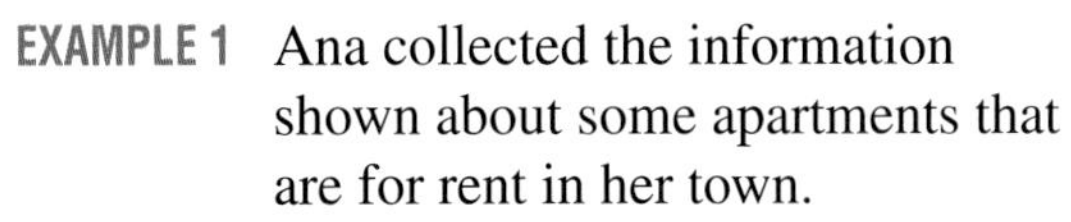

EXAMPLE 1 Ana collected the information shown about some apartments that are for rent in her town.

Monthly Rent	Number of Amenities
$ 600	5
1100	10
475	3
980	8
1050	9
450	3
675	6
350	1
570	6
700	7

QUESTIONS What is a linear model of the form $y = a + bx$ that will allow Ana to predict the number of amenities in an apartment in her town if she knows the monthly rental charge? Is there a good correlation between the monthly rental charge and the number of amenities?

SOLUTIONS

Use the linear regression function on your graphing calculator to find the line that best fits the data. The calculator also determines the value for r, the correlation coefficient, which is a measure of how well the line fits the data. If r is close to $+1$ or -1, the fit is very good. If r is close to 0, there is virtually no fit. Begin by entering the ordered pairs (x, y) of data, where x represents the rental charge and y represents the number of amenities in each apartment. Your calculator will display the following information.

$a = -1.414050978$
$b = 0.0103799295$
$r = 0.9556044872$

The linear equation that best relates the rental charge x and the number of amenities y is

$$y = -1.414050978 + 0.0103799295x$$

Since the value of r is close to $+1$, there is a very good correlation between monthly rental charge and number of amenities.

EXAMPLE 2 Ursula found an advertisement for an apartment that looks very appealing to her. The monthly rental charge for this apartment is listed at $625.

QUESTION Estimate the number of amenities in this apartment using the regression equation that you derived in Example 1.

INSTRUCTION

TEACHING THE LESSON

A natural follow-up to Example 1 is to obtain a scatter plot of the points entered in the graphing calculator. This will give a visual impression of the meaning of a high correlation. It can be seen that most of the points would be close to a straight line drawn through them.

A second activity would be to use the values for a and b to draw the regression line. Tracing along this line will give pairs that fit perfectly. This gives students another perspective of how the equation approximates the original pairs, even though perhaps none of the pairs satisfies the equation perfectly.

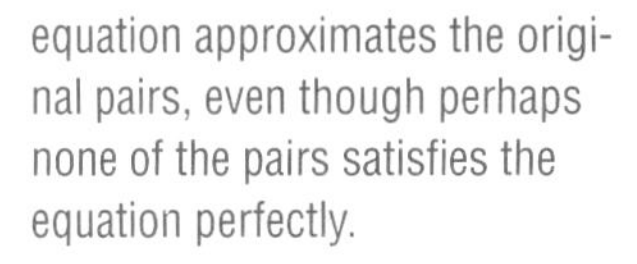

FOCUS ON ALGEBRA

The use of mathematical models and curve fitting to organize data to predict is a skill highly recommended and one that will be often used in statistics. *(NCTM Standard 10, p. 167.)*

COOPERATIVE LEARNING

Examples 1 and 4 can easily be done by pairs of students working together, checking entries and outcomes.

CRITICAL THINKING

Ask students what a correlation close to –1 might mean. Have them scatter points that they think would give this correlation and then think of a situation that might give a negative correlation. (**From ages 20 to 50, there might be a negative correlation between age of driver and numbers of car accidents.**)

COMMON ERRORS

Some students may get the impression that the value of *r* is associated with the slope of the best-fit line. Point out that *b* is the slope of the line that best represents the points, but that if *r* is close to 0, this line is not a good representation of the data. Some may want to know where *r* does come from. To answer this they will have to wait for statistics courses.

ESL STUDENTS

There are some important terms that are likely to be new to non-native speakers. Pay particular attention to: *sublet, access, default,* and *regression.*

TECHNOLOGY HINT

Different calculators and computer programs will have different capabilities regarding regression analysis, scatter diagrams, and curve fitting. Explain the possibilities, and as time permits, show some of the best features available.

SOLUTION

Substitute $x = 625$ into the equation from Example 1 and solve for y.

$$y = -1.414050978 + 0.0103799295x$$
$$y = -1.414050978 + 0.0103799295(625)$$
$$y = 5.07 \quad \text{To the nearest hundredth}$$

There should be approximately five amenities in this apartment.

EXAMPLE 3 Ana looked at an apartment with four amenities and a monthly rental charge of $800.

QUESTION Is this apartment fairly priced?

SOLUTION

Substitute $y = 4$ into the equation from Example 1 and solve for x.

$$y = -1.414050978 + 0.0103799295x$$
$$4 = -1.414050978 + 0.0103799295x$$
$$4 + 1.414050978 = 0.0103799295x$$
$$\frac{4 + 1.414050978}{0.0103799295} = x$$
$$522 = x \quad \text{To the nearest dollar}$$

This apartment appears to be significantly overpriced. A monthly rental charge of $522 seems more appropriate for this apartment.

EXAMPLE 4 Ana and Ursula are aware that there is a large variation in the number of apartments in each apartment building. They collected the information shown about some apartments that are for rent in their town.

Monthly Rent	Number of Apartments in Building
$800	12
750	180
500	160
550	8
375	210
400	2
400	40
675	50

QUESTION Is there a linear relationship between the monthly rental charge and the number of apartments in the building?

SOLUTION

Use the linear regression function on your graphing calculator to find the line that best fits the data. Begin by entering the ordered pairs (x, y) of data, where x represents the rental charge and y represents the number of amenities in each apartment. Your calculator will display the following information:

$$a = 110.9617225$$
$$b = -0.0507177033$$
$$r = -0.0988009201$$

Since the value of r is -0.10 to the nearest tenth, there is almost no linear relationship between monthly rental charge and number of apartments in the building.

TRY YOUR SKILLS

For Evelyn and Freda the most important factor in choosing an apartment is location. They want to be in walking distance from the campus because neither will have a car at college. Freda collected the information shown about eight two-bedroom rental apartments near school.

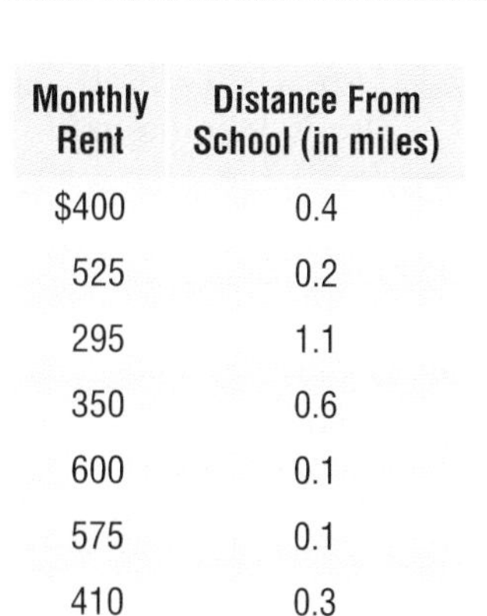

Monthly Rent	Distance From School (in miles)
$400	0.4
525	0.2
295	1.1
350	0.6
600	0.1
575	0.1
410	0.3
375	0.5

1. What is a linear model of the form $y = a + bx$ that will allow Freda to predict the distance from school of a rental apartment if she knows the monthly rental charge?
2. Is there a good correlation between the monthly rental charge and the distance from campus?
3. Estimate the distance from campus, to the nearest tenth of a mile, of an apartment that has a monthly rental charge of $435. 0.4 miles
4. Do you think that an apartment with a monthly rental charge of $590 that is located 0.7 mile from campus is fairly priced? no

GUIDED PRACTICE

Be sure to leave enough class time for student work and a step-by-step review of the *Try Your Skills* exercises. Students may think they understand the process as you explain it, but find it difficult to do the exercise by themselves.

RETEACHING

The best approach to reteaching is to assign tutors from the class to those students having difficulty with the concepts or calculator procedures. You may ask who needs help, who wants to tutor, and then allow students to pair themselves.

ENRICHMENT

Have students talk to a science teacher and then write a brief report on some data sets that give a nonlinear relationship. The relationship might be quadratic, exponential, or sinusoidal.

ADDITIONAL ANSWERS

1. $y = 1.573550568 - 0.0026312761x$
2. yes; $r = -0.884611342$

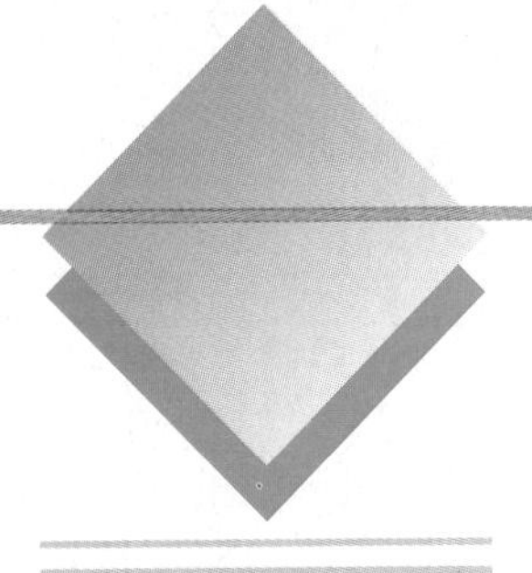

EXERCISE YOUR SKILLS

1. If you are about to rent an apartment, why should you read the lease very carefully before you sign it? A lease is usually a 1-year commitment.
2. In general, why would it be less desirable to live in an apartment building with a high rate of turnover than in one with a more stable population? See Additional Answers.
3. If you are considering renting an apartment in a certain building, why should you talk to the other tenants of the building before you make your final decision?
4. If you rent an apartment, why should you make a list of everything that is wrong with it before you sign the lease? Why should you ask the landlord to sign that list?

KEY TERMS

lease
real-estate agents
turnover

INDEPENDENT PRACTICE

ASSIGNMENT GUIDE

Exercises 1-4 review students' understanding of the steps to take in the process of renting an apartment.

Exercises 5-8, 9-12, and 13-15 offer three different sets of data for analysis. All students should do at least two of these exercise sets in order to reinforce the skills and concepts of the lesson. It would also be helpful to do a careful review of one of the data sets during class.

Rufus collected the information shown about apartments that are for rent in his town.

Monthly Rent	Number of Bedrooms
$ 900	3
1000	3
600	2
295	0*
450	1
1500	4
310	0*
425	1
575	2

*Studio or efficiency

5. What is a linear model of the form $y = a + bx$ that will allow Rufus to predict the number of bedrooms in an apartment in his town if he knows the monthly rental charge for the apartment?
6. Is there a good correlation between the monthly rental charge and the number of bedrooms? yes; $r = 0.9461001063$
7. Estimate the number of bedrooms in an apartment in Rufus' town that has a monthly rental charge of $500. Express your answer to the nearest whole number of bedrooms. 1
8. Do you think that an apartment with one bedroom and a monthly rental charge of $675 is fairly priced? no

Rakesh collected the information shown about apartments that are for rent in his town.

Monthly Rent	Area (in square feet)
$495	675
550	780
395	500
650	900
440	610
900	1030
975	1200
760	995

9. What is a linear model of the form $y = a + bx$ that will allow Rakesh to predict the area in an apartment in his town if he knows the monthly rental charge for the apartment?
10. Is there a good correlation between the monthly rental charge for the apartment and the area? yes; $r = 0.9769683519$

11. Estimate the area in square feet in an apartment in Rakesh's town that has a monthly rental charge of $500. 679 ft^2

12. Do you think that an apartment with 920 ft^2 and a monthly rental charge of $650 is fairly priced? yes

Gila collected the information shown about apartments that are for rent in her town.

Monthly Rent	Distance from Hardware Store (in miles)	Number of Bathrooms	Number of Closets
$1200	0.1	2.5	5
1100	4.1	2	4
400	0.2	1	1
375	1.0	1	1
400	1.9	1.5	2
700	3.1	2	3
1200	1.2	3	6
1575	1.8	3	6

13. Is there a good correlation between the monthly rental charge and the distance from the hardware store? no; $r = 0.145972874$

14. Is there a good correlation between the monthly rental charge and the number of bathrooms? yes; $r = 0.9289029843$

15. Is there a good correlation between the monthly rental charge and the number of closets? yes; $r = 0.9593668769$

CLOSURE

Ask four or five students to come to the chalkboard and show points that they think would give a correlation of 0.95, 0.6, 0, −0.6, −0.95. Review the scatter diagrams.

LESSON QUIZ

The following table shows the average rent paid by age.

Age	Rent
19	220
20	250
21	325
22	400
23	410
24	395
25	415

1. Use linear regression analysis to find r and the equation of the line that best fits the data. (**$r = 0.91$; $y = 34.29x - 409.29$**)
2. Is there a high correlation? (**Yes, 0.91 shows a strong linear relationship.**)

MIXED REVIEW

1-3 **1.** Jeremy is a real-estate agent who has sold a house for $179,500. He earns 4.5% commission on the first $100,000 and 6.5% commission on any amount over $100,000. How much did he earn from the sale of this house? $9667.50

6-2 **2.** Gladys has a balance of $1,455 on her MasterCard, which carries an APR of 17.5%. If she makes no further purchases with this card, how many months will it take her to pay off the balance? She can afford monthly payments of $170. 10

7-1 **3.** Use the Multiples-of-Salary Chart in the Reference Section to find the amount of life insurance that a 45-year-old man who has gross earnings of $65,000 would need in order to provide an income replacement of 60%. $390,000

9-3 **4.** Find the amount of the income tax refund or payment due to the nearest dollar for a family of two parents and three dependent children with wages of $44,000, $300 interest income, and dividends of $270. The family had medical expenses of $4,200, mortgage interest of $3,967 that they paid on their home, real estate taxes of $2100, state income tax payments of $1,680, and charitable contributions of $900. The husband made a tax-deductible contribution of $950 into his IRA. The wife did not work outside the home. The amount of withholding for the year was $5,985, and the filing status was married filing jointly. Refund: $2606

10-1 Mr. and Mrs. Anderson are purchasing a car for $13,000. Find the monthly payment under the terms indicated in Exercises 5 and 6.

5. 8.5% annual interest rate, 20% down, 4 years to pay $256.34

6. 11% annual interest rate, 15% down, 5 years to pay $240.25

12-2 In Exercises 7 and 8, find the total financed price T for a home that costs $80,000 with a down payment of $20,000. Use the formula $T = nM + d + p$, where n is the total number of monthly payments, M is the monthly payment, d is the down payment, and p is the cost of points paid on the loan.

7. Bank A provides a 7% 30-year mortgage with $1\frac{1}{2}$ points. $164,605.34

8. Bank B provides a $7\frac{1}{2}$% 30-year mortgage with no points. $171,030.33

13-1 **9.** If your gross annual income is $43,100, how much monthly rent can you afford? $828.85

12-1 **10.** The following table shows the correspondence between the prices of several homes and the floor area of those homes. Use a graphing calculator to find a linear model of the form $y = a + bx$ that best represents the data. See Additional Answers.

Floor Area (in square feet)	1510	1800	2230	3100	4500
Price (in thousands of dollars)	90	105	113	129	146

ADDITIONAL ANSWERS

10. $y = 70.70160176 + 0.0174651439x$ where x = floor area and y = price

13-3 CHOOSING A PLACE TO LIVE: SHOULD YOU BUY OR RENT?

QUICK REFERENCE

TEACHER'S RESOURCES AND ANSWER KEY

Lesson Quiz Answers 13-3, p. 88

Reteaching Activity Answers 13-3, p. 115

Enrichment Activity Answers 13-3, p. 116

EXTENSION ACTIVITIES

Reteaching Activity 13-3, p. 99

Enrichment Activity 13-3, p. 102

Yvette and Yvonne will be attending a large university in the Colorado mountains. Yvette wants to study architecture, and Yvonne will probably major in art history and secondary education. Since they are twins and are very close friends as well, they want to attend the same college.

The twins are now looking for a place to live in Colorado. Their parents are thinking about buying a condominium for the twins to live in rather than renting an apartment. There does not seem to be much to choose from in rental apartments in the resort and university town in the Colorado mountains.

If they buy a condominium, the twins' parents will make payments to build up their equity. If the twins rent an apartment, their parents will make rental payments every month but will have no equity invested after their daughters finish college. After their daughters attend the university, the parents can sell the condo. Alternatively, they can retain ownership of the condominium and use it as a vacation home in the summer and rent it to other university students during the school year.

Yvonne likes the idea of living in a condominium. She would like to learn to ski, but she may not be able to convince Yvette to learn too. Yvette is a serious student and plans to spend a lot of her time studying in the library.

OBJECTIVES: *In this lesson, we will help Yvette and Yvonne to:*

- *Investigate the financial aspects of condominium ownership.*
- *Compare the costs of renting or buying a particular home.*

ALGEBRA CONNECTIONS

How many calculator keystrokes does it take you to find each: the number of years it will take to double your money at (a) 6.5%, (b) 8%, (c) 9.5% interest?

HOUSING ALTERNATIVES

Rather than pay rent, Yvette's and Yvonne's parents prefer to buy their daughters' dwelling at school for the following reasons:

- The interest on their mortgage and their property taxes are tax deductible.*
- The money that they spend on mortgage payments provides them with equity in the property.

*If the dwelling is not the family's primary residence, the tax advantages may be diminished.

A condominium may be an apartment in a multiple-unit building, a townhouse in a development, or a semidetached living unit with common areas. A **condominium** (also called a *condo*) is a hybrid of individual ownership and common living. As in a house, when you own a condominium, you own the individual unit in which you live. A condo, however, also offers the advantages of shared living. All the condo owners in a particular building or development share the common costs of maintaining the building or development and its common areas. Condominium ownership involves (1) individual ownership of one unit and (2) joint ownership with the other unit owners of the facilities and common areas (yard, pool, patio, hallways, and so on). All of the unit owners are members of the **owners' association,** which is responsible for management of the building, grounds, and common areas. The purchase of a condominium means that the buyer automatically shares in the administration of the building and property. The condominium owner pays a monthly fee to cover the costs of operating and maintaining all of the common areas.

ADVANTAGES AND DISADVANTAGES OF CONDOMINIUMS

Advantages

- You build equity rather than just pay rent.
- You share costs of maintaining the building and common areas.
- You are not responsible for outside maintenance, such as lawn care and snow removal.
- The condominium may include use of a swimming pool, party room, or other recreational facility.

Disadvantages

- You cannot always control major maintenance expenses, such as roof repair.
- The owners' association may need to approve changes that affect common areas, such as painting the exterior of your unit.
- Resale of a condominium may be slower than resale of a single-family home.

Another housing alternative is a *mobile home.* A **mobile home** is a portable structure built on a chassis, designed to be used without a permanent foundation. It is usually parked in a permanent or long-term location. Unlike a recreational vehicle, it is designed as a year-round dwelling. In recent years mobile homes have grown in popularity among families with lower incomes. The mobile home owner must also have land on which to park the mobile home. Generally, the mobile home owner will rent space in a mobile home park and pay a monthly rental fee. The monthly rental fee for the land is likely to be less than the rental for an apartment. This is especially true in recreational and retirement areas or in other places where temporary housing is in demand. Ownership of a mobile home, therefore, combines features of ownership (the mobile home) and rental (the land).

ADVANTAGES AND DISADVANTAGES OF MOBILE HOMES

Advantages

- The cost is usually much lower than for other forms of housing.
- A mobile home can be moved to a new location.
- The costs to maintain a mobile home are low.
- The small, compact size of a mobile home makes it easy to keep clean and tidy.
- Without much yard space, there is little outdoor maintenance needed.

Disadvantages

- You cannot park mobile homes wherever you please—there are strict zoning regulations governing where you can put them.
- Living space in a mobile home is limited.
- There is usually little yard space around a mobile home.
- Mobile homes are not well-protected in storms because of lightweight construction.

ALGEBRA *REVIEW*

Evaluate.

1. 105% of 1500
   1575
2. 105% of (105% of 1500)
   1653.75
3. 98% of (98% of 975)
   936.39
4. $1450(1.08)^2$
   1691.28
5. $1450(1.08)^4$
   1972.71
6. $1450(1.08)^6$
   2300.97

Evaluate for *x* or *y*.

7. $x = \frac{98{,}000(0.01)(1 + 0.01)^{180}}{(1 + 0.01)^{180} - 1}$
   1176.16
8. $y = \frac{66{,}000(0.015)(1 + 0.015)^{360}}{(1 + 0.015)^{360} - 1}$
   994.68

Use the values for *x* and *y* from Exercises 7 and 8 to find *q* and *t*.

9. $q = 180x + 24{,}500.00$
   236,208.80
10. $t = 360y + 16{,}500.00$
    374,584.80

Yvonne's and Yvette's family have decided to buy a condominium for the twins. The students will not have to take care of maintaining a yard and will have the use of the common facilities. Yvette is looking forward to picking out some appropriate furniture for it. Yvonne is looking forward to the ski season.

Ask Yourself

1. What is a condominium?
   See Additional Answers.
2. What advantages are there to buying a condominium rather than renting an apartment?
   do not need to take care of grounds, still part of a community
3. Why do many people with lower incomes choose to buy a mobile home?
   The cost is usually much lower than for other forms of housing.

ADDITIONAL ANSWERS

1. A condominium is a multiple-unit dwelling where individuals own their own individual units and have joint ownership with other unit owners of the facilities and common areas.

SHARPEN YOUR SKILLS

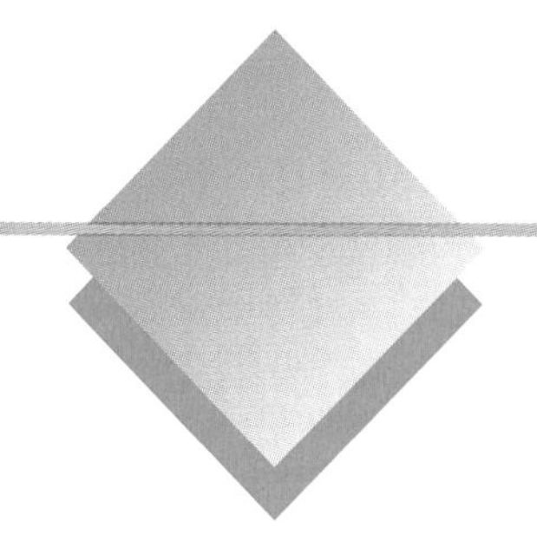

SKILL 1

EXAMPLE 1 Although Yvonne's and Yvette's parents have decided to buy a condominium, they are still interested in comparing the monthly cost of buying a condo with the monthly rental charge on an apartment. They have located an apartment that rents for \$480 per month and a condo with a purchase price of \$80,000.

QUESTIONS What is their monthly mortgage payment on the condo if they make a 20% down payment and borrow the remainder at 9% for 30 years? Which alternative would require a greater cash expenditure per month?

SOLUTIONS

If they make a down payment of 20%, they will borrow 80% of the purchase price.

$0.8(80{,}000) = 64{,}000$

$$M = \frac{Pr(1 + r)^n}{(1 + r)^n - 1}$$ Use the monthly payment formula.

$$M = \frac{64{,}000(0.0075)(1 + 0.0075)^{360}}{(1 + 0.0075)^{360} - 1}$$ $P = 64{,}000$; $r = (0.09 \div 12) = 0.0075$; $n = 30 \bullet 12 = 360$

$M = 514.96$ To the nearest hundredth

The monthly mortgage payment on the condo is \$514.96, which is \$514.96 − \$480 = \$34.96 more than the monthly rental charge on the apartment.

SKILL 2

EXAMPLE 2 Yvonne and Yvette will be at the university for 4 years. They can obtain only a one-year lease on the apartment in Example 1. When they renew the lease, the landlord will raise the rent by 5% each year.

QUESTION Are the monthly mortgage payments on the condo larger than all the monthly rent payments over the course of 4 years?

SOLUTION

The mortgage payments on the condominium are fixed at \$514.96 per month.

$4 \bullet 12 \bullet 514.96 = 24{,}718.08$

The mortgage payments for 4 years are \$24,718.08.

INSTRUCTION

TEACHING THE LESSON

This lesson reviews the formula for finding monthly payment. Students will have forgotten the formula, but once you have demonstrated it again they should find it easier to use than in the past. Stress efficiency with the calculator. Students should be able to enter the entire expression and find the answer without writing down any partial answers.

When covering Example 2, discuss the impact of inflation of income, expenses, and savings.

FOCUS ON ALGEBRA

In this lesson, students are shown how functions can be used to model real-world phenomena. Throughout the chapter different representations of relationships have been used: tables, verbal rules, equations, and graphs. *(NCTM Standard 6, p. 154)*

COOPERATIVE LEARNING

If time permits, small groups might be formed to use the calculator to experiment with exponents. Ask students to write down any rules, patterns, or unusual discoveries.

The monthly rent on the apartment increases 5% each year.

Year 1:	480.00
Year 2:	$480.00(1.05) = 504.00$
Year 3:	$480.00(1.05)^2 = 529.20$
Year 4:	$480.00(1.05)^3 = 555.66$

Rental payments for 4 years are calculated as follows:

$$12(480.00 + 504.00 + 529.20 + 555.66) = 24{,}826.32$$

The rental payments for 4 years are $24,826.32.

The rental payments are 24,826.32 − 24,718.08 = $108.24 more than the mortgage payments for 4 years.

Yvonne wonders why any college student would rent an apartment if the monthly mortgage payments over the course of 4 years are less. Yvette points out that the sum of the monthly mortgage payments is not the only factor to be considered when making a decision on whether to buy or rent. To buy a condominium, you must have enough money up front for a down payment and closing costs. In addition, your monthly costs are increased by the fee paid to the condominium association and property taxes paid to the town. Finally, moving from an apartment is much simpler than selling a condo when the student is ready to leave the area.

SKILL 3

CRITICAL THINKING

The lesson uses an annual interest rate of 9% which gives a monthly rate of 0.75%. A rate such as 7% does not give a terminating percent. Ask students how they could use the annual rate in the monthly payment formula so as to achieve a monthly payment amount as accurate as possible.

(Use $\frac{0.07}{12}$ in the formula.)

EXAMPLE 3 Sari found a condominium in which she wants to live. The owner will rent it to her for $1000 per month or sell it to her for $145,000.

QUESTION If she obtains a 15-year mortgage at 9% with a 20% down payment, will the total financed price or the total rental payments be greater over the 15 years? Assume that she can have a series of 3-year leases and that the rent increases by 10% at the end of each 3-year period.

SOLUTION

If she makes a down payment of 20%, she will borrow 80% of the purchase price.

$$0.8(145{,}000) = 116{,}000$$

$$M = \frac{Pr(1 + r)^n}{(1 + r)^n - 1}$$ Use the monthly payment formula.

$$M = \frac{116{,}000(0.0075)(1 + 0.0075)^{180}}{(1 + 0.0075)^{180} - 1}$$ $P = 116{,}000$; $r = (0.09 \div 12) = 0.0075$; $n = 15 \cdot 12 = 180$

$$M = 1176.55$$ To the nearest hundredth

Sari's monthly mortgage payment is \$1176.55. In the course of 15 years she will pay [ANS] (180) = \$211,778.86 in mortgage payments. Remember to use the calculation of the monthly mortgage payment to many decimal places. The down payment is 0.2(145,000) = \$29,000.

Total financed price of condo: 211,778.86 + 29,000 = \$240,778.86

The rent will increase by 10% at the end of each 3-year period.

Years 1–3:	1000
Years 4–6:	$1000(1.1) = 1100.00$
Years 7–9:	$1000(1.1)^2 = 1210.00$
Years 10–12:	$1000(1.1)^3 = 1331.00$
Years 13–15:	$1000(1.1)^4 = 1464.10$

In each 3-year period she will make 36 payments. For 15 years the calculations are as follows:

$$36(1000.00 + 1100.00 + 1210.00 + 1331.00 + 1464.10) = 219{,}783.60$$

The rental payments for 15 years would be \$219,738.60.

The total financed price of the condo is 240,778.86 − 219,783.60 = \$20,995.26 more than the rental payments for 15 years.

Although it appears that Sari will pay less if she rents, rather than buys, the condo, renting might not be the best choice for her. There are tax advantages to buying that will effectively increase her income. After 15 years she will own the condominium and will no longer have monthly mortgage payments, whereas if she rents, she will have to continue to pay rent. The value of the condo is likely to increase over 15 years, so she may be able not only to recover the total financed price, but also to make a profit. If she rents an apartment for 15 years, she will never be able to recover any of the rent payments she makes.

SKILL 4

EXAMPLE 4 Sari would like to estimate the price for which she will be able to sell her house at the end of 15 years.

QUESTION If the value of her condo increases by 5% each year, for what price will she be able to sell it at the end of 15 years?

SOLUTION

At the end of 15 years she will be able to sell her condo for

$$145{,}000(1.05)^{15} = \$301{,}444.59$$

Since the total financed price of her condo is \$240,778.86, she will make a profit of

$$301{,}444.59 - 240{,}778.86 = \$60{,}665.73$$

TECHNOLOGY HINT

Although they have already used the formula for monthly payments, some students will still find it difficult to enter. If necessary write out the keystrokes on the chalkboard.

RETEACHING

If students have trouble with the concept of interest and monthly payment, obtain a table of monthly payments from your bank and use it to show the variation with different interest rates.

ENRICHMENT

Have students enter the function for monthly payments into a graphing calculator in order to obtain a graph of interest rates and payments for a particular principal and length of time. Let students determine the ranges.

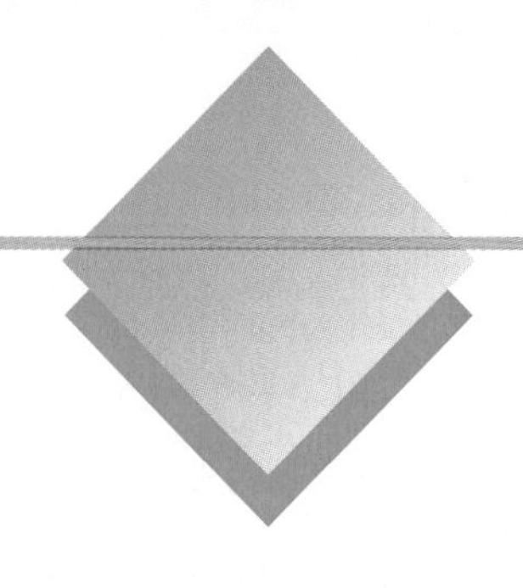

TRY YOUR SKILLS

Rico found a condominium in which he wants to live. The owner will rent it to him for $675 per month or sell it to him for $125,000.

1. Assume that he can have a series of 3-year leases and that the rent increases by 8% at the end of each 3-year period. What is the total of all rental payments for 30 years? $352,023.47
2. If he obtains a 30-year mortgage at 12% with a 20% down payment, what is the total financed price? $395,300.53
3. Will the total financed price or the total rental payments be greater over the 30 years? total financed price
4. Determine the value of Rico's condo at the end of 30 years if it increases in value by 4% each year. $405,424.69

GUIDED PRACTICE

While students are doing the *Try Your Skills* exercises, you can be available to give individual help as needed.

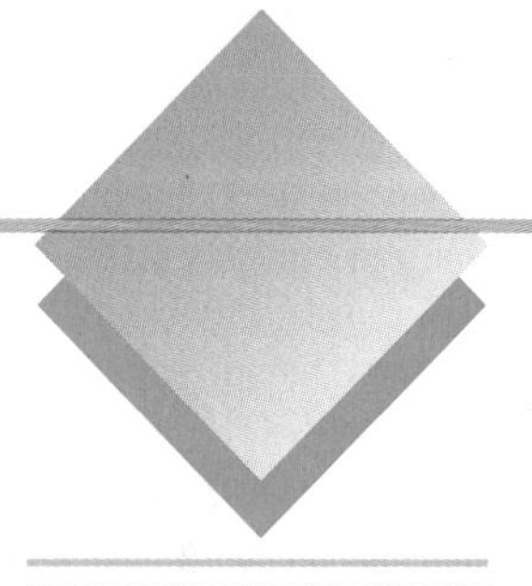

EXERCISE YOUR SKILLS

1. Why is it necessary to have an owners' association for condominium complexes? See Additional Answers.
2. Condominium owners' associations often have some strict rules that the individual home owners are obligated to obey. For example, the association may demand that each condo have the same exterior color. Why would associations have such rules?
3. If you had to choose between renting or buying a particular condo, which option would you choose? Why?

Compute the monthly payment for each condo if you can obtain a 30-year mortgage at 9% with 20% down for each. Compare your results with the rental charge for each condo, and state which alternative would require a greater cash expenditure per month.

4. Purchase price of $90,000; monthly rental charge of $750 $579.33; rental
5. Purchase price of $44,000; monthly rental charge of $220
6. Purchase price of $24,000; monthly rental charge of $230 $154.49; rental
7. Purchase price of $85,000; monthly rental charge of $435
8. Purchase price of $120,000; monthly rental charge of $1000
9. Purchase price of $140,000; monthly rental charge of $750

Assume that you can have a series of 3-year leases and that the rent increases by 9% at the end of each 3-year period.

10. In Exercise 4, will the total financed price or the total rental payments be greater over the 30 years? rental payments
11. In Exercise 5, will the total financed price or the total rental payments be greater over the 30 years? rental payments

KEY TERMS

condominium
mobile home
owners' association

ADDITIONAL ANSWERS

1. It is responsible for management of the grounds, building, and common areas.
2. so that the common areas, lawns, pools, etc. have pleasing scenery
3. renting to get acquainted with the community; buying if you are already a long-term resident of the area
5. $283.23; purchase
7. $547.14; purchase
8. $772.44; rental
9. $901.18; purchase

12. In Exercise 6, will the total financed price or the total rental payments be greater over the 30 years? rental payments
13. In Exercise 7, will the total financed price or the total rental payments be greater over the 30 years? rental payments
14. In Exercise 8, will the total financed price or the total rental payments be greater over the 30 years? rental payments
15. In Exercise 9, will the total financed price or the total rental payments be greater over the 30 years? rental payments
16. How much will the condominium in Exercise 4 be worth at the end of 30 years if it increases in value by 4% each year? $291,905.78
17. How much will the condominium in Exercise 5 be worth at the end of 30 years if it increases in value by 4% each year? $142,709.49
18. How much will the condominium in Exercise 6 be worth at the end of 30 years if it increases in value by 5% each year? $103,726.62
19. How much will the condominium in Exercise 7 be worth at the end of 30 years if it increases in value by 5% each year? $367,365.10
20. How much will the condominium in Exercise 8 be worth at the end of 30 years if it increases in value by 6% each year? $689,218.94
21. How much will the condominium in Exercise 9 be worth at the end of 30 years if it increases in value by 6% each year? $804,088.76

MIXED REVIEW

3-2 1. Use the Rule of 72 to determine how long it will take Gregory's $3200 savings account to double in value. It is growing at an annual rate of 4.5%. 16 years

5-4 2. You have a loan of $22,500.00 at an annual rate of 8% for 4 years for which you have been making monthly payments of $549.29. How much money will you save if you prepay it at the end of 37 months?

6-2 3. Determine the monthly payment that must be made to reduce a $1888.88 VISA balance to zero in 28 months. The card carries an APR of 13.9%.

8-1 4. How many shares of stock in SRB company can you purchase for $25,000 given the current price of the stock is $52 $\frac{1}{2}$ per share? Ignore commission costs. 476 shares

12-4 Calculate the tax savings for the following deductions. Assume that each family is in the 28% tax bracket and that their property tax is the given percentage of the assessed value of their home.

	Assessed Value	Property Tax	Amount of Property Tax	Mortgage Interest	Total Deductions	Tax Savings
5.	$100,000	2.0%	$2000	$ 8215	$10,215	$2860.20
6.	68,000	1.2%	816	4273	5,089	1424.92
7.	209,000	1.9%	3971	14562	18,533	5189.24

INDEPENDENT PRACTICE

ASSIGNMENT GUIDE

Exercises 1-3 review the information about condominiums and can provide the basis of a practical discussion.

Exercises 4-9 ask for the monthly payment under different conditions.

Exercises 10-15 and 16-21 refer back to Exercises 4-9, asking about total financing, and increase in value.

It is best to assign the exercises in groups of three from the above groups, for example, Exercises 4, 10, 16; 5, 11, 17, etc.

CLOSURE

Ask students to list the three factors that determine the monthly payment of a loan. Have them point to each of these and give examples in the monthly payment formula. Ask for an oral description of inflation and an explanation of how to find the future cost of a house if the value is increasing at an annual rate of 6%.

LESSON QUIZ

1. Find the monthly payment on a 25-year, $50,000 mortgage at an annual rate of 6%. (**$322.15**)
2. If a house now worth $65,000 increases in value at an annual rate of 3%, what will be its value after 10 years? Round your answer to the nearest hundred. (**$87,400**)

ADDITIONAL ANSWERS

2. $234.87
3. $79.38

CHAPTER 13 REVIEW

1. What are the advantages and disadvantages of sharing an apartment with another person? See Additional Answers.
2. What should you inspect in any apartment before renting it?
3. What are the advantages and disadvantages of owning, rather than renting, a condominium?

Determine how much rent you can afford for each gross annual income.

4. $62,900 5. $42,500 6. $63,100 7. $80,000
8. $120,000 9. $38,500 10. $54,000 11. $95,750

Determine the gross annual income that you must earn to be able to afford each monthly rent.

12. $785 $40,820 13. $1320 $68,640 14. $1095 $56,940 15. $840 $43,680
16. $560 $29,120 17. $490 $25,480 18. $870 $45,240 19. $390 $20,280

20. Jillian used a broker to find an apartment that rents for $635 per month. The broker charges a fee of 10% of one year's rent. When she signs the lease, she must pay a security deposit equal to two months' rent as well as the first and last month's rent. What is her total cash outlay for the year? How much will she have to pay before she moves into the apartment? $9652; $3302
21. Arnie pays $725 in rent and has monthly electric, water, and gas bills of $150, $75, and $90, respectively. He has the opportunity to move to another apartment in which utilities are included in the monthly rental fee of $990. His relocation bills will be approximately $75. Will he save money by moving to this apartment? See Additional Answers.

Room and board (food) cost $5800 in a dormitory on campus for the nine months of the school year.

22. Myrna would like to rent a furnished apartment near campus with her friend Lois. She found one that rents for $410 per month, including electricity, water, and gas. However, she must sign a lease for 12 months. She estimates that grocery expenses for two people will be about $300 per month. Should Myrna rent the apartment? Why or why not?
23. Erik is considering a furnished apartment exactly like the one that Myrna and Lois are considering. However, the rent is $460 per month and he will have no roommate. Should he rent the apartment? Why or why not?

ADDITIONAL ANSWERS

1. Advantages: shared expenses, companionship Disadvantages: lack of privacy
2. closets, sink, basement, toilets, shower
3. Advantages: interest on mortgage is tax deductible, mortgage payments provide equity Disadvantages: may be difficult to sell, requires large outlay of cash for purchase, extra fees to condo association, property taxes
4. $1209.62
5. $817.31
6. $1213.46
7. $1538.46
8. $2307.69
9. $740.38
10. $1038.46
11. $1841.35
21. yes; in $1\frac{1}{2}$ months moving costs will be covered
22. yes; rent = $4920 for the two students; food for 9 months = $2700; total = $7620 or $3810 each, even if they pay for food for 12 months, their expenses are $4260 each
23. no; rent = $5520; food = 9 • $150 = $1350; total = $6870; dorm is less expensive

David collected the following information about apartments that are for rent in his town.

Monthly Rent	Area (in square feet)
515	660
560	750
375	520
680	920
460	630
950	1095
995	1180
750	975

24. What is a linear model of the form $y = a + bx$ that will allow David to predict the area in square feet in an apartment in his town if he knows the monthly rental charge for the apartment? See Additional Answers.

25. Is there a good correlation between the monthly rental charge and the area?

26. Estimate the area in square feet in an apartment in David's town that has a monthly rental charge of $600. Express your answer to the nearest square foot. 778 ft^2

27. Do you think that an apartment with 780 ft^2 and a monthly rental charge of $575 is fairly priced? yes

Carmen and Bicki are going to be roommates at college. They have found an apartment that rents for $600 per month and a condominium with a purchase price of $85,000.

28. What is their monthly mortgage payment on the condo if they make a 20% down payment and borrow the remainder at 9% for 30 years? Which alternative would require a greater cash expenditure per month? $547.14; rent

29. Assume that Carmen and Bicki can have a series of 3-year leases and that the rent increases by 10% at the end of each 3-year period. Will the total financed price or the total rental payments be greater over the 30 years?

30. How much will Carmen and Bicki's condominium be worth at the end of 30 years if it increases in value by 6% each year? $488,196.75

31. Carmen and Bicki find another bank that will allow them to make a 25% down payment on the condominium and borrow the remainder at 8% for 15 years. What would be their monthly expenditure under this option? $609.23

32. Compare the total financed price in Question 28 to that in Question 31. Which is greater? **28:** $213,971.62 **31:** $130,911.08 Question 28 results in the greater price.

ADDITIONAL ANSWERS

24. $y = 156.903022 + 1.035908387x$

25. yes, $r = 0.9883815163$

29. rental payments

CHAPTER 13 TEST

Determine how much rent you can afford for each gross annual income.

1. $79,200 **2.** $34,500 **3.** $55,600 **4.** $69,600

Determine the gross annual income that you must earn to be able to afford each monthly rent.

5. $695 $36,140 **6.** $670 $34,840 **7.** $365 $18,980 **8.** $875 $45,500

9. Patricia used a broker to find an apartment that rents for $525 per month. The broker charges a fee of 16% of 1 year's rent. When she signs the lease, she must pay a security deposit equal to 1 month's rent, as well as the first and last month's rent. What is her total cash outlay for the year? How much will she have to pay before she moves into the apartment? $7833, $2583

ALTERNATIVE ASSESSMENT

Have students prepare a written report, with supporting data on three different linear relationships; one from science, one from social studies, and one from everyday living.

Leonard collected the following information about houses selling in his town.

Monthly Rent	Number of Amenities
$ 625	6
1140	10
500	3
990	8
1095	9
480	3
325	1
750	8

10. What is a linear model of the form $y = a + bx$ that will allow Leonard to predict the number of amenities in an apartment?

11. Is there a good correlation between the price and the number of amenities?

12. Estimate the number of amenities in an apartment with a monthly rental charge of $600. Round your answer to the nearest whole amenity. 5 amenities

ADDITIONAL ANSWERS

1. $1523.08
2. $663.46
3. $1069.23
4. $1338.46

10. $y = -1.56882617 + 0.0102541252x$
11. yes; $r = 0.9550274618$

Jared found a condominium in which he wants to live. The owner will rent it to him for $850 per month or sell it to him for $140,000.

13. If he obtains a 15-year mortgage at 9% with a 20% down payment, will the total financed price or the total rental payments be greater over the 15 years? (Assume that he can have a series of 3-year leases and that the rent increases by 8% at the end of each 3-year period.) financed price

14. Assuming the mortgage and lease conditions in Question 13, which alternative (renting or making monthly mortgage payments) would require a greater cash expenditure per month during the first 3-year lease? mortgage payment

15. Which alternative would require a greater cash expenditure per month during the fifth 3-year lease? rental payment

16. How much will Jared's condominium be worth at the end of 15 years if it increases in value by 4% each year? $252,132.09

CUMULATIVE REVIEW

1-2 **1.** You work 35 hours per week for 49 weeks. You get 3 weeks of paid vacation and 25% of your base salary in fringe benefits. You earn $4,850 per month. Determine your yearly earnings, including benefits.

4-3 **2.** The fixed costs for your business are $180 per week. The variable costs are $4.90 per item. What is the cost if you produce 95 items in 1 week?

4-3 **3.** Using the information in Exercise 2, determine your profit or loss for 1 week if you sell all the items for $8.75 each. Profit: $185.75

5-5 **4.** Determine the plan that would provide the lowest total financed price for a $20,595 car. Assume that the loan in Plan 1 is from a bank or other lending institution and that the loans in Plans 2 and 3 are from the car manufacturer's financing subsidiary. Plan 3

	Plan 1	Plan 2	Plan 3
Rebate	$2000	$1200	0
Rate	11.25%	8.9%	3.5%
Time	36 months	36 months	36 months
Loan Amount	$18,595	$19,395	$20,595
Monthly Payment	$610.98	$615.85	$603.48
Total Financed Price	$21,995.29	$22,170.73	$21,725.15

5-5 **5.** Determine which plan yields the highest profit to the company in Exercise 4.

7-1 **6.** Use the Multiples of Salary Chart in the Reference Section to find the amount of life insurance needed by a 35-year-old man who has gross earnings of $30,000 in order to provide 75% income replacement.

8-1 **7.** If you buy 850 shares of stock at a price of $23\frac{1}{4}$, what will the total cost of the stock be? Ignore commission costs. $19,762.50

12-4 **8.** A family that is in the 31% tax bracket pays $5,300 in mortgage interest and a property tax that is 1.42% of the assessed valuation of their home, which is $230,000. Find their possible income-tax savings. $2655.46

13-1 **9.** Determine how much monthly rent you can afford if your gross annual income is $55,500. $1067.31

13-3 Compute the monthly payment for each condo on which you can obtain a 30-year mortgage at 9% with 20% down. Compare your results with the rental charge for each condo and state which alternative would require a greater cash expenditure per month.

10. Purchase price of $125,800; monthly rental charge of $800 $809.77, purchase

11. Purchase price of $66,000; monthly rental charge of $600 $424.84, rental

12. If your condominium is currently worth $75,000 and it increases in value by 4% each year, how much will it be worth at the end of 20 years? $164,334.24

ADDITIONAL ANSWERS

1. $72,750
2. $645.50
5. Plan 2
6. $240,000

PROJECT 13–1: **Utilities**

Select one utility that is used in your home. Make telephone calls and visits to the utility's local office to answer the following questions.

1. How often does the utility company send bills to its customers?
2. How does the company measure its customers' usage? Does it have a meter in each customer's home?
3. In what units is each commodity measured? What is the cost per unit?
4. What state or local taxes are added to each bill?
5. Collect your family's bills for this utility for at least 3 months and find the average monthly usage and cost.

PROJECT 13–2: **Choosing an Apartment**

1. Find newspaper advertisements for 20 apartments that are for rent in your town.
2. For each apartment, find the number of square feet and the number of bedrooms.
3. Make a poster presenting the information that you gather.
4. Run a regression analysis to find a linear model of the form $y = a + bx$ that will allow you to predict the number of square feet in an apartment in your town if you know its monthly rental charge.
5. Is there a strong correlation between the number of square feet and the monthly rental charge?
6. Run a regression analysis to find a linear model of the form $y = a + bx$ that will allow you to predict the number of bedrooms in an apartment in your town if you know its monthly rental charge.
7. Is there a strong correlation between the number of bedrooms and the monthly rental charge?
8. Use your results from Exercise 4 to predict how many square feet are in an apartment in your town that rents for $600 per month.
9. Use your results from Exercise 6 to predict how many bedrooms are in an apartment in your town that rents for $450 per month.

Like distance, the absolute value of a number is always nonnegative. The absolute value of 5 can be thought of as the distance of 5 from 0 on the number line. There are two points that give this distance: 5 and -5. Therefore,

$$|5| = 5 \quad \text{and} \quad |-5| = 5$$

Solve each equation for *x*.

Example $|x + 1| - 3 = 5$

$|x + 1| = 8$ — Add 3 to each side.

$x + 1 = 8$ or $x + 1 = -8$ — Separate into two parts.

$x = 7$ or $x = -9$ — Check in original equation.

1. $|x| = 10$
10 or -10

2. $|x + 2| = 5$
3 or -7

3. $|x - 1| + 2 = 8$
7 or -5

4. $|x - 4| = 1$
5 or 3

5. $|x| + 1 = 5$
4 or -4

6. $|x| - 2 = 2.5$
4.5 or -4.5

An absolute value inequality has two conditions that must be satisfied.

Graph the solution of each inequality on a number line.

Examples **a.** $|x + 1| < 5$ **b.** $|x + 1| \geq 5$

The equation $|x + 1| = 5$ means $x + 1 = 5$ or $x + 1 = -5$.
Thus $x = 4$ and $x = -6$ are boundary points for each inequality.

a. For $|x + 1| < 5$, points inside these boundaries satisfy the inequality.

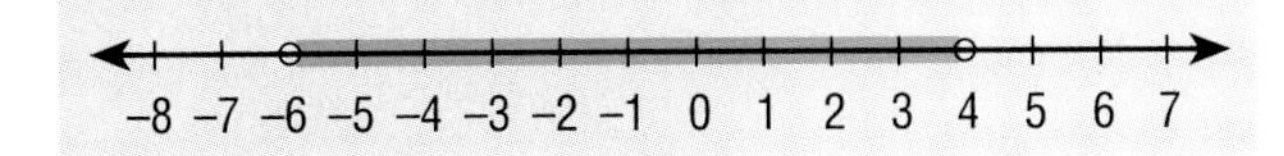

b. For $|x + 1| \geq 5$, points on and outside these boundaries satisfy the inequality.

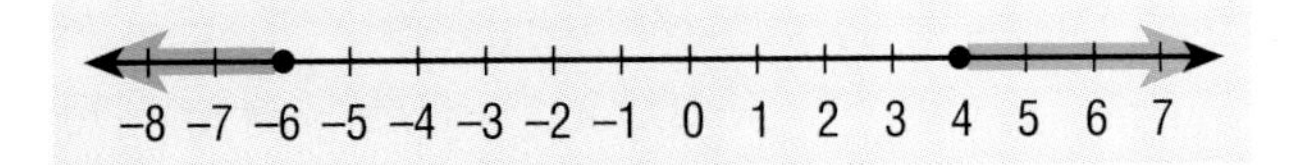

7. $|x| < 5$ **8.** $|x - 1| < 7$ **9.** $|x| > 2$ **10.** $|x + 1| \geq 3$

See Additional Answers.

Graph each equation on a coordinate system.

Example $y = |x|$

$y = x$ for $x \geq 0$
$y = -x$ for $x < 0$

Thus we have the graph shown.

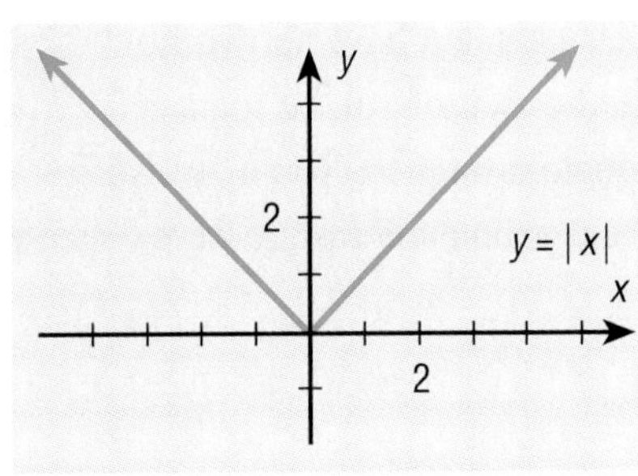

11. $y = -|x|$ **12.** $y = |x - 1|$

See Additional Answers.

13. $x = |y|$ **14.** $y = 2|x|$

Budgeting

IN THIS FINAL CHAPTER, WE WILL HELP the young people from the previous chapters use what they have learned in the construction of a budget to control their expenses. Especially expenses that recur, such as housing, food, and transportation.

Each individual and family has its own budgetary problems and its own solutions to those problems. Nevertheless, the solutions have common characteristics. The primary characteristic is that a budget is a means of taking control of your money rather than allowing it to take control of you.

This control is easiest to exert over monthly bills. However, you will need to make provisions for large or unexpected expenses. Earlier chapters have shown how some unexpected expenses can be managed by buying adequate insurance. You may also need to create and maintain a cash reserve.

Beyond paying your monthly bills, you will want to spend some of your money on personal needs and wants. An intelligently devised budget can help you to do this without damaging your economic health and security. It will even show you ways in which you can put aside some money for savings.

A budget should respect the needs and wishes of all family members. The adults of

the family bear the largest responsibility, since they have more experience and knowledge than the younger family members.

This chapter will concentrate on some aspects of consumer information and money management not covered in earlier chapters. It will also provide specific examples of sample budgets that you may be able to use as models for preparing your own budget.

14–1 Budgeting Housing Expenses

14–2 Food Budgets and Unit Prices

14–3 Living Within Your Budget

14-1 BUDGETING HOUSING EXPENSES

QUICK REFERENCE

TEACHER'S RESOURCES AND ANSWER KEY

Lesson Quiz Answers 14-1, p. 88
Reteaching Activity Answers 14-1, p. 116
Enrichment Activity Answers 14-1, p. 117

EXTENSION ACTIVITIES

Reteaching Activity 14-1, p. 103
Enrichment Activity 14-1, p. 106

People are always asking, "Where does the money go?" You have probably heard someone say something like, "Last week I had $50. Now it's gone, and I can't believe I spent it all." As you grow older, that $50 can become $100 or $500. To use money successfully, you must keep track of how you spend it.

Sylvia's parents learned the hard way. Her mother and father had a lot of disagreements about money before their divorce. Now Sylvia and her mother watch their expenses closely. Last year, Sylvia bought a dress on the layaway plan.

Sylvia is not sure how she will meet her expenses in college. She hopes that she will not have to drop out after a year or two as some of her friends did. Her father will help her, but she does not yet know just what all of her expenses will be. Further, she does not know how much she might need to earn herself to cover the expenses that her mother and father cannot pay.

Maria has had some experience in planning her finances by making car and insurance payments. As a result, she thinks that it must be much more complicated for her parents, who have to pay for a house, car, food, clothing, medical care, and even unexpected expenses. Maria and Sylvia both realize that among all the expenses that an individual or family has to budget carefully for, housing expenses are very important.

OBJECTIVES: *In this lesson, we will help Sylvia and her mother:*

- *Decide why a budget is needed.*
- *Study what goes into a budget.*
- *Learn how to calculate property taxes.*
- *Control monthly housing expenses.*

BUDGET GUIDELINES

A **budget** is a plan of action that shows how you will plan the distribution of your money to meet expenses and achieve financial goals. Sometimes Sylvia wishes she had never heard of budgets. First you have to have a plan. Then you have to write down all the amounts you spend and what they were for. Then you have to compare your expenditures with your plan. Then you have to figure out what went wrong. (Something always does seem to go wrong.) Sylvia's mother insists that without planning, they both would be worse off. In fact, Sylvia and her mother have worked out the following guidelines to help them budget their money.

- Have clear long-term and short-term goals.
- Recognize your standard of living and how it affects goals.
- See how goals change depending on your stage in life.
- Identify the alternatives when faced with decisions and choices.

FOCUS

ALGEBRA CONNECTIONS

If $y = 1.3x$ and $800 \leq x \leq 900$, what are the possible values of y? Budgeting is often not exact, but must be done within parameters. The equation and inequality above might represent the unit price of oil, oil consumption, and total cost. (**$1040 \leq y \leq 1170$**)

MAKING THE GUIDELINES WORK

Establishing Goals One of Sylvia's long-range goals is to graduate from college. She knows that she will have a much better chance of finding work that is satisfying and that pays well if she completes college.

Sylvia's mother is helping her understand that, to meet her long-term goals, she must be willing to work toward some short-term goals, such as the preparation of their monthly and yearly budgets. These budgets will allow them to save some of the money that Sylvia will need for college.

Standard of Living In discussions with their parents, both Sylvia and Maria have heard the expressions "standard of living," and "quality of life." A **standard of living** is the economic level at which an individual or family lives. People with a high standard of living can buy more of the things that most of us want.

Quality of life is another matter. It refers to the happiness and satisfaction that we find in our lives. This doesn't depend as much on money as on whether we are doing things that enrich our lives. For most of us, both standard of living and quality of life depend, in part, on whether we have adequate control over our income and spending.

Stage in Life Your stage in life plays a large part in your budgeting plans. Single people just starting out may spend much of their disposable income on entertainment and career clothes. A young family may have less disposable income because they need to spend more on insurance and medical care and to save for their children's education. A retired person living on a fixed income may not be able to afford luxury items.

Your stage in life can play a big role in determining your housing needs as well. A single person may live in a small apartment or share expenses with roommates in order to live in a city. Married couples with children will probably be more concerned about the school system available to them than couples without children. Elderly people may be more concerned about the ease of access in their homes, climate, public transportation, and the medical services available in their community.

Preferences Different people have different styles of living. Some do not want the responsibilities that come with home ownership. Of those who buy, some like old houses, some like new. Yvette and Yvonne want a condo with easy maintenance, a pool, and a health club. Sylvia's mother wants a small house in a clean, quiet neighborhood.

HOUSING EXPENSES

The more that Sylvia and her mother study budget guidelines, the more they notice how much those guidelines force them to concentrate on one particular aspect of their lives: their choice of housing. They already know that they should try not to spend more than 25 to 30 percent of their income on rent or mortgage payments. But what about other housing expenses? For example, if they decide to buy rather than rent, how much can they expect to pay in real-estate taxes? Whether they own or rent, they want to insure their belongings. If they own, they will have to have home insurance as well.

And what about **utilities**? Most home owners must pay for energy to regulate the temperature of their homes. This may mean heat in winter and air conditioning in summer. Other utility costs are for electricity, telephone, water and sewer, and regular monthly services. Some people might include cable TV in this category.

A home owner should also budget for home maintenance. This would include plumbing, electrical repairs, and perhaps landscaping. Sylvia and her mother know that maintenance expenses are not always predictable and may be higher if they buy an older home with appliances that will need to be replaced in a year or two. They will need to save to make sure they have money in the event of an expensive maintenance problem.

Of course, even utility expenses are not exactly the same every month. Extended use of the television, air conditioners, and other appliances can add considerably to your electric bill, and we all know how telephone costs vary according to use.

CRITICAL THINKING

Ask students to list some of the alternatives for an individual faced with a large unexpected expense. These might include: using savings, taking out a loan, cutting expenses, selling a house or car.

Ask Yourself

1. What is a budget?
   A plan of action that shows how you will spend your money.
2. Why is a budget useful?
   See Additional Answers.
3. What is the difference between standard of living and quality of life?
4. Why are housing expenses such an important factor in the preparation of a budget?
   Housing costs are important because they represent 25–30% of a family's or individual's income.

ALGEBRA *REVIEW*

Write each as an equation.

1. 6 is 5% of x
   $6 = 0.05x$
2. p is 10% of 80
   $p = (0.10)(80)$
3. 2.5 is x percent of 500
   $2.5 = 500x$
4. Find p in Exercise 2.
   $p = 8$
5. Find x in Exercise 3.
   $x = 0.5\%$

Write an equation and solve.

6. 50 is what percent of 75?
   $50 = 75x; x = 66\frac{2}{3}\%$
7. 2.8 is 5% of what number?
   $2.8 = 0.05x; x = 56$
8. What is 2.6% of 40?
   $x = (0.026)(40); x = 1.04$

Write a linear equation in two variables to express the relationship.

9.

cost (c)	\$8.00	\$14.00
tax (t)	\$0.28	\$ 0.49

$t = 0.035c$

10.

Selling price (e)	\$90	\$110
Sale price (a)	\$76.50	\$ 93.50

$e = 0.85a$

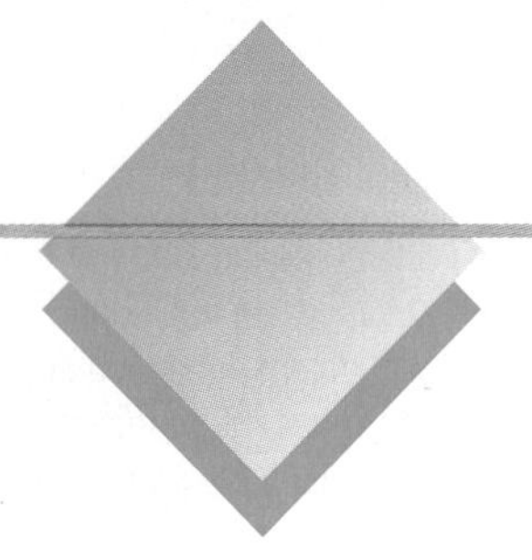

SHARPEN YOUR SKILLS

SKILL 1

The federal government with all its services is financed largely through taxes on the incomes of individuals. Many states tax incomes and also raise money through sales taxes. Local governments, that is, cities and towns, raise large amounts of money through taxes on property.

There are two concepts involved in property taxes: assessed value and mill rate. The **assessed value** is the amount that the local government assigns to a house or other property for tax purposes. Generally, this amount is less than the owner would receive if the house were sold. It is important that the assessed value of similar properties be consistent throughout the tax area.

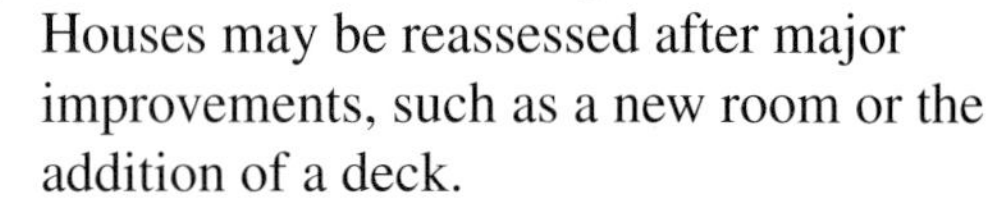

Houses may be reassessed after major improvements, such as a new room or the addition of a deck.

The **mill rate** is the amount of tax per thousand dollars of assessed value. In this respect it is somewhat like percent except that percent is in hundredths while mill rate is in thousandths.

EXAMPLE 1 Sylvia and her mother are considering buying a house that is assessed at $74,000. The tax rate in the town is 18.3 mills.

QUESTION What property taxes will Sylvia's mother have to pay?

SOLUTION

For every thousand dollars, Sylvia's mother will have to pay $18.30 in taxes.

$T = am$ where T = property tax
a = assessed value in thousands
m = mill rate

$$T = 74(18.3)$$
$$= 1354.20$$

The property tax is $1354.20 per year.

SKILL 2

The expenses you have for operating your home cannot be avoided, but they can be controlled. Heating bills, for example, can be budgeted in such a way that the oil company sends you a monthly bill rather than a bill each time a delivery is made. This flattens out the payments and helps you avoid surprises. A careful look at telephone bills will tell whether you are making more long-distance calls than you need. Many times you can send a message by mail and save money. Finally, if you compare electric bills from month to month, you will see whether you have saved money by conserving the use of electricity.

How can a home owner begin to keep track of utility and other operating expenses? The best way to start is to prepare a table like the one shown.

EXPENSES	
Bill Type	**Payment Amount**
Energy for heat or air conditioning	
Electricity for household appliances	
Water	
Telephone	
Home maintenance	

In the table above, home maintenance is not a utility expense. It refers to expenses such as repairs and removal of leaves in the fall and snow in the winter. Like a utility expense, home maintenance is a regular expense and should be included in a budget.

INSTRUCTION

TEACHING THE LESSON

Have students read the opening scenario showing the need for budgets.

Skill 1 introduces mill rate. Point out that mill rate is similar to percent. But percent means "parts out of 100," and mill rate means "parts out of 1000."

Since the value of a house is in thousands, it is convenient to move the decimal point in this number rather than to convert the mill rate to thousandths. The tax on a $50,000 home at 20 mills is found by multiplying: $50 \bullet 20 = 1000$. The tax is $1000.

The method of averaging explained in Skill 2 is itself sometimes called budgeting.

FOCUS ON ALGEBRA

As students approach the conclusion of the text, they should have matured not only in their ability to use algebra applied to financial situations. They should also have developed confidence in asking questions that require a mathematical analysis. *(NCTM Standard 1, p. 137)*

COMMON ERROR

Remind students to check every answer for correct placement of the decimal point. They should ask themselves whether the answer is reasonable in view of what they have learned and in the context of the problem. It is not reasonable that a house should sell for $75.63 or that a monthly mortgage payment should be $5.25.

TECHNOLOGY HINT

The spreadsheet used in Skill 2 is a type commonly used in both personal and business finances. Students should be able to write the formulas for each cell and explain how they are used.

EXAMPLE 2 Sylvia's mother remembers keeping a record of daily expenses on their recent trip and using the results to prepare a travel budget for her clients at her travel agency.

QUESTION How can Sylvia and her mother use a table of monthly housing expenses to prepare a budget for those expenses?

SOLUTION

They use a spreadsheet program to track their expenses for 4 months and then find the average of those expenses.

	A	B	C	D	E	F	G
1		Feb.	March	April	May	Total	Average
2	Heat/Air cond.	132	87	68	56	343	85.75
3	Electricity	50	48	49	52	199	49.75
4	Water	17	15	14	18	64	16.00
5	Telephone	38	45	62	19	164	41.00
6	Home Maintenance	62	17	40	0	119	29.75
7	Total	299	212	233	145	889	222.25

@SUM(B2..E2)

+F2/4

@SUM(B2..B6)

RETEACHING

If it is necessary to reteach mill rate and the calculation of property taxes, show how mill rate can be converted to thousandths while leaving the assessed value unchanged. For example, the tax on a house assessed at $65,000 with a mill rate of 18.5 is:

$65,000 • .0185 = $1202.50

On the basis of this 4-month record, Sylvia and her mother decide to budget $230 per month for utilities and home maintenance. They will pay particular attention to the energy budget, since the trial period did not include all of the winter months.

Using the ideas discussed in Chapter 12, they also plan to budget no more than about 28% of their family income, or about $600 a month, to cover the mortgage payments on the house they hope to buy. If they decide to rent instead of buy, then they will look for a house or apartment that rents for $600 per month. Then their entire housing budget, including the monthly mortgage payment or rent, will be $230 + $600, or $830.

TRY YOUR SKILLS

GUIDED PRACTICE

Have students write out the answers to the *Try Your Skills* exercises, then compare answers.

Write each decimal number as a mill rate.

1. 0.0183 18.3 mills **2.** 0.0200 20 mills **3.** 0.0016 1.6 mills **4.** 0.0102 10.2 mills

5. The mill rate in Smithtown is 23.4. What will be the yearly property tax on a house in Smithtown that has an assessed value of $65,000? $1521
6. Mr. and Mrs. Gant pay an annual property tax of $1827.50. Their house has an assessed value of $85,000. Find the mill rate. 21.5 mills
7. Construct a monthly housing budget for a family that spends the following amounts in a 4-month period: energy, $120, $108, $130, $126; electricity, $31, $28, $37, $40; water, $12, $13.50, $15.50, $21; telephone, $80, $36, $47, $51; home maintenance, $43, $72, $18, $10. See Additional Answers.

8. Suppose that the family of Exercise 7 has an annual income of $32,000. They wish to not spend more than 25% of their income on monthly rent. Find the total monthly housing budget of the family when the monthly rent is included. $926.42

EXERCISE YOUR SKILLS

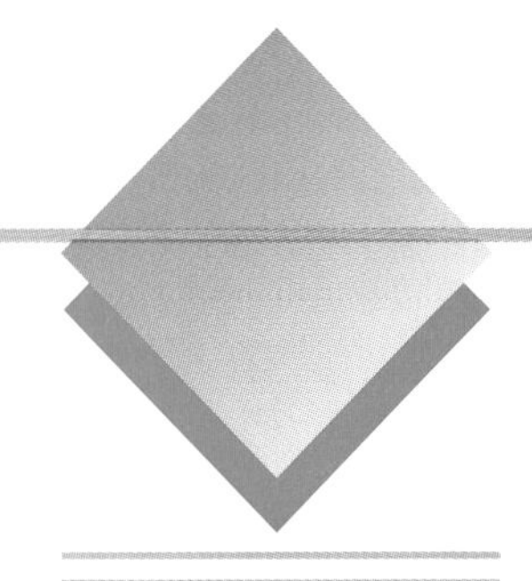

See Additional Answers.

1. Why do you think that the assessed value of a house is usually less than the amount that could actually be received from the sale of the house?
2. What factors might cause the mill rate of a community to rise or fall?
3. Do you think that 4 months is a satisfactory period of time to obtain a useful estimate of your monthly housing expenses? For which expenses would it be satisfactory and for which expenses would it not be satisfactory? See Additional Answers.
4. What will be the annual property tax on a house that is assessed at $78,000 with a mill rate of 25.7? $2004.60
5. What will be the quarterly payment of a property tax on a house with an assessed value of $41,000 and a mill rate of 13.6? $139.40

What is the mill rate of a house assessed at the following values with the given annual property tax?

6. $53,000; $678.40 12.8 mills
7. $160,000; $1328 8.3 mills
8. $112,000; $2430.40 21.7 mills
9. $99,550; $1582.85 15.9 mills

Construct a monthly budget for the housing expenses of each of the following families based on their actual expenditures over the indicated time periods.

10.

The McGrew Family	Sept.	Oct.	Nov.	Dec.	Total	Average
Heat/Air cond.	$71	$94	$99	$118	$382	$ 95.50
Electricity	34	55	40	48	177	44.25
Water	24	16	18	18	76	19.00
Telephone	34	32	59	74	199	49.75
Home Maintenance	9	28	43	40	120	30.00
Total	172	225	259	298	954	238.50

11.

The Lopez Family	June	July	Aug.	Sept.	Total	Average
Heat/Air cond.	$16	$58	$76	$ 70	$220	$ 55.00
Electricity	59	98	64	71	292	73.00
Water	19	21	30	28	98	24.50
Telephone	40	22	35	33	130	32.50
Home Maintenance	0	31	87	104	222	55.50
Total	134	230	292	306	962	240.50

KEY TERMS

assessed value
budget
mill rate
quality of life
standard of living
utilities

INDEPENDENT PRACTICE

ASSIGNMENT GUIDE

Exercises 1-3 review the concepts underlying property taxes and budgets.

Exercises 4-11 present applications similar to those in the lesson.

Exercises 12-14 include more extended applications.

ENRICHMENT

Have students write for literature on family budgeting. A librarian will be able to put them in touch with banks, insurance companies, and consumer advocate groups that offer such information. Students should work in groups to create a family budget for a family of four with a yearly income of $65,000.

INTERDISCIPLINARY INVESTIGATION

Economics: Have students look into public records of assessment and property tax rates in the town or city in which they live. Ask in particular for a report on what happens to assessment and the mill rate when a periodic reevaluation takes place.

CLOSURE

Have students explain in their own words the meaning of assessed value, property taxes, and mill rate.

Ask for a brief explanation of how to make a plan for monthly housing expenses.

You may wish to return to the *Ask Yourself* questions after working through the examples. These questions can help summarize the information.

LESSON QUIZ

1. Find the property tax on a home assessed at $55,000 with a mill rate of 14.7. (**$808.50**)
2. The following are 3-month totals.

Heat/Air Condit.	$327
Electricity	$187
Phone	$116

To the nearest dollar, what amount would you budget per month for each? (**Heat/Air Condit.: $109; Electricity: $62; Phone: $39**)

12. Construct a monthly budget for a family that has the following pattern of expenditures: energy, $1400 per year; electricity, $36 per month; water $70 every 3 months; telephone, $28 per month; home maintenance, $600 per year. See Additional Answers.

13. Suppose that the income of the McGrew family in Exercise 10 is $1600 per month and that they wish to budget no more than 25% of their income for monthly rent payments. Find the family's total monthly housing budget including monthly rent payments. $638.50

14. The Lopez family in Exercise 11 has an income of $40,000 a year and will not qualify for a mortgage if their mortgage payments are greater than 28% of their monthly income. Find the family's total monthly housing budget including mortgage payments. $1173.83

MIXED REVIEW

4-2 **1.** Alicia has her own business making figurines in her basement and selling them from her home. She sells the figurines at $35 each. Last year her weekly production was 10, not including two weeks that she took off for vacation. The cost of materials was $3.75 per figurine and 95% of the figurines were sold. What were her earnings for the year?

3-2 **2.** The island of Manhattan was bought from the Manhattan Indians by Peter Minuit, reportedly for the equivalent of $24. If instead of being used to buy Manhattan, the $24 had been invested at 6% compounded annually for 370 years (until the year 1996), do you think that the 1996 value of the investment would be worth more or less than the current value of the island of Manhattan? Explain your answer.

8-1 **3.** A stock's closing price was $52\frac{7}{8}$, $\frac{3}{4}$ of a point lower than yesterday's closing price. What was yesterday's closing price? $53\frac{5}{8}$

10-2 **4.** A car that was bought as a used car for $9000 depreciated over a 10-year period at the rate of 10% of the original price per year. What was the value of the car at the end of the seventh year? $2700

10-3 Brenda drives 10,000 miles a year. Last year her expenses for maintenance and repairs were $380 and her insurance was $1392. There are no loan payments on the car. Her cost of gasoline was $0.05 per mile. Find each of the following.

5. Her total annual operating expenses $2272

6. Her average cost per mile to operate the car $0.227

11-1 **7.** Use the US Mileage Chart in the Reference Section to find approximately how far Tulsa, Oklahoma, is from Boston, Massachusetts. 1537 mi

11-2 **8.** A motel charges $40 a night for a single individual and $50 for a couple. Mr. and Mrs. Price stayed at the motel. The charge for a video rental was $5.00 and the room tax was 6%. The couple rented two movies. Find the cost for them to stay at the motel for 3 nights. $169

14-2 FOOD BUDGETS AND UNIT PRICES

QUICK REFERENCE

TEACHER'S RESOURCES AND ANSWER KEY

Lesson Quiz Answers 14-2, p. 89

Reteaching Activity Answers 14-2, p. 116

Enrichment Activity Answers 14-2, p. 117

EXTENSION ACTIVITIES

Reteaching Activity 14-2, p. 104

Enrichment Activity 14-2, p. 107

When Olivia's parents were making plans to sell their house, they reviewed all of their finances. Since Olivia was interested, they decided that she was old enough to learn about household budgets. Olivia was surprised to discover how many decisions must be made when buying food and how much food costs. This doesn't mean that good food is necessarily expensive. Sometimes just the opposite is true. Some very nutritious foods are reasonably priced. Some prepared foods that contain a lot of fat or preservatives may be expensive.

Olivia not only studied the family's food budget, but also did some of the shopping, comparing prices and determining unit prices. She even began to feel like a parent when her little brother wanted to buy a lot of expensive snacks. She told him that the snacks were not good for him nor for their food budget.

Olivia's friend Jonathan helps keep the financial records for a local retail food store called Mrs. Harney's Pie Shop. The pies are not dessert pies but meat pies; they contain beef and chicken. Jonathan showed Olivia one reason why Mrs. Harney's unit costs are higher when she produces the pies in smaller quantities.

OBJECTIVES: *In this lesson, we will help Olivia and Jonathan to:*

- *Use proportions to determine unit costs based on retail prices.*
- *Develop a food budget for a family.*
- *Determine a business person's unit costs that incorporate the effect of fixed costs.*

WHAT IS THERE TO EAT?

There was a time when American families ate at least dinner together. Now many families are like Olivia's. Very often, some are working or busy at mealtimes, and the family seems to be going in all directions. So families don't have as many meals together as they once did.

Olivia was surprised to learn that her family spent more on food last year than on any other single item in their budget. They frequently eat out, and Olivia's father regularly brings home prepared meals. That may change with the purchase of a new home. The family will have to plan the budget more carefully.

To save on food, Olivia's family is trying to follow these guidelines:

- Make a weekly meal plan that includes nutritional values, personal preferences, preparation time, and cost.
- Watch for sales on products that are needed.
- Make a list and stick to it. Avoid impulse buying.
- Compare prices among brands and stores.
- Buy in quantity when possible.
- Use coupons.

As Olivia continued shopping, she became more interested in unit prices. The **unit price** is the price per unit of measurement, for example, \$1.25 per pound or \$0.23 per ounce. Unit prices are usually marked, sometimes on the item, often on the store shelf beneath the product. However, this is not the case in all stores or for all products.

CLOSURE

ALGEBRA CONNECTIONS

Solve each proportion for x.

$$\frac{0.62}{1} = \frac{1}{x}$$

$$\frac{1 \text{ kg}}{2.2 \text{ lb}} = \frac{x \text{ kg}}{1 \text{ lb}}$$

(**$x = 1.61$; $x = 0.45$**)

CRITICAL THINKING

Have students solve the following: If a certain herb costs 25 cents a gram, how many ounces can you buy for \$10? (**1.4 oz**)

To know what you are buying, it is necessary to know something about measurements. As you know, there are two widely used systems of measurement. The system that is most common in the United States uses inches, feet, and miles for distance; ounces and pounds for weight; and pints, quarts, and gallons for liquid volume. The **metric system** uses centimeters, meters, and kilometers for distance; grams and kilograms for mass; and liters for liquid measure. The table below shows some of the conversions between the two systems.

Common Conversion Factors
1 centimeter (cm) = 0.39 inches (in.)
1 meter (m) = 39.4 inches
1 kilometer (km) = 0.62 miles (mi)
1 gram (g) = 0.035 ounces (oz)
1 kilogram (kg) = 2.2 pounds (lb)
1 liter (L) = 1.06 quarts (qt)

Ask Yourself

1. What are three things to consider in preparing a food budget?
   Answers may vary.
2. What is an example of a nutritious food that is not expensive?
   Answers may vary.
3. What is an example of an expensive food that is not nutritious?
   Answers may vary.
4. What does unit price mean?
   price per unit of measurement
5. About how many grams are in an ounce?
   28.57 g
6. In many communities supermarkets are required to display unit prices, which is an additional cost for the supermarket. If they were allowed to do so, do you think that many supermarkets would discontinue unit pricing?
   yes

ALGEBRA *REVIEW*

Solve the proportion $\frac{a}{b} = \frac{x}{y}$ for

1. x $\frac{ay}{b}$ **2.** b $\frac{ay}{x}$ **3.** a $\frac{bx}{y}$

Solve the proportion $\frac{2}{x} = \frac{y}{7}$ for

4. x $\frac{14}{y}$ **5.** y $\frac{14}{x}$

6. For the proportion in Exercises 4 and 5, substitute values for x and make a table for x and y.
   Answers may vary.
7. Graph the points found in Exercise 6.
   Answers may vary.
8. Use a graphing calculator to graph the proportion in Exercises 4 and 5.
   See Additional Answers.
9. What is the difference between your graph and that of the graphing calculator?

The Pythagorean theorem states that in a right triangle, $c^2 = a^2 + b^2$, where c is the length of the hypotenuse and a and b are the lengths of the legs.

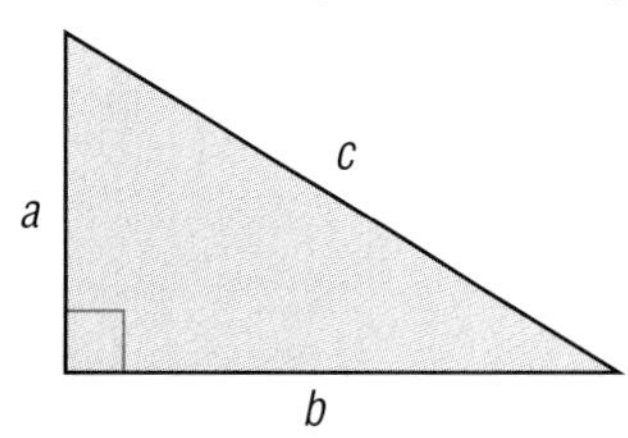

10. Find c if $a = 3$ and $b = 4$.
    5
11. Find b if $a = 5$ and $c = 13$.
    12
12. Find c if $a = 40$ and $b = 9$.
    41
13. Find a if $c = 25$ and $b = 7$.
    24

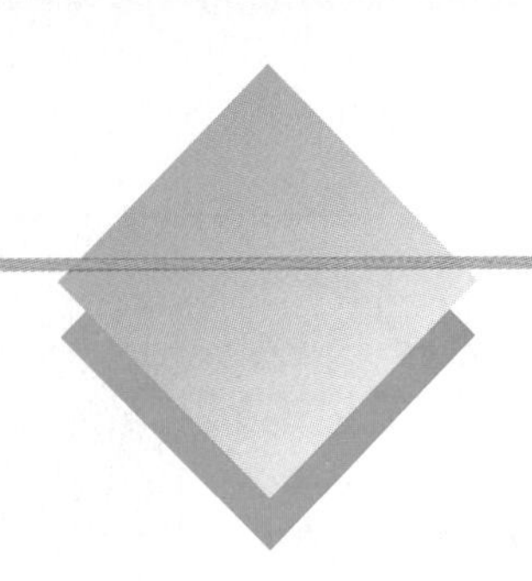

SHARPEN YOUR SKILLS

SKILL 1

Olivia went to the supermarket with her shopping list. She was surprised to find that some unit prices were not listed. Fortunately, she had her calculator.

EXAMPLE 1 Olivia found the following three sizes of Whole Grain Flakes cereal.

a. 12 oz for $1.98

b. 18 oz for $2.69

c. 24 oz for $3.39

QUESTION What is the unit price for each package?

SOLUTION

The unit price in this case is the price per ounce. You can think of this as a proportion problem. Remember that a proportion is an equality of two ratios. Use this method to find the unit price of each of the packages.

a. $\frac{1.98}{12} = \frac{u}{1}$ The price of 12 ounces ($1.98) divided by 12 equals the price of 1 ounce (u) divided by 1.

$u = 0.165$

The unit price is $0.165 per ounce.

b. $\frac{2.69}{18} = \frac{u}{1}$

$u = 0.149$

The unit price is $0.149 per ounce.

c. $\frac{3.39}{24} = \frac{u}{1}$

$u = 0.141$

The unit price is $0.141 per ounce.

The 24-ounce size has the lowest unit price. In terms of price, it is the best buy.

SKILL 2

Olivia has found a number of imported items that are measured in the metric system. In fact, she has developed an interest in comparing measurements. Now her friends ask her to do conversions.

EXAMPLE 2 A friend of Olivia's visiting from Argentina bought 12 pounds of potatoes and wondered how many kilograms that is.

QUESTION How would you convert 12 pounds to kilograms?

INSTRUCTION

TEACHING THE LESSON

Discuss with students how much of our eating is characterized by habit, but how we can change some of these habits if we wish.

The relationships between units of measure, especially ounces and grams are useful when planning a diet because nutritional information is often given in grams or milligrams, whereas our experiential knowledge of food is in ounces.

Work through the different conversions in both directions, and review the methods for solving proportions.

Skill 3 returns to the methods of fixed and variable costs in the production of a large number of items. These ideas have been covered several times before, so you may be able to have a student explain the example.

FOCUS ON ALGEBRA

The exercise material in the lesson is somewhat open ended. This gives students the opportunity to analyze data, formulate their own problems, and frame solutions. *(NCTM Standard 1, p. 137)*

COMMON ERROR

Some students may continue to mix up the terms in proportions. Remind them that keeping the units in the same relative position will help insure a correct proportion.

SOLUTION

To convert 12 pounds to kilograms, use the pound/kilogram conversion on one side of a proportion.

$$\frac{2.2 \text{ lb}}{1 \text{ kg}} = \frac{12 \text{ lb}}{t \text{ kg}}$$

$2.2t = 1 \bullet 12$, or 12   If $\frac{a}{b} = \frac{c}{d}$, then $ad = bc$.

$t = 12 \div 2.2$

$t = 5.5$ kg

12 pounds is equal to 5.5 kilograms

SKILL 3

In Chapter 4, you learned how a business calculates its variable costs by multiplying the number of items produced by its unit cost. For simplicity, it was assumed that the unit cost was constant. However, when a business needs to construct an accurate budget, it needs to account for the fact that the unit cost changes with the change in production volume.

Olivia's friend Jonathan is keeping the financial records at Mrs. Harney's Pie Shop. He taught Olivia something about unit costs from the seller's side. It is not only packaging that helps to determine unit cost but also quantity.

EXAMPLE 3 Including equipment, rent, salaries, and utilities, Mrs. Harney's Pie Shop has fixed expenses of $1250 a week. When fixed expenses are ignored, the unit cost is $2.30 per pie.

QUESTION Including fixed and unit costs, what is the unit cost for the pies in quantities of 1000, 2000, 3000, and 4000?

SOLUTION

The total cost is found by using the equation

$c = f + un$ where c is the total cost
f is the fixed cost
u is the unit cost per pie, not including fixed expenses
n is the number of pies produced

Substituting the given quantities gives the following values for the total cost.

$c = 1250 + 2.30(1000) = 3,550$
$c = 1250 + 2.30(2000) = 5,850$
$c = 1250 + 2.30(3000) = 8,150$
$c = 1250 + 2.30(4000) = 10,450$

The unit cost can be found by dividing each total cost by the given quantity.

$3,550 ÷ 1000 = $3.55 per pie for 1000 pies
$5,850 ÷ 2000 = $2.925 per pie for 2000 pies
$8,150 ÷ 3000 = $2.717 per pie for 3000 pies
$10,450 ÷ 4000 = $2.613 per pie for 4000 pies

You can see how the unit cost decreases as the number of pies increases.

AT-RISK STUDENTS

Have students collect labels and prices from cereal boxes. They can then compare the unit prices and nutritional information.

ESL STUDENTS

Students who have come from other countries will be more familiar with the metric system than Americans. You might ask such a student to describe metric speed limits and average weights or heights. This will give the student practice with English and give American students some first-hand information.

TECHNOLOGY HINT

Skill 3 can be further developed using the graphing calculator. Graph the function, $y = 1250 + 2.3x$. Using the trace feature, students can find many x and y values that satisfy the function. Compare and discuss the different ranges used by students.

RETEACHING

If some students still do not fully grasp the conversion relationships, show comparative units of length on the chalkboard or using the overhead projector. Make the point that, for example, 0.62 miles and 1 kilometer represent the same distance. That is why they are equal. Use this book as an example of something that weighs about 1 kilogram or 2.2 pounds.

ENRICHMENT

Have students prepare a menu for one day. For each item, they should include the weight, cost, number of calories, amount of fat, and whatever else they wish.

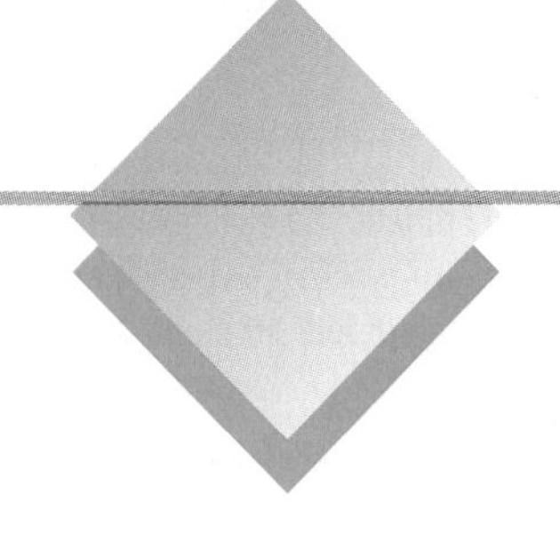

TRY YOUR SKILLS

1. Find the unit price of 8 ounces of cheese for $3.65. $0.456 per ounce
2. Find the unit price of 0.75 liters of skim milk for $0.60. $0.80 per liter
3. Find the better buy for ketchup: 14 oz for $1.65 or 20 oz for $2.15. 20 oz for $2.15
4. Convert 54 kilograms to pounds. 118.8 lb
5. Convert 17 ounces to grams. 485.714 grams

EXERCISE YOUR SKILLS

1. What might lead a family to overspend its food budget? Answers may vary.
2. What are three ways to economize when shopping? See Additional Answers.
3. Why are some foods more expensive than others? Answers may vary.

The following are the prices and sizes of products that you might find in a grocery store. For each one, use a proportion to calculate the unit price and decide which is the best buy in terms of cost. best buy circled

4.	Oil	8 oz $1.10	10 oz $1.29	**15 oz $1.89**
5.	Orange juice	6 oz $0.79	12 oz $1.29	**18 oz $1.79**
6.	Tuna	**6 oz $0.59**	13 oz $1.56	16 oz $1.76
7.	Raisins	**12 oz $1.98**	14 1-oz pkgs $3.04	16 oz $3.20
8.	Honey	4 oz $0.80	8 oz $0.96	**1 lb $1.33**

Convert each of the following.

9. 6.4 kilograms to pounds 14.08 lb
10. 2.4 ounces to grams 68.57 grams
11. 4 quarts to liters 3.77 liters
12. 1.5 pints to liters 0.708 liters

Mrs. Harney has expanded her business to include lasagna and manicotti take-out. Fixed costs are $940 per week for each of the dishes (including rent, salaries, utilities, and equipment). When fixed costs are ignored, the unit costs are $1.75 for each lasagna dinner and $1.45 for each manicotti dinner.

13. Allowing for the fixed costs, find the unit cost for lasagna for the quantities 500, 1000, 1500, and 2000. $3.63; $2.69; $2.38; $2.22
14. Allowing for the fixed costs, find the unit cost for manicotti for the same quantities. $3.33; $2.39; $2.08; $1.92

KEY TERMS

metric system
unit price

GUIDED PRACTICE

Have students do the *Try Your Skills* exercises and then ask for volunteers to explain them at the chalkboard.

INDEPENDENT PRACTICE

ASSIGNMENT GUIDE

Exercises 1-3 are best treated in discussion. Students may have widely different attitudes toward eating and nutrition.

Exercises 4-8 reinforce knowledge and skills on proportions, and ask about comparative shopping.

Exercises 9-12 are conversions between the customary and metric systems.

Exercises 13-14 review production costs as presented in Skill 3.

MIXED REVIEW

1-2 **1.** A company plans to hire a new production manager at a starting salary of $53,000 a year. Of the company's 260 operating days, the new manager will have 10 vacation days and 8 holidays. How much of the manager's salary is the company paying for non-working days?

5-1 **2.** Find the amount of money Caitlyn can borrow at an annual rate of 9.5% for 4 years. She can afford a monthly payment of $280. $11,145.11

9-1 **3.** Use the formula $t = 11{,}743.50 + 0.31(I - 51{,}900)$ to find the tax owed by a single taxpayer with a taxable income of $132,000. $36,574.50

12-1 Terry collected the following information about the amenities offered by several homes that are available for purchase. These included items such as fireplaces, security systems, and jacuzzis.

Number of amenities	7	7	5	3	2	1	1	1
Price, in thousands	$210	$234	$170	$160	$110	$100	$125	$96

4. Use a graphing calculator to represent these data by a linear model of the form $y = a + bx$. $y = 86.32 + 19.05x$

5. Use the linear model that you found in Exercise 4 to decide whether you think that $200,000 is a fair price for a house with six amenities.

12-2 **6.** Determine the total financed price on a $110,000 house that you plan to finance at 7.2% for 30 years with a 30% down payment. $221,160.09

14-1 Randolph Dillon has set a budget for himself of $620 a month but spent $648 during the month.

7. By how much was he over budget? $28

8. By what percent was he over budget? 4.5%

1-1 **9.** Randolph works from 3:00 P.M. until 6:00 P.M. 5 days per week. How many hours per week does he work? 15

2-2 **10.** How do you write $97.83 in words on a check? See Additional Answers.

4-1 **11.** Angela earns $125 in gross income each week. She is single and claims no exemptions. How much will her employer withhold for income tax?

13-2 Mr. and Mrs. Schwartz collected the following information about two-bedroom apartments facing south on different floors of a high-rise apartment building.

Floor	1	5	10	15	20	25	30
Monthly Rent	$380	$430	$440	$410	$470	$500	$510

12. What is a linear model of the form $y = a + bx$ that will allow the Schwartzes to predict the rent for any apartment in the same line?

13. Suppose that two apartments become available, one on the 17th floor for $430 and the other on the 26th floor for $505. Which apartment seems more attractively priced?

CLOSURE

Ask students to name the three most important factors to consider when preparing a menu or a food budget. For most people these would be: nutrition, cost, and taste. Priorities differ from person to person and from family to family.

LESSON QUIZ

1. Convert 8.7 pounds to kilograms. (**3.95 kg**)
2. Find the unit cost of 14 ounces of cereal at $3.15. (**22.5 cents an ounce**)

ADDITIONAL ANSWERS

1. $3669.23

5. yes; $y = 200.62$

10. Ninety-seven and 83/100 dollars

11. $12

12. $y = 386.92 + 4.072x$

13. 17th floor since predicted value is $456 while predicted value on 26th floor is $493

14-3 LIVING WITHIN YOUR BUDGET

QUICK REFERENCE

TEACHER'S RESOURCES AND ANSWER KEY

Lesson Quiz Answers 14-3, p. 89
Reteaching Activity
Answers 14-3, p. 117
Enrichment Activity
Answers 14-3, p. 117

EXTENSION ACTIVITIES

Reteaching Activity 14-3, p. 105
Enrichment Activity 14-3, p. 108

A budget is like a road map. Both can help you get where you want to go, but only if you use them and refer to them from time to time. Is this the road we planned to take? Will it take us where we want to go? Is this the amount we planned to spend? Are we staying within our budget? In spite of our efforts to live within our budgets, we sometimes find that "there is too much month left at the end of the money."

A budget is a plan of action that shows how you will spend and receive money over a specific time period. To budget well, you must first believe that budgeting is important. The young people and families in our stories have shown us and told us many of the benefits of budgeting. Here are some of their reminders.

- Budgets summarize your finances.
- Budgets encourage a sensible use of income.
- Budgets require an examination of goals, values, and priorities.
- Budgets help you get the maximum value from expenditures.
- Budgets help you live within your income.
- Budgets help you adjust to unforeseen expenses.
- Budgets help you prepare for income taxes.

In this lesson you will see what an overall family budget might look like and how a family might use it to review expenses and manage money effectively.

OBJECTIVES: *In this lesson, we will:*

- *Consider the reasons why a budget is useful.*
- *Review the categories of a typical family budget.*
- *Learn to make adjustments in a budget.*

A **cash shortage** means that you don't have enough money to cover some of your expenses. Cash shortages can occur in any family. Often they are due to large, irregular expenses such as automobile repairs. Sometimes a cash shortage occurs because a person loses his or her job. You can try to plan for irregular expenses by setting money aside in a special savings account.

The table shows typical budgets for three different levels of after-tax income.

	Level A	Level B	Level C
Total family budget	$14,808	$19,045	$27,181
Food	3,639	4,473	5,961
Housing	4,773	6,321	9,271
Clothing	891	1,183	2,036
Transportation	3,715	4,907	6,655
Health care	770	887	1,173
Entertainment	747	1,008	1,689
Personal care	153	196	299
Miscellaneous	120	$70	97

CLOSURE

ALGEBRA CONNECTIONS

Write an equation that could be used to solve any routine percent problem. Explain each term. Use the equation to solve the problem: $225 is 12% of a budget. How much is the entire budget? (***a*** **=** ***pb*****; amount = percent X base; $1875**)

CRITICAL THINKING

Ask students to name three instances when a budget would have to be revised. Have them explain why a change would be necessary and what might be involved.

Some families will need to include other items in their budget, such as life insurance on the principal wage earner.

The pie chart shows the results of a government survey on American spending decisions in a recent year.

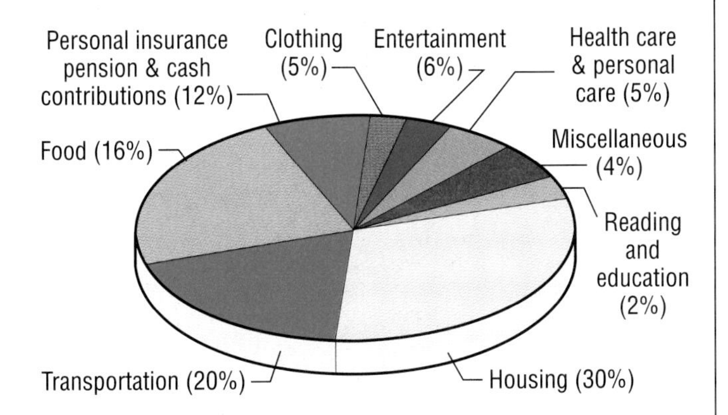

Ask Yourself

1. What are three reasons for preparing a budget?
   Answers may vary.
2. What are two expenditures that might cause a cash shortage?
   Answers may vary.
3. What are the three largest items in most family budgets?
   food, housing, clothing

ALGEBRA *REVIEW*

Housing	$ 600
Food	400
Transportation	450
Clothes	120
Miscellaneous	200
Savings	30
Total	1800

For a pie graph of the above budget, you can use the proportion

$$\frac{x}{1800} = \frac{d}{360}$$

to find the number of degrees d for each budgeted amount x.

How many degrees represent

1. Housing?
   120°
2. Transportation?
   90°
3. Food?
   80°
4. Clothes?
   24°
5. What budgeted amount requires a 40-degree pie slice?
   $200; Miscellaneous

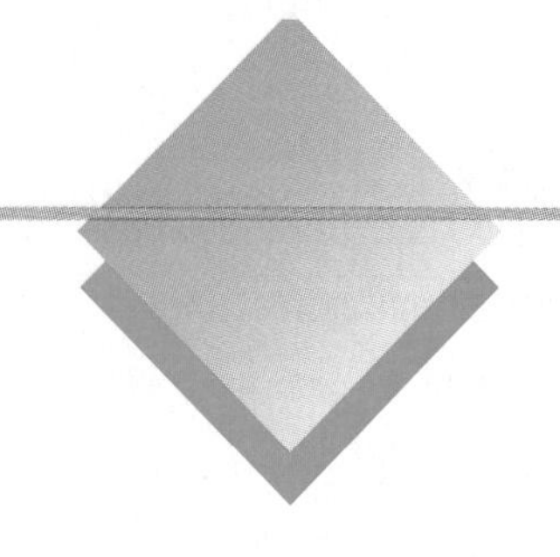

SHARPEN YOUR SKILLS

SKILL 1

Many people like to know not just how much they are spending in different budget categories, but also what the percentages are for the different categories. If one budget area such as food or housing takes too large a percentage of your money, this could be a reason why you have trouble with other categories.

INSTRUCTION

TEACHING THE LESSON

Explain the importance of looking at the percentages in a budget. Sometimes percentage can reveal why a budget does not work. For example, if a person is spending 60% of his or her take-home pay on housing, then it is unlikely that the other categories will be adequately covered.

EXAMPLE 1 Use the Level A budget on page 637.

QUESTION How would you find the percentages for the categories listed in the budget?

SOLUTION

The base for computing the percentages is the total budget, $14,808. The food amount is $3639.

What percent of $14,808 is $3639?

$$x \cdot 14{,}808 = 3639 \qquad \text{Let } x \text{ represent the percent.}$$

$$x = \frac{3639}{14{,}808}$$

$$x = 0.246$$

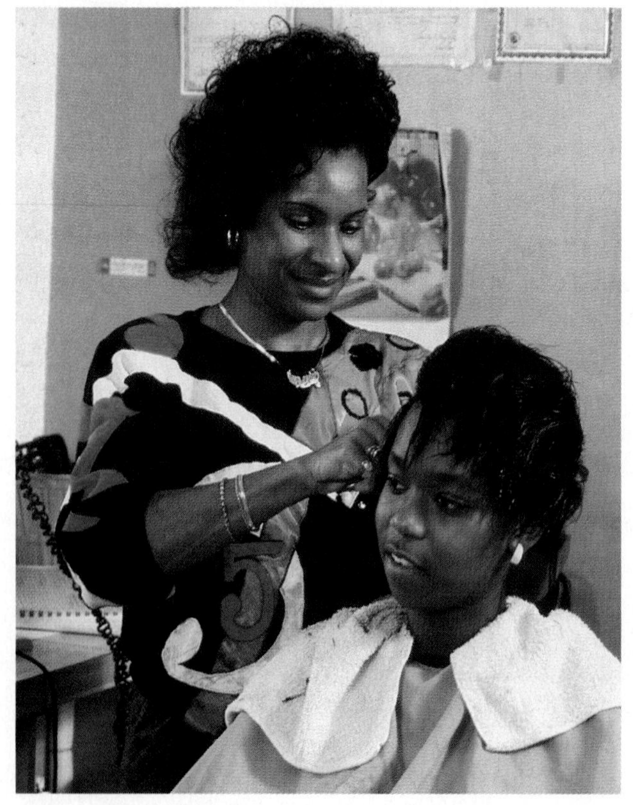

Food is about 25% of the total family budget. The rest of the percentages rounded to the nearest percent are given in the following table.

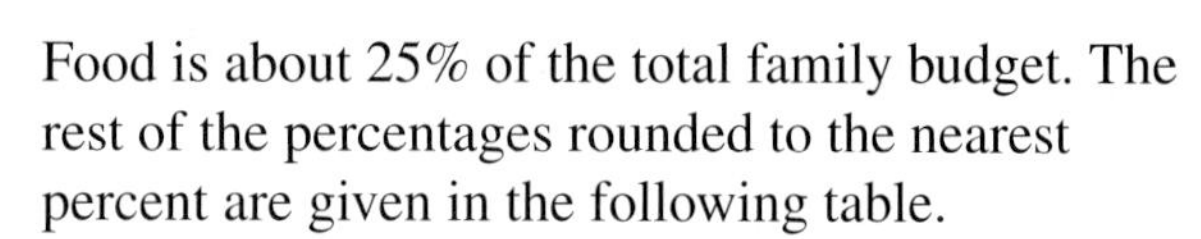

BUDGET CATEGORIES AND PERCENTAGES

	Budget	Percent
Total family budget	$14,808	100%
Food	3,639	25%
Housing	4,773	32%
Clothing	891	6%
Transportation	3,715	25%
Health care	770	5%
Entertainment	747	5%
Personal Care	153	1%
Miscellaneous	120	1%

It is important and useful to know what percent each category of expense represents compared to the entire budget. However, the real test of your budget is how well you stick to it. Are you spending more than you should or less than you are allowed?

SKILL 2

EXAMPLE 2 It is most important to keep track of your expenses and then, at the end of each month, determine whether you are over or under budget.

QUESTION Determine the amount by which the given actual expense is over or under the budgeted amount for the Level A budget on page 637.

SOLUTION

The third column shows the actual amount spent minus the amount budgeted. A negative amount means that the item is under budget. A positive amount means that the budgeted amount has been exceeded.

	Budget	Actual	Under (−)/Over (+)
Total family budget	$14,808	$15,111	+$303
Food	3,639	4,217	+ 578
Housing	4,773	4,792	+ 19
Clothing	891	1,003	+ 112
Transportation	3,715	3,419	− 296
Health Care	770	625	− 145
Entertainment	747	805	+ 58
Personal care	153	160	+ 7
Miscellaneous	120	90	− 30

TRY YOUR SKILLS

See Additional Answers.

A recently married couple, the Hollidays, have an annual take-home pay of $22,500. Use the given budget percentages to calculate the amount budgeted for the year in each category. Round the amounts to the nearest dollar.

1. Food, 22%
2. Housing, 28%
3. Clothing, 5%
4. Transportation, 20%
5. Health care, 4%
6. Entertainment, 7%
7. Personal care, 2%
8. Miscellaneous, 1%
9. Savings, 11%
10. The Hollidays expect to start having children next year. Therefore, they have decided to budget $320 a year for the premiums of a 5-year renewable term $150,000 insurance policy on Mr. Holliday's life. Can the family budget for life insurance in this category for savings? yes
11. Mr. Holliday, whose employer does not offer a retirement plan, would like to contribute $2000 each year to an IRA (Individual Retirement Arrangement). Can he afford to do this and also pay for the insurance premiums in the category for savings? yes

TECHNOLOGY HINT

As an extension to the lesson, have interested students prepare a three-month spreadsheet budget for a family or individual. Ask them to fill in the expenses and then tell a story about why the expenses were over or under budget.

FOCUS ON ALGEBRA

The circle graph, the budget tables, and the statements about budget are examples of the different ways in which mathematical ideas can be communicated. *(NCTM Standard 2, p. 140)*

COMMON ERRORS

Students should be continually reminded to double-check their over/under figures. It is easy to get signs mixed up.

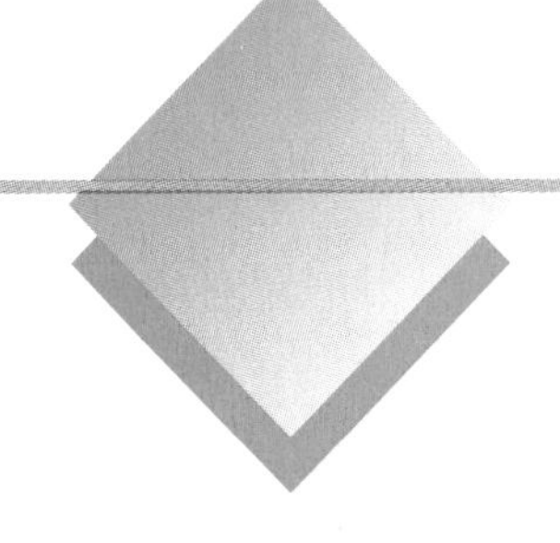

GUIDED PRACTICE

As a follow up to Skill 2, ask students to calculate the percent over or under for each category.

RETEACHING

If needed, draw up a fraction/decimal/percent table and have students find the missing amounts. Include a values less than 1% and greater than 100%.

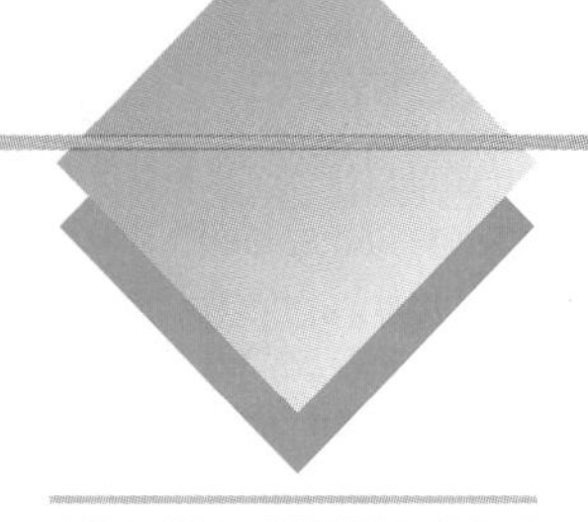

EXERCISE YOUR SKILLS

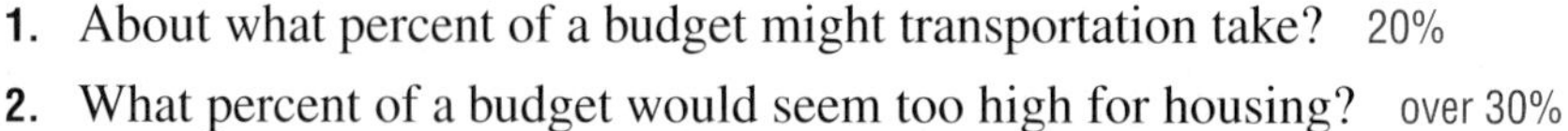

1. About what percent of a budget might transportation take? 20%
2. What percent of a budget would seem too high for housing? over 30%
3. Why is it helpful to find percentages as well as dollar amounts for budget categories? See Additional Answers.
4. How do you calculate the amount that you are over or under budget?

KEY TERM
cash shortage

INDEPENDENT PRACTICE

ASSIGNMENT GUIDE

Exercises 1-4 will allow you to discover whether students grasp the underlying budgetary concepts.

Exercises 5-22 involve percent, and over/under computations.

Exercise 23 asks students to think about budgets from a different perspective—that of a parent.

For each category in the Level B budget table, calculate the percent of the total to the nearest tenth of a percent.

		Budget	Percent of Total
5.	Total family budget	$19,045	100%
6.	Food	4,473	23.5%
7.	Housing	6,321	33.2%
8.	Clothing	1,183	6.2%
9.	Transportation	4,907	25.8%
10.	Health care	887	4.7%
11.	Entertainment	1,008	5.3%
12.	Personal care	196	1.0%
13.	Miscellaneous	70	0.4%

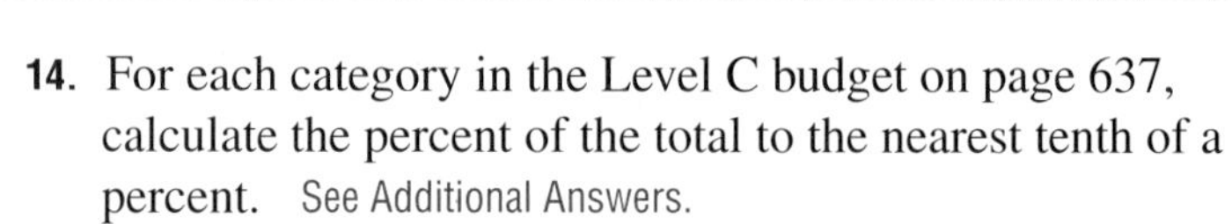

14. For each category in the Level C budget on page 637, calculate the percent of the total to the nearest tenth of a percent. See Additional Answers.

For each category in the Level C budget, calculate the amount over or under budget.

		Budget	Actual	Under(−)/Over(+)
	Total family budget	$27,181	$27,895	+$714
15.	Food	5,961	5,088	− 873
16.	Housing	9,271	9,271	0
17.	Clothing	2,036	2,874	+ 838
18.	Transportation	6,655	7,120	+ 465
19.	Health care	1,173	920	− 253
20.	Entertainment	1,689	2,145	+ 456
21.	Personal care	299	327	+ 28
22.	Miscellaneous	97	150	+ 53

23. List some of the differences in the ways in which a parent and a teenager think about a family budget. Prepare to discuss these with your classmates in small groups. Answers may vary.

MIXED REVIEW

2-2 **1.** How does the phrase "for deposit only" protect a person who is endorsing a check? It does not allow the check to be cashed.

3-1 **2.** A person saved $60 per week for 52 weeks in order to buy a one-year Certificate of Deposit (CD) that earned 6% a year. Find out how much was saved, how much interest is earned by the CD in 1 year, and the total amount of money in the CD account at the end of 1 year. $3120; $187.20; $3307.20

6-2 **3.** Arleigh wants to pay off the current balance of $3500 on his VISA card in $2\frac{1}{2}$ years. The annual interest rate is 17%. How large must the monthly payment be? $144.02

8-1 **4.** The annual dividend for Acme Enterprises is $1.20 per share. What additional information do you need in order to calculate the current rate of return on 500 shares of Acme Enterprises? price per share

9-3 **5.** Find the amount of the income tax refund or payment due for a family of two parents and one dependent child with wages of $32,000, taxable interest of $650, and dividends of $300. The family had medical expenses of $5200, mortgage interest of $2900 that they paid on their home, state income tax payments of $1080, and charitable contributions of $90. The wife made a tax-deferred contribution of $550 into her IRA. The amount of withholding tax for the year was $2401. payment due is $400

13-1 **6.** André is using a broker to find an apartment that rents for $480 per month. The broker charges a fee of one month's rent. At the time that he signs the lease, André must pay a security deposit equal to one-and-one-half of one month's rent, as well as the rent for the first month. How much will he have to pay before he moves into the apartment? $1680

8-3 **7.** How might a scatter plot of the 5-year performance of a mutual fund help you predict the fund's future performance? See Additional Answers.

12-3 A married couple has a combined take-home salary of $58,000. The couple is interested in buying a house with a 30-year mortgage but does not want to spend more than 25% of its take-home pay on the mortgage payments. The interest rate at the local bank is 7%.

8. What is the maximum amount that they should pay in monthly mortgage payments to the nearest dollar? $1208

9. With a 25% down payment, what is the maximum price to the nearest dollar that they can afford to pay for a house? $181,572 + $45,393 = $226,965

13-3 Compute the monthly payment for each condo, assuming that you can obtain a 30-year mortgage at 9% with 20% down for each condo. Compare your results with the rental charge for each condo and state which alternative would require a greater cash expenditure each month.

10. Purchase price of $60,000; monthly rental charge of $500

11. Purchase price of $100,000; monthly rental charge of $800

ENRICHMENT

Ask students to find other graphs showing expense budgets for families or individuals. Consumer magazines, newspapers, and public service literature are possible sources. Students should analyze and report on the similarities and differences in the distribution of expenses.

CLOSURE

This lesson can be used to initiate a general review. Some of the items on the budget can serve to review an entire chapter, for example, on car expenses, mortgage and home expenses, and travel.

LESSON QUIZ

1. If $720 is 6% of an annual budget and food takes 22% of that budget, in dollars how much is budgeted for food? (**$2640**)
2. For another budget, $325 is spent on food. This is $40 over budget. If food is budgeted at 20%, what is the amount of the total budget? (**$1425**)

ADDITIONAL ANSWERS

7. It might show a pattern, for example a linear pattern, that would suggest the fund's future performance.
10. $386.22; the rental option would be higher
11. $634.70; the rental option would be higher

CHAPTER 14 REVIEW

1. What will be the annual property tax on a house that is assessed at \$170,000 with a mill rate of 19.7? \$3349
2. What will be the quarterly payment of a property tax on a house with an assessed value of \$96,000 and a mill rate of 27.6? \$662.40

What is the mill rate of a house assessed at the following values?

3. \$87,000; \$974.40 11.2 mills
4. \$201,000; \$4482.30 22.3 mills
5. \$407,000; \$7692.30 18.9 mills
6. \$62,450; \$612.01 9.8 mills

7. Construct a monthly budget for the expenses incurred by the Matsuda family to operate their home based on actual expenditures over the time periods shown.

	Dec.	Jan.	Feb.	March	Total	Average
Heat	\$115	\$123	\$134	\$109	\$ 481	\$120.25
Electricity	47	42	56	39	184	46.00
Water	19	25	26	19	89	22.25
Telephone	56	82	38	50	226	56.50
Home maintenance	0	74	23	120	217	54.25
Total	237	346	277	337	1197	299.25

8. Construct a monthly operating budget for a family that has the following pattern of expenditures: energy, \$2600 per year; electricity, \$48 per month; water \$68 every 3 months; telephone, \$49 per month; home maintenance, \$725 per year. See Additional Answers.

For each item listed, calculate the unit price and decide which is the best buy in terms of cost. best buy circled

	A	B	C
9.	16 oz (circled)	12 oz	8 oz
	\$1.29 (circled)	\$1.19	\$0.89
10.	$14\frac{1}{2}$ oz (circled)	12 oz	8 oz
	\$1.59 (circled)	\$1.38	\$0.95
11.	6 oz	10 oz	12 oz (circled)
	\$0.95	\$1.49	\$1.69 (circled)
12.	16 oz	9 oz (circled)	12 oz
	\$2.15	\$1.15 (circled)	\$1.75

Convert each of the following. Round to the nearest hundredth.

13. 6 quarts to liters 5.66 liters
14. 7.5 ounces to grams 214.29 grams
15. 12 pounds to kilograms 5.45 kg
16. 3 kilograms to pounds 6.6 lb

17. Marshall Mills sells professionally made videos of local events such as graduations and weddings. For any event his fixed costs are $950 and his unit cost is $2.40 per cassette. Write an equation for his costs. $y = 950 + 2.4x$

Allowing for the fixed costs in Exercise 17, find the unit cost for the following number of video tapes sold for any event.

18. 20 $49.90 **19.** 75 $15.07 **20.** 100 $11.90 **21.** 130 $9.71

For each category in the budget table, calculate the percent of the total to the nearest tenth of a percent.

		Budget	Percent
22.	Total family budget	$1655	100.0%
23.	Food	410	24.8%
24.	Housing	455	27.5%
25.	Clothing	120	7.3%
26.	Transportation	385	23.3%
27.	Health care	70	4.2%
28.	Entertainment	60	3.6%
29.	Personal care	30	1.8%
30.	Miscellaneous	125	7.6%

For each of the following categories, calculate the amount over or under budget.

		Budget	Actual	Under(−)/Over(+)
31.	Total family budget	$1655	$1570	−$85
32.	Food	410	365	− 45
33.	Housing	455	455	0
34.	Clothing	120	150	+ 30
35.	Transportation	385	347	− 38
36.	Health care	70	25	− 45
37.	Entertainment	60	76	+ 16
38.	Personal care	30	52	+ 22
39.	Miscellaneous	125	100	− 25

Shauna has an annual take-home pay of $26,664. Use the given percentages to calculate the annual amount to the nearest dollar for each category.

40. Food, 21% $5599
41. Housing, 28% $7466
42. Car, 18% $4800
43. Clothing, 6% $1600
44. Health care, 10% $2666
45. Savings, 7% $1866
46. Entertainment, 8% $2133
47. Personal care, 2% $533

CHAPTER 14 TEST

1. Find the property tax on a house with a mill rate of 19.5 and an assessed value of $45,000. $877.50
2. Find the monthly payments in property taxes on a house assessed at $95,000 with a mill rate of 24.2. $191.58

Find the best buy based on unit price. best buy circled

	A	B	C
3.	(16 oz)	$14\frac{1}{2}$ oz	12 oz
	($1.39)	$1.29	$1.19
4.	1 quart	$\frac{1}{2}$ gallon	(gallon)
	$1.65	$2.98	($4.20)

ALTERNATIVE ASSESSMENT

Ask students to write material for a pamphlet advising young people as they prepare to buy their first house. The paper should include information about how much to spend, how to calculate the amount of an affordable mortgage, and how to calculate property taxes.

Have interested students interview a dietician about how he or she balances the demands of cost, nutrition, and good taste when preparing meals. Students should write a report based on their findings.

Perform each of the following conversions.

5. 85 grams to ounces 2.975 oz
6. 2.5 kilograms to pounds 5.5 lb

7. Judith Whitaker sells cookies. Her fixed costs are $80, and her unit cost is $0.16 per cookie. Write an equation for her costs. $y = 80 + 0.16x$

Allowing for the fixed costs in Test Item 7, find the unit cost for the following number of cookies baked.

8. 100 $0.96
9. 400 $0.36
10. 1000 $0.24

The Meyer family has a total annual take-home pay of $24,000. Use the given budget percentages to calculate the annual amount to the nearest dollar for each category.

11. Food, 22% $5280
12. Housing, 30% $7200
13. Car, 28% $6720
14. Clothing, 5% $1200
15. Insurance and pension, 11% $2640
16. Savings, 4% $960

For each category, calculate the amount over or under budget.

		Budget	Actual	Under(−)/Over(+)
17.	Total family budget	$2635	$2577	−$ 58
18.	Food	650	680	+ 30
19.	Housing	790	765	− 25
20.	Clothing	185	110	− 75
21.	Transportation	415	380	− 35
22.	Health care	85	412	+ 327
23.	Entertainment	225	50	− 175
24.	Personal care	35	30	− 5
25.	Miscellaneous	250	150	− 100

CUMULATIVE REVIEW

3-2 **1.** Devore has $25,500 in a CD that pays 4.8% interest, compounded quarterly. How much is her CD worth at the end of 2 years? $28,053.32

4-3 Your fixed costs for your business are $200 per week. Your unit cost is $3.15 per item.

2. What is the total cost of producing 100 items? $515

3. Each item sells for $5.75. How many must you sell to break even? 77

8-1 You have $90,000 to invest. Find the whole number of shares that you can buy in each of the companies below.

4. SmartMart at $73\frac{5}{8}$ 1222 **5.** Ocean Science Ind. at $33\frac{3}{4}$ 2666

9-1 Use the Tax Rate Schedule in the Reference Section to find the tax owed by each taxpayer. See Additional Answers.

6. A married person, filing separately, with $106,945 in taxable income

7. A head of household with $123,780 in taxable income $32,409.80

ADDITIONAL ANSWERS

6. $29,528.45

10-4 **8.** Use the Six-Month Basic Rate Schedule in Reference Section to find the 6-month premium for an insurance policy based on the following information: driver-rating factor, 2.25; collision deductible, $100; car-class rating, 25; comprehensive deductible, $50. $3577.50

12-3 **9.** The Li family has a monthly income of $3560. It can pay 28% of this income on mortgage payments. With a 20% down payment on a 30-year mortgage at 7.6% per year, what is the maximum amount that the family can pay for a house? $176,468.58

13-3 **10.** Use the Monthly Payment Formula to calculate the monthly mortgage payment for a condo that is selling for $79,500. Assume that the down payment is 20% and that the 30-year mortgage has an annual interest rate of 8%. Compare the rental costs of $600 per month with the monthly payments. $466.67 which is less than $600

6-7 **11.** Judy's monthly take-home pay is $2900. Credit card payments are $265. Determine, to the nearest tenth of a percent, how much Judy is spending on her credit payments. Is Judy in danger of credit overload? 9.1%; no

8-2 Find the purchase price of these shares of stock at the prices shown. Ignore the effect of commissions. Find the capital gain or capital loss at the current market value one year later by subtracting the purchase price from the current market value. Calculate the percent change.

			Original Purchase		Current Market			
	Company	**Number of Shares**	**Price/ Share**	**Total Price**	**Price/ Share**	**Total Value**	**Gain or Loss**	**Percent of Change**
12.	XYZ Inc	100	$59\frac{1}{2}$	$ 5950	76	$ 7600	$ 1650	27.7%
13.	Whiz Co	55	$46\frac{1}{4}$	2543.75	$38\frac{1}{2}$	2117.50	− 426.25	−16.8%
14.	AnimFrm	2,500	$6\frac{3}{8}$	15,937.50	$11\frac{1}{8}$	27,812.50	11,875	74.5%

PROJECT 14–1: **Electricity**

Electricity is delivered to almost every home, office, and factory in the United States. Power lines are often buried underground. But the costs of electricity are not invisible. We pay in several ways for the power that we use. First, we pay in hard-earned money that goes to the power companies for generating electricity. Second, we pay in the depletion of our natural resources such as coal, oil, and natural gas that may be burned to create electric power. Finally, we pay a price in the pollution that sometimes arises as fuels are converted to electric power.

1. Make an inventory of the electric appliances used in your home. List the kitchen appliances, radios, televisions, VCRs, lighting fixtures, computers, and so on. You will be surprised at how long your list becomes.
2. Find out the name of the company from which your household purchases electricity.
3. How much does the utility company charge its customers per kilowatt-hour? Does the company raise the rates at peak times of the day or year?
4. Find out from the utility company how it creates the electricity that it sells to you. What are the by-products of generating electricity by this method? What is the effect on our supplies of natural resources?
5. Find out how the electricity reaches your home from the place where it is generated.
6. Collect three or four monthly electricity bills for your household. Add the costs and divide to find the average monthly cost. Find out whether your usage is higher at certain seasons of the year. How do fluctuations in electricity use affect your household budget?

REFERENCE SECTION

The graphing calculator is used in many of the lessons in ***Mathematics of Money with Algebra.*** If you are just getting acquainted with this kind of calculator, the following specific instructions for the *TI-81* and the *Casio fx7700G* may be helpful. These directions should provide you with enough information to help you with other models of graphing calculators. If you need more information, be sure to consult your calculator's manual.

Notice that there is a label on each key and labels above each key. To use the labels above a key, press SHIFT or 2ND and then the key you want, or ALPHA and the key you want. The labels above the keys are the same color as the corresponding shift key. In this section, labels on the keys are shown in boxes ☐ and labels above the keys are shown in brackets [ ].

fx-7700G

AC is used to turn on the calculator and to clear the screen of the calculator. SHIFT [OFF] is used to turn off the calculator. EXE is used to execute a calculation.

TI-81

ON is used to turn on the calculator. CLEAR is used to clear the screen of the calculator. 2ND [OFF] is used to turn off the calculator. ENTER is used to enter a calculation.

Menus

Menus which give you several choices are used on most graphing calculators. Check the mode menu, which is used to set special features, before you begin to use your calculator.

fx-7700G

You make menu selections with the function keys F1 – F6. The label printed above each function key is active unless menu selections appear along the bottom of the screen.

The following should be displayed when you press MDISP.

```
      RUN/ COMP
  G-type: REC/CON
   angle: Deg
 display: Nrm1
```

To correct settings use:

MODE 1 for run (RUN).
MODE + for computation (COMP).
MODE SHIFT + for rectangular coordinates (REC).
MODE SHIFT 5 for connect (CON).
SHIFT [DISP] F3 EXE for norm 1 (Nrm 1).

TI-81

Menus appear on the screen when special keys such as MODE are pressed. Use the arrow keys ◄, ►, ▲, and ▼ to highlight your selection, then press ENTER.

The following should be displayed when you press MODE.

```
NORM Sci Eng
FLOAT 0123456789
Rad DEG
FUNCTION Param
CONNECTED Dot
SEQUENCE Simul
GRID OFF Grid on
RECT Polar
```

To make a correction, use the arrow keys to highlight the correct choice. Then press ENTER. To leave mode, press 2ND [QUIT].

Correcting Errors

If you make an error when entering a calculation, you can replay the calculation and use the insert and delete keys to make a correction. The arrow keys [◄], [►], [▲], and [▼] are used to position the cursor.

fx-7700G

Use the left arrow key [◄] to replay the previous line. Use the delete key [DEL] to delete a character at the cursor location. Use the insert key [SHIFT] [INS] to insert a character at the cursor location.

TI-81

Use the up arrow key [▲] to replay the previous line so you can edit it. Use the delete key [DEL] to delete a character at the cursor location. Use the insert key [INS] to insert a character at the cursor location.

Evaluating Expressions

You can use a graphing calculator to evaluate mathematical expressions. The graphing calculator uses the usual rules for the order of operations, with one exception. Implied multiplication will be calculated before any multiplication, division, addition, or subtraction that is shown by an operation symbol.

Setting the Range

You must set values for the portion of the graph that is displayed in the viewing screen. The range values define a viewing rectangle. The [RANGE] key is used on both the *fx-7700G* and the *TI-81* to set the viewing rectangle.

fx-7700G		TI-81
Range		Range
Xmin:-4.7	←Minimum value on *x*-axis →	Xmin=-4.8
max:4.7	←Maximum value on *x*-axis →	Xmax=4.7
scl:1.	←Scale used on *x*-axis →	Xscl=1
Ymin:-3.1	←Minimum value on *y*-axis →	Ymin=-3.1
max:3.1	←Maximum value on *y*-axis →	Ymax=3.2
scl:1.	←Scale used on *y*-axis →	Yscl=1
INIT		Xres=1

Graphing Equations

To graph an equation, the equation must be in the form $y = mx + b$. If necessary, solve the equation for y before inputting it into the calculator.

fx-7700G

To enter an equation, press [GRAPH].

```
Graph y=
```

Enter the equation using the key labeled [X,θ,T] for the variable x. Then press [EXE] to display the graph.

TI-81

To enter an equation, press [Y=].

$:Y_1 =$
$:Y_2 =$
$:Y_3 =$
$:Y_4 =$

Enter the equation using the key labeled [X|T] for the variable x. Then press [GRAPH] to display the graph.

Trace Feature

Solutions to an equation can be found by using the trace feature to identify coordinates of points on the graph. After pressing [TRACE], you will see a blinking cursor on the screen. Use [◀] and [▶] to move the trace cursor along the graph. The x- and y-coordinates of the point shown by the blinking cursor are displayed at the bottom of the screen. For the *fx7700G,* the trace label is above the [F1] key.

Graphing More than One Equation

You can graph more than one equation at a time on a graphing calculator.

fx-7700G

To graph more than one equation at the same time, press [SHIFT] [EXE] after entering the first equation. To graph $y = 3x$ and $y = x$, use:

[GRAPH] 3 [X, θ, T] [SHIFT] [EXE]
[GRAPH] [X, θ, T] [EXE]

To clear the screen before entering a new graph, press [F5] (cls) [EXE]. To switch between the graph screen and the text screen, press the [G↔T] key

TI-81

When you press [Y=] there is space for four equations to be entered, y_1, y_2, y_3, and y_4.

To clear the screen before entering a new graph, erase any previous equations. Press [Y=], then place the cursor over any character in the equation you want to erase, and press [CLEAR]. To keep the equation, but not have a graph drawn, place the cursor on the equals symbol, and press [ENTER]. To switch between the graph screen and the text screen, use [CLEAR] and [GRAPH].

Drawing a Scatter Plot

You can enter data and draw a scatter plot. First set the range.

fx-7700G

Set the mode menu as follows:
[MODE] [÷] for regression mode (REG).
[MODE] [4] for linear regression (LIN).
[MODE] [SHIFT] [1] for storing statistical data (STO).
[MODE] [SHIFT] [3] for drawing a statistical graph (DRAW).
Use [SHIFT] [,] and [F1] (DT) to enter the data. To enter (2, 3), use 2 [SHIFT] [,] 3 [F1] (DT). A dot is placed on the graph screen as each data point is entered.

TI-81

Pressing [2ND] [STAT] accesses the statistics menu.
To enter data, use [2ND] [STAT] [▶] [▶] [1] (Edit).
To leave the data screen, press [2ND] [QUIT].
To clear statistical data, use [2ND] [STAT] [▶] [▶] (Data) [2] (ClrStat) [ENTER].
To draw a scatter plot, use [2ND] [STAT] [▶] (Draw) [2] (Scatter) [ENTER].

Graphing a Linear Regression Equation

fx-7700G

Use [MODE] [÷] [F6] (REG) to get the linear regression menu for $y = a + bx$, where r is the correlation coefficient. To calculate a, b, and r, use [F1] (A) [EXE] [F2] (B) [EXE] [F3] (R) [EXE]. To graph the line of best fit, use [GRAPH] [SHIFT] [F4] (Line) [1] [EXE].

TI-81

To calculate a, b, and r for a linear regression $y = a + bx$ where r is the correlation coefficient, use [2ND] [STAT] [2] (LinReg) [ENTER].
To graph the line of best fit, use [Y=] [VARS] [▶] [▶] (LR) [4] (RegEQ) [GRAPH].

SINGLE Persons—**WEEKLY** Payroll Period

(For Wages Paid in 19--)

If the wages are–		And the number of withholding allowances claimed is—										
At least	But less than	0	1	2	3	4	5	6	7	8	9	10
		The amount of income tax to be withheld is—										
$0	**$50**	$0	$0	$0	$0	$0	$0	$0	$0	$0	$0	$0
50	**55**	1	0	0	0	0	0	0	0	0	0	0
55	**60**	1	0	0	0	0	0	0	0	0	0	0
60	**65**	2	0	0	0	0	0	0	0	0	0	0
65	**70**	3	0	0	0	0	0	0	0	0	0	0
70	**75**	4	0	0	0	0	0	0	0	0	0	0
75	**80**	4	0	0	0	0	0	0	0	0	0	0
80	**85**	5	0	0	0	0	0	0	0	0	0	0
85	**90**	6	0	0	0	0	0	0	0	0	0	0
90	**95**	7	0	0	0	0	0	0	0	0	0	0
95	**100**	7	1	0	0	0	0	0	0	0	0	0
100	**105**	8	1	0	0	0	0	0	0	0	0	0
105	**110**	9	2	0	0	0	0	0	0	0	0	0
110	**115**	10	3	0	0	0	0	0	0	0	0	0
115	**120**	10	4	0	0	0	0	0	0	0	0	0
120	**125**	11	4	0	0	0	0	0	0	0	0	0
125	**130**	12	5	0	0	0	0	0	0	0	0	0
130	**135**	13	6	0	0	0	0	0	0	0	0	0
135	**140**	13	7	0	0	0	0	0	0	0	0	0
140	**145**	14	7	1	0	0	0	0	0	0	0	0
145	**150**	15	8	1	0	0	0	0	0	0	0	0
150	**155**	16	9	2	0	0	0	0	0	0	0	0
155	**160**	16	10	3	0	0	0	0	0	0	0	0
160	**165**	17	10	4	0	0	0	0	0	0	0	0
165	**170**	18	11	4	0	0	0	0	0	0	0	0
170	**175**	19	12	5	0	0	0	0	0	0	0	0
175	**180**	19	13	6	0	0	0	0	0	0	0	0
180	**185**	20	13	7	0	0	0	0	0	0	0	0
185	**190**	21	14	7	1	0	0	0	0	0	0	0
190	**195**	22	15	8	1	0	0	0	0	0	0	0
195	**200**	22	16	9	2	0	0	0	0	0	0	0
200	**210**	23	17	10	3	0	0	0	0	0	0	0
210	**220**	25	18	11	5	0	0	0	0	0	0	0
220	**230**	26	20	13	6	0	0	0	0	0	0	0
230	**240**	28	21	14	8	1	0	0	0	0	0	0
240	**250**	29	23	16	9	2	0	0	0	0	0	0
250	**260**	31	24	17	11	4	0	0	0	0	0	0
260	**270**	32	26	19	12	5	0	0	0	0	0	0
270	**280**	34	27	20	14	7	0	0	0	0	0	0
280	**290**	35	29	22	15	8	2	0	0	0	0	0
290	**300**	37	30	23	17	10	3	0	0	0	0	0
300	**310**	38	32	25	18	11	5	0	0	0	0	0
310	**320**	40	33	26	20	13	6	0	0	0	0	0
320	**330**	41	35	28	21	14	8	1	0	0	0	0
330	**340**	43	36	29	23	16	9	2	0	0	0	0
340	**350**	44	38	31	24	17	11	4	0	0	0	0
350	**360**	46	39	32	26	19	12	5	0	0	0	0
360	**370**	47	41	34	27	20	14	7	0	0	0	0
370	**380**	49	42	35	29	22	15	8	2	0	0	0
380	**390**	50	44	37	30	23	17	10	3	0	0	0
390	**400**	52	45	38	32	25	18	11	5	0	0	0
400	**410**	53	47	40	33	26	20	13	6	0	0	0
410	**420**	55	48	41	35	28	21	14	8	1	0	0
420	**430**	56	50	43	36	29	23	16	9	2	0	0
430	**440**	58	51	44	38	31	24	17	11	4	0	0
440	**450**	59	53	46	39	32	26	19	12	5	0	0
450	**460**	61	54	47	41	34	27	20	14	7	0	0
460	**470**	64	56	49	42	35	29	22	15	8	1	0
470	**480**	67	57	50	44	37	30	23	17	10	3	0
480	**490**	70	59	52	45	38	32	25	18	11	4	0
490	**500**	73	60	53	47	40	33	26	20	13	6	0
500	**510**	75	63	55	48	41	35	28	21	14	7	1
510	**520**	78	66	56	50	43	36	29	23	16	9	2
520	**530**	81	68	58	51	44	38	31	24	17	10	4
530	**540**	84	71	59	53	46	39	32	26	19	12	5
540	**550**	87	74	61	54	47	41	34	27	20	13	7
550	**560**	89	77	64	56	49	42	35	29	22	15	8
560	**570**	92	80	67	57	50	44	37	30	23	16	10
570	**580**	95	82	70	59	52	45	38	32	25	18	11
580	**590**	98	85	73	60	53	47	40	33	26	19	13

(Continued on next page)

SINGLE Persons—**WEEKLY** Payroll Period

(For Wages Paid in 19--)

If the wages are–		And the number of withholding allowances claimed is—										
At least	But less than	0	1	2	3	4	5	6	7	8	9	10
		The amount of income tax to be withheld is—										
$590	**$600**	$101	$88	$75	$63	$55	$48	$41	$35	$28	$21	$14
600	**610**	103	91	78	66	56	50	43	36	29	22	16
610	**620**	106	94	81	68	58	51	44	38	31	24	17
620	**630**	109	96	84	71	59	53	46	39	32	25	19
630	**640**	112	99	87	74	61	54	47	41	34	27	20
640	**650**	115	102	89	77	64	56	49	42	35	28	22
650	**660**	117	105	92	80	67	57	50	44	37	30	23
660	**670**	120	108	95	82	70	59	52	45	38	31	25
670	**680**	123	110	98	85	72	60	53	47	40	33	26
680	**690**	126	113	101	88	75	63	55	48	41	34	28
690	**700**	129	116	103	91	78	65	56	50	43	36	29
700	**710**	131	119	106	94	81	68	58	51	44	37	31
710	**720**	134	122	109	96	84	71	59	53	46	39	32
720	**730**	137	124	112	99	86	74	61	54	47	40	34
730	**740**	140	127	115	102	89	77	64	56	49	42	35
740	**750**	143	130	117	105	92	79	67	57	50	43	37
750	**760**	145	133	120	108	95	82	70	59	52	45	38
760	**770**	148	136	123	110	98	85	72	60	53	46	40
770	**780**	151	138	126	113	100	88	75	63	55	48	41
780	**790**	154	141	129	116	103	91	78	65	56	49	43
790	**800**	157	144	131	119	106	93	81	68	58	51	44
800	**810**	159	147	134	122	109	96	84	71	59	52	46
810	**820**	162	150	137	124	112	99	86	74	61	54	47
820	**830**	165	152	140	127	114	102	89	77	64	55	49
830	**840**	168	155	143	130	117	105	92	79	67	57	50
840	**850**	171	158	145	133	120	107	95	82	69	58	52
850	**860**	173	161	148	136	123	110	98	85	72	60	53
860	**870**	176	164	151	138	126	113	100	88	75	62	55
870	**880**	179	166	154	141	128	116	103	91	78	65	56
880	**890**	182	169	157	144	131	119	106	93	81	68	58
890	**900**	185	172	159	147	134	121	109	96	83	71	59
900	**910**	187	175	162	150	137	124	112	99	86	74	61
910	**920**	190	178	165	152	140	127	114	102	89	76	64
920	**930**	193	180	168	155	142	130	117	105	92	79	67
930	**940**	196	183	171	158	145	133	120	107	95	82	69
940	**950**	199	186	173	161	148	135	123	110	97	85	72
950	**960**	202	189	176	164	151	138	126	113	100	88	75
960	**970**	205	192	179	166	154	141	128	116	103	90	78
970	**980**	208	194	182	169	156	144	131	119	106	93	81
980	**990**	211	197	185	172	159	147	134	121	109	96	83
990	**1,000**	214	200	187	175	162	149	137	124	111	99	86
1,000	**1,010**	217	203	190	178	165	152	140	127	114	102	89
1,010	**1,020**	220	206	193	180	168	155	142	130	117	104	92
1,020	**1,030**	224	210	196	183	170	158	145	133	120	107	95
1,030	**1,040**	227	213	199	186	173	161	148	135	123	110	97
1,040	**1,050**	230	216	202	189	176	163	151	138	125	113	100
1,050	**1,060**	233	219	205	192	179	166	154	141	128	116	103
1,060	**1,070**	236	222	208	194	182	169	156	144	131	118	106
1,070	**1,080**	239	225	211	197	184	172	159	147	134	121	109
1,080	**1,090**	242	228	214	200	187	175	162	149	137	124	111
1,090	**1,100**	245	231	217	203	190	177	165	152	139	127	114
1,100	**1,110**	248	234	220	206	193	180	168	155	142	130	117
1,110	**1,120**	251	237	223	209	196	183	170	158	145	132	120
1,120	**1,130**	255	241	227	213	199	186	173	161	148	135	123
1,130	**1,140**	258	244	230	216	202	189	176	163	151	138	125
1,140	**1,150**	261	247	233	219	205	191	179	166	153	141	128
1,150	**1,160**	264	250	236	222	208	194	182	169	156	144	131
1,160	**1,170**	267	253	239	225	211	197	184	172	159	146	134
1,170	**1,180**	270	256	242	228	214	200	187	175	162	149	137
1,180	**1,190**	273	259	245	231	217	203	190	177	165	152	139
1,190	**1,200**	276	262	248	234	220	206	193	180	167	155	142
1,200	**1,210**	279	265	251	237	223	209	196	183	170	158	145
1,210	**1,220**	282	268	254	240	226	212	198	186	173	160	148
1,220	**1,230**	286	272	258	244	230	216	202	189	176	163	151
1,230	**1,240**	289	275	261	247	233	219	205	191	179	166	153
1,240	**1,250**	292	278	264	250	236	222	208	194	181	169	156

$1,250 and over Use Table 1(a) for a **SINGLE person** on page 26. Also see the instructions on page 24.

MARRIED Persons—**WEEKLY** Payroll Period

(For Wages Paid in 19--)

If the wages are–		And the number of withholding allowances claimed is—										
At least	But less than	0	1	2	3	4	5	6	7	8	9	10
		The amount of income tax to be withheld is—										
$0	$125	$0	$0	$0	$0	$0	$0	$0	$0	$0	$0	$0
125	130	1	0	0	0	0	0	0	0	0	0	0
130	135	2	0	0	0	0	0	0	0	0	0	0
135	140	3	0	0	0	0	0	0	0	0	0	0
140	145	3	0	0	0	0	0	0	0	0	0	0
145	150	4	0	0	0	0	0	0	0	0	0	0
150	155	5	0	0	0	0	0	0	0	0	0	0
155	160	6	0	0	0	0	0	0	0	0	0	0
160	165	6	0	0	0	0	0	0	0	0	0	0
165	170	7	0	0	0	0	0	0	0	0	0	0
170	175	8	1	0	0	0	0	0	0	0	0	0
175	180	9	2	0	0	0	0	0	0	0	0	0
180	185	9	3	0	0	0	0	0	0	0	0	0
185	190	10	3	0	0	0	0	0	0	0	0	0
190	195	11	4	0	0	0	0	0	0	0	0	0
195	200	12	5	0	0	0	0	0	0	0	0	0
200	210	13	6	0	0	0	0	0	0	0	0	0
210	220	14	8	1	0	0	0	0	0	0	0	0
220	230	16	9	2	0	0	0	0	0	0	0	0
230	240	17	11	4	0	0	0	0	0	0	0	0
240	250	19	12	5	0	0	0	0	0	0	0	0
250	260	20	14	7	0	0	0	0	0	0	0	0
260	270	22	15	8	2	0	0	0	0	0	0	0
270	280	23	17	10	3	0	0	0	0	0	0	0
280	290	25	18	11	5	0	0	0	0	0	0	0
290	300	26	20	13	6	0	0	0	0	0	0	0
300	310	28	21	14	8	1	0	0	0	0	0	0
310	320	29	23	16	9	2	0	0	0	0	0	0
320	330	31	24	17	11	4	0	0	0	0	0	0
330	340	32	26	19	12	5	0	0	0	0	0	0
340	350	34	27	20	14	7	0	0	0	0	0	0
350	360	35	29	22	15	8	1	0	0	0	0	0
360	370	37	30	23	17	10	3	0	0	0	0	0
370	380	38	32	25	18	11	4	0	0	0	0	0
380	390	40	33	26	20	13	6	0	0	0	0	0
390	400	41	35	28	21	14	7	1	0	0	0	0
400	410	43	36	29	23	16	9	2	0	0	0	0
410	420	44	38	31	24	17	10	4	0	0	0	0
420	430	46	39	32	26	19	12	5	0	0	0	0
430	440	47	41	34	27	20	13	7	0	0	0	0
440	450	49	42	35	29	22	15	8	1	0	0	0
450	460	50	44	37	30	23	16	10	3	0	0	0
460	470	52	45	38	32	25	18	11	4	0	0	0
470	480	53	47	40	33	26	19	13	6	0	0	0
480	490	55	48	41	35	28	21	14	7	1	0	0
490	500	56	50	43	36	29	22	16	9	2	0	0
500	510	58	51	44	38	31	24	17	10	4	0	0
510	520	59	53	46	39	32	25	19	12	5	0	0
520	530	61	54	47	41	34	27	20	13	7	0	0
530	540	62	56	49	42	35	28	22	15	8	1	0
540	550	64	57	50	44	37	30	23	16	10	3	0
550	560	65	59	52	45	38	31	25	18	11	4	0
560	570	67	60	53	47	40	33	26	19	13	6	0
570	580	68	62	55	48	41	34	28	21	14	7	1
580	590	70	63	56	50	43	36	29	22	16	9	2
590	600	71	65	58	51	44	37	31	24	17	10	4
600	610	73	66	59	53	46	39	32	25	19	12	5
610	620	74	68	61	54	47	40	34	27	20	13	7
620	630	76	69	62	56	49	42	35	28	22	15	8
630	640	77	71	64	57	50	43	37	30	23	16	10
640	650	79	72	65	59	52	45	38	31	25	18	11
650	660	80	74	67	60	53	46	40	33	26	19	13
660	670	82	75	68	62	55	48	41	34	28	21	14
670	680	83	77	70	63	56	49	43	36	29	22	16
680	690	85	78	71	65	58	51	44	37	31	24	17
690	700	86	80	73	66	59	52	46	39	32	25	19
700	710	88	81	74	68	61	54	47	40	34	27	20
710	720	89	83	76	69	62	55	49	42	35	28	22
720	730	91	84	77	71	64	57	50	43	37	30	23
730	740	92	86	79	72	65	58	52	45	38	31	25

(Continued on next page)

MARRIED Persons—**WEEKLY** Payroll Period

(For Wages Paid in 19--)

If the wages are–		And the number of withholding allowances claimed is—										
At least	But less than	0	1	2	3	4	5	6	7	8	9	10
		The amount of income tax to be withheld is—										
$740	**$750**	$94	$87	$80	$74	$67	$60	$53	$46	$40	$33	$26
750	**760**	95	89	82	75	68	61	55	48	41	34	28
760	**770**	97	90	83	77	70	63	56	49	43	36	29
770	**780**	98	92	85	78	71	64	58	51	44	37	31
780	**790**	100	93	86	80	73	66	59	52	46	39	32
790	**800**	103	95	88	81	74	67	61	54	47	40	34
800	**810**	106	96	89	83	76	69	62	55	49	42	35
810	**820**	108	98	91	84	77	70	64	57	50	43	37
820	**830**	111	99	92	86	79	72	65	58	52	45	38
830	**840**	114	101	94	87	80	73	67	60	53	46	40
840	**850**	117	104	95	89	82	75	68	61	55	48	41
850	**860**	120	107	97	90	83	76	70	63	56	49	43
860	**870**	122	110	98	92	85	78	71	64	58	51	44
870	**880**	125	113	100	93	86	79	73	66	59	52	46
880	**890**	128	115	103	95	88	81	74	67	61	54	47
890	**900**	131	118	106	96	89	82	76	69	62	55	49
900	**910**	134	121	108	98	91	84	77	70	64	57	50
910	**920**	136	124	111	99	92	85	79	72	65	58	52
920	**930**	139	127	114	101	94	87	80	73	67	60	53
930	**940**	142	129	117	104	95	88	82	75	68	61	55
940	**950**	145	132	120	107	97	90	83	76	70	63	56
950	**960**	148	135	122	110	98	91	85	78	71	64	58
960	**970**	150	138	125	112	100	93	86	79	73	66	59
970	**980**	153	141	128	115	103	94	88	81	74	67	61
980	**990**	156	143	131	118	105	96	89	82	76	69	62
990	**1,000**	159	146	134	121	108	97	91	84	77	70	64
1,000	**1,010**	162	149	136	124	111	99	92	85	79	72	65
1,010	**1,020**	164	152	139	126	114	101	94	87	80	73	67
1,020	**1,030**	167	155	142	129	117	104	95	88	82	75	68
1,030	**1,040**	170	157	145	132	119	107	97	90	83	76	70
1,040	**1,050**	173	160	148	135	122	110	98	91	85	78	71
1,050	**1,060**	176	163	150	138	125	112	100	93	86	79	73
1,060	**1,070**	178	166	153	140	128	115	103	94	88	81	74
1,070	**1,080**	181	169	156	143	131	118	105	96	89	82	76
1,080	**1,090**	184	171	159	146	133	121	108	97	91	84	77
1,090	**1,100**	187	174	162	149	136	124	111	99	92	85	79
1,100	**1,110**	190	177	164	152	139	126	114	101	94	87	80
1,110	**1,120**	192	180	167	154	142	129	117	104	95	88	82
1,120	**1,130**	195	183	170	157	145	132	119	107	97	90	83
1,130	**1,140**	198	185	173	160	147	135	122	109	98	91	85
1,140	**1,150**	201	188	176	163	150	138	125	112	100	93	86
1,150	**1,160**	204	191	178	166	153	140	128	115	102	94	88
1,160	**1,170**	206	194	181	168	156	143	131	118	105	96	89
1,170	**1,180**	209	197	184	171	159	146	133	121	108	97	91
1,180	**1,190**	212	199	187	174	161	149	136	123	111	99	92
1,190	**1,200**	215	202	190	177	164	152	139	126	114	101	94
1,200	**1,210**	218	205	192	180	167	154	142	129	116	104	95
1,210	**1,220**	220	208	195	182	170	157	145	132	119	107	97
1,220	**1,230**	223	211	198	185	173	160	147	135	122	109	98
1,230	**1,240**	226	213	201	188	175	163	150	137	125	112	100
1,240	**1,250**	229	216	204	191	178	166	153	140	128	115	102
1,250	**1,260**	232	219	206	194	181	168	156	143	130	118	105
1,260	**1,270**	234	222	209	196	184	171	159	146	133	121	108
1,270	**1,280**	237	225	212	199	187	174	161	149	136	123	111
1,280	**1,290**	240	227	215	202	189	177	164	151	139	126	114
1,290	**1,300**	243	230	218	205	192	180	167	154	142	129	116
1,300	**1,310**	246	233	220	208	195	182	170	157	144	132	119
1,310	**1,320**	248	236	223	210	198	185	173	160	147	135	122
1,320	**1,330**	251	239	226	213	201	188	175	163	150	137	125
1,330	**1,340**	254	241	229	216	203	191	178	165	153	140	128
1,340	**1,350**	257	244	232	219	206	194	181	168	156	143	130
1,350	**1,360**	260	247	234	222	209	196	184	171	158	146	133
1,360	**1,370**	262	250	237	224	212	199	187	174	161	149	136
1,370	**1,380**	265	253	240	227	215	202	189	177	164	151	139
1,380	**1,390**	268	255	243	230	217	205	192	179	167	154	142
1,390	**1,400**	271	258	246	233	220	208	195	182	170	157	144

$1,400 and over Use Table 1(b) for a **MARRIED person** on page 26. Also see the instructions on page 24.

SINGLE Persons—**MONTHLY** Payroll Period

(For Wages Paid in 19--)

If the wages are–		And the number of withholding allowances claimed is—										
At least	But less than	0	1	2	3	4	5	6	7	8	9	10
		The amount of income tax to be withheld is—										
$0	**$210**	$0	$0	$0	$0	$0	$0	$0	$0	$0	$0	$0
210	**220**	1	0	0	0	0	0	0	0	0	0	0
220	**230**	2	0	0	0	0	0	0	0	0	0	0
230	**240**	4	0	0	0	0	0	0	0	0	0	0
240	**250**	5	0	0	0	0	0	0	0	0	0	0
250	**260**	7	0	0	0	0	0	0	0	0	0	0
260	**270**	8	0	0	0	0	0	0	0	0	0	0
270	**280**	10	0	0	0	0	0	0	0	0	0	0
280	**290**	11	0	0	0	0	0	0	0	0	0	0
290	**300**	13	0	0	0	0	0	0	0	0	0	0
300	**320**	15	0	0	0	0	0	0	0	0	0	0
320	**340**	18	0	0	0	0	0	0	0	0	0	0
340	**360**	21	0	0	0	0	0	0	0	0	0	0
360	**380**	24	0	0	0	0	0	0	0	0	0	0
380	**400**	27	0	0	0	0	0	0	0	0	0	0
400	**420**	30	1	0	0	0	0	0	0	0	0	0
420	**440**	33	4	0	0	0	0	0	0	0	0	0
440	**460**	36	7	0	0	0	0	0	0	0	0	0
460	**480**	39	10	0	0	0	0	0	0	0	0	0
480	**500**	42	13	0	0	0	0	0	0	0	0	0
500	**520**	45	16	0	0	0	0	0	0	0	0	0
520	**540**	48	19	0	0	0	0	0	0	0	0	0
540	**560**	51	22	0	0	0	0	0	0	0	0	0
560	**580**	54	25	0	0	0	0	0	0	0	0	0
580	**600**	57	28	0	0	0	0	0	0	0	0	0
600	**640**	61	32	3	0	0	0	0	0	0	0	0
640	**680**	67	38	9	0	0	0	0	0	0	0	0
680	**720**	73	44	15	0	0	0	0	0	0	0	0
720	**760**	79	50	21	0	0	0	0	0	0	0	0
760	**800**	85	56	27	0	0	0	0	0	0	0	0
800	**840**	91	62	33	3	0	0	0	0	0	0	0
840	**880**	97	68	39	9	0	0	0	0	0	0	0
880	**920**	103	74	45	15	0	0	0	0	0	0	0
920	**960**	109	80	51	21	0	0	0	0	0	0	0
960	**1,000**	115	86	57	27	0	0	0	0	0	0	0
1,000	**1,040**	121	92	63	33	4	0	0	0	0	0	0
1,040	**1,080**	127	98	69	39	10	0	0	0	0	0	0
1,080	**1,120**	133	104	75	45	16	0	0	0	0	0	0
1,120	**1,160**	139	110	81	51	22	0	0	0	0	0	0
1,160	**1,200**	145	116	87	57	28	0	0	0	0	0	0
1,200	**1,240**	151	122	93	63	34	5	0	0	0	0	0
1,240	**1,280**	157	128	99	69	40	11	0	0	0	0	0
1,280	**1,320**	163	134	105	75	46	17	0	0	0	0	0
1,320	**1,360**	169	140	111	81	52	23	0	0	0	0	0
1,360	**1,400**	175	146	117	87	58	29	0	0	0	0	0
1,400	**1,440**	181	152	123	93	64	35	5	0	0	0	0
1,440	**1,480**	187	158	129	99	70	41	11	0	0	0	0
1,480	**1,520**	193	164	135	105	76	47	17	0	0	0	0
1,520	**1,560**	199	170	141	111	82	53	23	0	0	0	0
1,560	**1,600**	205	176	147	117	88	59	29	0	0	0	0
1,600	**1,640**	211	182	153	123	94	65	35	6	0	0	0
1,640	**1,680**	217	188	159	129	100	71	41	12	0	0	0
1,680	**1,720**	223	194	165	135	106	77	47	18	0	0	0
1,720	**1,760**	229	200	171	141	112	83	53	24	0	0	0
1,760	**1,800**	235	206	177	147	118	89	59	30	0	0	0
1,800	**1,840**	241	212	183	153	124	95	65	36	6	0	0
1,840	**1,880**	247	218	189	159	130	101	71	42	12	0	0
1,880	**1,920**	253	224	195	165	136	107	77	48	18	0	0
1,920	**1,960**	259	230	201	171	142	113	83	54	24	0	0
1,960	**2,000**	269	236	207	177	148	119	89	60	30	1	0
2,000	**2,040**	280	242	213	183	154	125	95	66	36	7	0
2,040	**2,080**	291	248	219	189	160	131	101	72	42	13	0
2,080	**2,120**	302	254	225	195	166	137	107	78	48	19	0
2,120	**2,160**	314	260	231	201	172	143	113	84	54	25	0
2,160	**2,200**	325	270	237	207	178	149	119	90	60	31	2
2,200	**2,240**	336	281	243	213	184	155	125	96	66	37	8
2,240	**2,280**	347	292	249	219	190	161	131	102	72	43	14
2,280	**2,320**	358	304	255	225	196	167	137	108	78	49	20
2,320	**2,360**	370	315	261	231	202	173	143	114	84	55	26
2,360	**2,400**	381	326	271	237	208	179	149	120	90	61	32

(Continued on next page)

SINGLE Persons—**MONTHLY** Payroll Period

(For Wages Paid in 19--)

If the wages are–		And the number of withholding allowances claimed is—										
At least	But less than	0	1	2	3	4	5	6	7	8	9	10
		The amount of income tax to be withheld is—										
$2,400	**$2,440**	$392	$337	$282	$243	$214	$185	$155	$126	$96	$67	$38
2,440	**2,480**	403	348	294	249	220	191	161	132	102	73	44
2,480	**2,520**	414	360	305	255	226	197	167	138	108	79	50
2,520	**2,560**	426	371	316	261	232	203	173	144	114	85	56
2,560	**2,600**	437	382	327	272	238	209	179	150	120	91	62
2,600	**2,640**	448	393	338	283	244	215	185	156	126	97	68
2,640	**2,680**	459	404	350	295	250	221	191	162	132	103	74
2,680	**2,720**	470	416	361	306	256	227	197	168	138	109	80
2,720	**2,760**	482	427	372	317	262	233	203	174	144	115	86
2,760	**2,800**	493	438	383	328	273	239	209	180	150	121	92
2,800	**2,840**	504	449	394	339	285	245	215	186	156	127	98
2,840	**2,880**	515	460	406	351	296	251	221	192	162	133	104
2,880	**2,920**	526	472	417	362	307	257	227	198	168	139	110
2,920	**2,960**	538	483	428	373	318	263	233	204	174	145	116
2,960	**3,000**	549	494	439	384	329	275	239	210	180	151	122
3,000	**3,040**	560	505	450	395	341	286	245	216	186	157	128
3,040	**3,080**	571	516	462	407	352	297	251	222	192	163	134
3,080	**3,120**	582	528	473	418	363	308	257	228	198	169	140
3,120	**3,160**	594	539	484	429	374	319	265	234	204	175	146
3,160	**3,200**	605	550	495	440	385	331	276	240	210	181	152
3,200	**3,240**	616	561	506	451	397	342	287	246	216	187	158
3,240	**3,280**	627	572	518	463	408	353	298	252	222	193	164
3,280	**3,320**	638	584	529	474	419	364	309	258	228	199	170
3,320	**3,360**	650	595	540	485	430	375	321	266	234	205	176
3,360	**3,400**	661	606	551	496	441	387	332	277	240	211	182
3,400	**3,440**	672	617	562	507	453	398	343	288	246	217	188
3,440	**3,480**	683	628	574	519	464	409	354	299	252	223	194
3,480	**3,520**	694	640	585	530	475	420	365	311	258	229	200
3,520	**3,560**	706	651	596	541	486	431	377	322	267	235	206
3,560	**3,600**	717	662	607	552	497	443	388	333	278	241	212
3,600	**3,640**	728	673	618	563	509	454	399	344	289	247	218
3,640	**3,680**	739	684	630	575	520	465	410	355	301	253	224
3,680	**3,720**	750	696	641	586	531	476	421	367	312	259	230
3,720	**3,760**	762	707	652	597	542	487	433	378	323	268	236
3,760	**3,800**	773	718	663	608	553	499	444	389	334	279	242
3,800	**3,840**	784	729	674	619	565	510	455	400	345	290	248
3,840	**3,880**	795	740	686	631	576	521	466	411	357	302	254
3,880	**3,920**	806	752	697	642	587	532	477	423	368	313	260
3,920	**3,960**	818	763	708	653	598	543	489	434	379	324	269
3,960	**4,000**	829	774	719	664	609	555	500	445	390	335	280
4,000	**4,040**	840	785	730	675	621	566	511	456	401	346	292
4,040	**4,080**	851	796	742	687	632	577	522	467	413	358	303
4,080	**4,120**	863	808	753	698	643	588	533	479	424	369	314
4,120	**4,160**	875	819	764	709	654	599	545	490	435	380	325
4,160	**4,200**	888	830	775	720	665	611	556	501	446	391	336
4,200	**4,240**	900	841	786	731	677	622	567	512	457	402	348
4,240	**4,280**	913	852	798	743	688	633	578	523	469	414	359
4,280	**4,320**	925	864	809	754	699	644	589	535	480	425	370
4,320	**4,360**	937	877	820	765	710	655	601	546	491	436	381
4,360	**4,400**	950	889	831	776	721	667	612	557	502	447	392
4,400	**4,440**	962	901	842	787	733	678	623	568	513	458	404
4,440	**4,480**	975	914	854	799	744	689	634	579	525	470	415
4,480	**4,520**	987	926	866	810	755	700	645	591	536	481	426
4,520	**4,560**	999	939	878	821	766	711	657	602	547	492	437
4,560	**4,600**	1,012	951	890	832	777	723	668	613	558	503	448
4,600	**4,640**	1,024	963	903	843	789	734	679	624	569	514	460
4,640	**4,680**	1,037	976	915	855	800	745	690	635	581	526	471
4,680	**4,720**	1,049	988	928	867	811	756	701	647	592	537	482
4,720	**4,760**	1,061	1,001	940	879	822	767	713	658	603	548	493
4,760	**4,800**	1,074	1,013	952	892	833	779	724	669	614	559	504
4,800	**4,840**	1,086	1,025	965	904	845	790	735	680	625	570	516
4,840	**4,880**	1,099	1,038	977	916	856	801	746	691	637	582	527
4,880	**4,920**	1,111	1,050	990	929	868	812	757	703	648	593	538
4,920	**4,960**	1,123	1,063	1,002	941	881	823	769	714	659	604	549
4,960	**5,000**	1,136	1,075	1,014	954	893	835	780	725	670	615	560

$5,000 and over Use Table 4(a) for a **SINGLE person** on page 26. Also see the instructions on page 24.

MARRIED Persons—**MONTHLY** Payroll Period

(For Wages Paid in 19--)

If the wages are—		And the number of withholding allowances claimed is—										
At least	But less than	0	1	2	3	4	5	6	7	8	9	10
		The amount of income tax to be withheld is—										
$0	$520	$0	$0	$0	$0	$0	$0	$0	$0	$0	$0	$0
520	540	2	0	0	0	0	0	0	0	0	0	0
540	560	5	0	0	0	0	0	0	0	0	0	0
560	580	8	0	0	0	0	0	0	0	0	0	0
580	600	11	0	0	0	0	0	0	0	0	0	0
600	640	16	0	0	0	0	0	0	0	0	0	0
640	680	22	0	0	0	0	0	0	0	0	0	0
680	720	28	0	0	0	0	0	0	0	0	0	0
720	760	34	4	0	0	0	0	0	0	0	0	0
760	800	40	10	0	0	0	0	0	0	0	0	0
800	840	46	16	0	0	0	0	0	0	0	0	0
840	880	52	22	0	0	0	0	0	0	0	0	0
880	920	58	28	0	0	0	0	0	0	0	0	0
920	960	64	34	5	0	0	0	0	0	0	0	0
960	1,000	70	40	11	0	0	0	0	0	0	0	0
1,000	1,040	76	46	17	0	0	0	0	0	0	0	0
1,040	1,080	82	52	23	0	0	0	0	0	0	0	0
1,080	1,120	88	58	29	0	0	0	0	0	0	0	0
1,120	1,160	94	64	35	5	0	0	0	0	0	0	0
1,160	1,200	100	70	41	11	0	0	0	0	0	0	0
1,200	1,240	106	76	47	17	0	0	0	0	0	0	0
1,240	1,280	112	82	53	23	0	0	0	0	0	0	0
1,280	1,320	118	88	59	29	0	0	0	0	0	0	0
1,320	1,360	124	94	65	35	6	0	0	0	0	0	0
1,360	1,400	130	100	71	41	12	0	0	0	0	0	0
1,400	1,440	136	106	77	47	18	0	0	0	0	0	0
1,440	1,480	142	112	83	53	24	0	0	0	0	0	0
1,480	1,520	148	118	89	59	30	1	0	0	0	0	0
1,520	1,560	154	124	95	65	36	7	0	0	0	0	0
1,560	1,600	160	130	101	71	42	13	0	0	0	0	0
1,600	1,640	166	136	107	77	48	19	0	0	0	0	0
1,640	1,680	172	142	113	83	54	25	0	0	0	0	0
1,680	1,720	178	148	119	89	60	31	1	0	0	0	0
1,720	1,760	184	154	125	95	66	37	7	0	0	0	0
1,760	1,800	190	160	131	101	72	43	13	0	0	0	0
1,800	1,840	196	166	137	107	78	49	19	0	0	0	0
1,840	1,880	202	172	143	113	84	55	25	0	0	0	0
1,880	1,920	208	178	149	119	90	61	31	2	0	0	0
1,920	1,960	214	184	155	125	96	67	37	8	0	0	0
1,960	2,000	220	190	161	131	102	73	43	14	0	0	0
2,000	2,040	226	196	167	137	108	79	49	20	0	0	0
2,040	2,080	232	202	173	143	114	85	55	26	0	0	0
2,080	2,120	238	208	179	149	120	91	61	32	3	0	0
2,120	2,160	244	214	185	155	126	97	67	38	9	0	0
2,160	2,200	250	220	191	161	132	103	73	44	15	0	0
2,200	2,240	256	226	197	167	138	109	79	50	21	0	0
2,240	2,280	262	232	203	173	144	115	85	56	27	0	0
2,280	2,320	268	238	209	179	150	121	91	62	33	3	0
2,320	2,360	274	244	215	185	156	127	97	68	39	9	0
2,360	2,400	280	250	221	191	162	133	103	74	45	15	0
2,400	2,440	286	256	227	197	168	139	109	80	51	21	0
2,440	2,480	292	262	233	203	174	145	115	86	57	27	0
2,480	2,520	298	268	239	209	180	151	121	92	63	33	4
2,520	2,560	304	274	245	215	186	157	127	98	69	39	10
2,560	2,600	310	280	251	221	192	163	133	104	75	45	16
2,600	2,640	316	286	257	227	198	169	139	110	81	51	22
2,640	2,680	322	292	263	233	204	175	145	116	87	57	28
2,680	2,720	328	298	269	239	210	181	151	122	93	63	34
2,720	2,760	334	304	275	245	216	187	157	128	99	69	40
2,760	2,800	340	310	281	251	222	193	163	134	105	75	46
2,800	2,840	346	316	287	257	228	199	169	140	111	81	52
2,840	2,880	352	322	293	263	234	205	175	146	117	87	58
2,880	2,920	358	328	299	269	240	211	181	152	123	93	64
2,920	2,960	364	334	305	275	246	217	187	158	129	99	70
2,960	3,000	370	340	311	281	252	223	193	164	135	105	76
3,000	3,040	376	346	317	287	258	229	199	170	141	111	82
3,040	3,080	382	352	323	293	264	235	205	176	147	117	88
3,080	3,120	388	358	329	299	270	241	211	182	153	123	94
3,120	3,160	394	364	335	305	276	247	217	188	159	129	100
3,160	3,200	400	370	341	311	282	253	223	194	165	135	106

(Continued on next page)

MARRIED Persons—**MONTHLY** Payroll Period

(For Wages Paid in 19--)

If the wages are–		And the number of withholding allowances claimed is—										
At least	But less than	0	1	2	3	4	5	6	7	8	9	10
		The amount of income tax to be withheld is—										
$3,200	$3,240	$406	$376	$347	$317	$288	$259	$229	$200	$171	$141	$112
3,240	3,280	412	382	353	323	294	265	235	206	177	147	118
3,280	3,320	418	388	359	329	300	271	241	212	183	153	124
3,320	3,360	424	394	365	335	306	277	247	218	189	159	130
3,360	3,400	430	400	371	341	312	283	253	224	195	165	136
3,400	3,440	439	406	377	347	318	289	259	230	201	171	142
3,440	3,480	450	412	383	353	324	295	265	236	207	177	148
3,480	3,520	461	418	389	359	330	301	271	242	213	183	154
3,520	3,560	472	424	395	365	336	307	277	248	219	189	160
3,560	3,600	483	430	401	371	342	313	283	254	225	195	166
3,600	3,640	495	440	407	377	348	319	289	260	231	201	172
3,640	3,680	506	451	413	383	354	325	295	266	237	207	178
3,680	3,720	517	462	419	389	360	331	301	272	243	213	184
3,720	3,760	528	473	425	395	366	337	307	278	249	219	190
3,760	3,800	539	485	431	401	372	343	313	284	255	225	196
3,800	3,840	551	496	441	407	378	349	319	290	261	231	202
3,840	3,880	562	507	452	413	384	355	325	296	267	237	208
3,880	3,920	573	518	463	419	390	361	331	302	273	243	214
3,920	3,960	584	529	475	425	396	367	337	308	279	249	220
3,960	4,000	595	541	486	431	402	373	343	314	285	255	226
4,000	4,040	607	552	497	442	408	379	349	320	291	261	232
4,040	4,080	618	563	508	453	414	385	355	326	297	267	238
4,080	4,120	629	574	519	465	420	391	361	332	303	273	244
4,120	4,160	640	585	531	476	426	397	367	338	309	279	250
4,160	4,200	651	597	542	487	432	403	373	344	315	285	256
4,200	4,240	663	608	553	498	443	409	379	350	321	291	262
4,240	4,280	674	619	564	509	455	415	385	356	327	297	268
4,280	4,320	685	630	575	521	466	421	391	362	333	303	274
4,320	4,360	696	641	587	532	477	427	397	368	339	309	280
4,360	4,400	707	653	598	543	488	433	403	374	345	315	286
4,400	4,440	719	664	609	554	499	444	409	380	351	321	292
4,440	4,480	730	675	620	565	511	456	415	386	357	327	298
4,480	4,520	741	686	631	577	522	467	421	392	363	333	304
4,520	4,560	752	697	643	588	533	478	427	398	369	339	310
4,560	4,600	763	709	654	599	544	489	434	404	375	345	316
4,600	4,640	775	720	665	610	555	500	446	410	381	351	322
4,640	4,680	786	731	676	621	567	512	457	416	387	357	328
4,680	4,720	797	742	687	633	578	523	468	422	393	363	334
4,720	4,760	808	753	699	644	589	534	479	428	399	369	340
4,760	4,800	819	765	710	655	600	545	490	436	405	375	346
4,800	4,840	831	776	721	666	611	556	502	447	411	381	352
4,840	4,880	842	787	732	677	623	568	513	458	417	387	358
4,880	4,920	853	798	743	689	634	579	524	469	423	393	364
4,920	4,960	864	809	755	700	645	590	535	480	429	399	370
4,960	5,000	875	821	766	711	656	601	546	492	437	405	376
5,000	5,040	887	832	777	722	667	612	558	503	448	411	382
5,040	5,080	898	843	788	733	679	624	569	514	459	417	388
5,080	5,120	909	854	799	745	690	635	580	525	470	423	394
5,120	5,160	920	865	811	756	701	646	591	536	482	429	400
5,160	5,200	931	877	822	767	712	657	602	548	493	438	406
5,200	5,240	943	888	833	778	723	668	614	559	504	449	412
5,240	5,280	954	899	844	789	735	680	625	570	515	460	418
5,280	5,320	965	910	855	801	746	691	636	581	526	472	424
5,320	5,360	976	921	867	812	757	702	647	592	538	483	430
5,360	5,400	987	933	878	823	768	713	658	604	549	494	439
5,400	5,440	999	944	889	834	779	724	670	615	560	505	450
5,440	5,480	1,010	955	900	845	791	736	681	626	571	516	462
5,480	5,520	1,021	966	911	857	802	747	692	637	582	528	473
5,520	5,560	1,032	977	923	868	813	758	703	648	594	539	484
5,560	5,600	1,043	989	934	879	824	769	714	660	605	550	495
5,600	5,640	1,055	1,000	945	890	835	780	726	671	616	561	506
5,640	5,680	1,066	1,011	956	901	847	792	737	682	627	572	518
5,680	5,720	1,077	1,022	967	913	858	803	748	693	638	584	529
5,720	5,760	1,088	1,033	979	924	869	814	759	704	650	595	540
5,760	5,800	1,099	1,045	990	935	880	825	770	716	661	606	551

$5,800 and over Use Table 4(b) for a **MARRIED person** on page 26. Also see the instructions on page 24.

19-- Tax Table

If line 5 (Form 1040EZ), line 22 (Form 1040A), or line 37 (Form 1040) is–		And you are—			
At least	But less than	Single	Married filing jointly *	Married filing separately	Head of a household
		Your tax is—			
0	5	0	0	0	0
5	15	2	2	2	2
15	25	3	3	3	3
25	50	6	6	6	6
50	75	9	9	9	9
75	100	13	13	13	13
100	125	17	17	17	17
125	150	21	21	21	21
150	175	24	24	24	24
175	200	28	28	28	28
200	225	32	32	32	32
225	250	36	36	36	36
250	275	39	39	39	39
275	300	43	43	43	43
300	325	47	47	47	47
325	350	51	51	51	51
350	375	54	54	54	54
375	400	58	58	58	58
400	425	62	62	62	62
425	450	66	66	66	66
450	475	69	69	69	69
475	500	73	73	73	73
500	525	77	77	77	77
525	550	81	81	81	81
550	575	84	84	84	84
575	600	88	88	88	88
600	625	92	92	92	92
625	650	96	95	96	96
650	675	99	99	99	99
675	700	103	103	103	103
700	725	107	107	107	107
725	750	111	111	111	111
750	775	114	114	114	114
775	800	118	118	118	118
800	825	122	122	122	122
825	850	126	126	126	126
850	875	129	129	129	129
875	900	133	133	133	133
900	925	137	137	137	137
925	950	141	141	141	141
950	975	144	144	144	144
975	1,000	148	148	148	148
1,000					
1,000	1,025	152	152	152	152
1,025	1,050	156	156	156	156
1,050	1,075	159	159	159	159
1,075	1,100	163	163	163	163
1,100	1,125	167	167	167	167
1,125	1,150	171	171	171	171
1,150	1,175	174	174	174	174
1,175	1,200	178	178	178	178
1,200	1,225	182	182	182	182
1,225	1,250	186	186	186	186
1,250	1,275	189	189	189	189
1,275	1,300	193	193	193	193

If line 5 (Form 1040EZ), line 22 (Form 1040A), or line 37 (Form 1040) is–		And you are—			
At least	But less than	Single	Married filing jointly *	Married filing separately	Head of a household
		Your tax is—			
1,300	1,325	197	197	197	197
1,325	1,350	201	201	201	201
1,350	1,375	204	204	204	204
1,375	1,400	208	208	208	208
1,400	1,425	212	212	212	212
1,425	1,450	216	216	216	216
1,450	1,475	219	219	219	219
1,475	1,500	223	223	223	223
1,500	1,525	227	227	227	227
1,525	1,550	231	231	231	231
1,550	1,575	234	234	234	234
1,575	1,600	238	238	238	238
1,600	1,625	242	242	242	242
1,625	1,650	246	246	246	246
1,650	1,675	249	249	249	249
1,675	1,700	253	253	253	253
1,700	1,725	257	257	257	257
1,725	1,750	261	261	261	261
1,750	1,775	264	264	264	264
1,775	1,800	268	268	268	268
1,800	1,825	272	272	272	272
1,825	1,850	276	276	276	276
1,850	1,875	279	279	279	279
1,875	1,900	283	283	283	283
1,900	1,925	287	287	287	287
1,925	1,950	291	291	291	291
1,950	1,975	294	294	294	294
1,975	2,000	298	298	298	298
2,000					
2,000	2,025	302	302	302	302
2,025	2,050	306	306	306	306
2,050	2,075	309	309	309	309
2,075	2,100	313	313	313	313
2,100	2,125	317	317	317	317
2,125	2,150	321	321	321	321
2,150	2,175	324	324	324	324
2,175	2,200	328	328	328	328
2,200	2,225	332	332	332	332
2,225	2,250	336	336	336	336
2,250	2,275	339	339	339	339
2,275	2,300	343	343	343	343
2,300	2,325	347	347	347	347
2,325	2,350	351	351	351	351
2,350	2,375	354	354	354	354
2,375	2,400	358	358	358	358
2,400	2,425	362	362	362	362
2,425	2,450	366	366	366	366
2,450	2,475	369	369	369	369
2,475	2,500	373	373	373	373
2,500	2,525	377	377	377	377
2,525	2,550	381	381	381	381
2,550	2,575	384	384	384	384
2,575	2,600	388	388	388	388
2,600	2,625	392	392	392	392
2,625	2,650	396	396	396	396
2,650	2,675	399	399	399	399
2,675	2,700	403	403	403	403

If line 5 (Form 1040EZ), line 22 (Form 1040A), or line 37 (Form 1040) is–		And you are—			
At least	But less than	Single	Married filing jointly *	Married filing separately	Head of a household
		Your tax is—			
2,700	2,725	407	407	407	407
2,725	2,750	411	411	411	411
2,750	2,775	414	414	414	414
2,775	2,800	418	418	418	418
2,800	2,825	422	422	422	422
2,825	2,850	426	426	426	426
2,850	2,875	429	429	429	429
2,875	2,900	433	433	433	433
2,900	2,925	437	437	437	437
2,925	2,950	441	441	441	441
2,950	2,975	444	444	444	444
2,975	3,000	448	448	448	448
3,000					
3,000	3,050	454	454	454	454
3,050	3,100	461	461	461	461
3,100	3,150	469	469	469	469
3,150	3,200	476	476	476	476
3,200	3,250	484	484	484	484
3,250	3,300	491	491	491	491
3,300	3,350	499	499	499	499
3,350	3,400	506	506	506	506
3,400	3,450	514	514	514	514
3,450	3,500	521	521	521	521
3,500	3,550	529	529	529	529
3,550	3,600	536	536	536	536
3,600	3,650	544	544	544	544
3,650	3,700	551	551	551	551
3,700	3,750	559	559	559	559
3,750	3,800	566	566	566	566
3,800	3,850	574	574	574	574
3,850	3,900	581	581	581	581
3,900	3,950	589	589	589	589
3,950	4,000	596	596	596	596
4,000					
4,000	4,050	604	604	604	604
4,050	4,100	611	611	611	611
4,100	4,150	619	619	619	619
4,150	4,200	626	626	626	626
4,200	4,250	634	634	634	634
4,250	4,300	641	641	641	641
4,300	4,350	649	649	649	649
4,350	4,400	656	656	656	656
4,400	4,450	664	664	664	664
4,450	4,500	671	671	671	671
4,500	4,550	679	679	679	679
4,550	4,600	686	686	686	686
4,600	4,650	694	694	694	694
4,650	4,700	701	701	701	701
4,700	4,750	709	709	709	709
4,750	4,800	716	716	716	716
4,800	4,850	724	724	724	724
4,850	4,900	731	731	731	731
4,900	4,950	739	739	739	739
4,950	5,000	746	746	746	746

Continued on next page

* This column must also be used by a qualifying widow(er).

19-- Tax Table—*Continued*

If line 5 (Form 1040EZ), line 22 (Form 1040A), or line 37 (Form 1040) is—		And you are—			
At least	But less than	Single	Married filing jointly *	Married filing sepa-rately	Head of a house-hold
		Your tax is—			
5,000					
5,000	**5,050**	754	754	754	754
5,050	**5,100**	761	761	761	761
5,100	**5,150**	769	769	769	769
5,150	**5,200**	776	776	776	776
5,200	**5,250**	784	784	784	784
5,250	**5,300**	791	791	791	791
5,300	**5,350**	799	799	799	799
5,350	**5,400**	806	806	806	806
5,400	**5,450**	814	814	814	814
5,450	**5,500**	821	821	821	821
5,500	**5,550**	829	829	829	829
5,550	**5,600**	836	836	836	836
5,600	**5,650**	844	844	844	844
5,650	**5,700**	851	851	851	851
5,700	**5,750**	859	859	859	859
5,750	**5,800**	866	866	866	866
5,800	**5,850**	874	874	874	874
5,850	**5,900**	881	881	881	881
5,900	**5,950**	889	889	889	889
5,950	**6,000**	896	896	896	896
6,000					
6,000	**6,050**	904	904	904	904
6,050	**6,100**	911	911	911	911
6,100	**6,150**	919	919	919	919
6,150	**6,200**	926	926	926	926
6,200	**6,250**	934	934	934	934
6,250	**6,300**	941	941	941	941
6,300	**6,350**	949	949	949	949
6,350	**6,400**	956	956	956	956
6,400	**6,450**	964	964	964	964
6,450	**6,500**	971	971	971	971
6,500	**6,550**	979	979	979	979
6,550	**6,600**	986	986	986	986
6,600	**6,650**	994	994	994	994
6,650	**6,700**	1,001	1,001	1,001	1,001
6,700	**6,750**	1,009	1,009	1,009	1,009
6,750	**6,800**	1,016	1,016	1,016	1,016
6,800	**6,850**	1,024	1,024	1,024	1,024
6,850	**6,900**	1,031	1,031	1,031	1,031
6,900	**6,950**	1,039	1,039	1,039	1,039
6,950	**7,000**	1,046	1,046	1,046	1,046
7,000					
7,000	**7,050**	1,054	1,054	1,054	1,054
7,050	**7,100**	1,061	1,061	1,061	1,061
7,100	**7,150**	1,069	1,069	1,069	1,069
7,150	**7,200**	1,076	1,076	1,076	1,076
7,200	**7,250**	1,084	1,084	1,084	1,084
7,250	**7,300**	1,091	1,091	1,091	1,091
7,300	**7,350**	1,099	1,099	1,099	1,099
7,350	**7,400**	1,106	1,106	1,106	1,106
7,400	**7,450**	1,114	1,114	1,114	1,114
7,450	**7,500**	1,121	1,121	1,121	1,121
7,500	**7,550**	1,129	1,129	1,129	1,129
7,550	**7,600**	1,136	1,136	1,136	1,136
7,600	**7,650**	1,144	1,144	1,144	1,144
7,650	**7,700**	1,151	1,151	1,151	1,151
7,700	**7,750**	1,159	1,159	1,159	1,159
7,750	**7,800**	1,166	1,166	1,166	1,166
7,800	**7,850**	1,174	1,174	1,174	1,174
7,850	**7,900**	1,181	1,181	1,181	1,181
7,900	**7,950**	1,189	1,189	1,189	1,189
7,950	**8,000**	1,196	1,196	1,196	1,196

If line 5 (Form 1040EZ), line 22 (Form 1040A), or line 37 (Form 1040) is—		And you are—			
At least	But less than	Single	Married filing jointly *	Married filing sepa-rately	Head of a house-hold
		Your tax is—			
8,000					
8,000	**8,050**	1,204	1,204	1,204	1,204
8,050	**8,100**	1,211	1,211	1,211	1,211
8,100	**8,150**	1,219	1,219	1,219	1,219
8,150	**8,200**	1,226	1,226	1,226	1,226
8,200	**8,250**	1,234	1,234	1,234	1,234
8,250	**8,300**	1,241	1,241	1,241	1,241
8,300	**8,350**	1,249	1,249	1,249	1,249
8,350	**8,400**	1,256	1,256	1,256	1,256
8,400	**8,450**	1,264	1,264	1,264	1,264
8,450	**8,500**	1,271	1,271	1,271	1,271
8,500	**8,550**	1,279	1,279	1,279	1,279
8,550	**8,600**	1,286	1,286	1,286	1,286
8,600	**8,650**	1,294	1,294	1,294	1,294
8,650	**8,700**	1,301	1,301	1,301	1,301
8,700	**8,750**	1,309	1,309	1,309	1,309
8,750	**8,800**	1,316	1,316	1,316	1,316
8,800	**8,850**	1,324	1,324	1,324	1,324
8,850	**8,900**	1,331	1,331	1,331	1,331
8,900	**8,950**	1,339	1,339	1,339	1,339
8,950	**9,000**	1,346	1,346	1,346	1,346
9,000					
9,000	**9,050**	1,354	1,354	1,354	1,354
9,050	**9,100**	1,361	1,361	1,361	1,361
9,100	**9,150**	1,369	1,369	1,369	1,369
9,150	**9,200**	1,376	1,376	1,376	1,376
9,200	**9,250**	1,384	1,384	1,384	1,384
9,250	**9,300**	1,391	1,391	1,391	1,391
9,300	**9,350**	1,399	1,399	1,399	1,399
9,350	**9,400**	1,406	1,406	1,406	1,406
9,400	**9,450**	1,414	1,414	1,414	1,414
9,450	**9,500**	1,421	1,421	1,421	1,421
9,500	**9,550**	1,429	1,429	1,429	1,429
9,550	**9,600**	1,436	1,436	1,436	1,436
9,600	**9,650**	1,444	1,444	1,444	1,444
9,650	**9,700**	1,451	1,451	1,451	1,451
9,700	**9,750**	1,459	1,459	1,459	1,459
9,750	**9,800**	1,466	1,466	1,466	1,466
9,800	**9,850**	1,474	1,474	1,474	1,474
9,850	**9,900**	1,481	1,481	1,481	1,481
9,900	**9,950**	1,489	1,489	1,489	1,489
9,950	**10,000**	1,496	1,496	1,496	1,496
10,000					
10,000	**10,050**	1,504	1,504	1,504	1,504
10,050	**10,100**	1,511	1,511	1,511	1,511
10,100	**10,150**	1,519	1,519	1,519	1,519
10,150	**10,200**	1,526	1,526	1,526	1,526
10,200	**10,250**	1,534	1,534	1,534	1,534
10,250	**10,300**	1,541	1,541	1,541	1,541
10,300	**10,350**	1,549	1,549	1,549	1,549
10,350	**10,400**	1,556	1,556	1,556	1,556
10,400	**10,450**	1,564	1,564	1,564	1,564
10,450	**10,500**	1,571	1,571	1,571	1,571
10,500	**10,550**	1,579	1,579	1,579	1,579
10,550	**10,600**	1,586	1,586	1,586	1,586
10,600	**10,650**	1,594	1,594	1,594	1,594
10,650	**10,700**	1,601	1,601	1,601	1,601
10,700	**10,750**	1,609	1,609	1,609	1,609
10,750	**10,800**	1,616	1,616	1,616	1,616
10,800	**10,850**	1,624	1,624	1,624	1,624
10,850	**10,900**	1,631	1,631	1,631	1,631
10,900	**10,950**	1,639	1,639	1,639	1,639
10,950	**11,000**	1,646	1,646	1,646	1,646

If line 5 (Form 1040EZ), line 22 (Form 1040A), or line 37 (Form 1040) is—		And you are—			
At least	But less than	Single	Married filing jointly *	Married filing sepa-rately	Head of a house-hold
		Your tax is—			
11,000					
11,000	**11,050**	1,654	1,654	1,654	1,654
11,050	**11,100**	1,661	1,661	1,661	1,661
11,100	**11,150**	1,669	1,669	1,669	1,669
11,150	**11,200**	1,676	1,676	1,676	1,676
11,200	**11,250**	1,684	1,684	1,684	1,684
11,250	**11,300**	1,691	1,691	1,691	1,691
11,300	**11,350**	1,699	1,699	1,699	1,699
11,350	**11,400**	1,706	1,706	1,706	1,706
11,400	**11,450**	1,714	1,714	1,714	1,714
11,450	**11,500**	1,721	1,721	1,721	1,721
11,500	**11,550**	1,729	1,729	1,729	1,729
11,550	**11,600**	1,736	1,736	1,736	1,736
11,600	**11,650**	1,744	1,744	1,744	1,744
11,650	**11,700**	1,751	1,751	1,751	1,751
11,700	**11,750**	1,759	1,759	1,759	1,759
11,750	**11,800**	1,766	1,766	1,766	1,766
11,800	**11,850**	1,774	1,774	1,774	1,774
11,850	**11,900**	1,781	1,781	1,781	1,781
11,900	**11,950**	1,789	1,789	1,789	1,789
11,950	**12,000**	1,796	1,796	1,796	1,796
12,000					
12,000	**12,050**	1,804	1,804	1,804	1,804
12,050	**12,100**	1,811	1,811	1,811	1,811
12,100	**12,150**	1,819	1,819	1,819	1,819
12,150	**12,200**	1,826	1,826	1,826	1,826
12,200	**12,250**	1,834	1,834	1,834	1,834
12,250	**12,300**	1,841	1,841	1,841	1,841
12,300	**12,350**	1,849	1,849	1,849	1,849
12,350	**12,400**	1,856	1,856	1,856	1,856
12,400	**12,450**	1,864	1,864	1,864	1,864
12,450	**12,500**	1,871	1,871	1,871	1,871
12,500	**12,550**	1,879	1,879	1,879	1,879
12,550	**12,600**	1,886	1,886	1,886	1,886
12,600	**12,650**	1,894	1,894	1,894	1,894
12,650	**12,700**	1,901	1,901	1,901	1,901
12,700	**12,750**	1,909	1,909	1,909	1,909
12,750	**12,800**	1,916	1,916	1,916	1,916
12,800	**12,850**	1,924	1,924	1,924	1,924
12,850	**12,900**	1,931	1,931	1,931	1,931
12,900	**12,950**	1,939	1,939	1,939	1,939
12,950	**13,000**	1,946	1,946	1,946	1,946
13,000					
13,000	**13,050**	1,954	1,954	1,954	1,954
13,050	**13,100**	1,961	1,961	1,961	1,961
13,100	**13,150**	1,969	1,969	1,969	1,969
13,150	**13,200**	1,976	1,976	1,976	1,976
13,200	**13,250**	1,984	1,984	1,984	1,984
13,250	**13,300**	1,991	1,991	1,991	1,991
13,300	**13,350**	1,999	1,999	1,999	1,999
13,350	**13,400**	2,006	2,006	2,006	2,006
13,400	**13,450**	2,014	2,014	2,014	2,014
13,450	**13,500**	2,021	2,021	2,021	2,021
13,500	**13,550**	2,029	2,029	2,029	2,029
13,550	**13,600**	2,036	2,036	2,036	2,036
13,600	**13,650**	2,044	2,044	2,044	2,044
13,650	**13,700**	2,051	2,051	2,051	2,051
13,700	**13,750**	2,059	2,059	2,059	2,059
13,750	**13,800**	2,066	2,066	2,066	2,066
13,800	**13,850**	2,074	2,074	2,074	2,074
13,850	**13,900**	2,081	2,081	2,081	2,081
13,900	**13,950**	2,089	2,089	2,089	2,089
13,950	**14,000**	2,096	2,096	2,096	2,096

* This column must also be used by a qualifying widow(er).

Continued on next page

19-- Tax Table—*Continued*

If line 5 (Form 1040EZ), line 22 (Form 1040A), or line 37 (Form 1040) is—		And you are—			
At least	But less than	Single	Married filing jointly *	Married filing sepa-rately	Head of a house-hold
		Your tax is—			
14,000					
14,000	**14,050**	2,104	2,104	2,104	2,104
14,050	**14,100**	2,111	2,111	2,111	2,111
14,100	**14,150**	2,119	2,119	2,119	2,119
14,150	**14,200**	2,126	2,126	2,126	2,126
14,200	**14,250**	2,134	2,134	2,134	2,134
14,250	**14,300**	2,141	2,141	2,141	2,141
14,300	**14,350**	2,149	2,149	2,149	2,149
14,350	**14,400**	2,156	2,156	2,156	2,156
14,400	**14,450**	2,164	2,164	2,164	2,164
14,450	**14,500**	2,171	2,171	2,171	2,171
14,500	**14,550**	2,179	2,179	2,179	2,179
14,550	**14,600**	2,186	2,186	2,186	2,186
14,600	**14,650**	2,194	2,194	2,194	2,194
14,650	**14,700**	2,201	2,201	2,201	2,201
14,700	**14,750**	2,209	2,209	2,209	2,209
14,750	**14,800**	2,216	2,216	2,216	2,216
14,800	**14,850**	2,224	2,224	2,224	2,224
14,850	**14,900**	2,231	2,231	2,231	2,231
14,900	**14,950**	2,239	2,239	2,239	2,239
14,950	**15,000**	2,246	2,246	2,246	2,246
15,000					
15,000	**15,050**	2,254	2,254	2,254	2,254
15,050	**15,100**	2,261	2,261	2,261	2,261
15,100	**15,150**	2,269	2,269	2,269	2,269
15,150	**15,200**	2,276	2,276	2,276	2,276
15,200	**15,250**	2,284	2,284	2,284	2,284
15,250	**15,300**	2,291	2,291	2,291	2,291
15,300	**15,350**	2,299	2,299	2,299	2,299
15,350	**15,400**	2,306	2,306	2,306	2,306
15,400	**15,450**	2,314	2,314	2,314	2,314
15,450	**15,500**	2,321	2,321	2,321	2,321
15,500	**15,550**	2,329	2,329	2,329	2,329
15,550	**15,600**	2,336	2,336	2,336	2,336
15,600	**15,650**	2,344	2,344	2,344	2,344
15,650	**15,700**	2,351	2,351	2,351	2,351
15,700	**15,750**	2,359	2,359	2,359	2,359
15,750	**15,800**	2,366	2,366	2,366	2,366
15,800	**15,850**	2,374	2,374	2,374	2,374
15,850	**15,900**	2,381	2,381	2,381	2,381
15,900	**15,950**	2,389	2,389	2,389	2,389
15,950	**16,000**	2,396	2,396	2,396	2,396
16,000					
16,000	**16,050**	2,404	2,404	2,404	2,404
16,050	**16,100**	2,411	2,411	2,411	2,411
16,100	**16,150**	2,419	2,419	2,419	2,419
16,150	**16,200**	2,426	2,426	2,426	2,426
16,200	**16,250**	2,434	2,434	2,434	2,434
16,250	**16,300**	2,441	2,441	2,441	2,441
16,300	**16,350**	2,449	2,449	2,449	2,449
16,350	**16,400**	2,456	2,456	2,456	2,456
16,400	**16,450**	2,464	2,464	2,464	2,464
16,450	**16,500**	2,471	2,471	2,471	2,471
16,500	**16,550**	2,479	2,479	2,479	2,479
16,550	**16,600**	2,486	2,486	2,486	2,486
16,600	**16,650**	2,494	2,494	2,494	2,494
16,650	**16,700**	2,501	2,501	2,501	2,501
16,700	**16,750**	2,509	2,509	2,509	2,509
16,750	**16,800**	2,516	2,516	2,516	2,516
16,800	**16,850**	2,524	2,524	2,524	2,524
16,850	**16,900**	2,531	2,531	2,531	2,531
16,900	**16,950**	2,539	2,539	2,539	2,539
16,950	**17,000**	2,546	2,546	2,546	2,546

If line 5 (Form 1040EZ), line 22 (Form 1040A), or line 37 (Form 1040) is—		And you are—			
At least	But less than	Single	Married filing jointly *	Married filing sepa-rately	Head of a house-hold
		Your tax is—			
17,000					
17,000	**17,050**	2,554	2,554	2,554	2,554
17,050	**17,100**	2,561	2,561	2,561	2,561
17,100	**17,150**	2,569	2,569	2,569	2,569
17,150	**17,200**	2,576	2,576	2,576	2,576
17,200	**17,250**	2,584	2,584	2,584	2,584
17,250	**17,300**	2,591	2,591	2,591	2,591
17,300	**17,350**	2,599	2,599	2,599	2,599
17,350	**17,400**	2,606	2,606	2,606	2,606
17,400	**17,450**	2,614	2,614	2,614	2,614
17,450	**17,500**	2,621	2,621	2,621	2,621
17,500	**17,550**	2,629	2,629	2,629	2,629
17,550	**17,600**	2,636	2,636	2,636	2,636
17,600	**17,650**	2,644	2,644	2,644	2,644
17,650	**17,700**	2,651	2,651	2,651	2,651
17,700	**17,750**	2,659	2,659	2,659	2,659
17,750	**17,800**	2,666	2,666	2,666	2,666
17,800	**17,850**	2,674	2,674	2,674	2,674
17,850	**17,900**	2,681	2,681	2,681	2,681
17,900	**17,950**	2,689	2,689	2,692	2,689
17,950	**18,000**	2,696	2,696	2,706	2,696
18,000					
18,000	**18,050**	2,704	2,704	2,720	2,704
18,050	**18,100**	2,711	2,711	2,734	2,711
18,100	**18,150**	2,719	2,719	2,748	2,719
18,150	**18,200**	2,726	2,726	2,762	2,726
18,200	**18,250**	2,734	2,734	2,776	2,734
18,250	**18,300**	2,741	2,741	2,790	2,741
18,300	**18,350**	2,749	2,749	2,804	2,749
18,350	**18,400**	2,756	2,756	2,818	2,756
18,400	**18,450**	2,764	2,764	2,832	2,764
18,450	**18,500**	2,771	2,771	2,846	2,771
18,500	**18,550**	2,779	2,779	2,860	2,779
18,550	**18,600**	2,786	2,786	2,874	2,786
18,600	**18,650**	2,794	2,794	2,888	2,794
18,650	**18,700**	2,801	2,801	2,902	2,801
18,700	**18,750**	2,809	2,809	2,916	2,809
18,750	**18,800**	2,816	2,816	2,930	2,816
18,800	**18,850**	2,824	2,824	2,944	2,824
18,850	**18,900**	2,831	2,831	2,958	2,831
18,900	**18,950**	2,839	2,839	2,972	2,839
18,950	**19,000**	2,846	2,846	2,986	2,846
19,000					
19,000	**19,050**	2,854	2,854	3,000	2,854
19,050	**19,100**	2,861	2,861	3,014	2,861
19,100	**19,150**	2,869	2,869	3,028	2,869
19,150	**19,200**	2,876	2,876	3,042	2,876
19,200	**19,250**	2,884	2,884	3,056	2,884
19,250	**19,300**	2,891	2,891	3,070	2,891
19,300	**19,350**	2,899	2,899	3,084	2,899
19,350	**19,400**	2,906	2,906	3,098	2,906
19,400	**19,450**	2,914	2,914	3,112	2,914
19,450	**19,500**	2,921	2,921	3,126	2,921
19,500	**19,550**	2,929	2,929	3,140	2,929
19,550	**19,600**	2,936	2,936	3,154	2,936
19,600	**19,650**	2,944	2,944	3,168	2,944
19,650	**19,700**	2,951	2,951	3,182	2,951
19,700	**19,750**	2,959	2,959	3,196	2,959
19,750	**19,800**	2,966	2,966	3,210	2,966
19,800	**19,850**	2,974	2,974	3,224	2,974
19,850	**19,900**	2,981	2,981	3,238	2,981
19,900	**19,950**	2,989	2,989	3,252	2,989
19,950	**20,000**	2,996	2,996	3,266	2,996

If line 5 (Form 1040EZ), line 22 (Form 1040A), or line 37 (Form 1040) is—		And you are—			
At least	But less than	Single	Married filing jointly *	Married filing sepa-rately	Head of a house-hold
		Your tax is—			
20,000					
20,000	**20,050**	3,004	3,004	3,280	3,004
20,050	**20,100**	3,011	3,011	3,294	3,011
20,100	**20,150**	3,019	3,019	3,308	3,019
20,150	**20,200**	3,026	3,026	3,322	3,026
20,200	**20,250**	3,034	3,034	3,336	3,034
20,250	**20,300**	3,041	3,041	3,350	3,041
20,300	**20,350**	3,049	3,049	3,364	3,049
20,350	**20,400**	3,056	3,056	3,378	3,056
20,400	**20,450**	3,064	3,064	3,392	3,064
20,450	**20,500**	3,071	3,071	3,406	3,071
20,500	**20,550**	3,079	3,079	3,420	3,079
20,550	**20,600**	3,086	3,086	3,434	3,086
20,600	**20,650**	3,094	3,094	3,448	3,094
20,650	**20,700**	3,101	3,101	3,462	3,101
20,700	**20,750**	3,109	3,109	3,476	3,109
20,750	**20,800**	3,116	3,116	3,490	3,116
20,800	**20,850**	3,124	3,124	3,504	3,124
20,850	**20,900**	3,131	3,131	3,518	3,131
20,900	**20,950**	3,139	3,139	3,532	3,139
20,950	**21,000**	3,146	3,146	3,546	3,146
21,000					
21,000	**21,050**	3,154	3,154	3,560	3,154
21,050	**21,100**	3,161	3,161	3,574	3,161
21,100	**21,150**	3,169	3,169	3,588	3,169
21,150	**21,200**	3,176	3,176	3,602	3,176
21,200	**21,250**	3,184	3,184	3,616	3,184
21,250	**21,300**	3,191	3,191	3,630	3,191
21,300	**21,350**	3,199	3,199	3,644	3,199
21,350	**21,400**	3,206	3,206	3,658	3,206
21,400	**21,450**	3,214	3,214	3,672	3,214
21,450	**21,500**	3,225	3,221	3,686	3,221
21,500	**21,550**	3,239	3,229	3,700	3,229
21,550	**21,600**	3,253	3,236	3,714	3,236
21,600	**21,650**	3,267	3,244	3,728	3,244
21,650	**21,700**	3,281	3,251	3,742	3,251
21,700	**21,750**	3,295	3,259	3,756	3,259
21,750	**21,800**	3,309	3,266	3,770	3,266
21,800	**21,850**	3,323	3,274	3,784	3,274
21,850	**21,900**	3,337	3,281	3,798	3,281
21,900	**21,950**	3,351	3,289	3,812	3,289
21,950	**22,000**	3,365	3,296	3,826	3,296
22,000					
22,000	**22,050**	3,379	3,304	3,840	3,304
22,050	**22,100**	3,393	3,311	3,854	3,311
22,100	**22,150**	3,407	3,319	3,868	3,319
22,150	**22,200**	3,421	3,326	3,882	3,326
22,200	**22,250**	3,435	3,334	3,896	3,334
22,250	**22,300**	3,449	3,341	3,910	3,341
22,300	**22,350**	3,463	3,349	3,924	3,349
22,350	**22,400**	3,477	3,356	3,938	3,356
22,400	**22,450**	3,491	3,364	3,952	3,364
22,450	**22,500**	3,505	3,371	3,966	3,371
22,500	**22,550**	3,519	3,379	3,980	3,379
22,550	**22,600**	3,533	3,386	3,994	3,386
22,600	**22,650**	3,547	3,394	4,008	3,394
22,650	**22,700**	3,561	3,401	4,022	3,401
22,700	**22,750**	3,575	3,409	4,036	3,409
22,750	**22,800**	3,589	3,416	4,050	3,416
22,800	**22,850**	3,603	3,424	4,064	3,424
22,850	**22,900**	3,617	3,431	4,078	3,431
22,900	**22,950**	3,631	3,439	4,092	3,439
22,950	**23,000**	3,645	3,446	4,106	3,446

* This column must also be used by a qualifying widow(er).

Continued on next page

19-- Tax Table—*Continued*

If line 5 (Form 1040EZ), line 22 (Form 1040A), or line 37 (Form 1040) is—		And you are—			
At least	But less than	Single	Married filing jointly *	Married filing separately	Head of a household
		Your tax is—			
23,000					
23,000	**23,050**	3,659	3,454	4,120	3,454
23,050	**23,100**	3,673	3,461	4,134	3,461
23,100	**23,150**	3,687	3,469	4,148	3,469
23,150	**23,200**	3,701	3,476	4,162	3,476
23,200	**23,250**	3,715	3,484	4,176	3,484
23,250	**23,300**	3,729	3,491	4,190	3,491
23,300	**23,350**	3,743	3,499	4,204	3,499
23,350	**23,400**	3,757	3,506	4,218	3,506
23,400	**23,450**	3,771	3,514	4,232	3,514
23,450	**23,500**	3,785	3,521	4,246	3,521
23,500	**23,550**	3,799	3,529	4,260	3,529
23,550	**23,600**	3,813	3,536	4,274	3,536
23,600	**23,650**	3,827	3,544	4,288	3,544
23,650	**23,700**	3,841	3,551	4,302	3,551
23,700	**23,750**	3,855	3,559	4,316	3,559
23,750	**23,800**	3,869	3,566	4,330	3,566
23,800	**23,850**	3,883	3,574	4,344	3,574
23,850	**23,900**	3,897	3,581	4,358	3,581
23,900	**23,950**	3,911	3,589	4,372	3,589
23,950	**24,000**	3,925	3,596	4,386	3,596
24,000					
24,000	**24,050**	3,939	3,604	4,400	3,604
24,050	**24,100**	3,953	3,611	4,414	3,611
24,100	**24,150**	3,967	3,619	4,428	3,619
24,150	**24,200**	3,981	3,626	4,442	3,626
24,200	**24,250**	3,995	3,634	4,456	3,634
24,250	**24,300**	4,009	3,641	4,470	3,641
24,300	**24,350**	4,023	3,649	4,484	3,649
24,350	**24,400**	4,037	3,656	4,498	3,656
24,400	**24,450**	4,051	3,664	4,512	3,664
24,450	**24,500**	4,065	3,671	4,526	3,671
24,500	**24,550**	4,079	3,679	4,540	3,679
24,550	**24,600**	4,093	3,686	4,554	3,686
24,600	**24,650**	4,107	3,694	4,568	3,694
24,650	**24,700**	4,121	3,701	4,582	3,701
24,700	**24,750**	4,135	3,709	4,596	3,709
24,750	**24,800**	4,149	3,716	4,610	3,716
24,800	**24,850**	4,163	3,724	4,624	3,724
24,850	**24,900**	4,177	3,731	4,638	3,731
24,900	**24,950**	4,191	3,739	4,652	3,739
24,950	**25,000**	4,205	3,746	4,666	3,746
25,000					
25,000	**25,050**	4,219	3,754	4,680	3,754
25,050	**25,100**	4,233	3,761	4,694	3,761
25,100	**25,150**	4,247	3,769	4,708	3,769
25,150	**25,200**	4,261	3,776	4,722	3,776
25,200	**25,250**	4,275	3,784	4,736	3,784
25,250	**25,300**	4,289	3,791	4,750	3,791
25,300	**25,350**	4,303	3,799	4,764	3,799
25,350	**25,400**	4,317	3,806	4,778	3,806
25,400	**25,450**	4,331	3,814	4,792	3,814
25,450	**25,500**	4,345	3,821	4,806	3,821
25,500	**25,550**	4,359	3,829	4,820	3,829
25,550	**25,600**	4,373	3,836	4,834	3,836
25,600	**25,650**	4,387	3,844	4,848	3,844
25,650	**25,700**	4,401	3,851	4,862	3,851
25,700	**25,750**	4,415	3,859	4,876	3,859
25,750	**25,800**	4,429	3,866	4,890	3,866
25,800	**25,850**	4,443	3,874	4,904	3,874
25,850	**25,900**	4,457	3,881	4,918	3,881
25,900	**25,950**	4,471	3,889	4,932	3,889
25,950	**26,000**	4,485	3,896	4,946	3,896

If line 5 (Form 1040EZ), line 22 (Form 1040A), or line 37 (Form 1040) is—		And you are—			
At least	But less than	Single	Married filing jointly *	Married filing separately	Head of a household
		Your tax is—			
26,000					
26,000	**26,050**	4,499	3,904	4,960	3,904
26,050	**26,100**	4,513	3,911	4,974	3,911
26,100	**26,150**	4,527	3,919	4,988	3,919
26,150	**26,200**	4,541	3,926	5,002	3,926
26,200	**26,250**	4,555	3,934	5,016	3,934
26,250	**26,300**	4,569	3,941	5,030	3,941
26,300	**26,350**	4,583	3,949	5,044	3,949
26,350	**26,400**	4,597	3,956	5,058	3,956
26,400	**26,450**	4,611	3,964	5,072	3,964
26,450	**26,500**	4,625	3,971	5,086	3,971
26,500	**26,550**	4,639	3,979	5,100	3,979
26,550	**26,600**	4,653	3,986	5,114	3,986
26,600	**26,650**	4,667	3,994	5,128	3,994
26,650	**26,700**	4,681	4,001	5,142	4,001
26,700	**26,750**	4,695	4,009	5,156	4,009
26,750	**26,800**	4,709	4,016	5,170	4,016
26,800	**26,850**	4,723	4,024	5,184	4,024
26,850	**26,900**	4,737	4,031	5,198	4,031
26,900	**26,950**	4,751	4,039	5,212	4,039
26,950	**27,000**	4,765	4,046	5,226	4,046
27,000					
27,000	**27,050**	4,779	4,054	5,240	4,054
27,050	**27,100**	4,793	4,061	5,254	4,061
27,100	**27,150**	4,807	4,069	5,268	4,069
27,150	**27,200**	4,821	4,076	5,282	4,076
27,200	**27,250**	4,835	4,084	5,296	4,084
27,250	**27,300**	4,849	4,091	5,310	4,091
27,300	**27,350**	4,863	4,099	5,324	4,099
27,350	**27,400**	4,877	4,106	5,338	4,106
27,400	**27,450**	4,891	4,114	5,352	4,114
27,450	**27,500**	4,905	4,121	5,366	4,121
27,500	**27,550**	4,919	4,129	5,380	4,129
27,550	**27,600**	4,933	4,136	5,394	4,136
27,600	**27,650**	4,947	4,144	5,408	4,144
27,650	**27,700**	4,961	4,151	5,422	4,151
27,700	**27,750**	4,975	4,159	5,436	4,159
27,750	**27,800**	4,989	4,166	5,450	4,166
27,800	**27,850**	5,003	4,174	5,464	4,174
27,850	**27,900**	5,017	4,181	5,478	4,181
27,900	**27,950**	5,031	4,189	5,492	4,189
27,950	**28,000**	5,045	4,196	5,506	4,196
28,000					
28,000	**28,050**	5,059	4,204	5,520	4,204
28,050	**28,100**	5,073	4,211	5,534	4,211
28,100	**28,150**	5,087	4,219	5,548	4,219
28,150	**28,200**	5,101	4,226	5,562	4,226
28,200	**28,250**	5,115	4,234	5,576	4,234
28,250	**28,300**	5,129	4,241	5,590	4,241
28,300	**28,350**	5,143	4,249	5,604	4,249
28,350	**28,400**	5,157	4,256	5,618	4,256
28,400	**28,450**	5,171	4,264	5,632	4,264
28,450	**28,500**	5,185	4,271	5,646	4,271
28,500	**28,550**	5,199	4,279	5,660	4,279
28,550	**28,600**	5,213	4,286	5,674	4,286
28,600	**28,650**	5,227	4,294	5,688	4,294
28,650	**28,700**	5,241	4,301	5,702	4,301
28,700	**28,750**	5,255	4,309	5,716	4,309
28,750	**28,800**	5,269	4,316	5,730	4,320
28,800	**28,850**	5,283	4,324	5,744	4,334
28,850	**28,900**	5,297	4,331	5,758	4,348
28,900	**28,950**	5,311	4,339	5,772	4,362
28,950	**29,000**	5,325	4,346	5,786	4,376

If line 5 (Form 1040EZ), line 22 (Form 1040A), or line 37 (Form 1040) is—		And you are—			
At least	But less than	Single	Married filing jointly *	Married filing separately	Head of a household
		Your tax is—			
29,000					
29,000	**29,050**	5,339	4,354	5,800	4,390
29,050	**29,100**	5,353	4,361	5,814	4,404
29,100	**29,150**	5,367	4,369	5,828	4,418
29,150	**29,200**	5,381	4,376	5,842	4,432
29,200	**29,250**	5,395	4,384	5,856	4,446
29,250	**29,300**	5,409	4,391	5,870	4,460
29,300	**29,350**	5,423	4,399	5,884	4,474
29,350	**29,400**	5,437	4,406	5,898	4,488
29,400	**29,450**	5,451	4,414	5,912	4,502
29,450	**29,500**	5,465	4,421	5,926	4,516
29,500	**29,550**	5,479	4,429	5,940	4,530
29,550	**29,600**	5,493	4,436	5,954	4,544
29,600	**29,650**	5,507	4,444	5,968	4,558
29,650	**29,700**	5,521	4,451	5,982	4,572
29,700	**29,750**	5,535	4,459	5,996	4,586
29,750	**29,800**	5,549	4,466	6,010	4,600
29,800	**29,850**	5,563	4,474	6,024	4,614
29,850	**29,900**	5,577	4,481	6,038	4,628
29,900	**29,950**	5,591	4,489	6,052	4,642
29,950	**30,000**	5,605	4,496	6,066	4,656
30,000					
30,000	**30,050**	5,619	4,504	6,080	4,670
30,050	**30,100**	5,633	4,511	6,094	4,684
30,100	**30,150**	5,647	4,519	6,108	4,698
30,150	**30,200**	5,661	4,526	6,122	4,712
30,200	**30,250**	5,675	4,534	6,136	4,726
30,250	**30,300**	5,689	4,541	6,150	4,740
30,300	**30,350**	5,703	4,549	6,164	4,754
30,350	**30,400**	5,717	4,556	6,178	4,768
30,400	**30,450**	5,731	4,564	6,192	4,782
30,450	**30,500**	5,745	4,571	6,206	4,796
30,500	**30,550**	5,759	4,579	6,220	4,810
30,550	**30,600**	5,773	4,586	6,234	4,824
30,600	**30,650**	5,787	4,594	6,248	4,838
30,650	**30,700**	5,801	4,601	6,262	4,852
30,700	**30,750**	5,815	4,609	6,276	4,866
30,750	**30,800**	5,829	4,616	6,290	4,880
30,800	**30,850**	5,843	4,624	6,304	4,894
30,850	**30,900**	5,857	4,631	6,318	4,908
30,900	**30,950**	5,871	4,639	6,332	4,922
30,950	**31,000**	5,885	4,646	6,346	4,936
31,000					
31,000	**31,050**	5,899	4,654	6,360	4,950
31,050	**31,100**	5,913	4,661	6,374	4,964
31,100	**31,150**	5,927	4,669	6,388	4,978
31,150	**31,200**	5,941	4,676	6,402	4,992
31,200	**31,250**	5,955	4,684	6,416	5,006
31,250	**31,300**	5,969	4,691	6,430	5,020
31,300	**31,350**	5,983	4,699	6,444	5,034
31,350	**31,400**	5,997	4,706	6,458	5,048
31,400	**31,450**	6,011	4,714	6,472	5,062
31,450	**31,500**	6,025	4,721	6,486	5,076
31,500	**31,550**	6,039	4,729	6,500	5,090
31,550	**31,600**	6,053	4,736	6,514	5,104
31,600	**31,650**	6,067	4,744	6,528	5,118
31,650	**31,700**	6,081	4,751	6,542	5,132
31,700	**31,750**	6,095	4,759	6,556	5,146
31,750	**31,800**	6,109	4,766	6,570	5,160
31,800	**31,850**	6,123	4,774	6,584	5,174
31,850	**31,900**	6,137	4,781	6,598	5,188
31,900	**31,950**	6,151	4,789	6,612	5,202
31,950	**32,000**	6,165	4,796	6,626	5,216

* This column must also be used by a qualifying widow(er).

Continued on next page

19-- Tax Table—*Continued*

If line 5 (Form 1040EZ), line 22 (Form 1040A), or line 37 (Form 1040) is—		And you are—			
At least	But less than	Single	Married filing jointly *	Married filing sepa-rately	Head of a house-hold
		Your tax is—			
32,000					
32,000	**32,050**	6,179	4,804	6,640	5,230
32,050	**32,100**	6,193	4,811	6,654	5,244
32,100	**32,150**	6,207	4,819	6,668	5,258
32,150	**32,200**	6,221	4,826	6,682	5,272
32,200	**32,250**	6,235	4,834	6,696	5,286
32,250	**32,300**	6,249	4,841	6,710	5,300
32,300	**32,350**	6,263	4,849	6,724	5,314
32,350	**32,400**	6,277	4,856	6,738	5,328
32,400	**32,450**	6,291	4,864	6,752	5,342
32,450	**32,500**	6,305	4,871	6,766	5,356
32,500	**32,550**	6,319	4,879	6,780	5,370
32,550	**32,600**	6,333	4,886	6,794	5,384
32,600	**32,650**	6,347	4,894	6,808	5,398
32,650	**32,700**	6,361	4,901	6,822	5,412
32,700	**32,750**	6,375	4,909	6,836	5,426
32,750	**32,800**	6,389	4,916	6,850	5,440
32,800	**32,850**	6,403	4,924	6,864	5,454
32,850	**32,900**	6,417	4,931	6,878	5,468
32,900	**32,950**	6,431	4,939	6,892	5,482
32,950	**33,000**	6,445	4,946	6,906	5,496
33,000					
33,000	**33,050**	6,459	4,954	6,920	5,510
33,050	**33,100**	6,473	4,961	6,934	5,524
33,100	**33,150**	6,487	4,969	6,948	5,538
33,150	**33,200**	6,501	4,976	6,962	5,552
33,200	**33,250**	6,515	4,984	6,976	5,566
33,250	**33,300**	6,529	4,991	6,990	5,580
33,300	**33,350**	6,543	4,999	7,004	5,594
33,350	**33,400**	6,557	5,006	7,018	5,608
33,400	**33,450**	6,571	5,014	7,032	5,622
33,450	**33,500**	6,585	5,021	7,046	5,636
33,500	**33,550**	6,599	5,029	7,060	5,650
33,550	**33,600**	6,613	5,036	7,074	5,664
33,600	**33,650**	6,627	5,044	7,088	5,678
33,650	**33,700**	6,641	5,051	7,102	5,692
33,700	**33,750**	6,655	5,059	7,116	5,706
33,750	**33,800**	6,669	5,066	7,130	5,720
33,800	**33,850**	6,683	5,074	7,144	5,734
33,850	**33,900**	6,697	5,081	7,158	5,748
33,900	**33,950**	6,711	5,089	7,172	5,762
33,950	**34,000**	6,725	5,096	7,186	5,776
34,000					
34,000	**34,050**	6,739	5,104	7,200	5,790
34,050	**34,100**	6,753	5,111	7,214	5,804
34,100	**34,150**	6,767	5,119	7,228	5,818
34,150	**34,200**	6,781	5,126	7,242	5,832
34,200	**34,250**	6,795	5,134	7,256	5,846
34,250	**34,300**	6,809	5,141	7,270	5,860
34,300	**34,350**	6,823	5,149	7,284	5,874
34,350	**34,400**	6,837	5,156	7,298	5,888
34,400	**34,450**	6,851	5,164	7,312	5,902
34,450	**34,500**	6,865	5,171	7,326	5,916
34,500	**34,550**	6,879	5,179	7,340	5,930
34,550	**34,600**	6,893	5,186	7,354	5,944
34,600	**34,650**	6,907	5,194	7,368	5,958
34,650	**34,700**	6,921	5,201	7,382	5,972
34,700	**34,750**	6,935	5,209	7,396	5,986
34,750	**34,800**	6,949	5,216	7,410	6,000
34,800	**34,850**	6,963	5,224	7,424	6,014
34,850	**34,900**	6,977	5,231	7,438	6,028
34,900	**34,950**	6,991	5,239	7,452	6,042
34,950	**35,000**	7,005	5,246	7,466	6,056

If line 5 (Form 1040EZ), line 22 (Form 1040A), or line 37 (Form 1040) is—		And you are—			
At least	But less than	Single	Married filing jointly *	Married filing sepa-rately	Head of a house-hold
		Your tax is—			
35,000					
35,000	**35,050**	7,019	5,254	7,480	6,070
35,050	**35,100**	7,033	5,261	7,494	6,084
35,100	**35,150**	7,047	5,269	7,508	6,098
35,150	**35,200**	7,061	5,276	7,522	6,112
35,200	**35,250**	7,075	5,284	7,536	6,126
35,250	**35,300**	7,089	5,291	7,550	6,140
35,300	**35,350**	7,103	5,299	7,564	6,154
35,350	**35,400**	7,117	5,306	7,578	6,168
35,400	**35,450**	7,131	5,314	7,592	6,182
35,450	**35,500**	7,145	5,321	7,606	6,196
35,500	**35,550**	7,159	5,329	7,620	6,210
35,550	**35,600**	7,173	5,336	7,634	6,224
35,600	**35,650**	7,187	5,344	7,648	6,238
35,650	**35,700**	7,201	5,351	7,662	6,252
35,700	**35,750**	7,215	5,359	7,676	6,266
35,750	**35,800**	7,229	5,366	7,690	6,280
35,800	**35,850**	7,243	5,377	7,704	6,294
35,850	**35,900**	7,257	5,391	7,718	6,308
35,900	**35,950**	7,271	5,405	7,732	6,322
35,950	**36,000**	7,285	5,419	7,746	6,336
36,000					
36,000	**36,050**	7,299	5,433	7,760	6,350
36,050	**36,100**	7,313	5,447	7,774	6,364
36,100	**36,150**	7,327	5,461	7,788	6,378
36,150	**36,200**	7,341	5,475	7,802	6,392
36,200	**36,250**	7,355	5,489	7,816	6,406
36,250	**36,300**	7,369	5,503	7,830	6,420
36,300	**36,350**	7,383	5,517	7,844	6,434
36,350	**36,400**	7,397	5,531	7,858	6,448
36,400	**36,450**	7,411	5,545	7,872	6,462
36,450	**36,500**	7,425	5,559	7,886	6,476
36,500	**36,550**	7,439	5,573	7,900	6,490
36,550	**36,600**	7,453	5,587	7,914	6,504
36,600	**36,650**	7,467	5,601	7,928	6,518
36,650	**36,700**	7,481	5,615	7,942	6,532
36,700	**36,750**	7,495	5,629	7,956	6,546
36,750	**36,800**	7,509	5,643	7,970	6,560
36,800	**36,850**	7,523	5,657	7,984	6,574
36,850	**36,900**	7,537	5,671	7,998	6,588
36,900	**36,950**	7,551	5,685	8,012	6,602
36,950	**37,000**	7,565	5,699	8,026	6,616
37,000					
37,000	**37,050**	7,579	5,713	8,040	6,630
37,050	**37,100**	7,593	5,727	8,054	6,644
37,100	**37,150**	7,607	5,741	8,068	6,658
37,150	**37,200**	7,621	5,755	8,082	6,672
37,200	**37,250**	7,635	5,769	8,096	6,686
37,250	**37,300**	7,649	5,783	8,110	6,700
37,300	**37,350**	7,663	5,797	8,124	6,714
37,350	**37,400**	7,677	5,811	8,138	6,728
37,400	**37,450**	7,691	5,825	8,152	6,742
37,450	**37,500**	7,705	5,839	8,166	6,756
37,500	**37,550**	7,719	5,853	8,180	6,770
37,550	**37,600**	7,733	5,867	8,194	6,784
37,600	**37,650**	7,747	5,881	8,208	6,798
37,650	**37,700**	7,761	5,895	8,222	6,812
37,700	**37,750**	7,775	5,909	8,236	6,826
37,750	**37,800**	7,789	5,923	8,250	6,840
37,800	**37,850**	7,803	5,937	8,264	6,854
37,850	**37,900**	7,817	5,951	8,278	6,868
37,900	**37,950**	7,831	5,965	8,292	6,882
37,950	**38,000**	7,845	5,979	8,306	6,896

If line 5 (Form 1040EZ), line 22 (Form 1040A), or line 37 (Form 1040) is—		And you are—			
At least	But less than	Single	Married filing jointly *	Married filing sepa-rately	Head of a house-hold
		Your tax is—			
38,000					
38,000	**38,050**	7,859	5,993	8,320	6,910
38,050	**38,100**	7,873	6,007	8,334	6,924
38,100	**38,150**	7,887	6,021	8,348	6,938
38,150	**38,200**	7,901	6,035	8,362	6,952
38,200	**38,250**	7,915	6,049	8,376	6,966
38,250	**38,300**	7,929	6,063	8,390	6,980
38,300	**38,350**	7,943	6,077	8,404	6,994
38,350	**38,400**	7,957	6,091	8,418	7,008
38,400	**38,450**	7,971	6,105	8,432	7,022
38,450	**38,500**	7,985	6,119	8,446	7,036
38,500	**38,550**	7,999	6,133	8,460	7,050
38,550	**38,600**	8,013	6,147	8,474	7,064
38,600	**38,650**	8,027	6,161	8,488	7,078
38,650	**38,700**	8,041	6,175	8,502	7,092
38,700	**38,750**	8,055	6,189	8,516	7,106
38,750	**38,800**	8,069	6,203	8,530	7,120
38,800	**38,850**	8,083	6,217	8,544	7,134
38,850	**38,900**	8,097	6,231	8,558	7,148
38,900	**38,950**	8,111	6,245	8,572	7,162
38,950	**39,000**	8,125	6,259	8,586	7,176
39,000					
39,000	**39,050**	8,139	6,273	8,600	7,190
39,050	**39,100**	8,153	6,287	8,614	7,204
39,100	**39,150**	8,167	6,301	8,628	7,218
39,150	**39,200**	8,181	6,315	8,642	7,232
39,200	**39,250**	8,195	6,329	8,656	7,246
39,250	**39,300**	8,209	6,343	8,670	7,260
39,300	**39,350**	8,223	6,357	8,684	7,274
39,350	**39,400**	8,237	6,371	8,698	7,288
39,400	**39,450**	8,251	6,385	8,712	7,302
39,450	**39,500**	8,265	6,399	8,726	7,316
39,500	**39,550**	8,279	6,413	8,740	7,330
39,550	**39,600**	8,293	6,427	8,754	7,344
39,600	**39,650**	8,307	6,441	8,768	7,358
39,650	**39,700**	8,321	6,455	8,782	7,372
39,700	**39,750**	8,335	6,469	8,796	7,386
39,750	**39,800**	8,349	6,483	8,810	7,400
39,800	**39,850**	8,363	6,497	8,824	7,414
39,850	**39,900**	8,377	6,511	8,838	7,428
39,900	**39,950**	8,391	6,525	8,852	7,442
39,950	**40,000**	8,405	6,539	8,866	7,456
40,000					
40,000	**40,050**	8,419	6,553	8,880	7,470
40,050	**40,100**	8,433	6,567	8,894	7,484
40,100	**40,150**	8,447	6,581	8,908	7,498
40,150	**40,200**	8,461	6,595	8,922	7,512
40,200	**40,250**	8,475	6,609	8,936	7,526
40,250	**40,300**	8,489	6,623	8,950	7,540
40,300	**40,350**	8,503	6,637	8,964	7,554
40,350	**40,400**	8,517	6,651	8,978	7,568
40,400	**40,450**	8,531	6,665	8,992	7,582
40,450	**40,500**	8,545	6,679	9,006	7,596
40,500	**40,550**	8,559	6,693	9,020	7,610
40,550	**40,600**	8,573	6,707	9,034	7,624
40,600	**40,650**	8,587	6,721	9,048	7,638
40,650	**40,700**	8,601	6,735	9,062	7,652
40,700	**40,750**	8,615	6,749	9,076	7,666
40,750	**40,800**	8,629	6,763	9,090	7,680
40,800	**40,850**	8,643	6,777	9,104	7,694
40,850	**40,900**	8,657	6,791	9,118	7,708
40,900	**40,950**	8,671	6,805	9,132	7,722
40,950	**41,000**	8,685	6,819	9,146	7,736

* This column must also be used by a qualifying widow(er).

Continued on next page

19-- Tax Table—*Continued*

If line 5 (Form 1040EZ), line 22 (Form 1040A), or line 37 (Form 1040) is—		And you are—			
At least	But less than	Single	Married filing jointly *	Married filing separately	Head of a household
		Your tax is—			
41,000					
41,000	**41,050**	8,699	6,833	9,160	7,750
41,050	**41,100**	8,713	6,847	9,174	7,764
41,100	**41,150**	8,727	6,861	9,188	7,778
41,150	**41,200**	8,741	6,875	9,202	7,792
41,200	**41,250**	8,755	6,889	9,216	7,806
41,250	**41,300**	8,769	6,903	9,230	7,820
41,300	**41,350**	8,783	6,917	9,244	7,834
41,350	**41,400**	8,797	6,931	9,258	7,848
41,400	**41,450**	8,811	6,945	9,272	7,862
41,450	**41,500**	8,825	6,959	9,286	7,876
41,500	**41,550**	8,839	6,973	9,300	7,890
41,550	**41,600**	8,853	6,987	9,314	7,904
41,600	**41,650**	8,867	7,001	9,328	7,918
41,650	**41,700**	8,881	7,015	9,342	7,932
41,700	**41,750**	8,895	7,029	9,356	7,946
41,750	**41,800**	8,909	7,043	9,370	7,960
41,800	**41,850**	8,923	7,057	9,384	7,974
41,850	**41,900**	8,937	7,071	9,398	7,988
41,900	**41,950**	8,951	7,085	9,412	8,002
41,950	**42,000**	8,965	7,099	9,426	8,016
42,000					
42,000	**42,050**	8,979	7,113	9,440	8,030
42,050	**42,100**	8,993	7,127	9,454	8,044
42,100	**42,150**	9,007	7,141	9,468	8,058
42,150	**42,200**	9,021	7,155	9,482	8,072
42,200	**42,250**	9,035	7,169	9,496	8,086
42,250	**42,300**	9,049	7,183	9,510	8,100
42,300	**42,350**	9,063	7,197	9,524	8,114
42,350	**42,400**	9,077	7,211	9,538	8,128
42,400	**42,450**	9,091	7,225	9,552	8,142
42,450	**42,500**	9,105	7,239	9,566	8,156
42,500	**42,550**	9,119	7,253	9,580	8,170
42,550	**42,600**	9,133	7,267	9,594	8,184
42,600	**42,650**	9,147	7,281	9,608	8,198
42,650	**42,700**	9,161	7,295	9,622	8,212
42,700	**42,750**	9,175	7,309	9,636	8,226
42,750	**42,800**	9,189	7,323	9,650	8,240
42,800	**42,850**	9,203	7,337	9,664	8,254
42,850	**42,900**	9,217	7,351	9,678	8,268
42,900	**42,950**	9,231	7,365	9,692	8,282
42,950	**43,000**	9,245	7,379	9,706	8,296
43,000					
43,000	**43,050**	9,259	7,393	9,720	8,310
43,050	**43,100**	9,273	7,407	9,734	8,324
43,100	**43,150**	9,287	7,421	9,748	8,338
43,150	**43,200**	9,301	7,435	9,762	8,352
43,200	**43,250**	9,315	7,449	9,776	8,366
43,250	**43,300**	9,329	7,463	9,791	8,380
43,300	**43,350**	9,343	7,477	9,806	8,394
43,350	**43,400**	9,357	7,491	9,822	8,408
43,400	**43,450**	9,371	7,505	9,837	8,422
43,450	**43,500**	9,385	7,519	9,853	8,436
43,500	**43,550**	9,399	7,533	9,868	8,450
43,550	**43,600**	9,413	7,547	9,884	8,464
43,600	**43,650**	9,427	7,561	9,899	8,478
43,650	**43,700**	9,441	7,575	9,915	8,492
43,700	**43,750**	9,455	7,589	9,930	8,506
43,750	**43,800**	9,469	7,603	9,946	8,520
43,800	**43,850**	9,483	7,617	9,961	8,534
43,850	**43,900**	9,497	7,631	9,977	8,548
43,900	**43,950**	9,511	7,645	9,992	8,562
43,950	**44,000**	9,525	7,659	10,008	8,576

If line 5 (Form 1040EZ), line 22 (Form 1040A), or line 37 (Form 1040) is—		And you are—			
At least	But less than	Single	Married filing jointly *	Married filing separately	Head of a household
		Your tax is—			
44,000					
44,000	**44,050**	9,539	7,673	10,023	8,590
44,050	**44,100**	9,553	7,687	10,039	8,604
44,100	**44,150**	9,567	7,701	10,054	8,618
44,150	**44,200**	9,581	7,715	10,070	8,632
44,200	**44,250**	9,595	7,729	10,085	8,646
44,250	**44,300**	9,609	7,743	10,101	8,660
44,300	**44,350**	9,623	7,757	10,116	8,674
44,350	**44,400**	9,637	7,771	10,132	8,688
44,400	**44,450**	9,651	7,785	10,147	8,702
44,450	**44,500**	9,665	7,799	10,163	8,716
44,500	**44,550**	9,679	7,813	10,178	8,730
44,550	**44,600**	9,693	7,827	10,194	8,744
44,600	**44,650**	9,707	7,841	10,209	8,758
44,650	**44,700**	9,721	7,855	10,225	8,772
44,700	**44,750**	9,735	7,869	10,240	8,786
44,750	**44,800**	9,749	7,883	10,256	8,800
44,800	**44,850**	9,763	7,897	10,271	8,814
44,850	**44,900**	9,777	7,911	10,287	8,828
44,900	**44,950**	9,791	7,925	10,302	8,842
44,950	**45,000**	9,805	7,939	10,318	8,856
45,000					
45,000	**45,050**	9,819	7,953	10,333	8,870
45,050	**45,100**	9,833	7,967	10,349	8,884
45,100	**45,150**	9,847	7,981	10,364	8,898
45,150	**45,200**	9,861	7,995	10,380	8,912
45,200	**45,250**	9,875	8,009	10,395	8,926
45,250	**45,300**	9,889	8,023	10,411	8,940
45,300	**45,350**	9,903	8,037	10,426	8,954
45,350	**45,400**	9,917	8,051	10,442	8,968
45,400	**45,450**	9,931	8,065	10,457	8,982
45,450	**45,500**	9,945	8,079	10,473	8,996
45,500	**45,550**	9,959	8,093	10,488	9,010
45,550	**45,600**	9,973	8,107	10,504	9,024
45,600	**45,650**	9,987	8,121	10,519	9,038
45,650	**45,700**	10,001	8,135	10,535	9,052
45,700	**45,750**	10,015	8,149	10,550	9,066
45,750	**45,800**	10,029	8,163	10,566	9,080
45,800	**45,850**	10,043	8,177	10,581	9,094
45,850	**45,900**	10,057	8,191	10,597	9,108
45,900	**45,950**	10,071	8,205	10,612	9,122
45,950	**46,000**	10,085	8,219	10,628	9,136
46,000					
46,000	**46,050**	10,099	8,233	10,643	9,150
46,050	**46,100**	10,113	8,247	10,659	9,164
46,100	**46,150**	10,127	8,261	10,674	9,178
46,150	**46,200**	10,141	8,275	10,690	9,192
46,200	**46,250**	10,155	8,289	10,705	9,206
46,250	**46,300**	10,169	8,303	10,721	9,220
46,300	**46,350**	10,183	8,317	10,736	9,234
46,350	**46,400**	10,197	8,331	10,752	9,248
46,400	**46,450**	10,211	8,345	10,767	9,262
46,450	**46,500**	10,225	8,359	10,783	9,276
46,500	**46,550**	10,239	8,373	10,798	9,290
46,550	**46,600**	10,253	8,387	10,814	9,304
46,600	**46,650**	10,267	8,401	10,829	9,318
46,650	**46,700**	10,281	8,415	10,845	9,332
46,700	**46,750**	10,295	8,429	10,860	9,346
46,750	**46,800**	10,309	8,443	10,876	9,360
46,800	**46,850**	10,323	8,457	10,891	9,374
46,850	**46,900**	10,337	8,471	10,907	9,388
46,900	**46,950**	10,351	8,485	10,922	9,402
46,950	**47,000**	10,365	8,499	10,938	9,416

If line 5 (Form 1040EZ), line 22 (Form 1040A), or line 37 (Form 1040) is—		And you are—			
At least	But less than	Single	Married filing jointly *	Married filing separately	Head of a household
		Your tax is—			
47,000					
47,000	**47,050**	10,379	8,513	10,953	9,430
47,050	**47,100**	10,393	8,527	10,969	9,444
47,100	**47,150**	10,407	8,541	10,984	9,458
47,150	**47,200**	10,421	8,555	11,000	9,472
47,200	**47,250**	10,435	8,569	11,015	9,486
47,250	**47,300**	10,449	8,583	11,031	9,500
47,300	**47,350**	10,463	8,597	11,046	9,514
47,350	**47,400**	10,477	8,611	11,062	9,528
47,400	**47,450**	10,491	8,625	11,077	9,542
47,450	**47,500**	10,505	8,639	11,093	9,556
47,500	**47,550**	10,519	8,653	11,108	9,570
47,550	**47,600**	10,533	8,667	11,124	9,584
47,600	**47,650**	10,547	8,681	11,139	9,598
47,650	**47,700**	10,561	8,695	11,155	9,612
47,700	**47,750**	10,575	8,709	11,170	9,626
47,750	**47,800**	10,589	8,723	11,186	9,640
47,800	**47,850**	10,603	8,737	11,201	9,654
47,850	**47,900**	10,617	8,751	11,217	9,668
47,900	**47,950**	10,631	8,765	11,232	9,682
47,950	**48,000**	10,645	8,779	11,248	9,696
48,000					
48,000	**48,050**	10,659	8,793	11,263	9,710
48,050	**48,100**	10,673	8,807	11,279	9,724
48,100	**48,150**	10,687	8,821	11,294	9,738
48,150	**48,200**	10,701	8,835	11,310	9,752
48,200	**48,250**	10,715	8,849	11,325	9,766
48,250	**48,300**	10,729	8,863	11,341	9,780
48,300	**48,350**	10,743	8,877	11,356	9,794
48,350	**48,400**	10,757	8,891	11,372	9,808
48,400	**48,450**	10,771	8,905	11,387	9,822
48,450	**48,500**	10,785	8,919	11,403	9,836
48,500	**48,550**	10,799	8,933	11,418	9,850
48,550	**48,600**	10,813	8,947	11,434	9,864
48,600	**48,650**	10,827	8,961	11,449	9,878
48,650	**48,700**	10,841	8,975	11,465	9,892
48,700	**48,750**	10,855	8,989	11,480	9,906
48,750	**48,800**	10,869	9,003	11,496	9,920
48,800	**48,850**	10,883	9,017	11,511	9,934
48,850	**48,900**	10,897	9,031	11,527	9,948
48,900	**48,950**	10,911	9,045	11,542	9,962
48,950	**49,000**	10,925	9,059	11,558	9,976
49,000					
49,000	**49,050**	10,939	9,073	11,573	9,990
49,050	**49,100**	10,953	9,087	11,589	10,004
49,100	**49,150**	10,967	9,101	11,604	10,018
49,150	**49,200**	10,981	9,115	11,620	10,032
49,200	**49,250**	10,995	9,129	11,635	10,046
49,250	**49,300**	11,009	9,143	11,651	10,060
49,300	**49,350**	11,023	9,157	11,666	10,074
49,350	**49,400**	11,037	9,171	11,682	10,088
49,400	**49,450**	11,051	9,185	11,697	10,102
49,450	**49,500**	11,065	9,199	11,713	10,116
49,500	**49,550**	11,079	9,213	11,728	10,130
49,550	**49,600**	11,093	9,227	11,744	10,144
49,600	**49,650**	11,107	9,241	11,759	10,158
49,650	**49,700**	11,121	9,255	11,775	10,172
49,700	**49,750**	11,135	9,269	11,790	10,186
49,750	**49,800**	11,149	9,283	11,806	10,200
49,800	**49,850**	11,163	9,297	11,821	10,214
49,850	**49,900**	11,177	9,311	11,837	10,228
49,900	**49,950**	11,191	9,325	11,852	10,242
49,950	**50,000**	11,205	9,339	11,868	10,256

* This column must also be used by a qualifying widow(er).

Continued on next page

19-- Tax Table—*Continued*

If line 37 (Form 1040) is—		And you are—			
At least	But less than	Single	Married filing jointly *	Married filing sepa-rately	Head of a house-hold
		Your tax is—			
50,000					
50,000	**50,050**	11,219	9,353	11,883	10,270
50,050	**50,100**	11,233	9,367	11,899	10,284
50,100	**50,150**	11,247	9,381	11,914	10,298
50,150	**50,200**	11,261	9,395	11,930	10,312
50,200	**50,250**	11,275	9,409	11,945	10,326
50,250	**50,300**	11,289	9,423	11,961	10,340
50,300	**50,350**	11,303	9,437	11,976	10,354
50,350	**50,400**	11,317	9,451	11,992	10,368
50,400	**50,450**	11,331	9,465	12,007	10,382
50,450	**50,500**	11,345	9,479	12,023	10,396
50,500	**50,550**	11,359	9,493	12,038	10,410
50,550	**50,600**	11,373	9,507	12,054	10,424
50,600	**50,650**	11,387	9,521	12,069	10,438
50,650	**50,700**	11,401	9,535	12,085	10,452
50,700	**50,750**	11,415	9,549	12,100	10,466
50,750	**50,800**	11,429	9,563	12,116	10,480
50,800	**50,850**	11,443	9,577	12,131	10,494
50,850	**50,900**	11,457	9,591	12,147	10,508
50,900	**50,950**	11,471	9,605	12,162	10,522
50,950	**51,000**	11,485	9,619	12,178	10,536
51,000					
51,000	**51,050**	11,499	9,633	12,193	10,550
51,050	**51,100**	11,513	9,647	12,209	10,564
51,100	**51,150**	11,527	9,661	12,224	10,578
51,150	**51,200**	11,541	9,675	12,240	10,592
51,200	**51,250**	11,555	9,689	12,255	10,606
51,250	**51,300**	11,569	9,703	12,271	10,620
51,300	**51,350**	11,583	9,717	12,286	10,634
51,350	**51,400**	11,597	9,731	12,302	10,648
51,400	**51,450**	11,611	9,745	12,317	10,662
51,450	**51,500**	11,625	9,759	12,333	10,676
51,500	**51,550**	11,639	9,773	12,348	10,690
51,550	**51,600**	11,653	9,787	12,364	10,704
51,600	**51,650**	11,667	9,801	12,379	10,718
51,650	**51,700**	11,681	9,815	12,395	10,732
51,700	**51,750**	11,695	9,829	12,410	10,746
51,750	**51,800**	11,709	9,843	12,426	10,760
51,800	**51,850**	11,723	9,857	12,441	10,774
51,850	**51,900**	11,737	9,871	12,457	10,788
51,900	**51,950**	11,751	9,885	12,472	10,802
51,950	**52,000**	11,767	9,899	12,488	10,816
52,000					
52,000	**52,050**	11,782	9,913	12,503	10,830
52,050	**52,100**	11,798	9,927	12,519	10,844
52,100	**52,150**	11,813	9,941	12,534	10,858
52,150	**52,200**	11,829	9,955	12,550	10,872
52,200	**52,250**	11,844	9,969	12,565	10,886
52,250	**52,300**	11,860	9,983	12,581	10,900
52,300	**52,350**	11,875	9,997	12,596	10,914
52,350	**52,400**	11,891	10,011	12,612	10,928
52,400	**52,450**	11,906	10,025	12,627	10,942
52,450	**52,500**	11,922	10,039	12,643	10,956
52,500	**52,550**	11,937	10,053	12,658	10,970
52,550	**52,600**	11,953	10,067	12,674	10,984
52,600	**52,650**	11,968	10,081	12,689	10,998
52,650	**52,700**	11,984	10,095	12,705	11,012
52,700	**52,750**	11,999	10,109	12,720	11,026
52,750	**52,800**	12,015	10,123	12,736	11,040
52,800	**52,850**	12,030	10,137	12,751	11,054
52,850	**52,900**	12,046	10,151	12,767	11,068
52,900	**52,950**	12,061	10,165	12,782	11,082
52,950	**53,000**	12,077	10,179	12,798	11,096
53,000					
53,000	**53,050**	12,092	10,193	12,813	11,110
53,050	**53,100**	12,108	10,207	12,829	11,124
53,100	**53,150**	12,123	10,221	12,844	11,138
53,150	**53,200**	12,139	10,235	12,860	11,152
53,200	**53,250**	12,154	10,249	12,875	11,166
53,250	**53,300**	12,170	10,263	12,891	11,180
53,300	**53,350**	12,185	10,277	12,906	11,194
53,350	**53,400**	12,201	10,291	12,922	11,208
53,400	**53,450**	12,216	10,305	12,937	11,222
53,450	**53,500**	12,232	10,319	12,953	11,236
53,500	**53,550**	12,247	10,333	12,968	11,250
53,550	**53,600**	12,263	10,347	12,984	11,264
53,600	**53,650**	12,278	10,361	12,999	11,278
53,650	**53,700**	12,294	10,375	13,015	11,292
53,700	**53,750**	12,309	10,389	13,030	11,306
53,750	**53,800**	12,325	10,403	13,046	11,320
53,800	**53,850**	12,340	10,417	13,061	11,334
53,850	**53,900**	12,356	10,431	13,077	11,348
53,900	**53,950**	12,371	10,445	13,092	11,362
53,950	**54,000**	12,387	10,459	13,108	11,376
54,000					
54,000	**54,050**	12,402	10,473	13,123	11,390
54,050	**54,100**	12,418	10,487	13,139	11,404
54,100	**54,150**	12,433	10,501	13,154	11,418
54,150	**54,200**	12,449	10,515	13,170	11,432
54,200	**54,250**	12,464	10,529	13,185	11,446
54,250	**54,300**	12,480	10,543	13,201	11,460
54,300	**54,350**	12,495	10,557	13,216	11,474
54,350	**54,400**	12,511	10,571	13,232	11,488
54,400	**54,450**	12,526	10,585	13,247	11,502
54,450	**54,500**	12,542	10,599	13,263	11,516
54,500	**54,550**	12,557	10,613	13,278	11,530
54,550	**54,600**	12,573	10,627	13,294	11,544
54,600	**54,650**	12,588	10,641	13,309	11,558
54,650	**54,700**	12,604	10,655	13,325	11,572
54,700	**54,750**	12,619	10,669	13,340	11,586
54,750	**54,800**	12,635	10,683	13,356	11,600
54,800	**54,850**	12,650	10,697	13,371	11,614
54,850	**54,900**	12,666	10,711	13,387	11,628
54,900	**54,950**	12,681	10,725	13,402	11,642
54,950	**55,000**	12,697	10,739	13,418	11,656
55,000					
55,000	**55,050**	12,712	10,753	13,433	11,670
55,050	**55,100**	12,728	10,767	13,449	11,684
55,100	**55,150**	12,743	10,781	13,464	11,698
55,150	**55,200**	12,759	10,795	13,480	11,712
55,200	**55,250**	12,774	10,809	13,495	11,726
55,250	**55,300**	12,790	10,823	13,511	11,740
55,300	**55,350**	12,805	10,837	13,526	11,754
55,350	**55,400**	12,821	10,851	13,542	11,768
55,400	**55,450**	12,836	10,865	13,557	11,782
55,450	**55,500**	12,852	10,879	13,573	11,796
55,500	**55,550**	12,867	10,893	13,588	11,810
55,550	**55,600**	12,883	10,907	13,604	11,824
55,600	**55,650**	12,898	10,921	13,619	11,838
55,650	**55,700**	12,914	10,935	13,635	11,852
55,700	**55,750**	12,929	10,949	13,650	11,866
55,750	**55,800**	12,945	10,963	13,666	11,880
55,800	**55,850**	12,960	10,977	13,681	11,894
55,850	**55,900**	12,976	10,991	13,697	11,908
55,900	**55,950**	12,991	11,005	13,712	11,922
55,950	**56,000**	13,007	11,019	13,728	11,936
56,000					
56,000	**56,050**	13,022	11,033	13,743	11,950
56,050	**56,100**	13,038	11,047	13,759	11,964
56,100	**56,150**	13,053	11,061	13,774	11,978
56,150	**56,200**	13,069	11,075	13,790	11,992
56,200	**56,250**	13,084	11,089	13,805	12,006
56,250	**56,300**	13,100	11,103	13,821	12,020
56,300	**56,350**	13,115	11,117	13,836	12,034
56,350	**56,400**	13,131	11,131	13,852	12,048
56,400	**56,450**	13,146	11,145	13,867	12,062
56,450	**56,500**	13,162	11,159	13,883	12,076
56,500	**56,550**	13,177	11,173	13,898	12,090
56,550	**56,600**	13,193	11,187	13,914	12,104
56,600	**56,650**	13,208	11,201	13,929	12,118
56,650	**56,700**	13,224	11,215	13,945	12,132
56,700	**56,750**	13,239	11,229	13,960	12,146
56,750	**56,800**	13,255	11,243	13,976	12,160
56,800	**56,850**	13,270	11,257	13,991	12,174
56,850	**56,900**	13,286	11,271	14,007	12,188
56,900	**56,950**	13,301	11,285	14,022	12,202
56,950	**57,000**	13,317	11,299	14,038	12,216
57,000					
57,000	**57,050**	13,332	11,313	14,053	12,230
57,050	**57,100**	13,348	11,327	14,069	12,244
57,100	**57,150**	13,363	11,341	14,084	12,258
57,150	**57,200**	13,379	11,355	14,100	12,272
57,200	**57,250**	13,394	11,369	14,115	12,286
57,250	**57,300**	13,410	11,383	14,131	12,300
57,300	**57,350**	13,425	11,397	14,146	12,314
57,350	**57,400**	13,441	11,411	14,162	12,328
57,400	**57,450**	13,456	11,425	14,177	12,342
57,450	**57,500**	13,472	11,439	14,193	12,356
57,500	**57,550**	13,487	11,453	14,208	12,370
57,550	**57,600**	13,503	11,467	14,224	12,384
57,600	**57,650**	13,518	11,481	14,239	12,398
57,650	**57,700**	13,534	11,495	14,255	12,412
57,700	**57,750**	13,549	11,509	14,270	12,426
57,750	**57,800**	13,565	11,523	14,286	12,440
57,800	**57,850**	13,580	11,537	14,301	12,454
57,850	**57,900**	13,596	11,551	14,317	12,468
57,900	**57,950**	13,611	11,565	14,332	12,482
57,950	**58,000**	13,627	11,579	14,348	12,496
58,000					
58,000	**58,050**	13,642	11,593	14,363	12,510
58,050	**58,100**	13,658	11,607	14,379	12,524
58,100	**58,150**	13,673	11,621	14,394	12,538
58,150	**58,200**	13,689	11,635	14,410	12,552
58,200	**58,250**	13,704	11,649	14,425	12,566
58,250	**58,300**	13,720	11,663	14,441	12,580
58,300	**58,350**	13,735	11,677	14,456	12,594
58,350	**58,400**	13,751	11,691	14,472	12,608
58,400	**58,450**	13,766	11,705	14,487	12,622
58,450	**58,500**	13,782	11,719	14,503	12,636
58,500	**58,550**	13,797	11,733	14,518	12,650
58,550	**58,600**	13,813	11,747	14,534	12,664
58,600	**58,650**	13,828	11,761	14,549	12,678
58,650	**58,700**	13,844	11,775	14,565	12,692
58,700	**58,750**	13,859	11,789	14,580	12,706
58,750	**58,800**	13,875	11,803	14,596	12,720
58,800	**58,850**	13,890	11,817	14,611	12,734
58,850	**58,900**	13,906	11,831	14,627	12,748
58,900	**58,950**	13,921	11,845	14,642	12,762
58,950	**59,000**	13,937	11,859	14,658	12,776

* This column must also be used by a qualifying widow(er).

Continued on next page

19-- Tax Table—*Continued*

If line 37 (Form 1040) is— At least	But less than	And you are— Single	Married filing jointly *	Married filing separately	Head of a household
		Your tax is—			
59,000					
59,000	59,050	13,952	11,873	14,673	12,790
59,050	59,100	13,968	11,887	14,689	12,804
59,100	59,150	13,983	11,901	14,704	12,818
59,150	59,200	13,999	11,915	14,720	12,832
59,200	59,250	14,014	11,929	14,735	12,846
59,250	59,300	14,030	11,943	14,751	12,860
59,300	59,350	14,045	11,957	14,766	12,874
59,350	59,400	14,061	11,971	14,782	12,888
59,400	59,450	14,076	11,985	14,797	12,902
59,450	59,500	14,092	11,999	14,813	12,916
59,500	59,550	14,107	12,013	14,828	12,930
59,550	59,600	14,123	12,027	14,844	12,944
59,600	59,650	14,138	12,041	14,859	12,958
59,650	59,700	14,154	12,055	14,875	12,972
59,700	59,750	14,169	12,069	14,890	12,986
59,750	59,800	14,185	12,083	14,906	13,000
59,800	59,850	14,200	12,097	14,921	13,014
59,850	59,900	14,216	12,111	14,937	13,028
59,900	59,950	14,231	12,125	14,952	13,042
59,950	60,000	14,247	12,139	14,968	13,056
60,000					
60,000	60,050	14,262	12,153	14,983	13,070
60,050	60,100	14,278	12,167	14,999	13,084
60,100	60,150	14,293	12,181	15,014	13,098
60,150	60,200	14,309	12,195	15,030	13,112
60,200	60,250	14,324	12,209	15,045	13,126
60,250	60,300	14,340	12,223	15,061	13,140
60,300	60,350	14,355	12,237	15,076	13,154
60,350	60,400	14,371	12,251	15,092	13,168
60,400	60,450	14,386	12,265	15,107	13,182
60,450	60,500	14,402	12,279	15,123	13,196
60,500	60,550	14,417	12,293	15,138	13,210
60,550	60,600	14,433	12,307	15,154	13,224
60,600	60,650	14,448	12,321	15,169	13,238
60,650	60,700	14,464	12,335	15,185	13,252
60,700	60,750	14,479	12,349	15,200	13,266
60,750	60,800	14,495	12,363	15,216	13,280
60,800	60,850	14,510	12,377	15,231	13,294
60,850	60,900	14,526	12,391	15,247	13,308
60,900	60,950	14,541	12,405	15,262	13,322
60,950	61,000	14,557	12,419	15,278	13,336
61,000					
61,000	61,050	14,572	12,433	15,293	13,350
61,050	61,100	14,588	12,447	15,309	13,364
61,100	61,150	14,603	12,461	15,324	13,378
61,150	61,200	14,619	12,475	15,340	13,392
61,200	61,250	14,634	12,489	15,355	13,406
61,250	61,300	14,650	12,503	15,371	13,420
61,300	61,350	14,665	12,517	15,386	13,434
61,350	61,400	14,681	12,531	15,402	13,448
61,400	61,450	14,696	12,545	15,417	13,462
61,450	61,500	14,712	12,559	15,433	13,476
61,500	61,550	14,727	12,573	15,448	13,490
61,550	61,600	14,743	12,587	15,464	13,504
61,600	61,650	14,758	12,601	15,479	13,518
61,650	61,700	14,774	12,615	15,495	13,532
61,700	61,750	14,789	12,629	15,510	13,546
61,750	61,800	14,805	12,643	15,526	13,560
61,800	61,850	14,820	12,657	15,541	13,574
61,850	61,900	14,836	12,671	15,557	13,588
61,900	61,950	14,851	12,685	15,572	13,602
61,950	62,000	14,867	12,699	15,588	13,616
62,000					
62,000	62,050	14,882	12,713	15,603	13,630
62,050	62,100	14,898	12,727	15,619	13,644
62,100	62,150	14,913	12,741	15,634	13,658
62,150	62,200	14,929	12,755	15,650	13,672
62,200	62,250	14,944	12,769	15,665	13,686
62,250	62,300	14,960	12,783	15,681	13,700
62,300	62,350	14,975	12,797	15,696	13,714
62,350	62,400	14,991	12,811	15,712	13,728
62,400	62,450	15,006	12,825	15,727	13,742
62,450	62,500	15,022	12,839	15,743	13,756
62,500	62,550	15,037	12,853	15,758	13,770
62,550	62,600	15,053	12,867	15,774	13,784
62,600	62,650	15,068	12,881	15,789	13,798
62,650	62,700	15,084	12,895	15,805	13,812
62,700	62,750	15,099	12,909	15,820	13,826
62,750	62,800	15,115	12,923	15,836	13,840
62,800	62,850	15,130	12,937	15,851	13,854
62,850	62,900	15,146	12,951	15,867	13,868
62,900	62,950	15,161	12,965	15,882	13,882
62,950	63,000	15,177	12,979	15,898	13,896
63,000					
63,000	63,050	15,192	12,993	15,913	13,910
63,050	63,100	15,208	13,007	15,929	13,924
63,100	63,150	15,223	13,021	15,944	13,938
63,150	63,200	15,239	13,035	15,960	13,952
63,200	63,250	15,254	13,049	15,975	13,966
63,250	63,300	15,270	13,063	15,991	13,980
63,300	63,350	15,285	13,077	16,006	13,994
63,350	63,400	15,301	13,091	16,022	14,008
63,400	63,450	15,316	13,105	16,037	14,022
63,450	63,500	15,332	13,119	16,053	14,036
63,500	63,550	15,347	13,133	16,068	14,050
63,550	63,600	15,363	13,147	16,084	14,064
63,600	63,650	15,378	13,161	16,099	14,078
63,650	63,700	15,394	13,175	16,115	14,092
63,700	63,750	15,409	13,189	16,130	14,106
63,750	63,800	15,425	13,203	16,146	14,120
63,800	63,850	15,440	13,217	16,161	14,134
63,850	63,900	15,456	13,231	16,177	14,148
63,900	63,950	15,471	13,245	16,192	14,162
63,950	64,000	15,487	13,259	16,208	14,176
64,000					
64,000	64,050	15,502	13,273	16,223	14,190
64,050	64,100	15,518	13,287	16,239	14,204
64,100	64,150	15,533	13,301	16,254	14,218
64,150	64,200	15,549	13,315	16,270	14,232
64,200	64,250	15,564	13,329	16,285	14,246
64,250	64,300	15,580	13,343	16,301	14,260
64,300	64,350	15,595	13,357	16,316	14,274
64,350	64,400	15,611	13,371	16,332	14,288
64,400	64,450	15,626	13,385	16,347	14,302
64,450	64,500	15,642	13,399	16,363	14,316
64,500	64,550	15,657	13,413	16,378	14,330
64,550	64,600	15,673	13,427	16,394	14,344
64,600	64,650	15,688	13,441	16,409	14,358
64,650	64,700	15,704	13,455	16,425	14,372
64,700	64,750	15,719	13,469	16,440	14,386
64,750	64,800	15,735	13,483	16,456	14,400
64,800	64,850	15,750	13,497	16,471	14,414
64,850	64,900	15,766	13,511	16,487	14,428
64,900	64,950	15,781	13,525	16,502	14,442
64,950	65,000	15,797	13,539	16,518	14,456
65,000					
65,000	65,050	15,812	13,553	16,533	14,470
65,050	65,100	15,828	13,567	16,549	14,484
65,100	65,150	15,843	13,581	16,564	14,498
65,150	65,200	15,859	13,595	16,580	14,512
65,200	65,250	15,874	13,609	16,595	14,526
65,250	65,300	15,890	13,623	16,611	14,540
65,300	65,350	15,905	13,637	16,626	14,554
65,350	65,400	15,921	13,651	16,642	14,568
65,400	65,450	15,936	13,665	16,657	14,582
65,450	65,500	15,952	13,679	16,673	14,596
65,500	65,550	15,967	13,693	16,688	14,610
65,550	65,600	15,983	13,707	16,704	14,624
65,600	65,650	15,998	13,721	16,719	14,638
65,650	65,700	16,014	13,735	16,735	14,652
65,700	65,750	16,029	13,749	16,750	14,666
65,750	65,800	16,045	13,763	16,766	14,680
65,800	65,850	16,060	13,777	16,781	14,694
65,850	65,900	16,076	13,791	16,797	14,708
65,900	65,950	16,091	13,805	16,812	14,722
65,950	66,000	16,107	13,819	16,828	14,736
66,000					
66,000	66,050	16,122	13,833	16,843	14,750
66,050	66,100	16,138	13,847	16,859	14,764
66,100	66,150	16,153	13,861	16,874	14,778
66,150	66,200	16,169	13,875	16,890	14,792
66,200	66,250	16,184	13,889	16,905	14,806
66,250	66,300	16,200	13,903	16,921	14,820
66,300	66,350	16,215	13,917	16,936	14,834
66,350	66,400	16,231	13,931	16,952	14,848
66,400	66,450	16,246	13,945	16,967	14,862
66,450	66,500	16,262	13,959	16,983	14,876
66,500	66,550	16,277	13,973	16,998	14,890
66,550	66,600	16,293	13,987	17,014	14,904
66,600	66,650	16,308	14,001	17,029	14,918
66,650	66,700	16,324	14,015	17,045	14,932
66,700	66,750	16,339	14,029	17,060	14,946
66,750	66,800	16,355	14,043	17,076	14,960
66,800	66,850	16,370	14,057	17,091	14,974
66,850	66,900	16,386	14,071	17,107	14,988
66,900	66,950	16,401	14,085	17,122	15,002
66,950	67,000	16,417	14,099	17,138	15,016
67,000					
67,000	67,050	16,432	14,113	17,153	15,030
67,050	67,100	16,448	14,127	17,169	15,044
67,100	67,150	16,463	14,141	17,184	15,058
67,150	67,200	16,479	14,155	17,200	15,072
67,200	67,250	16,494	14,169	17,215	15,086
67,250	67,300	16,510	14,183	17,231	15,100
67,300	67,350	16,525	14,197	17,246	15,114
67,350	67,400	16,541	14,211	17,262	15,128
67,400	67,450	16,556	14,225	17,277	15,142
67,450	67,500	16,572	14,239	17,293	15,156
67,500	67,550	16,587	14,253	17,308	15,170
67,550	67,600	16,603	14,267	17,324	15,184
67,600	67,650	16,618	14,281	17,339	15,198
67,650	67,700	16,634	14,295	17,355	15,212
67,700	67,750	16,649	14,309	17,370	15,226
67,750	67,800	16,665	14,323	17,386	15,240
67,800	67,850	16,680	14,337	17,401	15,254
67,850	67,900	16,696	14,351	17,417	15,268
67,900	67,950	16,711	14,365	17,432	15,282
67,950	68,000	16,727	14,379	17,448	15,296

* This column must also be used by a qualifying widow(er).

Continued on next page

19-- Tax Table–*Continued*

If line 37 (Form 1040) is– At least	But less than	And you are — Single	Married filing jointly *	Married filing separately	Head of a household
		Your tax is —			
68,000					
68,000	**68,050**	16,742	14,393	17,463	15,310
68,050	**68,100**	16,758	14,407	17,479	15,324
68,100	**68,150**	16,773	14,421	17,494	15,338
68,150	**68,200**	16,789	14,435	17,510	15,352
68,200	**68,250**	16,804	14,449	17,525	15,366
68,250	**68,300**	16,820	14,463	17,541	15,380
68,300	**68,350**	16,835	14,477	17,556	15,394
68,350	**68,400**	16,851	14,491	17,572	15,408
68,400	**68,450**	16,866	14,505	17,587	15,422
68,450	**68,500**	16,882	14,519	17,603	15,436
68,500	**68,550**	16,897	14,533	17,618	15,450
68,550	**68,600**	16,913	14,547	17,634	15,464
68,600	**68,650**	16,928	14,561	17,649	15,478
68,650	**68,700**	16,944	14,575	17,665	15,492
68,700	**68,750**	16,959	14,589	17,680	15,506
68,750	**68,800**	16,975	14,603	17,696	15,520
68,800	**68,850**	16,990	14,617	17,711	15,534
68,850	**68,900**	17,006	14,631	17,727	15,548
68,900	**68,950**	17,021	14,645	17,742	15,562
68,950	**69,000**	17,037	14,659	17,758	15,576
69,000					
69,000	**69,050**	17,052	14,673	17,773	15,590
69,050	**69,100**	17,068	14,687	17,789	15,604
69,100	**69,150**	17,083	14,701	17,804	15,618
69,150	**69,200**	17,099	14,715	17,820	15,632
69,200	**69,250**	17,114	14,729	17,835	15,646
69,250	**69,300**	17,130	14,743	17,851	15,660
69,300	**69,350**	17,145	14,757	17,866	15,674
69,350	**69,400**	17,161	14,771	17,882	15,688
69,400	**69,450**	17,176	14,785	17,897	15,702
69,450	**69,500**	17,192	14,799	17,913	15,716
69,500	**69,550**	17,207	14,813	17,928	15,730
69,550	**69,600**	17,223	14,827	17,944	15,744
69,600	**69,650**	17,238	14,841	17,959	15,758
69,650	**69,700**	17,254	14,855	17,975	15,772
69,700	**69,750**	17,269	14,869	17,990	15,786
69,750	**69,800**	17,285	14,883	18,006	15,800
69,800	**69,850**	17,300	14,897	18,021	15,814
69,850	**69,900**	17,316	14,911	18,037	15,828
69,900	**69,950**	17,331	14,925	18,052	15,842
69,950	**70,000**	17,347	14,939	18,068	15,856
70,000					
70,000	**70,050**	17,362	14,953	18,083	15,870
70,050	**70,100**	17,378	14,967	18,099	15,884
70,100	**70,150**	17,393	14,981	18,114	15,898
70,150	**70,200**	17,409	14,995	18,130	15,912
70,200	**70,250**	17,424	15,009	18,145	15,926
70,250	**70,300**	17,440	15,023	18,161	15,940
70,300	**70,350**	17,455	15,037	18,176	15,954
70,350	**70,400**	17,471	15,051	18,192	15,968
70,400	**70,450**	17,486	15,065	18,207	15,982
70,450	**70,500**	17,502	15,079	18,223	15,996
70,500	**70,550**	17,517	15,093	18,238	16,010
70,550	**70,600**	17,533	15,107	18,254	16,024
70,600	**70,650**	17,548	15,121	18,269	16,038
70,650	**70,700**	17,564	15,135	18,285	16,052
70,700	**70,750**	17,579	15,149	18,300	16,066
70,750	**70,800**	17,595	15,163	18,316	16,080
70,800	**70,850**	17,610	15,177	18,331	16,094
70,850	**70,900**	17,626	15,191	18,347	16,108
70,900	**70,950**	17,641	15,205	18,362	16,122
70,950	**71,000**	17,657	15,219	18,378	16,136

If line 37 (Form 1040) is– At least	But less than	And you are — Single	Married filing jointly *	Married filing separately	Head of a household
		Your tax is —			
71,000					
71,000	**71,050**	17,672	15,233	18,393	16,150
71,050	**71,100**	17,688	15,247	18,409	16,164
71,100	**71,150**	17,703	15,261	18,424	16,178
71,150	**71,200**	17,719	15,275	18,440	16,192
71,200	**71,250**	17,734	15,289	18,455	16,206
71,250	**71,300**	17,750	15,303	18,471	16,220
71,300	**71,350**	17,765	15,317	18,486	16,234
71,350	**71,400**	17,781	15,331	18,502	16,248
71,400	**71,450**	17,796	15,345	18,517	16,262
71,450	**71,500**	17,812	15,359	18,533	16,276
71,500	**71,550**	17,827	15,373	18,548	16,290
71,550	**71,600**	17,843	15,387	18,564	16,304
71,600	**71,650**	17,858	15,401	18,579	16,318
71,650	**71,700**	17,874	15,415	18,595	16,332
71,700	**71,750**	17,889	15,429	18,610	16,346
71,750	**71,800**	17,905	15,443	18,626	16,360
71,800	**71,850**	17,920	15,457	18,641	16,374
71,850	**71,900**	17,936	15,471	18,657	16,388
71,900	**71,950**	17,951	15,485	18,672	16,402
71,950	**72,000**	17,967	15,499	18,688	16,416
72,000					
72,000	**72,050**	17,982	15,513	18,703	16,430
72,050	**72,100**	17,998	15,527	18,719	16,444
72,100	**72,150**	18,013	15,541	18,734	16,458
72,150	**72,200**	18,029	15,555	18,750	16,472
72,200	**72,250**	18,044	15,569	18,765	16,486
72,250	**72,300**	18,060	15,583	18,781	16,500
72,300	**72,350**	18,075	15,597	18,796	16,514
72,350	**72,400**	18,091	15,611	18,812	16,528
72,400	**72,450**	18,106	15,625	18,827	16,542
72,450	**72,500**	18,122	15,639	18,843	16,556
72,500	**72,550**	18,137	15,653	18,858	16,570
72,550	**72,600**	18,153	15,667	18,874	16,584
72,600	**72,650**	18,168	15,681	18,889	16,598
72,650	**72,700**	18,184	15,695	18,905	16,612
72,700	**72,750**	18,199	15,709	18,920	16,626
72,750	**72,800**	18,215	15,723	18,936	16,640
72,800	**72,850**	18,230	15,737	18,951	16,654
72,850	**72,900**	18,246	15,751	18,967	16,668
72,900	**72,950**	18,261	15,765	18,982	16,682
72,950	**73,000**	18,277	15,779	18,998	16,696
73,000					
73,000	**73,050**	18,292	15,793	19,013	16,710
73,050	**73,100**	18,308	15,807	19,029	16,724
73,100	**73,150**	18,323	15,821	19,044	16,738
73,150	**73,200**	18,339	15,835	19,060	16,752
73,200	**73,250**	18,354	15,849	19,075	16,766
73,250	**73,300**	18,370	15,863	19,091	16,780
73,300	**73,350**	18,385	15,877	19,106	16,794
73,350	**73,400**	18,401	15,891	19,122	16,808
73,400	**73,450**	18,416	15,905	19,137	16,822
73,450	**73,500**	18,432	15,919	19,153	16,836
73,500	**73,550**	18,447	15,933	19,168	16,850
73,550	**73,600**	18,463	15,947	19,184	16,864
73,600	**73,650**	18,478	15,961	19,199	16,878
73,650	**73,700**	18,494	15,975	19,215	16,892
73,700	**73,750**	18,509	15,989	19,230	16,906
73,750	**73,800**	18,525	16,003	19,246	16,920
73,800	**73,850**	18,540	16,017	19,261	16,934
73,850	**73,900**	18,556	16,031	19,277	16,948
73,900	**73,950**	18,571	16,045	19,292	16,962
73,950	**74,000**	18,587	16,059	19,308	16,976

If line 37 (Form 1040) is– At least	But less than	And you are — Single	Married filing jointly *	Married filing separately	Head of a household
		Your tax is —			
74,000					
74,000	**74,050**	18,602	16,073	19,323	16,990
74,050	**74,100**	18,618	16,087	19,339	17,004
74,100	**74,150**	18,633	16,101	19,354	17,018
74,150	**74,200**	18,649	16,115	19,370	17,032
74,200	**74,250**	18,664	16,129	19,385	17,048
74,250	**74,300**	18,680	16,143	19,401	17,063
74,300	**74,350**	18,695	16,157	19,416	17,079
74,350	**74,400**	18,711	16,171	19,432	17,094
74,400	**74,450**	18,726	16,185	19,447	17,110
74,450	**74,500**	18,742	16,199	19,463	17,125
74,500	**74,550**	18,757	16,213	19,478	17,141
74,550	**74,600**	18,773	16,227	19,494	17,156
74,600	**74,650**	18,788	16,241	19,509	17,172
74,650	**74,700**	18,804	16,255	19,525	17,187
74,700	**74,750**	18,819	16,269	19,540	17,203
74,750	**74,800**	18,835	16,283	19,556	17,218
74,800	**74,850**	18,850	16,297	19,571	17,234
74,850	**74,900**	18,866	16,311	19,587	17,249
74,900	**74,950**	18,881	16,325	19,602	17,265
74,950	**75,000**	18,897	16,339	19,618	17,280
75,000					
75,000	**75,050**	18,912	16,353	19,633	17,296
75,050	**75,100**	18,928	16,367	19,649	17,311
75,100	**75,150**	18,943	16,381	19,664	17,327
75,150	**75,200**	18,959	16,395	19,680	17,342
75,200	**75,250**	18,974	16,409	19,695	17,358
75,250	**75,300**	18,990	16,423	19,711	17,373
75,300	**75,350**	19,005	16,437	19,726	17,389
75,350	**75,400**	19,021	16,451	19,742	17,404
75,400	**75,450**	19,036	16,465	19,757	17,420
75,450	**75,500**	19,052	16,479	19,773	17,435
75,500	**75,550**	19,067	16,493	19,788	17,451
75,550	**75,600**	19,083	16,507	19,804	17,466
75,600	**75,650**	19,098	16,521	19,819	17,482
75,650	**75,700**	19,114	16,535	19,835	17,497
75,700	**75,750**	19,129	16,549	19,850	17,513
75,750	**75,800**	19,145	16,563	19,866	17,528
75,800	**75,850**	19,160	16,577	19,881	17,544
75,850	**75,900**	19,176	16,591	19,897	17,559
75,900	**75,950**	19,191	16,605	19,912	17,575
75,950	**76,000**	19,207	16,619	19,928	17,590
76,000					
76,000	**76,050**	19,222	16,633	19,943	17,606
76,050	**76,100**	19,238	16,647	19,959	17,621
76,100	**76,150**	19,253	16,661	19,974	17,637
76,150	**76,200**	19,269	16,675	19,990	17,652
76,200	**76,250**	19,284	16,689	20,005	17,668
76,250	**76,300**	19,300	16,703	20,021	17,683
76,300	**76,350**	19,315	16,717	20,036	17,699
76,350	**76,400**	19,331	16,731	20,052	17,714
76,400	**76,450**	19,346	16,745	20,067	17,730
76,450	**76,500**	19,362	16,759	20,083	17,745
76,500	**76,550**	19,377	16,773	20,098	17,761
76,550	**76,600**	19,393	16,787	20,114	17,776
76,600	**76,650**	19,408	16,801	20,129	17,792
76,650	**76,700**	19,424	16,815	20,145	17,807
76,700	**76,750**	19,439	16,829	20,160	17,823
76,750	**76,800**	19,455	16,843	20,176	17,838
76,800	**76,850**	19,470	16,857	20,191	17,854
76,850	**76,900**	19,486	16,871	20,207	17,869
76,900	**76,950**	19,501	16,885	20,222	17,885
76,950	**77,000**	19,517	16,899	20,238	17,900

* This column must also be used by a qualifying widow(er).

Continued on next page

19-- Tax Table—*Continued*

If line 37 (Form 1040) is— At least	But less than	And you are— Single	Married filing jointly *	Married filing sepa-rately	Head of a house-hold
		Your tax is—			
77,000					
77,000	**77,050**	19,532	16,913	20,253	17,916
77,050	**77,100**	19,548	16,927	20,269	17,931
77,100	**77,150**	19,563	16,941	20,284	17,947
77,150	**77,200**	19,579	16,955	20,300	17,962
77,200	**77,250**	19,594	16,969	20,315	17,978
77,250	**77,300**	19,610	16,983	20,331	17,993
77,300	**77,350**	19,625	16,997	20,346	18,009
77,350	**77,400**	19,641	17,011	20,362	18,024
77,400	**77,450**	19,656	17,025	20,377	18,040
77,450	**77,500**	19,672	17,039	20,393	18,055
77,500	**77,550**	19,687	17,053	20,408	18,071
77,550	**77,600**	19,703	17,067	20,424	18,086
77,600	**77,650**	19,718	17,081	20,439	18,102
77,650	**77,700**	19,734	17,095	20,455	18,117
77,700	**77,750**	19,749	17,109	20,470	18,133
77,750	**77,800**	19,765	17,123	20,486	18,148
77,800	**77,850**	19,780	17,137	20,501	18,164
77,850	**77,900**	19,796	17,151	20,517	18,179
77,900	**77,950**	19,811	17,165	20,532	18,195
77,950	**78,000**	19,827	17,179	20,548	18,210
78,000					
78,000	**78,050**	19,842	17,193	20,563	18,226
78,050	**78,100**	19,858	17,207	20,579	18,241
78,100	**78,150**	19,873	17,221	20,594	18,257
78,150	**78,200**	19,889	17,235	20,610	18,272
78,200	**78,250**	19,904	17,249	20,625	18,288
78,250	**78,300**	19,920	17,263	20,641	18,303
78,300	**78,350**	19,935	17,277	20,656	18,319
78,350	**78,400**	19,951	17,291	20,672	18,334
78,400	**78,450**	19,966	17,305	20,687	18,350
78,450	**78,500**	19,982	17,319	20,703	18,365
78,500	**78,550**	19,997	17,333	20,718	18,381
78,550	**78,600**	20,013	17,347	20,734	18,396
78,600	**78,650**	20,028	17,361	20,749	18,412
78,650	**78,700**	20,044	17,375	20,765	18,427
78,700	**78,750**	20,059	17,389	20,780	18,443
78,750	**78,800**	20,075	17,403	20,796	18,458
78,800	**78,850**	20,090	17,417	20,811	18,474
78,850	**78,900**	20,106	17,431	20,827	18,489
78,900	**78,950**	20,121	17,445	20,842	18,505
78,950	**79,000**	20,137	17,459	20,858	18,520
79,000					
79,000	**79,050**	20,152	17,473	20,873	18,536
79,050	**79,100**	20,168	17,487	20,889	18,551
79,100	**79,150**	20,183	17,501	20,904	18,567
79,150	**79,200**	20,199	17,515	20,920	18,582
79,200	**79,250**	20,214	17,529	20,935	18,598
79,250	**79,300**	20,230	17,543	20,951	18,613
79,300	**79,350**	20,245	17,557	20,966	18,629
79,350	**79,400**	20,261	17,571	20,982	18,644
79,400	**79,450**	20,276	17,585	20,997	18,660
79,450	**79,500**	20,292	17,599	21,013	18,675
79,500	**79,550**	20,307	17,613	21,028	18,691
79,550	**79,600**	20,323	17,627	21,044	18,706
79,600	**79,650**	20,338	17,641	21,059	18,722
79,650	**79,700**	20,354	17,655	21,075	18,737
79,700	**79,750**	20,369	17,669	21,090	18,753
79,750	**79,800**	20,385	17,683	21,106	18,768
79,800	**79,850**	20,400	17,697	21,121	18,784
79,850	**79,900**	20,416	17,711	21,137	18,799
79,900	**79,950**	20,431	17,725	21,152	18,815
79,950	**80,000**	20,447	17,739	21,168	18,830

If line 37 (Form 1040) is— At least	But less than	And you are— Single	Married filing jointly *	Married filing sepa-rately	Head of a house-hold
		Your tax is—			
80,000					
80,000	**80,050**	20,462	17,753	21,183	18,846
80,050	**80,100**	20,478	17,767	21,199	18,861
80,100	**80,150**	20,493	17,781	21,214	18,877
80,150	**80,200**	20,509	17,795	21,230	18,892
80,200	**80,250**	20,524	17,809	21,245	18,908
80,250	**80,300**	20,540	17,823	21,261	18,923
80,300	**80,350**	20,555	17,837	21,276	18,939
80,350	**80,400**	20,571	17,851	21,292	18,954
80,400	**80,450**	20,586	17,865	21,307	18,970
80,450	**80,500**	20,602	17,879	21,323	18,985
80,500	**80,550**	20,617	17,893	21,338	19,001
80,550	**80,600**	20,633	17,907	21,354	19,016
80,600	**80,650**	20,648	17,921	21,369	19,032
80,650	**80,700**	20,664	17,935	21,385	19,047
80,700	**80,750**	20,679	17,949	21,400	19,063
80,750	**80,800**	20,695	17,963	21,416	19,078
80,800	**80,850**	20,710	17,977	21,431	19,094
80,850	**80,900**	20,726	17,991	21,447	19,109
80,900	**80,950**	20,741	18,005	21,462	19,125
80,950	**81,000**	20,757	18,019	21,478	19,140
81,000					
81,000	**81,050**	20,772	18,033	21,493	19,156
81,050	**81,100**	20,788	18,047	21,509	19,171
81,100	**81,150**	20,803	18,061	21,524	19,187
81,150	**81,200**	20,819	18,075	21,540	19,202
81,200	**81,250**	20,834	18,089	21,555	19,218
81,250	**81,300**	20,850	18,103	21,571	19,233
81,300	**81,350**	20,865	18,117	21,586	19,249
81,350	**81,400**	20,881	18,131	21,602	19,264
81,400	**81,450**	20,896	18,145	21,617	19,280
81,450	**81,500**	20,912	18,159	21,633	19,295
81,500	**81,550**	20,927	18,173	21,648	19,311
81,550	**81,600**	20,943	18,187	21,664	19,326
81,600	**81,650**	20,958	18,201	21,679	19,342
81,650	**81,700**	20,974	18,215	21,695	19,357
81,700	**81,750**	20,989	18,229	21,710	19,373
81,750	**81,800**	21,005	18,243	21,726	19,388
81,800	**81,850**	21,020	18,257	21,741	19,404
81,850	**81,900**	21,036	18,271	21,757	19,419
81,900	**81,950**	21,051	18,285	21,772	19,435
81,950	**82,000**	21,067	18,299	21,788	19,450
82,000					
82,000	**82,050**	21,082	18,313	21,803	19,466
82,050	**82,100**	21,098	18,327	21,819	19,481
82,100	**82,150**	21,113	18,341	21,834	19,497
82,150	**82,200**	21,129	18,355	21,850	19,512
82,200	**82,250**	21,144	18,369	21,865	19,528
82,250	**82,300**	21,160	18,383	21,881	19,543
82,300	**82,350**	21,175	18,397	21,896	19,559
82,350	**82,400**	21,191	18,411	21,912	19,574
82,400	**82,450**	21,206	18,425	21,927	19,590
82,450	**82,500**	21,222	18,439	21,943	19,605
82,500	**82,550**	21,237	18,453	21,958	19,621
82,550	**82,600**	21,253	18,467	21,974	19,636
82,600	**82,650**	21,268	18,481	21,989	19,652
82,650	**82,700**	21,284	18,495	22,005	19,667
82,700	**82,750**	21,299	18,509	22,020	19,683
82,750	**82,800**	21,315	18,523	22,036	19,698
82,800	**82,850**	21,330	18,537	22,051	19,714
82,850	**82,900**	21,346	18,551	22,067	19,729
82,900	**82,950**	21,361	18,565	22,082	19,745
82,950	**83,000**	21,377	18,579	22,098	19,760

If line 37 (Form 1040) is— At least	But less than	And you are— Single	Married filing jointly *	Married filing sepa-rately	Head of a house-hold
		Your tax is—			
83,000					
83,000	**83,050**	21,392	18,593	22,113	19,776
83,050	**83,100**	21,408	18,607	22,129	19,791
83,100	**83,150**	21,423	18,621	22,144	19,807
83,150	**83,200**	21,439	18,635	22,160	19,822
83,200	**83,250**	21,454	18,649	22,175	19,838
83,250	**83,300**	21,470	18,663	22,191	19,853
83,300	**83,350**	21,485	18,677	22,206	19,869
83,350	**83,400**	21,501	18,691	22,222	19,884
83,400	**83,450**	21,516	18,705	22,237	19,900
83,450	**83,500**	21,532	18,719	22,253	19,915
83,500	**83,550**	21,547	18,733	22,268	19,931
83,550	**83,600**	21,563	18,747	22,284	19,946
83,600	**83,650**	21,578	18,761	22,299	19,962
83,650	**83,700**	21,594	18,775	22,315	19,977
83,700	**83,750**	21,609	18,789	22,330	19,993
83,750	**83,800**	21,625	18,803	22,346	20,008
83,800	**83,850**	21,640	18,817	22,361	20,024
83,850	**83,900**	21,656	18,831	22,377	20,039
83,900	**83,950**	21,671	18,845	22,392	20,055
83,950	**84,000**	21,687	18,859	22,408	20,070
84,000					
84,000	**84,050**	21,702	18,873	22,423	20,086
84,050	**84,100**	21,718	18,887	22,439	20,101
84,100	**84,150**	21,733	18,901	22,454	20,117
84,150	**84,200**	21,749	18,915	22,470	20,132
84,200	**84,250**	21,764	18,929	22,485	20,148
84,250	**84,300**	21,780	18,943	22,501	20,163
84,300	**84,350**	21,795	18,957	22,516	20,179
84,350	**84,400**	21,811	18,971	22,532	20,194
84,400	**84,450**	21,826	18,985	22,547	20,210
84,450	**84,500**	21,842	18,999	22,563	20,225
84,500	**84,550**	21,857	19,013	22,578	20,241
84,550	**84,600**	21,873	19,027	22,594	20,256
84,600	**84,650**	21,888	19,041	22,609	20,272
84,650	**84,700**	21,904	19,055	22,625	20,287
84,700	**84,750**	21,919	19,069	22,640	20,303
84,750	**84,800**	21,935	19,083	22,656	20,318
84,800	**84,850**	21,950	19,097	22,671	20,334
84,850	**84,900**	21,966	19,111	22,687	20,349
84,900	**84,950**	21,981	19,125	22,702	20,365
84,950	**85,000**	21,997	19,139	22,718	20,380
85,000					
85,000	**85,050**	22,012	19,153	22,733	20,396
85,050	**85,100**	22,028	19,167	22,749	20,411
85,100	**85,150**	22,043	19,181	22,764	20,427
85,150	**85,200**	22,059	19,195	22,780	20,442
85,200	**85,250**	22,074	19,209	22,795	20,458
85,250	**85,300**	22,090	19,223	22,811	20,473
85,300	**85,350**	22,105	19,237	22,826	20,489
85,350	**85,400**	22,121	19,251	22,842	20,504
85,400	**85,450**	22,136	19,265	22,857	20,520
85,450	**85,500**	22,152	19,279	22,873	20,535
85,500	**85,550**	22,167	19,293	22,888	20,551
85,550	**85,600**	22,183	19,307	22,904	20,566
85,600	**85,650**	22,198	19,321	22,919	20,582
85,650	**85,700**	22,214	19,335	22,935	20,597
85,700	**85,750**	22,229	19,349	22,950	20,613
85,750	**85,800**	22,245	19,363	22,966	20,628
85,800	**85,850**	22,260	19,377	22,981	20,644
85,850	**85,900**	22,276	19,391	22,997	20,659
85,900	**85,950**	22,291	19,405	23,012	20,675
85,950	**86,000**	22,307	19,419	23,028	20,690

* This column must also be used by a qualifying widow(er).

Continued on next page

19-- Tax Table—*Continued*

If line 37 (Form 1040) is—		And you are—				If line 37 (Form 1040) is—		And you are—				If line 37 (Form 1040) is—		And you are—			
At least	But less than	Single	Married filing jointly *	Married filing separately	Head of a household	At least	But less than	Single	Married filing jointly *	Married filing separately	Head of a household	At least	But less than	Single	Married filing jointly *	Married filing separately	Head of a household
		Your tax is—						Your tax is—						Your tax is—			
86,000						**89,000**						**92,000**					
86,000	**86,050**	22,322	19,433	23,043	20,706	**89,000**	**89,050**	23,252	20,349	23,973	21,636	**92,000**	**92,050**	24,182	21,279	24,903	22,566
86,050	**86,100**	22,338	19,447	23,059	20,721	**89,050**	**89,100**	23,268	20,364	23,989	21,651	**92,050**	**92,100**	24,198	21,294	24,919	22,581
86,100	**86,150**	22,353	19,461	23,074	20,737	**89,100**	**89,150**	23,283	20,380	24,004	21,667	**92,100**	**92,150**	24,213	21,310	24,934	22,597
86,150	**86,200**	22,369	19,475	23,090	20,752	**89,150**	**89,200**	23,299	20,395	24,020	21,682	**92,150**	**92,200**	24,229	21,325	24,950	22,612
86,200	**86,250**	22,384	19,489	23,105	20,768	**89,200**	**89,250**	23,314	20,411	24,035	21,698	**92,200**	**92,250**	24,244	21,341	24,965	22,628
86,250	**86,300**	22,400	19,503	23,121	20,783	**89,250**	**89,300**	23,330	20,426	24,051	21,713	**92,250**	**92,300**	24,260	21,356	24,981	22,643
86,300	**86,350**	22,415	19,517	23,136	20,799	**89,300**	**89,350**	23,345	20,442	24,066	21,729	**92,300**	**92,350**	24,275	21,372	24,996	22,659
86,350	**86,400**	22,431	19,531	23,152	20,814	**89,350**	**89,400**	23,361	20,457	24,082	21,744	**92,350**	**92,400**	24,291	21,387	25,012	22,674
86,400	**86,450**	22,446	19,545	23,167	20,830	**89,400**	**89,450**	23,376	20,473	24,097	21,760	**92,400**	**92,450**	24,306	21,403	25,027	22,690
86,450	**86,500**	22,462	19,559	23,183	20,845	**89,450**	**89,500**	23,392	20,488	24,113	21,775	**92,450**	**92,500**	24,322	21,418	25,043	22,705
86,500	**86,550**	22,477	19,574	23,198	20,861	**89,500**	**89,550**	23,407	20,504	24,128	21,791	**92,500**	**92,550**	24,337	21,434	25,058	22,721
86,550	**86,600**	22,493	19,589	23,214	20,876	**89,550**	**89,600**	23,423	20,519	24,144	21,806	**92,550**	**92,600**	24,353	21,449	25,074	22,736
86,600	**86,650**	22,508	19,605	23,229	20,892	**89,600**	**89,650**	23,438	20,535	24,159	21,822	**92,600**	**92,650**	24,368	21,465	25,089	22,752
86,650	**86,700**	22,524	19,620	23,245	20,907	**89,650**	**89,700**	23,454	20,550	24,175	21,837	**92,650**	**92,700**	24,384	21,480	25,105	22,767
86,700	**86,750**	22,539	19,636	23,260	20,923	**89,700**	**89,750**	23,469	20,566	24,190	21,853	**92,700**	**92,750**	24,399	21,496	25,120	22,783
86,750	**86,800**	22,555	19,651	23,276	20,938	**89,750**	**89,800**	23,485	20,581	24,206	21,868	**92,750**	**92,800**	24,415	21,511	25,136	22,798
86,800	**86,850**	22,570	19,667	23,291	20,954	**89,800**	**89,850**	23,500	20,597	24,221	21,884	**92,800**	**92,850**	24,430	21,527	25,151	22,814
86,850	**86,900**	22,586	19,682	23,307	20,969	**89,850**	**89,900**	23,516	20,612	24,237	21,899	**92,850**	**92,900**	24,446	21,542	25,167	22,829
86,900	**86,950**	22,601	19,698	23,322	20,985	**89,900**	**89,950**	23,531	20,628	24,252	21,915	**92,900**	**92,950**	24,461	21,558	25,182	22,845
86,950	**87,000**	22,617	19,713	23,338	21,000	**89,950**	**90,000**	23,547	20,643	24,268	21,930	**92,950**	**93,000**	24,477	21,573	25,198	22,860
87,000						**90,000**						**93,000**					
87,000	**87,050**	22,632	19,729	23,353	21,016	**90,000**	**90,050**	23,562	20,659	24,283	21,946	**93,000**	**93,050**	24,492	21,589	25,213	22,876
87,050	**87,100**	22,648	19,744	23,369	21,031	**90,050**	**90,100**	23,578	20,674	24,299	21,961	**93,050**	**93,100**	24,508	21,604	25,229	22,891
87,100	**87,150**	22,663	19,760	23,384	21,047	**90,100**	**90,150**	23,593	20,690	24,314	21,977	**93,100**	**93,150**	24,523	21,620	25,244	22,907
87,150	**87,200**	22,679	19,775	23,400	21,062	**90,150**	**90,200**	23,609	20,705	24,330	21,992	**93,150**	**93,200**	24,539	21,635	25,260	22,922
87,200	**87,250**	22,694	19,791	23,415	21,078	**90,200**	**90,250**	23,624	20,721	24,345	22,008	**93,200**	**93,250**	24,554	21,651	25,275	22,938
87,250	**87,300**	22,710	19,806	23,431	21,093	**90,250**	**90,300**	23,640	20,736	24,361	22,023	**93,250**	**93,300**	24,570	21,666	25,291	22,953
87,300	**87,350**	22,725	19,822	23,446	21,109	**90,300**	**90,350**	23,655	20,752	24,376	22,039	**93,300**	**93,350**	24,585	21,682	25,306	22,969
87,350	**87,400**	22,741	19,837	23,462	21,124	**90,350**	**90,400**	23,671	20,767	24,392	22,054	**93,350**	**93,400**	24,601	21,697	25,322	22,984
87,400	**87,450**	22,756	19,853	23,477	21,140	**90,400**	**90,450**	23,686	20,783	24,407	22,070	**93,400**	**93,450**	24,616	21,713	25,337	23,000
87,450	**87,500**	22,772	19,868	23,493	21,155	**90,450**	**90,500**	23,702	20,798	24,423	22,085	**93,450**	**93,500**	24,632	21,728	25,353	23,015
87,500	**87,550**	22,787	19,884	23,508	21,171	**90,500**	**90,550**	23,717	20,814	24,438	22,101	**93,500**	**93,550**	24,647	21,744	25,368	23,031
87,550	**87,600**	22,803	19,899	23,524	21,186	**90,550**	**90,600**	23,733	20,829	24,454	22,116	**93,550**	**93,600**	24,663	21,759	25,384	23,046
87,600	**87,650**	22,818	19,915	23,539	21,202	**90,600**	**90,650**	23,748	20,845	24,469	22,132	**93,600**	**93,650**	24,678	21,775	25,399	23,062
87,650	**87,700**	22,834	19,930	23,555	21,217	**90,650**	**90,700**	23,764	20,860	24,485	22,147	**93,650**	**93,700**	24,694	21,790	25,415	23,077
87,700	**87,750**	22,849	19,946	23,570	21,233	**90,700**	**90,750**	23,779	20,876	24,500	22,163	**93,700**	**93,750**	24,709	21,806	25,430	23,093
87,750	**87,800**	22,865	19,961	23,586	21,248	**90,750**	**90,800**	23,795	20,891	24,516	22,178	**93,750**	**93,800**	24,725	21,821	25,446	23,108
87,800	**87,850**	22,880	19,977	23,601	21,264	**90,800**	**90,850**	23,810	20,907	24,531	22,194	**93,800**	**93,850**	24,740	21,837	25,461	23,124
87,850	**87,900**	22,896	19,992	23,617	21,279	**90,850**	**90,900**	23,826	20,922	24,547	22,209	**93,850**	**93,900**	24,756	21,852	25,477	23,139
87,900	**87,950**	22,911	20,008	23,632	21,295	**90,900**	**90,950**	23,841	20,938	24,562	22,225	**93,900**	**93,950**	24,771	21,868	25,492	23,155
87,950	**88,000**	22,927	20,023	23,648	21,310	**90,950**	**91,000**	23,857	20,953	24,578	22,240	**93,950**	**94,000**	24,787	21,883	25,508	23,170
88,000						**91,000**						**94,000**					
88,000	**88,050**	22,942	20,039	23,663	21,326	**91,000**	**91,050**	23,872	20,969	24,593	22,256	**94,000**	**94,050**	24,802	21,899	25,523	23,186
88,050	**88,100**	22,958	20,054	23,679	21,341	**91,050**	**91,100**	23,888	20,984	24,609	22,271	**94,050**	**94,100**	24,818	21,914	25,539	23,201
88,100	**88,150**	22,973	20,070	23,694	21,357	**91,100**	**91,150**	23,903	21,000	24,624	22,287	**94,100**	**94,150**	24,833	21,930	25,554	23,217
88,150	**88,200**	22,989	20,085	23,710	21,372	**91,150**	**91,200**	23,919	21,015	24,640	22,302	**94,150**	**94,200**	24,849	21,945	25,570	23,232
88,200	**88,250**	23,004	20,101	23,725	21,388	**91,200**	**91,250**	23,934	21,031	24,655	22,318	**94,200**	**94,250**	24,864	21,961	25,585	23,248
88,250	**88,300**	23,020	20,116	23,741	21,403	**91,250**	**91,300**	23,950	21,046	24,671	22,333	**94,250**	**94,300**	24,880	21,976	25,601	23,263
88,300	**88,350**	23,035	20,132	23,756	21,419	**91,300**	**91,350**	23,965	21,062	24,686	22,349	**94,300**	**94,350**	24,895	21,992	25,616	23,279
88,350	**88,400**	23,051	20,147	23,772	21,434	**91,350**	**91,400**	23,981	21,077	24,702	22,364	**94,350**	**94,400**	24,911	22,007	25,632	23,294
88,400	**88,450**	23,066	20,163	23,787	21,450	**91,400**	**91,450**	23,996	21,093	24,717	22,380	**94,400**	**94,450**	24,926	22,023	25,647	23,310
88,450	**88,500**	23,082	20,178	23,803	21,465	**91,450**	**91,500**	24,012	21,108	24,733	22,395	**94,450**	**94,500**	24,942	22,038	25,663	23,325
88,500	**88,550**	23,097	20,194	23,818	21,481	**91,500**	**91,550**	24,027	21,124	24,748	22,411	**94,500**	**94,550**	24,957	22,054	25,678	23,341
88,550	**88,600**	23,113	20,209	23,834	21,496	**91,550**	**91,600**	24,043	21,139	24,764	22,426	**94,550**	**94,600**	24,973	22,069	25,694	23,356
88,600	**88,650**	23,128	20,225	23,849	21,512	**91,600**	**91,650**	24,058	21,155	24,779	22,442	**94,600**	**94,650**	24,988	22,085	25,709	23,372
88,650	**88,700**	23,144	20,240	23,865	21,527	**91,650**	**91,700**	24,074	21,170	24,795	22,457	**94,650**	**94,700**	25,004	22,100	25,725	23,387
88,700	**88,750**	23,159	20,256	23,880	21,543	**91,700**	**91,750**	24,089	21,186	24,810	22,473	**94,700**	**94,750**	25,019	22,116	25,740	23,403
88,750	**88,800**	23,175	20,271	23,896	21,558	**91,750**	**91,800**	24,105	21,201	24,826	22,488	**94,750**	**94,800**	25,035	22,131	25,756	23,418
88,800	**88,850**	23,190	20,287	23,911	21,574	**91,800**	**91,850**	24,120	21,217	24,841	22,504	**94,800**	**94,850**	25,050	22,147	25,771	23,434
88,850	**88,900**	23,206	20,302	23,927	21,589	**91,850**	**91,900**	24,136	21,232	24,857	22,519	**94,850**	**94,900**	25,066	22,162	25,787	23,449
88,900	**88,950**	23,221	20,318	23,942	21,605	**91,900**	**91,950**	24,151	21,248	24,872	22,535	**94,900**	**94,950**	25,081	22,178	25,802	23,465
88,950	**89,000**	23,237	20,333	23,958	21,620	**91,950**	**92,000**	24,167	21,263	24,888	22,550	**94,950**	**95,000**	25,097	22,193	25,818	23,480

* This column must also be used by a qualifying widow(er).

Continued on next page

19-- Tax Table—*Continued*

If line 37 (Form 1040) is— At least	But less than	And you are— Single	Married filing jointly *	Married filing sepa-rately	Head of a house-hold
		Your tax is—			
95,000					
95,000	**95,050**	25,112	22,209	25,833	23,496
95,050	**95,100**	25,128	22,224	25,849	23,511
95,100	**95,150**	25,143	22,240	25,864	23,527
95,150	**95,200**	25,159	22,255	25,880	23,542
95,200	**95,250**	25,174	22,271	25,895	23,558
95,250	**95,300**	25,190	22,286	25,911	23,573
95,300	**95,350**	25,205	22,302	25,926	23,589
95,350	**95,400**	25,221	22,317	25,942	23,604
95,400	**95,450**	25,236	22,333	25,957	23,620
95,450	**95,500**	25,252	22,348	25,973	23,635
95,500	**95,550**	25,267	22,364	25,988	23,651
95,550	**95,600**	25,283	22,379	26,004	23,666
95,600	**95,650**	25,298	22,395	26,019	23,682
95,650	**95,700**	25,314	22,410	26,035	23,697
95,700	**95,750**	25,329	22,426	26,050	23,713
95,750	**95,800**	25,345	22,441	26,066	23,728
95,800	**95,850**	25,360	22,457	26,081	23,744
95,850	**95,900**	25,376	22,472	26,097	23,759
95,900	**95,950**	25,391	22,488	26,112	23,775
95,950	**96,000**	25,407	22,503	26,128	23,790
96,000					
96,000	**96,050**	25,422	22,519	26,143	23,806
96,050	**96,100**	25,438	22,534	26,159	23,821
96,100	**96,150**	25,453	22,550	26,174	23,837
96,150	**96,200**	25,469	22,565	26,190	23,852
96,200	**96,250**	25,484	22,581	26,205	23,868
96,250	**96,300**	25,500	22,596	26,221	23,883
96,300	**96,350**	25,515	22,612	26,236	23,899
96,350	**96,400**	25,531	22,627	26,252	23,914
96,400	**96,450**	25,546	22,643	26,267	23,930
96,450	**96,500**	25,562	22,658	26,283	23,945
96,500	**96,550**	25,577	22,674	26,298	23,961
96,550	**96,600**	25,593	22,689	26,314	23,976
96,600	**96,650**	25,608	22,705	26,329	23,992
96,650	**96,700**	25,624	22,720	26,345	24,007
96,700	**96,750**	25,639	22,736	26,360	24,023
96,750	**96,800**	25,655	22,751	26,376	24,038
96,800	**96,850**	25,670	22,767	26,391	24,054
96,850	**96,900**	25,686	22,782	26,407	24,069
96,900	**96,950**	25,701	22,798	26,422	24,085
96,950	**97,000**	25,717	22,813	26,438	24,100
97,000					
97,000	**97,050**	25,732	22,829	26,453	24,116
97,050	**97,100**	25,748	22,844	26,469	24,131
97,100	**97,150**	25,763	22,860	26,484	24,147
97,150	**97,200**	25,779	22,875	26,500	24,162
97,200	**97,250**	25,794	22,891	26,515	24,178
97,250	**97,300**	25,810	22,906	26,531	24,193
97,300	**97,350**	25,825	22,922	26,546	24,209
97,350	**97,400**	25,841	22,937	26,562	24,224
97,400	**97,450**	25,856	22,953	26,577	24,240
97,450	**97,500**	25,872	22,968	26,593	24,255
97,500	**97,550**	25,887	22,984	26,608	24,271
97,550	**97,600**	25,903	22,999	26,624	24,286
97,600	**97,650**	25,918	23,015	26,639	24,302
97,650	**97,700**	25,934	23,030	26,655	24,317
97,700	**97,750**	25,949	23,046	26,670	24,333
97,750	**97,800**	25,965	23,061	26,686	24,348
97,800	**97,850**	25,980	23,077	26,701	24,364
97,850	**97,900**	25,996	23,092	26,717	24,379
97,900	**97,950**	26,011	23,108	26,732	24,395
97,950	**98,000**	26,027	23,123	26,748	24,410
98,000					
98,000	**98,050**	26,042	23,139	26,763	24,426
98,050	**98,100**	26,058	23,154	26,779	24,441
98,100	**98,150**	26,073	23,170	26,794	24,457
98,150	**98,200**	26,089	23,185	26,810	24,472
98,200	**98,250**	26,104	23,201	26,825	24,488
98,250	**98,300**	26,120	23,216	26,841	24,503
98,300	**98,350**	26,135	23,232	26,856	24,519
98,350	**98,400**	26,151	23,247	26,872	24,534
98,400	**98,450**	26,166	23,263	26,887	24,550
98,450	**98,500**	26,182	23,278	26,903	24,565
98,500	**98,550**	26,197	23,294	26,918	24,581
98,550	**98,600**	26,213	23,309	26,934	24,596
98,600	**98,650**	26,228	23,325	26,949	24,612
98,650	**98,700**	26,244	23,340	26,965	24,627
98,700	**98,750**	26,259	23,356	26,980	24,643
98,750	**98,800**	26,275	23,371	26,996	24,658
98,800	**98,850**	26,290	23,387	27,011	24,674
98,850	**98,900**	26,306	23,402	27,027	24,689
98,900	**98,950**	26,321	23,418	27,042	24,705
98,950	**99,000**	26,337	23,433	27,058	24,720
99,000					
99,000	**99,050**	26,352	23,449	27,073	24,736
99,050	**99,100**	26,368	23,464	27,089	24,751
99,100	**99,150**	26,383	23,480	27,104	24,767
99,150	**99,200**	26,399	23,495	27,120	24,782
99,200	**99,250**	26,414	23,511	27,135	24,798
99,250	**99,300**	26,430	23,526	27,151	24,813
99,300	**99,350**	26,445	23,542	27,166	24,829
99,350	**99,400**	26,461	23,557	27,182	24,844
99,400	**99,450**	26,476	23,573	27,197	24,860
99,450	**99,500**	26,492	23,588	27,213	24,875
99,500	**99,550**	26,507	23,604	27,228	24,891
99,550	**99,600**	26,523	23,619	27,244	24,906
99,600	**99,650**	26,538	23,635	27,259	24,922
99,650	**99,700**	26,554	23,650	27,275	24,937
99,700	**99,750**	26,569	23,666	27,290	24,953
99,750	**99,800**	26,585	23,681	27,306	24,968
99,800	**99,850**	26,600	23,697	27,321	24,984
99,850	**99,900**	26,616	23,712	27,337	24,999
99,900	**99,950**	26,631	23,728	27,352	25,015
99,950	**100,000**	26,647	23,743	27,368	25,030
100,000 or over—use tax rate schedules					

* This column must also be used by a qualifying widow(er).

19--
Tax Rate Schedules

Caution: *Use **only** if your taxable income (Form 1040, line 37) is $100,000 or more. If less, use the **Tax Table.** Even though you cannot use the tax rate schedules below if your taxable income is less than $100,000, all levels of taxable income are shown so taxpayers can see the tax rate that applies to each level.*

Schedule X—Use if your filing status is **Single**

If the amount on Form 1040, line 37, is: *Over—*	*But not over—*	Enter on Form 1040, line 38	*of the amount over—*
$0	$21,450	.... **15%**	**$0**
21,450	51,900	**$3,217.50 + 28%**	**21,450**
51,900	....	**11,743.50 + 31%**	**51,900**

Schedule Y-1—Use if your filing status is **Married filing jointly** or **Qualifying widow(er)**

If the amount on Form 1040, line 37, is: *Over—*	*But not over—*	Enter on Form 1040, line 38	*of the amount over—*
$0	$35,800	.... **15%**	**$0**
35,800	86,500	**$5,370.00 + 28%**	**35,800**
86,500	....	**19,566.00 + 31%**	**86,500**

Schedule Y-2—Use if your filing status is **Married filing separately**

If the amount on Form 1040, line 37, is: *Over—*	*But not over—*	Enter on Form 1040, line 38	*of the amount over—*
$0	$17,900	.... **15%**	**$0**
17,900	43,250	**$2,685.00 + 28%**	**17,900**
43,250	....	**9,783.00 + 31%**	**43,250**

Schedule Z—Use if your filing status is **Head of household**

If the amount on Form 1040, line 37, is: *Over—*	*But not over—*	Enter on Form 1040, line 38	*of the amount over—*
$0	$28,750	.... **15%**	**$0**
28,750	74,150	**$4,312.50 + 28%**	**28,750**
74,150	....	**17,024.50 + 31%**	**74,150**

19-- Standard Deduction Tables

Caution: If you are married filing a separate return and your spouse itemizes deductions, or if you are a dual-status alien, you cannot take the standard deduction even if you were 65 or older or blind.

Table 20-1. Standard Deduction Chart for Most People*

If Your Filing Status is:	Your Standard Deduction Is:
Single	$3,600
Married filing joint return or Qualifying widow(er) with dependent child	6,000
Married filing separate return	3,000
Head of household	5,250

* DO NOT use this chart if you were 65 or older or blind, OR if someone can claim you as a dependent.

Table 20-2. Standard Deduction Chart for People Age 65 or Older or Blind*

Check the correct number of boxes below. Then go to the chart.

You	65 or older ☐	Blind ☐
Your spouse, if claiming spouse's exemption	65 or older ☐	Blind ☐

Total number of boxes you checked ☐

If Your Filing Status is:	And the Number in the Box Above is:	Your Standard Deduction is:
Single	1	$4,500
	2	5,400
Married filing joint return or Qualifying widow(er) with dependent child	1	6,700
	2	7,400
	3	8,100
	4	8,800
Married filing separate return	1	3,700
	2	4,400
	3	5,100
	4	5,800
Head of household	1	6,150
	2	7,050

* If someone can claim you as a dependent, use the worksheet in Table 20-3, instead.

Table 20-3. Standard Deduction Worksheet for Dependents*

If you were 65 or older or blind, check the correct number of boxes below. Then go to the worksheet.

You	65 or older ☐	Blind ☐
Your spouse, if claiming spouse's exemption	65 or older ☐	Blind ☐

Total number of boxes you checked ☐

1. Enter your **earned income** (defined below). If none, go on to line 3	**1.**_______
2. Minimum amount	**2.** $600
3. Compare the amounts on lines 1 and 2. Enter the **larger** of the two amounts here	**3.**_______
4. Enter on line 4 the amount shown below for your filing status. • Single, enter $3,600 • Married filing separate return, enter $3,000 • Married filing jointly or Qualifying widow(er) with dependent child, enter $6,000 • Head of household, enter $5,250	**4.**_______
5. Standard deduction. **a.** Compare the amounts on lines 3 and 4. Enter the **smaller** of the two amounts here. If under 65 and not blind, stop here. This is your standard deduction. Otherwise, go on to line 5b	**5a.**_______
b. If 65 or older or blind, multiply $900 ($700 if married or qualifying widow(er) with dependent child) by the number in the box above. Enter the result	**5b.**_______
c. Add lines 5a and 5b. This is your standard deduction for 1992.	**5c.**_______

Earned income *includes wages, salaries, tips, professional fees, and other compensation received for personal services you performed. It also includes any amount received as a scholarship that you must include in your income.*

* Use this worksheet ONLY if someone can claim you as a dependent.

ORGANIZATION OF THE FEDERAL RESERVE SYSTEM

Seattle
Helena
Minneapolis
Portland
Salt Lake City
San Francisco
Los Angeles
Denver
Omaha
Kansas City
Oklahoma City
El Paso
Dallas
San Antonio
Houston
St. Louis
Little Rock
Memphis
Louisville
Nashville
New Orleans
Birmingham
Atlanta
Jacksonville
Miami
Chicago
Detroit
Cincinnati
Cleveland
Pittsburg
Buffalo
Boston
New York
Philadelphia
Baltimore
WASHINGTON
Board of Governors
Richmond
Charlotte
(1) (2) (3) (4) (5) (6) (7) (8) (9) (10) (11) (12)

LEGEND

— BOUNDARIES OF FEDERAL RESERVE DISTRICTS

— BOUNDARIES OF FEDERAL RESERVE BRANCH TERRITORIES

★ BOARD OF GOVERNORS OF THE FEDERAL RESERVE SYSTEM

■ CITY WHERE FEDERAL RESERVE BANK IS LOCATED

● CITY WHERE A BRANCH OF FEDERAL RESERVE BANK IS LOCATED

(3) FEDERAL RESERVE DISTRICT NUMBER. THIS NUMBER APPEARS ON THE CURRENCY ISSUED BY THE FEDERAL RESERVE BANK IN THE DISTRICT

Hypothetical Credit-Scoring Table

Fill out your credit profile by answering the nine questions below in Table 1. Circle the one response that applies to you, and then find your total score by adding up the points you got for each response. The points are found in the lower right-hand corner of each box. (For example: if you are 25 years old, you get 5 points.) Once you've totaled your score, look at Table 2 to find out how good a credit "bet" you may be.

1.

1.	age?	under 25 12	25–29 5	30–34 0	35–39 1	40–44 18	45–49 22	50 or over 31
2.	time at address?	less than 1 yr. 9	1yr. 0	2–3 yrs. 5	4–5 yrs. 0	6–9 yrs. 5	10 yrs. or more 21	
3.	age of auto?	none 0	0–1 yrs. 12	2 yrs. 16	3–4 yrs. 13	5–7 yrs. 3	8 yrs. or more 0	
4.	monthly auto payment?	none 18	less than $125 6	$126–$150 1	$151–$199 4	$200 or more 0		
5.	housing cost?	less than $274 0	$275–$399 10	$400 or more 12	owns clear 12	lives with relatives 24		
6.	checking and savings accounts	both 15	checking only 2	savings only 2	neither 0			
7.	finance company reference	yes 0	no 15					
8.	major credit cards?	none 0	1 5	2 or more 15				
9.	ratio of debt to income?	no debts 41	1%–5% 16	6%–15% 20	16% or over 0			

2.

A lender using this scoring table selects a cutoff point from a table like this, which gauges how likely applicants are to repay loans.

Total Score	Probability of Repayment
90	89 in 100
95	91 in 100
100	92 in 100
105	93 in 100
110	94 in 100
115	95 in 100
120	95.5 in 100
125	96 in 100
130	96.25 in 100

Source: Federal Reserve Board. Developed by Fair, Isaac, and Co., Inc. Modified to update.

ACCUMULATED CASH VALUE OF $100,000 WHOLE LIFE POLICY AGE OF ISSUE: 25					
Year	**Person's Age**	**Cash Value**	**Year**	**Person's Age**	**Cash Value**
1	25	$ 0	11	35	$10,187
2	26	700	12	36	11,501
3	27	1500	13	37	12,860
4	28	2300	14	38	14,246
5	29	3100	15	39	15,667
6	30	4020	16	40	17,094
7	31	5158	17	41	18,555
8	32	6349	18	42	20,014
9	33	7538	19	43	21,563
10	34	8898	20	44	23,197

ACCUMULATED CASH VALUE OF $100,000 WHOLE LIFE POLICY						
Age at Issue	**After Paying Premiums for:**					
	5 years	**10 years**	**15 years**	**17 years**	**20 years**	**At Age 65**
20	$ 2,212	$ 7,105	$12,840	$15,371	$19,438	$57,716
25	3,100	8,898	15,667	18,555	23,197	55,890
30	3,999	11,027	18,837	22,098	27,339	53,533
35	5,151	13,369	22,203	25,868	31,759	50,453
40	6,367	15,921	26,016	30,001	36,407	46,426
45	7,737	18,715	29,992	34,586	41,181	41,181
50	9,210	21,707	34,080	38,556	45,751	34,080
55	10,842	24,827	37,980	42,562	49,928	24,827

MULTIPLES-OF-SALARY CHART								
	Current Age							
	25 Years		**35 Years**		**45 Years**		**55 Years**	
Current Gross Earnings	**75%**	**60%**	**75%**	**60%**	**75%**	**60%**	**75%**	**60%**
$ 7,500	4.0	3.0	5.5	4.0	7.5	5.5	6.5	4.5
9,000	4.0	3.0	5.5	4.0	7.5	5.5	6.5	4.5
15,000	4.5	3.0	6.5	4.5	8.0	6.0	7.0	5.5
23,500	6.5	4.5	8.0	5.5	8.5	6.5	7.5	5.5
30,000	7.5	5.0	8.0	6.0	8.5	6.5	7.0	5.5
40,000	7.5	5.0	8.0	6.0	8.0	6.0	7.0	5.5
65,000	7.5	5.5	7.5	6.0	7.5	6.0	6.5	5.0

COMPARISON TABLE FOR TERM AND WHOLE LIFE PREMIUMS			
Policy Face Value is $100,000			
Age	Five-Year Renewable Term	Whole Life	First-Year Difference
20	$205	$ 775	$ 570
25	207	918	711
30	218	1112	894
35	254	1374	1120
40	363	1729	1366
45	562	2127	1565
50	878	2689	1811

EXPECTED DEATHS PER 100,000 ALIVE AT SPECIFIED AGE		
Age	Expected Deaths Within 1 Year	Expected to be Alive in 1 Year
15	63	99,937
16	79	99,921
17	91	99,909
18	99	99,901
19	103	99,897
20	106	99,894
21	110	99,890
22	113	99,887
23	115	99,885
24	117	99,883
25	118	99,882
26	120	99,880
27	123	99,877
28	127	99,873
29	132	99,868
45	315	99,685
46	341	99,659
47	371	99,629
48	405	99,595
49	443	99,557

Moonbeam	Dealer Cost	Sticker Price
Model A, 4-door wagon, 8 cylinder	$12,725	$14,722
Model B, 4-door sedan, 6 cylinder	14,062	16,295
Model C, 4-door sedan, 4 cylinder	11,053	12,755
Options		
Conventional spare tire	$ 62	$ 73
Air conditioning	695	817
Electronic climate control	850	1000
Antilock brake system	838	985
Rear window defroster	136	160
AM/FM radio and cassette player	132	155
Rear-facing third seat	132	155
Cruise control	178	210
Stripe, painted	51	61

AUTOMOBILE INSURANCE, SIX-MONTH BASIC RATE SCHEDULE						
Car Class Rating	Collision Deductible			Comprehensive Deductible		
	$100	$250	$500	$50	$250	$500
1–10	$ 500	$ 430	$ 370	$140	$130	$110
11–20	820	750	690	350	310	280
21–30	1140	1040	940	450	410	370
31–40	1280	1190	1100	670	610	560

DRIVER RATING FACTOR (Part I)
Category: Married Youths

Age	Sex	Owner or Usual Driver	Driver Training	Drive to Work	Rating Factor
16–17	F	Yes	Yes	Yes	2.40
				No	2.25
			No	Yes	2.60
				No	2.45
		No	Yes	Yes	1.90
				No	1.75
			No	Yes	2.05
				No	1.90
	M	Yes	Yes	Yes	2.45
				No	2.30
			No	Yes	2.70
				No	2.55
		No	Yes	Yes	2.45
				No	2.30
			No	Yes	2.70
				No	2.55
18–21	F	Yes	Yes	Yes	1.95
				No	1.80
			No	Yes	2.10
				No	1.95
		No	Yes	Yes	1.55
				No	1.40
			No	Yes	1.65
				No	1.50
	M	Yes	Yes	Yes	2.00
				No	1.85
			No	Yes	2.15
				No	2.00
		No	Yes	Yes	2.00
				No	1.85
			No	Yes	2.15
				No	2.00

DRIVER RATING FACTOR (Part II)
Category: Unmarried Youths

Age	Sex	Owner or Usual Driver	Driver Training	Drive to Work	Rating Factor
16–17	F	Yes	Yes	Yes	2.40
				No	2.25
			No	Yes	2.60
				No	2.45
		No	Yes	Yes	1.90
				No	1.75
			No	Yes	2.05
				No	1.90
	M	Yes	Yes	Yes	3.75
				No	3.60
			No	Yes	4.20
				No	4.05
		No	Yes	Yes	2.60
				No	2.45
			No	Yes	2.85
				No	2.70
18–21	F	Yes	Yes	Yes	1.95
				No	1.80
			No	Yes	2.10
				No	1.95
		No	Yes	Yes	1.55
				No	1.40
			No	Yes	1.65
				No	1.50
	M	Yes	Yes	Yes	3.00
				No	2.85
			No	Yes	3.35
				No	3.20
		No	Yes	Yes	2.15
				No	2.00
			No	Yes	2.30
				No	2.15

DRIVER RATING FACTOR (Part III)
Category: All Adults

Age	Sex	Owner or Usual Driver	Drive to Work	Rating Factor
21–24	F	Yes	Yes	1.50
			No	1.35
		No	Yes	1.40
			No	1.25
	M	Yes	Yes	2.25
			No	2.10
		No	Yes	1.55
			No	1.40
25–29	Both	Yes	Yes	1.65
			No	1.50
30–49	Both	Yes	Yes	1.25
			No	1.00
50–64	Both	Yes	Yes	1.15
			No	0.90
65 +	Both	Yes	Yes	1.05
			No	0.80

United States Mileage Chart

	Atlanta, GA	Boston, MA	Cheyenne, WY	Chicago, IL	Cincinnati, OH	Cleveland, OH	Dallas, TX	Denver, CO	Des Moines, IA	Detroit, MI	Indianapolis, IN	Kansas City, MO	Louisville, KY	Memphis, TN	Milwaukee, WI	Minneapolis, MN	New Orleans, LA	Omaha, NE	Philadelphia PA	Pittsburgh, PA	Portland, OR	St. Louis, MO	Salt Lake City, UT	San Francisco, CA	Seattle, WA	Toledo, Ohio	Tulsa, OK	Washington, DC	Wichita, KS
Albuquerque, NM	1381	2172	517	1281	1372	1560	638	417	977	1525	1266	782	1301	1010	1319	1190	1134	858	1899	1619	1371	1038	604	1115	1440	1469	645	1824	593
Amarillo, TX	1097	1897	511	1043	1096	1285	358	423	742	1269	991	547	1019	726	1084	975	850	643	1624	1344	1655	756	888	1399	1724	1210	361	1549	350
Atlanta, GA		1037	1442	674	440	672	795	1398	870	699	493	798	382	371	761	1068	479	986	741	687	2601	541	1878	2496	2618	640	772	608	903
Austin, TX	919	1911	994	1110	1083	1327	193	906	877	1315	1037	682	982	615	1184	1129	517	837	1615	1367	2069	823	1302	1748	2138	1256	450	1482	548
Birmingham, AL	150	1165	1347	642	465	709	645	1286	787	724	475	697	364	246	728	1006	342	898	869	741	2505	465	1781	2366	2535	665	647	736	778
Boston, MA	1037		1907	963	840	628	1748	1949	1280	695	906	1391	941	1296	1050	1368	1507	1412	296	561	3046	1141	2343	3095	2976	739	1537	429	1587
Charleston, SC	289	929	1722	877	603	730	1072	1678	1150	842	696	1078	591	660	964	1282	720	1266	633	666	2881	821	2158	2785	2890	783	1061	500	1192
Cheyenne, WY	1442	1907		954	1174	1279	869	100	627	1211	1068	650	1161	1101	987	788	1361	495	1678	1390	1159	901	436	1188	1228	1176	765	1611	583
Chicago, IL	674	963	954		287	335	917	996	327	266	181	499	292	530	87	405	912	459	738	452	2083	289	1390	2142	2013	232	683	671	696
Cleveland, OH	672	628	1279	335	244		1159	1321	652	170	294	779	101	468	374	692	786	693	567	287	2333	340	1610	2467	2348	111	925	346	975
Columbus, OH	533	735	1235	308	108	139	1028	1229	618	192	171	656	209	576	395	713	894	750	462	182	2391	406	1671	2423	2321	133	802	387	852
Dallas, TX	795	1748	869	917	920	1159		781	684	1143	865	489	819	452	991	936	496	644	1452	1204	2009	630	1242	1753	2078	1084	257	1319	365
Denver, CO	1398	1949	100	996	1164	1321	781		669	1253	1058	600	1120	1040	1029	841	1273	537	1691	1411	1238	857	504	1235	1307	1218	681	1616	509
Des Moines, IA	870	1280	627	327	571	652	684	669		584	465	195	566	599	361	252	978	132	1051	763	1786	333	1063	1815	1749	549	443	984	392
Detroit, MI	699	695	1211	266	259	170	1143	1253	584		278	743	360	713	353	671	1045	716	573	287	2349	513	1647	2399	2279	59	909	506	940
Flagstaff, AZ	1704	2495	757	1604	1695	1883	961	657	1300	1848	1589	1105	1624	1333	1642	1481	1457	1171	2222	1942	1241	1361	511	792	1347	1792	968	2147	916
Harrisburg, PA	700	373	1579	639	468	314	1383	1592	952	474	534	1019	569	931	726	1044	1142	1084	102	189	2722	769	2015	2767	2652	415	1165	107	1215
Indianapolis, IN	493	906	1068	181	106	294	865	1058	465	278		485	111	435	268	586	796	587	633	353	2227	235	1504	2256	2194	219	631	558	681
Jackson, MS	391	1406	1257	742	655	899	404	1169	809	914	646	644	554	212	824	1036	178	845	1110	939	2401	495	1646	2157	2470	855	527	977	708
Kansas City, MO	798	1391	650	499	591	779	489	600	195	743	485		520	451	537	447	806	201	1118	838	1809	257	1086	1835	1839	687	248	1043	197
Knoxville, TN	193	911	1372	527	253	485	837	1328	800	512	346	728	241	385	614	932	596	916	615	511	2531	471	1808	2510	2540	453	786	482	871
Louisville, KY	382	941	1161	292	101	345	819	1120	566	360	111	520		367	379	697	685	687	668	388	2320	263	1597	2349	2305	301	659	582	710
Mackinaw City, MI	935	916	1291	387	495	439	1261	1341	673	284	460	864	562	880	368	508	1247	805	842	556	2128	651	1691	2443	2058	328	1047	775	1061
Miami, FL	655	1504	2097	1329	1095	1264	1300	2037	1525	1352	1148	1448	1037	997	1416	1723	856	1641	1208	1200	3256	1196	2532	3053	3273	1293	1398	1075	1529
Minneapolis, MN	1068	1368	788	405	692	740	936	841	252	671	586	447	697	826	332		1214	357	1143	857	1678	552	1186	1940	1608	637	695	1076	644
New Orleans, LA	479	1507	1361	912	786	1030	496	1273	978	1045	796	806	685	390	994	1214		1007	1211	1070	2505	673	1738	2249	2574	986	647	1078	816
Norfolk, VA	540	558	1764	831	604	508	1329	1758	1141	666	700	1162	642	877	918	1236	1019	1273	263	384	2914	905	2200	2952	2844	607	1278	188	1352
Pierre, SD	1361	1726	434	763	1050	1098	943	518	492	1029	944	592	1055	1043	690	394	1394	391	1501	1215	1353	824	823	1575	1283	995	760	1434	578
Pittsburgh, PA	687	561	1390	452	287	129	1204	1411	763	287	353	838	388	752	539	857	1070	895	288		2535	588	1826	2578	2465	228	964	221	1034
Portland, ME	1139	106	1986	1042	942	707	1850	2028	1359	775	1001	1486	1043	1398	1129	1447	1609	1491	398	663	3125	1236	2422	3174	3055	818	1632	531	1682
Portland, OR	2601	3046	1159	2083	2333	2418	2009	1238	1786	2349	2227	1809	2320	2259	2010	1678	2505	1654	2821	2535		2060	767	636	172	2315	1913	2754	1739
San Antonio, TX	983	1988	1027	1187	1160	1404	270	939	954	1392	1114	759	1059	692	1261	1206	550	914	1692	1444	2086	900	1319	1737	2155	1333	527	1559	625
San Francisco, CA	2496	3095	1188	2142	2362	2467	1753	1235	1815	2399	2256	1835	2349	2125	2175	1940	2249	1683	2866	2578	636	2089	752		808	2364	1760	2799	1695
Seattle, WA	2618	2976	1228	2013	2300	2348	2078	1307	1749	2279	2194	1839	2305	2290	1940	1608	2574	1638	2751	2465	172	2081	836	808		2245	1982	2684	1808
Tulsa, OK	772	1537	765	683	736	925	257	681	443	909	631	248	659	401	757	695	647	387	1264	984	1913	396	1172	1760	1982	850		1189	182
Washington, DC	608	429	1611	671	481	346	1319	1616	984	506	558	1043	582	867	758	1075	1078	1116	133	221	2754	793	2047	2799	2684	447	1189		1239
Wichita, KS	903	1587	583	696	787	975	365	509	392	940	681	197	710	532	734	644	815	298	1314	1034	1739	447	1003	1695	1808	884	182	1239	

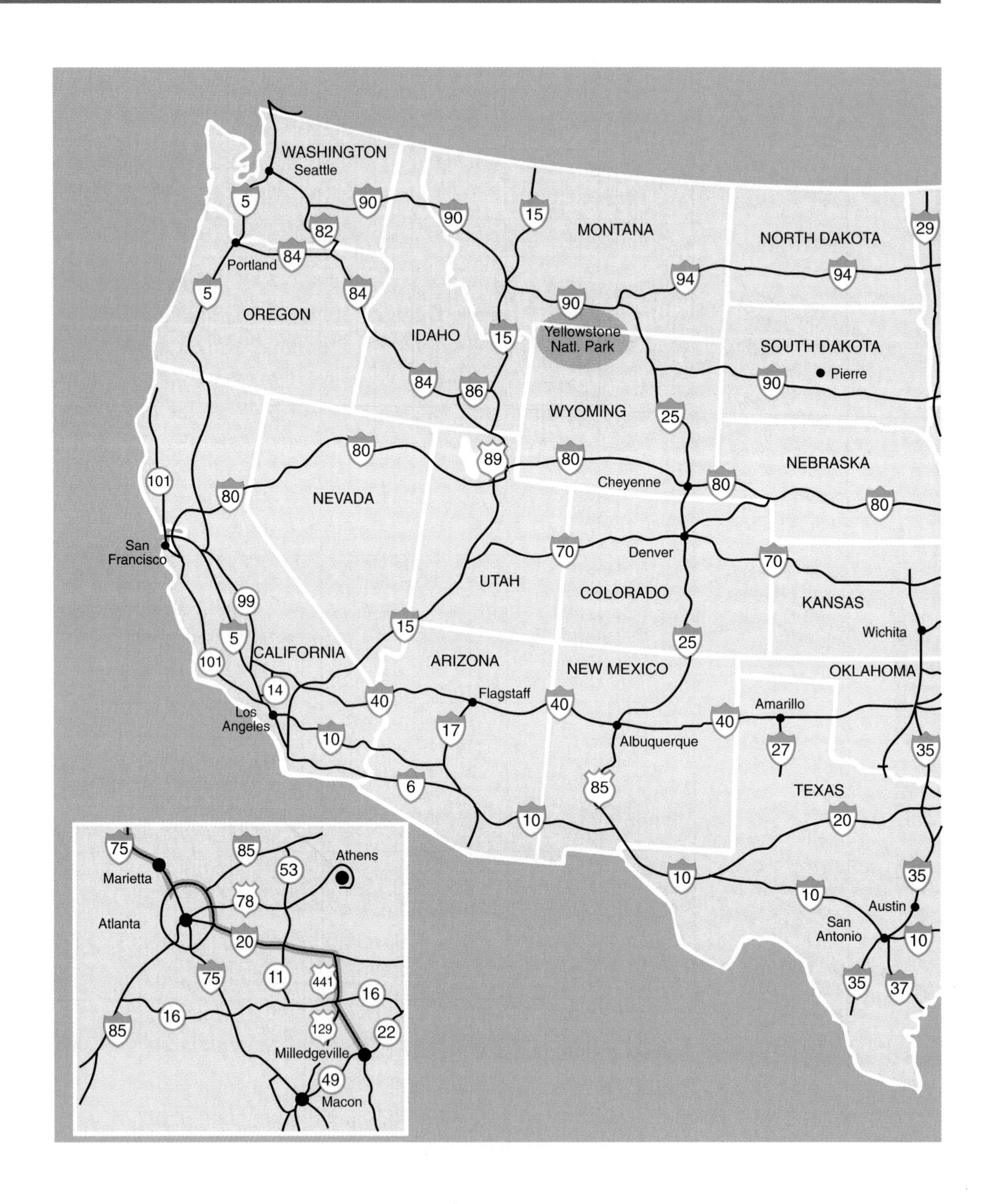

WASHINGTON
Seattle
Portland
OREGON
IDAHO
MONTANA
NORTH DAKOTA
SOUTH DAKOTA
Pierre
Yellowstone Natl. Park
WYOMING
Cheyenne
NEBRASKA
NEVADA
San Francisco
UTAH
Denver
COLORADO
KANSAS
Wichita
CALIFORNIA
ARIZONA
NEW MEXICO
OKLAHOMA
Flagstaff
Amarillo
Los Angeles
Albuquerque
TEXAS
Austin
San Antonio
Marietta
Atlanta
Athens
Milledgeville
Macon

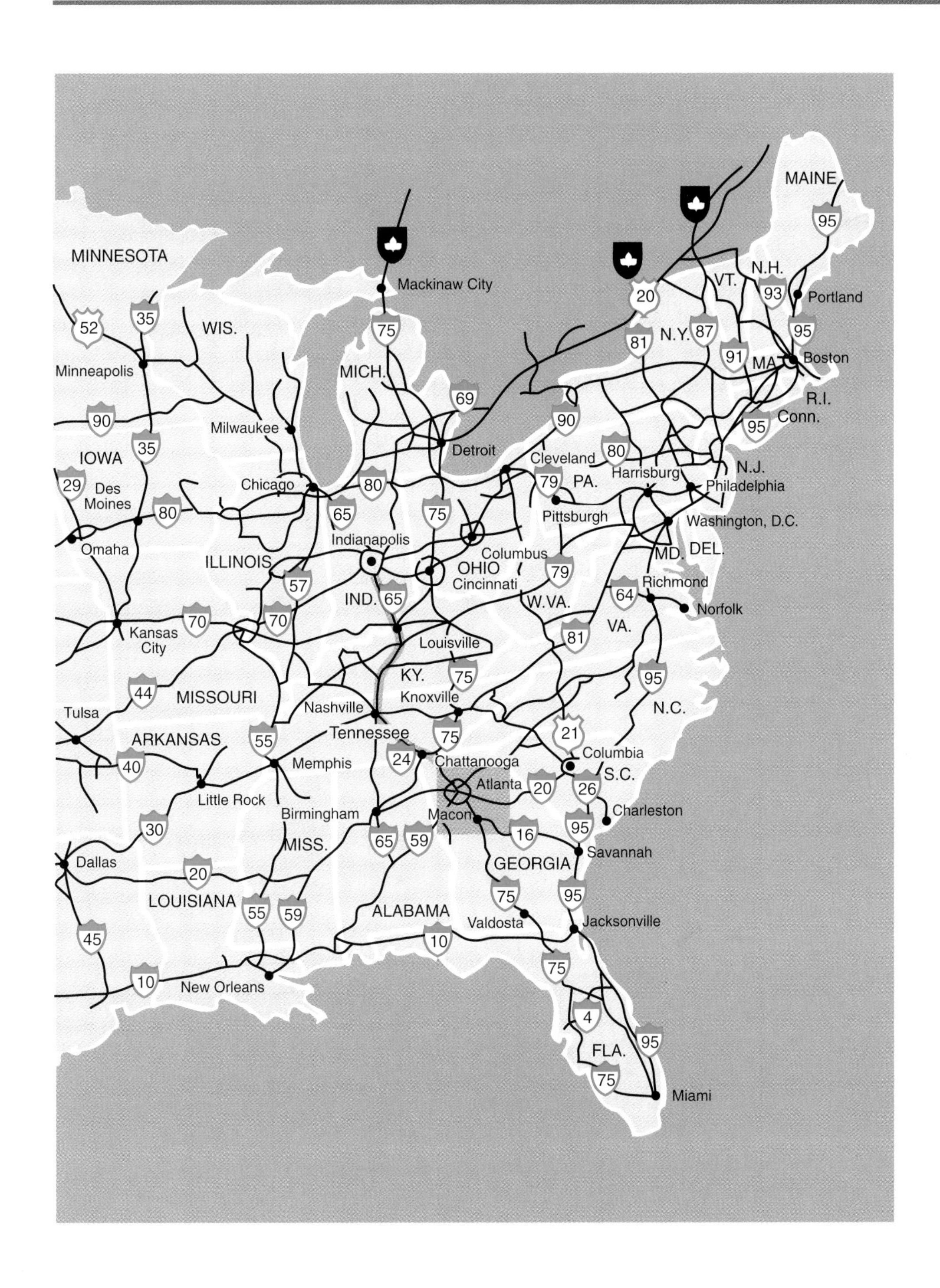
MAINE
95
N.H.
VT.
93
Portland
95
Boston
MA
91
N.Y.
87
81
20
Mackinaw City
MINNESOTA
52
35
WIS.
75
MICH.
Minneapolis
69
R.I.
Conn.
95
90
90
Milwaukee
35
IOWA
Detroit
80
Cleveland
N.J.
Harrisburg
Philadelphia
29
Des Moines
Chicago
80
79
PA.
80
65
75
Pittsburgh
Washington, D.C.
Omaha
Indianapolis
ILLINOIS
Columbus
MD.
DEL.
OHIO
79
Cincinnati
57
Richmond
IND.
65
64
W.VA.
Norfolk
70
70
Kansas City
VA.
81
Louisville
KY.
75
95
44
MISSOURI
Knoxville
N.C.
Tulsa
Nashville
Tennessee
21
ARKANSAS
55
75
Columbia
40
Memphis
24
Chattanooga
S.C.
Atlanta
20
26
Little Rock
Birmingham
Macon
Charleston
30
16
95
MISS.
65
59
Savannah
Dallas
GEORGIA
20
LOUISIANA
75
95
55
59
ALABAMA
Valdosta
Jacksonville
45
10
75
10
New Orleans
4
95
FLA.
75
Miami

Time Zone World Map

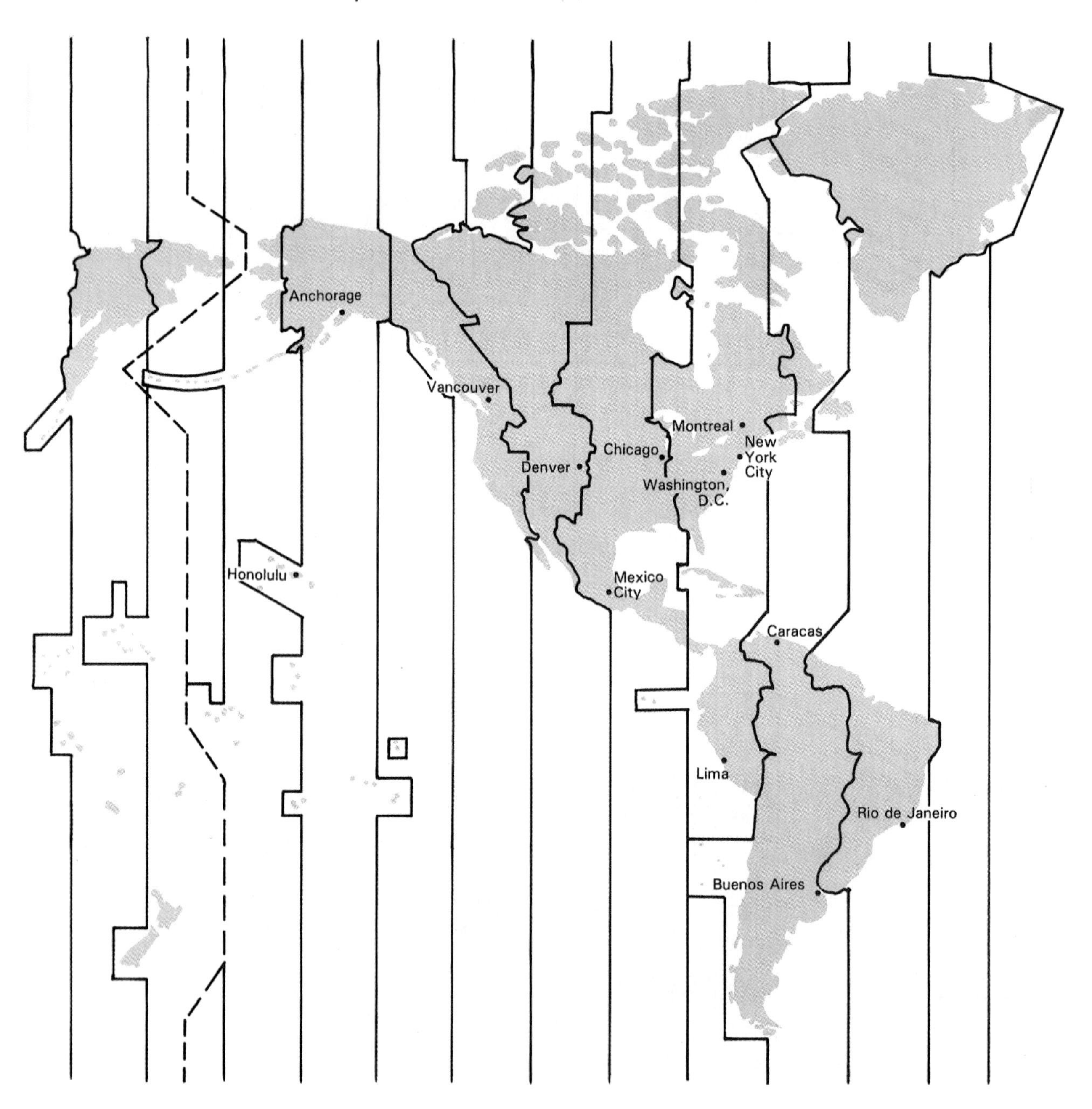

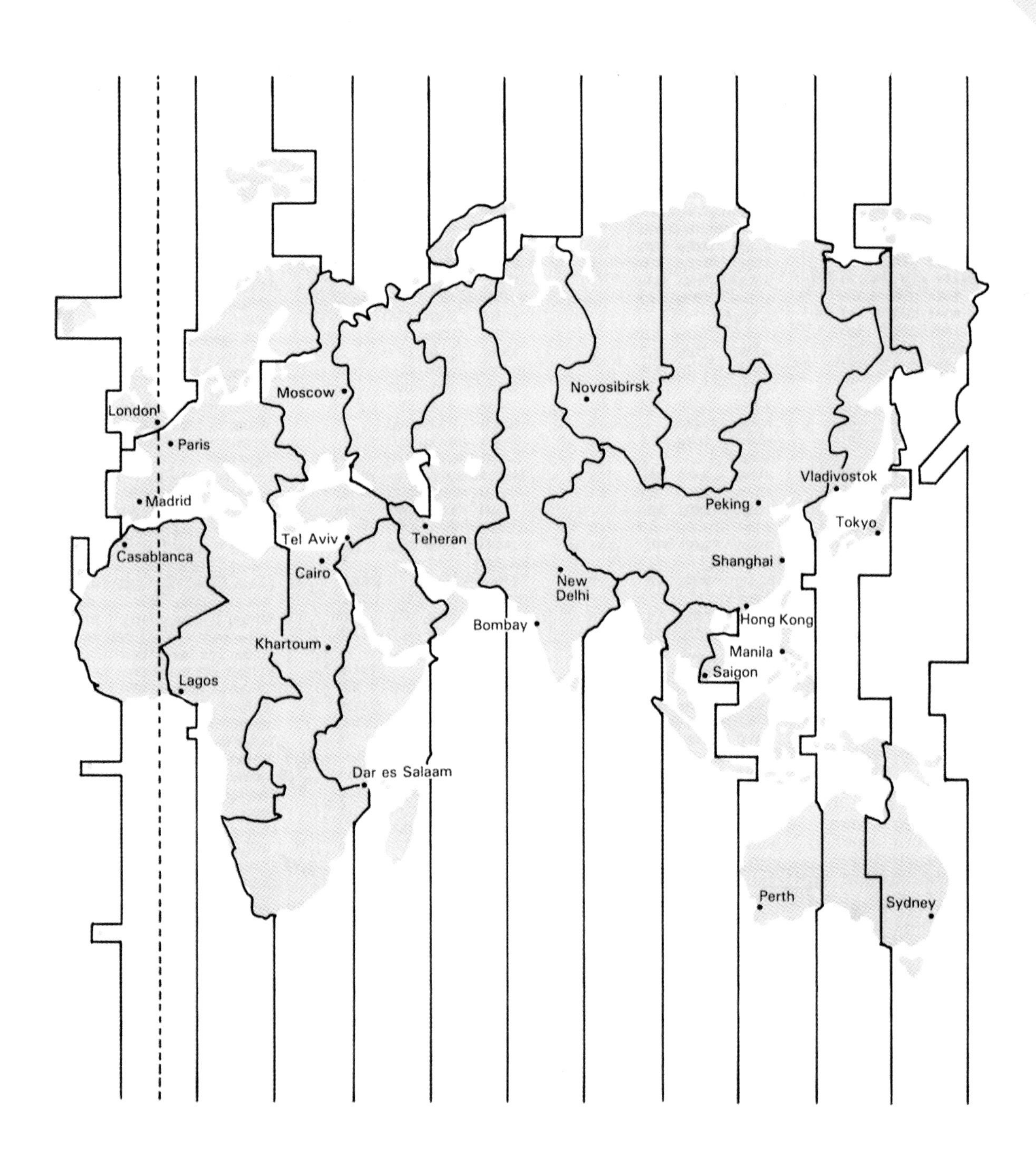
London
Paris
Madrid
Casablanca
Lagos
Moscow
Tel Aviv
Cairo
Khartoum
Dar es Salaam
Teheran
Bombay
New Delhi
Novosibirsk
Peking
Shanghai
Hong Kong
Manila
Saigon
Vladivostok
Tokyo
Perth
Sydney

WINDWARD AIRLINES SCHEDULE

Leave	Arrive	Stops/Via	Rmks
From Chicago, IL			
To Muscle Shoals, AL 491 mi			
1:00p0	5:10p	ATL	L
4:55p0	9:00p	ATL	D X6
To Myrtle Beach, SC 743 mi			
8:10a0	1:18p	ATL	B
9:54a0	3:15p	ATL	S
Eff. May 15			
1:00p0	6:59p	ATL	L
6:44p0	11:45p	ATL	D X6
To Naples, FL 1136 mi			
8:15a0	1:48p	MCO	B
2:10p0	7:25p	MCO	S
To Nashville, TN 401 mi			
5:30a0	8:45a	CVG	S X67
6:15a0	9:24a	CVG	S X67
7:40aM	10:55a	CVG	X67
9:57a0	12:59p	CVG	X67
11:35aM	2:50p	CVG	X67
3:20pM	6:55p	CVG	X6
3:50p0	6:55p	CVG	
5:35p0	8:43p	CVG	X6
5:40pM	8:43p	CVG	X67
Eff. May 1			
To Nassau, Bah 1301 mi			
8:10a0	1:55p	ATL	
To New Orleans, LA 831 mi			
6:15a0	10:14a	CVG	B/S X67
8:10a0	12:11p	ATL	B
9:25aM	1:35p	CVG	L6
Eff. May 6			
9:57a0	1:35p	CVG	L
10:18a0	2:45p	DFW	X67
11:35a0	3:55p	ATL	L
1:00p0	5:20p	ATL	L
3:14p0	7:20p	ATL	S6
3:20pM	7:45p	CVG	DX6
3:50p0	7:45p	CVG	D
6:44p0	10:35p	ATL	D
8:45p0	12:59a	DFW	S
To New York, NY/ Newark, NJ 734 mi			
L-LaGuardia; J-Kennedy; E-Newark			
7:40aM	12:25pL	CVG	X67
11:35aM	4:40pL	CVG	X67
11:35aM	4:55pE	CVG	X67
3:20pM	7:45pE	CVG	D X6
5:35p0	10:20pL	CVG	
5:40pM	10:20pL	CVG	X67
Eff. May 1			
To Norfolk/Virginia Beach/ Williamsburg, VA 707 mi			
6:30a0	12:29p	ATL	B X7
9:54a0	3:10p	ATL	S
11:35aM	3:55p	CVG	X67
1:00p0	6:25p	ATL	L
5:35p0	9:40p	CVG	
5:40pM	9:40p	CVG	X67
To Oakland, CA 1843 mi			
Also see San Francisco, San Jose			
8:00a0	12:05p	SLC	B
11:45a0	3:45p	SLC	
3:10p0	7:10p	SLC	D
6:15p0	9:55p	SLC	
To Oklahoma City, OK 695 mi			
8:30a0	12:35p	DFW	B
10:18a0	2:15p	DFW	X67
12:10p0	4:15p	DFW	X7
3:50p0	9:25p	DFW	S
8:45p0	12:35a	DFW	S
To Ontario, CA 1707 mi			
Also see Los Angeles, Burbank, Long Beach and Orange County			
8:00a0	11:55a	SLC	B
8:30a0	1:10p	DFW	
11:45a0	3:25p	SLC	
12:10p0	4:20p	DFW	X7
3:50p0	8:10p	DFW	
6:15p0	9:45p	1	
To Orange County, CA 1732 mi			
Also see Los Angeles, Burbank, Long Beach and Ontario			
6:15a0	12:05p	CVG	X67
8:00a0	12:05p	SLC	B
8:30a0	1:15p	DFW	
9:25aM	3:40p	CVG	6
Eff. May 6			
11:45a0	3:40p	SLC	
3:10p0	7:35p	SLC	S
3:50p0	8:10p	DFW	
To Orlando, FL 995 mi			
5:30a0	10:00a	CVG	B X67
8:15a0	11:42a		B
9:25aM	2:35p	CVG	L6
Eff. May 6			
9:45a0	2:54p	ATL	S
9:57a0	2:35p	CVG	L
11:35a0	5:00p	ATL	L
2:10p0	5:37p	0	S
3:14p0	8:29p	ATL	S X6
Disc. Apr. 30			
3:14p0	8:29p	ATL	S
Eff. May 1			
3:14p0	8:29p	ATL	S6
3:20pM	8:52p	CVG	DX6
3:50p0	8:52p	CVG	D
4:55p0	9:44p	ATL	D
6:44p0	11:35p	ATL	D
7:59p0	12:55a	ATL	S
To Palm Springs, CA 1658 mi			
8:00a0	11:45a	SLC	B
Disc. Apr. 30			
11:45a0	3:45p	SLC	
3:10p0	8:25p	SLC	S
6:15p0	10:25p	SLC	X5
To Panama City, FL 812 mi			
6:30a0	10:15a	ATL	B X7
8:10a0	12:05p	ATL	B
9:54a0	1:35p	ATL	S
11:35a0	3:29p	ATL	L
1:00p0	5:07p	ATL	L
3:14p0	6:59p	ATL	S X6
Disc. Apr. 30			
3:14p0	6:59p	ATL	S
Eff. May 1			
3:14p0	6:59p	ATL	S6
4:55p0	8:45p	ATL	D
6:44p0	10:30p	ATL	D X6
To Paris, Fra 4154 mi			
Orly Airport			
11:35aM	7:25a	CVG	X67
Eff. May 8; aircraft change enroute			
11:35aM	7:25a	CVG	X67
Disc. May 5; aircraft change enroute			
1:00p0	7:25a	ATL	
To Pasco/Richland/ Kennewick, WA 1579 mi			
8:00a0	11:40a	SLC	B
3:10p0	7:33p	SLC	S
6:15p0	9:40p	SLC	
To Pensacola, FL 787 mi			
8:10a0	11:53a	ATL	B
11:35a0	3:12p	ATL	L
3:14p0	6:50p	ATL	S X6
Disc. Apr. 30			
3:14p0	6:50p	ATL	S
Eff. May 1			
3:14p0	6:50p	ATL	S6
6:44p0	10:20p	ATL	D
To Philadelphia, PA 669 mi			
6:15a0	11:50a	CVG	B/SX67
11:35aM	4:23p	CVG	X67
5:35p0	10:05p	CVG	
5:40pM	10:05p	CVG	X67
Eff. May 1			
To Phoenix, AZ 1445 mi			
6:15a0	9:50a	CVG	X67
8:00a0	11:30a	SLC	B
8:30a0	12:15p	DFW	
12:10p0	3:35p	DFW	X7
3:20pM	7:39p	CVG	X6
3:50p0	7:22p	DFW	
3:50p0	7:39p	CVG	
6:15p0	9:45p	SLC	
6:44p0	10:40p	DFW	
To Pittsburgh, PA 403 mi			
7:40aM	11:45a	CVG	X67
11:35aM	3:50p	CVG	X67
5:35p0	9:40p	CVG	
5:40pM	9:40p	CVG	X67
Eff. May 1			
To Pocatello, ID 1268 mi			
8:00a0	12:55p	SLC	B X67
11:45a0	4:59p	SLC	X6
3:10p0	6:45p	SLC	S
6:15p0	10:25p	SLC	
To Portland, ME 894 mi			
11:35aM	6:30p	CVG	X67
To Portland, OR 1749 mi			
6:15a0	11:00a	CVG	X67
8:00a0	12:00n	SLC	B
11:45a0	3:35p	SLC	
12:10p0	6:19p	DFW	X7
3:10p0	7:05p	SLC	D
3:20pM	10:20p	CVG	X6
6:15p0	10:20p	SLC	
To Raleigh/Durham, NC 636 mi			
5:30a0	9:20a	CVG	B/SX67
6:15a0	12:43p	CVG	B/SX67
6:30a0	11:30a	ATL	B X7
7:40aM	12:43p	CVG	X67
9:54a0	2:55p	ATL	S
1:00p0	5:57p	ATL	L
3:20pM	7:50p	CVG	S X6
3:50p0	7:50p	CVG	S
4:55p0	9:50p	ATL	D
To Reno, NV/Tahoe 1680mi			
8:00a0	11:30a	SLC	B
11:45a0	4:27p	SLC	
3:10p0	6:20p	1	S
6:15p0	9:21p	SLC	
To Richmond/ Williamsburg, VA 632 mi			
6:15a0	11:35a	CVG	B/SX67
6:30a0	11:25a	ATL	B X7
7:40aM	11:35a	CVG	X67
9:54a0	3:05p	ATL	S
11:35aM	3:55p	CVG	X67
1:00p0	6:40p	ATL	L
5:35p0	9:50p	CVG	
5:40pM	9:50p	CVG	X67
Eff. May 1			
To Roanoke, VA 520 mi			
5:30a0	9:50a	CVG	S X67
6:15a0	11:45a	CVG	S X67
7:40aM	11:45a	CVG	X67
11:35aM	3:50p	CVG	X367
11:35aM	4:25p	CVG	3
5:35p0	9:55p	CVG	X6
5:40pM	9:55p	CVG	X67
Eff. May 1			
To Rock Springs, WY 1099 mi			
11:45a0	4:30p	SLC	
To Sacramento, CA 1790 mi			
8:00a0	11:50a	SLC	B
11:45a0	3:40p	SLC	
3:10p0	6:40p	SLC	S
6:15p0	9:50p	SLC	
To St. George, UT 1414 mi			
6:15p0	10:37p	SLC	X67
To Salt Lake City, UT 1257mi			
7:40aM	1:55p	CVG	X67
8:00a0	10:09a	0	B
8:30a0	1:36p	DFW	
9:25aM	2:14p	CVG	
Eff. May 6			
11:35aM	5:28p	CVG	S X67
11:45a0	1:57p	0	
3:10p0	5:21p	0	S
3:20pM	8:34p	CVG	X6
6:15p0	8:25p	0	
6:44p0	11:41p	DFW	
To San Angelo, TX 1011 mi			
8:30a0	12:58p	DFW	B
12:10p0	4:30p	DFW	X7
To San Antonio, TX 1040 mi			
8:30a0	12:38p	DFW	B
10:18a0	2:25p	DFW	X67
12:10p0	4:05p	DFW	X7
3:50p0	7:50p	DFW	S
8:45p0	12:40a	DFW	S
To San Diego, CA 1729 mi			
6:15a0	10:20a	CVG	X67
8:00a0	11:50a	SLC	B
Disc. Apr. 30			
8:00a0	11:50a	SLC	B
Eff. May 1			
8:30a0	12:55p	DFW	
11:45a0	3:30p	SLC	
12:10p0	5:25p	DFW	W X7
3:20pM	9:11p	CVG	X6
3:50p0	7:40p	DFW	
6:15p0	10:00p	SLC	
6:44p0	11:05p	DFW	
To San Francisco, CA 1854 mi			
Also see Oakland and San Jose			
6:15a0	11:15a	CVG	X67
8:00a0	12:00n	SLC	B
8:30a0	1:20p	DFW	
11:45a0	3:45p	SLC	C
12:10p0	4:45p	DFW	X7
3:10p0	7:10p	SLC	D
3:20pM	8:35p	CVG	X6
3:50p0	8:35p	CVG	
3:50p0	8:50p	DFW	
6:15p0	10:00p	SLC	
6:44p0	11:45p	DFW	
To San Jose, CA 1837 mi			
Also see San Francisco and Oakland			
7:40aM	3:45p	CVG	X67
8:00a0	12:10p	SLC	B

Measurement Equivalents and Conversion Table

U.S. Customary System

Length
1 inch
12 inches = 1 foot
36 inches = 3 feet = 1 yard

Volume or Capacity
1 pint = 1/2 quart = 1/8 gallon
2 pints = 1 quart = 1/4 gallon
4 pints = 2 quarts = 1/2 gallon
8 pints = 4 quarts = 1 gallon

Dry Measure
1 pint
2 pints = 1 quart
16 pints = 8 quarts = 1 peck
64 pints = 32 quarts = 4 pecks = 1 bushel

Weight
1 ounce (oz)
16 ounces = 1 pound (lb)
2000 pounds = 1 ton

Conversion Factors

Metric System	U.S. Customary System		
Meters	**Yards**	**Inches**	
1	1.094	39.37	
0.914	1	36	
Centimeters	**Inches**	**Feet**	
1	0.394	0.0328	
2.54	1	0.0833	
30.48	12	1	
Kilometers	**Miles**		
1	0.621		
1.609	1		
Grams	**Ounces**	**Pounds**	
1	0.035	0.0022	
28.35	1	0.0625	
453.59	16	1	
1000	35.274	2.205	
Kilograms	**Ounces**	**Pounds**	
1	35.274	2.205	
0.028	1	0.0625	
0.454	16	1	
Liters	**Pints**	**Quarts**	**Gallons**
1	2.114	1.057	0.264
0.473	1	0.5	0.125
0.946	2	1	0.25
3.785	8	4	1

Metric System

Length
1 meter (m) = 100 cm = 1000 mm

1 millimeter (mm)	= 0.001	m
1 centimeter (cm)	= 0.01	m
1 decimeter (dm)	= 0.1	m
1 dekameter (dkm)	= 10	m
1 hectometer (hm)	= 100	m
1 kilometer (km)	= 1000	m

Volume or Capacity
1 liter (L) = 100 cL = 1000 mL

1 milliliter (mL)	= 0.001	L
1 centiliter (cL)	= 0.01	L
1 deciliter (dL)	= 0.1	L
1 dekaliter (dkL)	= 10	L
1 hectoliter (hL)	= 100	L
1 kiloliter (kL)	= 1000	L

Weight
1 gram (g) = 100 cg = 1000 mg

1 milligram (mg)	= 0.001	g
1 centigram (cg)	= 0.01	g
1 decigram (dg)	= 0.1	g
1 dekagram (dkg)	= 10	g
1 hectogram (hg)	= 100	g
1 kilogram (kg)	= 1000	g

Common Conversion Factors
1 centimeter = 0.39 inches
1 meter = 39.4 inches
1 kilometer = 0.62 miles
1 gram = 0.035 ounces
1 kilogram = 2.20 pounds
1 liter = 1.06 quarts

Each formula is identified by lesson number.

1–1 Earnings Formula

$E = rh + t$

where E = earnings, r = hourly rate, h = number of hours, and t = tips

1–3 Commission Formula

$C = prn$

where C = commission, p = price of one item, r = commission rate, and n = number of items sold

1–3 Earnings with Commission Formula

$E = s + rt$

where E = earnings, s = monthly salary, r = commission rate, and t = total sales for the month

1–3 Piece Rate Formula

$E = rn$

where E = earnings, r = the piece or item rate, and n = the number of items

1–4 FICA taxes—Social Security and Medicare

If annual income is less than $57,600, then the deduction for FICA taxes is 7.65% of gross pay.

1–4 Take-home Pay

$T = g - (w + p)$

where T = take-home pay, g = gross pay, w = amount withheld for income taxes, and f = deduction for FICA taxes

2–1 Amount Earned or Charged for Checks

$A = br - cn - s$

where A = amount earned or charged, b = monthly average balance, r = monthly interest rate, n = number of checks, c = charge for each check, and s = service charge

3–1 Simple Interest Formula

$i = prt$

where i = interest, p = principal, r = interest rate, and t = time

3–2 Compound Interest Formula

$B = p(1 + r)^n$

where B = balance, p = original principal, r = interest rate for the time period, and n = total number of time periods

3–2 Rule of 72

$$\frac{72}{\text{Annual interest rate} \cdot 100} = \text{years to double}$$

3–3 Multiplier Effect

$$S = \frac{a}{1 - r}$$

where S = maximum amount of money a demand deposit can create, a = initial loan amount, and $1 - r$ = reserve requirement

$$M = \frac{D + S}{D}$$

where M = multiplier, D = original demand deposit, and S = maximum amount of money a demand deposit can create

4–1 FICA taxes

Social Security: 6.2% of all income at or under $57,600

Medicare: 1.45% of all income at or under $135,000

4–3 Cost Function

$$c = un + f$$

where c = total cost, u = unit cost, n = number of units, and f = fixed cost

4–3 Revenue Function

$$r = sn$$

where r = revenue, s = selling price per unit, or unit price, and n = number of units sold

4–3 Profit Function

$$p = r - c$$

where p = profit, r = revenue, and c = total cost

5–1 Monthly Payment Formula

$$M = \frac{Pr(1 + r)^n}{(1 + r)^n - 1}$$

where M = monthly payment, P = amount of loan, r = monthly interest rate, and n = number of payment periods

5–1 Amount Formula

$$P = \frac{M[(1 + r)^n - 1]}{r(1 + r)^n}$$

where P = amount of loan, r = monthly interest rate, n = number of payment periods, and M = monthly payment

5–3 Loans With Down Payment Formulas

$$D = rC$$

where D = down payment, r = percent for down payment, and C = cost

$$P = C - D$$

where P = loan amount, C = cost, and D = down payment

$$T = nM + D$$

where T = total amount, n = number of payments, M = monthly payment, and D = down payment

5–4 Amortization Schedule Formulas

$$I_1 = rL \qquad I_2 = rB_1$$
$$R_1 = P - I_1 \qquad R_2 = P - I_2$$
$$B_1 = L - R_1 \qquad B_2 = B_1 - R_2$$

where L = loan amount, r = monthly interest rate, P = payment amount, I_1 = interest due at end of month 1, R_1 = loan reduction at end of month 1, and B_1 = balance at end of month 1

5–4 Prepayment Formula

$$A = \frac{M[1 - (1 + r)^{-q}]}{r}$$

where M = monthly payment, r = monthly interest rate, q = number of remaining payment periods and A = prepayment amount

5–4 Rule of 78

The buyer pays a portion of the yearly interest equal to

$\frac{12}{78}$ in month 1

$\frac{11}{78}$ in month 2

$\frac{10}{78}$ in month 3

⋮

$\frac{1}{78}$ in month 12

6–1 Effective Interest Rate Formula

$$i_{\text{eff}} = \left(1 + \frac{i}{12}\right)^{12} - 1$$

where i_{eff} = effective interest rate, and i = APR (annual percentage rate)

6–2 Time-to-Pay-Off Formula

$$n = \frac{\log\left(\frac{M}{M - Pr}\right)}{\log(1 + r)}$$

where P = amount of loan, r = monthly interest rate, M = monthly payment, and n = number of payment periods

6–2 Monthly Payment Formula for Paying Off Balance

$$M = \frac{Pr(1 + r)^n}{(1 + r)^n - 1}$$

where P = current balance, r = monthly interest rate, n = number of payment periods, and M = monthly payment

6–5 Average Daily Balance Formula

$$b = \frac{s}{d}$$

where b = average daily balance, s = sum of the daily balances, and d = total number of days in the billing cycle

7–1 Replacement Life Insurance Formula

$$R = mS$$

where R = required replacement insurance, m = multiple from the table, and S = original gross salary or wage

7–2 Probability of an Event

$$P(E) = \frac{m}{n}$$

where $P(E)$ = the probability of an event E, m = the number of times the event occurs, and n = the number of all possible outcomes

7–2 Expected Value Formula

$$E = P_1v_1 + P_2v_2$$

where v_1 and v_2 are values and P_1 and P_2 are the corresponding probabilities

7–2 Profit on Insurance

$$P = R - B - C$$

where P = profit, R = revenue received as premiums, B = benefits paid out, and C = costs or expenses

7–3 Future Value of a Periodic Investment Formula

$$A = \frac{p[(1 + r)^n - 1]}{r}$$

where A = the future value of the investment, p = the investment made at the end of each period, r = the interest rate for the period, and n = the number of periods

8–2 Commission Cost Formula

$$c = npr$$

where c = cost of commission, n = number of shares, p = price of one share, and r = commission rate

8–2 Rate-of-Change Formula

$$r = \frac{|P_n - P_o|}{P_o}$$

where r = rate of increase or decrease, P_o = original price, and P_n = new price

9–1 Formulas for Tax Rate Schedule X (for single taxpayers)

i. $t = 0.15I$
if $I \leq 21{,}450$

ii. $t = 3{,}217.50 + 0.28(I - 21{,}450)$
if $21{,}450 < I \leq 51{,}900$

iii. $t = 11{,}743.50 + 0.31(I - 51{,}900)$
if $51{,}900 < I$

where I = taxable income and t = tax on the income

9–1 Formulas for Tax Rate Schedule Y-1 (for married filing jointly or qualifying widow or widower)

i. $t = 0.15I$
if $I \leq 35{,}800$

ii. $t = 5{,}370.00 + 0.28(I - 35{,}800)$
if $35{,}800 < I \leq 86{,}500$

iii. $t = 19{,}566.00 + 0.31(I - 86{,}500)$
if $86{,}500 < I$

where I = taxable income and t = tax on the income

10–3 Average Miles per Gallon

$$a = m \div g$$

where a = average miles per gallon, m = total miles driven, and g = number of gallons purchased

10–3 Average Cost per Mile

$$b = c \div m$$

where b = average cost per mile, c = total cost, and m = total miles driven

10–4 Annual Insurance Premium Formula

$$p = 2rb$$

where p = the annual premium paid to the company, r = the driver rating factor, and b = the six month basic rate for the car

11–1 Number of Gallons Needed

$$g = \frac{d}{a}$$

where g = number of gallons needed, d = total distance in miles, and a = average miles per gallon

11–1 Total Cost for Gas

$$t = gc$$

where t = total cost for gas, g = number of gallons, and c = average cost per gallon

12–4 Periodic Investment Formula

$$p = \frac{Ar}{(1 + r)^n - 1}$$

where p = periodic investment, A = future value of periodic investment, r = periodic interest rate, and n = number of periods

14–2 Common Conversion Factors

1 centimeter (cm) = 0.39 inches (in.)
1 meter (m) = 39.4 inches
1 kilometer (km) = 0.62 miles (mi)
1 gram (g) = 0.035 ounces (oz)
1 kilogram (kg) = 2.2 pounds (lb)
1 liter (l) = 1.06 quarts (qt)

SELECTED ANSWERS

CHAPTER 1

Algebra Refresher

1. 10.5
3. 11.8
5. 379.5
7. 5
9. 28.8
11. 35,175
13. $x = 11$
15. $t = 1377$
17. $x = 75$

Lesson 1–1

Algebra Review

1. 29
3. 137.5
5. 167.8
7. 120.5

Try Your Skills

1. $342.75
3. $397.38
5. Weekly: $714
Monthly: $3,094
Yearly: $37,128
7. Hourly: $17.50
Monthly: $3,033.33
Yearly: $36,400

Exercise Your Skills

1. Hourly employees are paid for exact hours worked including overtime. Salaried employees are paid a flat rate no matter how many hours they work.
3. Answers may vary. (Ex.: to complete assigned projects)
5. Answers may vary. (Ex.: 10–20 hours)
7. $339.75
9. $291.50
11. $393.75
13. $394.38
15. $327.25
17. $465.63
19. Weekly: $428
Monthly: $1,854.67
Yearly: $22,256
21. Hourly: $7.50
Monthly: $1,300
Yearly: $15,600
23. Hourly: $11.25
Monthly: $1,950
Yearly: $23,400
25. Hourly: $19.23
Weekly: $769.23
Monthly: $3333.33

Lesson 1–2

Algebra Review

1. 4.8
3. 12
5. 200
7. 175

Try Your Skills

1. **a.** $1698.00 **b.** $1200.00
3. **a.** $5954.50 **b.** $3807.69
5. **a.** $42,372.50; $41,680; $40,990 **b.** A
7. $3933.88

Exercise Your Skills

1. Answers may vary.
3. Answers may vary. (Ex.: What do you like most about your job? What skills do you use? What training did you need?
5. **a.** $1518 **b.** $600
7. **a.** $4542.00 **b.** $3033.33
9. **a.** $9273 **b.** $3675
11. **a.** $10,398.00 **b.** $4,766.67
13. **a.** $29,280; $33,000; $30,675 **b.** B

Mixed Review

1. $406.00
3. $464
5. Weekly: $250
Monthly: $1,083.33
Yearly: $13,000
7. Hourly: $21.92
Weekly: $876.92
Yearly: $45,600

Lesson 1–3

Algebra Review

1. $x = 2$
3. $a = 3$
5. $x = 2028$
7. $12.5x$
9. $10b$
11. $14y + 5$

Try Your Skills

1. **a.** \$15.08 **b.** $y = 1.5075x$
   c. \$15.08; \$27.14; \$36.18; \$48.24
3. **a.** \$8.23 **b.** $y = 1.64625x$
   c. \$8.23; \$19.76; \$24.69; \$29.63
5. \$718.75; \$1193.75
7. \$690
9. 20 memberships

Exercise Your Skills

1. Since commission is a percent of sales, you will earn more.
3. No. Next month's sales will probably be back to average or below.
5. **a.** \$6.75
   b. $y = 0.375x$
   c. \$6.75; \$10.50; \$14.25; \$18.00
7. **a.** \$11.39
   b. $y = 0.75905x$
   c. \$11.39; \$18.98; \$26.57; \$34.16
9. \$877.61
11. \$3497
13. $E = 0.07t + 2400$; $E = 0.10t$, $t = \$80{,}000$
15. \$620

Mixed Review

1. \$450
3. **a.** \$5460.00 **b.** \$4000.00
5. **a.** \$8047.50 **b.** \$3807.69

Lesson 1–4

Algebra Review

1. 0.05
3. 0.075
5. 0.0765
7. 30%
9. 150%
11. 6
13. 969.6
15. 376.38

Try Your Skills

1. \$1930.40
3. \$2850.90
5. \$1182.08
7. \$1710.00
9. \$2168.92

	Income Tax Withholding	FICA Withholding	Take-Home Pay
11.	\$286	\$183.60	\$1930.40
13.	289	260.10	2850.90
15.	0	97.92	1182.08
17.	137	153.00	1710.00
19.	186	195.08	2168.92

Exercise Your Skills

1. Answers may vary. (Ex.: no one likes tax increases, tax laws affect many people)
3. The workers in the future
5. Retirement income will not cover living expenses.
7. \$2040.25
9. \$1384.25
11. \$1355.25
13. \$3305.92
15. \$2873.25

	Income Tax Withholding	FICA Withholding	Take-Home Pay
17.	268	191.25	2040.75
19.	1	114.75	1384.25
21.	30	114.75	1355.25
23.	542	318.75	3305.92
25.	359	267.75	2873.25

Mixed Review

1. \$568.75
3. \$3885
5. 75%; $1.50n \bullet 5 = 150 + 5n$; $n = 60$; 60 is 75% of 80.
7. \$398

Chapter 1 Review

1. Hourly. You would be paid for overtime.
3. Deductions are subtracted.
5. \$354.50
7. \$1116.00
9. \$144.38
11. \$51,806
13. \$10,215; \$837.63
15. \$2,925; \$204.75
17. \$544
19. \$400
21. \$193; \$114.75; \$1192.25
23. \$83; \$132.44; \$1515.76
25. **a.** \$3237.75 **b.** \$1988.46

Cumulative Review

1. $520
3. $250
5. No. Some benefits, such as health insurance, are the same for all employees.
7. $12,500
9. $4851.00
11. $494; $228.74; $2267.26
13. $213; $168.30; $1818.70

CHAPTER 2

Algebra Refresher

1. (5, −3)
3. (2.5, 1.5)
5. (4, 3)
7. (2, 1)
9. (3, 0)

Lesson 2–1

Algebra Review

1. 3, 7.5, 21
3. −3, −7.5, −21; they are opposites.
5. −4, −8.5, −22; they are one less than the opposites.
7. −2, 0, 2
9. 67.7, 2.7, −1.3

Try Your Skills

1. $0.02 earned
3. $3.68 charged
5. $1.28 earned
7. $2.00; $1.80; $0; +$0.20
   $2.16; $0.40; $0; +$1.76
   $1.40; $1.20; $0; +$0.20

Exercise Your Skills

1. $A = br - cn - s$
3. Information is concise and easy to compare.
5. Writing large numbers of checks
9. C3, F3, G3
11. cn
13. $b = A3$, $r = B3$, $c = D3$, $A = G3$
15. $3.25; $3.00; $0.41; $0; $649.84
    $3.13; $3.00; $0.59; $0; $624.54
    $2.40; $3.00; $0.45; $0; $478.95
    $0.95; $3.00; $0.70; $0; $187.25
17. $A = br$
19. 0.4%
21. Bank S: $y = 4.4 - 0.07x$
    Bank T: $y = 2 - 0.025x$
    $x = 53.3$
    The account at Bank S will be more expensive if fewer than 53 checks are written and less expensive if more than 53 checks are written.

Mixed Review

1. $275.50
3. $3600
5. 20%
7. B5 = 0.0765*B4
9. B8 = B4 − B7

Lesson 2–2

Algebra Review

1. $y = 2x$
3. $y = 2x + 1$
5. $y = -2x + 1$
7. $y = 2x$
9. $y = 3x - 0.5$

Try Your Skills

1. Balance: $400.00
3. $196.50
5. $195.87

Exercise Your Skills

1. So you will not forget to enter it later
3. How much money remains in the account
9. 478.50
11. 428.50
13. 409.24
15. 279.20
17. 518.58
19. 367.15
21. 468.65
23. 419.65
25. 306.65
27. 301.65
29. 221.35
31. 311.85
33. 288.57
35. 276.52
37. 252.41
39. 145.70
41. Answers may vary. (Ex.: For deposit only. Latoya S. Marshall.)

43. Answers may vary. (Ex.: Pay to the order of Judy Bruns. L. Marshall.

Mixed Review

1. $14.10
3. $446.50
5. $258
7. $5625
9. $0.39 charged

Lesson 2–3

Algebra Review

1. $x = 2$
3. $x = 4$
5. $x = 1$
7. $x = 3$
9. $x = 4$

Try Your Skills

1. deposits not recorded
3. check fees, service charges, ATM withdrawal
5. $23.88; $40.00
7. $88.72

Exercise Your Skills

1. Answers may vary. (Ex.: convenient, open 24 hours)
3. additional deposits made after statement prepared, outstanding checks
5. 11
7. $140.22
9. $2.30
11. $326.68
13. $570.56

Mixed Review

1. $225.25
3. $1158
5. $0.20 charged
7. $4.35
9. $250.70
11. $39.76

Chapter 2 Review

1. $0.42
3. $423.57
5. Ridgewood Savings Bank: $y = 6.5 - 0.10x$
   Second National Bank: $y = 3.5 - 0.04x$
   $x = 50$
   The account at Ridgewood Savings Bank will be more expensive if fewer than 50 checks are written and less expensive if more than 50 checks are written.
7. $455.78
   369.83
   331.30
   323.85
   309.09
   255.19
   360.07
   343.79
9.

Beginning Balance	Number of Checks	Cost of Checks	Extra Charges	Service Charges	Interest Earned	New Balance
$155	11	$0.25	0	$5.00	$0.70	$150.45
278	17	0.38	0	5.00	1.25	273.87
505	21	0.47	$0.15	0	2.27	506.65
789	27	0.61	1.05	0	3.55	790.89

Cumulative Review

1. $64.00
3. $112.70
5. $9500
7. $154.38
9. $1.00
11. Answers may vary.
13.

Amount	Balance
	95.16
24.65	−24.65
	70.51

CHAPTER 3

Algebra Refresher

1. $11a + 12$
3. $x + 7y$
5. $4x + 5.97$
7. $31a - 20b$
9. $-x^2 + x + 1.06$

Lesson 3–1

Algebra Review

1. $A = 18.2$
3. $s = 20.6$
5. $i = 3375$
7. $B = 780.63$

Try Your Skills

1. $520; $18.20; $538.20
3. $1000; $35.00; $1035.00
5. $1500; $52.50; $1552.50
7. 17; $340
9. 19; $342
11. 10; $350
13. 6 wk
15. 20 wk

Exercise Your Skills

1. The bank has use of the funds for a guaranteed amount of time.
3. to avoid penalties for withdrawing the funds
5. $2400; $96.00; $2496.00
7. 10; $150
9. 6; $150
11. 14; $210
13. 8; $200
15. 10; $150
17. 6; $150
19. 8 wk
21. 4 wk
23. 5 wk; 5 wk; 3 wk.

Mixed Review

1. $13,000
3. $3.20; $3.50; $0.42; $799.28
5. $0.66; $3.50; $0.24; $162.92
7. $152.48
9. $183.75

Lesson 3–2

Algebra Review

1. x^4
3. x^8
5. $15y^8$
7. a^6
9. 15,609.06

Try Your Skills

1. First Period: $60.00; $1060.00
   Second Period: $63.60; $1123.60
3. $328.10
5. $55,093.29
7. 48 years

Exercise Your Skills

1. 12
3. 3
5. 1.5%; 3
7. $15.23; $1030.23
9. $15.69; $1061.37
11. $16.16; $1093.45
13. $16.65; $1126.50
15. $169.02
17. $2770.17; slightly higher interest rates yield much higher interest over time
19. 36 y
21. 21.6 y
23. 14 y
25. 16 y
27. Robert:

$$2000(1.06)^{41} = \$21,805.72$$

Amos:

$$2000(1.06)^{30} = 11,486.98$$
$$2000(1.06)^{20} = 6,414.27$$
$$2000(1.06)^{10} = 3,581.70$$
$$\$21,482.95$$

Mixed Review

1. $627.75; $1527.75
3. $80 deposit
5. $189.63
7. Bank A; Bank B is a better deal only if the number of checks reaches 188 checks a month.

Lesson 3–3

Algebra Review

1. yes; 0.5
3. no
5. yes; 0.04
7. yes; 0.07

Try Your Skills

1.

Level	Demand Deposit	Required Reserve	Loans and Investments
1	$500.00	$100.00	$400.00
2	400.00	80.00	320.00
3	320.00	64.00	256.00
4	256.00	51.20	204.80
5	204.80	40.96	163.84
6	163.84	32.77	131.07
7	131.07	26.21	104.86
8	104.86	20.97	83.89
9	83.89	16.78	67.11
10	67.11	13.42	53.69
		Total after 10 levels	$1785.26

3.

Level	Demand Deposit	Required Reserve	Loans and Investments
1	$500.00	$125.00	$375.00
2	375.00	93.75	281.25
3	281.25	70.31	210.94
4	210.94	52.74	158.20
5	158.20	39.55	118.65
6	118.65	29.66	88.99
7	88.99	22.25	66.74
8	66.74	16.69	50.05
9	50.05	12.51	37.54
10	37.54	9.39	28.15
		Total after 10 levels	$1415.51

5. $2,000 **7.** $1,500
9. 5 **11.** 4
13. The greater the reserve requirement, the less extra money generated.

Exercise Your Skills

1. No, the information is transmitted electronically.
3. A financial institution creates extra money or credit from excess reserves.
5.

Level	Demand Deposit	Required Reserve	Loans and Investments
1	$500.00	$75.00	$425.00
2	425.00	63.75	361.25
3	361.25	54.19	307.06
4	307.06	46.06	261.00
5	261.00	39.15	221.85
6	221.85	33.28	188.57
7	188.57	28.29	160.28
8	160.28	24.04	136.24
9	136.24	20.44	115.80
10	115.80	17.37	98.43
		Total after 10 levels	$2275.48

7.

Level	Demand Deposit	Required Reserve	Loans and Investments
1	$12,500.00	$1875.00	$10,625.00
2	10,625.00	1593.75	9,031.25
3	9,031.25	1354.69	7,676.56
4	7,676.56	1151.48	6,525.08
5	6,525.08	978.76	5,546.32
6	5,546.32	831.95	4,714.37
7	4,714.37	707.16	4,007.21
8	4,007.21	601.08	3,406.13
9	3,406.13	510.92	2,895.21
10	2,895.21	434.28	2,460.93
		Total after 10 levels	$56,888.06

9.

Level	Demand Deposit	Required Reserve	Loans and Investments
1	$12,500.00	$2500.00	$10,000.00
2	10,000.00	2000.00	8,000.00
3	8,000.00	1600.00	6,400.00
4	6,400.00	1280.00	5,120.00
5	5,120.00	1024.00	4,096.00
6	4,096.00	819.20	3,276.80
7	3,276.80	655.36	2,621.44
8	2,621.44	524.29	2,097.15
9	2,097.15	419.43	1,677.72
10	1,677.72	335.54	1,342.18
		Total after 10 levels	$44,631.29

11. $2833.33
13. $70,833.33
15. $50,000.00
17. 6.7
19. 6.7
21. 5
23. the greater the reserve requirement, the less extra money generated
25. $5250

Mixed Review

1. $30,600 **3.** $200; 4 wk
5. $180; 9 wk **7.** $19,087.31
9. $368.66 **11.** 12 y
13. 6 y **15.** $1380

Chapter 3 Review

1. $180
3. $1170
5. 11; $165
7. 4; $200
9. $1300.00; $39.00; $1339.00
11. $3400.00; $102.00; $3502.00
13. $50.00; $2050.00
15. $52.53; $2153.78
17. $25.00; $2025.00
19. $25.63; $2075.94
21. $26.27; $2128.16
23. $26.93; $2181.69
25. $2207.63; $2208.97; They only differ by one cent.
27. 283.41 pounds
29. 24 y
31. 9 y
33.

Level	Demand Deposit	Required Reserve	Loans and Investments
1	$2000.00	$400.00	$1600.00
2	1600.00	320.00	1280.00
3	1280.00	256.00	1024.00
4	1024.00	204.80	819.20
5	819.20	163.84	655.36
		Total after 10 levels	$5378.56

35.

Level	Demand Deposit	Required Reserve	Loans and Investments
1	$11,500.00	$1955.00	$9545.00
2	9,545.00	1622.65	7922.35
3	7,922.35	1346.80	6575.55
4	6,575.55	1117.84	5457.71
5	5,457.71	927.81	4529.90
		Total after 10 levels	$34,030.51

37. \$8000
39. \$56,147.06
41. 5
43. 5.9
45. \$3644.44

Cumulative Review

1. a. \$6742.50 **b.** \$4500.00
3. \$4340
5. \$2.38
7. \$14,116.50
9. Check Number Date To: (name of recipient) For: (purpose of payment) Amount: \$243.50 Balance: \$739.11
11. 18 y
13. 6.9 y

CHAPTER 4

Lesson 4–1

Algebra Review

1. $y = 2x - 1; m = 2$
3. $y = -x + 1; m = -1$
5. $y = x + 3; m = 1$
7. $y = 6.50x$
9. $m = 7.25$

Try Your Skills

1. \$137.50
3. \$10.52
5. \$72.00
7. \$5.51

Exercise Your Skills

1. Answers may vary. (Ex.: good management, good market, quality product)
3. (1) efforts and skills; (2) workers
5.

Employee	Gross Pay	Income Tax Withholding	FICA Withholding	Total Deductions	Take-Home Pay
Kendra	105.00	9.00	8.03	17.03	87.97
Paulo	73.80	0.00	5.65	5.65	68.15
Michael	176.90	19.00	13.53	32.53	144.37
Total	335.70	28.00	27.21	55.12	300.49

7. Week ending 4/4

Employee	Gross Pay	Income Tax Withholding	FICA Withholding	Total Deductions	Take-Home Pay
Catlyn	60.00	2.00	4.59	6.59	53.41
Sara	90.00	0.00	6.89	6.89	83.11
Joleen	208.00	23.00	15.91	38.91	169.09
Hernando	87.36	0.00	6.68	6.68	80.68
Total	445.36	25.00	34.07	59.07	386.29

Week ending 4/11

Employee	Gross Pay	Income Tax Withholding	FICA Withholding	Total Deductions	Take-Home Pay
Catlyn	52.50	1.00	4.02	5.02	47.48
Sara	126.00	5.00	9.64	14.64	111.36
Joleen	128.00	12.00	9.79	21.79	106.21
Hernando	62.40	0.00	4.77	4.77	57.63
Total	368.90	18.00	28.22	46.22	322.68

Week ending 4/18

Employee	Gross Pay	Income Tax Withholding	FICA Withholding	Total Deductions	Take-Home Pay
Catlyn	78.75	4.00	6.02	10.02	68.73
Sara	112.50	3.00	8.61	11.61	100.89
Joleen	128.00	12.00	9.79	21.79	106.21
Hernando	187.20	14.00	14.32	28.32	158.88
Total	506.45	33.00	38.74	71.74	434.71

Week ending 4/25

Employee	Gross Pay	Income Tax Withholding	FICA Withholding	Total Deductions	Take-Home Pay
Catlyn	120.00	11.00	9.18	20.18	99.82
Sara	81.00	0.00	6.20	6.20	74.80
Joleen	192.00	22.00	14.69	36.69	155.31
Hernando	124.80	4.00	9.55	13.55	111.25
Total	517.80	37.00	39.62	76.62	441.18

9.

Employee	Exemptions	Hourly Rate	Hours Worked	Gross Pay	Income Tax Withholding	FICA Withholding	Take-Home Pay
Adikes	1	\$7.85	10	\$ 78.50	\$ 0.00	\$ 6.01	\$ 72.49
Carney	1	5.30	18	95.40	1.00	7.30	87.10
Inez	0	6.78	27	183.06	20.00	14.00	149.06
Sun-Li	0	9.36	17	159.12	16.00	12.17	130.95
Total				516.08	37.00	39.48	439.60

11. When the total gross salary has reached $135,000, there are no FICA taxes for the remainder of the year.
13. $4824.00
15. 5528.70
17. $135,000

Mixed Review

1. $225.60
3. $75
5. 10 wk
7. $1071.91
9. $760

Lesson 4–2

Algebra Review

1. $x = 10$
3. $x = 6$
5. $x = 20$
7. $x = -1.5$
9. $x = -\frac{1}{3}$
11. $x = 4$
13. $m = 5$
15. $m = 3$

Try Your Skills

1. $60.00
3. $7.00
5. $7.00
7. $12.00

Exercise Your Skills

1. labor, advertising, energy and transportation, packaging
3. Yes, probably a lot
5. Labor: $202.50; Materials: $208.00; Advertising, energy, and transportation costs: $32.50; Cost: $443.00
7. Labor: $240.00; Materials: $200.00, $20.00; Packaging: $10.00; Advertising, energy, and transportation costs: $32.50; Cost: $502.50
9. Labor: $180.00; Materials: $200.00, $40.00; Advertising, energy, and transportation costs: $32.50; Cost: $452.50
11. Labor: $250.00; Materials: $150.00, $20.00; Packaging: $7.50; Advertising, energy, and transportation costs: $32.50; Cost: $460.00
13. Labor: $80.00; Materials: $215.00; Advertising, energy, and transportation costs: $32.50; Cost: $327.50
15. Labor: $157.50; Materials: $175.00, $25.00; Advertising, energy, and transportation costs: $32.50; Cost: $390.00
17. lower price, enhance quality of product
19. decrease price, find ways to decrease costs

Mixed Review

1. $55.77
3. $415.07
5. $8349.79
7.

Level	Demand Deposit	Required Reserve	Loans and Investments
1	$1800	$270.00	$1530.00
2	1530	229.50	1300.50

9. $530.03
11. $5528.70
13. 48 y
15. 27 y
17. 5.9

Lesson 4–3

Algebra Review

1. $c = 12g$
3. $c = hn$
5. $c = e \div 15$
7. Answers may vary. r = sales revenue, p = unit price, n = number units sold; $r = pn$
9. Answers may vary. x = unit price of first item, y = unit price of second item, a = average unit price; $a = (x + y) \div 2$

Try Your Skills

1. $c = 6.77(23) + 282 = \$437.71$
3. $4.20
5. $209.40
7. $p = 3.8n - 75$
9. +A4*B4+C4
11. +F4−D4

Exercise Your Skills

1. competition, demand
3. The cost is greater than the revenue.
5. It can drive some companies out of business. It can create unfair high prices.
7. $502.50; $732.50; $962.50
9. $500.00; $1000.00; $1500.00
11. ($2.50); $267.50; $537.50
13.

Unit Cost	Number Produced	Fixed Cost	Total Cost	Unit Price	Revenue	Profit (Loss)
$1.20	200	$212.50	$452.50	$2.00	$400.00	($52.50)
1.20	300	212.50	572.50	2.00	600.00	27.50
1.20	400	212.50	692.50	2.00	800.00	107.50

15. $c = 3.55n + 282.50$; $r = 10n$; $p = 6.45n - 282.50$

19.

Unit Cost	Number Produced	Fixed Cost	Total Cost	Unit Price	Revenue	Profit (Loss)
$2.75	100	$112.50	$387.50	$4.00	$ 400.00	$ 12.50
2.75	200	112.50	662.50	4.00	800.00	137.50
2.75	300	112.50	937.50	4.00	1200.00	262.50

21. $c = 4n + 190$; $r = 7.5n$; $p = 3.5n - 190$

Mixed Review

1. $5.00 per hour
3. $204.58
5. To: Farley Smith; For: VCR repair; Balance: $363.50
7. To: Abe's Supermarket; For: groceries; Balance: $668.73
9. 9 memberships

Lesson 4–4

Algebra Review

1. $y = -x + 10$; $m = -1$
3. $y = \frac{2}{3}x + 2$; $m = \frac{2}{3}$
7. (10, −2)
9. (7.18, 0.53)

Try Your Skills

1. $c = 1.04n + 242.5$
3. 253 stickers

Exercise Your Skills

1. cost function and revenue function
3. $c = 1.20n + 212.50$
5. 266 pennants
7. $r = 4n$
9. 90 cans
11. $r = 9n$
15. 25 book bags
17. $c = 2.30n + 272.50$
21. $p = 2.7n - 272.50$
23. 173 towels

Mixed Review

1. 12%
3.

Level	Demand Deposit	Required Reserve	Loans and Investments
1	$5000	$1000	$4000
2	4000	800	3200

5. 5
7. $172.50

Lesson 4–5

Algebra Review

3. (3, 2)
7. (0, 6), (4, 4), (4, 6)

Try Your Skills

1. (4, 8)
3. (4, 8), (12, 8), (12, 4)
5. A (7.5, 45) represents the least number of hours (7.5 hours) you can work in a week to earn $45.
   B (10, 45) represents the lowest rate you would accept (45 ÷ 10 = $4.50).
   C (10, 60) represents the most you can earn ($60).

Exercise Your Skills

1. $x \leq 200$; $y \leq 400$; $x + y \geq 300$
3. A(0, 300); B(200, 100); C(200, 400); D (0, 400)
5. $r = 2x + 2y$
7. 200 pennants; 400 stickers
9. $x \geq 4000$; $y \geq 6000$; $x + y \leq 15{,}000$
11. A(4000, 6000); B(9000, 6000); C(4000, 11,000)
13. $c = 3.55x + 6.40y$
17. 40 mugs. 10 organizers
19. $x \geq 40$; $y \geq 50$; $x \leq 80$; $y \leq 2x$; $x + y = 210$
21. A(40, 50); B(80, 50); C(80, 130); D(70, 140), E(40, 80)
23. $c = 2.5x + 9y$
25. $p = 1.5x + y$
29. 5 pairs of earrings; 10 necklaces
31. 30 pairs of earrings; 15 necklaces

Mixed Review

1. 25
3. $1146.74
5. $13,150.13
7. $9.41; $94.59
9. $15.38; $157.62
11. $50

Chapter 4 Review

1. $127.30
3. $930.70
5. $5.00; $9.74; $112.56
7. $183.00; $71.20; $676.50
9. $162.90
11. $253.70
13. profit: $71.70
15. $117.50; $2.27; $162.90; $7.00; $140.00; ($22.90)
17. $117.50; $2.27; $253.70; $7.00; $420.00; $166.30
19. $r = 7n$

23. 25 kittens
25. Answers may vary. (Ex.: higher prices, higher production, lower costs, and so on)
27. $c = 2.27x + 3y$
29. $p = 4.73x + 4.5y$
33. 40 kittens and 10 puppies
35. 40 kittens and 20 puppies

Cumulative Review

1. $192.88
3. $19,972.68
5. $79.81
7. Answers may vary.
9. $43.75
11. $22,000
13. Loss: $10

CHAPTER 5

Algebra Refresher

1. $\frac{1}{64}$
3. $\frac{1}{531,441}$
5. 4096
7. $\frac{x^2z^{11}}{y^7}$
9. $\frac{x^{12}}{y^4z^{14}}$
11. 0.136
13. 0.0078
15. 16
17. 22.01
19. 6702.88

Lesson 5–1

Algebra Review

1. 0.89
3. 500
5. 1.03
7. 4800
9. 4.34
11. $\frac{y}{(1-r)^2}$
13. $\frac{3rq^{24}}{z(q+2)^{24}}$

Try Your Skills

1. $198.01; $11,880.72
3. $96.80; $76.09; $63.74
5. $161.34; $126.81; $106.24
7. $225.87; $177.54; $148.73
9. $4648.69; $5914.22; $7059.81

Exercise Your Skills

1. People do not like the sound of being "in debt."
3. The loan enables someone to acquire a home or a car long before he or she could save the full amount needed to purchase it.
5. repossess the item(s) bought with the credit
7. $323.27; $15,517.17
9. $217.45; $7,828.15
11. $146.24; $8,774.20
13. $252.51; $12,120.48
15. $200.72; $7,225.98
17. $134.99; $8,099.26
19. $318.96; $15,310.08
21. $585.44; $21,075.78
23. $393.71; $23,622.84
25. $309.66; $14,863.54
27. $293.56; $10,568.00
29. $197.42; $11,845.16
31. $318.00; $248.85; $207.58
33. $381.60; $298.62; $249.10
35. $445.20; $348.39; $290.62
37. $9434.04 for 3 years; $12,055.43 for 4 years; $14,452.01 for 5 years

Mixed Review

1. $2.54
3. $8164.22
5. 9 wk
7. $948
9. 78 items

Lesson 5–2

Algebra Review

1. 174.46
3. $z = 6978.40$
5. $s = 278.40$
7. $x = 44.70$
9. $z = 5659.61$
11. 48(359.28) is smaller.

Try Your Skills

3. $3,187.06; $687.06
5. $126.81; $6,087.02; $1087.02
7. $242.00; $8,712.14; $1212.14
9. $159.35; $9,561.17; $2061.17
11. $253.63; $12,174.04; $2174.04
13. Total cost is $2061.17 versus $808.09. You might consider total cost as a factor in your choice.

Exercise Your Skills

1. You could find yourself with excessive debt.
3. The stress and worry can undermine one's health.
5. $8,292.96; $1792.96
7. $317.81; $11,441.14; $1941.14
9. $213.73; $12,823.82; $3323.82
11. $159.48; $7,655.04; $1655.04
13. $401.44; $14,451.97; $2451.97
15. $269.98; $16,198.52; $4198.52
17. $465.15; $22,327.20; $4827.20
19. $389.73; $14,030.45; $2380.45
21. $262.10; $15,726.06; $4076.06
23. $233.24; $11,195.50; $2420.50
25.

Loan Amount	Number of Years	Monthly Payment	Total Payment	Total Cost
$20,000	5	$400.76	$24,045.54	$4045.54
20,000	6	345.80	24,897.76	4897.76
20,000	7	306.77	25,768.30	5768.30
20,000	8	277.68	26,657.03	6657.03
20,000	9	255.22	27,563.79	7563.79
20,000	10	237.40	28,488.42	8488.42

Mixed Review

1. $31.06
3. $238.75
5. 9 y

Lesson 5–3

Algebra Review

1. $1110
3. $1290
5. 2160
7. $z = 3600(0.6)$

Try Your Skills

1. $114.40; $4118.24; $106.48; $3833.34; $284.90
3. $62.40; $2246.31; $58.08; $2090.91; $155.40
5. $2490; $9,960; $80.20; $26,550.19; $1762.52

Exercise Your Skills

1. the income left after paying for basic needs
3. The financial institution makes higher profits.
5. $83.04; $2989.29; $79.50; $2861.98; $127.31
7. $33.21; $1195.72; $31.80; $1144.79; $50.93
9. $83.04; $2989.29; $1750.00; $58.13; $2842.50; $146.79
11. $33.21; $1195.72; $700.00; $23.25; $1137.00; $58.72
13. $8400.00; $33,600; $45,198.38; $46,304.47
15. $8400.00; $33,600; $46,276.65; $47,773.16
17. $8400.00; $33,600; $47,374.93; $49,277.21
19. 5 y; 8%; 10% down

Mixed Review

1. $21,500
3. $446.77 − 57.49 = 389.28
5. $29,900
7. The average-balance method; the minimum-balance method will result in extra charges even if your balance drops below the limit for one day.
9. 13 items

Lesson 5–4

Algebra Review

1. 1.25%
3. 44
5. 0.06
7. 19
9. 1.0016
11. 450
13. 3688.87

Try Your Skills

1. $1963.03
3. $46.97; $9.63; $37.34; $1888.54
5. $1515.26
7. $20.77

Exercise Your Skills

1. When the prepayment penalty is less than the interest.
3. It is a schedule that lists the interest portion of monthly loan payments.
5. 2; $231.92; $62.21; $169.71; $9161.70
7. 4; $231.92; $59.94; $171.98; $8818.88
9. 6; $231.92; $57.64; $174.28; $8471.47
11. 8; $231.92; $55.31; $176.61; $8119.42
13. 10; $231.92; $52.94; $178.98; $7762.65
15. 12; $231.92; $50.55; $181.37; $7401.11
17. $2937.82
19. $26,028.08
21. $40.38

Mixed Review

1. **a.** $3346.67; **b.** $2964.53; loan **a** is greater
3. $9566.82
5. $895
7. $x + y \leq 160$; $x \geq 40$; $x \leq 100$; $y > 60$
9. (40, 60), (40, 120), (100, 60)

Lesson 5–5

Algebra Review

1. 7.578
3. 1.572
5. y
7. $48y$

Try Your Skills

1. $33,600; $34,600
   $812.41; $765.85
   $40,064.17; $38,995.73; $36,760.69
   Plan 3
3. $86.27

Exercise Your Skills

1. Answers may vary
3. Some lending institutions will give you a better deal than others.
5. $10,500; $11,200; $12,000
   $353.79; $354.86; $359.65
   $12,736.31; $12,774.79; $12,947.43
   Plan 1
7. $6800; $7550; $7800
   $172.47; $183.43; $183.18
   $8278.35; $8804.78; $8792.79
   Plan 1
9. 0; $1884.09; $1198.72
   0; $600.00; $1200.00
   0; $2484.09; $2398.72
   Plan 2
11. 0; $1548.10; $990.82
    0; $300.00; $1500.00
    0; $1848.10; $2490.82
    Plan 3
13. 0; $1699.73; $861.78
    0; $1000.00; $2000.00
    0; $2699.73; $2861.78
    Plan 3
15. $719.10
17. $599.75

Mixed Review

1. $484.50
3. $70.38
5. $31
7. $2156
9. $75.10
11. $r = 4n$

Chapter 5 Review

1. $326.19
3. $52.87
5. $1766.95; $530,084.40
7. $33,600.54
9. $8431.20
11. $1096.77; $39,483.82; $3096.61
13. $709.67; $42,580.43; 0
15. $7,036.14; $6,575.98; $480.16
17. $13,305.26; $12,244.60; $1060.66
19. $936; $8424; $11,386.69
21. $2808; $6552; $10,936.31; $450.38
23. $773.79
25. $16.15
27. 0; $2918.71; $1194.62
    0; $500.00; $1750.00
    0; $3418.71; $2944.62
    Plan 2
29. No; sometimes you cannot afford the higher monthly payments.
31. Since the unpaid balance decreases each month, the interest due also decreases.

Cumulative Review

1. $4.00
3. $17,605
5. 16 years
7. $1546.68 - 597.99 = 948.69$
9. $5594.88
11. $29,877.31
13. $21.15
15. The amount of the loan is lower and therefore the interest payments are less.

CHAPTER 6

Algebra Refresher

1. 15
3. 22,765.04
5. 766,584
7. 27,399.68
9. d: {7, 3, 2}; r: {3, 7, 4}
11. Yes; each member of the domain is paired with a different member of the range.
13. 272
15. 2
17. 4
19. January: 6.89%; February: 8.41%; March: 9.85%; April: 8.96%; May: 7.01%; June: 8.76%; July: 9.67%; August: 9.00%; September: 8.53%; October: 7.10%; November: 8.24%; December: 7.58%

Lesson 6–1

Algebra Review

1. 0.125
3. 1.19
5. 3330%
7. 76%
9. 0.13

Try Your Skills

1. $0.00; $0.00; $2300.00
3. $2048.00; $0.00; $20.48; $275.00; $1793.48
5. 12.68%

Exercise Your Skills

1. the amount of time that elapses before you must pay interest on the purchase
3. VISA is a credit card with a monthly payment whereas American Express is a charge card with the entire balance expected on receipt of the bill.
5. $3000.00; $0.00; $40.00; $325.00; $2715.00
7. $0.00; $3215.00
9. $2907.24; $0.00; $33.68; $345.00; $2595.92
11. $2280.99; $0.00; $26.42; $345.00; $1962.41
13. $1640.14; $0.00; $19.00; $345.00; $1314.14
15. $984.36; $0.00; $11.40; $345.00; $650.76
17. $313.30; $0.00; $3.63; $316.93; $0.00
19. 14.37%; 1.13%
21. 17.23%; 1.33%
23. 13.1%
25. 15.68%

Mixed Review

1. $650.67
3. $8121.43
5. 16 wk
7. $150
9. $195.89
11. $r = 10x + 20y$
13. $540

Lesson 6–2

Algebra Review

1. d: {9, 3}; r: {2, 8}
3. No; the domain value of 9 is paired with range values of 2 and 8.

Try Your Skills

1. 2.44
3. 0.16
5. 10 mo
7. 9 mo
9. $1137.18

Exercise Your Skills

1. After 7 months, there would still be a small balance.
3. 2.22
5. 0.50
7. 15 mo
9. 18 mo
11. 18 mo
13. 15 mo
15. 23 mo
17. 19 mo
19. $118.53
21. $54.66
23. $46.64

Mixed Review

1. $3039.77
3. $43.35 per wk
5. 24 y
7. 39 items

Lesson 6–3

Algebra Review

1. 0.139
3. 2.45
5. 45,600%
7. 9%
9. $25.60
11. $797.98
13. 15%
15. 25
17. $52,000

Try Your Skills

1. $12.44; $1007.44; $50.00
3. $921.41; $11.52; $932.93; $47.00
5. $47.00
7. $906.44; $11.33; $917.77; $92.00
9. $752.09; $9.40; $761.49; $76.00
11. $3.51
13. $902.46; $6.77; $909.23; $91.00
15. $742.37; $5.57; $747.94; $75.00
17. $17.55

Exercise Your Skills

1. Answers may vary.
3. the balance due decreases, thus the interest on the balance also decreases
5. $366.00; $6.41; $372.41; $37.00
7. $307.28; $5.38; $312.66; $31.00

9. $257.59; $4.51; $262.10; $26.00
11. $216.23; $3.78; $220.01; $22.00
13. $181.48; $3.18; $184.66; $20.00
15. $147.54; $2.58; $150.12; $20.00
17. $407.00; $20.00
19. $373.77; $6.54; $380.31; $20.00
21. $346.62; $6.07; $352.69; $20.00
23. $318.51; $5.57; $324.08; $20.00
25. $289.40; $5.06; $294.46; $20.00
27. $259.26; $4.54; $263.80; $20.00
29. $68.07
33. $367.64; $37.00
35. $300.95; $3.01; $303.96; $30.00
37. $248.70; $2.49; $251.19; $25.00
39. $205.45; $2.05; $207.50; $21.00
41. $168.37; $1.68; $170.05; $20.00
43. $131.55; $1.32; $132.87; $20.00
45. $24.25

Mixed Review

1. **a.** $4753.14 **b.** $3608.78
   Loan **b** has the greater total cost.
3. $145.86
5. 10 mo

Lesson 6–4

Algebra Review

1. 14.00
3. 3051.76
5. 7,112,448.00
7. 231.00
9. 15,616.26
11. 494.64

Try Your Skills

1. $56.00
3. $533.17; $8.00; $541.17; $54.00
5. $521.47; $7.82; $529.29; $53.00
7. $455.58; $6.83; $462.41; $46.00
9. $545.00; $8.18; $553.18; $55.00
11. $549.12; $8.24; $557.36; $56.00
13. $95.12

	Month	Purchases	Balance	Interest	Amount Owed	Payment
15.	2	$86.35	593.42	8.90	602.32	60.00
17.	4	40.00	535.45	8.03	543.48	54.00
19.	6	22.80	539.75	8.10	547.85	55.00
21.	8	97.88	557.27	8.36	565.63	57.00
23.	10	31.00	564.38	8.47	572.85	57.00
25.	12	67.00	598.58	8.98	607.56	61.00

Exercise Your Skills

1. Answers may vary.
3. Creditors cannot cancel a divorced or widowed person's credit unless the income drops dramatically.
5. $799.79; $12.00; $811.79; $81.00
7. $667.75; $10.02; $677.77; $68.00
9. $715.72; $10.74; $726.46; $73.00
11. $682.26; $10.23; $692.49; $69.00
13. $653.22; $9.80; $663.02; $66.00
15. $667.98; $10.02; $678.00; $68.00
17. $147.00
19. $1210.27; $18.15; $1228.42; $123.00
21. $1223.51; $18.35; $1241.86; $124.00
23. $1021.63; $15.32; $1036.95; $104.00
25. $851.94; $12.78; $864.72; $86.00
27. $1043.11; $15.65; $1058.76; $106.00
29. $204.44

	Month	Purchases	Balance	Interest	Amount Owed	Payment
31.	2	0.00	799.79	12.00	811.79	81.00
33.	4	0.00	686.05	10.29	696.34	70.00
35.	6	158.80	730.54	10.96	741.50	74.00
37.	8	125.00	734.51	11.02	745.53	75.00
39.	10	0.00	715.27	10.73	726.00	73.00
41.	12	123.00	719.80	10.80	730.60	73.00

	Month	Purchases	Balance	Interest	Amount Owed	Payment
43.	1	0.00	$1450.63	$ 21.76	$1472.39	$147.00
45.	3	0.00	1224.50	18.37	1242.87	124.00
47.	5	0.00	1249.39	18.74	1268.13	127.00
49.	7	0.00	1042.25	15.63	1057.88	106.00
51.	9	15.00	884.16	13.26	897.42	90.00
53.	11	45.00	1096.69	16.45	1113.14	111.00

55. Total Interest: $209.56

Mixed Review

1. $1981.35
3. $79.30
5. $182,438.25; Total cost for 15 years = $358,753.78
   Total cost for 30 years = $541,192.03
7. $p = 0.2x + 0.25y$
9. $1775 when $x = 2000$ and $y = 5500$
11. $864
13. The option of Exercise 12 is better than the cost of the installment option, $794.76.

Lesson 6–5

Algebra Review

1. 51.4
3. 12.47
5. 290.73

Try Your Skills

1. 1; $567.00
15; $8505.00
30; $18,060.00
a. $602.00; **b.** $7.53; **c.** $574.53
3. 1; $580.00
$580.00; 5; $2900.00
$640.00; 1; $640.00
$640.00; 12; $7680.00
a. $641.94; **b.** $11.23; **c.** $651.23

Exercise Your Skills

1. to verify the customer's right to use the card
3. 1; $1850.50
23; $42,561.50
a. $1872.63; **b.** $28.09; **c.** $1878.59
5. 15; $27,193.05
1; $1724.87
15; $25,873.05
a. $1767.45; **b.** $26.51; **c.** $1751.38
7. 20; $33,878.00
1; $1614.90
10; $16,149.00
a. $1665.87; **b.** $24.99; **c.** $1639.89
9. 18; $28,599.30
1; $1517.85
12; $18,214.20
a. $1559.08; **b.** $23.39; **c.** $1541.24
11. 13; $19,443.19
1; $1431.63
16; $22,908.08
a. $1459.36; **b.** $21.89; **c.** $1453.52
13. $0.33
15. $0.64
17. $0.77
19. $0.62
21. $0.42
23. 18; $15,750.00
1; $785.00
$785; 6; $4710.00
$920.00; 1; $920.00
$920.00; 5; $4600.00
a. $863.39; **b.** $10.79; **c.** $930.79
25. 8; $10,200.00
1; $1500.00
$1500.00; 8; $12,000.00
$1325.00; 1; $1325.00
$1325.00; 12; $15,900.00
a. $1364.17; **b.** $17.05; **c.** $1342.05

Mixed Review

1. $1052.01
3. $104.04
5. $45.00
7. $200.92

Lesson 6–6

Algebra Review

1. 98
3. 83
5. Tom: B
7. Let S represent the test score, x represent the number of questions 1–10 answered correctly, y represent the number of questions 11–14 answered correctly, and z represent the number of questions 15–16 answered correctly. $S = 5x + 7y + 11z + 10$, if the bonus question is answered correctly and $S = 5x + 7y + 11z$, if the bonus question is not answered correctly.

Try Your Skills

1. 5
3. 13
5. 24
7. 15
9. 41
11. 96.25%

Exercise Your Skills

1. to secure credit for a large loan
3. Answers may vary.
5. 22; 21; 3; 18; 10; 15; 15; 15; 0; Score: 119; 95%
7. 18; 21; 12; 18; 12; 15; 15; 15; 20;
Score: 146; greater than 96.25%
9. 31; 21; 13; 18; 10; 15; 15; 15; 16;
Score: 154; greater than 96.25%
11. 12; 21; 12; 0; 0; 15; 15; 5; 16; Score: 96; 91%

Mixed Review

1. $1706.43
3. $778.22 - 105.89 = 672.33$
5. $20,832

Lesson 6–7

Algebra Review

1. $54.45
3. $1370.88
5. 0.78%

7. 37,500
9. 435.29

Try Your Skills

1. 21.6%; Yes, they are over the 20% limit.

Exercise Your Skills

1. Answers may vary.
3. Answers may vary.
5. 55.8%; Yes, they are in credit overload.
7. 17%; They are not in credit overload.
9. 11.4%; no
11. 18.3%; no

Mixed Review

1. 10 months
3. 18.62%
5. $8349.79
7. $129.40
9. $4.85
13. $3.90 added

Chapter 6 Review

1.

Month	Previous Balance	New Charges	Finance Charges	Payment Received	New Balance
1	0.00	$2315.00	0.00	0.00	$2315.00
2	$2315.00	0.00	$24.89	$250.00	2089.89
3	2089.89	0.00	22.47	250.00	1862.36
4	1862.36	0.00	20.02	250.00	1632.38
5	1632.38	0.00	17.55	250.00	1399.93
6	1399.93	0.00	15.05	250.00	1164.98
7	1164.98	0.00	12.52	250.00	927.50
8	927.50	0.00	9.97	250.00	687.47
9	687.47	0.00	7.39	250.00	444.86
10	444.86	0.00	4.78	250.00	199.64
11	199.64	0.00	2.15	201.79	0.00
12	0.00	0.00	0.00	0.00	0.00

3. 11 mo

5.

Month	Balance	Interest	Amount Owed	Payment
1	$875.00	$8.75	$883.75	$44
2	839.75	8.40	848.15	42
3	806.15	8.06	814.21	41
4	773.21	7.73	780.94	39

7.

Month	Balance	Interest	Amount Owed	Payment
1	$875.00	$8.75	$883.75	$88
2	795.75	7.96	803.71	80
3	723.71	7.24	730.95	73
4	657.95	6.58	664.53	66

9. $2.41

11.

Month	New Charges	Balance	Interest	Amount Owed	Payment
1	0.00	$980.12	$14.70	$994.82	$ 99.00
2	$ 87.35	983.17	14.75	997.92	100.00
3	0.00	897.92	13.47	911.39	91.00
4	0.00	820.39	12.31	832.70	83.00
5	25.50	775.20	11.63	786.83	79.00
6	100.00	807.83	12.12	819.95	82.00
7	53.33	791.28	11.87	803.15	80.00
8	0.00	723.15	10.85	734.00	73.00

13. $381.39; $6.67
15. $778.06; $9.73
17. 19.1%; No, they are not in credit overload but are approaching the 20% limit.

Cumulative Review

1. $355.25
3. $67,500
5. One hundred ten and $\frac{23}{100}$
7. $608.00
9. $3996
11. $1800
13. Plan 3; Plan 1: $18,260.93; Plan 2: $18,085.16; Plan 3: $17,776.15

CHAPTER 7

Algebra Refresher

1. {1, 3, 5}
3. {1, 2, 3, 4}
5. 0.5
7. $0.\overline{6}$
9. 0
11. $0 < P < 1$

Lesson 7–1

Algebra Review

1. 0.5
3. 0.125
5. 0.5

Try Your Skills

1. $487,500
3. $280,000
5. $918
7. $2748
9. $6381

Exercise Your Skills

1. to replace lost income
3. No, she probably does not support the family with income.
5. Answers may vary.
7. $200,000
9. $195,000
11. $105,750
13. $90,000
15. $556
17. $2593.50
19. $6722.50
21. Term; $447; $2049; $4527.50

Mixed Review

1. $453.75
3. $1380
5. $797.60; $5.98; $803.58; $80.00
7. $104.20

Lesson 7–2

Algebra Review

1. 0.458
3. 0.167
5. 0.583
7. 0.583
9. 0

Try Your Skills

1. $2.50
3. 0.00094
5. $E = 0.99906x + 0.00094(x - 60{,}000)$
7. $72,000; $1000x = 200(1000) - 1.03(100{,}000) - 25(1000)$
9. $767,750; $5000x = 350(5000) - 6.35(135{,}000) - 25(5000)$

Exercise Your Skills

1. The older a person becomes the more likely death may be.
3. anyone who is alive past his/her life expectancy
5. life expectancy changes over the years
7. $4.00
9. $0.40
11. $88.50
13. $380.00
15. $717.50
17. $195.50
19. $295
21. $146.25
23. no; $13 reduction
25. $133,333.33

Mixed Review

1. Five hundred and $\frac{00}{100}$
3. 30% down payment; the less owed, the less interest owed
5. $7.50; $1007.50; $50.00
7. $14.68
9. $355.25
11. 18 mo
13. $6000
15. $84

Lesson 7–3

Algebra Review

1. 3.03
3. 4183.63
5. 102,320.24
7. 204,887.17

Try Your Skills

1. $91,523.93
3. term: $8280; whole: $36,720
5. $4020
7. $17,094
9. $218
11. $894
13. $3999

Exercise Your Skills

1. no taxes paid on income invested
3. early withdrawal
5. $214,438.05
7. term: $7630; whole: $38,920
9. A person cannot borrow from an annuity.
11. $6537.92
13. IRA
15. when it is unlikely he/she will need to borrow the money in the future

Mixed Review

1. $9000
3. compound interest
5. 16.9%
7. $2160

Chapter 7 Review

1. to protect dependent family members
3. Premiums are lower than renewable term.
5. $240,000
7. $459
9. $726
11. 0.00094
13. $121.60
15. $34
17. $45,426.37
19. $61,830
21. $1366
23. $6367

Cumulative Review

1. $7351.50
3. $8500
5. $92.39
7.

Month	Balance	Interest	Amount Owed	Payment
1	$690.00	$12.08	$702.08	$70.00
2	632.08	11.06	643.14	64.00
3	579.14	10.13	589.27	59.00

CHAPTER 8

Algebra Refresher

1. good
3. excellent
5. $y = -3.07x + 1.19$
7. $y = 1.27x + 4.25$
9. $y = 0.98x + 0.66$

Lesson 8–1

Algebra Review

1. $x = 62.35$
3. $x = 345.38$
5. $x = 6.125$ or $6\frac{1}{8}$

Try Your Skills

1. $49,998.00
3. 5479; $49,995.88
5. 51
7. $651.70

Exercise Your Skills

1. If the company is not successful, part or all of the investment could be lost.
3. schools, hospitals, streets, and so on
5. 5,633; $49,992.88
7. 555; $49,950.00
9. 671; $49,989.50
11. The bond would be a safer, long-term investment if he is a beginning investor.
13. LukInd
15. MamaMi
17. $16\frac{1}{2}$
19. $64

Mixed Review

1. $300; $15,600
3. $360; $18,720
5. $9.37; $633.87; $63
7. $605.81; $9.09; $614.90; 61
9. $4800
11. $1223.19
13. $3105.06

Lesson 8–2

Algebra Review

1. 7
3. 2
5. $15
7. $30x$

Try Your Skills

1. $6337.50
3. $1462.50
5. $95.06; $6432.56
7. $21.94; $1484.44
9. $105 loss
11. 15.79% loss
13.

r	0	0.005	0.01	0.02	0.03	0.04	0.05	0.06	0.07	0.08
t	2000	2010	2020	2040	2060	2080	2100	2120	2140	2160

Exercise Your Skills

1. without patience, money may foolishly be lost
3. Answers may vary.
5. $2,203.75; $3,547.50; $1343.75 gain; 61%
7. $120,000.00; $109,500.00; $10,500 loss; 8.8%
9. $3,061.25; $5,347.50; $2286.25 gain; 74.7%
11. $256.28; $13,070.03
13. $255; $17,255
15. 0%; −7%
$1080; 10%; 2%
$1080; −5%; −12%
$1080; 30%; 20%
$1080; 75%; 62%
$1080; 130%; 113%
$1080; 220%; 196%
$1080; 305%; 275%

Mixed Review

1. \$6227
3. \$1469.33
5. \$199.08
7. \$402.83
9. \$317.04

Lesson 8–3

Algebra Review

1. c
3. a

Try Your Skills

3. (0, 10)
5. c
7. \$20.70

Exercise Your Skills

1. to ensure sound trading practices
3. if the broker will sell only one stock or fund, makes promises for a quick, sure profit, has inside information, and urges you to buy before "prices go up"
7. $y = 2.98x + 19.97$; good, $r = 0.9797$
9. \$48.28
11. \$52.75
13. \$34.84
15. The line of best fit is accurate in predicting stock prices.

Mixed Review

1. \$6445.25
3. \$9,960; \$39,840; \$320.80; \$106,200.74; \$7050.09
5. Helps to determine the credit risk of a borrower in an objective way that does not discriminate.
7. Chuck Lewis (Charles Lewis)
9. \$4513.70
11. \$700.00

Lesson 8–4

Algebra Review

1. 12
3. 5
5. 3

Try Your Skills

1. 5.4%
3. $\frac{10,000}{113.6} = \frac{x}{100}$

   $x = \$8,802.82$
5. 4%
7. 0.85

Exercise Your Skills

1. to measure inflation or deflation
3. Food; explanations may vary.
5. \$691.50
7. \$35.18
9. 5.4%
11. 7.5%
13. 18.87 compared to 22.3
17. 80%

Mixed Review

1. \$443.95; \$375; (\$68.95)
3. \$775
5. \$1777.50
9. 0.01671

Chapter 8 Review

1. through careful investments
3. It is a way to have a stake in many different corporations, rather than risking all your money on the future of just one or two corporations.
5. The auction method
7. Stocks are not instant profit and take time to develop.
9. 2622; \$19,992.75
11. 133; \$19,950
13. \$46,250; \$75,625; \$29,375
15. \$108,750; \$222,500; \$113,750
19. \$23.14
21. close; model is \$27.916 and actual is \$30.00

Cumulative Review

1. \$239.15
3. \$1.50
5. 12 wk
7. \$427.50; \$9427.50
   \$447.81; \$9875.31
9. \$2467.50; \$25,370; \$22,902.50
11. 14.93%

CHAPTER 9

Algebra Refresher

1. The number -1.5 is half way between -1 and -2.
3. The number $-\frac{8}{3}$ is a little less than half way between -3 and -2.
5. The number $\sqrt{2}$ is a little less than half way between 1 and 2.

Lesson 9–1

Algebra Review

1. 840
3. 3214.50
5. 11,740.70
7. 19,137.00

Try Your Skills

1. $309
3. $339
5. $1,654
7. $8,030
9. $t = 0.15I$; $336.75
11. $t = 11{,}743.50 + 0.31(I - 51{,}900)$; 27,894.50
13. $t = 11{,}743.50 + 0.31(I - 51{,}900)$; 36,574.50
15. $t = 11{,}743.50 + 0.31(I - 51{,}900)$; 57,654.50

Exercise Your Skills

1. Income tax is the largest revenue-producing tax the government has.
3. No; instead of the government telling us what we owe, we inform them
5. $11,217
7. $1,526
9. $35,686.00
11. $9,154.62
13. Taxpayer #11
15. all three
17. $18,660; $12,760; $1,916; $2,040; +$124
19. $28,536; $24,936; $4,191; $4,572; +$381
21. $42,000; $3,960; −$64
23. $63,600; $53,000; $10,193; $10,260; +$67
25. $21,360; $9,210; $1,384; $1,764; +380
27. $38,508; $24,058; $3,611; $3,444; −$167
29. $30,940.50

Mixed Review

1. $528
3. $29; $21.96; $236.04
5.

Level	Demand Deposit	Required Reserve	Loans and Investments
1	$1,200.00	$240.00	$ 960.00
2	960.00	192.00	768.00
3	768.00	153.60	614.40
4	614.40	122.88	491.52
5	491.52	98.30	393.22
		Total after 5 levels:	$3,227.14

7. $4,800
9. $1245.80
11. $196.00

Lesson 9–2

Try Your Skills

1. Refund: $225
3. Refund: $388
5. Tax owed: $114

Exercise Your Skills

1. Not everyone has the same income, number of dependents, investments, and so on.
3. It is easier than documenting your itemized deductions, and it saves the average person more money.
5. Refund: $501
7. Refund: $79
9. Tax owed: $169
11. Refund: $367
13. Refund: $597
15. Refund: $151

Mixed Review

1. $303.28; $18,196.52
3. 5-year term; $1565
5. 120; 95.5%
7. 6%

Lesson 9–3

Algebra Review

1. 330
3. 7750
5. 950

Try Your Skills

1. +$4000
3. None
5. +$4300
7. no effect

Exercise Your Skills

1. The IRS and tax preparers can make mistakes; it is good to be informed just in case.
3. When the taxpayers' deductions are larger than the standard deduction.
5. $11,054
7. $72,275
9. Refund: $945.00
11. Line 27
13. $222
15. No. The standard deduction would be higher.

Mixed Review

1. 5 mo
3. $2120
5. 2666; $49,987.50
7. $14,533
9. $820
11. $4781.95

Chapter 9 Review

1. how you submit your claim form—single, married filing jointly, married filing separately, head of household
3. It lowers the amount of taxes.
5. $19,729
7. $7,022
9. $2,929
11. $26,231.00
13. $28,742.04
15. $11,280; $0; $0; $60; +$60
17. $53,400; $38,950; $7176; $8928; +$1752
19. Refund: $823
21. Refund: $1134
23. Adjusted gross income: $64,035; Itemized deductions: $11,475; Tax owed: $79

Cumulative Review

1. A: $47,270; B: $53,340; C: $48,018
3. $314
5. $1413.73
7. $1387.71
9. $1,163,652.10 − $398,041.76 = $765,610.34
11. $22,721

CHAPTER 10

Algebra Refresher

1. mean: 77.7; median: 78; mode: 78
3. mean: 630; median: 320; mode: 220
5. mean: 80.2; median: 82; mode: 75

Lesson 10–1

Algebra Review

1. $1129.40
3. $105.06
5. 40%
7. $4905
9. $4876

Try Your Skills

1. $2432
3. $2081.25
5. 0.007
7. $14,623.11

Exercise Your Skills

1. 15.7%
3. 15.4%
5. 17.6%
7. 17.4%
9. $15,203; $17,637
11. $12,067; $13,948
13. $429.24
15. $15,732.99
17. $17,204
19. $228.10
21. 6% APR for 6 years with 20% down
23. In Exercise 19, although the high down payment decreased the amount financed, the increased term of the loan increased the total finance charge. The less expensive financing is in Exercise 20 because the term of the loan is decreased to 4 years. You would choose the financing in Exercise 19 if you could not afford a high monthly payment. If you could afford the higher monthly payments, it would be better to choose the financing in Exercise 20, and lower the total finance charge.
25. $145.86

Mixed Review

1. $3506.38
3. 72%
5. $17,141.34
7. $y = 0.998x + 8.942$
9. $18,250

Lesson 10–2

Algebra Review

5. 1154.39
7. 9.5
9. 5.4

Try Your Skills

1. $396.40
3. $319.60
5. $8000
7. $2364.34

Exercise Your Skills

1. For a new car, depreciation is greater and begins immediately after you purchase the car. For a used car, depreciation is more gradual.
3. $620.61
5. $518.61
11. 3 years
13. $60
17. $6000
19. Dena will need to choose 10% down at 6.5% annual interest over 5 years to make her monthly payments as low as possible.
21. $3200

Mixed Review

1. $62.40
3. $3.56
5. $14,666.75
7. $273

Lesson 10–3

Algebra Review

1. $6\frac{2}{3}$
3. 8.75 minutes per mile
5. 13 miles per gallon
7. (7, 15)

Try Your Skills

1. 22.7 mpg
3. $455.83
5. $376.95

Exercise Your Skills

1. the miles per gallon affect the annual operating cost
3. in order to put money aside to pay for the repairs needed
5. 417; 22.7; $0.060
7. 151; 15.0; 0.093
9. 211; 19.4; 0.068
11. $100.55
13. $0.499
17. 1428.6
19. Mark's car
21. lease

Mixed Review

1. $2000
3. $54,750
5. Refund: $396

Lesson 10–4

Algebra Review

1. 945
3. 5280
5. p decreases
7. 228,000
9. −7501.31

Try Your Skills

1. 2.10
3. $p = 2rb$; p represents the annual insurance premium; r represents the driver rating factor; b represents the six-month basic rate for the car

Exercise Your Skills

1. Any driver might have an accident.
3. The insured person pays the deductible amount of the bill resulting from the accident before the insurance company begins to pay. The premium is the amount one pays for the insurance coverage.
5. 1.65
7. 2.40
9. 1.25
11. $2340
13. $1152
15. $550
17. $610.60
19. expenses paid out (claims): number of people paying premiums; and overhead costs

Mixed Review

1. $4200
3. 10%
5. $1801.63
7. $4171.15
9. itemizing; $9770

Chapter 10 Review

1. factors that raise monthly payments: higher interest, longer term, higher amount financed
   factors that lower monthly payments: lower interest, shorter term, smaller amount financed
3. loan payments followed by insurance costs
5. the driver rating factor, the car being insured, coverage and deductible
7. 20%
9. $2922.50; 19.8%
11. $408.87
13. Interest $67.20; Principal $341.67
15. $1680
17. $4704
19. $0.05 per mile
21. $1431; $119.25
27. 3.35
29. $2835
31. 1.65

Cumulative Review

1. $790
3. $43.33
5. $14.31
7. $28,499
9. $1564
11. $3600
13. **b** and **d**

CHAPTER 11

Algebra Refresher

1. $y = -2x - 3$; $y = -2x$; $y = -2x + 4$; $m = -2$

Lesson 11–1

Algebra Review

1. $g = 20$
3. $a = 10.71$
5. $c = 1.21$
7. $(a + b) \div 2$
9. d

Try Your Skills

1. North on 59, East on 10
3. 2010 miles
5. 18 gallons
7. $41.14
9. Answers may vary. (Ex.: $g = d \div a$; g = gallons of gas, d = total distance in miles, a = average miles per gallon)

Exercise Your Skills

1. Answers may vary.
3. Answers may vary.
5. east on 16; north on 95
7. northeast on 30; east on 40
9. 681 mi
11. 916 mi
13. 695 mi
15. 26.3; $33.93
17. 28.4; $35.22
19. 92.7; $125.15
21. Answers may vary.
23. Answers may vary.

Mixed Review

1.

Checks/Deposits	Amount	Balance
To: Grocery Mart		$778.22
For: Groceries	$105.89	− 105.89
		672.33

3. $16.55
5. $327.50
7. $7271

Lesson 11–2

Algebra Review

1. 11.2
3. 10.0
5. $c + 1$
7. (15, 75)

Try Your Skills

1. $50.06
3. $4.73, $5.44, $18.53

Exercise Your Skills

1. $181.44
3. $331.76
5. $85.95; $21.49
7. $124.70; $79.85; $93.17; $139.38; $437.10; $109.28
9. Answers may vary.
11. $y = 7500 + 180x$
13. 20 people
15. loss: $1800
17. 50
19. Gomezes want dinner prices to cover costs; patrons want an inexpensive dinner.
21. Lower prices may lead to more patrons; if so, revenues may increase.
23. $y = 445x$
25. Yes, if more than 41 people take the tour.

Mixed Review

1. $568.75
3. $502.25
5. Loan **b**; **a.** $16,753.14; **b.** $18,608.78
7. $11,000
9. 42 years
11. 24 years
13. $86,062.50
15. $900

Lesson 11–3

Algebra Review

1. 100 mi
3. 1400 mi

Try Your Skills

1. 8:15 A.M.
3. Yes
5. $625 is an expensive flight, so Ramón would likely save money by driving, given he has a car to use.

Exercise Your Skills

1. 9:54 A.M. or 11:35 A.M.
3. Midway
5. If she's traveling on the weekend, the second flight is not available.
7. Car travel costs include gas, lodging and meals; air travel costs include ticket and transportation to and from airport.
9. The ticket is probably not refundable so you can't change your plans.
11. 8:30 A.M.
13. 6:15 P.M.
15. 381 min
17. Central
19. Mountain
21. Mountain
23. 1749 mi
25. $415.40

Mixed Review

1. $20.95; $1696.95; $170
3. $1391.04; $17.39; $1408.43; $141

5. $16,430
7. 616; 31.6; $0.045
9. $630.75
11. $4560.83
13. $4016.25

Chapter 11 Review

1. faster
3. unexpected events
5. often less expensive, more flexible
7. 20 east, 55 south, 10 east
9. 20 east, 95 northeast
11. 30 northeast, 40 east, 55 north, 57 north
13. 1315 mi
15. 1448 mi
17. $81.14, $104.59, $89.35, $41.22
19. $341.28
21. $196.56
23. $85.28; $21.32
25. $130.30; $124.84; $91.13; $79.93; $426.20; $106.55
27. $y = 450x$
29. 15 people
31. $1600
33. pacific
35. central
37. 8:30 A.M.
39. 230 min
41. 1187 mi
43. $347.16

Cumulative Review

1. $8.50/h; $340/wk; $1473.33/mo
3. $220/wk; $953.33/mo; $11,440/y
5. $7045.45
7. 1026
9. $805
11. $319
13. 19
15. $5549
17. 27%

CHAPTER 12

Algebra Refresher

1. $x^2 + x - 2$
3. $x^2 + 6x + 9$
5. $x^2 - 4x - 5$
7. $(x - 1)(x + 2)$
9. $(x - 1)(x - 3)$
11. $(x + 5)(x - 4)$
13. $x = -4$ or $x = -1$
15. $x = 3$ or $x = -2$

Lesson 12–1

Algebra Review

1. 184.97
3. 10.78
5. −44.60
7. 268.20
9. 1449.78

Try Your Skills

1. $y = -4.74 + 0.0763x$
5. yes

Exercise Your Skills

1. Answers may vary.
3. Answers may vary.
5. yes; $r = -0.9873$
7. 19 years
9. $y = -1.8969 + 0.0151x$
13. no; $220,000 would be fair; $x = 223.11$
15. yes; $r = -0.9627$
17. about 1; $y = 1.059$

Mixed Review

1. $864.57
3. 17.8%
5. 12.1%
7. $218.40

Lesson 12–2

Algebra Review

1. 1988

Try Your Skills

1. $659,520
3. $48,296
5. $448,294; $24,548
7. $472,842; $0

Exercise Your Skills

1. down payment, closing costs, mortgage, additional costs
3. between 2 and 4 times
5. 2.5:1
7. $40,000; $160,000; $545,481
9. $40,000; $160,000; $588,538
11. $60,000; $140,000; $502,296
13. $60,000; $140,000; $539,971
15. $30,000; $90,000; $267,740
17. $30,000; $90,000; $290,698
19. $30,000; $90,000; $194,311
21. $30,000; $90,000; $184,816
23. lowest: 25% down, 8% interest, 15 years; highest: 20% down, 9% interest, 30 years
25. Plan C

Mixed Review

1. $1.19
3. $21.01; $1221.51; $122
   $1099.51; $19.24; $1118.75; $112
   $1006.75; $17.62; $1024.37; $102
5. 763
7. 16.2%

Lesson 12–3

Algebra Review

1. 4000
3. 1424
5. $540
7. 2222.22
9. $\frac{5ab^{36}}{z(c+2)^{36}}$

Try Your Skills

1. $115,149.42
3. yes; 20% is $28,787.36

Exercise Your Skills

1. the higher the interest rate, the higher the total cost of the loan
3. federal government
5. $700; $86,997.31; $108,746.64
7. $980; $121,796.23; $152,245.29
9. $742; $92,217.14; $115,271.43
11. $1468.60; $182,520.35; $228,150.44
13. $978.60; $121,622.23; $152,027.79
15. $1617.56; $201,033.37; $251,291.71
17. no; $160,000
19. yes
21. yes

Mixed Review

1. 33 wk
3. $456,238.19
5. $52.42
7. $469

Lesson 12–4

Algebra Review

1. 0.227
3. 0.0131
5. $7019.44
7. 95,244.28
9. 18.83

Try Your Skills

1. yes
3. $120.67
5. $2025; $17,901; $5012.28

Exercise Your Skills

1. Answers may vary.
3. document with photos and contact the insurance agent before beginning to clean
5. $1025
7. yes
9. $75
11. $11.91
13. $22.54
15. $5.47
17. $10.92
19. $23.18
21. $21.57
23. $3360; $13,985; $3915.80
25. $2562.50; $18,706.25; $5237.75
27. $867; $8,899.50; $2491.86

Mixed Review

1. 14 years
3. $880
5. 1615 mi
7. $116.60

Chapter 12 Review

1. Answers may vary.
3. 28%
7. 4
9. $534,332; $64,206
11. $576,890; $21,648
13. $896; $122,110.01; $152,637.51
15. $924; $125,925.95; $157,407.44
17. $1736; $236,588.15; $295,735.19
19. $665; $90,628.52; $113,285.65
21. $628.32; $85,629.65; $107,037.06
23. no; $125,000
25. $107.23
27. $2180.50; $19,168.50; $5367.18

Cumulative Review

1. $178.52
3. $21.70; $1261.70; $126
5. $1039.57; $18.19; $1057.76; $106
7. $240,000
9. 491.6 mi; 28.3 mi/gal; $0.04/mi

CHAPTER 13

Algebra Refresher

1. 25
3. 8
5. 5
7. 2
9. −2, 2
11. $\frac{9}{2}$

Lesson 13–1

Algebra Review

1. $2146.56 **3.** $7936.50
5. 375 **7.** F

Try Your Skills

1. $721.15
3. $437.50
5. $15,340
7. Yes. The lease is for $4800; food costs $4200; 9000 ÷ 2 − $4500. Jacob will save $1500 over a 12-month period by living in the apartment. Even if they leave the apartment after 9 months, their grocery expenses will only be $3150, lowering their apartment costs further.

Exercise Your Skills

1. sharing expenses, companionship
3. Answers may vary. (Ex.: rent—do not have to worry about maintenance)
5. $423.08
7. $569.23
9. $867.21
11. $346.15
13. $36,400
15. $40,300
17. $50,700
19. $33,540
21. $7979; $2929
23. Yes, in 2 months he will recover his moving cost and start saving $108/mo.
25. no; rent is $3480, food for 9 months is $1800; total: $5280

Mixed Review

1. $13,809.99 **3.** $16,637.50
5. 6:15 A.M. **7.** $1281
9. 16.4%; no

Lesson 13–2

Algebra Review

1. 34.31
3. −138.84
5. $y = 2x + 3$
7. $y = 150x + 100$
y = cost; x = number of rooms
9. $970.00

Try Your Skills

1. $y = 1.573550568 - 0.0026312761x$
3. 0.4 miles

Exercise Your Skills

1. A lease is usually a 1-year commitment.
3. Tenants will give you an inside view of the way the apartment building is run.
5. $y = -0.4737640598 + 0.0033466353x$
7. 1
9. $y = 141.1439994 + 1.076640466x$
11. 679 ft^2
13. no; $r = 0.145972874$
15. yes; $r = 0.9593668769$

Mixed Review

1. $9667.50 **3.** $390,000
5. $256.34 **7.** $164,605.34
9. $828.85

Lesson 13–3

Algebra Review

1. 1575 **3.** 936.39
5. 1972.71 **7.** 1176.16
9. 236,208.80

Try Your Skills

1. $352,023.47 **3.** total financed price

Exercise Your Skills

1. It is responsible for management of the grounds, building, and common areas.
3. renting to get acquainted with the community; buying if you are already a long-term resident of the area
5. $283.23; purchase
7. $547.14; purchase
9. $901.18; purchase
11. rental payments
13. rental payments
15. rental payments
17. $142,709.49
19. $367,365.10
21. $804,088.76

Mixed Review

1. 16 years **3.** $79.38
5. $2000; $10,215; $2860.20
7. $3971; $18,533; $5189.24

Chapter 13 Review

1. Advantages: shared expenses, companionship
   Disadvantages: lack of privacy
3. Advantages: interest on mortgage is tax deductible, mortgage payments provide equity
   Disadvantages: may be difficult to sell, requires large outlay of cash for purchase, extra fees to condo association, property taxes
5. \$817.31
7. \$1538.46
9. \$740.38
11. \$1841.35
13. \$68,640
15. \$43,680
17. \$25,480
19. \$20,280
21. yes; in $1\frac{1}{2}$ months moving costs will be covered
23. no; rent = \$5520; food = 9 • \$150 = \$1350; total = \$6870; dorm is less expensive
25. yes, $r = 0.9883815163$
27. yes
29. rental payments
31. \$609.23

Cumulative Review

1. \$72,750
3. Profit: \$185.75
5. Plan 2
7. \$19,762.50
9. \$1067.31
11. \$424.84, rental

CHAPTER 14

Algebra Refresher

1. 10 or −10
3. 7 or −5
5. 4 or −4

Lesson 14–1

Algebra Review

1. $6 = 0.05x$
3. $2.5 = 500x$
5. $x = 0.5\%$
7. $2.8 = 0.05x; x = 56$
9. $t = 0.035c$

Try Your Skills

1. 18.3 mills
3. 1.6 mills
5. \$1521
7.

	Month 1	Month 2	Month 3	Month 4	Total	Average
Energy	\$120	\$108	\$130	\$126	\$ 484	\$121.00
Electricity	31	28	37	40	136	34.00
Water	12	13.50	15.50	21	62	15.50
Telephone	80	36	47	51	214	53.50
Home Maint.	43	72	18	10	143	35.75
Total	286	257.50	247.50	248	1039	259.75

Exercise Your Skills

1. It is determined by the local government and not by the owner.
3. It is satisfactory for expenses not affected by seasons such as telephone and water but not for heating/air conditioning.
5. \$139.40
7. 8.3 mills
9. 15.9 mills
11. \$220; \$55.00
    \$292; \$73.00
    \$98; \$24.50
    \$130; \$32.50
    \$222; \$55.50
    \$134; \$230; \$292; \$306; \$962; \$240.50
13. \$638.50

Mixed Review

1. \$14,750
3. $53\frac{5}{8}$
5. \$2272
7. 1537 mi

Lesson 14–2

Algebra Review

1. $\frac{ay}{b}$
3. $\frac{bx}{y}$
5. $\frac{14}{x}$
9. graphing calculator shows more points
11. 12
13. 24

Try Your Skills

1. \$0.456 per ounce
3. 20 oz for \$2.15
5. 485.714 grams

Exercise Your Skills

1. Answers may vary.
3. Answers may vary.
5. 18 oz; $1.79
7. 12 oz; $1.98
9. 14.08 lb
11. 3.77 liters
13. $3.63; $2.69; $2.38; $2.22

Mixed Review

1. $3669.23
3. $36,574.50
5. yes; $y = 200.62$
7. $28
9. 15
11. $12
13. 17th floor since predicted value is $456 while predicted value on 26th floor is $493

Lesson 14–3

Algebra Review

1. 120°
3. 80°
5. $200; Miscellaneous

Try Your Skills

1. $4950
3. $1125
5. $900
7. $450
9. $2475
11. yes

Exercise Your Skills

1. 20%
3. If one budget area takes too large a percentage of the income, it may mean trouble in other areas.
5. 100%
7. 33.2%
9. 25.8%
11. 5.3%
13. 0.4%
15. −$873
17. +$838
19. −$253
21. +$28
23. Answers may vary.

Mixed Review

1. It does not allow the check to be cashed.
3. $144.02
5. payment due is $400
7. It might show a pattern, for example a linear pattern, that would suggest the fund's future performance.
9. $181,572 + $45,393 = $226,965
11. $634.70; the rental option would be higher

Chapter 14 Review

1. $3349
3. 11.2 mills
5. 18.9 mills
7. $481; $120.25
$184; $46.00
$89; $22.25
$226; $56.50
$217; $54.25
$237; $346; $277; $337; $1197; $299.25
9. 16 oz; $1.29
11. 12 oz; $1.69
13. 5.66 liters
15. 5.45 kg
17. $y = 950 + 2.4x$
19. $15.07
21. $9.71
23. 24.8%
25. 7.3%
27. 4.2%
29. 1.8%
31. −$85
33. $0
35. −$38
37. +$16
39. −$25
41. $7466
43. $1600
45. $1866
47. $533

Cumulative Review

1. $28,053.32
3. 77
5. 2666
7. $32,409.80
9. $176,468.58
11. 9.1%; no
13. $2543.75; $2117.50; −$426.25; −16.8%

GLOSSARY

add-on costs Expenses that add to the price of a product as it goes through processing, marketing, and distribution. Examples include labor, packaging, advertising, energy, and transportation.

adjustable-rate mortgage (ARM) Mortgage loan with an interest rate that can be adjusted up or down during the life of the loan

adjusted gross income Total of income earned as well as income from interest and other sources minus IRA deductions

advance purchase Paying ahead of time; if you pay for an airline ticket ahead of time you may get a lower price.

advertising Materials and activities used to promote the sales of a product

all-purpose bank cards Credit cards widely accepted by businesses that may not have their own charge accounts; examples include MasterCard and VISA.

amortize To pay off; a mortgage is usually amortized over a period from 15 to 30 years

amortization schedule A list showing the amounts of principal and interest that are paid with each monthly payment of a loan

annual percentage rate (APR) The percent cost of credit on a yearly basis

annuity Investment plan that provides income upon retirement

appraised value The approximate selling price of a house as determined by an independent agent

assessed value Amount that a local government assigns to a house or other property for tax purposes

assumability clause Statement in a mortgage allowing a new buyer to take over the mortgage from the specified holder

automated teller machines (ATMs) Computerized machines that perform banking services quickly and automatically at convenient locations and during nonbanking hours

average-balance account Checking account in which a fee is charged if the average balance for the month falls below a stated minimum

average daily balance The average of daily balances over the month, found by figuring the unpaid balance for each day of the billing cycle, adding these balances together, and dividing the total by the number of days

bait-and-switch Unethical sales practice in which one product is used to attract customers and another, more expensive product is then urged on them

balloon mortgage A form of mortgage in which the monthly payments are low, but there is a large final payment

bankruptcy The condition in which an individual or business declares the inability to pay its debts; assets must be sold in return for a discharge from outstanding debts

bar graph A graph in which the lengths of bars represent amounts

base salary Wages earned regardless of commissions or piece-work rates

bear market Condition in which stock prices are falling

beneficiary The person designated to receive the death benefit in a life insurance policy

blank endorsement Your signature on the back of a check with no additional message

bodily injury liability Insurance that covers injuries to others in an automobile accident caused by the insured person

bond A loan to a government or corporation for which the holder receives interest during the life of the bond, and the full amount at the maturity date

book value Standard estimate of the resale value of a car

break-even point (*also* **break-even value**) The point at which cost and revenue are equal

broken line graph A graph in which line segments connecting points represent data

broker Salesperson who specializes in buying and selling stocks and bonds

budget A plan showing categories and projected amounts of expenses or income

bull market Condition in which stock prices are rising

canceled checks Checks, drawn on an account, that have cleared through the bank and been returned to the account holder for record keeping

cap A limit to the total amount of increase allowable in an adjustable-rate mortgage

capital gain Increase in the market value of a security or other asset; increase in the value of property during the time that it is owned

cash discount A lower price given to a customer who pays cash instead of using credit

cash shortage Insufficient funds to cover expenses

cash-value insurance Insurance plus savings plan; one of two main types of life insurance

cell In a spreadsheet, the intersection of a row and column

certificate of deposit (CD) Special form of savings in which a fixed amount of money is deposited for a specified length of time

charge card A card that allows you to make purchases for later billing

check register A record of checks written, fees charged, and deposits made in a checking account

checks Forms filled out and signed by the owner of a checking account authorizing the bank to pay the amount shown to the person designated

closing costs Charges and fees associated with the transfer of ownership of a home to a buyer

collateral Something of value offered as assurance that a loan will be repaid; (*See* **security**)

collection agency A firm hired by creditors to collect overdue debts or repossess items not paid for

collision Insurance that pays for damage to the insured person's car in an accident

commercial banks Banks offering a wide variety of services in addition to savings accounts: checking accounts, credit cards, loans, financial counseling, safe deposit boxes, traveler's checks, money orders and transfers, and trust and investment services

commission Money paid a salesperson for each item sold; a percentage of the cost of the item. Also, broker's fee for buying and selling securities

commodity An item that has not yet been enhanced, refined, or packaged to become a consumer good

common logarithm function Function of the form $y = \log_{10} x$, where y represents the exponent to which 10 must be raised to obtain x.

common ratio The ratio between two consecutive terms in a geometric series

common stock Stock that entitles the owner to receive dividends and vote for the company's board of directors

competition Rivalry among sellers for consumers' purchases, or rivalry among producers when seeking the rights for production

complement In probability, if E is an event then "E does not occur" or "not E" is the complement of E

compound interest Interest paid on the principal and also on previously earned interest, assuming that the interest is left in the account

compounded quarterly Interest calculated and paid four times a year, every three months

compounded semiannually Interest calculated and paid twice a year, every six months

comprehensive physical damage Insurance that pays for damage to a car that results from fire, falling objects, theft, storm, flood, earthquake, mischief, flying objects, and vandalism

condominium Living space in which individual units are privately owned and common spaces are jointly owned

connection Change of planes during a trip

constraints Conditions that limit business activities

Consumer Price Index (CPI) An economic tool for comparing prices from one era to another

consumers Purchasers of goods and users of services; all the members of a society who need to buy things

corporate bond Bond issued by a private company to raise money for plant expansion or other operations

corporation A business made up of a number of owners who have shares of stock in the company

cosign Parents or others can cosign a loan with a borrower; this makes the cosigner responsible for repaying the loan if the borrower does not

cost-per-check account Checking accounts that charge a small fee for every check that clears during the month

coverage The items and dollar amounts for which an insurance company will compensate a policy holder; homeowner's insurance coverage may include dwelling, personal property, loss of use, personal liability, and medical payments

credit A form of debt that occurs when cash, goods, or services are provided in exchange for a promise to pay at a future date

credit card A small plastic card that identifies the holder and gives the holder privileges of making purchases on credit; the holder is not required to pay the total outstanding bill every month, but to pay a minimum required amount

credit counseling Advice about how to solve credit problems

credit limit The maximum value a credit card holder can charge on an account; once the balance is partly paid back the holder can again make purchases up to the limit

credit rating An indication of a person's ability to secure goods, services, and money in return for the promise to pay

credit risk A customer who is unlikely to be able to pay for purchases made on credit is referred to as a poor credit risk

credit unions Not-for-profit savings and lending financial institutions in which all members belong to the same social or professional organization

credit-scoring systems Award points for various factors to determine credit worthiness

creditor Business, bank, or individual who extends credit

debit card A card that enables the card holder to pay through a direct deduction from a checking account

decreasing term insurance Kind of term insurance in which the death benefits decrease over time and the premium is lower than for renewable term insurance

deductible The portion of a loss that the insured person pays; the insurance company pays the costs above the deductible

deduction **1.** An amount of money subtracted from your gross pay for such items as FICA taxes, income taxes, insurance premiums, and union dues **2.** An amount that you may deduct from your gross income to find the amount on which you owe taxes; state and local taxes, interest paid on a mortgage, and charitable contributions are deductible

deferred payment price The amount including interest that you pay for an item you buy on installment

deflation A general decrease in prices

demand deposits Checks

dependent A person supported by another; a tax payer's dependents include children and spouse supported by the tax payer

deposit ticket A form used by banks for deposits to accounts

depreciation The difference between the original purchase price of a car or other asset, and its resale price at a given time

direct flight A plane flight that has one or more stops but you will not change planes

disability insurance Benefits paid through social security to those who can no longer work because of injury or illness

discount broker A broker in securities who charges much lower commissions than a full-service broker, but provides no investment advice

discretionary income Income remaining after you pay for basic needs such as food, clothing, and shelter

diversify Acquiring a balanced variety of securities rather than owning too much of any one; diversification cuts down the risk of investment

dividend The part of the profits of a corporation that each stockholder receives

down payment The amount of a purchase you make in cash before taking out a loan for the remainder

drawer The person who writes and signs a check

driver rating factor The multiple applied to a base premium to determine the insurance costs for an individual; in general, the factor for a young person is higher than for an older person, so the younger person pays higher premiums

easy-money policy An easy-money policy results when the reserve requirement is lowered and money is thus freed

effective interest rate Interest rate when the result of compounding is included

efficient Working well; a business is efficient when it is producing a product with a minimum of energy, expense, and waste.

endorsement Signature and directions to the bank on the back of a check

endowment policy Type of cash-value life insurance which builds up value quickly through high premiums

entrepreneurship Putting together the ingredients of a business, including making plans, hiring workers, obtaining raw materials, and obtaining equipment

Equal Credit Opportunity Act Law that prohibits discrimination in granting credit on the basis of sex, marital status, race, color, religion, age, or national origin

equity The difference between what a home, car, or other object is worth and what the buyer still owes on it in the form of a mortgage or loan

excess reserves Money remaining from a deposit after the required reserves have been subtracted; money available for investment or loans by a bank

exemption A tax deduction taken for dependents

expected value The amount of money to be won or lost in the long run

exponential regression equation The equation of the form $y = ab^x$ that best fits a set of points

face value of a bond The original loan amount that appears on the bond

face value of the policy The amount of the death benefit in a life insurance policy

Fair Credit Billing Act Law that allows people to preserve their good credit rating while settling a dispute with a store or credit card company

Fair Credit Reporting Act Law giving people the right to examine their own credit files and have incorrect information removed

Fair Debt Collection Practices Act Law prohibiting violence, the threat of violence, harassment, and certain other methods in the collection of debts

Federal Housing Administration loans Loans in which federal insurance is provided

Federal Reserve note Main type of U.S. currency

Federal Reserve System A system of banks under the direction of a board of governors that directs the purchases and sales by the reserve banks of federal government securities and other obligations.

FICA Federal Insurance Contributions Act (*See* **Social Security**)

FICA deduction An amount withheld from a paycheck and remitted to the federal government as part of the FICA tax

FICA tax A percentage of income paid to the federal government by employees and employers to pay for pensions and Medicare

finance charge **1.** A charge levied on the unpaid balance of a credit account **2.** The amount over the loan amount that you pay to a lending institution for a loan; also called interest

financing Borrowing money needed for a purchase

financial responsibility laws Laws requiring drivers to pay, either with insurance or with personal funds, for damage to other people or to property

fixed costs A company's costs that remain constant over a period of time; may include labor, transportation, advertising, energy, and so on

fixed-rate mortgage (FRM) A mortgage paid off in equal monthly payments on a regular schedule over the years of the loan; (*Compare* **adjustable rate mortgage**)

Form 1040 Federal income tax form that is used when you are itemizing deductions and/or reporting self-employment income

Form 1040A Basic form for calculating and filing individual or joint federal income tax

Form 1040 EZ Simplified federal tax form for single individuals with no dependents

Form W-2 Form from an employer that provides a record of wages earned as well as federal, state, local, and other taxes withheld during the past year

Form W-4 Filled out at the beginning of employment, the form provides the IRS with information used to determine how much of the worker's wages to withhold for income tax

401(k) plan A pension plan approved by the IRS through which an employer can make tax-sheltered contributions

free checking account Has no minimum balance requirement or service charges

fringe benefits Benefits offered to workers in addition to wages or salaries; these include paid vacations, life and health insurance, and retirement plans

full endorsement Your signature on the back of a check made out to you, with directions to the bank to transfer the check to someone else

full service broker Broker offering reports on companies and advice about buying and selling securities

future value of a periodic investment The value, at a given time in the future, of money periodically invested in an interest-bearing account

garnishment To hold part of a debtor's earnings in order to pay a debt; a legal method of debt collection

geometric series A sum of terms in which the ratio of two consecutive terms is constant

grace period The time between when a bill is received and when interest is charged

graduated commission Commission plan in which a salesperson receives a higher rate cf commission for sales above a certain amount

gross pay Total pay before deductions have been taken

group life insurance Insurance purchased through the place of employment, covering employees as a group

health insurance Hospital and medical insurance; for senior citizens, Medicare

homeowner's insurance Insurance covering a dwelling and personal property against damage or loss from fire and lightning, theft, storm, explosion, and so on. Mortgage lenders typically require the homeowner to take out this insurance

hourly rate Amount of money paid for each hour of work

income The money a person receives from work or investments

income security Guarantee that income will continue

individual income tax Tax on an individual's earnings from wages, salary, tips, interest, rents, dividends, and capital gains; in the U.S. the largest revenue producing tax for the federal government. The U.S. income tax is based on ability to pay, voluntary compliance, and paying-as-you-earn through withholding

Individual Retirement Arrangement (IRA) An account in which money is invested for retirement and taxes are deferred until retirement

inflation A general increase in prices

installment loan Borrowed money that is paid back in installments, often monthly; a common way to buy expensive items

interest **1.** The amount over the loan amount that you pay to a lending institution for a loan; also called finance charge. **2.** Money paid to you by a financial institution for the privilege of using your money to make investments and loans to others.

interest due Interest to be paid, found by multiplying the previous month's unpaid balance by the monthly interest rate

interest rate The percent that is the basis for interest earned or paid

interstate highways Major, controlled access highways linking cities in the U.S.

inventory Items produced or purchased and held for resale; stock on hand

investing Purchasing securities, real estate, or some other item with the hope that it will increase in value

item rate *See* **piece rate**

itemized deductions Amounts listed on Schedule A of Form 1040 and deducted from the adjusted gross income

joint return Income tax form completed by a married couple showing combined incomes and deductions

Keogh plan A tax-sheltered pension plan, similar to an IRA, for self-employed persons

labor Work; the use of the efforts and skills of workers in manufacturing a product or extending a service

late payment penalties Possible extra charge imposed if payment on a credit account, loan, or mortage is late

lease Rental agreement between renter and landlord

leasing An arrangement under which a car is hired for a monthly fee

legal tender Money that by law must be accepted for paying debts and taxes

liability legal responsibility for the results of an accident

life insurance A contract to pay a specified amount of money to a designated person upon the death of the insured

linear programming A mathematical method for planning within given constraints

limited payment life insurance Cash-value life insurance which accumulates value more quickly than ordinary life because higher premium payments are made over a shorter period of time

liquidation bankruptcy Outright bankruptcy; an extreme measure in which a debtor who cannot pay must sell most assets and then is left free of debt; the person cannot reapply for credit for seven years.

linear regression equation The equation, $y = ax + b$, of the line that best fits a given set of points

liquidity The ease and speed with which you can withdraw money from deposit

loan sharks Illegal money lenders who charge extremely high interest for short-term cash loans and may use violence to collect the debt

loss region On a graph displaying revenue and cost, the region between the lines and below the break-even point

low-balance account Checking account in which a fee is charged if the balance falls below the minimum any time during the month

low-load insurance policy An insurance policy with a very low commission

maintainance fee *See* **service charge**

management Leadership in a company; those people who make decisions affecting labor, production, advertising, distribution, and so on

market basket List of goods and services that are tracked in the CPI

market economy An economic system in which profit is the incentive

market rate The interest rate currently charged by lending institutions such as banks and credit unions

markup The difference between the price at which a retailer buys an item and sells it

maximize To find the greatest number possible within given constraints

mean Average, found by adding scores and dividing by the total number of scores

median The middle score when scores are ordered from least to greatest

medical payments Insurance coverage for medical costs resulting from an automobile accident

Medicare A federally funded health insurance program for people over 65

metric system A standard system of measurement based on the decimal system. Some basic units are the meter, gram, and liter

mileage chart Chart showing distances between cities or towns

miles per gallon The average number of miles that a car drives on one gallon of gas

mill rate Amount of tax per thousand dollars of assessed value

minimize To find the least number possible within given constraints

minimum-balance account Checking account that requires you to keep a minimum amount on deposit

minimum monthly payment The least amount that a credit card holder may pay on a monthly bill

mobile home A portable living space designed to be used without a permanent foundation, intended as a year-round dwelling

mode In a set of scores, the score that appears most frequently

money market account An account with restricted access in exchange for a higher rate of interest

money supply The amount of money (coins, currency, and demand deposits) that is in circulation; this amount is regulated by the Federal Reserve System

monthly interest rate Percent used to calculate interest for one month, the annual percentage rate divided by 12

mortgage Loan for the purpose of buying property; usually paid in monthly payments of principal and interest, over a period of from 15 to 30 years

multiplier effect In the multiplier effect, money is loaned or invested over and over again, by different institutions, each institution keeping the required portion in reserve but passing on the larger portion for additional investment or loan making

multi-purpose travel and entertainment cards Charge cards often used for travel and entertainment for which an annual fee is charged and for which the total balance must be paid each month; examples include American Express, Diners Club, and Carte Blanche

municipal bond Issued by states, cities, counties, school districts, or other governmental bodies to raise money for schools, hospitals, streets, and so on

mutual fund A means of pooling funds with many other investors to acquire a wide variety of stocks, bonds, and other kinds of investments

nonrefundable A purchase is nonrefundable when you cannot return it for money

non-stop flight A plane flight that does not make any stops between two points

note reduction An amount found by subtracting the interest due from the monthly payment

NOW account Negotiable Order of Withdrawal account; checking account in which the balance earns interest but which may charge a service or per-check fee if the balance falls below a required minimum

objective function The equation used to find a maximum or minimum in linear programming

odd lot A stock trade of fewer than 100 shares

odometer The counter on the dashboard of a car that shows the number of miles the car has been driven

outstanding checks Checks that have been written but have not yet cleared through the bank

over-the-counter market A national network of dealers and brokers in securities

overtime hours Hours worked in addition to regular hours; pay for overtime is usually higher than for regular hours

owner's association The group of owners of individual units in a condominium; responsible for maintenance of building, grounds, and common areas

passbook or regular savings account An account that allows deposits and withdrawals at any time

pay-as-you-earn The payroll withholding feature of the income tax laws

payment number The number of months that have elapsed since an amount was borrowed

pawnbroker A person who will make a cash loan for about 40 percent of the value of an item left with the lender

payee The person or organization to whom a check is written

piece rate Pay based on the number of items produced

points One-time charge by the lending institution at the time of obtaining a mortgage; each point is equal to one percent of the mortgage value

preferred stock Stock which entitles the owner to receive dividends ahead of common stock owners, but which does not usually entitle the owner to vote for the board of directors

premium An amount of money paid on a regular basis for an insurance policy

prepayment clause Statement allowing a mortgage holder to make additional payments toward the principal

prepayment penalty Charge that may be imposed when a person pays back the full amount of a loan ahead of the payment schedule

prime interest rate The interest rate that large banks give their best customers

principal An amount of money invested or borrowed

probability of an event $P(\text{probability}) = \frac{m}{n}$ where m is the number of times the event occurs; n is the number of all possible outcomes.

producers Those who manufacture goods

profit The amount of money a business makes above the cost of the products or services it produces

profit region On a graph displaying revenue and cost, the region between the lines and above the break-even point

progressive tax schedule The means through which higher income is taxed at a higher rate

property damage liability Insurance that pays for damage done to other people's property

push money Money paid to a salesperson by a manufacturer to sell its goods ahead of others

quality of life The degree of satisfaction that people experience in their lives

rate of interest *See* **interest rate**

real-estate agents Individuals licensed by the state to help people find apartments

real wages The value of wages in terms of the goods and services that can be purchased. In times of inflation, real wages may fall while the dollar amount of wages rises

rebate An amount paid back to an individual purchasing a car or other item, either in cash or as a reduction of the purchase price.

reconciliation The process of finding the correct balance in a checking account by comparing the bank statement with the check register

refund An amount of money sent back to a taxpayer when the amount withheld is greater than the total payments due

regular hours The hours an employee is required to work each week

regular savings accounts An account that allows deposits and withdrawals at any time

renewable convertible term insurance Life insurance covering a person for a period of time such as one, five, or ten years and can be renewed without a medical examination

rent Money paid to a landlord, usually by the month, for the use of an apartment or house

rent-to-own plan A purchase plan in which you pay in monthly installments to rent an item; later you can choose to buy the item, applying the payments already made to the purchase price.

replacement cost Insurance coverage which compensates owners for lost or damaged items of personal property at the amount it would cost to replace them

required reserve Percentage of deposits a bank must hold in reserve to repay depositors; the bank may not use this amount for loans or investments

reservation A hotel's or restaurant's agreement to hold a room or a table for the person making the reservation; sometimes a deposit, or part of the price, is required to hold the reservation

restrictive endorsement Your signature on the back of a check with the words "for deposit only," protects the check against being cashed by someone else

revenue Money received for the sale of goods and services

round lot A stock trade of 100 shares

Rule of 72 Method for determining the time it will take for money invested to double at given interest rates: Divide 72 by the annual interest rate times 100, the quotient is the doubling time in years; for 9%, $72 \div 9 = 8$; an investment will double in eight years

Rule of 78 Rule used by lending institutions to determine payment for a buyer who purchases merchandize on credit and wishes to prepay

salary Weekly, monthly, or yearly rate of pay

savings banks Banks offering a variety of services, including checking and savings accounts, and real estate loans

savings and loan associations Banking institutions that specialize in lending money for home purchases and construction

scatter plot A graph in which a finite number of points represent data

secondary roads Well-marked federal and state highways that pass through towns

Securities and Exchange Commission (SEC) An independent federal agency whose purpose is to prevent unsound stock selling practices

security Something of value offered as assurance that a loan will be repaid; also collateral

securities stocks and bonds

security deposit Money, often equal to one month's rent, given to the landlord when a lease is signed; the deposit is refunded when the renter leaves the property in good condition

self-employment income Income earned by people working for themselves—such as physicians, farmers, writers, and consultants

service charge (*also* **maintainance fee**) A fee charged by a bank for a checking account

shares Units of an investment such as "100 shares of stock"

shareholders Those who own a company by owning shares of its stock

simple interest Interest paid one time a year at the end of the year on the total balance in a savings account

Simplified Employee Pension Plan (SEP) A plan that allows a person to contribute to an IRA without all of the usual IRA rules and limitations

single-purpose credit cards Credit cards issued by companies and businesses for use in purchasing their products and services; examples include oil companies and department stores

Social Security Compulsory federal insurance program which provides a level of income security for workers who have retired, surviving dependents of workers, and others

spreadsheet program A computer program that arranges data in rows and columns so that calculations can be made.

standard deduction One of the amounts you are permitted to deduct from your income to determine your taxable income; alternatively you may choose to itemize deductions for such things as charitable contributions, medical expenses, and interest paid on a mortgage loan

standard of living Economic level at which an individual or family lives; the affordable goods and services owned or used by an individual

statement Record of a month's activities in a checking account, credit card or other similar account sent monthly to the holder of the account

sticker price The price shown in a car window in a show room; usually higher than the dealer actually expects to receive

stocks Shares in the ownership of a corporation

sum of an infinite geometric series If a geometric series having an infinite number of terms has a sum, that is the sum of the series. For example, $1 + \frac{1}{2} + \frac{1}{4} + \frac{1}{8} + \ldots = 2$

survivors insurance Money for living expenses paid to survivors of workers who would have been eligible for social security

take-home pay Amount of money a worker receives after deductions have been made for social security, income tax, and other purposes

tax liability The amount of income tax you owe

tax preparation software A computer program used to compute taxes

Tax Rate Schedules Formulas from which taxes are calculated on the basis of fixed amounts and percentages for given incomes

tax shelter An investment or retirement plan that reduces or defers the payment of taxes

tax tables Charts that indicate the income tax owed based on your taxable income

taxable income Gross income minus personal exemptions (for self and dependents) and the standard deduction or itemized deductions; the amount you use to look up your taxes in the tax table

technology The development of new machines or labor-saving methods; technology requires better trained workers to fill new types of positions

term insurance Life insurance that pays death benefits for a stated number of years; one of two major kinds; (*Compare* **whole life**)

tight-money policy A tight-money policy results when the reserve requirement is higher so that less money is available for investments and loans

time zone A region within which all places have the same time

title Document stating the ownership of a car

total financed price Total purchase price of a house plus the financing costs

total income All income from wages, tips, interest, dividends, and other sources

total payment Amount of income tax to be paid

trade-in Selling a used car to the dealer from whom one purchases a new car

Truth in Lending Act Law that requires creditors to tell customers the exact cost of buying on credit

turnover Rate at which people move out and new people move into apartments or other rental units

uninsured and underinsured motorist protection Insurance that pays for damage to your car if you are in an accident caused by a driver who had little or no bodily injury insurance

unit cost The cost to produce one item

unit price The price per standard unit of a product; the cost used to compare different size packages of the product, such as cost per ounce, per pound, per square foot, and so on

U.S. Savings Bonds Bonds issued and backed by the U.S. government

universal life insurance Gives whole life protection and permits savings

unpaid balance The amount of borrowed money remaining to be paid

utilities Suppliers of heat, water, electricity, and natural gas to households; customers pay periodic bills, usually monthly, for these services. The products themselves are also called utilities

variable costs A company's costs that vary depending upon the number of items produced; may include materials, packaging, and the like

Veterans Administration loans Loans with special privileges and insurance given to veterans who served during wars or at other specified times

wage Hourly or daily rate of pay

Wage Earner Plan—Chapter XIII A plan under Chapter XIII of the Federal Bankruptcy Act in which people can restructure their debt without declaring bankruptcy

warranty The guarantee to keep a car in good repair; the period during which the guarantee is in force

whole life insurance Life insurance in which premiums are paid for your entire life

INDEX

In this index, *See* cross-references indicate the word under which page numbers are listed. *See also* cross-references indicate the page numbers of related topics or more detailed breakdowns of a topic.

PHOTO ACKNOWLEDGMENTS

CHAPTER 1
pp. 2-3: Photo by Ian Crysler; **p. 4:** © Stephen Whalen/Picture Perfect USA, Inc.; **p. 6:** © Campbell & Boulanger/Picture Perfect USA, Inc.; **p. 12:** © SuperStock, Inc.; **p. 14:** © Charles Gupton/Stock, Boston; **p. 16:** © C. Orrico/SuperStock, Inc.; **p. 23:** © SuperStock, Inc.; **p. 24:** Location courtesy of Northland Volkswagen/Cincinnati, Ohio; **p. 30:** Jeff Greenberg, Photographer; **p. 35:** © C. J. Allen/Stock, Boston; **p. 36:** © Gala/SuperStock, Inc.; **p. 37:** © Charles Gupton/Stock, Boston.

CHAPTER 2
p. 51: top: © FPG International Corp./J. McNee, **bottom:** © FPG International Corp./Ron Chapple; **p. 52:** Photo by Jim Whitmer; **p. 53:** © Bob Daemmrich/Stock, Boston; **p. 56:** © Rick McClain/Picture Perfect USA, Inc.; **p. 63:** © Bob Daemmrich/Stock, Boston; **p. 67:** © Warren Morgan/Westlight; **p. 70:** © R. Chen/SuperStock, Inc.; **p. 73:** © David R. Frazier/Tony Stone Images; **p. 74:** © Poulides & Thatcher/Tony Stone Images; **p. 79:** © Bob Daemmrich/Stock, Boston.

CHAPTER 3
pp. 90-91: bottom: © Comstock, Inc./Michael Thompson; **p. 92:** © Billy E. Barnes/Picture Perfect USA, Inc.; **p. 93:** © Walter Hodges/Westlight; **p. 97:** © G. Brettnacher/SuperStock, Inc.; **p. 102:** © H. Richard Johnston/Tony Stone Images; **p. 103:** © Melanie Carr/Picture Perfect USA, Inc.; **p. 111:** © SuperStock, Inc.; **p. 114:** Photo by Jim Whitmer; **p. 123:** © P. Cantor/SuperStock, Inc.

CHAPTER 4
p. 133: Photos by Ian Crysler; **p. 134:** © W. Cody/Westlight; **p. 135:** © Richard Pasley/Stock, Boston; **p. 140:** © SuperStock, Inc.; **p. 143:** © Robert Frerck/Tony Stone Images; **p. 147:** © Gene Stein/Westlight; **p. 151:** © Charles Gupton/Stock, Boston; **p. 152:** © Nik Wheeler/Westlight; **p. 158:** © Wendt WorldWide; **p. 161:** © Digital Art/Westlight; **p. 163:** © Bob Daemmrich/Stock, Boston; **p. 169:** © SuperStock, Inc.; **p. 175:** © Lawrence Manning/Westlight.

CHAPTER 5
p. 185: Photo by David Michael Allen; **p. 186:** Jeff Greenberg, Photographer; **p. 187:** © J. Trotter/SuperStock, Inc.; **p. 190:** © Wendt WorldWide; **p. 196:** © Jim Richardson/Westlight; **p. 197:** © Mark E. Gibson; **p. 205:** © Warren Morgan/Westlight; **p. 206:** © Don Smetzer/Tony Stone Images; **p. 210:** © Guy Motil/Westlight; **p. 214:** © William Johnson/Stock, Boston; **p. 216:** © Graham Pym/Picture Perfect USA, Inc.; **p. 222:** © Robert E. Daemmrich/Tony Stone Images; **p. 223:** © Spencer Grant/Stock, Boston; **p. 225:** © Andy Sacks/Tony Stone Images; **p. 226:** Location courtesy of Chevrolet; **p. 233:** © SuperStock, Inc.

CHAPTER 6
pp. 242–243: Photo by David Michael Allen; **p. 244:** Location courtesy of Lazarus; **p. 245:** © William Johnson/Stock, Boston; **p. 247:** © Mark E. Gibson; **p. 253:** © Walter Hodges/Westlight; **p. 257:** © Digital Art/Westlight; **p. 259:** © Charles Thatcher/Tony Stone Images; **p. 260:** © S. William/SuperStock, Inc.; **p. 265:** Photo by Jim Whitmer; **p. 271:** © Graham Lawrence/Picture Perfect USA, Inc.; **p. 272:** © Tony Stone Images; **p. 280:** © Michael Neveux/Westlight; **p. 281:** © Mark E. Gibson; **p. 291:** © John Maher/Stock, Boston; **p. 292:** © Mark E. Gibson; **p. 295:** © Mark E. Gibson; **p. 299:** Photo by Jim Whitmer; **p. 301:** © Ellis Herwig/Stock, Boston.

CHAPTER 7
p. 319: Photo by David Michael Allen; **p. 320:** © K. Coppieters/SuperStock, Inc.; **p. 321:** © Steve Chenn/Westlight; **p. 327:** © Wendt WorldWide; **p. 328:** Photo by Jim Whitmer; **p. 330:** © John Fortunato/Tony Stone Images; **p. 333:** © Ong & Associates/SuperStock, Inc.; **p. 336:** © SuperStock, Inc.; **p. 340:** Photo by Jim Whitmer; **p. 341:** Photo by Jim Whitmer; **p. 344:** © Gary A. Bartholomew/Westlight.

CHAPTER 8
p. 357: © Comstock, Inc.; **p. 358:** © Walter Hodges/Westlight; **p. 359:** © SuperStock, Inc./P. Cantor; **p. 361:** © Wendt WorldWide; **p. 368:** © Greg Pease/Tony Stone Images; **p. 369:** © R. Llewellyn/SuperStock, Inc.; **p. 374:** © Frank Siteman/Stock, Boston; **p. 379:** © Tangent/Picture Perfect USA, Inc.; **p. 380:** © Tony Stone Images; **p. 382:** © SuperStock, Inc.; **p. 388 :** © FPG International Corp./Telegraph Colour Library; **p. 389:** © Tony Stone Images; **p. 392:** © SuperStock, Inc.

CHAPTER 9
p. 405: © FPG International Corp./Dick Luria; **p. 406:** © Robert E. Daemmrich/Tony Stone Images; **p. 413:** © Bill Horsman/Stock, Boston; **p. 419:** © Adamsmith Productions/Westlight; **p. 420:** © Mark E. Gibson/Picture Perfect USA, Inc.; **p. 430:** © Walter Hodges/Westlight; **p. 434:** © SuperStock, Inc.; **p. 435:** © Walter Hodges/Westlight; **p. 437:** © Steve Chenn/Westlight.

CHAPTER 10
p. 453: © Donovan Reese/Tony Stone Images; **p. 454:** Photo by Jim Whitmer; **p. 455:** © SuperStock, Inc.; **p. 457:** © Donald Johnston/Tony Stone Images; **p. 463:** © Rhoda Sidney/Stock, Boston; **p. 464:** © Bill Ross/Westlight; **p. 475:** © Robert Landau/Westlight; **p. 477:** © Joseph Pobereskin/Tony Stone Images; **p. 478:** © Dick Wade/Picture Perfect USA, Inc.; **p. 484:** © Mark E. Gibson; **p. 485:** © Eric Neurath/Stock, Boston; **p. 487:** © Tim Andrew/Picture Perfect USA, Inc.

CHAPTER 11

pp. 498–499: Photo by David Michael Allen; **p. 500:** © Joseph Nettis/Stock, Boston; **p. 505:** © SuperStock, Inc.; **p. 506:** © Bob Daemmrich/Stock, Boston; **p. 509:** © Bob Daemmrich/Stock, Boston; **p. 511:** © Melanie Carr/Picture Perfect USA, Inc.; **p. 513:** © John W. Warden/Picture Perfect USA, Inc.; **p. 517:** © Rainer Grosskopf/Tony Stone Images; **p. 520:** © Mark Segal/Tony Stone Images; **p. 523:** Photo courtesy of Delta Air Lines, Inc.

CHAPTER 12

p. 534: © J. Whitmer/SuperStock, Inc.; **p. 535:** © Wendt WorldWide; **p. 540:** © Ron Slenzak/Westlight; **p. 543:** © Walter Hodges/Westlight; **p. 544:** © Barbara Filet/Tony Stone Images; **p. 546:** © Laima Druskis/Stock, Boston; **p. 557:** © Bob Daemmrich/Stock, Boston; **p. 558:** © Dave Moore/Picture Perfect USA, Inc.; **p. 563:** Photo by Jim Whitmer; **p. 566:** © Wendt WorldWide; **p. 567:** © Gary Irving/Tony Stone Images.

CHAPTER 13

pp. 582–583: background: © Russell Thompson/Picture Perfect USA, Inc., **sign:** Photo by David Michael Allen; **p. 584:** © Doug Wilson/Westlight; **p. 585:** © Chris Andrews/Stock, Boston; **p. 587:** © Bill Gallery/Stock, Boston; **p. 594:** © Frank Siteman/Stock, Boston; **p. 599:** Photo by Jim Whitmer; **p. 603:** © SuperStock, Inc.; **p. 604:** © W. Woodworth/SuperStock, Inc.; **p. 605:** © Lionel Delevingne/Stock, Boston.

CHAPTER 14

p. 620: © Eric Curry/Westlight; **p. 621:** © Seth Resnick/Stock, Boston; **p. 622:** © Walter Urie/Westlight; **p. 624:** © Wendt WorldWide; **p. 625:** © Dean Abramson/Stock, Boston; **p. 630:** © Lawrence Migdale/Stock, Boston; **p. 638:** © Lawrence Migdale/Stock, Boston; **p. 640:** © Bob Daemmrich/Stock, Boston.

CHAPTER 1

Lesson 1–1

Exercise Your Skills

1. Hourly employees are paid for exact hours worked including overtime. Salaried employees are paid a flat rate no matter how many hours they work.
2. Answers may vary. (Ex.: to meet an important deadline)
3. Answers may vary. (Ex.: to complete assigned projects)
4. Answers may vary. (Ex.: A baseball team makes a large profit.)
5. Answers may vary. (Ex.: 10–20 hours)

Lesson 1–4

Ask Yourself

1. Five broad economic functions of our government are: (1) the upkeep and funding of national parks, and (2) public highways, (3) redistribution of income to care for the needy, (4) regulation of certain industries such as communications, and (5) the establishment and preservation of the legal framework for the protection of the rights of citizens.
2. Answers may vary. (Ex.: Social Security, Medicare, programs for disabled veterans, disaster relief, various welfare services, nation's defense)

Try Your Skills

	Income Tax Withholding	FICA Withholding	Take-Home Pay
11.	$286	$183.60	$1930.40
12.	0	53.55	646.45
13.	289	260.10	2850.90
14.	653	335.07	3391.93
15.	0	97.92	1182.08
16.	720	351.90	3528.10
17.	137	153.00	1710.00
18.	109	168.94	1930.39
19.	186	195.08	2168.92
20.	172	127.50	1367.17

Exercise Your Skills

1. Answers may vary. (Ex.: no one likes tax increases, tax laws affect many people)
2. To help protect citizens from economic insecurity

	Income Tax Withholding	FICA Withholding	Take-Home Pay
16.	$292	$201.96	$2146.04
17.	268	191.25	2040.75
18.	699	361.08	3659.92
19.	1	114.75	1384.25
20.	379	306.00	3315.00
21.	30	114.75	1355.25
22.	287	229.50	2483.50
23.	542	318.75	3305.92
24.	742	359.55	3598.45
25.	359	267.75	2873.25

CHAPTER 2

Lesson 2–1

Exercise Your Skills

2. maintain the account, provide records, honor checks
3. Information is concise and easy to compare.
4. Answers may vary. (Ex.: other services, convenience, low service charges)
6. Low-balance account charges fee even if account falls below minimum only one day in a month. Average-balance account can drop to zero as long as customer deposits enough money to bring the average balance for the month to the minimum required.
8. A3, B3, D3, E3
9. C3, F3, G3

Lesson 2–2

Try Your Skills

2.–6.

CHECK NUMBER	DATE	CHECKS/DEPOSITS	AMOUNT	BALANCE
101	2/2	TO: Josh Harmon		400.00
		FOR: Electrical Work	175.60	−175.60
102	2/3	TO: Sunrise Shop		224.40
		FOR: Jacket	27.90	−27.90
103	2/5	TO: Bright Spot		196.50
		FOR: Lamp	46.23	−46.23
	2/6	TO: Deposit		150.27
		FOR:	45.60	+45.60
104	2/6	TO: Garden Center		195.87
		FOR: Plants	19.25	−19.25
		TO:		176.62

7.

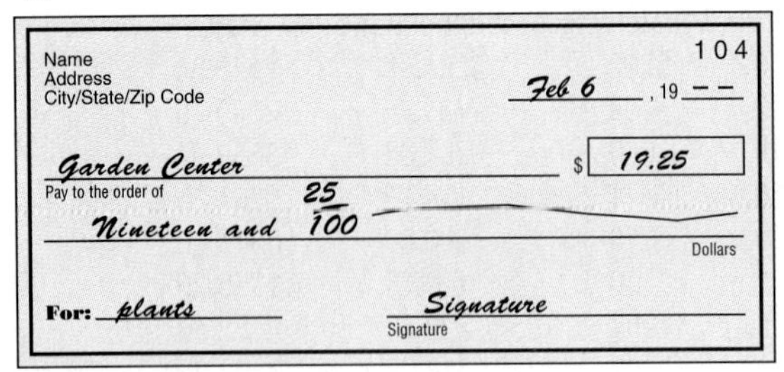
Name 104
Address
City/State/Zip Code Feb 6, 19 – –

Garden Center $ 19.25
Pay to the order of
Nineteen and 25/100 Dollars

For: plants Signature
Signature

Exercise Your Skills

4.

Eldorado Golf Tournament 351
Eldorado Golf Club
Inland, IN 47304 March 4, 19 – –
Telephone 555-7902

Mark Louis $ 240,000.00
Pay to the order of
Two hundred forty thousand and 00/100 Dollars

Inland Bank
15 Commercial Street
Inland, IN 47304

For: 2nd place Signature
Signature

5.

Eldorado Golf Tournament 352
Eldorado Golf Club
Inland, IN 47304 March 4, 19 – –
Telephone 555-7902

Peggy Race $ 150,000.00
Pay to the order of
One hundred fifty thousand and 00/100 Dollars

Inland Bank
15 Commercial Street
Inland, IN 47304

For: 3rd place Signature
Signature

6.

Eldorado Golf Tournament 353
Eldorado Golf Club
Inland, IN 47304 March 4, 19 – –
Telephone 555-7902

Pablo Chosa $ 75,000.00
Pay to the order of
Seventy five thousand and 00/100 Dollars

Inland Bank
15 Commercial Street
Inland, IN 47304

For: 4th place Signature
Signature

7.

Eldorado Golf Tournament 354
Eldorado Golf Club
Inland, IN 47304 March 4, 19 – –
Telephone 555-7902

Tina Marin $ 50,000.00
Pay to the order of
Fifty thousand and 00/100 Dollars

Inland Bank
15 Commercial Street
Inland, IN 47304

For: 5th place Signature
Signature

9. 478.50
10. 453.50
11. 428.50
12. 423.50
13. 409.24
14. 394.20
15. 279.20
16. 590.39
17. 518.58
18. 407.32
19. 367.15
20. 567.15
21. 468.65
22. 444.65
23. 419.65
24. 369.65
25. 306.65
26. 426.65
27. 301.65
28. 276.65
29. 221.35
30. 211.85
31. 311.85
32. 306.85
33. 288.57
34. 284.07
35. 276.52
36. 261.52
37. 252.41
38. 157.41
39. 145.70

Chapter 2 Review

9.

Beginning Balance	Number of Checks	Cost of Checks	Extra Charges	Service Charges	Interest Earned	New Balance
$155	11	$0.25	0	$5.00	$0.70	$150.45
278	17	0.38	0	5.00	1.25	273.87
505	21	0.47	$0.15	0	2.27	506.65
789	27	0.61	1.05	0	3.55	790.89

CHAPTER 3

Lesson 3–1

Try Your Skills

13.–16.

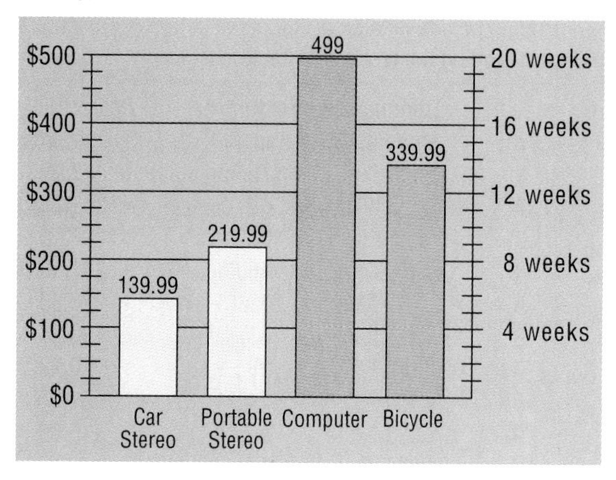

Exercise Your Skills

1. The bank has use of the funds for a guaranteed amount of time.
2. to help secure the funds for other depositors
3. to avoid penalties for withdrawing the funds

19.–21.

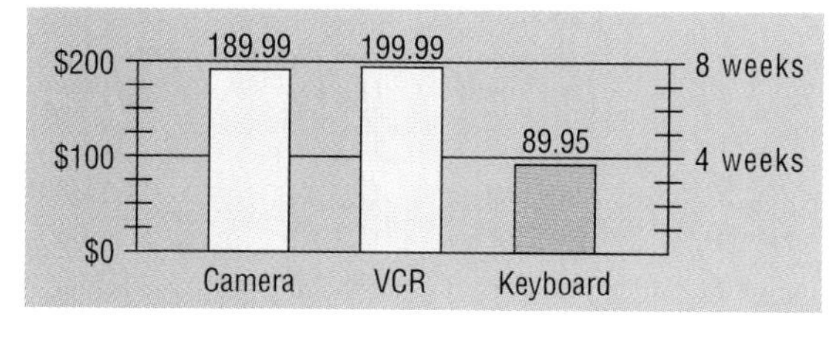

22.–24.

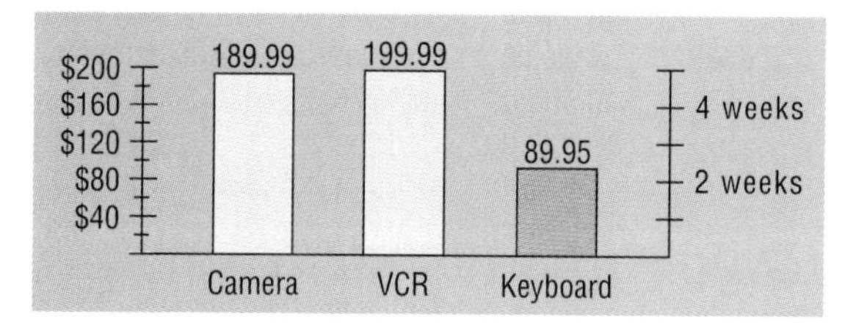

Lesson 3–3

Ask Yourself

1. Washington, Oregon, California, Idaho, Utah, Nevada, Arizona, Hawaii, Alaska
2. To act as a central bank; to serve as a bank for the U.S. government; to supervise financial institutions; to regulate and manage the nation's money supply.
3. Money that by law must be accepted for payments of debts and taxes.
4. Encourages financial institutions to borrow short-term money from the Federal Reserve banks at relatively low interest rates. With low rates, businesses and wage earners will be encouraged to borrow money and buy more goods.
5. A specified percent of a customer's deposit which must be deposited with the Federal Reserve System.
6. A tight-money policy reduces the multiplier effect; an easy-money policy increases the multiplier effect.
7. The multiplier effect is the creation of new money or credit which occurs from the required reserves of the Federal Reserve System.
8. The Federal Reserve System consists of 12 Federal Reserve banks, 25 branch banks, a board of governors, and the Federal Open Market Committee.

Try Your Skills

1.

Level	Demand Deposit	Required Reserve	Loans and Investments
1	$500.00	$100.00	$ 400.00
2	400.00	80.00	320.00
3	320.00	64.00	256.00
4	256.00	51.20	204.80
5	204.80	40.96	163.84
6	163.84	32.77	131.07
7	131.07	26.21	104.86
8	104.86	20.97	83.89
9	83.89	16.78	67.11
10	67.11	13.42	53.69
		Total after 10 levels	$1785.26

2.

Level	Demand Deposit	Required Reserve	Loans and Investments
1	$2500.00	$500.00	$2000.00
2	2000.00	400.00	1600.00
3	1600.00	320.00	1280.00
4	1280.00	256.00	1024.00
5	1024.00	204.80	819.20
6	819.20	163.84	655.36
7	655.36	131.07	524.29
8	524.29	104.86	419.43
9	419.43	83.89	335.54
10	335.54	67.11	268.43
		Total after 10 levels	$8926.25

3.

Level	Demand Deposit	Required Reserve	Loans and Investments
1	$500.00	$125.00	$ 375.00
2	375.00	93.75	281.25
3	281.25	70.31	210.94
4	210.94	52.74	158.20
5	158.20	39.55	118.65
6	118.65	29.66	88.99
7	88.99	22.25	66.74
8	66.74	16.69	50.05
9	50.05	12.51	37.54
10	37.54	9.39	28.15
		Total after 10 levels	$1415.51

4.

Level	Demand Deposit	Required Reserve	Loans and Investments
1	$2500.00	$625.00	$1875.00
2	1875.00	468.75	1406.25
3	1406.25	351.56	1054.69
4	1054.69	263.67	791.02
5	791.02	197.76	593.26
6	593.26	148.32	444.94
7	444.94	111.24	333.70
8	333.70	83.43	250.27
9	250.27	62.57	187.70
10	187.70	46.93	140.77
		Total after 10 levels	$7077.60

5. $2,000
6. $10,000
7. $1,500
8. $7,500
9. 5
10. 5
11. 4
12. 4

Exercise Your Skills

1. No, the information is transmitted electronically.
2. Too much money causes inflation; too little money causes economic activity to diminish.
3. A financial institution creates extra money or credit from excess reserves.
4. by changing the reserve requirement

5.

Level	Demand Deposit	Required Reserve	Loans and Investments
1	$500.00	$75.00	$ 425.00
2	425.00	63.75	361.25
3	361.25	54.19	307.06
4	307.06	46.06	261.00
5	261.00	39.15	221.85
6	221.85	33.28	188.57
7	188.57	28.29	160.28
8	160.28	24.04	136.24
9	136.24	20.44	115.80
10	115.80	17.37	98.43
		Total after 10 levels	$2275.48

6.

Level	Demand Deposit	Required Reserve	Loans and Investments
1	$2500.00	$375.00	$2125.00
2	2125.00	318.75	1806.25
3	1806.25	270.94	1535.31
4	1535.31	230.30	1305.01
5	1305.01	195.75	1109.26
6	1109.26	166.39	942.87
7	942.87	141.43	801.44
8	801.44	120.22	681.22
9	681.22	102.18	579.04
10	579.04	86.86	492.18
		Total after 10 levels	$11,377.58

7.

Level	Demand Deposit	Required Reserve	Loans and Investments
1	$12,500.00	$1875.00	$10,625.00
2	10,625.00	1593.75	9,031.25
3	9,031.25	1354.69	7,676.56
4	7,676.56	1151.48	6,525.08
5	6,525.08	978.76	5,546.32
6	5,546.32	831.95	4,714.37
7	4,714.37	707.16	4,007.21
8	4,007.21	601.08	3,406.13
9	3,406.13	510.92	2,895.21
10	2,895.21	434.28	2,460.93
		Total after 10 levels	$56,888.06

8.

Level	Demand Deposit	Required Reserve	Loans and Investments
1	$12,500.00	$2250.00	$10,250.00
2	10,250.00	1845.00	8,405.00
3	8,405.00	1512.90	6,892.10
4	6,892.10	1240.58	5,561.52
5	5,651.52	1017.27	4,634.25
6	4,634.25	834.17	3,800.08
7	3,800.08	684.01	3,116.07
8	3,116.07	560.89	2,555.18
9	2,555.18	459.93	2,095.25
10	2,095.25	377.15	1,718.10
		Total after 10 levels	$49,117.55

9.

Level	Demand Deposit	Required Reserve	Loans and Investments
1	$12,500.00	$2500.00	$10,000.00
2	10,000.00	2000.00	8,000.00
3	8,000.00	1600.00	6,400.00
4	6,400.00	1280.00	5,120.00
5	5,120.00	1024.00	4,096.00
6	4,096.00	819.20	3,276.80
7	3,276.80	655.36	2,621.44
8	2,621.44	524.29	2,097.15
9	2,097.15	419.43	1,677.72
10	1,677.72	335.54	1,342.18
		Total after 10 levels	$44,631.29

10.

Level	Demand Deposit	Required Reserve	Loans and Investments
1	$12,500.00	$3125.00	$9375.00
2	9,375.00	2343.75	7031.25
3	7,031.25	1757.81	5273.44
4	5,273.44	1318.36	3955.08
5	3,955.08	988.77	2966.31
6	2,966.31	741.58	2224.73
7	2,224.73	556.18	1668.55
8	1,668.55	417.14	1251.41
9	1,251.41	312.85	938.56
10	938.56	234.64	703.92
		Total after 10 levels	$35,388.25

11. $2833.33
12. $14,166.67
13. $70,833.33
14. $56,944.44
15. $50,000.00
16. $37,500.00
17. 6.7
18. 6.7
19. 6.7
20. 5.6
21. 5
22. 4
23. the greater the reserve requirement, the less extra money generated

Chapter 3 Review

8.

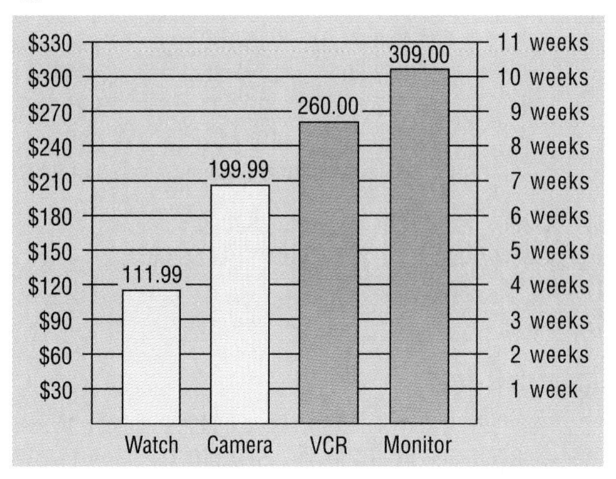

33.

Level	Demand Deposit	Required Reserve	Loans and Investments
1	$2000.00	$400.00	$1600.00
2	1600.00	320.00	1280.00
3	1280.00	256.00	1024.00
4	1024.00	204.80	819.20
5	819.20	163.84	655.36
		Total after 5 levels	$5378.56

34.

Level	Demand Deposit	Required Reserve	Loans and Investments
1	$1000.00	$250.00	$ 750.00
2	750.00	187.50	562.50
3	562.50	140.63	421.87
4	421.87	105.47	316.40
5	316.40	79.10	237.30
		Total after 5 levels	$2288.07

35.

Level	Demand Deposit	Required Reserve	Loans and Investments
1	$11,500.00	$1955.00	$9545.00
2	9,545.00	1622.65	7922.35
3	7,922.35	1346.80	6575.55
4	6,575.55	1117.84	5457.71
5	5,457.71	927.81	4529.90
		Total after 5 levels	$34,030.51

36.

Level	Demand Deposit	Required Reserve	Loans and Investments
1	$25,000.00	$3750.00	$21,250.00
2	21,250.00	3187.50	18,062.50
3	18,062.50	2709.38	15,353.12
4	15,353.12	2302.97	13,050.15
5	13,050.15	1957.52	11,092.63
		Total after 5 levels	$78,808.40

37. $8000 **38.** $3000

39. $56,147.06 **40.** $141,666.67

41. 5 **42.** 4

43. 5.9 **44.** 6.7

Chapter 3 Test

	12.	13.	14.	15.
1 y	$ 9,095.00	$ 9,105.41	$ 9,110.80	$ 9,114.47
2 y	9,731.65	9,753.95	9,765.50	9,773.35
3 y	10,412.87	10,448.67	10,467.23	10,479.87
4 y	11,141.77	11,192.88	11,219.40	11,237.46

18.

Level	Demand Deposit	Required Reserve	Loans and Investments
1	$3000.00	$480.00	$2520.00
2	2520.00	403.20	2116.80
3	2116.80	338.69	1778.11
4	1778.11	284.50	1493.61
5	1493.61	238.98	1254.63
		Total after 5 levels	$9163.15

Chapter 3 Cumulative Review

9. Check Number Date To: (name of recipient) For: (purpose of payment) Amount: $243.50 Balance: $739.11

CHAPTER 4

Algebra Refresher

1.

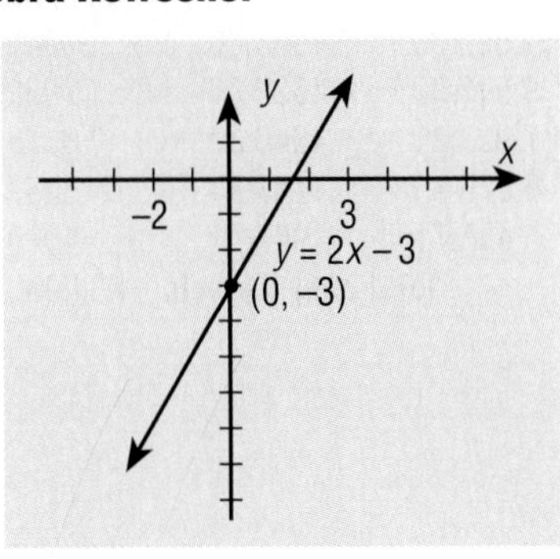

2.

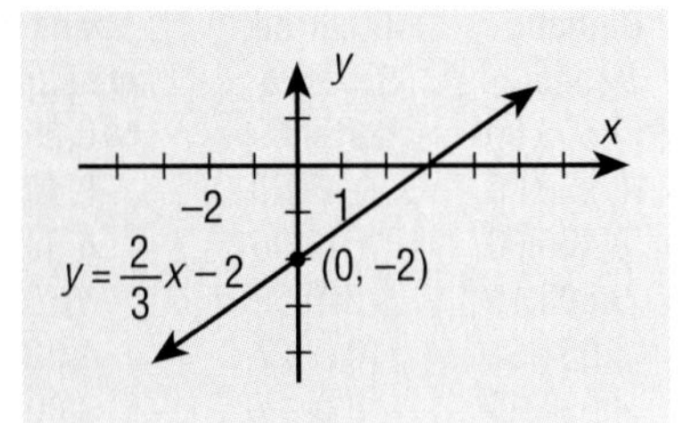

3.

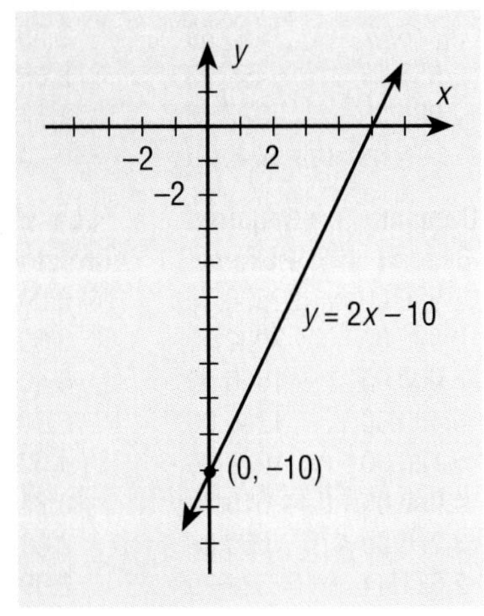

4.

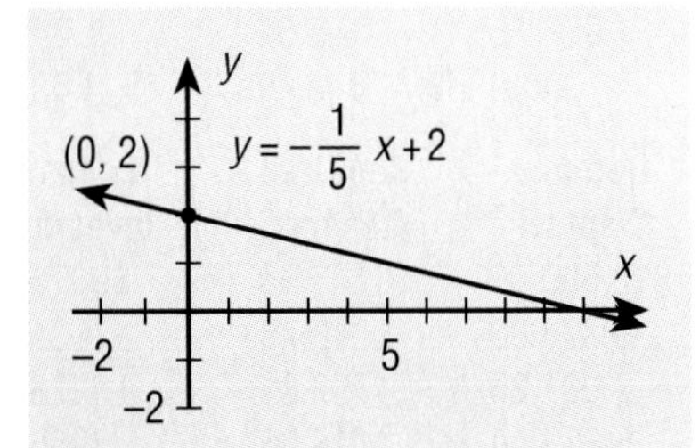

5.

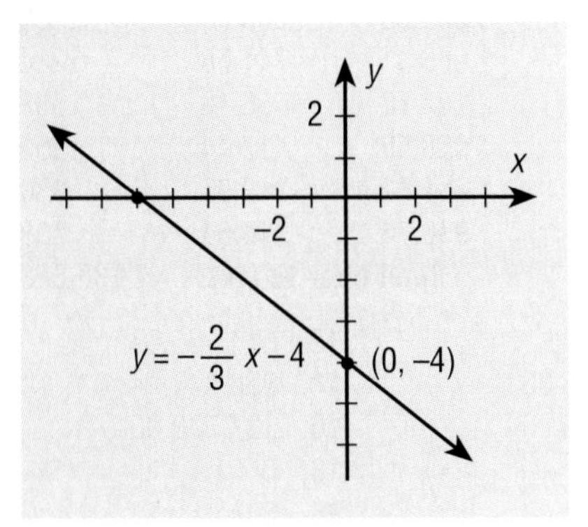

6.

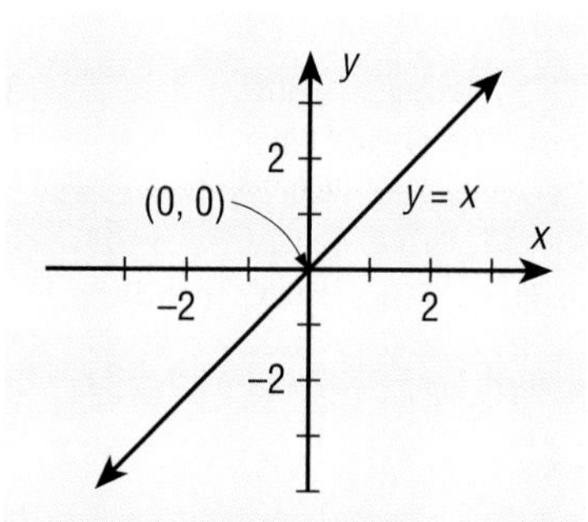

7.

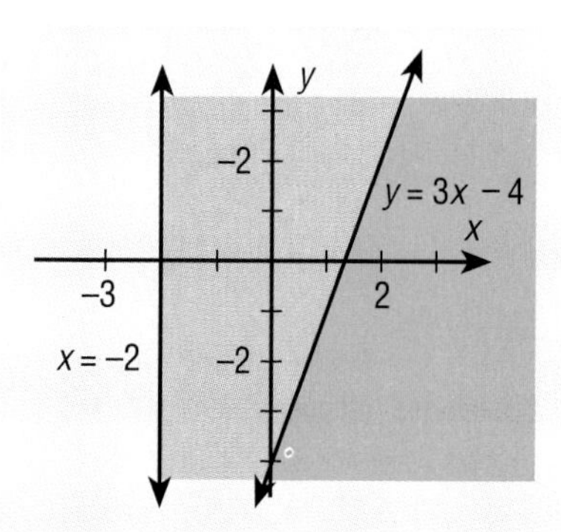

8.

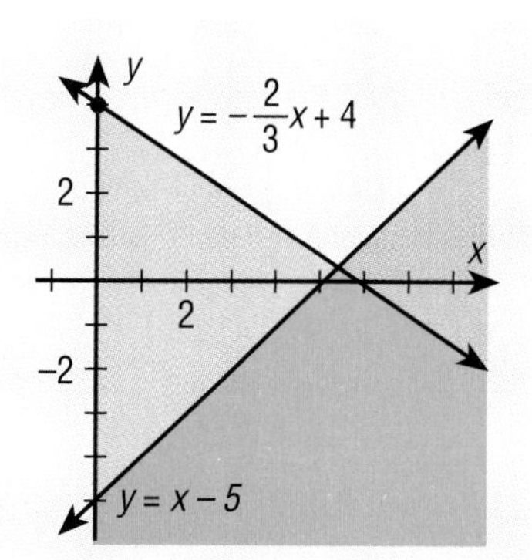

9.

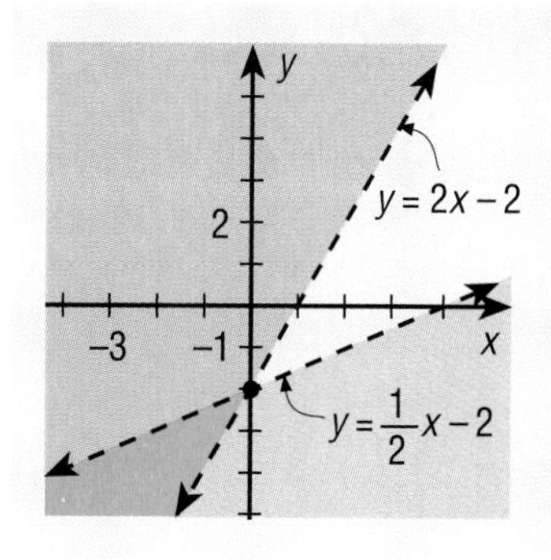

10.

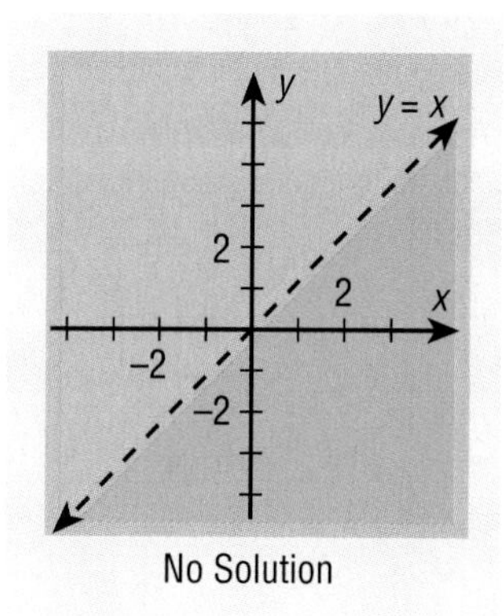

No Solution

Lesson 4–1

Algebra Review

1.

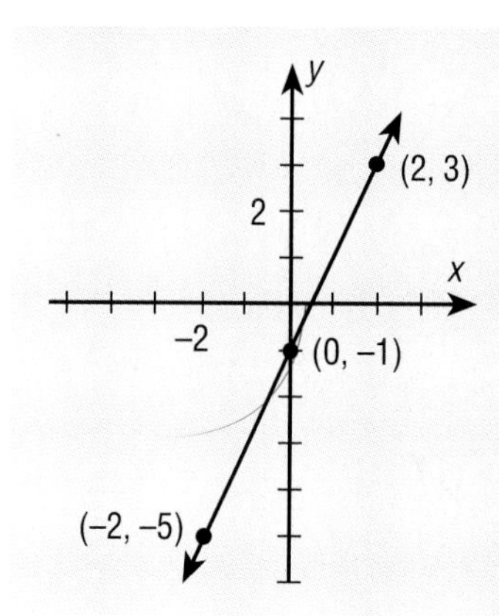

2.

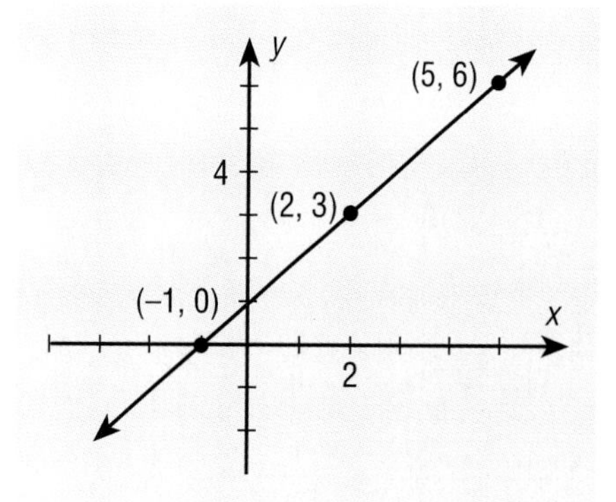

3.

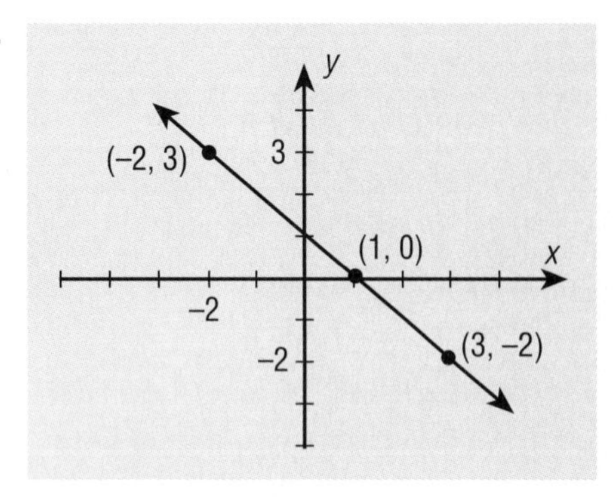

4.

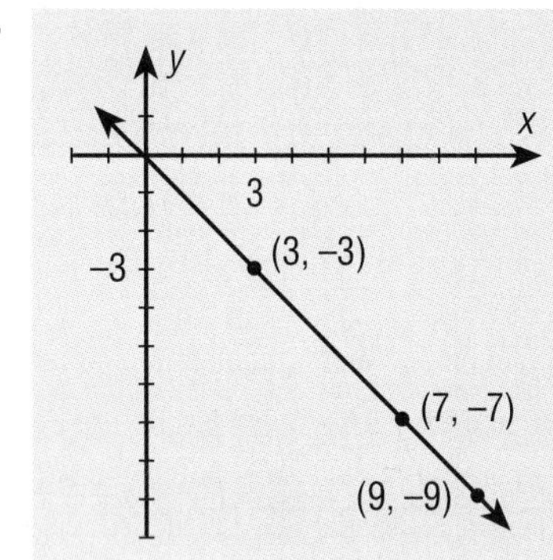

5.

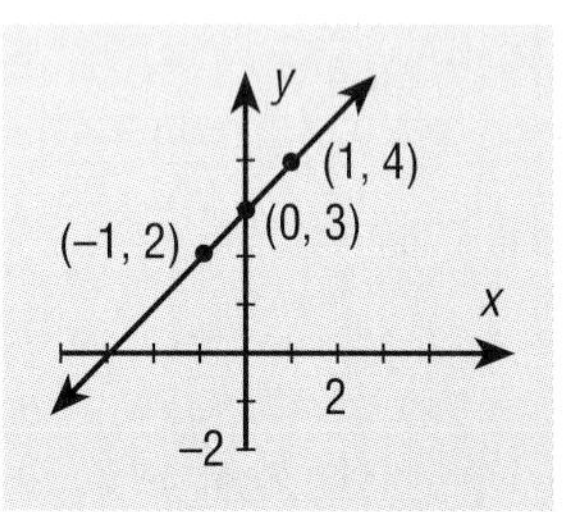

6.

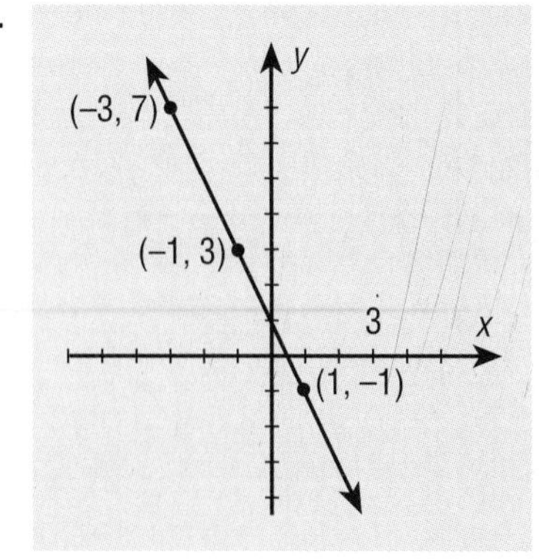

Exercise Your Skills

5.

Employee	Gross Pay	Income Tax Withholding	FICA Withholding	Total Deductions	Take-Home Pay
Kendra	105.00	9.00	8.03	17.03	87.97
Paulo	73.80	0.00	5.65	5.65	68.15
Michael	176.90	19.00	13.53	32.53	144.37
Total	335.70	28.00	27.21	55.12	300.49

6. Week ending 4/4

Employee	Exemptions	Marital Status	Hourly Rate	Hours Worked	Gross Pay
Catlyn	0	Single	$3.75	16	$ 60.00
Sara	1	Single	4.50	20	90.00
Joleen	0	Single	8.00	26	208.00
Hernando	1	Single	6.24	14	87.36
				Total	445.36

Week ending 4/11

Employee	Exemptions	Marital Status	Hourly Rate	Hours Worked	Gross Pay
Catlyn	0	Single	$3.75	14	$ 52.50
Sara	1	Single	4.50	28	126.00
Joleen	0	Single	8.00	16	128.00
Hernando	1	Single	6.24	10	62.40
				Total	368.90

Week ending 4/18

Employee	Exemptions	Marital Status	Hourly Rate	Hours Worked	Gross Pay
Catlyn	0	Single	$3.75	21	$ 78.75
Sara	1	Single	4.50	25	112.50
Joleen	0	Single	8.00	16	128.00
Hernando	1	Single	6.24	30	187.20
				Total	506.45

Week ending 4/25

Employee	Exemptions	Marital Status	Hourly Rate	Hours Worked	Gross Pay
Catlyn	0	Single	$3.75	32	$120.00
Sara	1	Single	4.50	18	81.00
Joleen	0	Single	8.00	24	192.00
Hernando	1	Single	6.24	20	124.80
				Total	517.80

7. Week ending 4/4

Employee	Gross Pay	Income Tax Withholding	FICA Withholding	Total Deductions	Take-Home Pay
Catlyn	60.00	2.00	4.59	6.59	53.41
Sara	90.00	0.00	6.89	6.89	83.11
Joleen	208.00	23.00	15.91	38.91	169.09
Hernando	87.36	0.00	6.68	6.68	80.68
Total	445.36	25.00	34.07	59.07	386.29

Week ending 4/11

Employee	Gross Pay	Income Tax Withholding	FICA Withholding	Total Deductions	Take-Home Pay
Catlyn	52.50	1.00	4.02	5.02	47.48
Sara	126.00	5.00	9.64	14.64	111.36
Joleen	128.00	12.00	9.79	21.79	106.21
Hernando	62.40	0.00	4.77	4.77	57.63
Total	368.90	18.00	28.22	46.22	322.68

Week ending 4/18

Employee	Gross Pay	Income Tax Withholding	FICA Withholding	Total Deductions	Take-Home Pay
Catlyn	78.75	4.00	6.02	10.02	68.73
Sara	112.50	3.00	8.61	11.61	100.89
Joleen	128.00	12.00	9.79	21.79	106.21
Hernando	187.20	14.00	14.32	28.32	158.88
Total	506.45	33.00	38.74	71.74	434.71

Week ending 4/25

Employee	Gross Pay	Income Tax Withholding	FICA Withholding	Total Deductions	Take-Home Pay
Catlyn	120.00	11.00	9.18	20.18	99.82
Sara	81.00	0.00	6.20	6.20	74.80
Joleen	192.00	22.00	14.69	36.69	155.31
Hernando	124.80	4.00	9.55	13.55	111.25
Total	517.80	37.00	39.62	76.62	441.18

8. Employee: Catlyn

Week Ending	Gross Pay	Income Tax Withholding	FICA Withholding	Take-Home Pay
4/4	$ 60.00	$ 2.00	$ 4.59	$ 53.41
4/11	52.50	1.00	4.02	47.48
4/18	78.75	4.00	6.02	68.73
4/25	120.00	11.00	9.18	99.82
Total	311.25	18.00	23.81	269.44

Employee: Sara

Week Ending	Gross Pay	Income Tax Withholding	FICA Withholding	Take-Home Pay
4/4	$ 90.00	$ 0.00	$ 6.89	$ 83.11
4/11	126.00	5.00	9.64	111.36
4/18	112.50	3.00	8.61	100.89
4/25	81.00	0.00	6.20	74.80
Total	409.50	8.00	31.34	370.16

Employee: Joleen

Week Ending	Gross Pay	Income Tax Withholding	FICA Withholding	Take-Home Pay
4/4	$208.00	$23.00	$15.91	$169.09
4/11	128.00	12.00	9.79	106.21
4/18	128.00	12.00	9.79	106.21
4/25	192.00	22.00	14.69	155.31
Total	656.00	69.00	50.18	536.82

Employee: Hernando

Week Ending	Gross Pay	Income Tax Withholding	FICA Withholding	Take-Home Pay
4/4	$ 87.36	$ 0.00	$ 6.68	$ 80.68
4/11	62.40	0.00	4.77	57.63
4/18	187.20	14.00	14.32	158.88
4/25	124.80	4.00	9.55	111.25
Total	461.76	18.00	35.32	408.44

9.

Employee	Exemptions	Hourly Rate	Hours Worked	Gross Pay	Income Tax Withholding	FICA Withholding	Take-Home Pay
Adikes	1	$7.85	10	$ 78.50	$ 0.00	$ 6.01	$ 72.49
Carney	1	5.30	18	95.40	1.00	7.30	87.10
Inez	0	6.78	27	183.06	20.00	14.00	149.06
Sun-Li	0	9.36	17	159.12	16.00	12.17	130.95
Total				516.08	37.00	39.48	439.60

Lesson 4–2

Algebra Review

13.

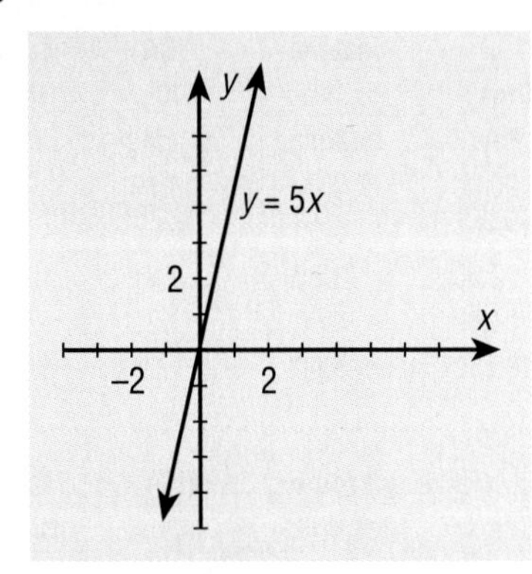

14.

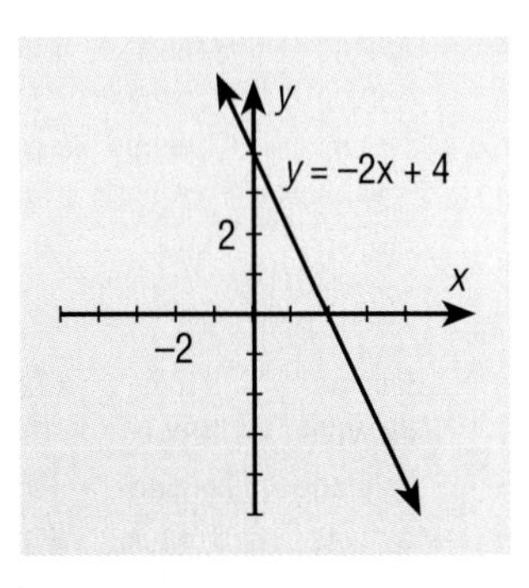

15.

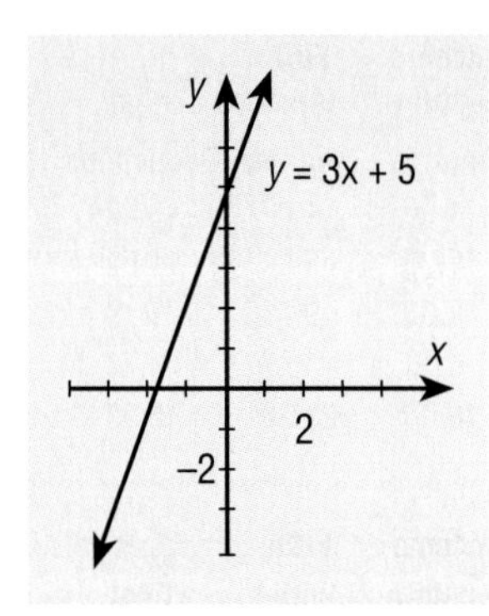

16.

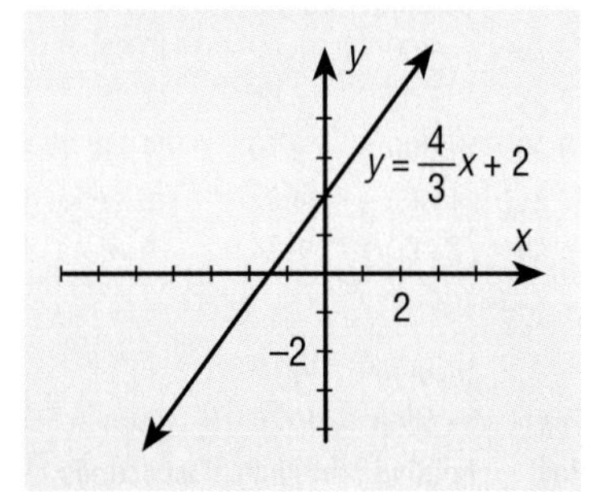

Mixed Review

	Level	Demand Deposit	Required Reserve	Loans and Investments
6.	1	$2400	$480.00	$1920.00
	2	1920	384.00	1536.00
7.	1	1800	270.00	1530.00
	2	1530	229.50	1300.50
8.	1	1800	450.00	1350.00
	2	1350	337.50	1012.50

Lesson 4–3

Exercise Your Skills

13.

Unit Cost	Number Produced	Fixed Cost	Total Cost	Unit Price	Revenue	Profit (Loss)
$1.20	200	$212.50	$452.50	$2.00	$400.00	($52.50)
1.20	300	212.50	572.50	2.00	600.00	27.50
1.20	400	212.50	692.50	2.00	800.00	107.50

14.

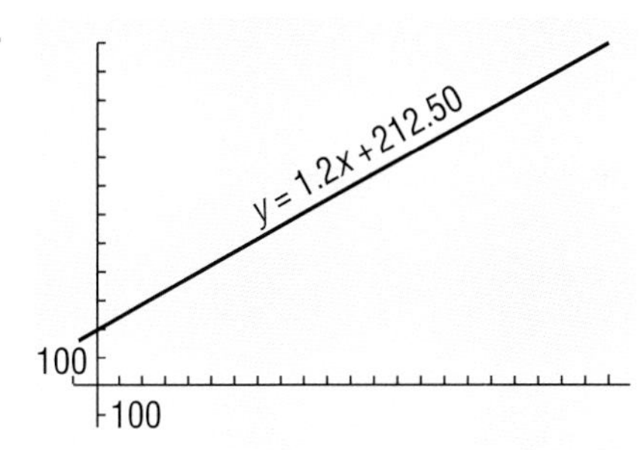

16.

Unit Cost	Number Produced	Fixed Cost	Total Cost	Unit Price	Revenue	Profit (Loss)
$3.55	25	$282.50	$371.25	$10.00	$250.00	($121.25)
3.55	50	282.50	460.00	10.00	500.00	40.00
3.55	75	282.50	548.75	10.00	750.00	201.25

17.

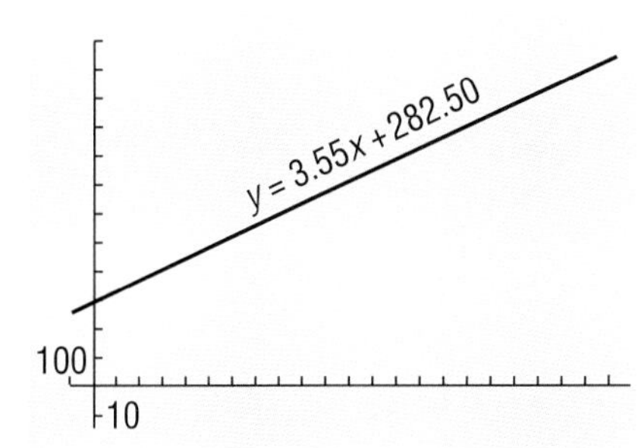

19.

Unit Cost	Number Produced	Fixed Cost	Total Cost	Unit Price	Revenue	Profit (Loss)
$2.75	100	$112.50	$387.50	$4.00	$ 400.00	$ 12.50
2.75	200	112.50	662.50	4.00	800.00	137.50
2.75	300	112.50	937.50	4.00	1200.00	262.50

20.

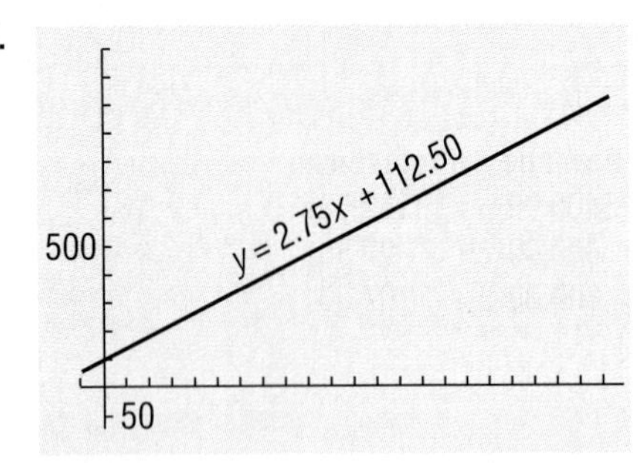

22.

Unit Cost	Number Produced	Fixed Cost	Total Cost	Unit Price	Revenue	Profit (Loss)
$4.00	25	$190.00	$290.00	$7.50	$187.50	($102.50)
4.00	50	190.00	390.00	7.50	375.00	(15.00)
4.00	75	190.00	490.00	7.50	562.50	72.50

23.

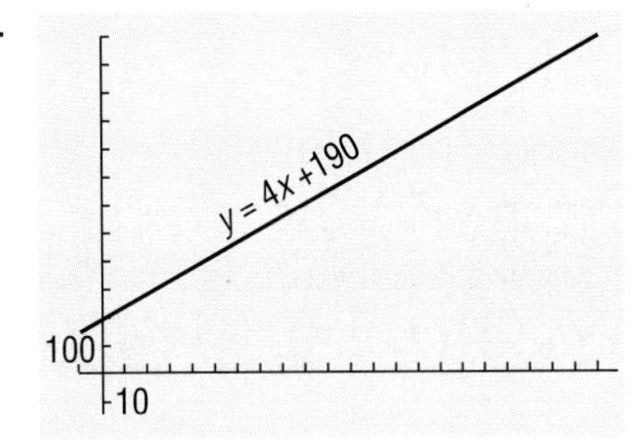

Lesson 4–4

Algebra Review

5.

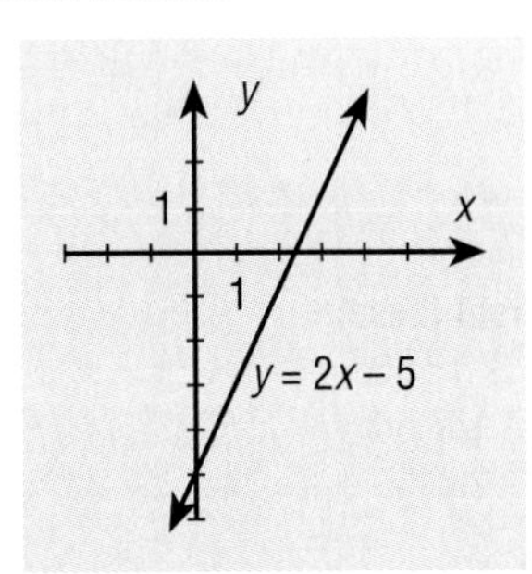

6.

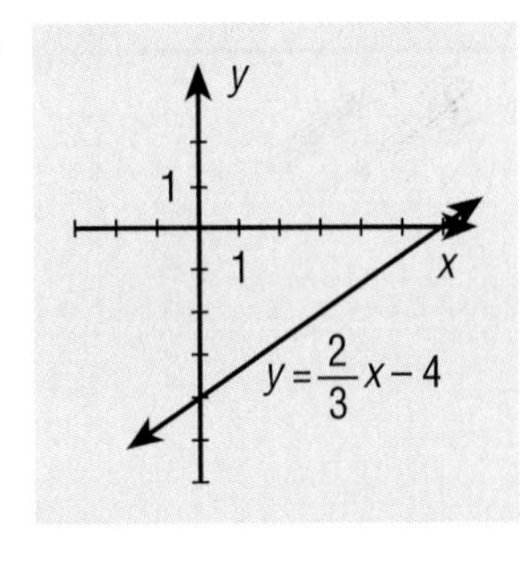

Try Your Skills

4.

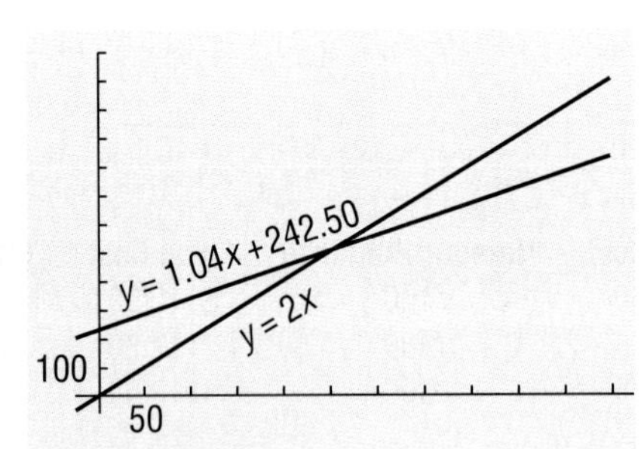

Exercise Your Skills

2. It tells how many units must be sold for revenue to cover costs.

8.

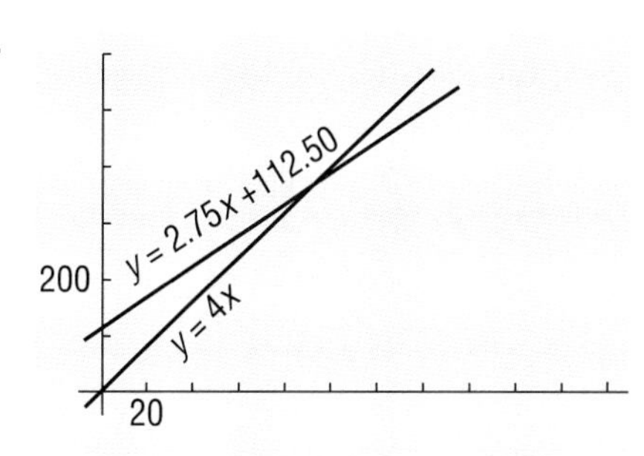

13.

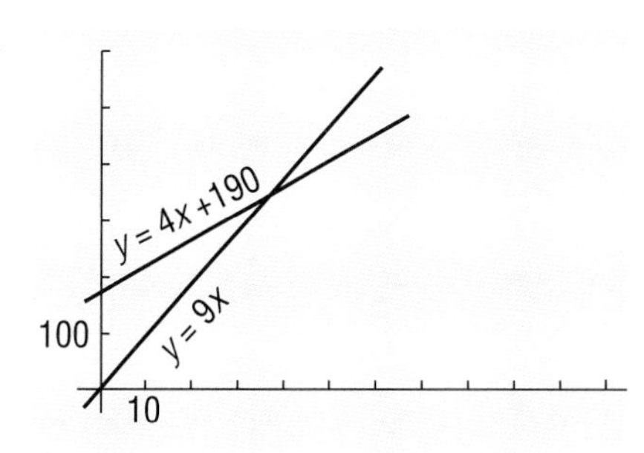

19.

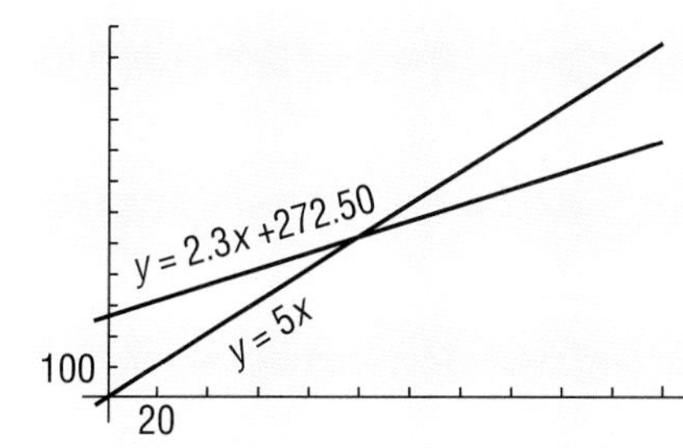

Mixed Review

3.

Level	Demand Deposit	Required Reserve	Loans and Investments
1	$5000	$1000	$4000
2	4000	800	3200

Lesson 4–5

Algebra Review

1.

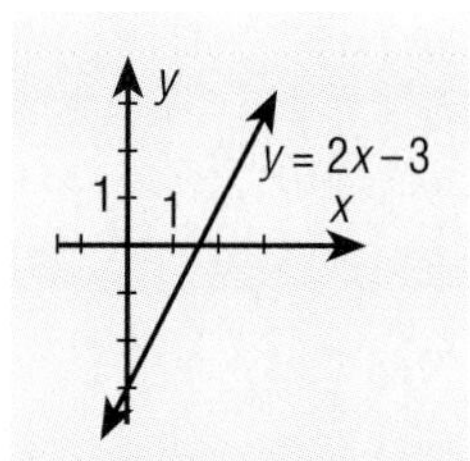

2.

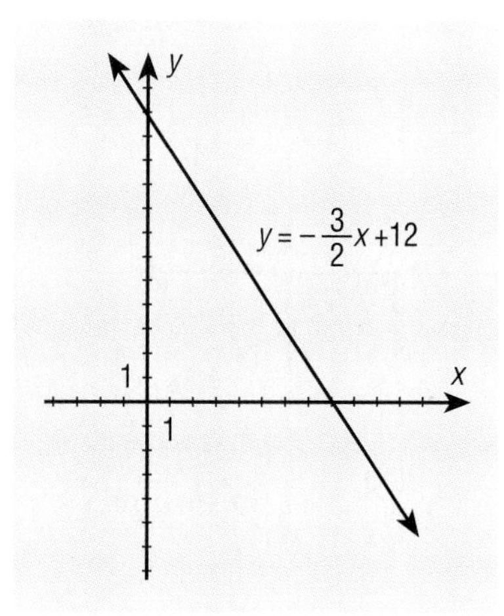

5.

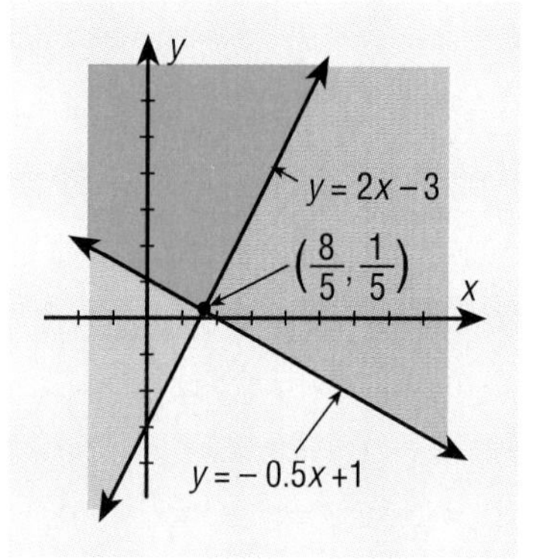

6.

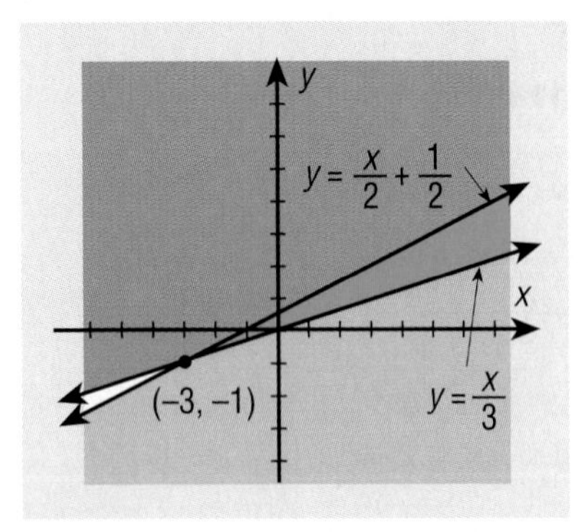

7.

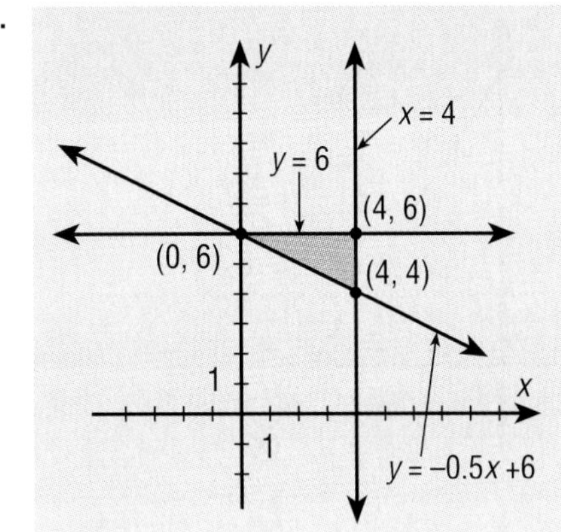

Try Your Skills

2.

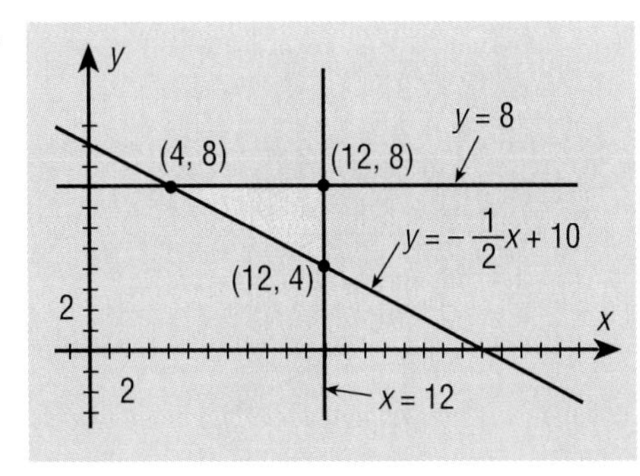

5.

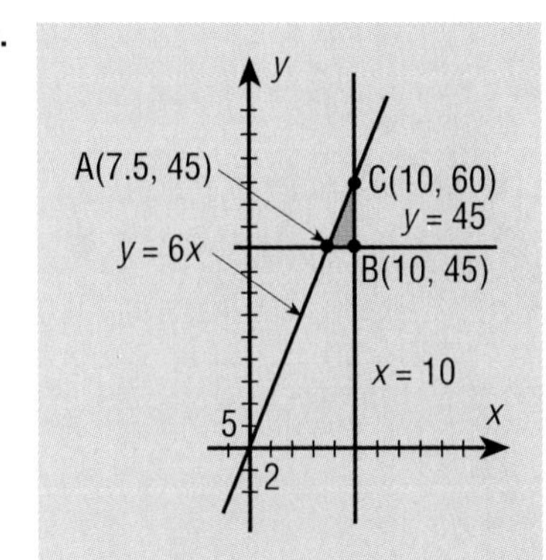

A (7.5, 45) represents the least number of hours (7.5 hours) you can work in a week to earn $45.

B (10, 45) represents the lowest rate you would accept (45 ÷ 10 = $4.50).

C (10, 60) represents the most you can earn ($60).

Exercise Your Skills

2.

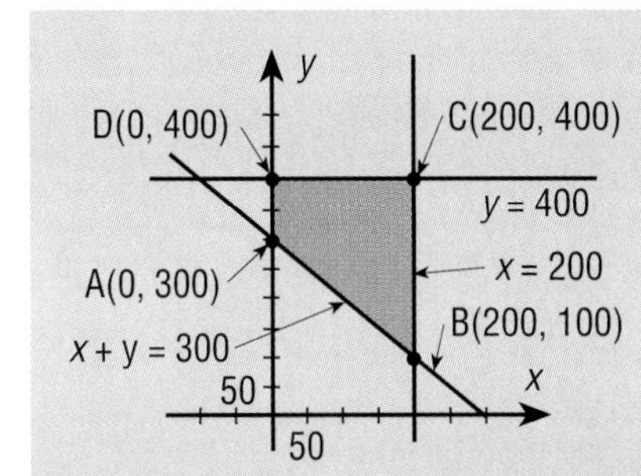

10.

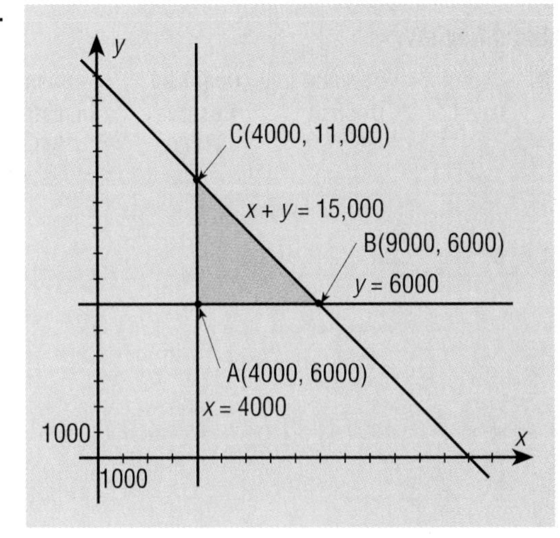

15.

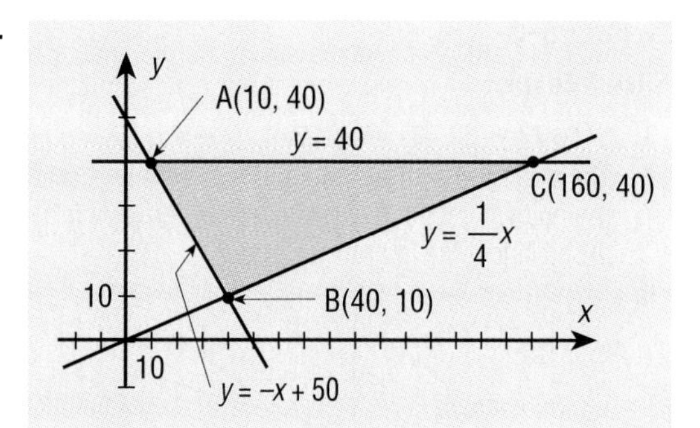

20.

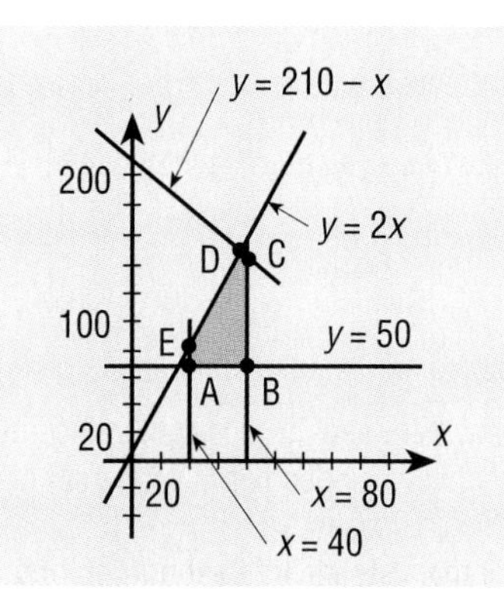

27.

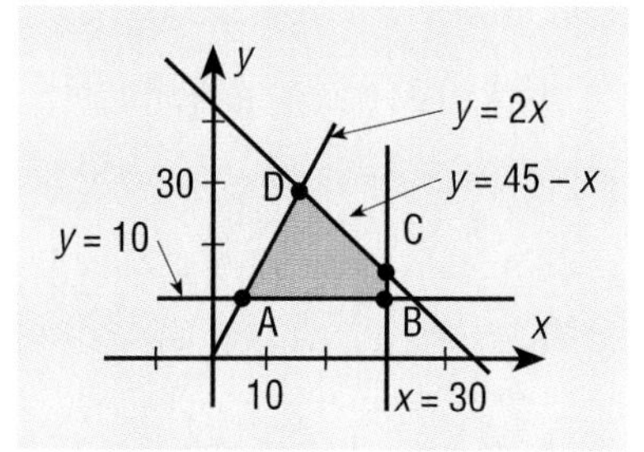

Chapter 4 Review

21. and 22.

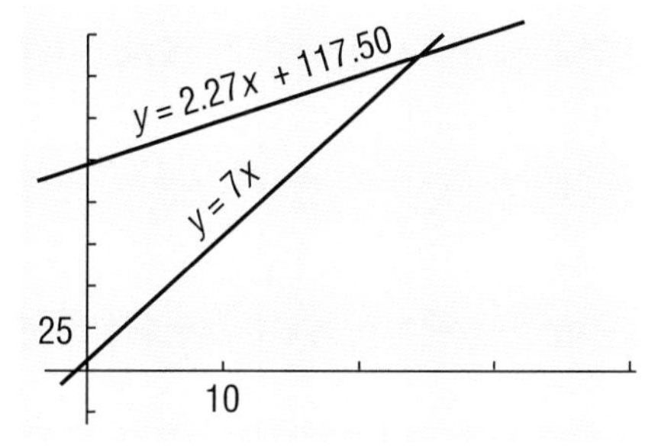

31.

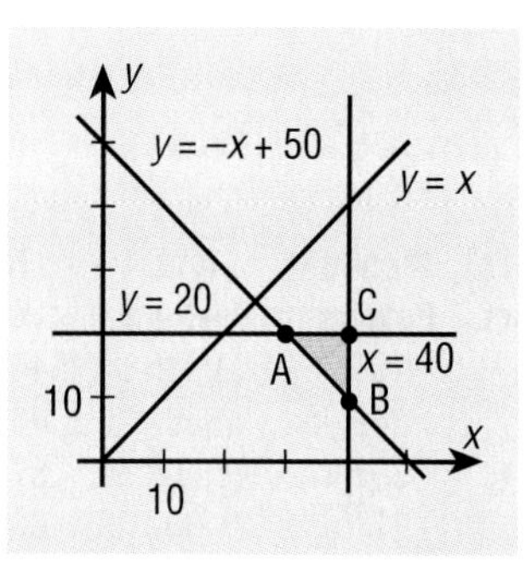

Chapter 4 Test

16.

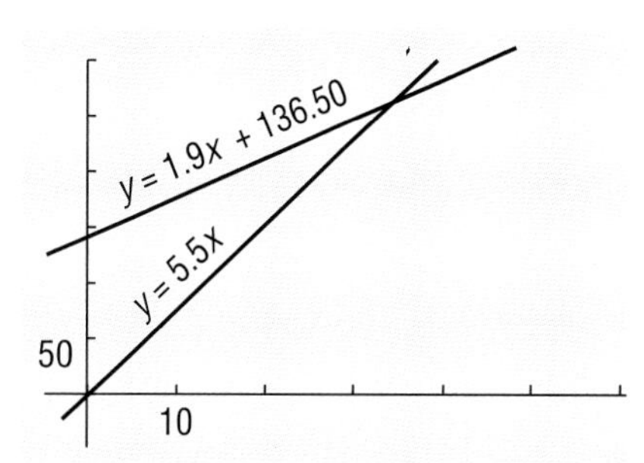

18.

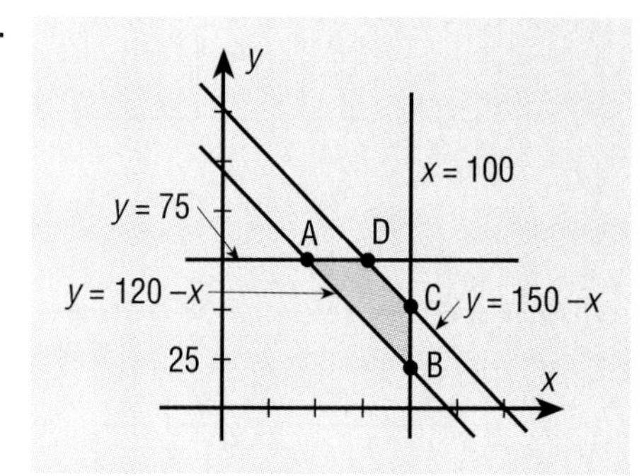

A(45, 75); B(100, 20); C(100, 50); D(75, 75)

Constraints: $x \le 100$
$y \le 75$
$x + y \ge 120$
$x + y \le 150$

Cost function: $c = 1.9x + 2.25y$

Revenue function: $r = 5x + 6y$

Profit function: $p = 3.1x + 3.75y$

CHAPTER 5

Lesson 5–2

Exercise Your Skills

25.

Loan Amount	Number of Years	Monthly Payment	Total Payment	Total Cost
$20,000	5	$400.76	$24,045.54	$4045.54
20,000	6	345.80	24,897.76	4897.76
20,000	7	306.77	25,768.30	5768.30
20,000	8	277.68	26,657.03	6657.03
20,000	9	255.22	27,563.79	7563.79
20,000	10	237.40	28,488.42	8488.42

26.

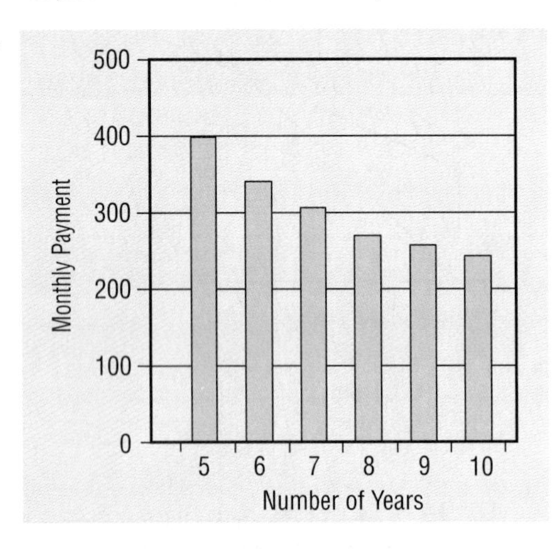

27.

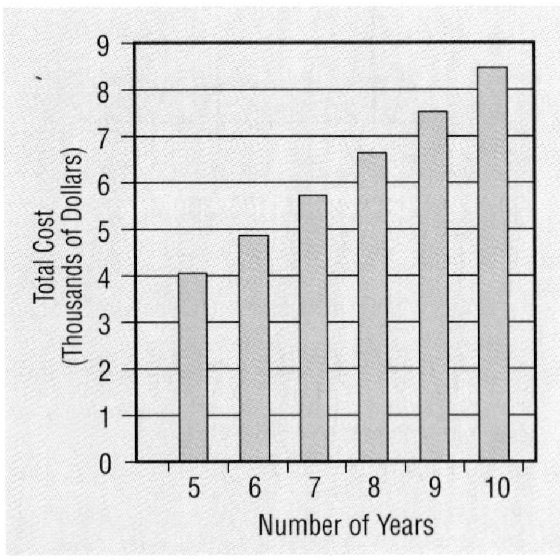

Lesson 5–3

Mixed Review

3. $446.77 − 57.49 = 389.28
7. The average-balance method; the minimum-balance method will result in extra charges even if your balance drops below the limit for one day.
8.

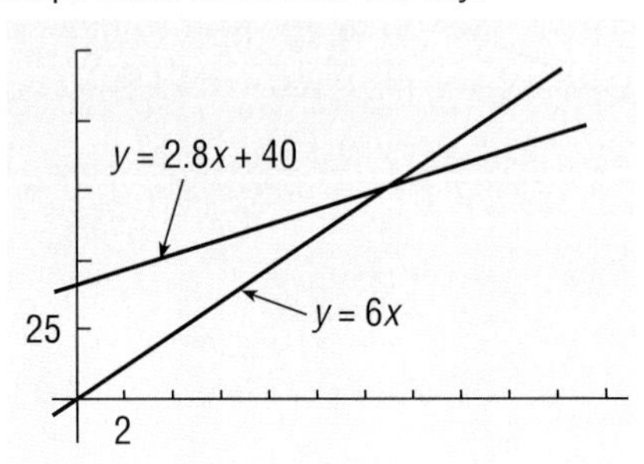

Lesson 5–4

Exercise Your Skills

1. When the prepayment penalty is less than the interest.
2. Since the unpaid balance decreases each month, the interest due also decreases.
3. It is a schedule that lists the interest portion of monthly loan payments.
16. $1204.26
17. $2937.82
18. $9511.37
19. $26,028.08

Mixed Review

1. **a.** $3346.67; **b.** $2964.53; loan **a** is greater
3. $9566.82
7. $x + y \leq 160$; $x \geq 40$; $x \leq 100$; $y > 60$
8.

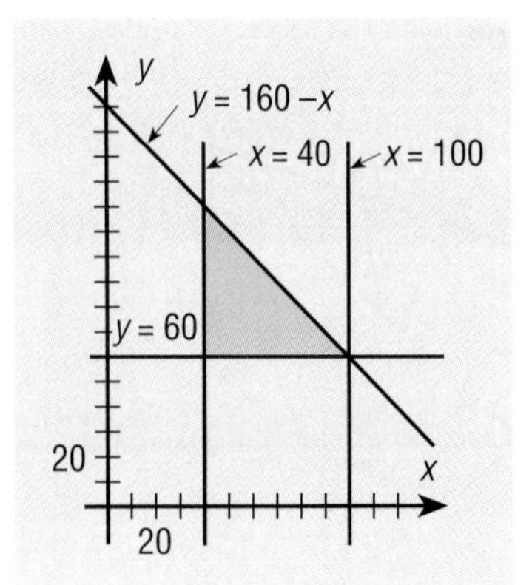

Lesson 5–5

Ask Yourself

1. When there is a cash purchase the merchant is not subjected to pay 4–7% of the purchase price to a credit card company.

Chapter 5 Review

18.

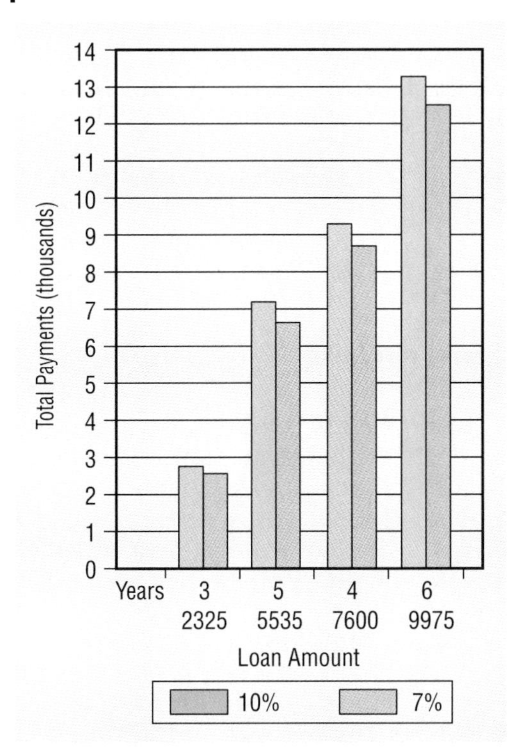

22.

Payment Number	Payment Amount	Interest Due	Note Rebate	Unpaid Balance
1	$382.50	$53.13	$329.37	$8170.63
2	382.50	51.07	331.43	7839.20
3	382.50	49.00	333.50	7505.70
4	382.50	46.91	335.59	7170.11
5	382.50	44.81	337.69	6832.42
6	382.50	42.70	339.80	6492.62

Chapter 5 Test

4.

Number of Years	Monthly Payment	Total Payment	Savings from 5 Years
3	$332.14	$11,957.15	$1389.52
4	263.34	12,640.24	706.43
5	222.44	13,346.67	

8.

Percent Down	Down Payment	Loan Amount	Total Amount	Savings Over 10% Down
10	$ 655	$5895	$7256.36	
20	1310	5240	7177.88	$ 78.48
30	1965	4585	7099.39	156.97

9.

Payment Number	Payment Amount	Interest Due	Note Reduction	Unpaid Balance
1	$109.50	$18.00	$91.50	$3508.50
2	109.50	17.54	91.96	3416.54

11.

	Plan 1	Plan 2	Plan 3
Loan amount	$25,550	$26,500	$27,500
Monthly payment	682.28	670.84	632.06
Total financed price	32,749.27	32,200.15	30,338.91

12.

	Plan 1	Plan 2	Plan 3
Interest	0	$5700.15	$2838.91
Rebate not paid	0	950.00	1950.00
Profit	0	6650.15	4788.91

CHAPTER 6

Lesson 6–1

Exercise Your Skills

1. the amount of time that elapses before you must pay interest on the purchase
2. The annual percentage rate is simple interest. The effective interest rate is higher because the interest is compounded.
3. VISA is a credit card with a monthly payment whereas American Express is a charge card with the entire balance expected on receipt of the bill.

Lesson 6–2

Exercise Your Skills

1. After 7 months, there would still be a small balance.

7.	15 mo	**8.**	13 mo	**9.**	18 mo
10.	6 mo	**11.**	18 mo	**12.**	27 mo
13.	15 mo	**14.**	17 mo	**15.**	23 mo
16.	14 mo	**17.**	19 mo	**18.**	16 mo

Lesson 6–3

Exercise Your Skills

31.

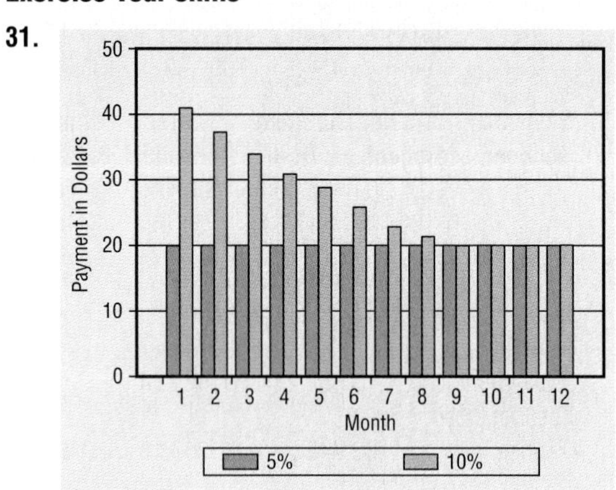

46.

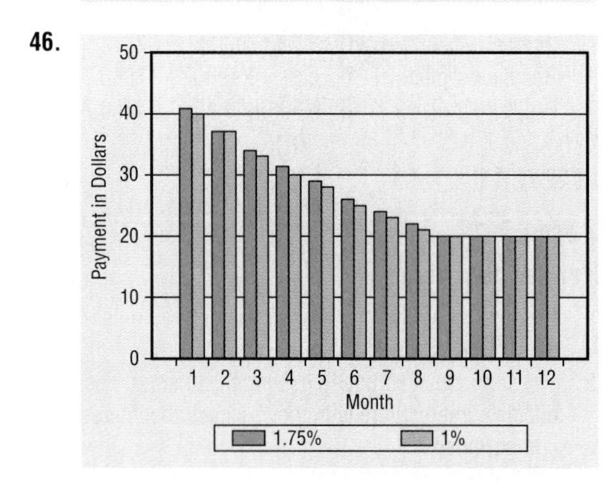

Lesson 6–4

Try Your Skills

	Month	Purchases	Balance	Interest	Amount Owed	Payment
14.	1	0.00	$554.75	$ 8.32	$563.07	$56.00
15.	2	$86.35	593.42	8.90	602.32	60.00
16.	3	0.00	542.32	8.13	550.45	55.00
17.	4	40.00	535.45	8.03	543.48	54.00
18.	5	75.99	565.47	8.48	573.95	57.00
19.	6	22.80	539.75	8.10	547.85	55.00
20.	7	10.00	502.85	7.54	510.39	51.00
21.	8	97.88	557.27	8.36	565.63	57.00
22.	9	75.00	583.63	8.75	592.38	59.00
23.	10	31.00	564.38	8.47	572.85	57.00
24.	11	66.00	581.85	8.73	590.58	59.00
25.	12	67.00	598.58	8.98	607.56	61.00
26.			Total Interest	$100.79		

Exercise Your Skills

	Month	Purchases	Balance	Interest	Amount Owed	Payment
30.	1	0.00	$875.66	$ 13.13	$888.79	$89.00
31.	2	0.00	799.79	12.00	811.79	81.00
32.	3	$ 20.00	750.79	11.26	762.05	76.00
33.	4	0.00	686.05	10.29	696.34	70.00
34.	5	0.00	626.34	9.40	635.74	64.00
35.	6	158.80	730.54	10.96	741.50	74.00
36.	7	0.00	667.50	10.01	677.51	68.00
37.	8	125.00	734.51	11.02	745.53	75.00
38.	9	112.00	782.53	11.74	794.27	79.00
39.	10	0.00	715.27	10.73	726.00	73.00
40.	11	0.00	653.00	9.80	662.80	66.00
41.	12	123.00	719.80	10.80	730.60	73.00
42.			Total Interest	$133.12		

	Month	Purchases	Balance	Interest	Amount Owed	Payment
43.	1	0.00	$1450.63	$ 21.76	$1472.39	$147.00
44.	2	$ 15.00	1340.39	20.11	1360.50	136.00
45.	3	0.00	1224.50	18.37	1242.87	124.00
46.	4	249.00	1367.87	20.52	1388.39	139.00
47.	5	0.00	1249.39	18.74	1268.13	127.00
48.	6	0.00	1141.13	17.12	1158.25	116.00
49.	7	0.00	1042.25	15.63	1057.88	106.00
50.	8	0.00	951.88	14.28	966.16	97.00
51.	9	15.00	884.16	13.26	897.42	90.00
52.	10	344.00	1151.42	17.27	1168.69	117.00
53.	11	45.00	1096.69	16.45	1113.14	111.00
54.	12	68.00	1070.14	16.05	1086.19	109.00
55.			Total Interest	$209.56		

Chapter 6 Review

1.

Month	Previous Balance	New Charges	Finance Charges	Payment Received	New Balance
1	0.00	$2315.00	0.00	0.00	$2315.00
2	$2315.00	0.00	$24.89	$250.00	2089.89
3	2089.89	0.00	22.47	250.00	1862.36
4	1862.36	0.00	20.02	250.00	1632.38
5	1632.38	0.00	17.55	250.00	1399.93
6	1399.93	0.00	15.05	250.00	1164.98
7	1164.98	0.00	12.52	250.00	927.50
8	927.50	0.00	9.97	250.00	687.47
9	687.47	0.00	7.39	250.00	444.86
10	444.86	0.00	4.78	250.00	199.64
11	199.64	0.00	2.15	201.79	0.00
12	0.00	0.00	0.00	0.00	0.00

5.

Month	Balance	Interest	Amount Owed	Payment
1	$875.00	$8.75	$883.75	$44
2	839.75	8.40	848.15	42
3	806.15	8.06	814.21	41
4	773.21	7.73	780.94	39

7.

Month	Balance	Interest	Amount Owed	Payment
1	$875.00	$8.75	$883.75	$88
2	795.75	7.96	803.71	80
3	723.71	7.24	730.95	73
4	657.95	6.58	664.53	66

11.

Month	New Charges	Balance	Interest	Amount Owed	Payment
1	0.00	$980.12	$14.70	$994.82	$ 99.00
2	$ 87.35	983.17	14.75	997.92	100.00
3	0.00	897.92	13.47	911.39	91.00
4	0.00	820.39	12.31	832.70	83.00
5	25.50	775.20	11.63	786.83	79.00
6	100.00	807.83	12.12	819.95	82.00
7	53.33	791.28	11.87	803.15	80.00
8	0.00	723.15	10.85	734.00	73.00

Chapter 6 Test

4.

Month	Balance	Interest	Amount Owed	Payment
1	$795.00	$9.94	$804.94	$40
2	764.94	9.56	774.50	39
3	735.50	9.19	744.69	37
4	707.69	8.85	716.54	36

6.

Month	Balance	Interest	Amount Owed	Payment
1	$795.00	$9.94	$804.94	$80
2	724.94	9.06	734.00	73
3	661.00	8.26	669.26	67
4	602.26	7.53	609.79	61

Chapter 6 Cumulative Review

12.

Month	Balance	Interest	Amount Owed	Payment
1	$540.00	$8.10	$548.10	$55
2	493.10	7.40	500.50	50
3	450.50	6.76	457.26	46

CHAPTER 7

Lesson 7–3

Exercise Your Skills

2. No taxes paid on interest until used during retirement. When taxes are paid, they are less than they would have been if paid earlier.

Mixed Review

4.

Month	Purchase	Balance	Interest	Amount Owed	Payment
1	0	$720.00	$10.80	$730.80	$73.00
2	$125	782.80	11.74	794.54	79.00
3	0	715.54	10.73	726.27	73.00

Chapter 7 Cumulative Review

7.

Month	Balance	Interest	Amount Owed	Payment
1	$690.00	$12.08	$702.08	$70.00
2	632.08	11.06	643.14	64.00
3	579.14	10.13	589.27	59.00

CHAPTER 8

Algebra Refresher

1.

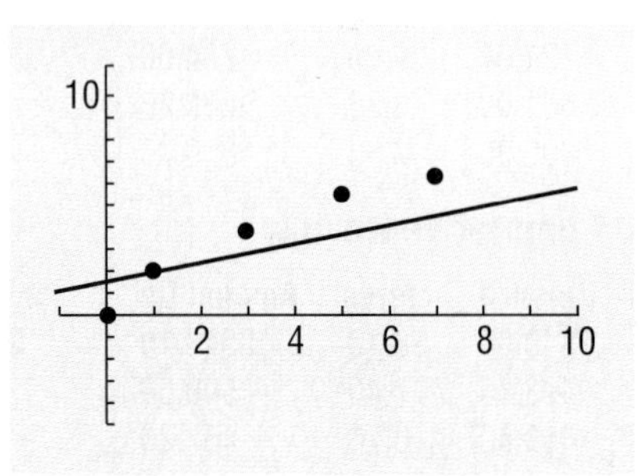

2.

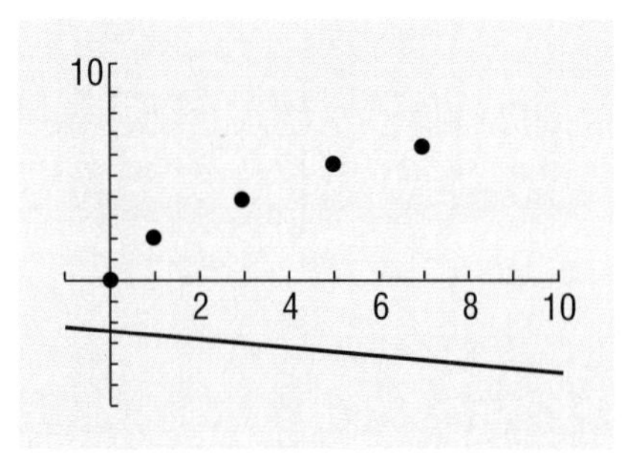

3.

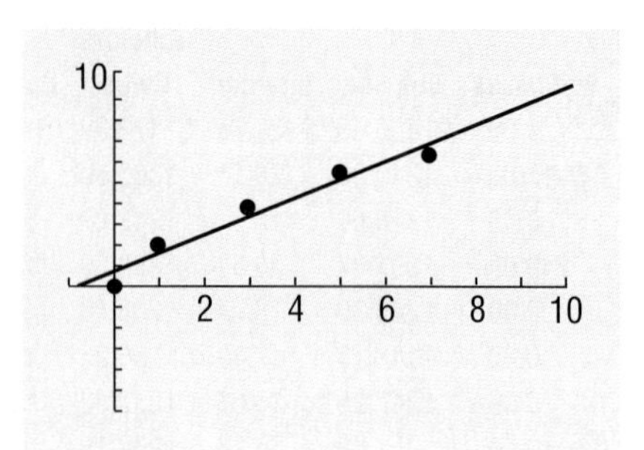

Lesson 8–2

Try Your Skills

12.

r	0	0.005	0.01	0.02	0.03	0.04	0.05	0.06	0.07	0.08
t	500	502.50	505	510	515	520	525	530	535	540

13.

r	0	0.005	0.01	0.02	0.03	0.04	0.05	0.06	0.07	0.08
t	2000	2010	2020	2040	2060	2080	2100	2120	2140	2160

Exercise Your Skills

16. and 17.

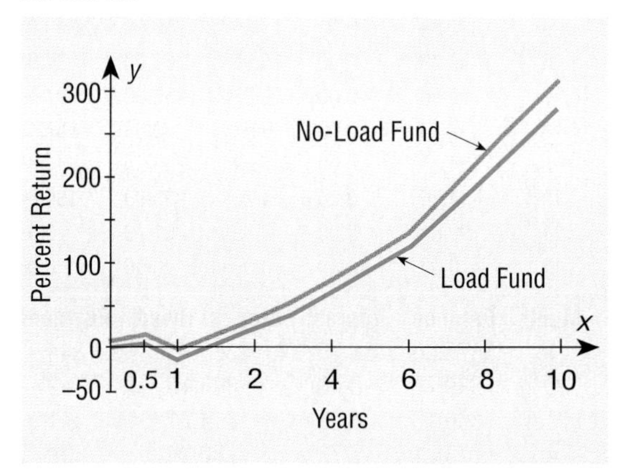

Lesson 8–3

Try Your Skills

1.

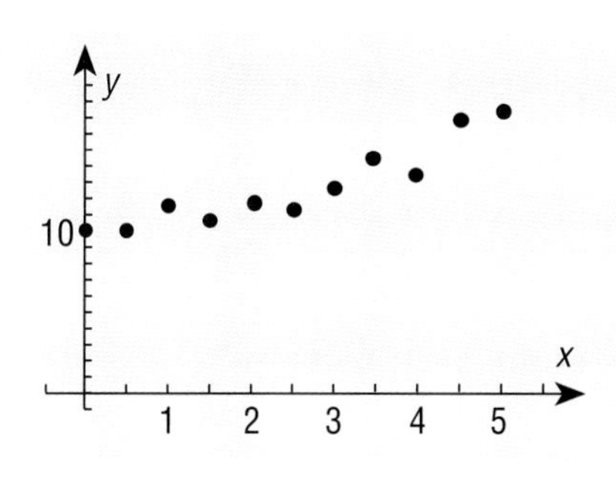

2.

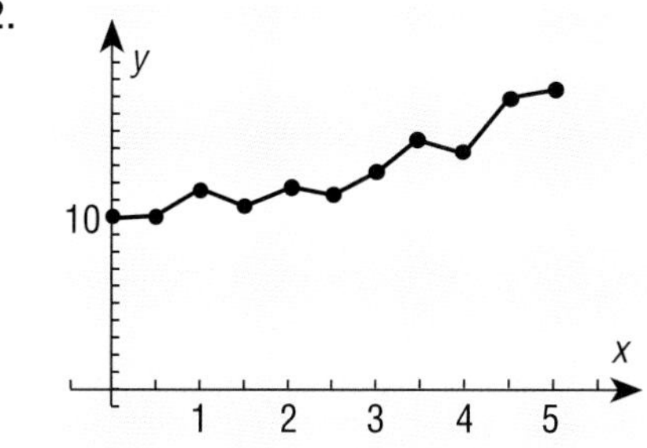

Exercise Your Skills

4.

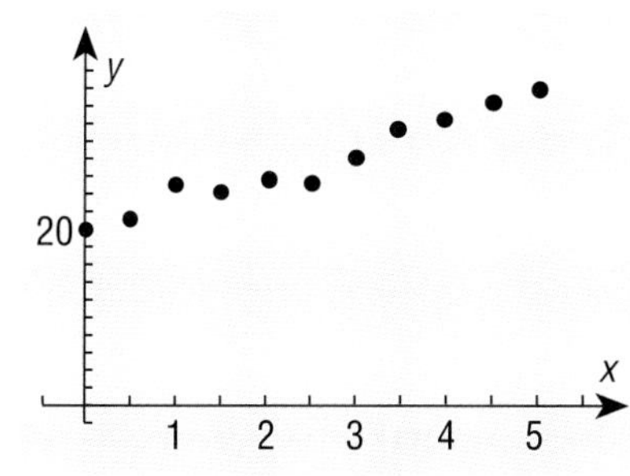

5.

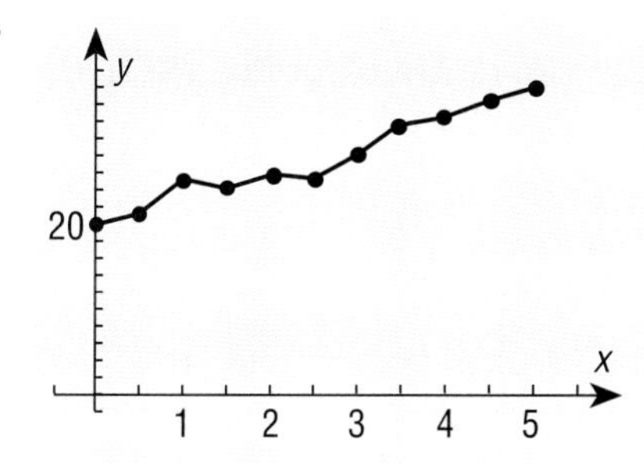

Lesson 8–4

Exercise Your Skills

15.

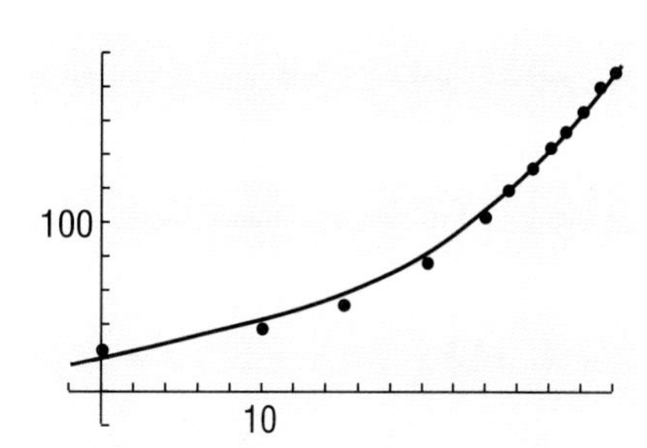

Mixed Review

	Check Number	Date	Checks/Deposits	Amount	Balance
6.	201	9/27	To: Aaron Jones		450.00
			For: Car Repair	$158.50	−158.50
7.		9/29	Deposit		291.50
				$130.00	+130.00
8.	202	9/29	To: Superior Supermarket		421.50
			For: Groceries	$87.63	−87.63
					333.87

Chapter 8 Review

16.

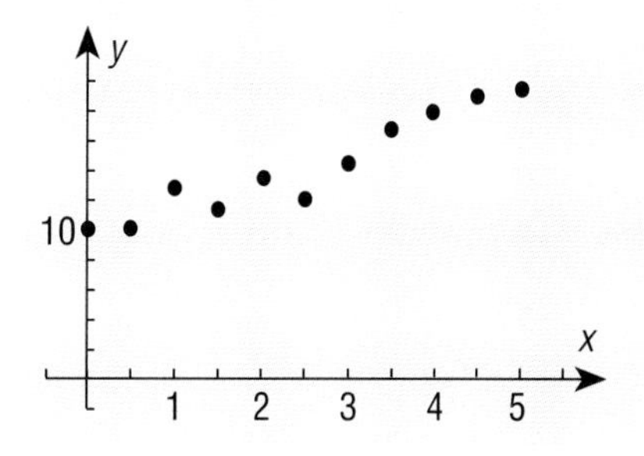

17.

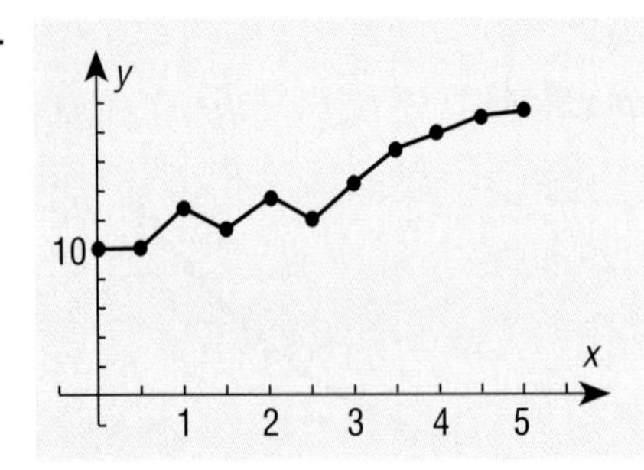

Chapter 8 Test

13.

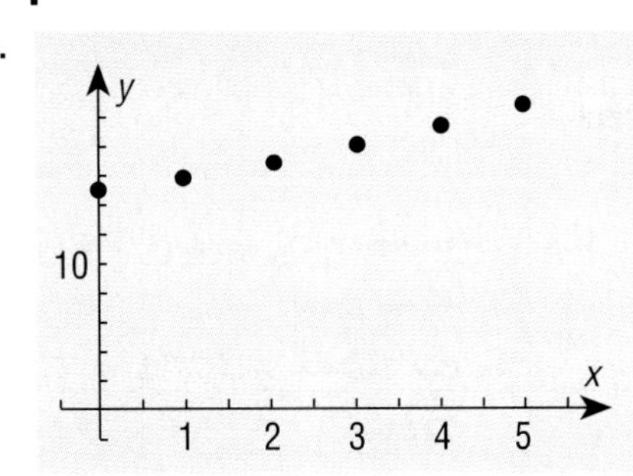

14.

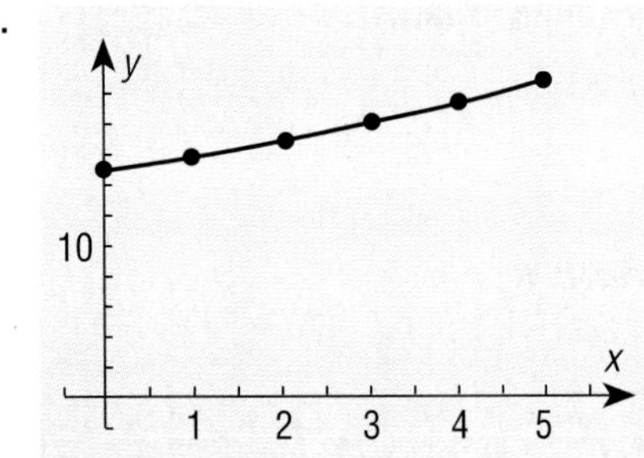

CHAPTER 9

Algebra Refresher

1.

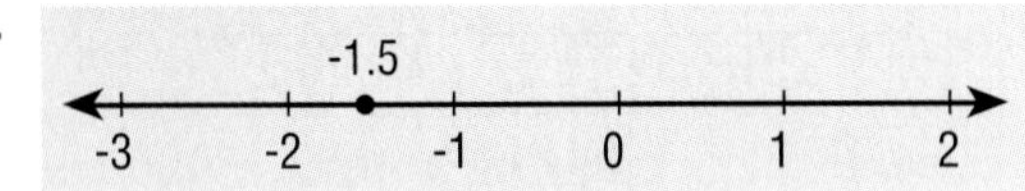

The number -1.5 is half way between -1 and -2.

2.

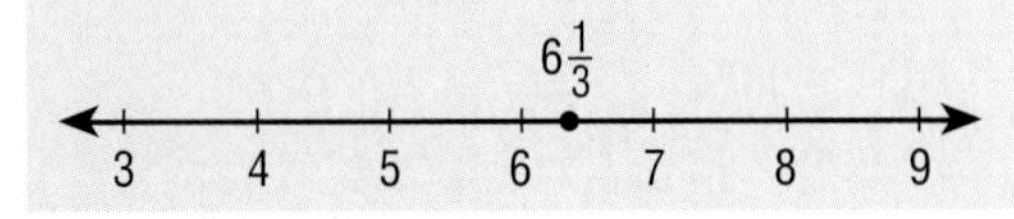

The number $6\frac{1}{3}$ is a little less than half way between 6 and 7.

3.

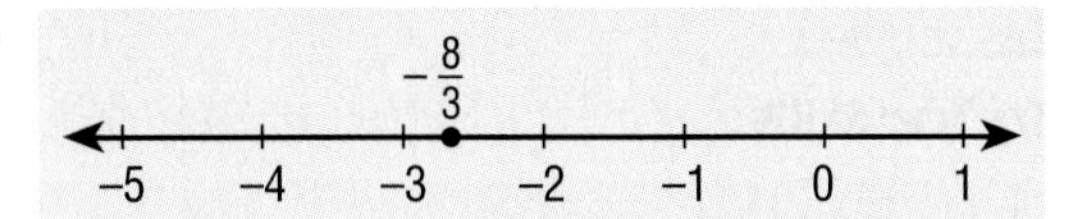

The number $-\frac{8}{3}$ is a little less than half way between -3 and -2.

4.

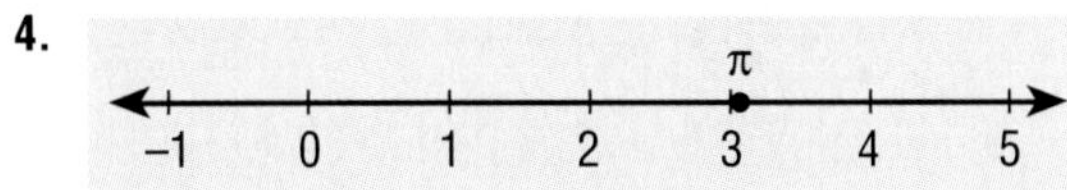

The number π is a little more than 3.

5.

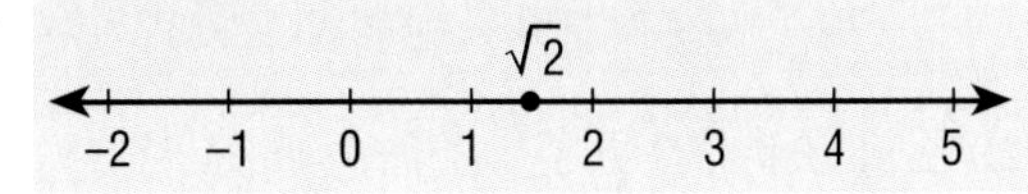

The number $\sqrt{2}$ is a little less than half way between 1 and 2.

6.

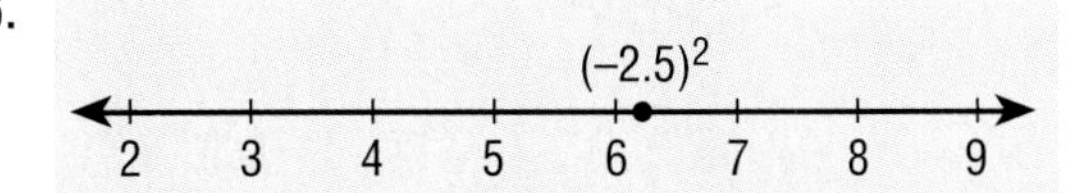

The number for the expression $(-2.5)^2$ is a fourth of the way between 6 and 7.

7.

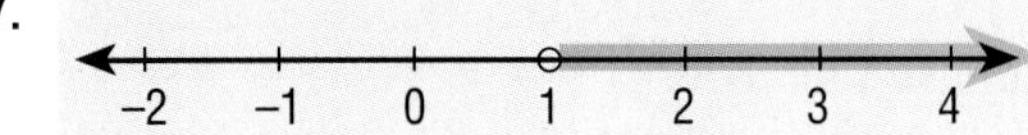

8.

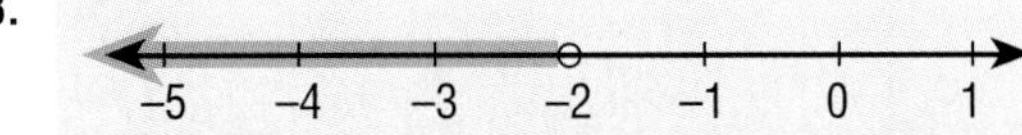

9.

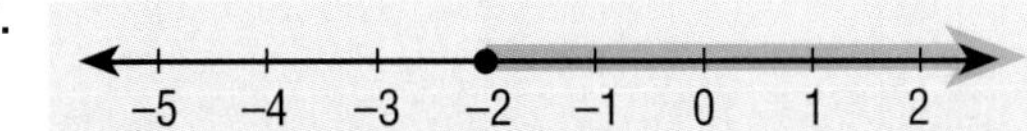

10.

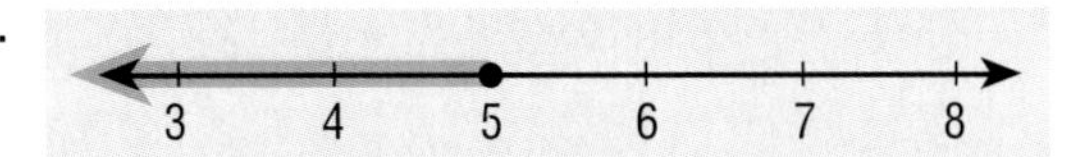

11.

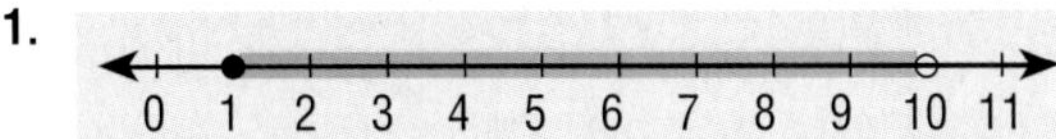

12.

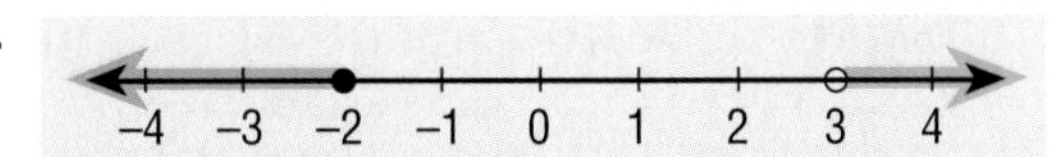

13.

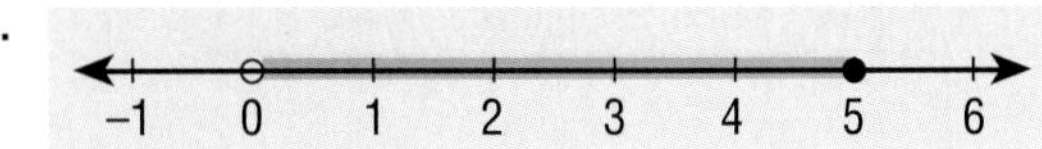

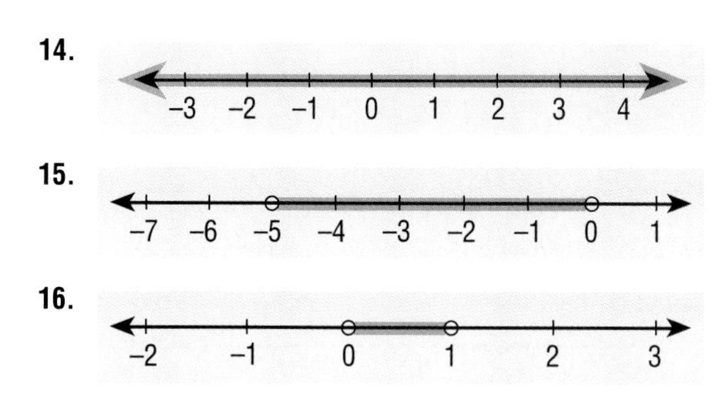

Lesson 9–1

Mixed Review

5.

Level	Demand Deposit	Required Reserve	Loans and Investments
1	$1,200.00	$240.00	$ 960.00
2	960.00	192.00	768.00
3	768.00	153.60	614.40
4	614.40	122.88	491.52
5	491.52	98.30	393.22
		Total after 5 levels:	$3,227.14

6.

Level	Demand Deposit	Required Reserve	Loans and Investments
1	$10,000	$2,000.00	$ 8,000.00
2	8,000	1,600.00	6,400.00
3	6,400	1,280.00	5,120.00
4	5,120	1,024.00	4,096.00
5	4,096	819.20	3,276.80
		Total after 5 levels:	$26,892.80

Lesson 9–2

Ask Yourself

2. **a.** be a relative of the filer, or a member of the filer's household

b. be a citizen or resident of the U.S.

c. have gross income of less than $2300 unless he/she is under 19 and a child of the filer, or unless he/she is under 24 and qualifies as a student and is a child of the filer

d. have received more than half of his/her support from the filer

3. tips, taxable scholarships, wages, fellowship grants or any other taxable income from investments or savings ($400 or less)

Algebra Review

1.

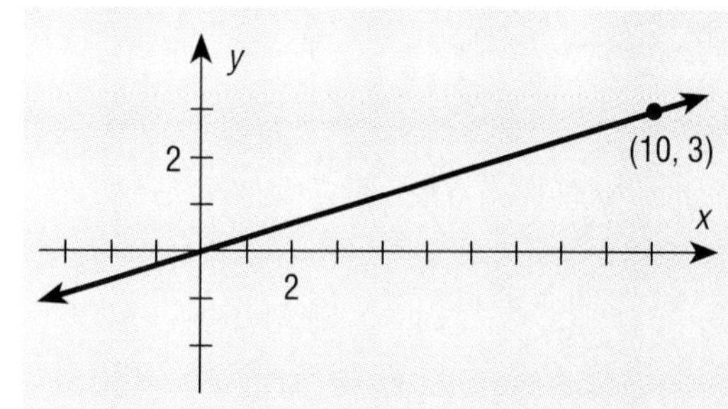

2.

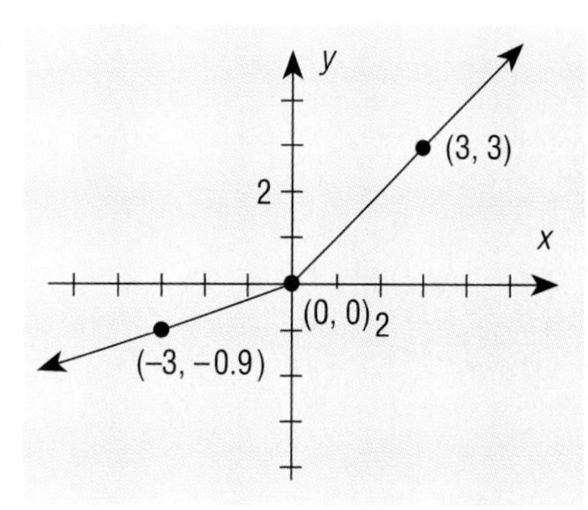

3.

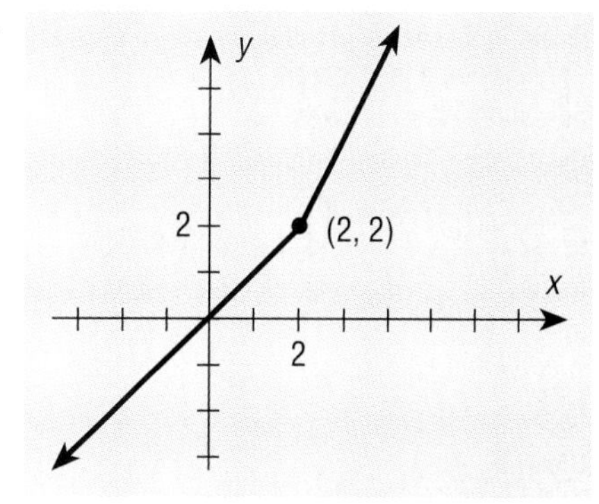

4.

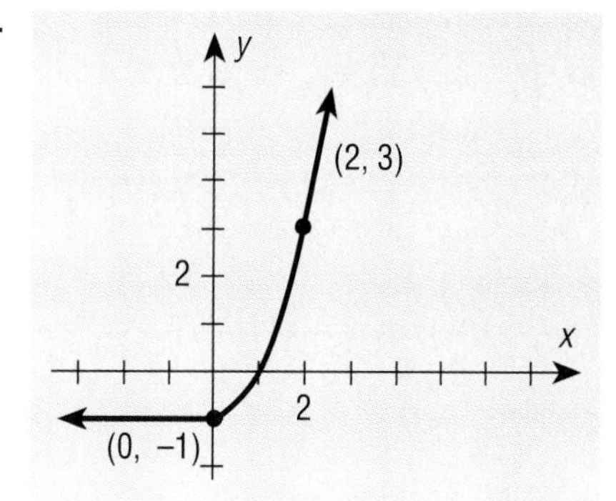

5.

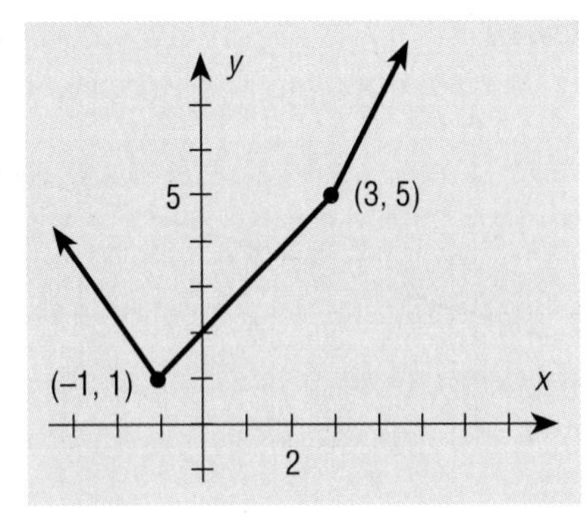

6.

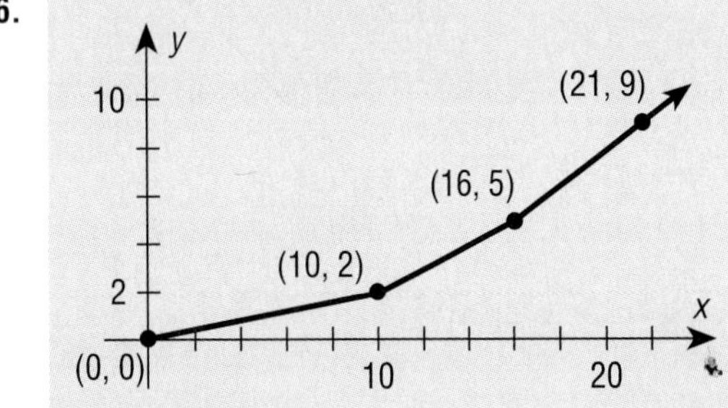

Exercise Your Skills

1. Not everyone has the same income, number of dependents, investments, and so on.
2. It often results in lower taxes.
3. It is easier than documenting your itemized deductions, and it saves the average person more money.
4. Refund: $340
5. Refund: $501
6. Refund $661
7. Refund: $79
8. Refund: $1911

Lesson 9–3

Algebra Review

6.

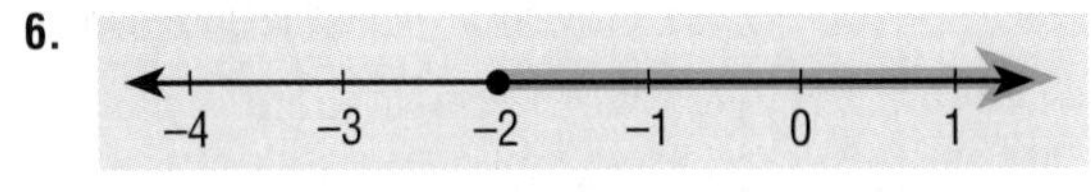

7.

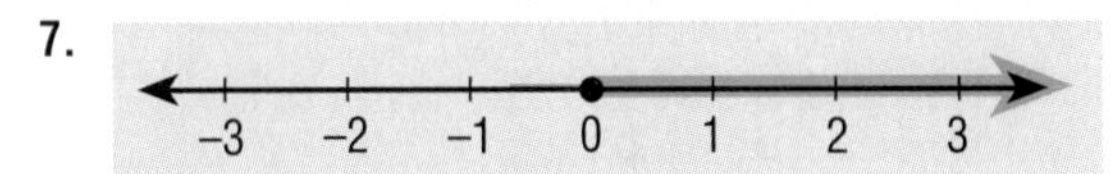

8.

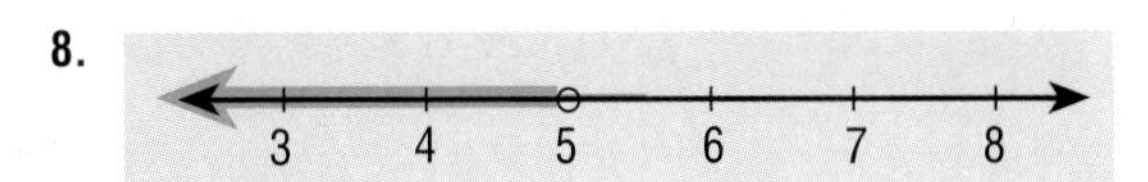

9.

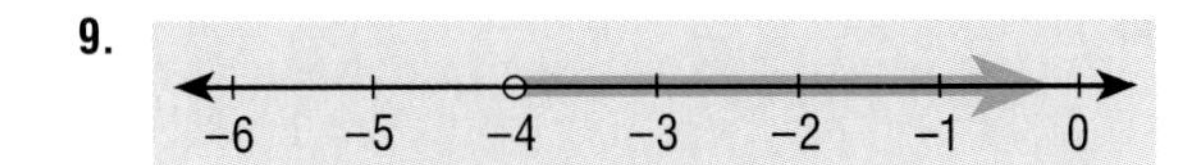

10.

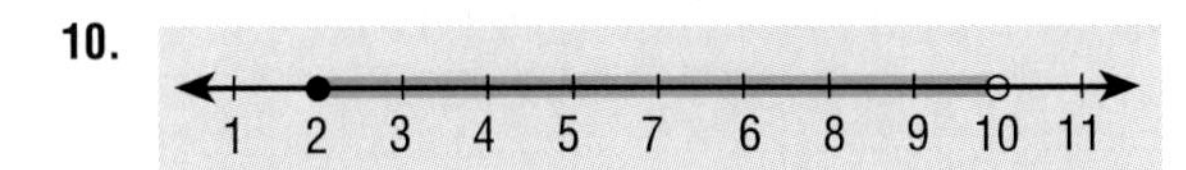

11.

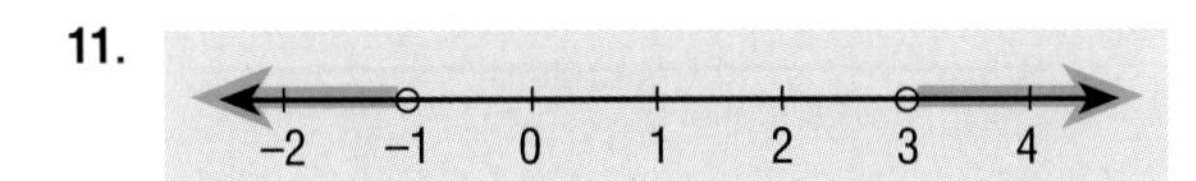

12.

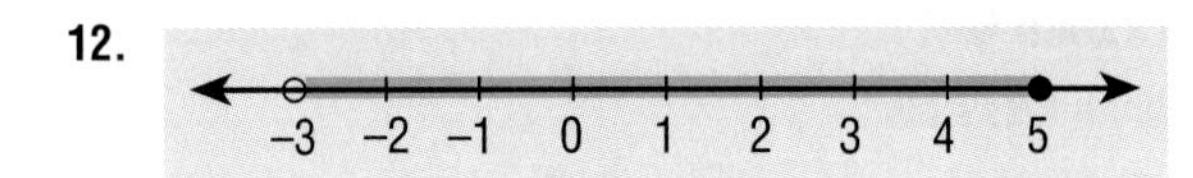

13.

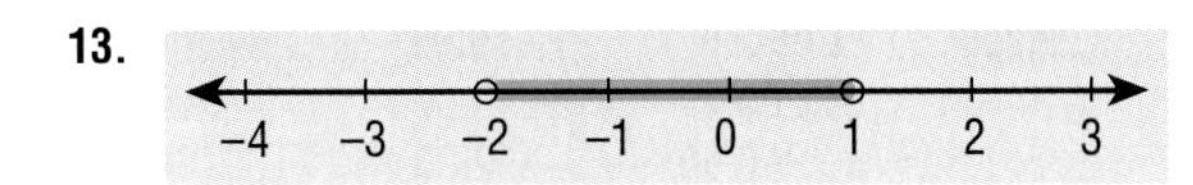

14.

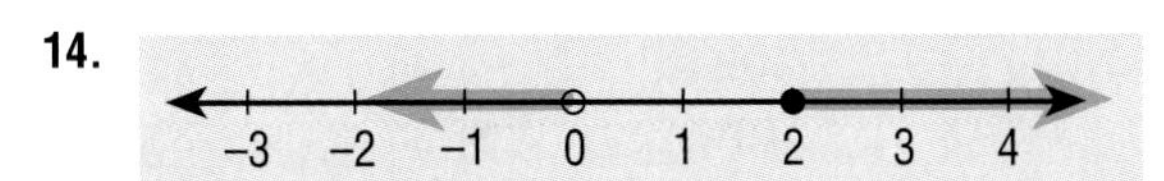

Exercise Your Skills

1. The IRS and tax preparers can make mistakes; it is good to be informed just in case.
2. A CPA is better prepared and trained for difficult work.

CHAPTER 10

Lesson 10–2

Algebra Review

1.

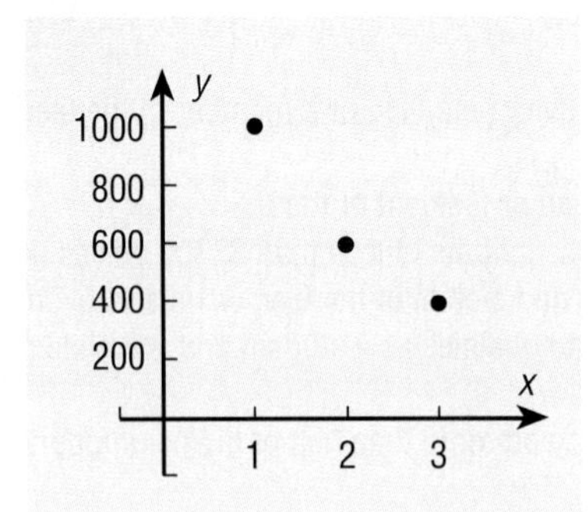

2.

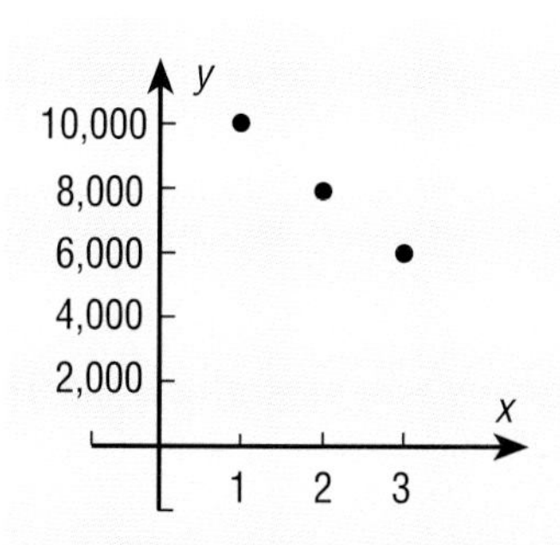

3.

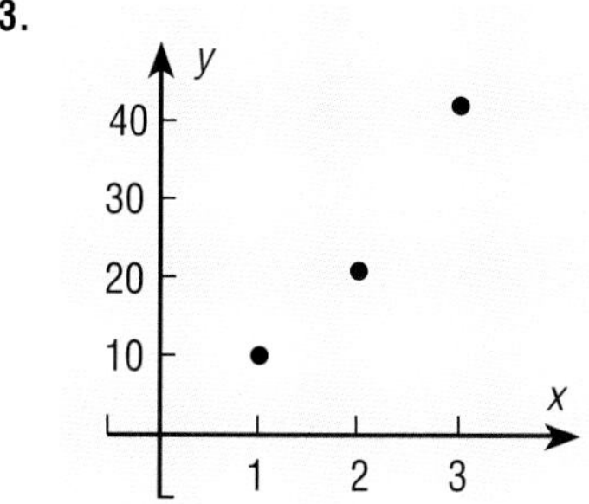

Exercise Your Skills

7.–8.

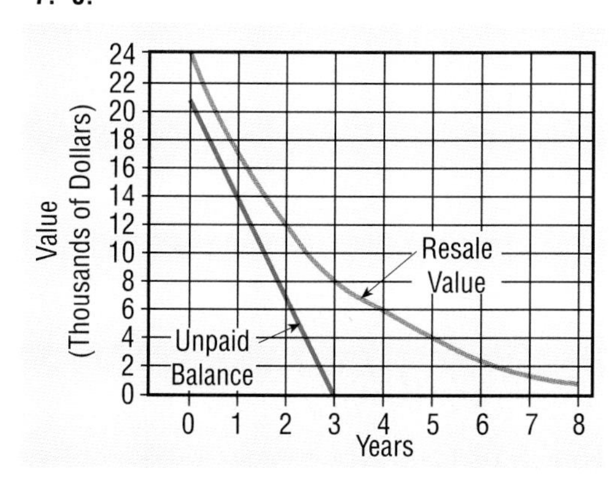

11.

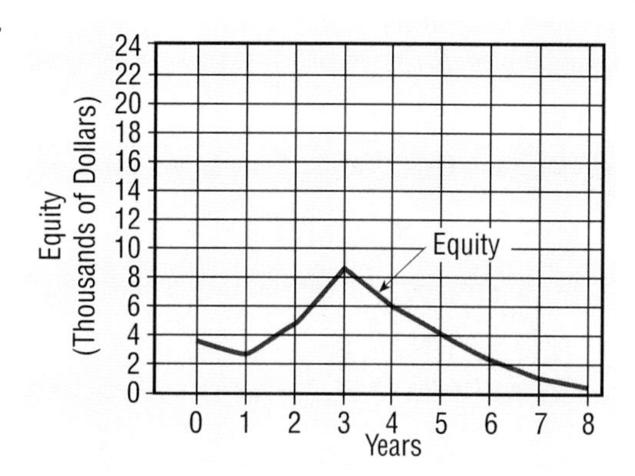

15.–16.

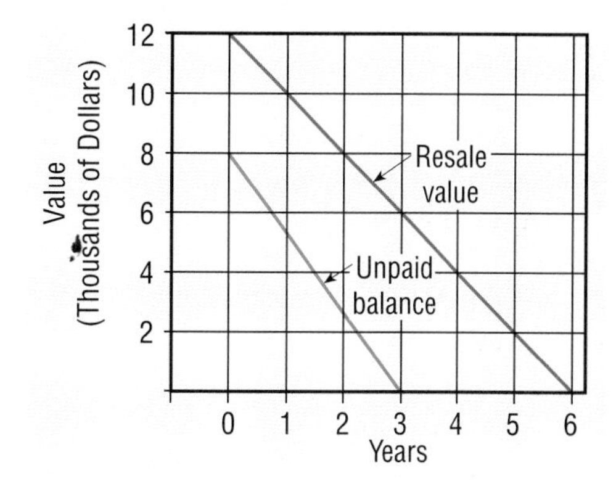

Lesson 10–3

Ask Yourself

2. Maintenance refers to the things that should be done to a car at certain intervals (every so many miles) to maintain the car in good condition. Repairs are unscheduled work that needs to be done on the car to keep it running.
4. Leasing is an arrangement by which a car is rented for a monthly fee under a contract that extends for several years.

Exercise Your Skills

15.

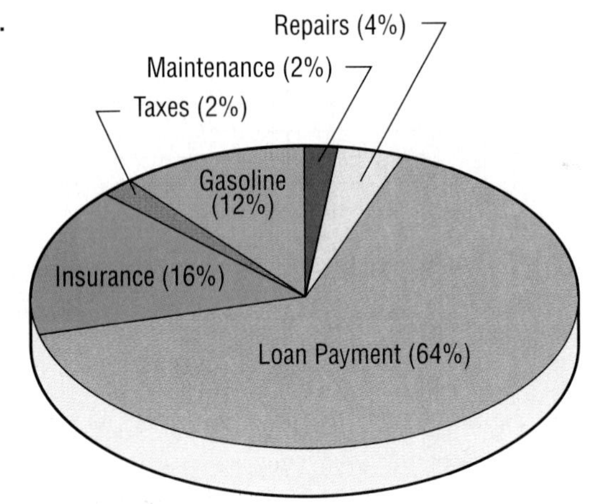

18.

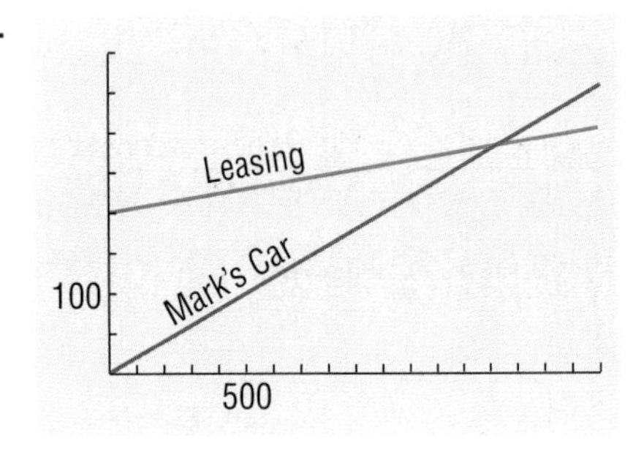

23.

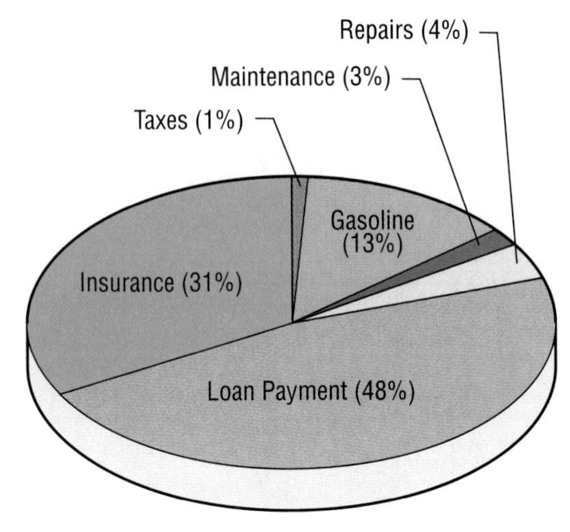

25.

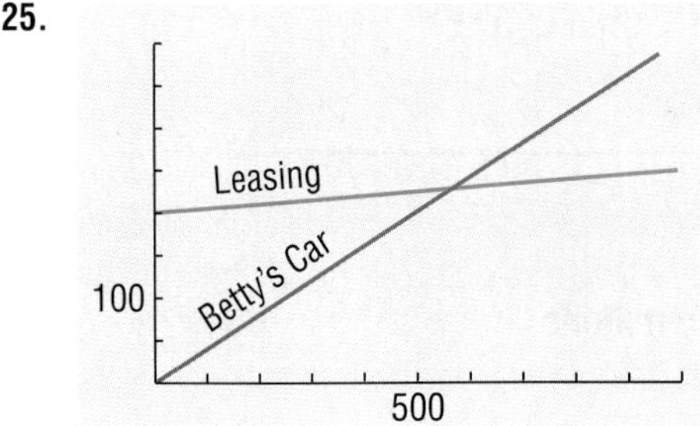

Chapter 10 Review

16.

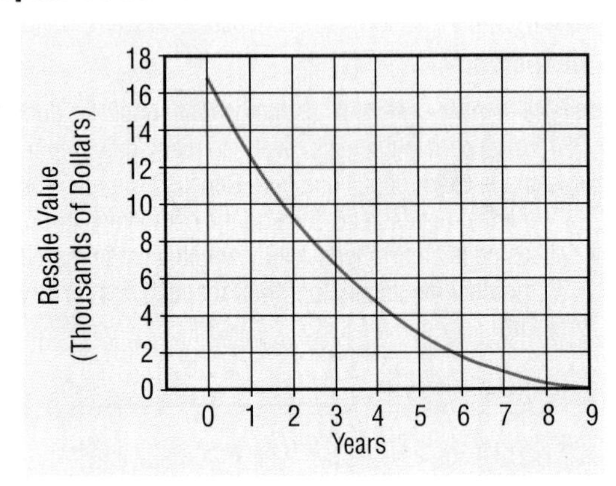

Chapter 10 Test

8.

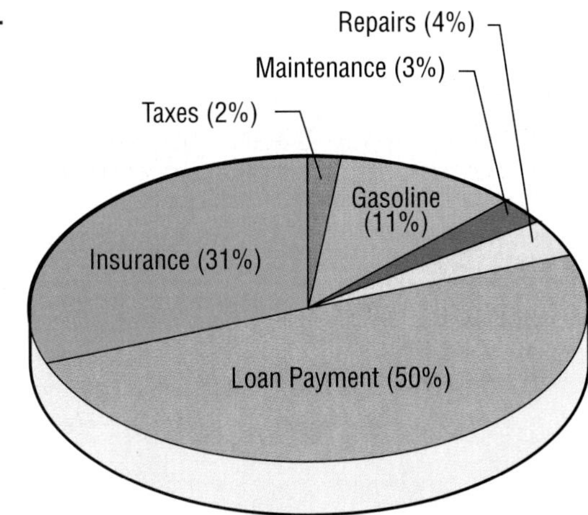

CHAPTER 11

Algebra Refresher

3.

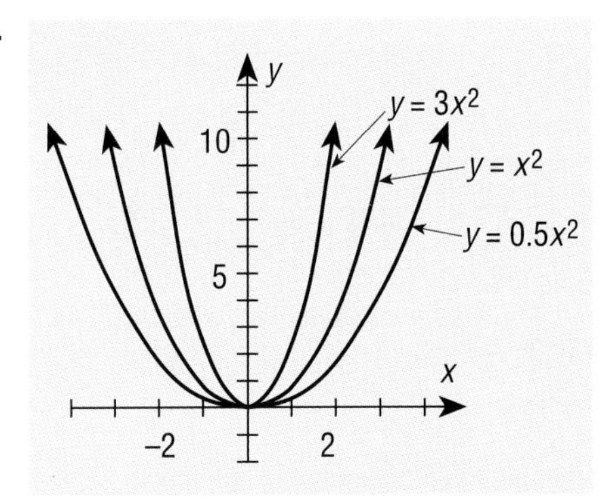

4.

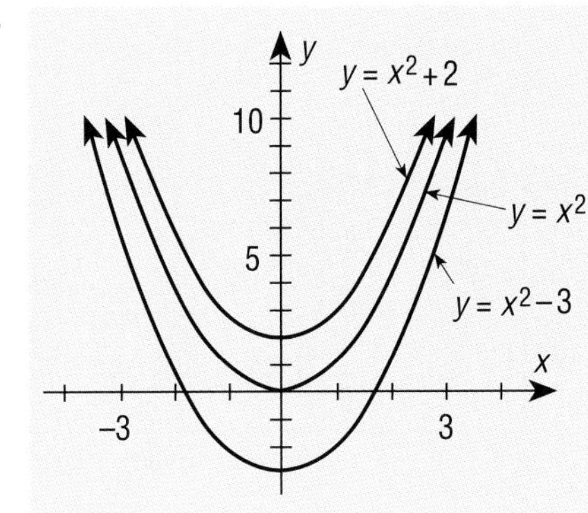

Lesson 11–1

Ask Yourself

1. If destination is your only concern, you would take interstate highways; if you want to see what an area is like, you would take secondary roads.

Mixed Review

1. Checks/Deposits	Amount	Balance
To: Grocery Mart		$778.22
For: Groceries	$105.89	− 105.89
		672.33

Lesson 11–2

Exercise Your Skills

12.

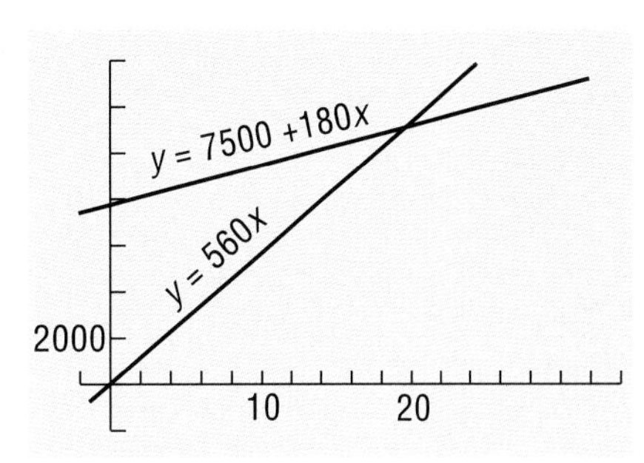

19. Gomezes want dinner prices to cover costs; patrons want an inexpensive dinner.
20. If the same number of people purchase meals, revenue would go down.
21. Lower prices may lead to more patrons; if so, revenues may increase.
22. Advertise a special with a lower price and see how many more patrons come as a result.
23. $y = 445x$
24. $y = 9100 + 225x$

Lesson 11–3

Algebra Review

5.

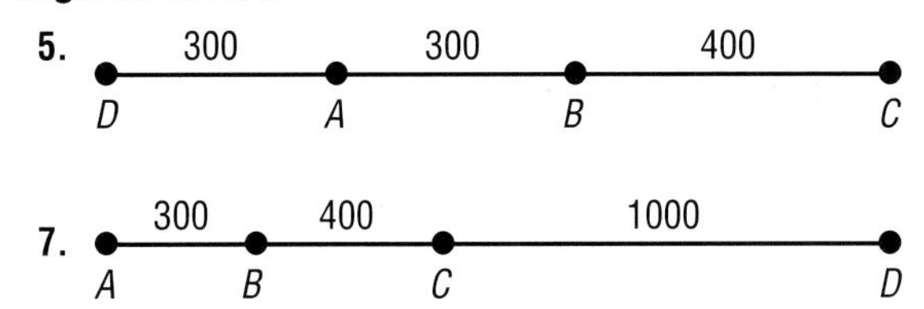

7.

Chapter 11 Review

28.

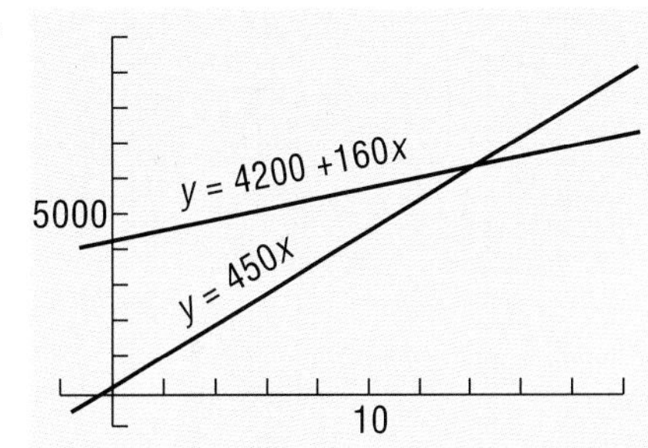

Chapter 11 Test

10.

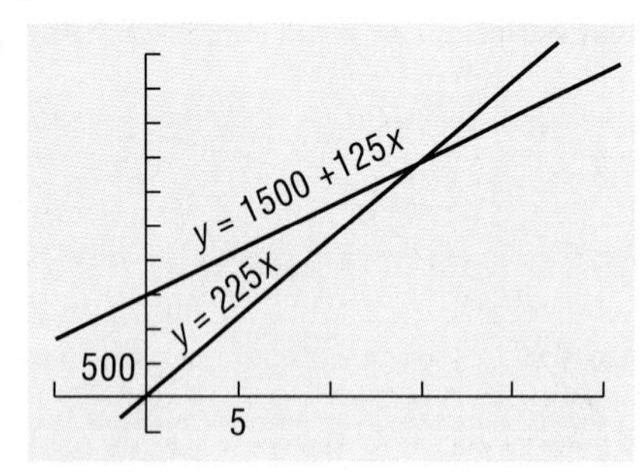

16.

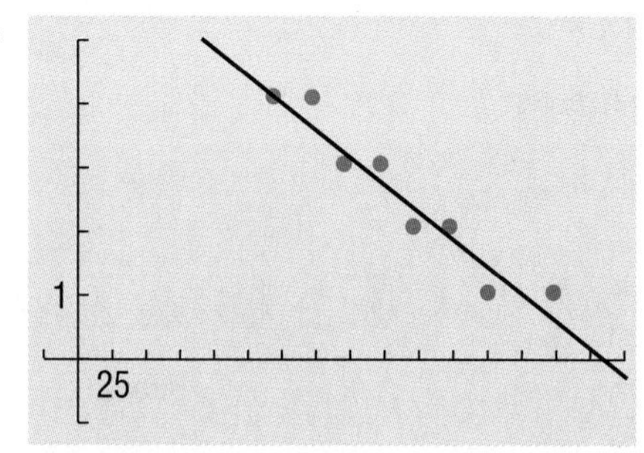

CHAPTER 12

Lesson 12–1

Try Your Skills

3.

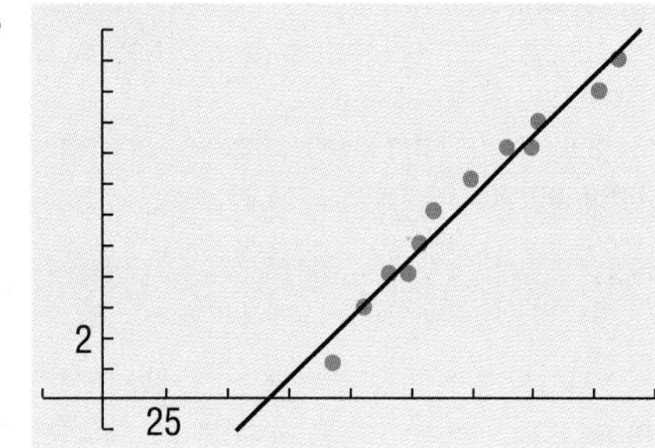

Exercise Your Skills

6.

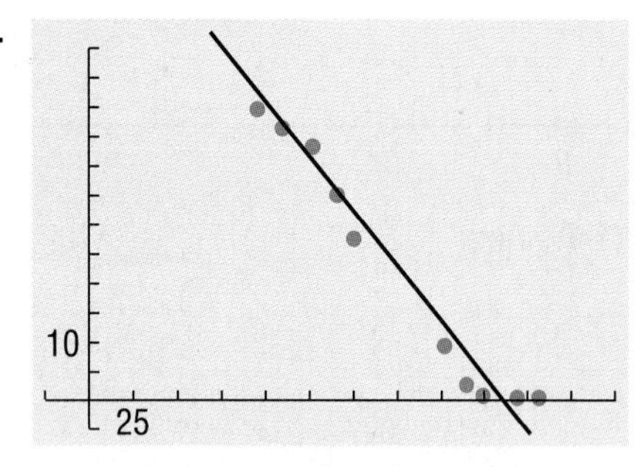

11.

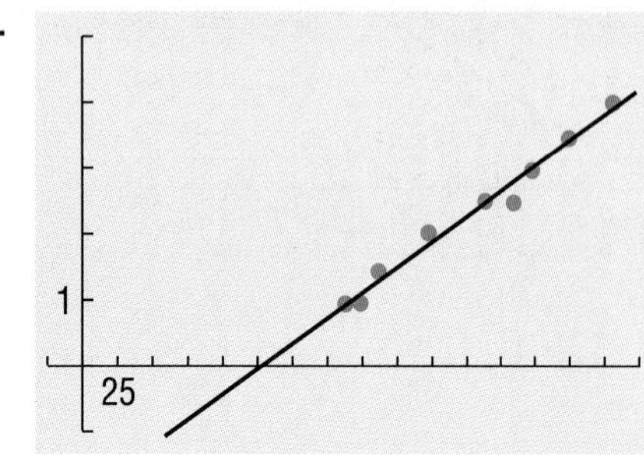

Lesson 12–2

Algebra Review

3.

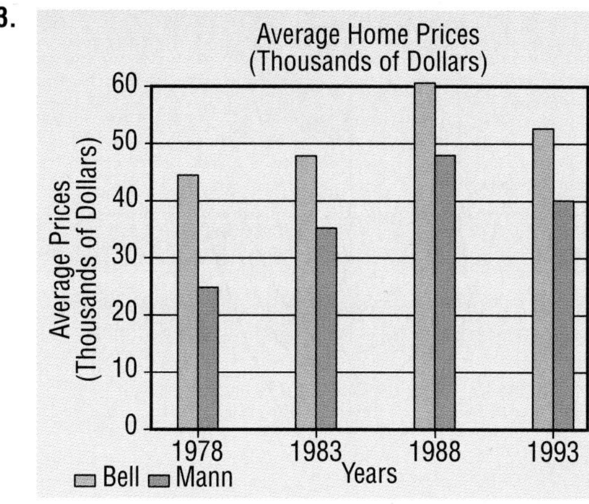

Exercise Your Skills

24.

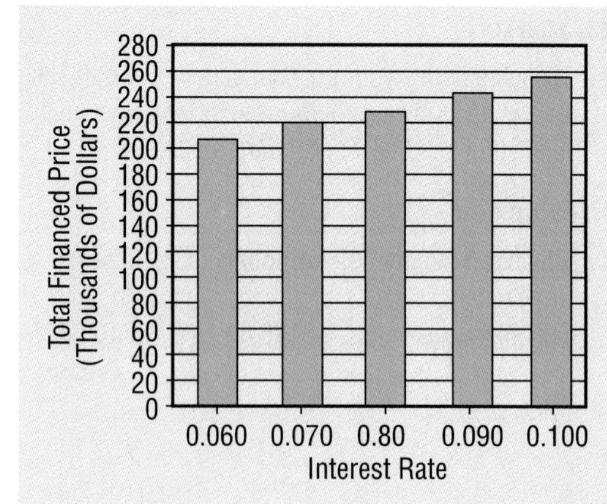

Chapter 12 Review

5.

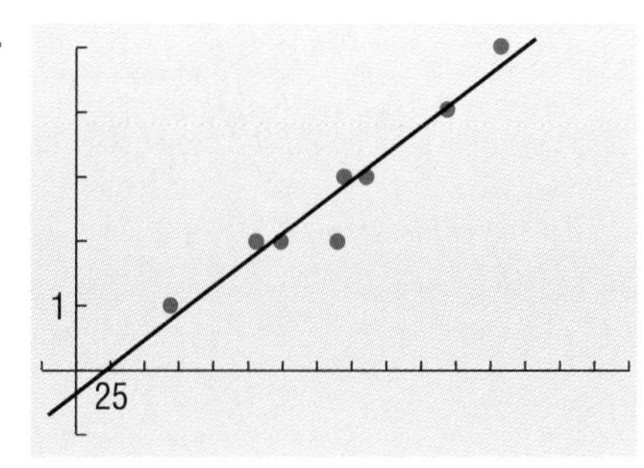

Chapter 12 Test

3.

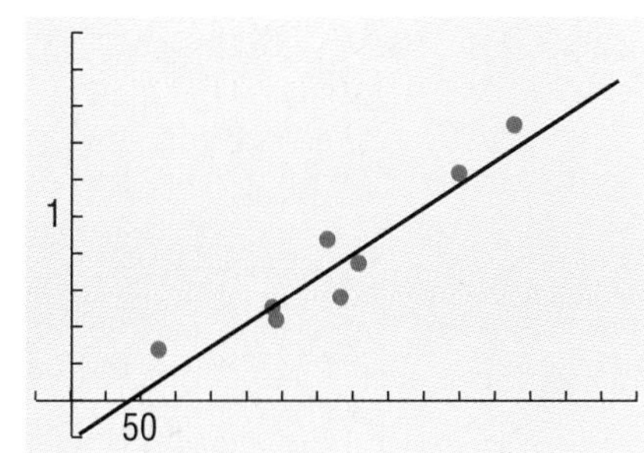

CHAPTER 13

Lesson 13–1

Exercise Your Skills

4. \$1278.85
5. \$423.08
6. \$673.08
7. \$569.23
8. \$379.81
9. \$867.21
10. \$651.92
11. \$346.15

Lesson 13–2

Algebra Review

7. $y = 150x + 100$
   y = cost; x = number of rooms
10. $y = x + 120$
   y = heating cost; x = number of square feet

Ask Yourself

3. A lease is an agreement between the landlord and tenant. The lease imposes a financial obligation on the tenant, the person renting the apartment.

Exercise Your Skills

2. People who are staying longer in an apartment are more apt to take care of the area.
3. Tenants will give you an inside view of the way the apartment building is run.
4. Make a list and have it signed so you are not liable for the damage that existed before you moved in. Otherwise, when you move out, your security deposit may be used to repair damages, since you are responsible for the condition of the apartment.
5. $y = -0.4737640598 + 0.0033466353x$
9. $y = 141.1439994 + 1.076640466x$

CHAPTER 14

Algebra Refresher

7.

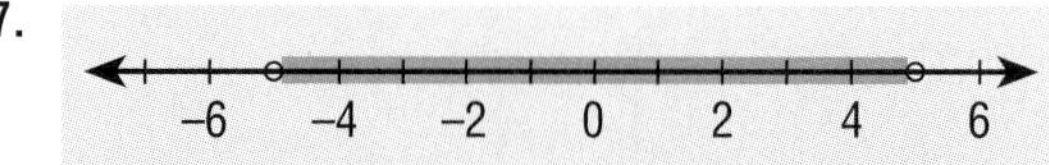

8.

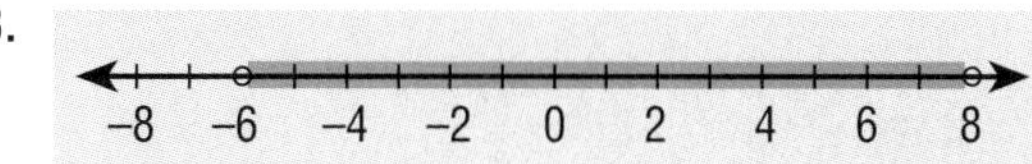

9.

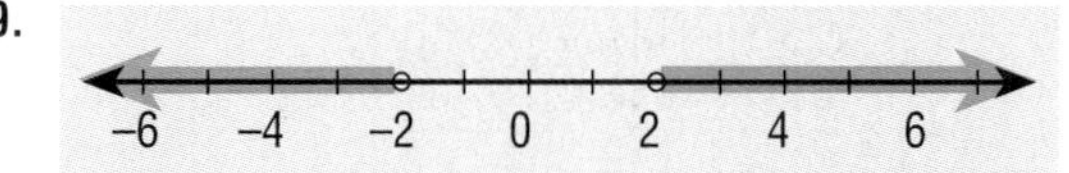

10.

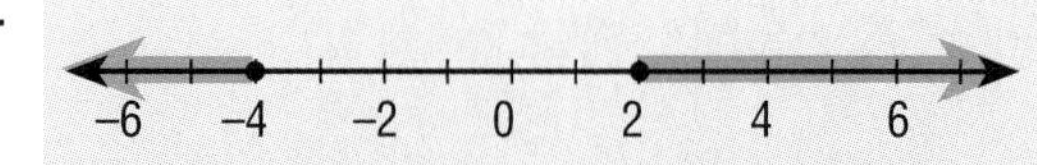

11.

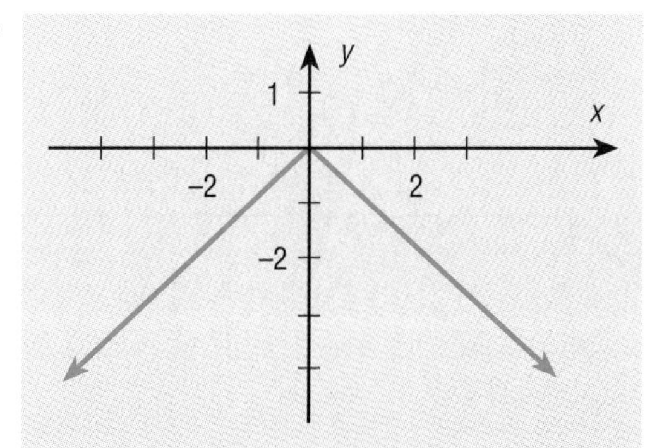

12.

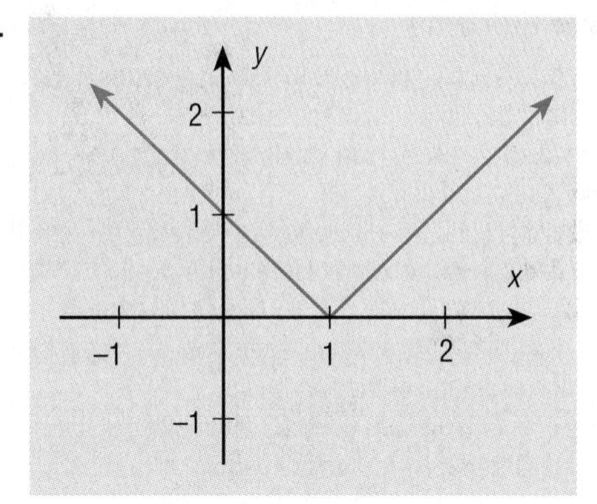

13.

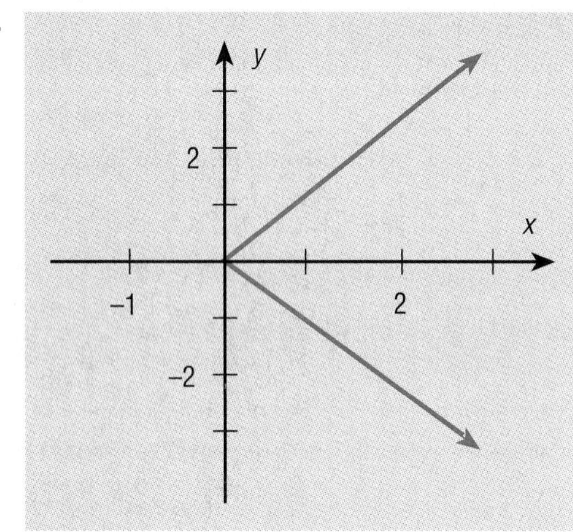

14.

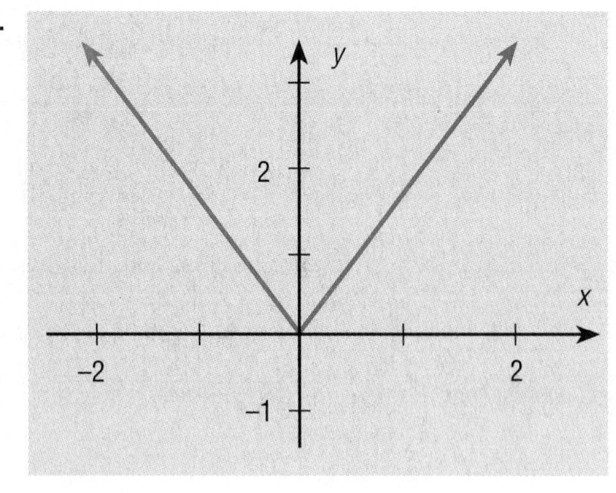

Lesson 14–1

Ask Yourself

2. Answers may vary. (Ex.: helps you establish goals; recognize standard of living and its effect on goals; helps you make choices)
3. Standard of living is the economic level at which an individual or family lives. Quality of life refers to the happiness and satisfaction that we find in our lives.

Try Your Skills

7.

	Month 1	Month 2	Month 3	Month 4	Total	Average
Energy	$120	$108	$130	$126	$ 484	$121.00
Electricity	31	28	37	40	136	34.00
Water	12	13.50	15.50	21	62	15.50
Telephone	80	36	47	51	214	53.50
Home Maint.	43	72	18	10	143	35.75
Total	286	257.50	247.50	248	1039	259.75

Exercise Your Skills

1. It is determined by the local government and not by the owner.
2. Answers may vary. (Ex.: school levy)
3. It is satisfactory for expenses not affected by seasons such as telephone and water but not for heating/air conditioning.

12.

	Monthly Budget
Energy	$116.67
Electricity	36.00
Water	23.33
Telephone	28.00
Home Maintenance	50.00
Total	$254.00

Mixed Review

1. $14,750
2. The $24 would have grown to over $55 billion (55,384,000,000). The investment will be worth more unless Manhattan is worth more than about $90 per square foot.

Lesson 14–2

Algebra Review

8.

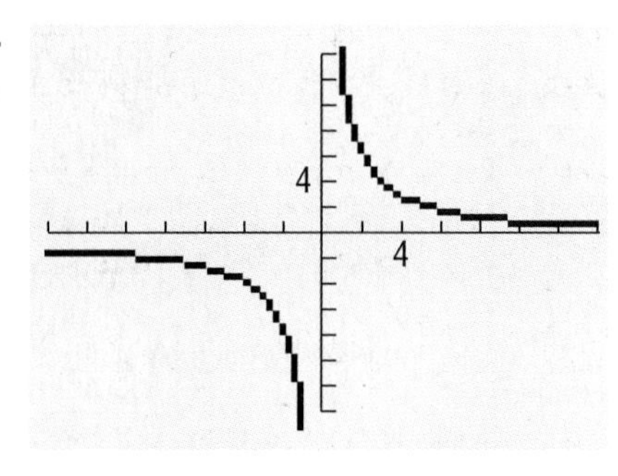

9. graphing calculator shows more points

Exercise Your Skills

2. Answers may vary. (Ex.: watch for sales, buy in quantity, use a list)

Lesson 14–3

Try Your Skills

1. $4950	2. $6300	3. $1125
4. $4500	5. $900	6. $1575
7. $450	8. $225	9. $2475

Exercise Your Skills

3. If one budget area takes too large a percentage of the income, it may mean trouble in other areas.
4. actual amount less budgeted amount

14.

	Budget	Percent of Total
Total family budget	$27,181	100%
Food	5,961	21.9%
Housing	9,271	34.1%
Clothing	2,036	7.5%
Transportation	6,655	24.5%
Health care	1,173	4.3%
Entertainment	1,689	6.2%
Personal care	299	1.1%
Miscellaneous	97	0.4%

Chapter 14 Review

8.

	Monthly Budget
Energy	$216.67
Electricity	48.00
Water	22.67
Telephone	49.00
Home Maintenance	60.42
Total	$396.76

ALGEBRA TOPICS in Algebra Refreshers and Algebra Reviews

Algebra Refresher 1 Simplify expressions, solve linear equations
Algebra Review 1–1 Evaluate expressions
Algebra Review 1–2 Solve proportions
Algebra Review 1–3 Solve linear equations, simplify expressions
Algebra Review 1–4 Find percents

Algebra Refresher 2 Solve systems of equations by substitution and by the addition method
Algebra Review 2–1 Make a table of values for an equation
Algebra Review 2–2 Determine an equation from a table of values
Algebra Review 2–3 Solve linear equations

Algebra Refresher 3 Commutative, associative, and distributive properties, definition of subtraction
Algebra Review 3–1 Use formulas
Algebra Review 3–2 Use exponents
Algebra Review 3–3 Find a common ratio between two terms of a series

Algebra Refresher 4 Graph linear equations and inequalities; slope
Algebra Review 4–1 Determine equations from data
Algebra Review 4–2 Graph linear equations
Algebra Review 4–3 Write equations from verbal information
Algebra Review 4–4 Solve systems of equations
Algebra Review 4–5 Solve systems of equations

Algebra Refresher 5 Properties of exponents
Algebra Review 5–1 Evaluate expressions with exponents, solve formulas with exponents
Algebra Review 5–2 Evaluate formulas with exponents using calculator memory
Algebra Review 5–3 Percent problems, evaluate expressions by factoring
Algebra Review 5–4 Evaluate expressions with negative exponents
Algebra Review 5–5 Compare quantities with exponents

Algebra Refresher 6 Order of operations, relations, functions, domain, range
Algebra Review 6–1 Decimal/percent conversion
Algebra Review 6–2 Domain and range
Algebra Review 6–3 Solve problems involving percent
Algebra Review 6–4 Evaluate expressions with exponents
Algebra Review 6–5 Find the average of scores with different frequencies
Algebra Review 6–6 Interpret information
Algebra Review 6–7 Solve problems involving percent

Algebra Refresher 7 Probability and events
Algebra Review 7–1 Geometric probability
Algebra Review 7–2 Determine probabilities
Algebra Review 7–3 Evaluate formulas with exponents

Algebra Refresher 8 Scatter plot, line of best fit, linear regression equation, correlation coefficient
Algebra Review 8–1 Solve linear equations
Algebra Review 8–2 Absolute value
Algebra Review 8–3 Determine the equation that best fits a scatter plot
Algebra Review 8–4 Solve proportions

Algebra Refresher 9 Compound Inequalities
Algebra Review 9–1 Evaluate expressions
Algebra Review 9–2 Graph functions defined by more than one rule
Algebra Review 9–3 Graph compound inequalities on the number line

Algebra Refresher 10 Mean, median, mode
Algebra Review 10–1 Write and solve equations involving percent
Algebra Review 10–2 Determine scale for graphs, evaluate expressions with exponents
Algebra Review 10–3 Solve problems involving average, solve systems of equations
Algebra Review 10–4 Evaluate formulas

Algebra Refresher 11 Families of linear and quadratic graphs, changing parameters
Algebra Review 11–1 Evaluate formulas, write equations from written information
Algebra Review 11–2 Solve problems involving averages, solve systems of equations by graphing
Algebra Review 11–3 Solve problems involving distances and logic

Algebra Refresher 12 Operations with polynomials, factoring
Algebra Review 12–1 Solve linear equations
Algebra Review 12–2 Interpret information from tables, draw a bar graph
Algebra Review 12–3 Solve equations involving percent, exponents, and radicals
Algebra Review 12–4 Express percents as decimals, evaluate expressions with exponents

Algebra Refresher 13 Solve radical equations, solve rational equations
Algebra Review 13–1 Evaluate expressions with percent, test the truth of inequalities
Algebra Review 13–2 Write an equation from a table of values, solve cost problems
Algebra Review 13–3 Evaluate expressions with exponents

Algebra Refresher 14 Absolute value equations, inequalities, and graphs
Algebra Review 14–1 Write equations from information
Algebra Review 14–2 Solve proportions, use the Pythagorean theorem
Algebra Review 14–3 Find degrees in a pie graph